KT-528-606

The Material World

Rodney Cotterill was a native of the Isle of Wight, and he was educated at University College London, Yale and the Cavendish Laboratory in Cambridge, where he was also a member of Emmanuel College. He researched and taught initially in physics and materials science, subsequently in biophysics, and ultimately in neuroscience, and he was one of the pioneers of computer simulation. He spent periods at Argonne National Laboratory, near Chicago, the University of Tokyo, and the Quantum Protein Centre at the Technical University of Denmark, near Copenhagen. Professor Cotterill was a Fellow of the Royal Danish Academy of Sciences and Letters, and a Member of its præsidium, until his death in June 2007.

The Material World

RODNEY COTTERILL

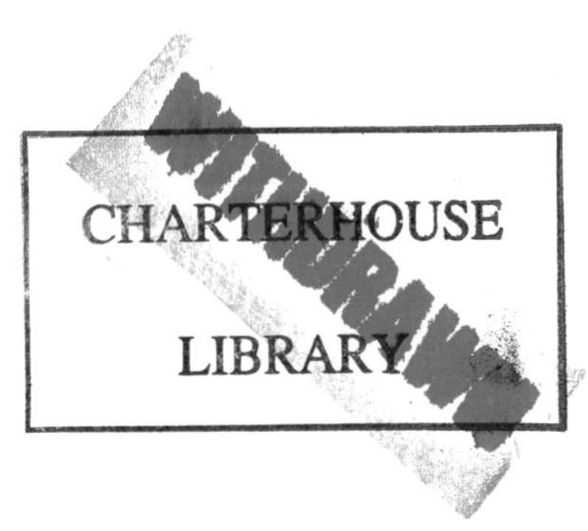

CAMBRIDGE UNIVERSITY PRESS
Cambridge, New York, Melbourne, Madrid, Cape Town, Singapore, São Paulo, Delhi

Cambridge University Press
The Edinburgh Building, Cambridge CB2 8RU, UK

Published in the United States of America by Cambridge University Press, New York

www.cambridge.org
Information on this title: www.cambridge.org/9780521451475

First published 2008

Printed and bound in Singapore by Imago

A catalogue record for this publication is available from the British Library

ISBN 978-0-521-45147-5 hardback

In honour – and *awe* – of the polymaths of bygone days – of the likes of Aristotle, Francis Bacon, Roger Bacon, Robert Boyle, René Descartes, Thomas Edison, Michael Faraday, Benjamin Franklin, Galileo Galilei, Nicholaas Hartsoeker, Hermann von Helmholtz, Christiaan Huygens, Antoine Lavoisier, Gottfried Leibnitz, John Locke, John Tyndall, Thomas Willis, Christopher Wren and Thomas Young – and not the least of Leonardo da Vinci, Robert Hooke, Isaac Newton and Niels Steensen.

With special thanks to Bjørn G Nielsen, Henrik Georg Bohr, Jens Ulrik Madsen, Claus Helix Nielsen and Anetta Claussen, all of whom were instrumental in completing this new edition.

International praise for Rodney Cotterill's

The Material World

'In undertaking to give an account of matter in all its aspects, from the individual atom to the living organism, Rodney Cotterill has done what very few would dare. This is a book that should be in every university science department, every technological firm and every sixth-form college. Let me commend a magnificent achievement.'

Sir Brian Pippard FRS *Interdisciplinary Science Reviews*

'The unity at depth between physics and chemistry, between crystal defects and restriction enzymes, has never been served so well in a popular and visually engaging book.'

Philip Morrison *Scientific American*

'The beautifully produced *The Material World* by Rodney Cotterill should allow teachers and students alike to move away from the narrow traditional education patterns and adopt a broader, more thoughtful approach to science education.'

Sir Tom Blundell FRS *New Scientist*

'This is a grand tour of the sciences. Beautifully written, with a fine poetic turn of phrase, it will be an outstanding source of reference for high school, college and university students, and should find a market among science teachers, science buffs, and young scientists who might be encouraged to range farther afield during their own research careers.'

Graeme O'Neill in *The Canberra Times*

'An instant classic.'

Robert Cahn FRS *Contemporary Physics*

'A godsend to anyone who knows little or no physics, chemistry or biology but would like a picture of the world at the microscopic and atomic levels.'

Carol Rasmussen *New York Library Journal*

'Will it not soon be impossible for one person to grasp the breadth of the topics we call science? Not yet. After several delightful hours with Rodney Cotterill's *The Material World*, I can confirm that at least one polymath survives.'

Peter Goodhew *The Times Educational Supplement*

Contents

Preface

In 1855, Heinrich Geissler (1815–1879) devised a vacuum pump based on a column of mercury which functioned as a piston. He and his colleague Julius Plücker (1801–1868) used it to remove most of the air from a glass tube into which two electrical leads had been sealed, and they used this simple apparatus to study electrical discharges in gases. Their experiments, and related ones performed around the same time by Michael Faraday (1791–1867) and John Gassiot (1797–1877), probed the influence of magnetic and electric fields on the glow discharge, and it was established that the light was emitted when 'negative rays' struck the glass of the tube. The discharge tube underwent a succession of design modifications, in the hands of William Crookes (1832–1919), Philipp Lenard (1862–1947) and Jean Perrin (1870–1942), and this field of activity culminated with the discovery of X-rays by Wilhelm Röntgen (1845–1923), in 1895, and of the electron two years later, by Joseph (J.J.) Thomson (1856–1940). These landmarks led, respectively, to investigations of the atomic arrangements in matter and explanations of the forces through which the atoms interact. The repercussions were felt throughout the physical and life sciences. Jens (J. F.) Willumsen (1863–1958) captured the spirit of these pioneering efforts in his 'A Physicist', painted in 1913.

I have written this book for people interested in science who seek a broader picture than is normally found between a single pair of covers. It presents a non-mathematical description of the physics, chemistry and biology of nature's materials. Although it has been written primarily as a self-contained review for the non-specialist, it can also be used as an introduction, to precede more detailed studies of specific substances and phenomena.

The general approach adopted in the text has been to proceed from the relatively simple to the more complex. After the Prologue, which provides the book with a cosmic backdrop, the story starts with single atoms and the groups of atoms known as molecules, and it continues with the cooperative properties of large numbers of atoms. The nature and consequences of symmetry in crystals have been allocated their own chapter, as have those departures from regularity that play a key role in the behaviour of materials. Such imperfections are presented in the context of the inorganic domain, but they have counterparts in the mutations and variations inevitably present in living organisms. Water and the Earth's minerals are so important to our environment that they too have been discussed in separate chapters, and the balance of the book is broadly divided into three parts, each covered in several chapters: inorganic materials, organic non-biological materials, and biological materials. The final chapter is devoted exclusively to the mind.

Nature is not aware of the boundaries between our scientific disciplines, and the areas obscured by professional demarcations are potentially as rich as any established field of scientific inquiry. Some of the most exciting developments in recent times have indeed come through exploration of these twilight zones. Around the middle of the twentieth century, for example, physics impinged upon biology and produced the revolution we now know as molecular biology. It has already changed our approach to the body, and it may ultimately influence our attitude to the mind; if Alexander Pope were alive today, he might feel that the proper study of mankind should involve proteins, lipids and nucleic acids. The individual chapters of this book admittedly reflect current academic divisions, and could seem to reinforce them. I hope that the inclusion of such wide variety in a single volume will be perceived as compensation, and that the unusual association of apparently disparate topics might even reveal new avenues for investigation.

It would be gratifying if the book's provision of a global view could help alleviate the scarcity of time that bedevils so many professionals these days.

The pressures of securing adequate funding have never been greater, and one of the chief casualties has been the time that could have been spent in the library. Those who would have us focus on a narrow range of prestigious projects are no lovers of the diversity so vital to the scientific enterprise. In an era dominated by buzz-words, it is worth bearing in mind that there is hardly a word in this book's index that does not give a buzz to someone or other.

Stand still you ever-moving spheres of heaven, that time may cease, and midnight never come.

Christopher Marlowe (1564–1593)
Faustus

Prologue

There is broad scientific agreement that the *Universe* came into existence about 13 700 000 000 years ago, as a result of the largest-ever explosion. In the briefest of instants, the explosion created all the *matter* and *energy* that has ever existed, and although matter and energy are interconvertible, as Albert Einstein's *special theory of relativity* tells us, their sum has remained constant ever since. Fred Hoyle referred to this cataclysmic event as the *big bang* – facetiously in fact, because he advocated the now-defunct rival idea of *continuous creation*. In 1948, Hoyle, Hermann Bondi and Thomas Gold had put forward the idea that the Universe is in a *steady state*, that it had no beginning, and that it will have no end. There is now strong evidence that the big bang did take place, as first surmised by Georges Lemaître in 1927, and thus that the Universe certainly did have a beginning. It remains a moot point, however, as to whether it will have an end.

One might speculate as to what was present in the cosmos before the primordial explosion, but Stephen Hawking has argued that such a question would be incorrectly posed, and therefore futile. According to his theory, the big bang created *time* itself, and it thus coincided with what could be called a *pole in time*. Just as it would be meaningless to ask which direction is North when one is standing on the North Pole – pointing upward would merely indicate the direction of increasing altitude – so is it meaningless to ask what came before the big bang.

The huge explosion created a prodigious number of *elementary particles* of matter, which immediately started to fly apart from one another, as do gas molecules in the more familiar type of explosion. But the moving particles were travelling through the vacuum of *space*, and they have continued to do so ever since. The Universe has always been expanding, therefore, but that is not to say that it is featureless and inert. On the contrary, the various *fundamental forces* – of which there are just four – have contrived to cause clustering of particles on a number of very different scales of distance. The shortest-ranged of these forces, the so-called *weak nuclear force* and *strong nuclear force*, created *nuclear matter*, the most common example being the tiny clumps of it present in *atomic nuclei*, while the longer-ranged *electromagnetic force* was responsible for the formation of *atoms* themselves.

The *gravitational force* is by far the weakest of the four, but given the great amount of mass and time at its disposal, it was sufficient to gradually produce the very large clusters of atoms we now call heavenly bodies. The composition of some of these is such that the interconversion predicted by Einstein's theory comes into play, and *thermonuclear reactions* convert matter into vast amounts of energy. These bodies are *stars*, and the one closest to

It was natural for early astronomers to assume that the various heavenly bodies revolve around the Earth, the latter remaining fixed in position. This picture looks like a sixteenth century woodcut at first glance, and it seems to show a person succumbing to curiosity regarding what lies beyond the firmament of fixed stars, as well as the Sun and Moon. It was actually produced fairly recently, however, by Camille Flammarion (1842–1925), an enthusiastic popularizer of science. Published in his book *L'Atmosphère*, in 1888, it is a tongue-in-cheek depiction of a medieval missionary's claim to have found the point at which the sky touches the earth.

Un missionnaire du moyen âge raconte qu'il avait trouvé le point où le ciel et la Terre se touchent...

us – *the Sun* – is so bright that it is visible during the day. Indeed, it *creates* the day! It provides us with light and heat. All the other stars are so far away that their radiations are visible only during the night. Our Sun is surrounded by a number of smaller bodies – *the planets* – which do not generate light, but they reflect it and this makes most of them visible to the naked eye during darkness. Some planets have even smaller bodies revolving around *them*, and the larger of these are known as *moons*. Planets have also been discovered revolving around some of the other stars, and there has been much speculation about whether they too support living organisms. There has indeed been speculation about whether life existed at earlier times on other planets in our own *solar system*, *Mars* in particular.

Solar systems are not the largest manifestations of gravity at work. It transpires that such systems are themselves clustered into what are known as *galaxies*, and the galaxy of which our own solar system is a member contains approximately 10 000 000 000 stars. It is a typical example of a *spiral galaxy*. It has a central stellar cluster and the remainder of its stars are distributed among a number of 'arms', which spiral out from the centre. Our own solar system is located in one of the arms, and this is why we see more of the other stars when we look in a certain direction at the night sky. We call this dense region the *Milky Way*, which is indeed the name given to our galaxy itself. Certain other stars in our galaxy are so close that they appear particularly bright, and are thus easily identifiable. The night sky seems to be a two-dimensional dome – the *firmament* – but the various stars really lie at different distances. This causes some of them to appear grouped together as *constellations*, which acquired names reflecting their fanciful similarities to familiar objects and animals. There is a seven-star constellation variously

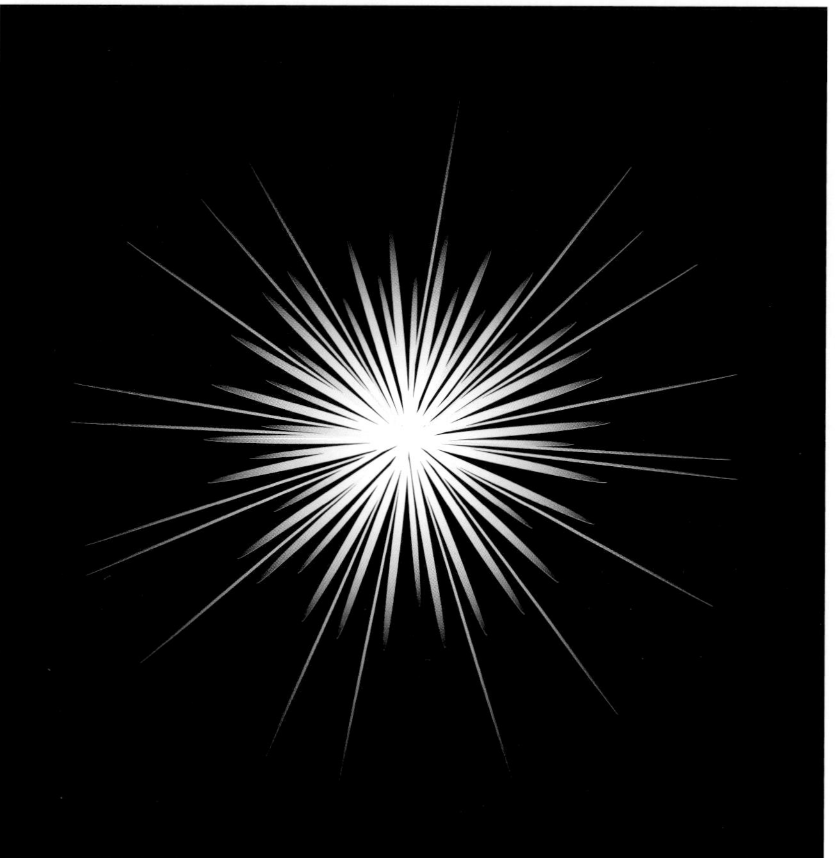

There is general scientific agreement that the Universe was created about fourteen thousand million years ago, as a result of the largest-ever explosion. Fred Hoyle (1915–2001), an advocate of the rival theory of continuous creation, referred to this cataclysmic event as the *big bang* – humorously, because he did not believe that it ever took place. The picture shown here is, of course, just an artists's impression. In reality, there would have been nothing to see because the wavelengths involved were much shorter than those of the visible spectrum.

known as the Great Bear (*Ursa major*), the Plough, the Drinking Gourd or the Big Dipper, for example. Early observers of the firmament divided it into twelve more or less equally spaced zones, each containing a constellation, the number twelve being roughly compatible with the approximately twelve full moons in each year. These constellations become prominent at various times of the year, always in the same sequence, and gave rise to the *signs of the zodiac*. This early rationalization in *astronomy* was subsequently embraced by *astrology*, according to which date of birth dictates traits in a person's character.

Galaxies lying farther away from us naturally appear smaller, and are visible as entire objects. The apparent density of matter – stars and the material from which they are being formed – is then so large that the distant galaxy appears as a luminous cloud, or *nebula*. A famous example is the nebula that seems to lie in the constellation of Andromeda, but is really very much farther away. The typical galaxy is an extremely large structure; the Milky Way measures about 1 000 000 000 000 000 000 kilometres in diameter. A good idea of the vastness of the Universe is provided by the fact that it contains more than 1 000 000 000 galaxies. If we wished to write down the total mass of the Universe, in kilograms, we would have to write 1 followed by forty-one zeros. Such large numbers are more conveniently written in the short-hand *decimal exponent notation*, that mass then becoming simply 10^{41} kg. We will take advantage of this notation in the rest of the book.

Energy

Energy is the capacity for doing *work*, and it comes in various forms, which are always associated with *matter*, the one exception being *radiant energy*. The most fundamental distinction among the various forms is between *potential energy* and *kinetic energy*, the latter being associated with *motion*. But even this neat categorization is blurred by the fact that all matter in the Universe is perpetually in motion through space. Potential energy thus refers to the capacity for doing work that stems from the *relative* positions of particles of matter. The blacksmith's raised hammer, when held stationary, has unvarying potential energy relative to the target object it will subsequently strike, even though both hammer and target are hurtling through space at the same velocity. The various forms of energy are interconvertible, the conversion *always* requiring the presence of matter, even when one of the forms is radiant energy. *Electrical energy* is a form of potential energy that derives from the electrical charges – some positive, some negative – on interacting particles of matter, and *chemical energy* is one manifestation of this. *Nuclear energy* is produced by somewhat similar interactions between the particles in atomic nuclei. *Thermal energy* stems from the motions of the particles in a solid, liquid or gas, and is thus a form of kinetic energy. The heat produced by the blacksmith's hammer when it strikes its target is a familiar example. The unit of energy is the *joule* (named for James Joule, 1818–1889).

The remarkably weak *gravitational force* really comes into its own over very large distances because of its long-ranged nature; in fact, gravitational attraction appears to stretch out toward infinity. The resulting forces between the heavenly bodies profoundly influence their motions. Indeed, they also cause them. A body's *state of motion* is said to remain constant either if it is at rest or if its velocity – that is to say its speed along a given direction – remains unchanged. As Robert Hooke, and more rigorously Isaac Newton, argued, the application of a force changes the state of motion, speeding it up or slowing it down according to whether the force lies in a direction that promotes acceleration or retardation. In our own solar system, for example, the gravitational force between the massive central Sun and a given planet causes the latter to be constantly attracted toward the former (and vice versa, but let us ignore that minor factor). This does not send the planet crashing into the Sun, however, because it also has a velocity that lies tangential to its *orbit*, and this produces the counterbalancing *centrifugal force* that permits stability to prevail. Our familiarity with *artificial satellites* permits us to appreciate the underlying principles, so we are not surprised when such a device does indeed come crashing down to Earth if its speed falls below the necessary value.

Orbital stability is in fact a relative thing, and one must always ask *For how long?* One theory of the origin of the Earth's single Moon runs as follows. There was formerly an additional planet, *Theia*, which had a mass comparable to that of Mars and an orbit close to that of the Earth. The consequent perturbation of Theia's motion caused its orbit to become chaotic, and the two planets ultimately collided, with the generation of so much energy that both bodies melted. Within about an hour, the remains of the totally disrupted Theia were hurled into an irregular orbit around the now-molten Earth, with Theia's iron core located about 50 000 km from the Earth's centre. This situation too was unstable, and the huge lump of iron crashed into the Earth's surface, within about a day after the initial impact. It penetrated deep within our planet's mass and became permanently incorporated in it. Gravitational attraction then ensured that the remnant debris gradually condensed to produce our Moon, initially in an orbit lying at just 22 000 km,

Three of the bodies in our solar system can come between the Earth and the Sun: Mercury, Venus and our own Moon. The latter can thereby cause a total eclipse if it is suitably positioned, the eclipse otherwise merely being partial. (Leonardo da Vinci, 1452–1519, demonstrated the underlying principle, using simple geometry.) Eclipses by Mercury and Venus are only ever partial because the angle subtended at the Earth's surface by either of these planets is very much smaller than that subtended by the Sun. Partial eclipses by Mercury are relatively frequent, because that planet's frequency of revolution about the Sun is relatively large; there was a so-called *Mercury passage* in 2003, for example. The rarest of all types of eclipse is the *Venus passage;* such partial eclipses occur in pairs – about eight years apart – less than once a century. The Venus passage shown here took place on 8 June 2004, and it was the first such event since 1882. No living person had ever before seen such a thing. The passage took six hours and twenty minutes, and the next three passages will occur in 2012 (6 June), 2117 and 2125. Astronomers of earlier times used such passages to check their calculations of planetary sizes and distances.

but now in a stable one at 384 400 km. There is much evidence to support this remarkable theory. The iron-depleted Moon (radius 1738 km) has a density of only 3.34 g/cm^3 (grams per cubic centimetre), whereas the Earth (mean radius 6368 km) has a density of 5.52 g/cm^3. It must be stressed, however, that this fascinating story is nevertheless just conjecture.

On the cosmic scale, then, gravitational influences make motion the norm. *Everything* is perpetually moving! The Moon revolves around the Earth at a speed of about 3600 km/h (kilometres per hour), while the Earth's speed around the Sun (roughly 1.5×10^8 km away) is about 10^5 km/h. And our galaxy is rotating too. This is why its arms do not merely radiate out from the centre like the straight spokes of a wheel; they are a series of vast spirals, and this gives the Milky Way the overall form of a gigantic vortex. It is because of this rotation that the Earth has an additional speed, relative to the Universe, of almost 10^6 km/h. Such relative motions were a constant source of wonderment among the scientists of former times, and some museums can boast an antique *orrery*, in which the planetary movements around the Sun are duplicated in a small-scale clockwork model. Modern counterparts of such devices are to be seen in today's *planetariums*, which additionally reproduce stellar movements.

Gravity also plays a key role in the ultimate fate of each star. Stars vary considerably in size, and also in the surface temperature generated by the thermonuclear reactions in their interiors. These temperatures lie in the approximate range 3000 to 30 000 °C, our Sun being on the cool side with its roughly 6000 °C. This nevertheless makes it hotter than the type of star known as a *red giant*, but it is colder than a so-called *white dwarf*. A white dwarf has the equivalent of a solar mass within the size of the Earth, so it is a million times denser than the Sun. The radiation generated by those thermonuclear reactions produces a force on the constituent particles that counterbalances the inwardly directed gravity. As the reactions weaken, over thousands of millions of years, gravity gradually predominates, pulling the

This typical example of a spiral galaxy was photographed with the Very Large Telescope at the European Southern Observatory on September 21, 1998. It has been given the designation NGC 1232, and it is located about 20 degrees south of the celestial equator, in the constellation of Eridanus, which means the river. Its distance from the Earth is about 100 million light-years. This picture is actually a composite of three individual exposures in the ultra violet, blue and red wave regions, respectively.

particles into an ever-decreasing radius. (This is a simplification of the actual situation, but it is adequate for our purposes.) Various scenarios are then possible. Some heavy stars become mechanically unstable, throbbing with alternating increases and decreases of radius until there is a huge explosion known as a *supernova*. The average-sized star is ultimately shrunk by gravity down to a stable and very dense sphere, and it becomes a white dwarf. But if the initial mass is sufficiently large, the collapse proceeds further and the end product is a *neutron star*. *Pulsars* are believed to be rapidly rotating neutron stars which emit radiation along specific directions, this being detectable on Earth as periodic bursts of energy. A neutron star has the equivalent of a solar mass within a sphere of radius 10 km, so it is 10^9 times denser than a white dwarf and 10^{15} denser than the Sun. Even this does not exhaust the possibilities. If the initial mass is somewhat larger still, gravity shrinks the star into a *black hole*, from which nothing can escape – not even light itself! We can understand how this situation arises by recalling that the speed of an artificial satellite must increase with decreasing distance from the Earth's surface, if it is to remain in stable orbit. If the Earth had a larger mass, the required speed would increase. The concentration of mass in a black hole is so large that even a particle of light, which travels at the fastest possible speed, cannot stay in orbit. During recent years, it has been discovered that there is a black hole located at the very centre of our own galaxy. It is several million times more massive than our Sun. It is a sobering thought that this black hole had been in existence for millions of years before our species emerged.

The metric system
In order to avoid spelling out large and small numbers, it is more convenient to use the decade power notation. Examples are: 1000 = 10^3; 1 000 000 000 = 10^9; 1/1000 000 = 0.000 001 = 10^{-6}.

Decimal multiples for certain powers have been given standard prefixes and symbols. These are indicated in the following table.

Factor	Prefix	Symbol
10	deca-	da
10^2	hecto-	h
10^3	kilo-	k
10^6	mega-	M
10^9	giga-	G
10^{12}	tera-	T
10^{15}	peta-	P
10^{18}	exa-	E
10^{21}	zetta-	Z
10^{24}	yotta-	Y
10^{-1}	deci-	d
10^{-2}	centi-	c
10^{-3}	milli-	m
10^{-6}	micro	µ
10^{-9}	nano-	n
10^{-12}	pico-	p
10^{-15}	femto-	f
10^{-18}	atto-	a
10^{-21}	zepto-	z
10^{-24}	yocto-	y

How do we know all these things? The key appears in this Prologue's first sentence; our knowledge is scientific. But what *is* science, that we should place such confidence in it? *Science*, according to John Ziman's admirably straightforward definition, is nothing more than *public knowledge*. And the word public indicates that we must limit ourselves to knowledge about which there is some sort of consensus. At the dawn of mankind's independence from the other mammals, this consensus would have existed only on the local scale of the tribe, and it must have been dominated by the immediate needs of sheer survival. The world of those early humans was surely an inhospitable place. Some of the animals being killed for food were ferocious and the hunt must occasionally have gone wrong. External stimuli were very much to the fore, even before humans were able to apply names to such threats – even before there was any form of language. Non-animal stimuli would also have been forcing themselves on the anxious human psyche. Lightning, thunder, storm-force winds, flash floods, landslides, forest fires, the Northern and Southern Lights, earthquakes, tidal waves and volcanoes must all have been terrifying.

An early feature of the consensus was the threat posed by things not immediately visible. A predator would often be heard in the undergrowth before it sprang into view. This inevitably extrapolated to threats detectable by the senses which happened to remain latent. In the most spectacular cases, such as those we have just identified, the latency stretched beyond the lifespan of the individual, and this must have been the harbinger of mankind's preoccupation with the supernatural; the seeds of *religion* were surely being planted very early in mankind's history. The influence would have been particularly strong when the unseen thing could be likened to something familiar. Lightning looks as if it could be a scaled-up version of the sparks generated by struck flint, for example, so this paved the way for Thor's hammer and anvil. And when that god's voice bellowed, there were claps of thunder. The tendency to anthropomorphize things is also seen in the idea that earthquakes were caused by the footfalls of unseen giants. High winds, similarly, were generated when these huge beings expelled air. One can see such wisdom depicted in ancient maps, human figures with puckered lips and puffed-out cheeks frequently being located at the four corners. Such myths may now seem humorous, but we should bear in mind that the general acceptance of unseen influences would subsequently lead to what was surely mankind's greatest intellectual achievement: the realization that there are things – objects and agencies – that lie beyond the reach of the unaided senses. This massive step forward was later to provide the dynamo for much of our industry, and it was going to revolutionize medicine by making us appreciate that illness can be caused by things invisible to the naked eye. But the more immediate result was the emergence of mankind's aspiration to emulate the unseen gods' immortality – to share their apparently eternal life.

It ought to be stressed that such a desire was – and still is – entirely consistent with one of the animal kingdom's most basic urges: the survival instinct. In species not possessing consciousness, this instinct influences behaviour in an automatic fashion. In mankind, it is augmented by the power of reason, but there was – and again still is – an understandable tendency to hold those beliefs which provide most comfort. And here we begin to glimpse the great parting of the ways, because science has no truck with comfort. When the evidence has accumulated, and been duly evaluated, the resulting new

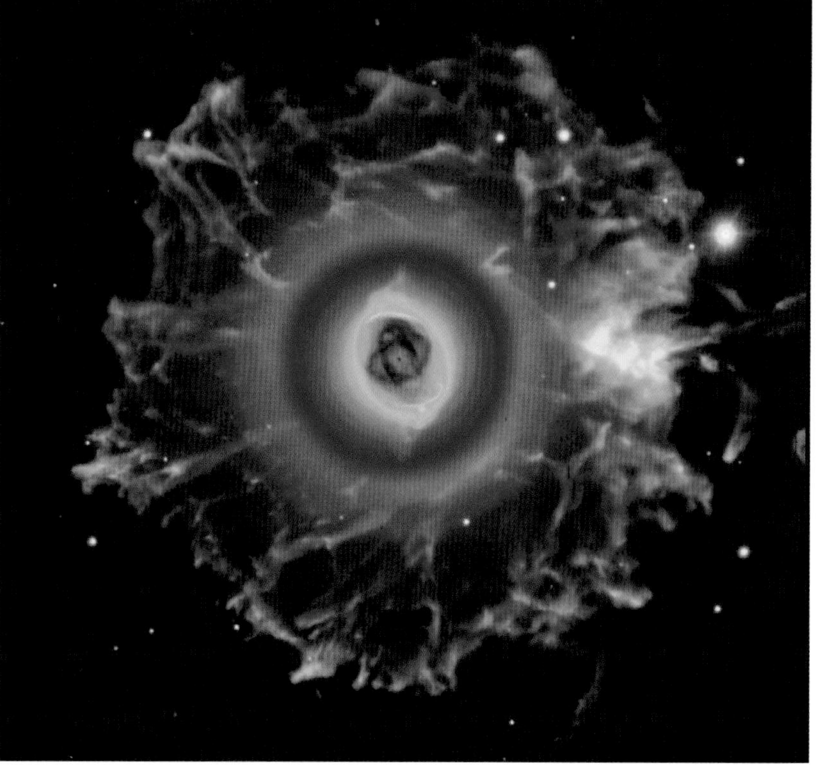

Some galaxies are so distant that their individual stars cannot be seen with the naked eye. Instead, the entire galaxy is seen as a *nebula*: a diffuse cloud of stars and the matter of which newer stars are being created. The examples shown here were photographed at the European Southern Observatory, the one at the top being the well-known Crab Nebula in the constellation of Taurus. The equally famous object below this is the Cat's Eye Nebula, designated NGC 6543. It is an example of a planetary nebula, and it was imaged by R. Corradi in the spectral emission lines of ionized oxygen (the green and blue colours) and nitrogen (red). This is the deepest image ever taken of the Cat's Eye, which is five light-years in diameter. A planetary nebula is a transitory object observed when a star is evolving toward the white-dwarf state, the outer envelope being lost while the remnant core becomes the white dwarf. The majority of them have a mass that is about 0.6 that of our Sun. White dwarfs are essentially dead stars that have ceased to produce energy; they simply radiate what energy they have accumulated, while cooling down.

Forces and motion

In 1679, Isaac Newton (1642–1727) received a letter from Robert Hooke (1635–1703) in which Hooke reiterated the gist of the remarkable paragraph with which he had concluded his *Attempt to Prove the Motion of the Earth* (1674). It correctly identified – for the first time ever – the dynamical components of orbital motion. Hooke had realized that such motion is a consequence of the continuous deflection of a moving body from its (straight) tangential path by a force that attracts it toward the object it is orbiting. In the case of an orbiting moon, for example, the moon is the body while the attracting object is the planet it is orbiting. Hooke had indeed glimpsed the concept of universal gravitation. Until that time, Newton's investigation of the issue had penetrated no farther than rather vague ideas about centrifugal force, but now he set about a thorough mathematical analysis of the underlying physics, and the upshot was enunciation of his three *laws of motion*, namely: (1) a body continues in its state of motion – that is to say it either remains at rest or carries on moving along a straight path at constant speed – unless it is acted upon by a force; (2) the rate of change of a body's *momentum* (the product of its mass and velocity) is proportional to the magnitude of the applied force, and it takes place in the direction of that force; and (3) to every action there is an equal and oppositely directed reaction (*action* being formally defined as the product of work and time, while *work* is the product of force and the distance moved in the direction of that force). These fundamental laws of physics appeared in Newton's *Principia mathematica* (1687), but he was regrettably loathe to give Hooke any real credit for the decisive inspiration. He preferred to cite his celebrated observation of an apple (the moving body – here commemorated by a postage stamp) as it fell from a tree (here reproduced in William Stukeley's sketch dated 1840), having been attracted by the gravitational force due to the Earth. Newton claimed that this historic event took place in the garden of Woolsthorpe, his birthplace, here appropriately photographed while adorned with a rainbow. Newton had correctly perceived that such a spectrum is the result of the angular separation of the various wavelengths (then merely called colours) due to their differential refraction. This explanation was far superior to the one offered in Hooke's *Micrographia* (1665).

consensus must be accepted irrespective of whether it provides solace or poses a threat. The salvaging fact is that knowledge cannot, in itself, ever be a bad thing. Lack of knowledge, on the other hand, is frequently downright dangerous.

A major subsequent step was the willingness, on the part of a few individuals, to reflect on phenomena which, though sometimes frightening, were not obviously related to survival. This must have come much later, because such leisurely contemplation is something of a luxury. The night sky would

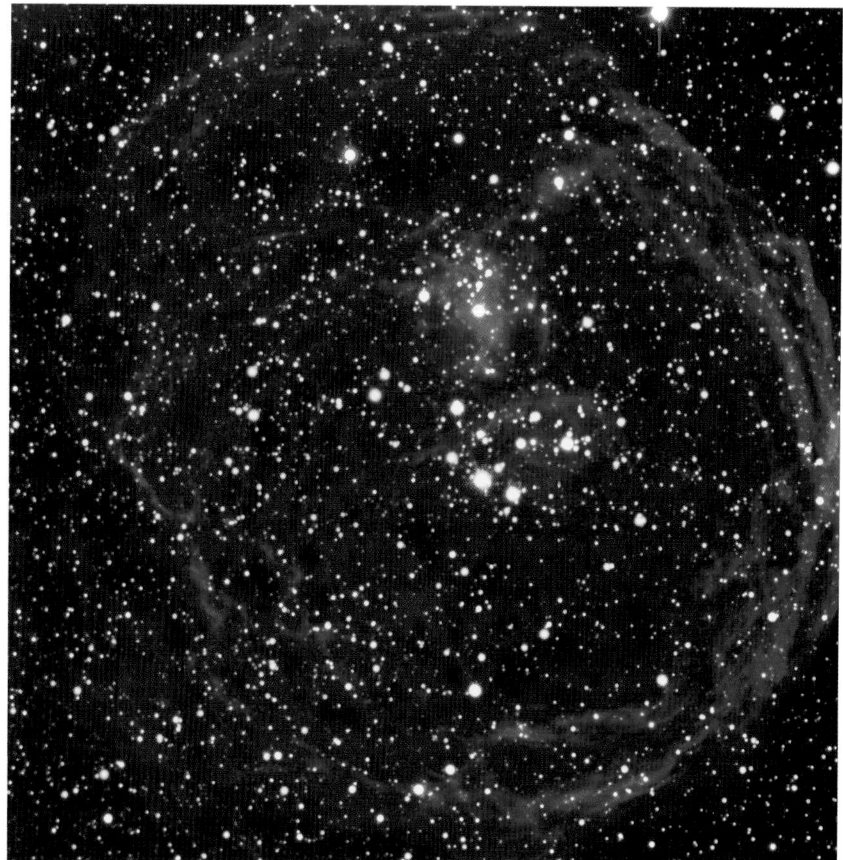

The typical star can remain stable for many millions of years, the inward pull of gravity on its constituent elementary particles being counterbalanced by the outward pressure of the radiation created by the star's thermonuclear reactions. Ultimately, however, those reactions slow down and gravity provokes collapse to a smaller radius. Exactly what then happens is determined primarily by the star's initial size, and one possibility is a series of unstable oscillations which terminate in a colossal explosion known as a *supernova*. The example shown here is designated N70, and it is located in what is known as the Large Magellanic Cloud. It lies only about 180 000 light-years from our solar system, the only known galaxy that is closer to us being the so-called Sagittarius Dwarf.

have been an obvious target of conjecture, by dint of its size. It is relatively easy to locate the planets that are visible to the naked eye, that is to say Mercury, Venus, Mars, Jupiter and Saturn; one has merely to study the sky just after sunset or just before sunrise. When suitably positioned with respect to the Earth, those planets are the first bodies to appear at dusk, and the last to disappear at dawn, because their proximity to Earth makes them relatively bright. And it must soon have been clear that, unlike the stars in the firmament, they do not have fixed positions. Indeed, the word *planet* comes from the Greek for wandering star. A much less frequent feature of the night sky, and thus more mysterious, was the occasional appearance of a *comet*, and there was a natural tendency to accord such an event special significance. The 'star' said to have guided the Three Wise Men to Jesus may have been the comet later associated with the name of Edmond Halley. Comets are now believed to be composites of dust and frozen water – one could call them dirty snowballs – whose infrequent visibility is attributable to their highly elongated orbits around the Sun; they spend most of their time in the outermost reaches of the solar system. Eclipses of the Moon are more frequent, and its entrance into the shadow of the Earth causes it to take on a deep red glow, which perhaps was likened to blood, given that our close neighbour's surface features contrive to give the impression of a 'Man in the Moon'. Total eclipses of the Sun are much rarer, in any given location, and most people would never have observed one. As the present author can verify, the experience is both awe-inspiring and deeply humbling.

Around the turn of the Millennium, it became clear that there is a *black hole* located right at the centre of our own galaxy, its mass being several million times larger than that of the Sun. It is located in the vicinity of the radio source designated SgrA*, and it lies about 24 000 light-years from Earth. The upper picture shows a region 3.2 light-years wide, recorded in the infrared part of the spectrum, the black hole lying roughly at the centre. The lower picture shows that central region magnified, the point source marked S2 being the star whose orbit revealed the presence of the black hole. That orbit around the black hole obeys the rules of motion first enunciated for the motion of the planets around our Sun by Johannes Kepler (1571–1630), as do those of other stars in the vicinity. The photographs were made with the Very Large Telescope, a European facility built at the top of Mount Paranal in the north of Chile.

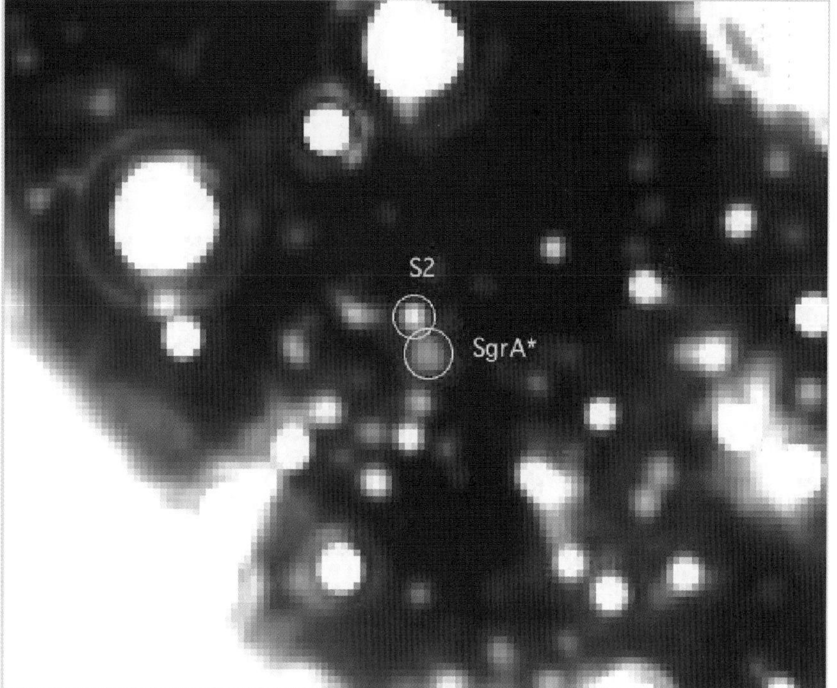

Lightning sometimes causes forest fires, as shown in this example that occurred near Grenoble in France. The resulting blaze continued for several days. Such large-scale upheavals must have appeared terrifying to early humans and they probably gave rise to myths and early religious awakenings.

All these observations could be called passive, however; they were not producing knowledge that one could deem useful. The consensus really started to win its spurs when it began to generate predictions. Just when it occurred to people that heavenly bodies are spheres rather than discs or points is difficult to say. We know that early humans used fires for rendering meat more digestible, and when a person is close to the blaze only one side of the roughly spherical head is illuminated. Similarly, the side of the Moon visible when it is not 'full' always points toward the Sun, or to the place where the Sun has recently set, or will soon rise. So the appropriate conclusion was there for the asking, even if it was not being drawn. It would certainly have been useful to know that mariners are *not* in danger of falling over the edge of an assumed flat Earth when they sail far from land. Yet here too the clues – such as the fact that the hull of a departing ship disappears first and the mast-top last – were originally being missed. Was there perhaps an independent thinker who had the insight to fold and crease a sheet of papyrus to create a straight edge, hold it against the sea's horizon, and notice the curvature?

But the idea of a spherical Earth had certainly established itself toward the end of the third century BC, because we then find the Greek mathematician Eratosthenes calculating its diameter by comparing the Sun's midsummer illumination within two deep Egyptian wells in Syene (the present Aswan) and, about 800 km to the North, Alexandria. The beams reached the very bottom of the well in Syene, this being possible because Syene lay close to the Tropic of Cancer (see below), whereas they penetrated only a fraction of the way down the North side of the wall of the Alexandrian well, that well's bottom thus lying in shadow. Eratosthenes then used simple geometry to produce a remarkably good estimate of the Earth's diameter. Ironically, Alexandria was the home of Claudius Ptolemaeus (Ptolemy), who, about 350 years later, propounded a *geocentric view of creation*. This was hardly a bold hypothesis, given that the heavens do indeed appear to rotate about

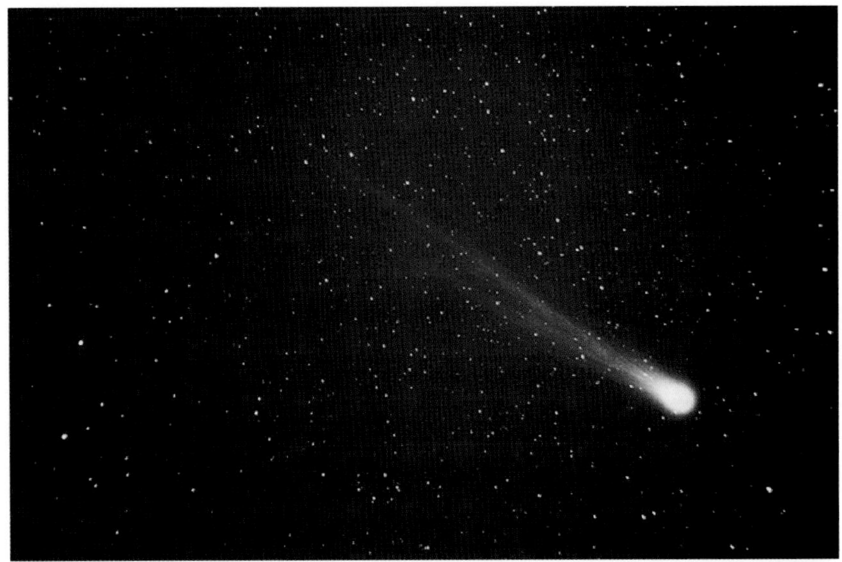

Comets are aggregates of dust and frozen water which revolve around the Sun in highly elongated elliptical orbits. They thus spend most of the time well away from the Earth, and are visible only after intervals that can last many decades. The example shown here is the comet discovered by Edmund Halley (1656–1742), who was the first person to predict the return of such an object. Its closest approach to the Sun is 88 million km and its greatest distance is 5281 million km. Mention of this comet first appeared in Chinese writings in 466 BC, and some believe it to have been the 'star' that guided the Three Wise Men to Jesus. Its appearance in 1066 horrified the English, who took it to foreshadow invasion by William of Normandy; as shown in the lower picture, the celestial harbinger is depicted in the Bayeux tapestry. The photograph of *Halley's comet* was taken in 1986, seventy-six years after its previous visit to our vicinity.

our world; not only the Sun but also the Moon and stars dutifully rise in the East and set in the West, while the Earth seems to stand steady as a rock.

The understanding of the world must have seemed complete at that juncture. Ptolemy's geocentric principle harmonized with mankind's self-centred view of himself, and this in turn was in accord with the deep-rooted tenets of religion. Our earthly mortality was clearly not negotiable, but there was the consoling promise of posthumous ascent to the heavens, to join the invisible but omnipresent gods. The one factor that was to threaten this complacency was the very thing that has always provided the well-spring of science: mankind's innate *curiosity*. If the present author's view is correct, this inquisitiveness is shared by all mammals and possibly by some bird species as well. As will be argued in the final chapter, they all possess *consciousness*, though their levels of *intelligence* differ widely. There continued to be people who asked searching questions of their environment, therefore, and slowly but surely the neat picture of things began to crumble.

The position of the questioning person was initially precarious and vulnerable. The status quo was based on a symbiotic pact between the religious establishment and those in power, the latter providing patronage and the former reciprocating with guarantees that privileges enjoyed in this world would be duplicated in the next. Learning, based as it was on expensive parchment and laborious copying by hand, was the monopoly of these two groups. Dogma dominated over discussion, faith over free thinking. The catalysts of change were increasing urban prosperity, increasing availability of easily affordable paper from Egypt, and the consequent realization of mechanical printing's potential. And although it would be an exaggeration to call them turncoats, some of the first people to break out of the religious straightjacket were actually inhabitants of monasteries; they looked beyond rite and rote, and added reason. The most prominent of these was the Franciscan monk Roger Bacon who, in the thirteenth century, began to look for logical explanations of such familiar phenomena as the concentrating of the Sun's rays by a liquid-filled glass sphere and the magnifying effect of a hemispherical piece of glass. In doing so, he stumbled on the principle of refraction, which is the change in a light ray's direction when it passes from one medium to another. He can appropriately be credited with a flight of fancy, indeed, because he also foresaw ascents in lighter-than-air balloons.

A major milestone was Nicolas Copernicus's conjecture, in the sixteenth century, that the Sun, and *not* the Earth, is the heavenly body around which the others revolve. But this was just a *theory*, which the church elders probably felt they could ignore. The science of the time had another string to its bow, however: *detailed observation*. And detailed observation found one of its most exquisite practitioners in the person of Tycho Brahe, who set about measuring the positions and relative movements of the stars and planets with previously unimagined precision. Fate even rewarded his care with a bonus: the discovery of a supernova in 1572. His assistant, Johannes Kepler, used some of the observations to rationalize planetary motion. He was able to show that Brahe's data are consistent with each planet moving around the Sun in an elliptical orbit, rather than a circular one, with the Sun at one of the ellipse's foci; that a line drawn between the planet and the Sun sweeps out equal areas of the ellipse in equal times; and that the square of the planet's period of revolution about the Sun is proportional to the cube of its orbit's characteristic distances, such as the distance of its closest approach to the Sun. Quantitative details such as these were precisely what Isaac Newton needed, several decades later, when he focussed his mathematical genius on the movements of the celestial spheres, and administered the *coup de grace* to Ptolemy's theory. The new understanding soon led to the first prediction, by Edmond Halley, of the return of a comet.

Science was coming of age, and its most ambitious architect, Francis Bacon, set his sights on nothing less than a complete restructuring of the scientific endeavour. He took stock of the situation in the various disciplines, compiling lists of what remained to be clarified in each of them. He advocated that *deduction* – explanation of the particular in terms of the general – be augmented by *induction* – in which the general is arrived at via examples of the particular. And in 1620 he noted that the east coast of the Americas bears a striking resemblance to the west coast of Europe and Africa, thereby paving the way for the subsequent theory of continental drift. Above all, he stressed the importance of *experiment*; if anyone deserves to be called the founder of empirical science it is Bacon. Sadly, his final experiment indirectly killed him, though the experiment itself was successful. He caught a severe chill when stuffing the decapitated, disemboweled carcass of a chicken with snow, in an effort to preserve the meat, and he failed to recover from the ensuing pneumonia.

Given that the education offered by universities in that era was heavily slanted toward theology, Newton was probably wise in confining his doubts concerning the numerical validity of the scriptures to his private diaries. Galileo Galilei, whose own terrestrial experiments and telescopic observations were lending massive support to the emerging understanding of mechanics and gravitation, was less cautious, and he came into public conflict with the Church. He spent the balance of his life essentially under house arrest, though privately maintaining his defiance. The depth of the wound felt by the Church can be gauged from the fact that he was not posthumously pardoned until about 350 years later. Once again, we should bear in mind how natural – how *biological*, one could say – religion's resistance to these revolutionary ideas really was. The instinct for survival had now gained the authority of the written page. We may be tempted to look back on medieval inquisitions as tribunals of fiends, but to the majority of people in those days they were more akin to teams of doctors, fighting to save the afterlife.

The Earth's axis of rotation is not at right angles to its plane of revolution about the Sun, and it is this tilt that gives us our seasons. Without them, the fact of the revolution would have been much less obvious. Midsummer, in the Northern Hemisphere, is the time when the axis maximally points toward the Sun, and midwinter occurs just six months later, when the pointing is maximally away from the Sun. The opposite is true, naturally, for the Southern Hemisphere. There are two times each year, symmetrically positioned between these extremes, when the Sun crosses over the *equator* and there is equal day and night all over the world: the *equinoxes*. The angle of tilt is, of course, equal to the latitudes (positive and negative) of the *Tropic of Cancer* and the *Tropic of Capricorn*. Only between these two latitudes is the Sun ever directly overhead. These geometrical considerations gave Ole Rømer an astonishingly simple idea, in 1675, for measuring the *speed of light*. By using the eclipses of some of the planet Jupiter's moons as a sort of celestial stopwatch, he compared the timing of these events when the Earth was at two diametrically opposite places in its orbit around the Sun, one closer to Jupiter and the other more remote from it. He detected a discrepancy of about a quarter of an hour between the two timings, and argued that the difference was caused by the light having to travel across the diameter of the Earth's orbit. He was thereby able to calculate the speed of light, which turned out to be approximately 3×10^5 km/s (kilometres per second). This is of course a huge speed compared with anything we normally encounter in our daily lives. The *light-year* is the distance travelled through space by a light pulse in one year, namely about 10^{13} km, and it is widely used when describing astronomical distances. The diameter of the Milky Way, for example, is about 10^5 light-years.

It is high time we considered another important facet of science: the *accidental discovery*. And we could hardly choose a more appropriate example than something that happened during a lecture given in 1820 by Hans Christian Ørsted. His demonstrations that day included the influence of the Earth's magnetic field on a compass needle and the flow of electric current through a conductor when this is connected across the poles of a battery. The compass happened to be lying near the conductor, and Ørsted was surprised to note that every time current was flowing through the latter the needle was deflected. Until that moment, electricity and magnetism had always been regarded as two unrelated phenomena. Ørsted had accidentally discovered that they are actually different aspects of one and the same thing: *electromagnetism*. A magnetic field is produced when a current flows through a conductor, and this will influence any other magnet in the vicinity. Conversely, a moving magnet generates a current in a nearby conductor. This discovery soon led to the production of devices now found in just about every home, factory and transport vehicle, and in many other places as well: *electrical motors* and *dynamos*. But the astonishing thing is that these ubiquitous workhorses were only minor products of Ørsted's discovery. Its deeper significance was that it would soon reveal the nature of light itself! This was accomplished by James Clerk Maxwell, who produced a mathematical description of the propagation of *electromagnetic waves*.

Maxwell's equations apply to all wavelengths and frequencies, not just to those corresponding to *visible light*, and this fact was soon seized upon by Heinrich Hertz, who produced the first *radio waves* in 1886. These travel through a vacuum at the same speed as light itself, so delays analogous to those that apply to light occur when radio signals are transmitted to distant satellites, for example. It became apparent that there is a continuous *spectrum* of electromagnetic waves. At wavelengths somewhat shorter than those used in radio, there is the *microwave radiation* used in radar, while the *infrared waves* that familiarly produce heat lie at even shorter wavelengths. The visible part of the spectrum comes next and progressing through ever-decreasing wavelengths we find the *ultraviolet waves*, *X-rays*, and ultimately *gamma rays*. All these radiations are manifestations of the same basic mechanism that Maxwell had uncovered, and they all propagate through a vacuum at the same speed. Here indeed were agencies lying beyond the immediate grasp of the human senses. And in one particular case, the agency permitted science to look back across almost the entire history of the Universe and to come tantalizingly close to catching a glimpse of the big bang itself.

As was mentioned at the start of this Prologue, the primordial explosion generated both matter and energy, and in 1950 George Gamow argued that they would initially have been in strong mutual interaction. But after about 300 000 years had elapsed since the big bang, Gamow's analysis suggested, this interaction would have weakened markedly, and thereafter matter and energy were essentially unaffected by each other. This stage in the Universe's development is now called the *epoch of last scattering*, and Gamow estimated that the wavelength of the radiation then existing has been stretched out, during the intervening period to the present day, until it now corresponds to the microwave region of the electromagnetic spectrum. This corresponds to a cooling of the Universe from about 3000 °C, at the time of the above epoch, to about −270 °C at present. In other words, Gamow concluded that remnants of the aftermath of the big bang should still be propagating through the Universe as radiation: the *cosmic microwave background*. Once again, an accidental discovery provided science with unexpected progress.

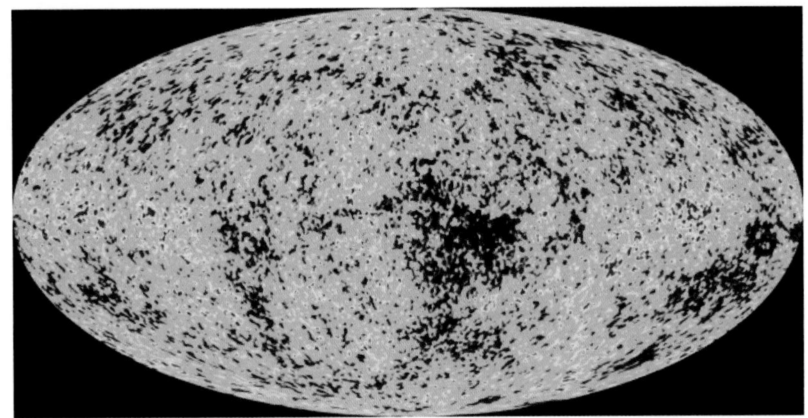

The strongest support for the big bang theory comes from observations of the red shift, which shows that the Universe is perpetually expanding, and the *cosmic microwave background*. The existence of the latter was predicted in 1950 by George Gamow (1904–1968) and it was observed accidentally fifteen years later by Arno Penzias and Robert Wilson. This 'picture' of the radiation is somewhat misleading because microwaves lie well removed from the visible region of the electromagnetic spectrum. But the whole-sky image does give a good idea of the radiation's remarkable uniformity. It was taken in early 2003 with the Wilkinson Microwave Anisotropy Probe, which permits an angular resolution of 0.2 degrees, the telescope being located in an orbit more than a million kilometres from the Earth. The observations indicate that the first stars in the Universe were formed a mere 200 million years or so after the big bang.

In 1965, Arno Penzias and Robert Wilson were calibrating a radio horn they had constructed for communicating with artificial satellites when they discovered that they were detecting radiation with a wavelength of 7.35 cm – microwaves – irrespective of which part of the sky their horn was pointing toward. The picture we have of creation thereby received pivotal endorsement.

It is remarkable that our understanding of the heavens was so advanced compared with that of substances here on Earth. One sees in retrospect that this stemmed from the weakness of the gravitational force; its consequences were thus relatively easy to divine. The objects that surround us owe their physical properties to the much stronger electromagnetic force, the discovery of which soon began to unlock the secrets of terrestrial materials, both animate and inanimate. Superficially, the actors on this microscopic stage were just miniature counterparts of the celestial spheres: the atoms whose existence had been predicted centuries earlier. And the *atom* itself seemed to be nothing other than a tiny counterpart of our solar system, with one type of elementary particle, the *electron*, revolving in orbit about a sphere later discovered to be composed of two other types of elementary particle, *protons* and *neutrons*. But then came the first of two shocks that rocked science back on its heels. The mechanical rules governing the motions of the elementary particles – the rules of *quantum mechanics* – were clearly not those revealed by Isaac Newton's analysis. They were altogether more strange, and one even had to give up any hope of knowing exactly where a given particle was at any given time. It is for this reason that one can validly use terms such as particle, wave and ray as if they are interchangeable; they are indeed.

The second shock was just as profound, because it indicated that our knowledge of celestial mechanics also needed sweeping revision. As mentioned earlier, large bodies such as the Earth move through space at prodigious speeds. The question arose as to whether these could be measured experimentally, by methods analogous to those used for what are regarded as high speeds here on our planet. An obvious example of the latter is *sound*, which is *not* an electromagnetic radiation; it travels through still air at 332 m/s at 0 °C. A person travelling at 100 m/s toward a source of sound would observe that the sound waves were passing at 432 m/s, and the same would be true if it were the source that were travelling toward the observer. This change in the observed speed of sound could thus be used to measure the relative speed between the sound source and the observer. (By the same reasoning, a person who travelled at 332 m/s *away* from a sound source would hear nothing.) Do similar rules apply to electromagnetic waves, as they travel through a cosmic medium analogous to air, namely the *ether* conjectured to pervade all of space? (This ether must not be confused with the chemical substance – ether – bearing the same name.)

That question is what Albert Michelson and Edward Morley attempted to answer in 1887, by comparing what happened to light waves travelling in two mutually perpendicular directions, one of these being roughly parallel to the direction of the Earth's revolution around the Sun while the other was transverse to that path. The experiment involved observations of what is known as an *interference pattern*. If a stone falls into water, a set of circular waves is set up, radiating outward from the point where the stone entered. If two stones simultaneously fall into water, the resulting two sets of waves will overlap, producing an interference pattern that is stationary, even though the individual waves are moving. If the stones fall into the water at slightly different times, the resulting pattern is still stationary but it is displaced from the one produced by the simultaneous entry of the stones. Measurement of the distance of displacement can be used to determine the time lapse between the entries of the two stones; the displacement of the pattern thus functions as a sort of stop-watch. This was the essence of the approach employed by Michelson and Morley. They were surprised to obtain a null result, which indicated that the ether does not exist. There is apparently something peculiar about light in that it does not obey the additive rule of velocities clearly applicable in the case of sound.

We should pause here and take a more detailed look at what Robert Hooke was saying in a letter to Isaac Newton, in 1679. In effect, he was stressing the vital difference between *speed* and *velocity*. As we noted earlier, speed is just a quantity, irrespective of direction – it is what is known as a *scalar*. Velocity, on the other hand, is speed in a given direction – it is what is known as a *vector*. When a smaller body is in a stable orbit around a larger body, as in the case of the Earth moving around the Sun, the speed of the smaller body is constant but its velocity is perpetually changing, because its *direction* is perpetually changing. Although this may sound remarkable, the smaller body is constantly accelerating toward the larger, but it never gets closer to the latter because its forward movement acts to counteract this. In order to bring out a deep truth about motion, we need to consider a simpler situation here on Earth in which the *velocity* is constant, that is to say one in which both speed and direction remain the same. This would be true of an elevator travelling up or down at a constant rate, for example. The same would be true of a train travelling at a fixed speed along a long stretch of perfectly straight and perfectly smooth rail. Under these conditions, a passenger would find that the usual laws of mechanics still apply, despite the movement; a ball thrown against a wall or the floor would bounce back in just the same way as it does when the elevator or train is standing still. And we could imagine making the example more complicated by thinking of a billiard table being installed in one of the train's carriages; the multiple collisions between the billiard balls would still obey the usual laws, provided the train continued to move at constant speed in the same direction.

In fact – and here is the importance of this imaginary exercise – *there would be no way that a ball-throwing or billiard-playing passenger could know that the elevator or train was moving*! It is true that the train passenger could make that discovery through the simple expedient of looking out of the window, but even then, he or she might prefer to believe that the countryside was moving past the stationary train, precisely as the people of bygone days regarded the heavens as moving around the stationary Earth. In fact, in the years following Einstein's great breakthrough, it became popular for waggish passengers to ask railway officials when the next station would arrive at the train, rather than the other way around. Hooke and Newton deserve great credit for having perceived these truths in an age when the fastest mode of travel was the horse-drawn carriage, which made for a bumpy ride even at the best of times. Those gentlemen were both Fellows of the Royal Society of London, and there were pressures even in those days for scientists to produce 'useful' results; the monarch of that time, King William III, demanded of the Society's Fellows that they strive to improve the performance of such vehicles.

Let us make a final visit to that moving train and imagine jumping from it. In movies, such hazardous behaviour is usually indulged in by desperadoes who, for obvious reasons, do not wish to wait and alight at the next station. But these people know that a jump is more dangerous the greater is the train's speed. And even if they have no understanding of the underlying theory, they instinctively sense that they will not land at the place the train is passing the instant they make their leap; they realize that they will have the train's forward velocity when they jump. Similarly, if one wished to hit a tree with a ball thrown from the train, one would have to adjust one's aim accordingly, to allow for the train's velocity being added to that of the ball. The remarkable thing is that such addition of velocities does not apply to light, though there is still the relatively trivial need to make allowances for the fact that the speed of light is not infinite.

This has been a long digression, but it was desirable because we needed to appreciate what the null result of the *Michelson–Morley experiment* really implied. It was showing them and their colleagues that one simply cannot obtain information about movement at constant velocity by performing experiments that probe the validity of the usual physical laws. Let us now return to that famous experiment and consider more of the details. It involved two sets of light waves that were passing back and forth between mirrors located at the ends of two rigid rods of equal length, mutually arranged at right angles, and a third mirror located at the rods' intersection. That third mirror was special in that its layer of deposited silver was only partial, this thus letting the two sets of light waves interfere, in a manner analogous to what was described earlier in the case of waves on a water surface. A light pulse directed at the partially silvered mirror therefore gave rise to two half-strength pulses, one travelling toward each of the two other mirrors. And those mirrors reflected the pulses back along the directions from whence they came, toward the partially silvered mirror, and set up the interference pattern.

The key point to bear in mind is that although these light pulses were always travelling at the same speed, the mirrors would have moved during the pulses' journeys. So we have to explain why Michelson and Morley nevertheless obtained a null result. In the case of the rod and mirror lined up with the direction of the Earth's motion, the light waves would have been travelling alternately with and against the latter, but it can be shown that these two modes would not exactly compensate for each other; there would nevertheless be a net shortening of the round-trip time compared to what would have been observed had the apparatus been stationary in space. The waves travelling between the centrally located partial mirror and the transversely positioned mirror, on the other hand, would actually have traced out a zig-zag path, because of the Earth's motion, and this too would have produced a net shortening of the round-trip time. But simple geometrical considerations show that this transverse-case shortening does not equal the longitudinal-case shortening. There would thus be a temporal mismatch, and we noted earlier that this should be detectable by the resulting shift of the interference pattern. But Michelson and Morley detected *no* such shift. The question was: *why*?

We need to pause again and consider another problem that had arisen. Maxwell's equations, then only twenty or so years old, did not possess the symmetry expected of the reciprocity between magnetism and electricity observable empirically. Thus whereas the equations predicted that an electric field is generated in the vicinity of a moving magnet, and that this will produce a current in a nearby stationary conductor, they did not predict that the same thing will happen if the conductor is moving and the magnet is stationary. Joseph Larmor and Hendrik Lorentz independently discovered mathematical modifications to the spatial and temporal dimensions that would correct for this inadequacy. Length is contracted in the direction of motion but not in the transverse direction, they found, and time becomes dilated because of the motion. Such modifications are known as *transformations*, and the remarkable fact is that the *Lorentz contraction*, as it is now called, is precisely what is required to produce the null result of the Michelson–Morley experiment. The longitudinal contraction, the lack of a transverse contraction, and the *time dilation* are collectively referred to as the *Lorentz transformation*.

Given that Albert Einstein's famous paper of 1905 established his *special theory of relativity*, its title seems rather tame: 'On the electrodynamics of moving bodies'. But the fact is that he too had been troubled by the apparent inadequacies of Maxwell's equations. Following a suggestion by Henri Poincaré, Einstein proposed that any law of physics remaining unchanged by the Lorentz transformation can be regarded as a faithful description of what happens in Nature, but not otherwise. This indicated that Newton's laws of motion required revision. It is true that the modifications are negligible at speeds we encounter terrestrially, but Einstein was looking for rules that would apply even near the speed of light itself. He imagined trying to read the time from a clock face while travelling away from it, for example. If the speed of travel were slow, the correct time would be read of course. But in the limiting case of travel at the speed of light, the time would appear perpetually frozen at a specific moment. This is in agreement with the Lorentz transformation. And the Einstein analysis produced other surprises. Mass increases as the speed increases, such that it would become infinite if the speed of light could ever be attained. Another peculiar result concerned *failure of simultaneity at a distance*. Suppose that a person travelling in a space ship wished to synchronize two clocks lying at opposite ends of the vehicle. This could be achieved by precisely locating the mid-point between the clocks, and then by simultaneously dispatching from that point two light pulses that were to accomplish the synchronizing. But a person observing this attempt while not travelling with the space ship would judge it a failure, because he would note that one of the pulses was travelling in the same direction as the ship while the other pulse was travelling in the opposing direction. The person aboard the space ship, on the other hand, would be unaware of any limitation in his concept of simultaneity because, as we have seen, he would have no way of determining that he was moving. The upshot of all this, as first suggested by Hermann Minkowski, was that the two variables previously assumed to be mutually independent, namely *space* and *time*, are merely different components in a single composite, that is to say *space-time*. And another surprising corollary is that one has to be very careful about what one means by the term *straight line*, when the underlying *frame of reference* is moving.

The most spectacular result in Einstein's special theory of relativity is its discovery that mass and energy can be converted into each other. This relationship is captured in what is surely the most famous of all equations, $E = mc^2$. And in stating that the amount of energy that can be created from a given mass is equal to the mass multiplied by the square of the speed of light, it indicated that enormous amounts of energy would be liberated if the appropriate type of reaction could be provoked. Ernest Rutherford, discoverer of the atomic nucleus, was sceptical about this prediction and he went on record as saying that the idea of gaining energy from the atom was preposterous. The production of the first *atomic bomb*, less than a decade later, served as a reminder that the word *never* is dangerous when applied to scientific predictions. It has often been said that science lost its innocence when that bomb was produced, but the event also served to underline the seriousness of the scientific endeavour, and the reliability of its predictions when observations have been correctly interpreted.

Einstein's famous relationship is really a consequence of something embodied in Maxwell's analysis, and of what is known as the *conservation of momentum, momentum* being mass multiplied by velocity. Maxwell showed that the momentum inherent in radiation is equal to its energy divided by the velocity of light, so when an object such as an elementary particle emits radiation, momentum can only be conserved if there is a simultaneous decrease in the particle's mass. In his *general theory of relativity*, Einstein extended his analysis to include the effects of acceleration, rather than just of constant velocity. Isaac Newton had attributed the fall of his famous apple to gravity, and Galileo Galilei had found that objects with different masses dropped from the Tower of Pisa accelerate at the same rate, if the small disturbances due to air resistance are ignored. But once again Einstein imagined what would happen if one were moving at the same pace as the moving object. To take a rather dramatic example, let us return to that elevator we considered earlier, and suppose that the cable breaks while we are descending from the top of a skyscraper. Ignoring the small effect of friction, the elevator would go into free fall, and if we dropped an apple in order to observe its acceleration toward the elevator's floor we would be surprised to find that gravity had been suddenly switched off! Such weightlessness has of course become quite familiar in this age of artificial satellites. In other words, gravity and acceleration are also a pair of parameters which have an intimate mutual relationship. One remarkable prediction of Einstein's general theory was that the path taken by a light beam travelling toward us from a distant star would show a slight deflection if it passed sufficiently close to the Sun en route. This proved to be the case.

We should now return to the question of the Universe's expansion, which has apparently continued unabated since the big bang. In the 1920s, studying the light spectra emitted by distant suns using an instrument known as a *spectroscope*, Edwin Hubble had noticed a shift toward longer wavelengths, this effect increasing with increasing distance from our own solar system. (The determination of distance had become much easier because of Hubble's own felicitous discovery of stars called *Cepheid variables*, which all happen to emit the same amount of light in a given time; measurement of their apparent brightness thus provided a direct indication of their distance.) This is reminiscent of the effect observed when a moving vehicle passes us: the sound of its motor or whistle is higher pitched when it is approaching, compared with what is heard when the vehicle is stationary, whereas the pitch drops when it is moving away from us. Christian Doppler had analyzed the terrestrial version of the effect in the middle of the nineteenth century, and the spectroscopic observations were showing that it has a cosmic-scale counterpart. But, as noted above, the observed spectral modification was always toward longer wavelength, irrespective of which part of the night sky Hubble was aiming his spectroscope at. This *gravitational red shift* thus indicated that every other sun was moving away from our own, and at a speed that increased the farther the other sun was from us. The situation could be compared with what is observed during the production of currant buns and muffins. When the dough is expanding because of the action of the yeast, every currant is moving away from every other currant, with a speed that is proportional to its distance from any specific currant.

The techniques used to detect distant galaxies and measure their red shifts has long since been extended to other wavelengths, and radio astronomy in particular has established that there are galaxies 5×10^9 light-years away that are receding from us at about half the speed of light. Indeed, there are bodies lying still farther away – about 10^{10} light-years – which are increasing their distances from us at about 90% of the speed of light. These are the very powerful radio-wave emitters known as *quasars*. It is a sobering thought that radio telescopes aimed at such objects are looking simultaneously out over vast distances and back over extremely long times. Indeed, they are looking more than half way back to the big bang itself.

At present, the observable Universe is a staggering 10^{23} km in diameter. But what of its future? Will it go on expanding forever? The situation is complicated by the possibility that most of its mass might not be visible to us at any detectable wavelength. There have been strong hints of the existence of what is known as *dark matter*, and some people believe that its mass far exceeds that of the matter we are able to detect. Perhaps the gravitational influence of this ghostly substance plays an important role in galaxy formation. Even more importantly, it could ultimately provide the brake that would slow down the Universe's expansion to a full stop, reverse it, and then initiate a process of shrinkage that would culminate in what could be called a *big crunch*. Those enthusiastic about salvaging indefinite existence might be inclined to take heart at this scenario; perhaps such a big crunch would be followed by a new big bang, and so on, ad infinitum. But there would be no possibility of any meaningful continuity at the level of life-supporting molecules, given the enormous compaction that would have to intervene between successive crunches and bangs.

This does not unduly bother those who subscribe to the *dualism* championed by René Descartes. They believe that the soul can continue its existence in space, independent of any supporting matter, and some believe in a cosmic intelligence. But even this venerable issue has latterly entered the limelight in the scientific endeavour, though its move from the wings to centre stage required a number of preliminaries. First, there was the realization by Leonardo da Vinci, and subsequently by Robert Hooke and Niels Steensen, that there were earlier life forms on Earth that subsequently died out. Then there was Charles Darwin's theory of the inter-species competition which decreed that such demise is a natural consequence of time's passing. And Gregor Mendel – another monk, but this time of the Augustine order – discovered the mechanism on which the competition is based. Scientists in the physical disciplines, meanwhile, were discovering the ultimate goal of the competition, namely limited energy resources. As became clear through the important discipline of *thermodynamics*, that unavoidable competition is a consequence of processes played out at the molecular level, processes on a scale too small to permit tampering by human fingers!

In a sense, traditional religion has been focussing on the wrong goal. Its province is really interactions between people; anyone who doubts that should contemplate the majority of the Ten Commandments. A huge web of interconnected facts has now placed the ideas of the physical and biological sciences in an impregnable position; these issues are now part of the consensus of public knowledge, and are not negotiable. We unwittingly endorse them every time we use a kitchen appliance, board a plane or visit the doctor. Philip Anderson famously noted that 'More is different', by which he meant that although the individual scientific disciplines can be arranged in a hierarchy, with mathematics figuring as the most fundamental category, this emphatically does not imply that everything is reducible to mathematics. Every step to a higher level of complexity, he argued, brings in new phenomena that could not have been predicted from our knowledge, however complete, of what lies below. According to Anderson's persuasive view, physics is not just applied mathematics, chemistry is not just applied physics, and biology is not just applied chemistry. And one could continue upward into even greater complexity; physiology is not just applied biology, and psychology is not just applied physiology. Indeed, sociology is not just applied psychology, and sociology, which deals with the above-mentioned interactions between people, is certainly a most difficult subject. The juggernaut of the physical and biological sciences now has its own momentum. Those lower echelons of the hierarchy are driven forward by an inner logic and it is difficult to imagine them ever again being slowed down. If Aristotle returned to Earth today and sat in on any university lecture in the physical or biological sciences, he would feel totally lost. But place Socrates among people discussing issues in the humanities and one would see him smile knowingly.

The question remains, however, as to the nature of the mind. Those who do not subscribe to the ideas of an immaterial soul and a cosmic consciousness must show how mind emerges solely from matter. And this too requires a decision as to which level in the hierarchy is likely to hold the key. Some – Roger Penrose being prominent among their number – feel that one can confine oneself to the level of certain molecules. Others – Julian Jaynes being a leading member of that school – look to the other extreme of sociology and maintain that consciousness and the mind are byproducts of the language through which people interact. In favouring the intermediate level of anatomy and physiology, the present author is merely following the lead set by Niels Steensen. A contemporary of the dualism-espousing Descartes, Steensen was stressing the importance of the brain's white matter, which we now know is composed of the fibres through which nerve cells communicate, at a time when most people were not even aware of the grey matter. It is remarkable that Steensen, having helped to establish geology, crystallography, anatomy and physiology, suddenly dropped all scientific activity and became a priest. He was recently beatified, and appears to be well on the way to actual sainthood. Could it be that he looked so deeply into Nature's workings that he was appalled by the prospect of even the human mind being just another manifestation of the material world?

1

I can trace my ancestry back to a protoplasmal primordial atomic globule. Consequently, my family pride is something inconceivable. I can't help it. I was born sneering.

William S. Gilbert (1836–1911)
The Mikado

Solo atoms: *electrons, nuclei and quanta*

Our world is a material world. Everything we see, touch, taste or smell is composed of one or more materials. Even our hearing depends on the interaction of the eardrums with the gaseous form of that familiar, though invisible, material known as air. Interest in the matter that makes up our environment has grown with the realization that it can be brought under control. At the dawn of civilization this control was of a rather rudimentary type. The earliest use of such materials as stone, wood, and the bones and skin of animals involved relatively minor alterations to these substances, and they were naturally taken for granted. Just which physical phenomena led to speculation about the nature of materials remains a matter of conjecture. The stimulation probably came from observations of simple modifications of state, such as the irreversible change caused by the burning of wood and the reversible changes between the solid, liquid, and gaseous forms of substances like water.

The great diversity of form and behaviour seen in the material world stems from the wide range of *chemical composition*. Although chalk and cheese, for example, are both composed of *atoms*, the atoms are of different kinds and are combined in different ways. Strictly speaking, several hundred different types of atom are now known to exist and they are referred to as *isotopes*. More significantly, these isotopes are distributed amongst 90 naturally occurring, and about a score of artificially produced, atomic species known as *elements*. It is the combination of atoms of different elements that produces *chemical compounds*. The idea that matter is composed of large numbers of minute particles is a surprisingly old one. It seems to have appeared first with the speculations, around 400 BC, of the Greek philosopher Leucippus and his student Democritus, and the word atom comes directly from the Greek word *atoma*, which means indivisible.

The atomic idea languished for over 2000 years, and the next real advances came through the chemical knowledge acquired as a byproduct of the painstaking, though futile, endeavours of the *alchemists*. Just when the modern era started can be debated. A good choice would be 1661, the year of publication of a book by Robert Boyle in which he discussed the concept of pure substances, the above-mentioned elements, of which more complicated substances were composed. Boyle was especially interested in the use of colour as a means of distinguishing between different elements, and it was he who developed the *flame test* that was later to become so important in connection with *atomic spectra*. The formation of compounds was put on a quantitative basis by Antoine Lavoisier and Joseph Priestley, who carefully weighed the ingredients before and after a *chemical reaction*

ELEMENTS

Element	W.t	Element	W.t
Hydrogen.	1	Strontian	46
Azote	5	Barytes	68
Carbon	5.4	Iron	50
Oxygen	7	Zinc	56
Phosphorus	9	Copper	56
Sulphur	13	Lead	90
Magnesia	20	Silver	190
Lime	24	Gold	190
Soda	28	Platina	190
Potash	42	Mercury	167

The foundations of modern atomic theory were laid by John Dalton (1766–1844) in 1808. His symbols for the elements, shown here on one of his listings of atomic weights, seem to emphasize atomicity by their use of circles.

had been made to occur by heating. The experiments variously involved the combination of oxygen and a metal such as tin, lead, or mercury. In Lavoisier's case, two elements were made to combine to form a compound, while Priestley drove the reaction in the other direction. In both types of experiment it was found that the weight of the compound equalled the weight of its constituent elements, and it was therefore concluded that a chemical reaction leads to no net loss or gain of matter; that matter is in fact indestructible. The significance of weight was first fully appreciated by John Dalton who, in 1808, put forward the idea that the defining characteristic of an element is the weight of one of its atoms. By assuming that all atoms of a given element have the same weight, and by permitting different elements to associate with one another only in certain proportions,

Mass and weight
Although the terms mass and weight are often used as if they were interchangeable, they are not equivalent. Mass, the more fundamental quantity, is a measure of the force required to impart a given acceleration to an object. It is thus an expression of the amount of matter present. Weight is actually a force of attraction, and it depends on the local gravitational acceleration. The mass of a given object is not altered when it is transferred from the Earth's surface to the Moon, but its weight decreases.

he was able to explain the weights of certain simple compounds. Through his elegant analysis Dalton had in fact demonstrated the existence of atoms as individual entities, and it is interesting to note that even his choice of symbols for the different elements seemed to emphasize atomicity. In the seventeenth century gold, mercury, and lead were denoted by ☉, ☿, and ♄, respectively, but after Dalton the standard symbols for those same elements were ◎, ❂, and Ⓛ.

Elements are the letters of the chemical alphabet, while compounds can be compared with words. Using some or all of the letters a, r, and t, once or several times, we can form a variety of words such as rat, art, tart, and so on. Similarly, using some or all of the elements carbon, hydrogen and oxygen we can make such compounds as sugar, alcohol, benzene, and candle wax. The analogy is not perfect, however, because whereas words are limited to arrangements along a single line, compounds form three-dimensional patterns in space. Moreover, the alphabet of elements is divided into not just two classes, as with consonants and vowels, but into a whole series of classes. And the rules which govern the making of compounds, the construction of chemical words, are intimately connected with the very existence of these classes of elements.

By the latter part of the nineteenth century, so many elements had been isolated and identified (a total of 65, for instance, by the year 1870), that the stage was set for the next great advance in *atomic theory*. The breakthrough was achieved independently by several individuals, who had remarkably different degrees of success in getting their revolutionary ideas accepted. The important step was based on the recognition of a sort of family likeness between certain groups of elements. An example of such a family is the *alkali metals*, lithium, sodium, potassium, rubidium, and caesium, all of which had been discovered by the year 1861. The compounds produced by combining any one of these metals with chlorine resemble one another to a remarkable extent. They are all colourless and dissolve readily in water. The crystals of their solid forms also have the same shape, with the same geometrical relationships between their surface faces. Indeed, it is now known that the actual *crystal structures* of the compounds formed between the four lightest members of the alkali metals with chlorine are identical. Similar family likenesses had been noticed in the groups known as *alkaline earths* (beryllium, magnesium, calcium, strontium, and barium), the *halogens* (fluorine, chlorine, bromine, and iodine), the *noble metals* (copper, silver, and gold), and so on. In 1865, John Newlands showed that if the elements are listed in ascending order of their *atomic weights*, members of any particular chemical family appear at regular intervals. Every eighth element belongs to the same family, and this was the basis of his *law of octaves*. The proposal met with a scornful reaction, and one of his colleagues ridiculed Newlands by suggesting that he might alternatively have listed the elements in alphabetical order of their symbols. A similar conjecture by Dmitri Mendeleev, in 1869, won greater support, particularly from Lothar Meyer in 1871, and it is Mendeleev's name which is now generally associated with the *periodic table of the elements*. He recognized the importance of leaving gaps in the pattern for elements which had not yet been discovered. This enabled him to predict, with remarkable accuracy, the existence of germanium, which was to occupy a gap in the family that starts with carbon and silicon. His name for this missing element was eka-silicon.

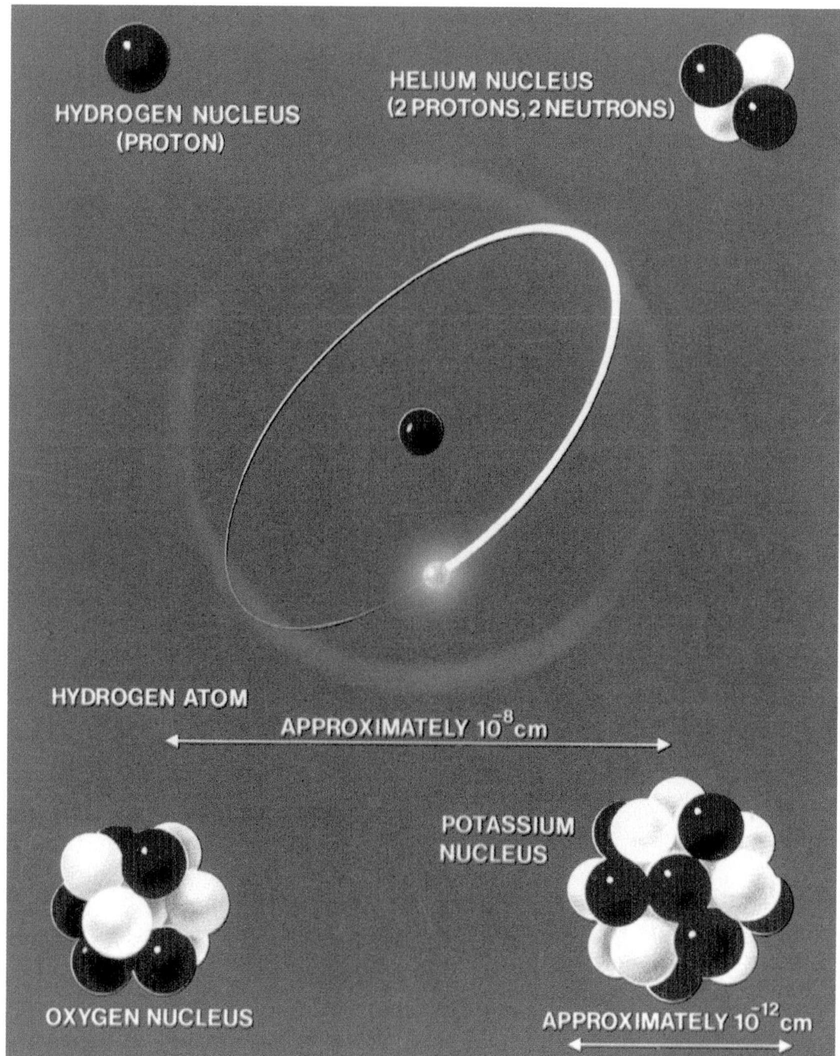

The hydrogen atom consists of one negatively charged electron in orbit around a single positively charged proton. The nuclei of all other elements are heavier, and they contain both neutrons and protons. The number of the latter determines the atomic number, and this also equals the number of electrons in an electrically neutral atom.

With the approach of the twentieth century, the underlying cause of the arrangement of the elements in the periodic table became a major challenge. The generally held view of atoms was that expressed by Isaac Newton, who envisaged them as 'hard, impenetrable, movable particles', but evidence began to accumulate which showed the atom to have an internal structure. It was ultimately to transpire that this structure varies from one element to another. Several of the key discoveries owed much to a simple device developed by William Crookes. This was a glass tube into which wires had been sealed at both ends. When most of the air is pumped out of such a tube, and the wires are connected to a high-voltage electrical supply, a *fluorescent glow* appears at the inner surface of the glass. This discharge appears when the pressure falls below about one hundredth of an atmosphere (0.01 atm), and it is now widely exploited in illumination and advertizing. Crookes demonstrated that the glow is caused by an emission from the wire having the negative polarity: the cathode. Such *cathode rays*, as they became called, are themselves invisible; they yield light only through their influence on matter. In 1895, Wilhelm Röntgen discovered that cathode rays give rise to other

Units

Système International d'Unités is an internationally agreed coherent system of units which is now in use for all scientific purposes. The seven basic units are as follows:

Property	Unit	Symbol
length	metre	m
mass	kilogram	kg
time	second	s
electric current	ampere	A
temperature	kelvin	K
amount of substance	mole	mol
luminous intensity	candela	cd

The secondary units derived from these basic quantities include those given below:

Property	Unit	Symbol
frequency	hertz	Hz
force	newton	N
energy	joule	J
power	watt	W
pressure	pascal	Pa
luminous flux	lumen	lm
illuminance	lux	lx
electric charge	coulomb	C
electric potential	volt	V
electric capacitance	farad	F
electric resistance	ohm	Ω
electric conductance	siemens	S
inductance	henry	H
magnetic flux	weber	Wb
magnetic flux density	tesla	T

invisible rays when they strike a solid target. These new rays were capable of passing through moderate thicknesses of matter such as human flesh. Being unable to identify the nature of the emissions, Röntgen called them *X-rays*. They are now a familiar medical aid, of course. Shortly thereafter, Henri Becquerel discovered that potassium uranyl sulphate continuously emits rays, some of which are even more penetrating than X-rays. We will return to these spontaneous emissions, known as *radioactivity*, after considering another epoch-making discovery afforded by Crookes's device.

Joseph (J. J.) Thomson and John Townsend modified a Crookes tube by making the anode – the conductor with the positive polarity – in the form of a plate with a small hole. Some of the cathode rays, which had been shown to travel in straight lines, passed through the hole and hit a fluorescent glass plate. The latter, which was the ancestor of today's television screen, enabled Thomson and Townsend to study the influence of electric and magnetic fields on the cathode rays. They were thereby able first to show that the rays are particles bearing a negative electrical charge, a result independently obtained by Jean Perrin, and later to measure the ratio of their charge to their mass. Following G. Johnstone Stoney, who was the first to postulate the existence of discrete units of electrical charge, Thomson called the particles *electrons*. When Robert Millikan later made an absolute determination of their charge (by ingeniously balancing the electrical and gravitational forces on a charged oil drop), it was possible to calculate the electron's mass and show that it is about 2000 times smaller than that of the smallest atom, hydrogen. The first subatomic particle had been discovered; atoms did indeed have an internal structure.

Several years later, Henry Moseley went on to show that the number of electrons in each element's atoms is identical to that element's *atomic number*; one for hydrogen, two for helium, three for lithium, and so on. But there was a more immediate problem: electrons are negatively charged but atoms are electrically neutral. There must also be positive charges, but it was not clear how either they or the electrons were arranged within the atom. Thomson himself imagined the electrons dotted in a positive ball, like

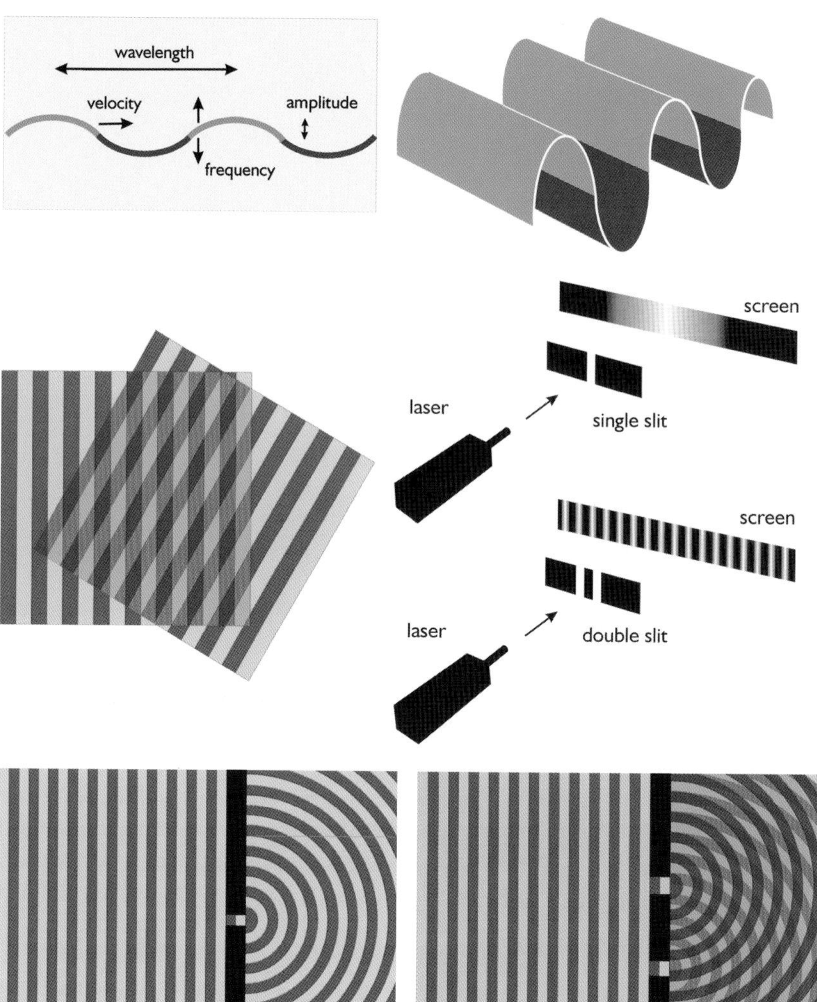

The defining characteristic of *wave motion* is its periodic variation of amplitude with time and position. At any instant, a wave consists of a series of crests (indicated by the green colour) and valleys (shown in red). The wave nature of light was originally proposed by Christiaan Huygens (1629–1695), and it was first demonstrated by Thomas Young (1773–1829), in 1801, who used a double-slit device and obtained an *interference pattern* by making the two waves mutually overlap. This took the form of alternating light bands, at the positions where the two wave trains reinforced one another, and dark bands, where they cancelled. The basis of such reinforcement and cancellation is indicated in the middle-left picture, which involves ideally flat wave fronts.

raisins in a cake, while Philipp Lenard believed the positive and negative charges to be grouped in pairs. It is particularly instructive to consider the inspired guess of Hantaro Nagaoka, who, in 1904, saw the atom as being a minute replica of the planet Saturn, which is so heavy that it can attract and hold its rings in orbit. He might also have compared the atom to our solar system, in which the central mass is the Sun, around which rotate the relatively small planets Mercury, Venus, Earth, Mars, and so on. The outermost planet, Pluto, is roughly six thousand million kilometres from the Sun. This makes the solar system about a hundred thousand million million million times larger than a single atom. The diameter of the smallest atom, that of the element hydrogen, is approximately one ten thousand millionth of a metre, while the diameter of the *nucleus* that lies at its centre is smaller by a factor a hundred thousand than this. But it is not only size which separates the cosmic and atomic concepts. Nagaoka's conjecture, brilliant as it was, suffered from a fundamental inadequacy, which we must soon discuss. Before this, however, we should consider how it was possible to establish that the atom does have a rather open structure, reminiscent of the solar system. This brings us back to radioactivity.

Atomic-scale units
Although the *Système International d'Unités* is now the only official set, some other units still find widespread use, primarily because they involve conveniently small numbers at the atomic scale. An example is the ångström unit (symbol Å), which is 10^{-10} m. Named for Anders Ångström (1814–1874), it is still popular because atomic diameters are approximately 1 Å. Similarly, the dalton, named after John Dalton (1766–1844), is convenient for molecular masses because the mass of a hydrogen atom is approximately one of these units.

Following Becquerel's original observations, Pierre and Marie Curie investigated uranium by studying one of its ores: *pitchblende*. They discovered it to contain traces of a second element, which they called radium; it too emitted radiation, and more intensely than uranium itself. Intriguingly, the radium seemed to lose weight as it gave off its rays; atoms were apparently not as permanent as had been assumed. It soon became clear that radioactivity manifested itself in at least three forms of emission, which were dubbed alpha, beta, and gamma. *Alpha rays* can travel through about 6 cm of air, at atmospheric pressure, or through about one page of this book. The beta variety is about a hundred times more penetrating, it requiring about half a millimetre (0.5 mm) of aluminium foil to reduce its intensity to half. *Gamma rays*, discovered by Paul Villard, can pass through many centimetres of even a dense metal such as lead. By probing their motions under the influence of a magnetic field, it was found that alpha rays are positively charged and that *beta rays* are negatively charged, while gamma rays possess no charge. Ernest Rutherford and his colleagues used an approach similar to that of Thomson and Townsend, and showed the alpha variety to consist of particles having four times the mass of a hydrogen atom and two units of positive charge; they are in fact identical to helium atoms that have lost their two electrons. It had become clear that radioactive rays are emissions of *subatomic particles*, and Rutherford realized their potential as probes of *atomic structure*.

We can now appreciate how it was possible to establish that the atom is a rather open entity. The critical experiments were carried out by Rutherford and his colleagues Hans Geiger and Ernest Marsden, in the period 1907–1911. They fired a beam of alpha particles at a thin gold foil and found that most of them passed right through this target. The much smaller electrons present in the gold clearly offered only minor resistance. But about one alpha particle in 10 000 was deflected through a large angle, some of these probing projectiles actually rebounding towards the alpha-particle source. As Rutherford noted, *it was like firing an artillery shell at a piece of tissue paper and having it come back and hit you*. The atom was clearly a tenuous structure with most of its mass concentrated in a heavy nucleus at the centre. It is interesting to note that the path of a comet can show somewhat similar behaviour, entering the solar system and leaving in the reverse direction, after being deflected by the massive central body: the Sun. But the *forces* involved are quite different; in this celestial analogy we have *gravitational attraction* whereas the alpha-particle rebounds were caused by *electrical repulsion* because the nucleus too bears a positive charge.

Once it had been established that an atom consists of small negatively charged electrons moving around a far more massive and positively charged nucleus, it was apparent that the students of atomic structure had been posed a difficult problem. The theory of moving electrical charges, which had been founded by James Maxwell a few decades earlier, predicted that an electron must radiate energy if it is diverted from a straight path. Just as the oscillating electrical charge in a radio antenna gives off *radiant energy*, so should the orbiting electrons in an atom gradually radiate their energy away and spiral in towards the nucleus. How could atoms be stable? What is it that keeps the electrons moving in their orbits around the nucleus? And there was another experimental fact that had never been explained. It concerned the spectra of *optical radiation*: the light emitted from atoms when they are in what is known as the *excited state*. If a piece of string is dipped first into a concentrated solution of table salt and then held in a naked flame, it is

Planck's constant
Max Planck (1858–1947) put forward his law of electromagnetic radiation in 1900. It states that the energy of such radiation is composed of discrete amounts known as *quanta*. The magnitude of one *quantum* is the product of Planck's constant (symbol h), 6.626×10^{-34} Js (joule seconds), and the frequency of the radiation. Typical frequencies of infrared (heat), visible, and ultraviolet radiations are 10^{14} Hz, 10^{15} Hz, and 10^{16} Hz, respectively. The corresponding quantum energies are about 6.6×10^{-20} J, 6.6×10^{-19} J, and 6.6×10^{-18} J. The stronger types of bonds between atoms have energies in the range 2×10^{-19} to 10^{-18} J. Under certain conditions, therefore, these bonds are vulnerable to modification by single quanta of ultraviolet and even visible radiation. Ultraviolet radiation is capable of altering the structure of some polymers, including those found in human skin. Heat quanta can achieve this only if they can be made to act cooperatively.

observed to emit a bright yellow light. If a barium salt is used, a light green colour is seen, and with a strontium salt the colour is a beautiful red. If the emitted light is analyzed, by making it pass through a prism, it is found that the *spectral lines* are very sharp and uniquely associated with the element in question. They constitute, so to speak, the element's fingerprint. These two problems are related. If an electron spiralled in towards the nucleus, the energy of the emitted radiation would equal the change in the energy of the electron in its motion around the nucleus. Since this energy would change as the distance between electron and nucleus decreased, the anticipated spiralling towards the nucleus should cause emission of radiation with an ever-changing energy. Now the colour of emitted light is determined by its energy, so the collapsing atom should emit light of a constantly changing hue. This conflicts with the salted string and flame experiment. Since only specific colours are emitted from atoms of a given element, only certain changes in the energy of motion of the electrons appear to be permitted. Of all possible *orbits* of an electron around the nucleus only certain special ones occur in practice. This appears to make the atomic domain quite different from the motion of an artificial satellite around the Earth. The small booster rockets attached to these devices can be used to vary the orbit continuously, and there seems to be no obvious limit to the fineness with which the orbit can be adjusted. The truth is that the discrete nature of energy also applies to macroscopic systems, but its consequences are much too small to be observable.

The distinction between continuous and discrete energy levels is so important that the point is worth illustrating further. An object can have both *kinetic energy* and *potential energy*, the former being associated with its motion and the latter with its position relative to a *force field*. Consider the force of *gravity*. The potential energy of a body depends on its height above the surface of the Earth. Thus a person standing on a chair has a higher potential energy than one of the same *mass* standing on the ground in the same location. If the person on the chair jumps off, part of the potential energy will be converted into kinetic energy as he or she falls towards the ground. During the descent, the person's height above ground level continually decreases, so the potential energy diminishes. A simple illustration of the difference between discrete and continuous changes of potential energy is afforded by the playground slide. One mounts this by climbing a ladder and then slides down the polished surface so as to return to the ground. The ladder provides only a discrete number of positions above ground level at which the climber can remain stationary and have only potential energy. These are the rungs, and they are associated with stability. The slide, on the other hand, represents a continuous set of potential energy states, and it is associated with kinetic energy and a lack of stability.

It was Niels Bohr who first realized that only certain discrete electron orbits can be stable, and in order to appreciate what led him to this great breakthrough, in 1913, it will be necessary to make a rather wide digression. In 1666, Isaac Newton had advanced a theory that light consists of particles which move in straight lines, at a velocity that depends on both their colour and the medium through which they are travelling. This view held sway until Thomas Young, in 1801, performed an experiment in which light emerging from one slit was made to illuminate two other slits, beyond which was placed a screen. One might expect this arrangement to produce two lines of light on the screen, one corresponding to each of the two secondary slits.

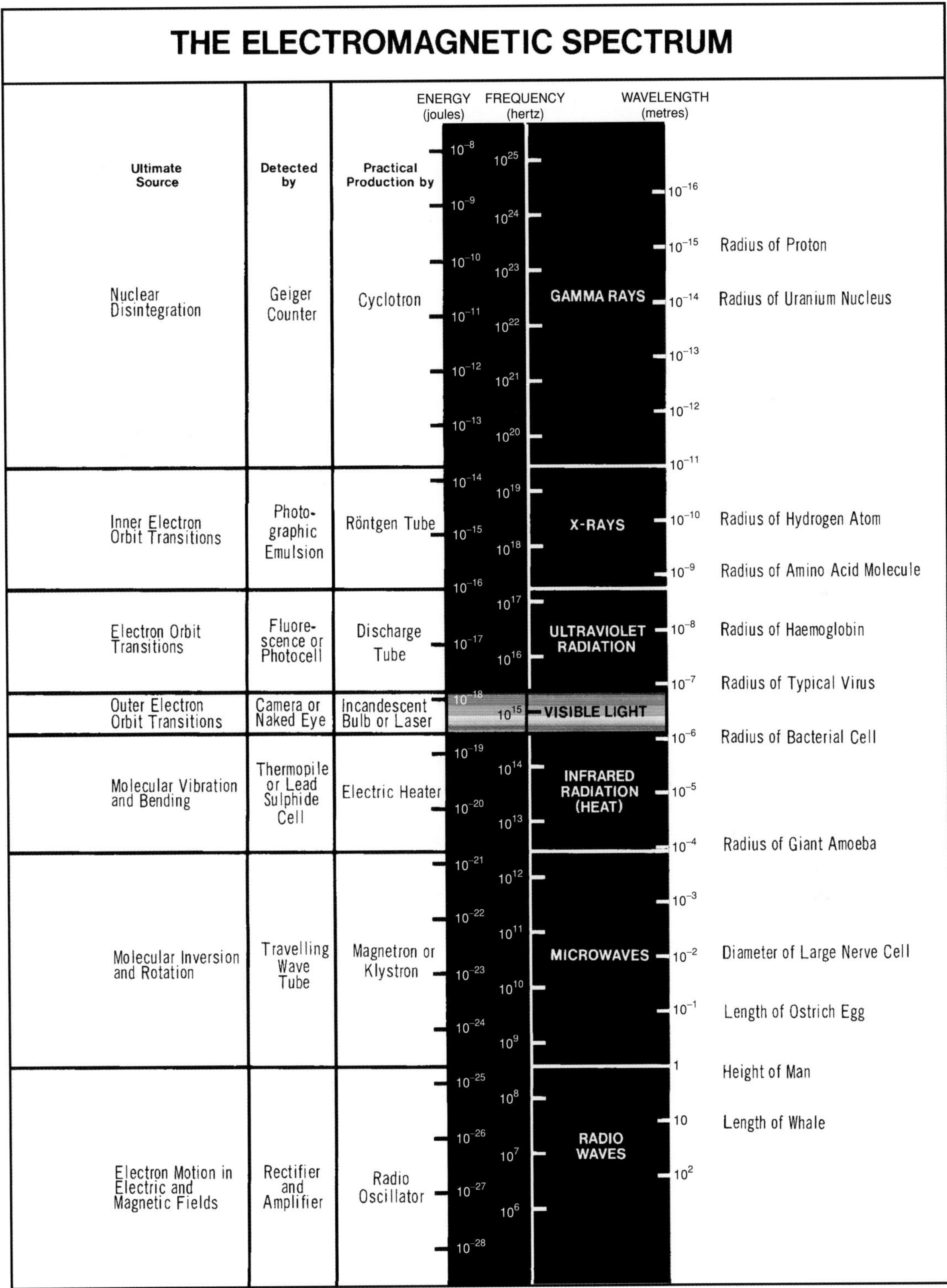

THE ELECTROMAGNETIC SPECTRUM
Ultimate Source
Detected by
Practical Production by
ENERGY (joules)
FREQUENCY (hertz)
WAVELENGTH (metres)
Nuclear Disintegration
Geiger Counter
Cyclotron
GAMMA RAYS
Inner Electron Orbit Transitions
Photographic Emulsion
Röntgen Tube
X-RAYS
Electron Orbit Transitions
Fluorescence or Photocell
Discharge Tube
ULTRAVIOLET RADIATION
Outer Electron Orbit Transitions
Camera or Naked Eye
Incandescent Bulb or Laser
VISIBLE LIGHT
Molecular Vibration and Bending
Thermopile or Lead Sulphide Cell
Electric Heater
INFRARED RADIATION (HEAT)
Molecular Inversion and Rotation
Travelling Wave Tube
Magnetron or Klystron
MICROWAVES
Electron Motion in Electric and Magnetic Fields
Rectifier and Amplifier
Radio Oscillator
RADIO WAVES
10^{-8} 10^{-9} 10^{-10} 10^{-11} 10^{-12} 10^{-13} 10^{-14} 10^{-15} 10^{-16} 10^{-17} 10^{-18} 10^{-19} 10^{-20} 10^{-21} 10^{-22} 10^{-23} 10^{-24} 10^{-25} 10^{-26} 10^{-27} 10^{-28}
10^{25} 10^{24} 10^{23} 10^{22} 10^{21} 10^{20} 10^{19} 10^{18} 10^{17} 10^{16} 10^{15} 10^{14} 10^{13} 10^{12} 10^{11} 10^{10} 10^{9} 10^{8} 10^{7} 10^{6}
10^{-16}
10^{-15} Radius of Proton
10^{-14} Radius of Uranium Nucleus
10^{-13}
10^{-12}
10^{-11}
10^{-10} Radius of Hydrogen Atom
10^{-9} Radius of Amino Acid Molecule
10^{-8} Radius of Haemoglobin
10^{-7} Radius of Typical Virus
10^{-6} Radius of Bacterial Cell
10^{-5}
10^{-4} Radius of Giant Amoeba
10^{-3}
10^{-2} Diameter of Large Nerve Cell
10^{-1} Length of Ostrich Egg
1 Height of Man
10 Length of Whale
10^{2}

All radiations can be described in terms of waves, and many familiar varieties, such as radio waves and heat, are different manifestations of the same basic type: *electromagnetic radiation*. Only a small part of the electromagnetic spectrum falls in the visible region.

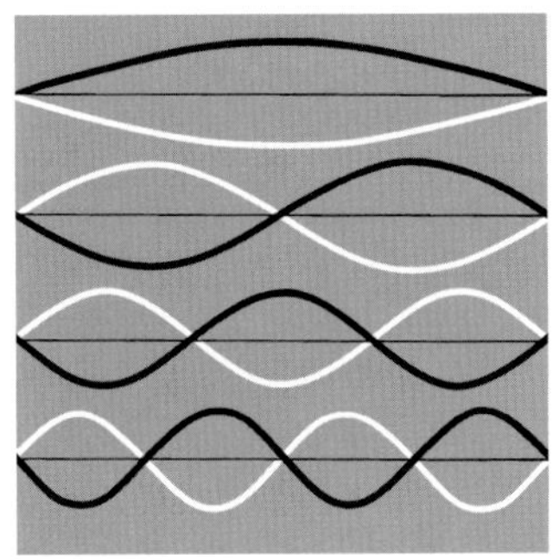

The vibrations of a system can be characterized by the number and positions of the *nodes* (i.e. locations of zero vibrational amplitude). In a one-dimensional system such as a guitar string, these nodes are points. For *the fundamental*, which corresponds to the lowest possible note or frequency, there are just two nodes; one at each end. For the first, second, and third *harmonics*, there are three, four, and five nodes respectively, and the frequency increases as the wavelength becomes progressively shorter.

Instead, one sees a series of alternating light and dark bands, known as *interference fringes*. Young recognized that this is similar to what is observed if one throws two stones simultaneously into water. A series of waves, with successive peaks and valleys, is generated by each stone. Where the two wave systems overlap, one sees secondary peaks of double height where two primary peaks coincide, secondary valleys of double depth where two primary valleys coincide, and cancellation where a primary peak coincides with a primary valley. Young's fringes thus led him to conclude that light is a *wave motion*, in direct contradiction to Newton's theory. This dichotomy was not resolved until over a hundred years later. Then, in a remarkable series of advances during the early decades of the twentieth century, the two different views were reconciled and incorporated into modern *atomic theory*.

Twentieth-century physics was born, appropriately enough, in the year 1900, with the pioneering hypothesis of Max Planck. To appreciate his achievement, another digression is necessary. We must consider what is expected of radiation consisting of waves, as advocated by Young, and begin by defining the characteristic properties of a wave. The *wavelength* is the distance between two successive peaks, or wave crests. If a stone is thrown into water near a floating cork, the cork is observed to bob up and down at a certain frequency, and this is simply the *wave frequency*. The range of the fluctuation in the cork's height, from the mean level to the extreme, is the *wave amplitude*. A wave also has velocity. This is the rate at which a surf-rider travels, for example. Now it turns out that all radiation can be described in terms of waves. The things that distinguish the different types of radiation are the wavelength, the frequency, which is inversely related to wavelength, the *wave velocity*, and the way in which the radiation propagates. *Radio and radar waves* have wavelengths in excess of 1 mm (millimetre), while heat (*infrared*) radiation had a wavelength between 1 mm and 1 μm (micrometre). For visible and *ultraviolet* light the range is from 1 μm to about 1 nm (nanometre), while for X-rays the limits are from 1 nm to 10 pm (picometres). Gamma rays have wavelengths that lie between about ten picometres and 10 fm (femtometres). Finally, there are *cosmic rays* with wavelengths less than 10 fm. The entire spectrum is continuous, one type changing smoothly to the next at these limits. All of these are basically manifestations of the same phenomenon, known as *electromagnetic radiation*, which travels at a velocity of 300×10^6 m/s. Electromagnetic radiation can even travel through a vacuum, and this distinguishes it from sound, which is passed along by atoms. If an artificial satellite exploded in space, we might see a flash, but there would be no bang.

Consider the possible vibrations of a string on an instrument such as a guitar. It is held fixed at either end, and when it vibrates, the amplitude at the ends is zero. These fixed points are the nut and the bridge, and they are referred to as *nodes*. If the string is plucked at its middle, the position of maximum amplitude lies exactly halfway between nut and bridge, and the note played is known as the fundamental. If the string is depressed at its middle and plucked halfway between that point and the bridge, a new tone is heard lying just one octave higher than the original note. This is the first harmonic, and its wavelength is only half that of the *fundamental*. In this case only half the string is in vibration. A skilled guitar player can, however, remove the first finger so quickly after the string is plucked that the *first harmonic* is played with the entire string in motion. There

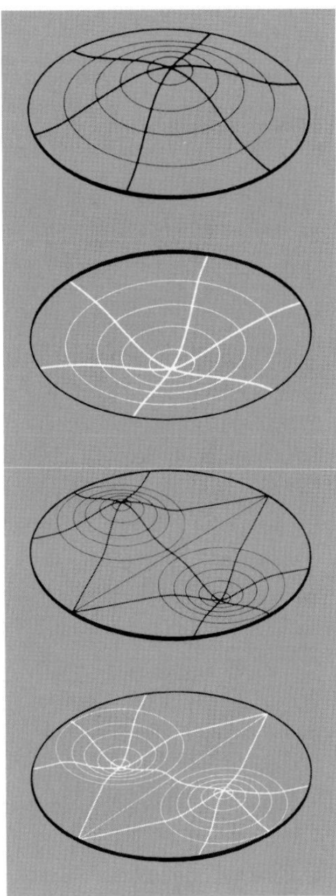

In a two-dimensional system, such as a drum skin, the circumference constitutes a node, and this is the only node for the fundamental vibration (top two diagrams). For the first harmonic (bottom two diagrams) there is an extra node lying along a diameter, and higher harmonics are associated with further nodal lines.

are now three nodes: one at the nut, one at the bridge, and one halfway between.

Lord Rayleigh (John William Strutt) and James Jeans studied the properties of radiation at the end of the nineteenth century. The essence of their analysis hinged on the fact that a vibrating string has two halves, four quarters, eight eighths, and so on. Since the intensity of emitted sound is related to the number of vibrating segments, we should expect the volume of sound to rise with increasing pitch. More generally, this means that intensity should rise with increasing frequency, and therefore with decreasing wavelength. Up to a point, this is exactly what is observed. When a piece of metal is heated, it glows visibly, giving off light energy. The analysis predicts that it should emit a lower intensity of infrared radiation, because infrared wavelengths are longer than visible wavelengths. This is indeed the case. By the same token, the intensity of ultraviolet emission should be greater than the intensity of the visible emission, and the X-ray and gamma-ray intensities should be greater still. It is here that the Rayleigh–Jeans theory failed. It encountered what became known as the *ultraviolet catastrophe*. There is a marked falling away of intensity for the ultraviolet and all lower wavelengths. This was a catastrophe for a theory, but not for living organisms; a red-hot poker would otherwise have been a lethal weapon, emitting X-rays and gamma rays.

Max Planck resolved the difficulty by assuming that radiant energy is not continuous and that it comes in discrete amounts; it too has its atomicity. These packets of energy became known as *quanta* and he postulated that their energy is proportional to the frequency. Planck showed that although the Rayleigh–Jeans analysis is adequate at low frequency, it becomes increasingly difficult for a heated solid to concentrate sufficient energy to emit quanta when the frequency becomes large. The intensity therefore falls off, this becoming quite noticeable when the ultraviolet range is reached. In suggesting that light energy comes in packets, Planck had reverted to the Newtonian corpuscular theory, and the issue appeared to be clinched, a few years later, by Albert Einstein's explanation of the *photo-electric effect*. If a metallic surface is illuminated with radiation, it gives off electrons, producing a detectable current. This is the case only if the wavelength of the radiation is less than a certain value, that is to say only if its energy is greater than a certain level. This threshold energy is related to the strength of the bond by which each electron is attached to the surface. Planck's packets of energy had been shown to exist, and Einstein had added the important point that the quantum remains a packet as it travels.

As attractive as Planck's *quantum theory* was, the wave experiments of Young – subsequently extended to the case of electrons by Clinton Davisson, Lester Germer and George Thomson – were not to be denied. The *wave nature of light* can be readily demonstrated by, for example, observing the fringes when a distant street lamp is viewed through the fine mesh of a bird's feather or an umbrella. The vital reconciliation between the two opposing viewpoints was established in 1923 by Louis de Broglie, who suggested that all matter simultaneously has both wave and particle properties. The implication of this *wave–particle duality* is that radiation and matter are equivalent. Because all radiation is associated with a frequency and hence an energy, as noted earlier, these ideas fitted in nicely with the epoch-making discovery by Albert Einstein, that matter and energy are equivalent. Those two familiar principles, the indestructibility of matter and the conservation of energy, are

The stability of atoms, and the sharp spectral lines of their electromagnetic wave emissions, were explained by Niels Bohr (1885–1962) in 1913. Using the quantum ideas of Max Planck, he argued that only certain discrete orbits are possible, and showed that each spectral line corresponds to the jumping of an electron from one orbit to another.

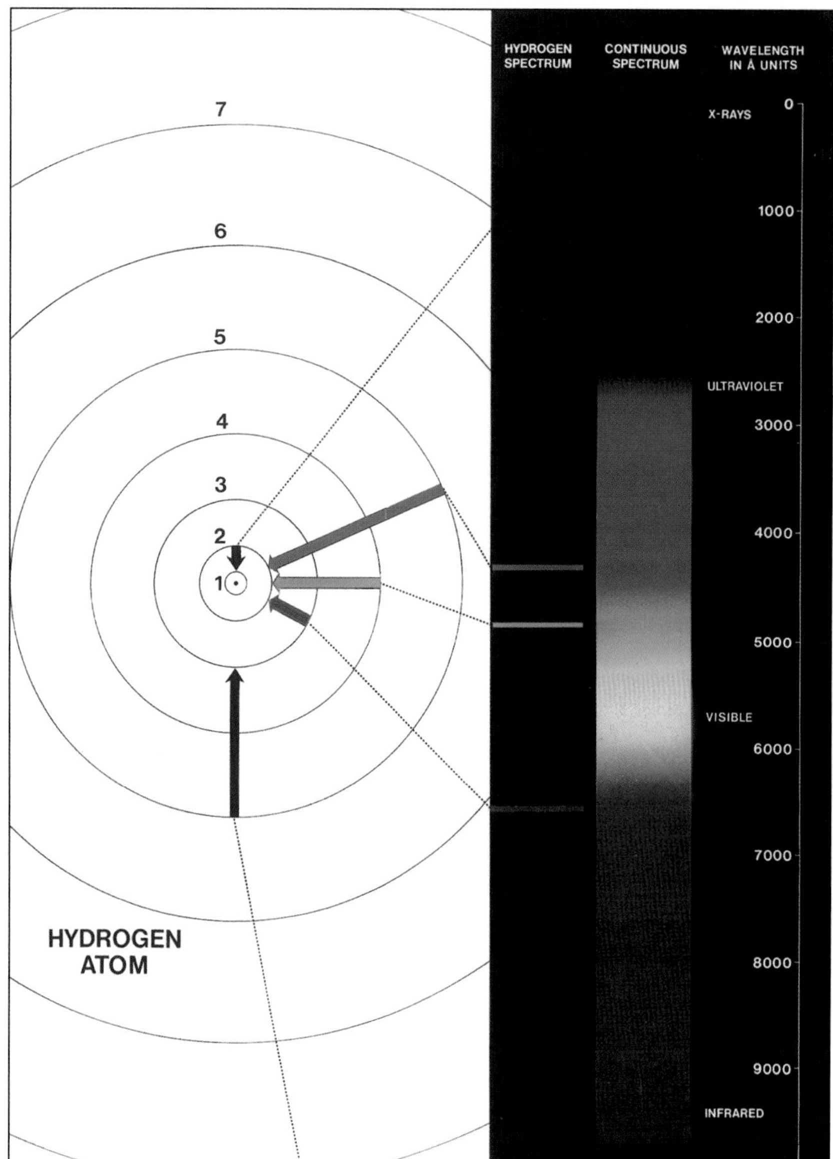

not strictly observed. They are merely manifestations of the same physical law, which stipulates that the total amount of matter and energy remains constant.

The acceptance of wave–particle duality requires far more than a disposition for compromise. It demands no less than a fundamental change in our way of looking at nature. If each elementary packet of matter is simultaneously a particle and a wave, we might wonder about the size of the particle and the wavelength of the wave. The duality actually precludes precise replies to either of these questions. Consider first the case in which the wave aspect is emphasized. Accurate measurement of wavelength requires as many waves as possible, in order to get good statistics, but the number of peaks and valleys is limited by the particle's size. In other words, the particle aspect limits the accuracy with which the wavelength can be determined, and the smaller the particle, the greater is the uncertainty. Conversely, if

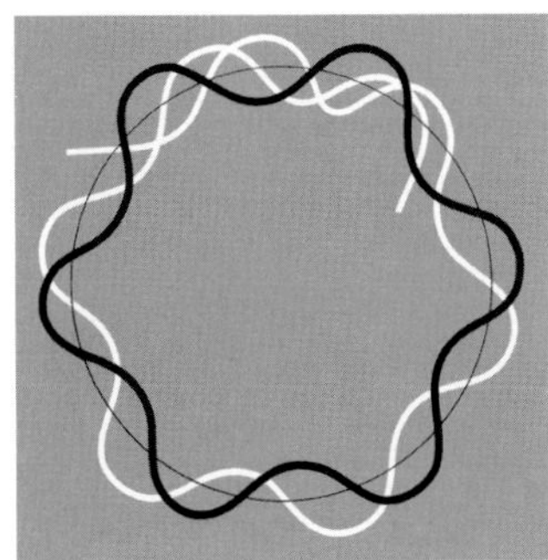

Louis de Broglie (1892–1987) attempted to reconcile the Bohr model of atomic orbits with wave–particle duality by suggesting that only those orbits which comprise a whole number of wavelengths around their orbital paths are permitted.

the particle aspect is stressed, the wave aspect imposes uncertainty in position because of its undulations of amplitude. It was Werner Heisenberg, in 1927, who first recognized this fundamental uncertainty in our ability to define the position and wavelength simultaneously; the more accurately we try to measure the one quantity, the greater is the uncertainty that must be tolerated in the other.

The capstone to these developments was provided by Max Born, who argued that the correct interpretation of the *wave intensity* (i.e. the square of the wave amplitude – which is positive, of course) is that it is a measure of the probability of finding the corresponding particle at the position in question. Thus although we must give up any hope of being able to make precise statements about a particle, its behaviour is not governed solely by whim. The picture of matter became a statistical one, the actual statistics being determined by the prevailing conditions. The situation is not unlike that which confronts the supplier of ready-made clothing to a chain of department stores. He does not need to know the chest, waist, hip, and arm measurements of any individual. The statistics of these measurements in the population as a whole are sufficient. Having to forego the possibility of making exact statements about a single particle is not really a disadvantage, because experimental situations invariably involve vast numbers of these entities.

We can now return to the question of atomic structure and the theory of Niels Bohr. He realized that the packaging of energy into Planck's quanta automatically leads to stable electron orbits in an atom. And he made a bold assumption: when an electron is orbiting around a nucleus it does not radiate energy, even though it is describing a curved path. He further assumed that, instead of a continuous series of states, only certain *discrete energy levels* are available to an electron, and these are separated by amounts equivalent to a single energy quantum. An electron can be excited from one state into a higher state by absorbing such a quantum, and it will decay back to the lower state by re-emitting a quantum of the same energy. Many of these quantum emissions give light in the visible region, and because they have definite energies they correspond to a particular colour. These are indeed the *sharp spectral lines* referred to earlier. An atom containing several electrons arranged in different energy levels can be made to emit a quantum by the forcible ejection of one of the electrons in an inner, low energy, level. This can be achieved in a target that is bombarded with electrons, and it makes an electron in the adjacent higher level unstable against decay with associated quantum emission. The emitted quantum in this case is an *X-ray*. The subsequent advances, by Louis de Broglie and Erwin Schrödinger, brought modifications to Bohr's model. The de Broglie theory required replacement of the point-like electron by a wave, and successively higher energy states permitted increasingly higher numbers of nodes to be accommodated in the corresponding orbit. One can imagine this arrangement as bending the guitar string, referred to earlier, into a circle and making the two ends coincide. Finally, Schrödinger replaced these simple waves by a series of diffuse clouds. These are to be regarded as three-dimensional waves and, as will be seen later, they too have characteristic nodes. Following the suggestion of Max Born, the intensity of these clouds is a measure of the *electron probability density*.

It is not difficult to see what role the uncertainty principle plays in smearing out a well-defined electron orbit into the cloud required by Schrödinger's

Quantum numbers

The microscopic state of a system is specified by *quantum numbers*. For an electron within an atom, four such numbers are required. These are the *principal quantum number, n*, which defines the energy level or shell that the electron occupies; the *azimuthal quantum number, l*, which determines the shape and multiplicity of the orbit within the shell; the *magnetic quantum number*, m_l, which determines the orientation of that orbit with respect to a strong magnetic field; and the *spin quantum number*, m_s, which determines the direction of the electron's spin in a magnetic field.

The principal quantum number can be any positive integer: 1, 2, 3, 4 . . . The azimuthal number can have any positive integer value up to and including $n-1$. The magnetic numbers are positive or negative integers lying in the range $-l$ to $+l$, while the spin quantum number can only be $+1/2$ or $-1/2$.

As an example of the use of these rules, we find that there are eight distinct combinations for the state $n = 2$. These are

$l = 0$	$m_l = 0$	$m_s = 1/2$
$l = 0$	$m_l = 0$	$m_s = 1/2$
$l = 1$	$m_l = 1$	$m_s = 1/2$
$l = 1$	$m_l = 1$	$m_s = -1/2$
$l = 1$	$m_l = 0$	$m_s = 1/2$
$l = 1$	$m_l = 0$	$m_s = -1/2$
$l = 1$	$m_l = -1$	$m_s = 1/2$
$l = 1$	$m_l = -1$	$m_s = -1/2$

quantum treatment. When an electron is at a relatively large distance from the nucleus, say half a nanometre, the two particles mutually attract. This *electrostatic force* makes the electron move towards the nucleus at an ever-increasing speed. As the distance between nucleus and electron decreases, the position of the latter becomes more precisely defined, and its kinetic energy must rise by the uncertainty principle. A balancing of the potential and kinetic energies occurs for electron–nucleus distances of the order of a tenth of a nanometre, or one *ångström unit*, and this is therefore the approximate radius of the atom. To illustrate the point further, let us compare an orbiting electron with an artificial Earth satellite. One could measure the position of the latter by making it reflect radar waves. Because the satellite is a relatively massive object, any recoil due to the radar beam would be quite negligible. Now suppose we try to do a similar thing with the orbiting electron. Radiation could in principle be reflected from it, in order to establish its position. But the energy of the radiation would not be negligible compared with the kinetic energy of the electron itself. In trying to locate the electron we would run the risk of changing its energy. This is the general problem that confronts any attempt at simultaneous measurement of two properties, such as energy and position, of particles having atomic dimensions, and it is the epitome of the Heisenberg *uncertainty principle*. At the macroscopic level accessible to our senses, such effects are far too small to be observed. A defeated tennis player cannot invoke Heisenberg's principle as an excuse; the uncertainty in position of a ball travelling at 50 m/s is a mere 10^{-34} m.

The atomic weight invoked by Newlands and Mendeleev and the atomic number rationalized by Moseley are not the same thing. Both rise systematically as we proceed through the *periodic table*, and the atomic number is a direct indication of an element's position in the table, but weight and

In the quantum mechanical model of the atom proposed by Erwin Schrödinger (1887–1961) in 1926, the electrons appear neither as particles nor waves, but as both simultaneously. As shown by Max Born (1882–1970), the probability of finding an electron at a given position, with respect to the nucleus, is determined by the wave intensity. This, in turn, is related to the wave amplitude, the form of which is governed by three-dimensional nodal surfaces.

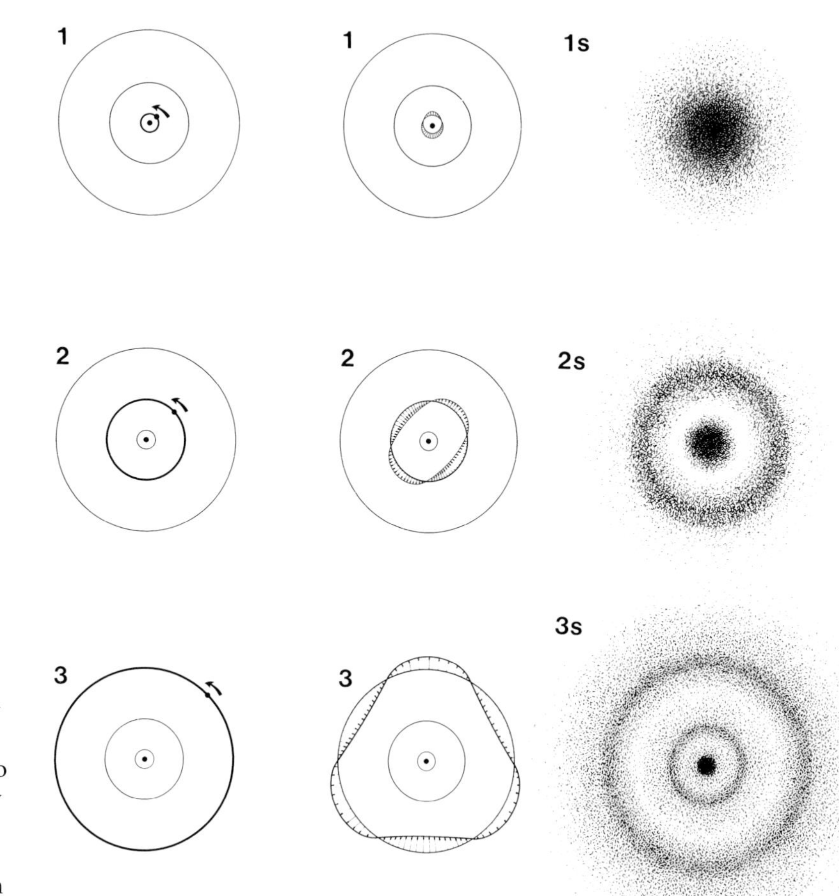

number are not simply related. To understand the disparity, we must return to the question of radioactivity, and to the studies of Rutherford and his colleagues in 1919. They fired *alpha particles*, from a radium source, through a tube containing pure nitrogen gas, and checked for possible emissions with the usual fluorescent screen. Particles were being given off which travelled farther than the alpha variety, and the influence on their trajectories of a magnetic field showed them to be equivalent to the positively charged nuclei of hydrogen, now known as *protons*. But no hydrogen had been present in the tube. The protons were apparently being knocked out of the nitrogen nuclei. The experiments of Becquerel and the Curies had shown that certain special nuclei spontaneously disintegrate; it now became clear that stable nuclei can also be induced to break up. And the original nitrogen nuclei were thereby converted to carbon nuclei; the ancient dream of alchemists had been realized: the *artificial transmutation* of one element into another. The nucleus clearly had its own composite structure and the difference between atomic weight and atomic number was soon explained; the balance between weight (mass) and charge is made up by *neutrons*.

The atomic nucleus is thus composed of a number of elementary particles known as protons and neutrons. Just as with that other particle, the electron,

According to Bohr's theory, the electrons in an unexcited atom are in their (lowest-energy) ground state. In 1923, Satyendra Bose (1894–1974) and Albert Einstein (1879–1955) predicted that there should be an analogous ground state for a group of atoms, but calculations indicated that such a *Bose–Einstein condensate* would exist only at temperatures below about 100 billionths of a degree above the absolute zero temperature (see Chapter 3). In 1995, research teams led by Eric Cornell and Carl Wieman, using about 2000 rubidium atoms, and Wolfgang Ketterle, using about 10 000 000 sodium atoms, succeeded in producing such condensates. Both teams cooled their group of atoms first by retarding the individual thermal motions with laser beams and then by letting the most energetic atoms evaporate, while the remainder of the group was confined in a magnetic trap. (The latter cooling is similar to that observed in a cup of beverage, due to the escape of the hottest atoms in the form of steam.) The sequence shown here indicates the velocity distribution in a 200 × 270 micrometre cloud of rubidium atoms just before (left, at a temperature of 400 nanokelvin) and just after (centre, at 200 nK) condensation, the right-hand picture (at 50 nK) showing the nearly pure condensate formed after further evaporation. (The colour indicates the number of atoms at each velocity, red and white corresponding to the fewest and the most, respectively.) The central peak, where the atoms are most densely packed and where they are barely moving, is not infinitely sharp because that would violate the uncertainty principle enunciated by Werner Heisenberg (1901–1976). The wave functions of the individual atoms in a condensate are all in phase with one another.

all protons are identical, and so are all neutrons. Protons and neutrons have approximately the same mass, 1.7×10^{-24} g, roughly 2000 times that of the electron. The diameters of the proton and neutron can also be regarded as being essentially equal, the value being about 10^{-14} m. The diameter of the electron must be even smaller than this, but we have already seen that its exact size is not a particularly significant concept. The important difference between the proton and the neutron, which was discovered by James Chadwick in 1932, is that the former carries a positive electric charge, 1.6×10^{-19} C (coulombs), whereas the latter is electrically neutral. The

Radioactive isotopes

All atoms of the same element have the same *atomic number*, and therefore the same number of protons in their nuclei. But these nuclei may contain varying quantities of neutrons and thus differ in *mass number*, in which case they are referred to as different *isotopes* of the element. Some isotopes are unstable, and are said to be *radioactive*, their nuclei being subject to spontaneous disintegration, which is accompanied by the emission of one or more types of radiation. The most common emission is that of *beta particles*, and it occurs either when a neutron in the nucleus is converted to a proton and an electron, or when a proton is converted to a neutron, a *positron* (i.e. a positively charged electron), and a *neutrino*. In these cases, the respective beta emissions consist of electrons and positrons, and they clearly produce a change in the atomic number by one unit while leaving the mass number unaltered. *Alpha particles* comprise two neutrons and two protons, and they are emitted by certain isotopes of the heavier elements. When this occurs, the atomic number of the resulting (daughter) nucleus is two units less than that of the parent, while the mass number is reduced by four units. *Gamma rays* are emitted in conjunction with alpha or beta decay if the daughter nucleus is formed in the excited state (i.e. a state other than that corresponding to the lowest possible energy).

Radioactive isotopes find widespread use in medicine, technology, the physical and life sciences, and even in archaeology. In biology, for example, they are useful as tracers that can be employed as markers of specific chemical components. In the following list of common examples, the standard convention is used, the mass number appearing as a superscript and the atomic number as a subscript. The *half-value period*, or *half-life*, is the time taken for the activity of the isotope to decay to half its original value.

Isotope	Emission	Half-value period
${}^{3}_{1}H$	beta	12.26 years
${}^{14}_{6}C$	beta	5730 years
${}^{22}_{11}Na$	beta + gamma	2.60 years
${}^{32}_{15}P$	beta	14.28 days
${}^{35}_{16}S$	beta	87.9 days
${}^{42}_{19}K$	beta + gamma	12.36 hours
${}^{59}_{26}Fe$	beta + gamma	45.6 days
${}^{82}_{35}Br$	beta + gamma	35.34 hours
${}^{99}_{43}Tc$	beta + gamma	6.0 hours
${}^{125}_{53}I$	gamma	60.2 days
${}^{203}_{80}Hg$	beta + gamma	46.9 days

The case of ${}^{125}_{53}I$ is special in that it involves capture, by the nucleus, of an orbiting electron, and no beta ray is emitted.

Industrial application of radioactive isotopes includes thickness measurement of such products as metallic foil and sheet, paper, fabrics, and glass. Isotopes are also used in various measuring instruments, and they provide the source of ionization in fire-prevention equipment that is based on smoke detection. Another common application is the radiographic inspection of welded joints. The following list includes some of the most common industrial isotopes.

Isotope	Emission	Half-value period
${}^{60}_{27}Co$	beta + gamma	5.26 years
${}^{137}_{55}Cs$	beta + gamma	30.0 years
${}^{192}_{77}Ir$	beta + gamma	74.2 days
${}^{241}_{95}Am$	alpha + gamma	458 years

In radiocarbon dating, the age of objects having a biological origin is estimated by measuring their content of ${}^{14}_{6}C$. This is done by comparing the radioactive strength of a known mass of the object with that of a standard source. In 1946, Willard Libby (1908–1980) predicted that ${}^{14}_{6}C$ should be produced by the impact of cosmic rays on the Earth's atmospheric nitrogen. A very small proportion of the (stable) ${}^{14}_{7}N$ nuclei absorb a neutron and emit a proton, thereby becoming radioactive ${}^{14}_{6}C$ nuclei which decay with a half-value period of 5730 years. At the beginning of the twentieth century, the ratio of ${}^{14}_{6}C$ to the stable isotope ${}^{12}_{6}C$ was about 1.5×10^{-12} to 1, but this has been changed subsequently by a few per cent due to the burning of coal and oil. Some of the radioactive carbon atoms are absorbed by plants during photosynthesis, and thus find their way into most living things. When the organism dies it ceases to acquire ${}^{14}_{6}C$, and the radioactivity gradually diminishes. A measurement of the residual radioactivity therefore permits a determination of the age of an organism's remains. The method has been calibrated against specimens of wood taken from specific annual rings of trees felled at a known date. Checked against such ancient wood as that found in the tombs of the Pharaohs, the method has been found to be fairly reliable, and is now archaeology's most accurate dating technique.

negative charge of the electron is equal in magnitude to the charge of the proton: it is -1.6×10^{-19} C. Protons and neutrons are referred to collectively as *nucleons*. It is found that groups of up to about 250 of these nucleons can cluster together to form stable nuclei. For larger clusters the *inter-nucleon forces*, which will be discussed later, are insufficient to stop the nucleus from breaking apart. More specifically, a nucleus can be stable if it contains less than about 100 protons, and less than about 150 neutrons, there being the further requirement that the number of neutrons and protons must be very roughly equal. The nucleus of a helium atom, and thus also an alpha particle, comprises two protons and two neutrons.

All nuclei of atoms of the same chemical element contain an identical number of protons. This number must equal the number of electrons, in an electrically neutral atom; it is thus the *atomic number* of the element. Nuclei having the same number of protons but different numbers of neutrons are referred to as *isotopes* of the element in question. The total number of nucleons is called the *mass number*. This might appear to imply that the mass of an atomic nucleus is simply the mass number times the mass of a nucleon, but that would ignore the Einstein principle alluded to earlier. The point is that the energy with which the nucleons in a nucleus are bound together is itself equivalent to a certain amount of mass. When a number of nucleons come together to form a nucleus, therefore, one can imagine a fraction of their individual masses as being converted into a *binding energy*, and the nuclear mass is thereby lowered. The binding energy per nucleon is greatest for nuclei with mass number about 50. Nuclei with masses in excess of this have a lower stability because of the electrical repulsion between their protons, and at sufficiently large nuclear mass this manifests itself in *nuclear fission*: the splitting of the nucleus into two smaller, and more stable, nuclei with an associated release of energy. For nuclei with mass numbers below

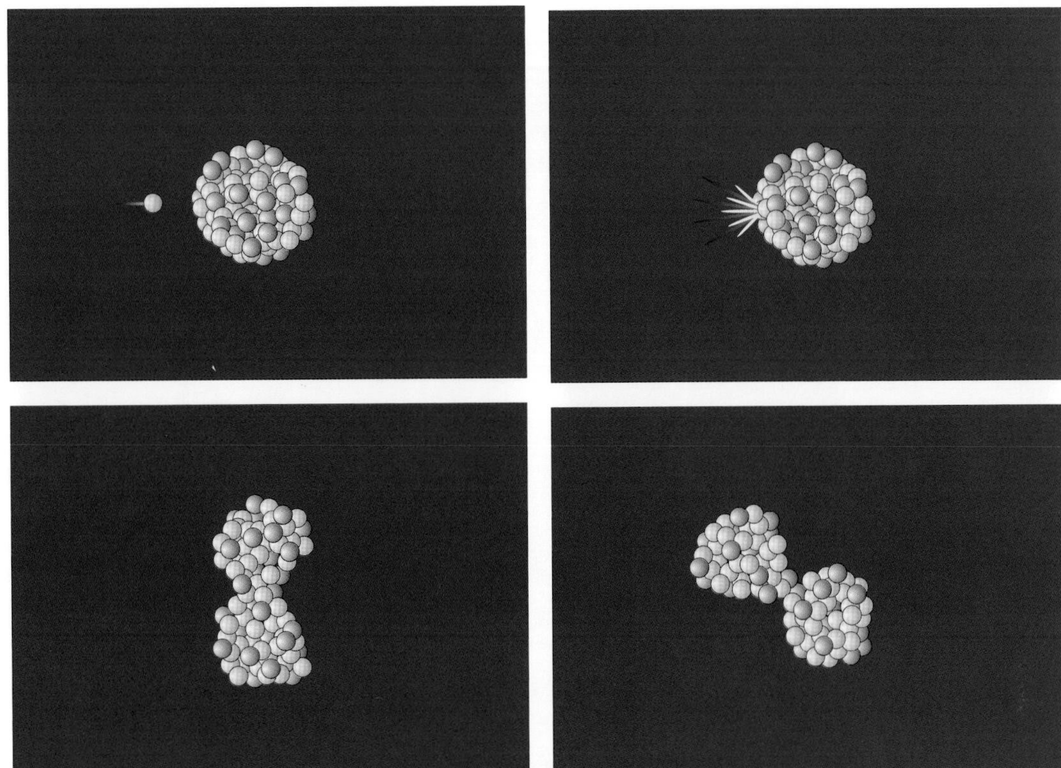

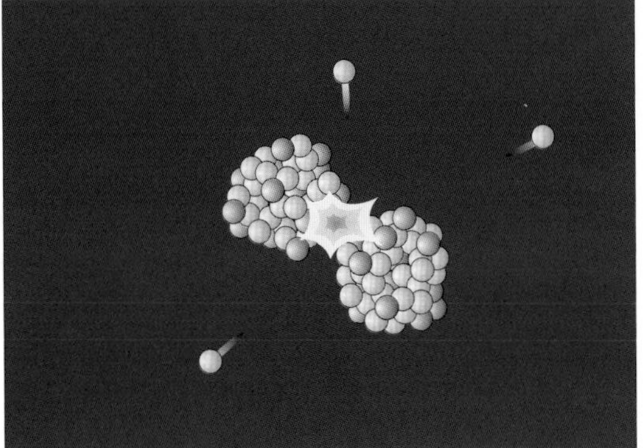

The fission of a uranium nucleus by a slow neutron, as originally observed in experiments performed in 1938 by Otto Hahn (1879–1968) and Fritz Strassmann (1902–1980), and explained shortly thereafter by Otto Frisch (1904–1979) and Lise Meitner (1878–1968). The original nucleus is deformed by the impact and, after a series of oscillations, becomes unstable against the mutual repulsion of its protons. The nucleus then breaks up into two smaller nuclei, known as fission fragments, and several fast-moving neutrons. As originally perceived by Christian Møller (1904–1981), a chain reaction can be obtained in a sufficiently dense packing of uranium atoms if the ejected neutrons are slowed down to fission-producing velocities by a moderator.

50 there is also a decreased binding energy per nucleon, this being due to the fact that such a large fraction of the nucleons are forced to lie at the nuclear surface, where they cannot make so many nucleon–nucleon bonds. In this case, greater stability can be achieved by consolidation of two smaller nuclei into a single larger nucleus: the energy-producing process of *nuclear fusion*.

The smallest atom is that of the element hydrogen. Its nucleus consists of a single proton, around which a lone electron is in orbit. Because the mass of an electron is negligible compared with that of a proton, the mass of the hydrogen atom is approximately one nucleon mass. It is customary to use the mass of the nucleon as the unit of atomic mass, so the mass of the hydrogen atom is approximately unity. This might lead one to expect all atomic masses to be roughly whole numbers, with small departures from exactly integer masses due to the binding-energy effect discussed earlier.

The High Flux Isotope Reactor at Oak Ridge National Laboratory, Tennessee. The blue glow is known as Cerenkov radiation – in honour of its discoverer, Pavel Cerenkov (1904–1990) – and is caused by particles that are moving faster than light photons would move in the same medium, i.e. the light water moderator.

But this is not the whole story. Looking at a table of atomic masses, we find that most elements have non-integer values. Copper, for example, has an atomic mass of 63.54. The reason for such fractioned values is that most elements can exist in a number of different stable *isotopes*. Thus the element copper, which has 29 protons, has isotopes with 34 and 36 neutrons. The corresponding total numbers of nucleons are, respectively, 63 and 65. These isotopes have different probabilities of occurrence in our terrestrial environment. The isotope containing 63 nucleons (atomic mass 62.9298) occurs with a probability of 69.09%, whereas for the isotope with 65 nucleons (atomic mass 64.9278) the figure is 30.91%. Averaging over the various isotopes of an element thus produces a non-integer atomic mass. Joseph Thomson had come to this conclusion in 1912, and confirmation was first provided by Francis Aston, who used the mass spectrograph he had invented

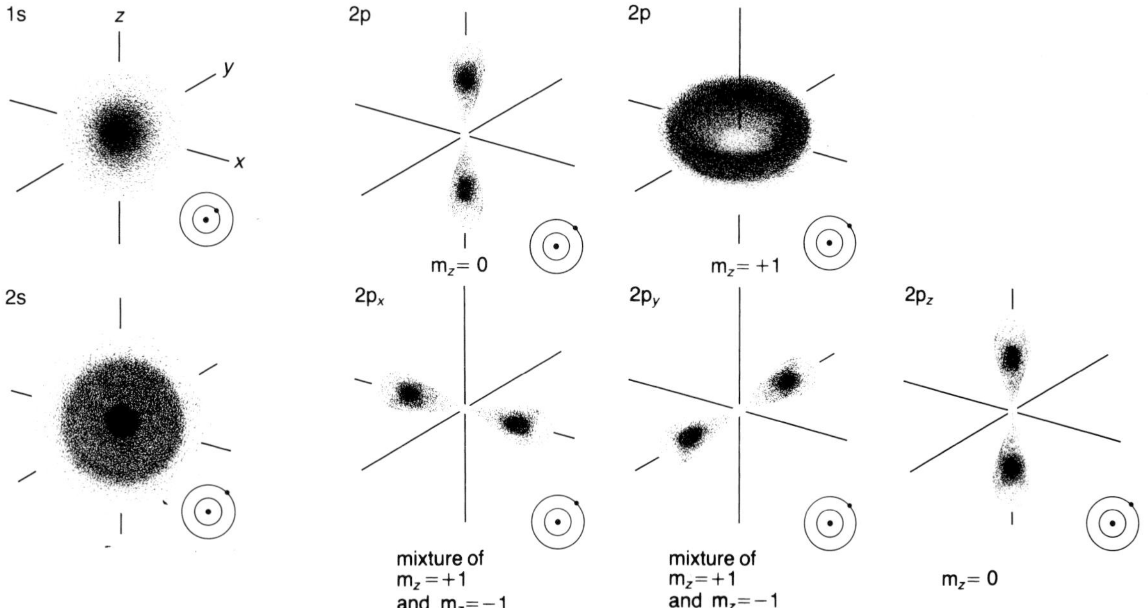

The lowest energy (ground) state for a hydrogen atom, the *1s* state, is characterized by a spherically-symmetric wave intensity with a single spherical nodal surface at infinity. The lowest energy excited state is the *2s*, and this has an additional spherical nodal surface centred on the nucleus. The *2p* states, which have slightly higher energy, have nodal surfaces which pass through the nucleus. There are three of them, corresponding to the three possible values of the magnetic quantum number m_l. States with lobes extending along one of the Cartesian axes can be obtained by mixing the three *2p* states in the correct proportions. The insets in this and later diagrams indicate schematically the equivalent Bohr orbits.

to demonstrate that there are two isotopes of neon: one with an atomic mass of 20 and the other of 22.

We still have the question of why nucleons are bound to one another. The attraction between the proton and the electron in a hydrogen atom is understandable because they have electrical charges of opposite sign. In the nucleus there are only positive charges, and one type of nucleon, the neutron, has no charge at all. The answer to this riddle was supplied by Hideki Yukawa, in 1935, and once again it involved a radical change of attitude. He suggested that nucleons are bound together through the act of constantly exchanging a particle between themselves. The idea seems less eccentric when one considers the simple analogy of two children playing catch with a ball. They soon discover that there is a certain distance between them which leads to the most comfortable game. If they are too far apart, too much effort must be expended in throwing, while if they are too close, they will have insufficient time for catching. The particle postulated by Yukawa to be the ball in this process was calculated to have a mass about ten times smaller than that of a nucleon. It was subsequently discovered and is now known as a *pi-meson* or *pion*. We have already encountered equivalence between particles and waves, and between matter and energy. To these we can now add equivalence between forces and particles.

Because it involves particle exchange, the nucleon–nucleon force is short-ranged and it is saturated, which means that it can be shared only among a limited number of nucleons. There is also the repulsion between protons, and because this repulsion is electric and therefore acts over much larger distances, the instability of all but a few selected isotopes is not really surprising. The modern theory of the nucleus, which is due to Aage Bohr, Ben Mottelson, and James Rainwater, again reveals a picture of duality; in some ways it acts like a drop of liquid, but other properties can only be explained on the basis of quantized waves. In 1933, Irène and Frédéric Joliot-Curie discovered that bombardment of stable isotopes with subatomic particles can lead to phenomena other than disintegration; such isotopes can also be

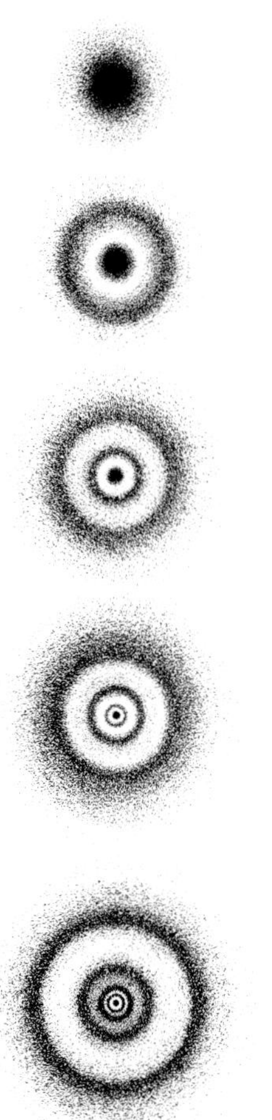

The outermost *s* states of the noble gas atoms, shown in cross-section. These are (from top to bottom) the *1s* state of helium, the *2s* state of neon, the *3s* of argon, *4s* of krypton, and *5s* of xenon. Notice that the principal quantum number equals the number of (spherical) nodal surfaces: one for helium, two for neon, and so on. The complete electron complement of xenon, for example, includes all the lower *s* states (i.e. *1s*, *2s*, *3s*, and *4s*) and also all the lower *p* states (*2p*, *3p*, etc.). The *4s* state of xenon is not the same as that of krypton because the nuclear charges of these two elements are different, so the forces on the respective electrons are not the same.

provoked into radioactivity. In the mildest cases, the nuclei are merely shaken by the incoming particle and the subsequent pulsations emit energy in the form of gamma rays. In the more violent cases, single nucleons, or even groups of nucleons such as the alpha particle, are liberated. These particle emissions have already been referred to by the collective term *radioactivity*. They manifest the usual uncertainty that must be tolerated at the atomic level. Data on the rate of disintegration of atomic nuclei are analogous to the actuarial tables of the life-insurance company. The irrelevance of the fate of a particular individual is not a sign of callous indifference on the part of such a company. It is simply a reflection of the fact that statistical information is all that is required to run the business. Similarly, although it is not possible to predict exactly when any particular atomic nucleus will disintegrate, one can make precise calculations of just what fraction of the atoms in a large sample will decay in a given time. The time required for half the atoms of a radioactive isotope to break up is referred to as the *half-value period* and also, unfortunately, by the misnomer *half-life*.

One mode of nuclear decay has come to have special significance. It was discovered by Otto Hahn and Fritz Strassmann in 1938, and it involved hitting the nuclei of uranium atoms with slow neutrons. Subsequent chemical analysis detected the presence of barium, even though there had been none of that element present in the target before the experiment. Now the atomic weight of barium is only about half that of uranium, and the inescapable conclusion was that some of the uranium nuclei had been split in two. The observation of this process, the *nuclear fission* referred to earlier, was about as surprising as would have been the cleaving of a loaf of bread by hitting it with a crumb. Otto Frisch and his aunt, Lise Meitner, realized that fission is the result of combined effects of the repulsion between the protons and the distortion of the nucleus due to the neutron's impact, and they also saw that the process would involve a large release of energy because of the great strength of nuclear forces. The fissioning of the uranium nucleus also produces several free neutrons. The energy with which the latter are ejected is much higher than that required to provoke further fissions in other nuclei, but they can readily be slowed down to the right energy by making them collide with the light atoms of what is known as a *moderator*, such as graphite. This is the principle of the *chain reaction* exploited in the *nuclear reactor*, the first example of which was put into operation in 1942 by Enrico Fermi and his colleagues.

The significance for the theory of atomic structure of the discoveries described so far is clear, but we should also contemplate their impact on society at large. The discovery of the electron led to rapid advances in radio techniques and later to television. It also gave birth to *solid state electronics*, and to the *transistor* and modern *computer circuitry*. In the early 1930s, new microscopes were developed that were based on electron waves rather than light waves. The smallest detail that can be perceived with any microscope is dictated by the wavelength of the rays employed. In an *electron microscope* – invented in 1931 by Max Knoll and Ernst Ruska – this can be comparable to atomic diameters, so it is clear that such an instrument has a much higher resolution than its more common optical counterpart. Electron microscopes have, consequently, greatly increased our knowledge of materials at the microscopic level. Similarly, radioactivity has provided us with a unique tool. Used to label particular chemical components, it is widely employed in biochemical detective work, as we shall see in later chapters.

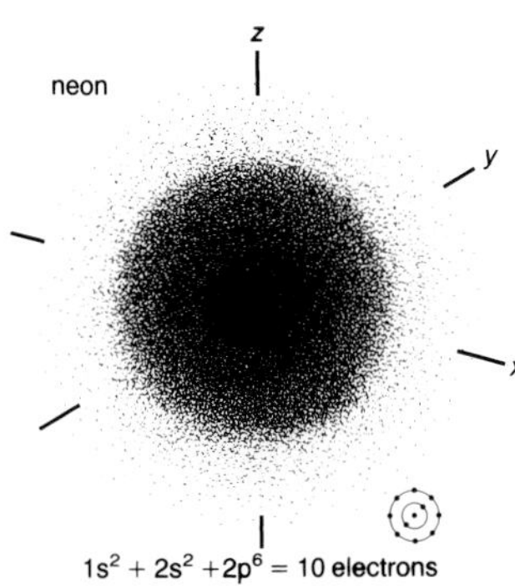

The electron probability density in the neutral neon atom comprises contributions from the *1s*, *2s*, and *2p* states. Because the combined wave intensity of the individual states is spherically symmetric for a complete shell, the electron distribution is non-directional in neon, and in all other neutral noble gas atoms.

We can now return to the question of atomic structure and consider the proton and electron in hydrogen. Because the charges of these two particles are equal in magnitude but opposite in sign, the atom as a whole is electrically neutral. Indeed, overall electrical neutrality is looked upon as being the natural state of all isolated atoms. This means that if the nucleus has 10 protons, as in the case of neon, there will be 10 electrons deployed in orbits around the nucleus. Similarly, a neutral atom of gold has 79 protons in its nucleus (and anywhere between 81 and 92 neutrons) and 79 orbiting electrons. If one electron is removed from a neutral gold atom, 78 will be left in orbit, and there will be a single positive proton charge on the nucleus that is not balanced out. Such an uncompensated atom is said to be *ionized*. Doubly and triply ionized atoms result from the removal of a second and third electron, and so on. *Ionized atoms* will be discussed in the next chapter in connection with the ways in which groups of atoms get together to form the condensed (liquid and solid) states of matter.

We turn finally to the actual electronic structure of atoms, as dictated by the mechanical rules peculiar to the quantum domain. Schrödinger showed that the variation of electron probability density is characterized by nodes. The situation is thus similar to a (one-dimensional) guitar string and a (two-dimensional) drum skin, in which variations in vibrational amplitude are determined by the nodal points and lines. In three dimensions there are nodal surfaces, at which the electron probability density is zero. One thing can be stated immediately: because an electron must be located somewhere or other in space, there will be a nodal surface at infinity. In practice, the electron probability density falls essentially to zero within a few tenths of a nanometre, so it is a good approximation to imagine a spherical nodal surface surrounding each nucleus at roughly this distance. In the case of the vibrating guitar string, we can adequately describe the situation by stating the number of nodal points: two for the fundamental, three for the first harmonic, and so on. The vibrations of the two-dimensional drum skin are more complicated, and a single number is insufficient to uniquely characterize the pattern of nodal points and lines. For electron orbits around a nucleus it is found that a complete description requires no less than four different numbers, which are known as *quantum numbers*.

The need for as many as four numbers to describe the pattern of electron clouds in an atom seems less surprising when we consider the variety of possible symmetries in three dimensions. The plant world offers such examples as the concentric spherical layers of the onion, the crescent-shaped segments of the orange, and the numerous conical divisions of the blackberry. Three of the four quantum numbers can be looked upon as a shorthand recipe for slicing up the space around the nucleus. The *principal quantum number* is equal to the total number of nodal surfaces, both planar and spherical, and it essentially determines the energy of the electron. The greater the number of nodes, the smaller is the corresponding electron wavelength and, by the uncertainty principle, the larger is the energy. The *azimuthal quantum number* determines the number of nodal planes passing through the centre of the atom. If it is zero, the electron probability density is spherically symmetrical and takes the form of a fuzzy ball-shaped cloud around the nucleus. A zero azimuthal number combined with principal numbers of two or more gives a set of concentric spheres which do indeed resemble an onion. If the azimuthal number is not zero, the electron probability clouds are concentrated out along certain directions, like the petals of a flower. This

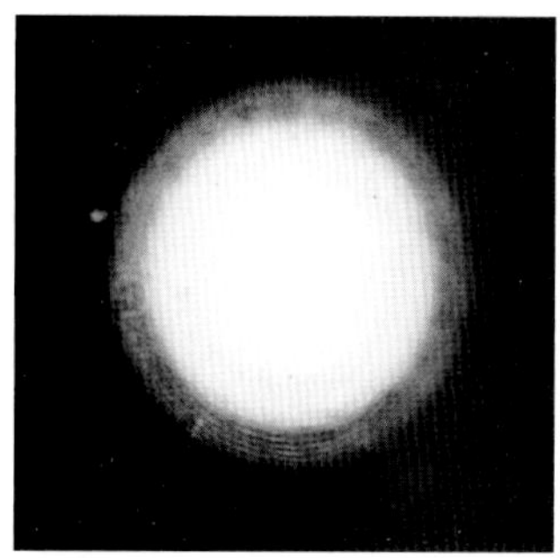

The quantum theory of atomic structure permits one to calculate the probability that an electron will be located at a given position relative to the atomic nucleus, but it cannot make precise predictions about the location of any specific particle. In this picture of the electron distribution in argon, taken by holography in an electron microscope, the variation of photographic intensity is proportional to the electron probability density. The magnification is approximately one thousand million times.

is the origin of *directional bonding* between atoms, which will be discussed in the next chapter. The third member of the quartet is the *magnetic quantum number*. It determines the number of different possible orientations of the non-spherical electron probability clouds. Orientation is important only in relation to externally imposed directionality such as that of a *magnetic field*, and hence the name of this third quantum number. So far, only the orbital motion of an electron around the considerably heavier nucleus has been considered. But the electron can also spin about a line passing through itself, in much the same way as the Earth daily spins about its axis while performing its annual motion around the Sun. It is the direction of the Earth's spin that dictates that the Sun rises in the East. If it had the opposite spin, the Sun would rise in the West. The spinning motion of the electron gives rise to the fourth and last of the quantum numbers. It simply determines the direction in which the electron is spinning about its own axis, and it can thus have only two different values. *Electron spin* was discovered by Samuel Goudsmit and George Uhlenbeck in 1925.

The energy of an electron when it occupies an orbit around an atomic nucleus is determined primarily by the principal quantum number, and the higher this number is, the greater is the energy. Let us again consider an atom of the lightest element, hydrogen. It has a single electron spinning and rotating about the simplest of atomic nuclei, a single proton. The lowest energy state, known as the *ground state*, corresponds to the lowest possible principal quantum number: one. The electron probability cloud then has the fuzzy sphere form. If a hydrogen atom is bathed in light of the correct colour, it can absorb a quantum packet of light energy, also known as a *photon*, and its electron will jump into a higher energy state. The atom will have become excited, and the electron probability cloud is inevitably changed. One possibility is that it maintains its spherical symmetry but acquires one or more extra layers, like an apple becoming an onion. Alternatively, the probability cloud can develop lobes which fan out from the nucleus like arms reaching into the surrounding space.

Let us consider the common situation in which a single atom contains a number of electrons. Until now we have encountered nothing that would prevent all electrons from seeking the lowest energy level and therefore entering the same cloud-like probability orbit around the nucleus. It turns out, however, that this is not possible. Such a situation would be in violation of Wolfgang Pauli's *exclusion principle*, expounded in 1925, which states that no two electrons can have exactly the same set of values for the four quantum numbers. As more and more electrons are added to an atom, they are forced to occupy quantum states of ever-increasing energy. One can compare the energy levels of an atom with the apartments in a high-rise residential building and with everyone trying to live as close to the ground as possible.

We can now discuss the atomic structure of different elements. Mendeleev's periodic table arranges these in a systematic fashion according to increasing atomic number. We have already seen that when the nucleus contains a single proton, the atomic number is one, and the element is hydrogen. A nucleus containing two protons gives helium. For three protons we have lithium, for four protons we have beryllium, and so on. In the lowest energy state for the single electron in hydrogen, the probability density has the spherical distribution in space described above. Now in

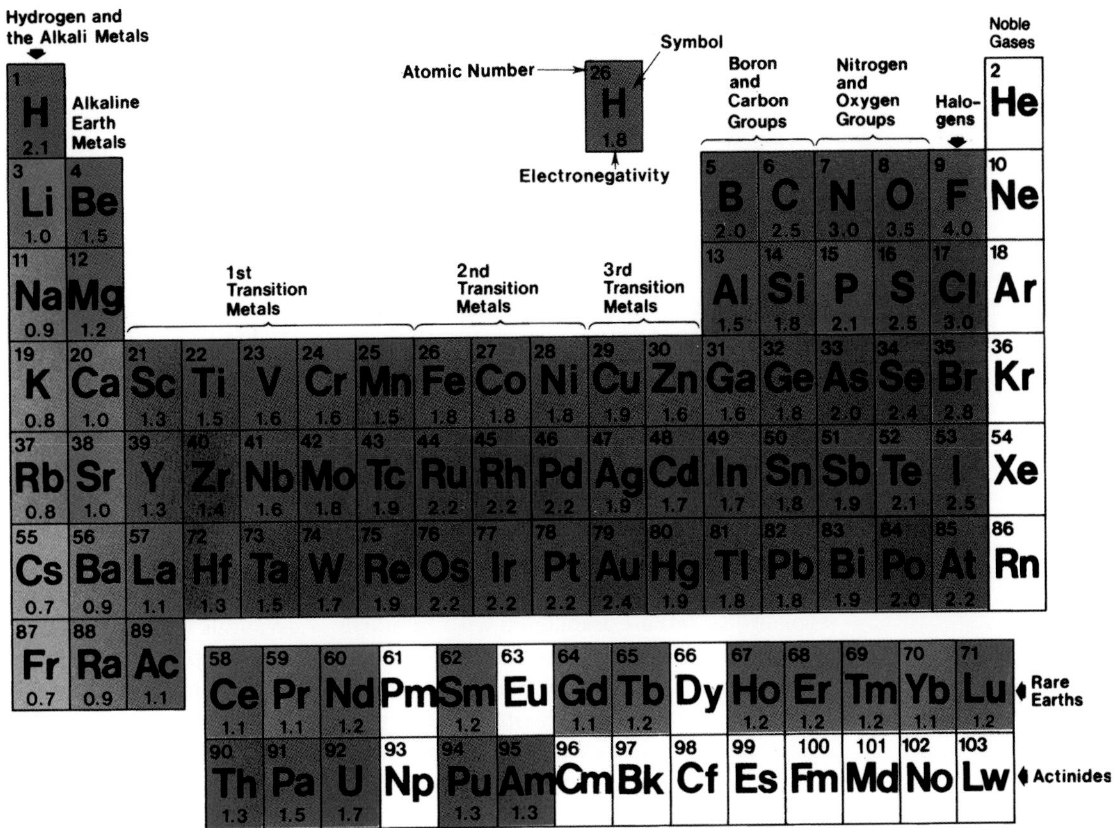

The periodic table of the elements. The order of filling up the electron shells begins with *1s, 2s, 2p, 3s, 3p*, but the sequence then goes *4s, 3d, 4p, 5s, 4d, 5p*, and this produces the transition groups. The sequence is further disrupted in the rare earths.

the absence of external stimuli it turns out that a change in direction of electron spin does not change the energy. When a second electron is present, therefore, as in helium, this too can occupy the lowest energy state as long as it has the opposite spin of the first electron. In such a case, where a single energy level can be occupied by more than one electron, the level is said to be *degenerate*. Because of spin, all atomic energy levels have a degeneracy of at least two: the apartments in our atomic residential building are designed for couples.

Because there are only two different *spin quantum numbers*, the possibilities for putting electrons into the lowest energy state are already exhausted with helium. In lithium, the Pauli exclusion principle requires that for the third electron another of the four quantum numbers also be changed. This means a probability cloud with a symmetry different from that of the first and second electrons. Lithium's third electron is therefore forced to enter a level with a higher energy. At this point it is convenient to introduce the abbreviated nomenclature usually employed in atomic physics to denote the different orbital states. The ones with the simplest form, the spherical fuzzy clouds with various numbers of layers, are called s states, while increasing numbers of cleavages through the centre give the p, d, and f states. These designations are derived from the sets of *spectral lines* to which the different quantum states correspond, namely the **s**harp, **p**rincipal, **d**iffuse, and **f**ine, the initial letter being used in each case. The two lowest energy electrons in the lithium atom occupy 1s orbitals. The number indicates the value of the principal quantum number, which should not be confused with the principal spectral lines. We have to decide what to do with lithium's third electron. It turns out that the next lowest energy is the 2s. Hence lithium has two electrons with opposing spins in the 1s level and one in the 2s. This situation is usually written in the shorthand form $1s^2 2s^1$.

In beryllium the fourth electron can also occupy the 2s level provided its spin opposes that of the electron already in that level. Beryllium's electron configuration is thus $1s^2 2s^2$, and both the 1s and 2s levels are now saturated. For boron, therefore, a new level must be used. It is the 2p, and it is the lowest level to display the non-spherical lobe structure described earlier. Because the azimuthal quantum number can now take on three different values, the 2p level can hold up to six electrons. These are systematically added in the succession of elements stretching from boron, $1s^2 2s^2 2p^1$, to neon, $1s^2 2s^2 2p^6$. The element having eleven electrons, sodium, must have its eleventh electron in a new level, and this is the 3s. States having the same principal quantum number are said to occupy the same *shell*, and the letters s, p, and so on, are associated with *sub-shells*.

Magnesium, with its twelve electrons, has two in the 3s state, while the next element, aluminium, has two electrons in the 3s and one in the 3p state. We see that an interesting pattern begins to emerge as electrons are systematically added, progressing through the table of elements. At regular intervals there is an element for which the highest energy level has just its correct complement of electrons. This corresponds to an orbit becoming saturated with electrons or, as it is usually put, a *shell of electrons* becoming complete. Elements for which this is the case are remarkable for their low chemical activity, and they are known as the *inert, or noble, gases*. All elements with an atomic number just one above an inert gas have a single electron in an otherwise empty shell. These are the alkali metals, and they all have similar chemical properties. Indeed it is found in general that elements having the same number of electrons in their outermost shell show similar chemical behaviour. This is the origin of the regularity that led Newlands and Mendeleev to their independent formulations of the periodic table. The explanation of the systematic variation of chemical properties through the orbital structure of the different elements was one of the crowning achievements of twentieth-century science: it united physics and chemistry.

Summary

The idea that matter ultimately consists of microscopic fundamental particles was already familiar in ancient Greek philosophy, and the word atom comes from the Greek term for indivisible. Twenty-three centuries passed before the nature of atoms was clarified, and they were found to be surprisingly open structures, superficially reminiscent of our solar system. The relatively massive nucleus, containing protons and neutrons, bears a positive electric charge, while the much smaller orbiting electrons are negatively charged. Atomic diameters are typically a one hundred millionth of a centimetre, and this microscopic domain is governed by its own set of

The periodic table of the elements

Element	Chemical symbol	Electron configuration	Element	Chemical symbol	Electron configuration	Element	Chemical symbol	Electron configuration
hydrogen	H	Is^1	krypton	Kr	$[Ar]3d^{10}4s^2 4p^6$	lutetium	Lu	$[Xe]4f^{14}5d^1 6s^2$
helium	He	Is^2	rubidium	Rh	$[Kr]5s^1$	hafnium	Hf	$[Xe]4f^{14}5d^2 6s^2$
lithium	Li	$Is^2 2s^1$	strontium	Sr	$[Kr]5s^2$	tantalum	Ta	$[Xe]4f^{14}5d^3 6s^2$
beryllium	Be	$Is^2 2s^2$	yttrium	Y	$[Kr]4d^1 5s^2$	tungsten	W	$[Xe]4f^{14}5d^4 6s^2$
boron	B	$Is^2 2s^2 2p^1$	zirconium	Zr	$[Kr]4d^2 5s^2$	rhenium	Re	$[Xe]4f^{14}5d^5 6s^2$
carbon	C	$Is^2 2s^2 2p^2$	niobium	Nb	$[Kr]4d^4 5s^1$	osmium	Os	$[Xe]4f^{14}5d^6 6s^2$
nitrogen	N	$Is^2 2s^2 2p^3$	molybdenum	Mo	$[Kr]4d^5 5s^1$	iridium	Ir	$[Xe]4f^{14}5d^7 6s^2$
oxygen	O	$Is^2 2s^2 2p^4$	technetium	Tc	$[Kr]4d^5 5s^2$	platinum	Pt	$[Xe]4f^{14}5d^9 6s^1$
fluorine	F	$Is^2 2s^2 2p^5$	ruthenium	Ru	$[Kr]4d^7 5s^1$	gold	Au	$[Xe]4f^{14}5d^{10}6s^1$
neon	Ne	$Is^2 2s^2 2p^6$	rhodium	Rh	$[Kr]4d^8 5s^1$	mercury	Hg	$[Xe]4f^{14}5d^{10}6s^2$
sodium	Na	$[Ne]\ 3s^1$	palladium	Pd	$[Kr]4d^{10}5s^0$	thallium	Tl	$[Xe]4f^{14}5d^{10}6s^2 6p^1$
magnesium	Mg	$[Ne]\ 3s^2$	silver	Ag	$[Kr]4d^{10}5s^1$	lead	Pb	$[Xe]4f^{14}5d^{10}6s^2 6p^2$
aluminium	Al	$[Ne]3s^2 3p^1$	cadmium	Cd	$[Kr]4d^{10}5s^2$	bismuth	Bi	$[Xe]4f^{14}5d^{10}6s^2 6p^3$
silicon	Si	$[Ne]3s^2 3p^2$	indium	In	$[Kr]4d^{10}5s^2 5p^1$	polonium	Po	$[Xe]4f^{14}5d^{15}6s^2 6p^4$
phosphorus	P	$[Ne]3s^2 3p^3$	tin	Sn	$[Kr]4d^{10}5s^2 5p^2$	astatine	At	$[Xe]4f^{14}5d^{10}6s^2 6p^5$
sulphur	S	$[Ne]3s^2 3p^4$	antimony	Sb	$[Kr]4d^{10}5s^2 5p^3$	radon	Rn	$[Xe]4f^{14}5d^{10}6s^2 6p^6$
chlorine	Cl	$[Ne]3s^2 3p^5$	tellurium	Te	$[Kr]4d^{10}5s^2 5p^4$	francium	Fr	$[Rn]\ 7s^1$
argon	Ar	$[Ne]3s^2 3p^6$	iodine	I	$[Kr]4d^{10}5s^2 5p^5$	radium	Ra	$[Rn]\ 7s^2$
potassium	K	$[Ar]4s^1$	xenon	Xe	$[Kr]4d^{10}5s^2 5p^6$	actinium	Ac	$[Rn]6d^1 7s^2$
calcium	Ca	$[Ar]4s^2$	aesium	Cs	$[Xe]6s^1$	thorium	Th	$[Rn]5f^0 6d^2 7s^2$
scandium	Sc	$[Ar]3d^1 4s^2$	barium	Ba	$[Xe]6s^2$	protoactinium	Pa	$[Rn]5f^2 6d^1 7s^2$
titanium	Ti	$[Ar]3d^2 4s^2$	lanthanum	La	$[Xe]5d^1 6s^2$	uranium	U	$[Rn]5f^3 6d^1 7s^2$
vanadium	V	$[Ar]3d^3 4s^2$	cerium	Ce	$[Xe]4f^2 5d^0 6s^2$	neptunium	Np	$[Rn]5f^4 6d^1 7s^2$
chromium	Cr	$[Ar]3d^5 4s^1$	praseodymium	Pr	$[Xe]4f^3 5d^0 6s^2$	plutonium	Pu	$[Rn]5f^6 6d^0 7s^2$
manganese	Mn	$[Ar]3d^5 4s^2$	neodymium	Nd	$[Xe]4f^4 5d^0 6s^2$	americium	Am	$[Rn]5f^7 6d^0 7s^2$
iron	Fe	$[Ar]3d^6 4s^2$	promethium	Pm	$[Xe]4f^5 5d^0 6s^2$	curium	Cm	$[Rn]5f^7 6d^1 7s^2$
cobalt	Co	$[Ar]3d^7 4s^2$	samarium	Sm	$[Xe]4f^6 5d^0 6s^2$	berkelium	Bk	$[Rn]5f^9 6d^0 7s^2$
nickel	Ni	$[Ar]3d^5 4s^2$	europium	Eu	$[Xe]4f^7 5d^0 6s^2$	californium	Cf	$[Rn]5f^{10}6d^0 7s^2$
copper	Cu	$[Ar]3d^{10}4s^1$	gadolinium	Gd	$[Xe]4f^7 5d^1 6s^2$	einsteinium	Es	$[Rn]5f^{11}6d^0 7s^2$
zinc	Zn	$[Ar]3d^{10}4s^2$	terbium	Tb	$[Xe]4f^9 5d^0 6s^2$	fermium	Fm	$[Rn]5f^{12}6d^0 7s^2$
gallium	Ga	$[Ar]3d^{10}4s^2 4p^1$	dysprosium	Dy	$[Xe]4f^{10}5d^0 6s^2$	mendelevium	Md	$[Rn]5f^{13}6d^0 7s^2$
germanium	Ge	$[Ar]3d^{10}4s^2 4p^2$	holmium	Ho	$[Xe]4f^{11}5d^0 6s^2$	nobelium	No	$[Rn]5f^{14}6d^0 7s^2$
arsenic	As	$[Ar]3d^{10}4s^2 4p^3$	erbium	Er	$[Xe]4f^{12}5d^0 6s^2$	lawrencium	Lw	$[Rn]5f^{14}6d^1 7s^2$
selenium	Se	$[Ar]3d^{10}4s^2 4p^4$	thulium	Tm	$[Xe]4f^{13}5d^0 6s^2$			
bromine	Br	$[Ar]3d^{10}4s^2 4p^5$	ytterbium	Yb	$[Xe]4f^{14}5d^0 6s^2$			

Electron configuration and the periodic table of the elements
Pauli's exclusion principle – named for Wolfgang Pauli (1900–1958) – requires that no two electrons in an atom have the same values of the four quantum numbers. Because these numbers (see the first table) dictate the electron configuration, this means that each element must be characterized by a unique pattern of electron probability clouds around its nuclei. In the second table, for the first eleven elements, the principal (n) and azimuthal (l) quantum numbers are indicated, the latter being denoted by a standard letter (s, p, etc.). No distinction is made between states with various magnetic quantum numbers, m_l within a group having the same value of l. Because the two possible values of the spin quantum number, m_s, refer to opposite spin directions, these can be indicated by arrows, as shown.

Electron configuration

Symbol	1s	2s	2p	3s
H	↑			
He	↑↓			
Li	↑↓	↑		
Be	↑↓	↑↓		
B	↑↓	↑↓	↑	
C	↑↓	↑↓	↑ ↑	
N	↑↓	↑↓	↑ ↑ ↑	
O	↑↓	↑↓	↑↓↑ ↑	
F	↑↓	↑↓	↑↓↑↓↑	
Ne	↑↓	↑↓	↑↓↑↓↑↓	
Na	↑↓	↑↓	↑↓↑↓↑↓	↑

mechanical rules; energy also comes in discrete packets known as quanta, and matter is simultaneously particle-like and wave-like. One has to forego a precise definition of particle location and replace it with a statistical description. The number of electrons in an electrically neutral atom is the defining characteristic of an element, and quantum theory explains the arrangement of the periodic table, which is the basis of chemical behaviour.

2

The force of nature could no farther go; to make a third she join'd the former two.
John Dryden (1631–1700)
Lines Under Portrait of Milton

Atomic duets: *the chemical bond*

Most substances are mixtures of two or more elements, and *chemical compounds* are found to have definite compositions. They are also found to consist of *molecules*, which are the smallest particles that can exist in the free state and still retain the characteristics of the substance. Thus the *chemical formula* for *water* is always H_2O, indicating that its molecules contain two hydrogen atoms and one oxygen. (The subscript 1 is not written explicitly; H_2O implies H_2O_1.) The formula H_2O_2, in which the relative proportions of hydrogen and oxygen are different from those in water, belongs to a different chemical compound, the common bleaching agent *hydrogen peroxide*. Similarly, *sodium chloride*, also known as common salt, has the formula NaCl, and not Na_2Cl or $NaCl_3$. This shows that the *forces* that hold atoms together in groups must have certain characteristics which determine the rules of combination. The occurrence of *chemical reactions* reveals that the forces can be overcome, allowing the atoms to move apart and establish bonds with other atoms. A simple example of such an event is the reaction

$$Zn + 2HCl \rightarrow ZnCl_2 + H_2$$

in which the addition of *hydrochloric acid* to zinc produces zinc chloride and hydrogen, which is liberated as a gas. This demonstrates that the binding forces between zinc and chlorine atoms are stronger than those which hold hydrogen to chlorine. The reaction provides another example of the existence of numerical *rules of combination*, since the chemical formula for zinc chloride is $ZnCl_2$ and not simply ZnCl. It also implies conservation of matter: atoms are neither created nor destroyed during a chemical reaction. The arrow indicates the direction in which the chemical reaction proceeds. If two opposing arrows are used, i.e. $\leftrightarrows$, the reaction takes place readily in either direction, and it is said to be *reversible*. A single arrow indicates that the reaction is *irreversible* under the prevailing conditions.

The Greek philosopher Empedocles, around 500 BC, suggested that chemical changes are brought about by something akin to emotional relationships between things. Thus two substances which love each other will unite to form a third, while decomposition is simply the separation of two substances with a mutual hate. On the basis of the above chemical reaction, we would have to conclude that chlorine is capable of a fickle change of heart. The modern approach to reactions is based on knowledge of atomic structure, and this leads quite logically to the fundamental rules of *chemical composition*.

An early hint of the nature of chemical combination came from a puzzling result obtained by Joseph Gay-Lussac. He found that when one volume of

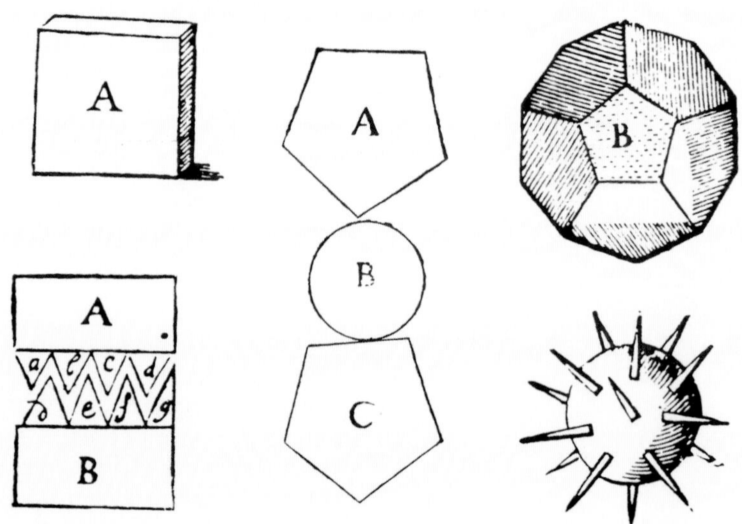

Among the early speculations regarding the cohesion of materials, those appearing in *Principes de Physique* (1696), by Nicolas Hartsoeker (1656–1725) are particularly noteworthy because of their resourceful fantasy. The particles in a refractory material were believed to be rectangular (top left), while those of a fusible material were thought to have less pointed corners (top right). Mercury's spherical particles could soften gold by penetrating between the latter's polyhedral particles (centre). That iron is hard when cold but malleable when hot was explained by the separation of interlocking bonds upon heating (lower left). Corrosive sublimate (lower right) was believed to be a composite material, one component of which provided the burr-like bonding arms.

oxygen and one volume of nitrogen are made to react, not one but two volumes of nitric oxide are produced. John Dalton, believing the atoms in a gas to be separate entities, found this result inexplicable. The correct answer was provided by Amedeo Avogadro, in 1811, and it involved two assumptions, both of which had far-reaching implications. One was that equal volumes of gases, at the same temperature and pressure, contain equal numbers of particles. The other assumption was that the particles, which are now recognized as the molecules referred to earlier, consist of chemically combined groups of atoms. Dalton's incorrect version of the Gay-Lussac reaction was

$$N + O \rightarrow NO$$

whereas the correct form

$$N_2 + O_2 \rightarrow 2NO$$

embodies both Avogadro's assumptions and Gay-Lussac's result. Although the existence of molecules had been established, there remained the question of the forces by which they are held together.

The terms force and energy were used rather loosely in Chapter 1. In order to appreciate the main points of the present chapter, we must delve deeper into these concepts, and a simple example will prove useful. We noted in the previous chapter that the potential energy of a person standing on a chair is greater than that of the same person standing on the floor. This excess potential energy can be converted to kinetic energy by jumping off the chair. The kinetic energy is then changed to other energy forms, such as sound and heat, when the floor is reached. Another familiar example of the conversion of potential to kinetic energy is the generation of *hydroelectricity*, when the water descending from a reservoir turns the shaft of a turbine. In general, *energy* is defined as the capacity for doing work, and it exists in such varied forms as potential, kinetic, electrical, thermal, chemical, mechanical, nuclear, and radiant energy. Energy can readily be transformed from one form to another. The definition of *force* is that it is an external agency capable

A system is said to be *stable* when it is at its condition of absolute minimum energy, and *metastable* when it is at a local but higher minimum. These states are separated by one or more *energy barriers*. As with the stable and metastable conditions, the peak of the energy barrier corresponds to a situation of zero net force, but the system is *unstable* against an arbitrarily small displacement in either direction.

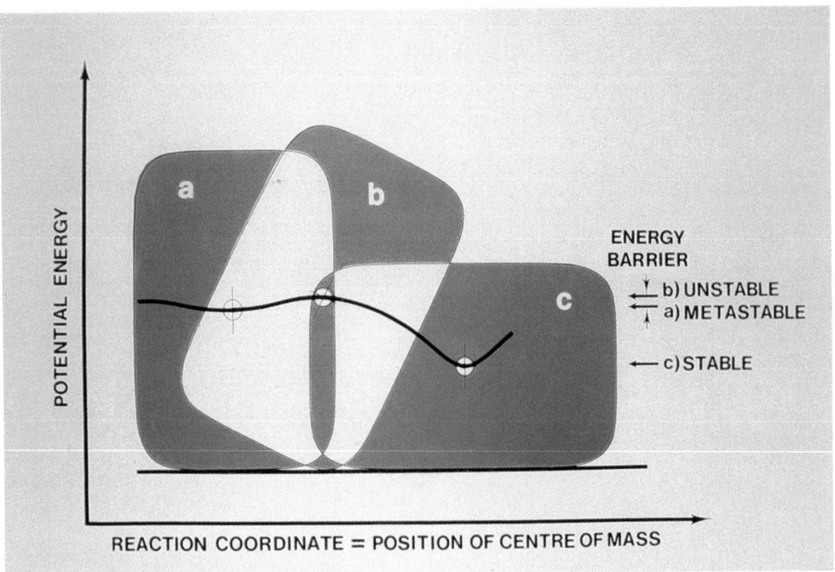

of altering the state of rest or motion of an object, and we see that this implies acceleration. Indeed, Isaac Newton's famous law states that force is simply the product of mass and acceleration.

Let us consider a car on a road that consists of a series of hills, valleys, and level segments. If the vehicle is on the flat and is stationary, an *equilibrium* situation prevails. The downward force due to the weight of the car is exactly counterbalanced by the upward reactive force of the road, which acts through the tyres. There are no lateral, or sideways, forces. If the engine is started and the vehicle put into gear, there is first an acceleration, and the vehicle subsequently achieves a constant velocity when the push of the engine is exactly counterbalanced by the frictional forces acting on the tyres and bearings. Now let us consider the same vehicle located on a hill between two valleys, and assume that the frictional forces are negligibly small. If the car is precisely at the crest of the hill and is stationary, it is in equilibrium, because there are again no lateral forces, while the vertical forces are exactly balanced. This is referred to as *unstable equilibrium* because the smallest displacement in either direction causes the car to accelerate as it moves downhill. The acceleration is greatest where the hill is steepest. Just as when the person jumps from the chair, there is conversion of potential energy to the kinetic energy of motion, the amount of potential energy being directly related to the height of the hill. Lateral force is related to slope, and equilibrium prevails when the slope is zero. The bottom of a valley is also a position of equilibrium because here too there is no slope. But in this case the equilibrium is said to be *stable*; a small displacement gives rise to a force that tends to return the car to the bottom of the valley. Absolute potential energy is usually less important than differences in potential energy. If there are several valleys with different depths, the lowest is said to be associated with stability, and all the others with *metastability*. When the vehicle is in a metastable valley, it can lower its energy by travelling to a lower valley, but only through the temporary investment of energy to get it over the intervening hill, or hills. The use of a car in this analogy rather than a ball, for instance, gives the issue directionality. Thus two hills of equal slope will

The modern view of interatomic forces dates from 1824, with the conjecture by Ludwig Seeber (1793–1855) that stability must result from a balance between attractive and repulsive forces. The force corresponding to any particular interatomic separation is given by the slope of the interatomic potential, and it is thus zero when that potential is at a minimum.

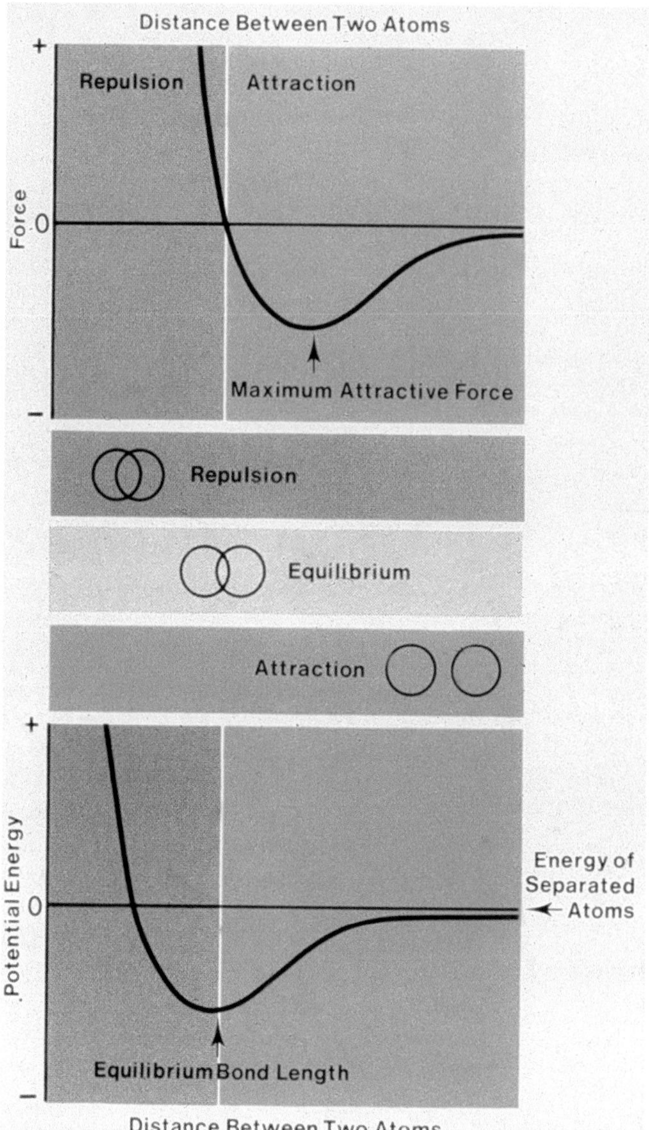

have opposite effects if one is encountered in the downhill direction while the other is approached uphill. The downhill slope could be said to attract the car towards its destination while the uphill slope repels it away from that goal. Two potential energy valleys separated by a hill is a frequently encountered situation in the interactions between atoms, and the hill is referred to as an *energy barrier*, the word potential being implicit in this term.

In the previous chapter only individual atoms were considered; interactions between atoms were neglected. Even in a gas the atoms are usually grouped into molecules. In a reasonably rarefied gas these molecules behave as independent particles, interacting with one another only during collisions. But in the condensed forms of matter, which include all liquid and solid materials, interactions between atoms are of paramount importance. Experience tells us quite a lot about the forces between atoms. Atmospheric air has a rather low density, so the distance from a molecule to its closest

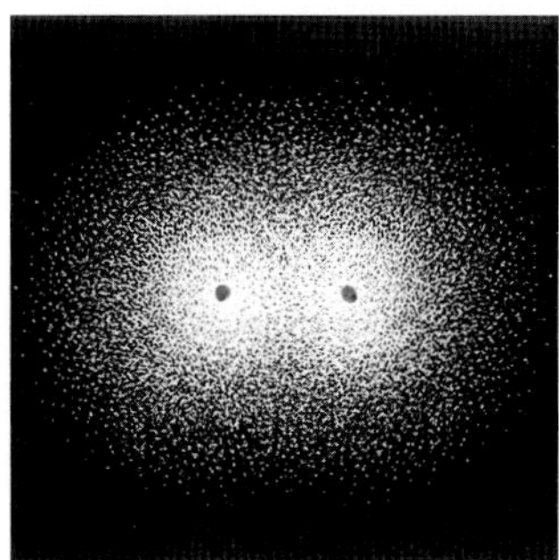

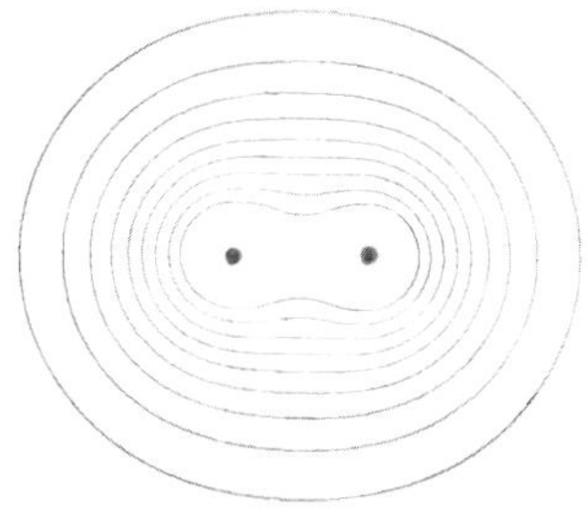

In the hydrogen molecule the two electrons have opposite spins, and they are thus paired. The final electron configuration is not simply equal to the superposition of two individual hydrogen electron orbitals (top), there being an increased tendency for the electrons to lie between the two nuclei (bottom). The contours of constant electron density (middle) are shown for the equilibrium situation, in which the nuclei are separated by a distance of 70 picometres. The large dots indicate the positions of the hydrogen nuclei (i.e. the protons), but their size is shown greatly exaggerated.

neighbours is relatively large, as was first surmised by Robert Hooke. In air at atmospheric pressure the average value of this distance is about 20 nm, which is approximately 100 times greater than the corresponding distance in a liquid or solid. The air is nevertheless stable, and at normal temperatures it shows no tendency to condense into a liquid. We are therefore safe in assuming that under atmospheric conditions the attractive forces between the molecules are insufficient to make them form large clusters. But these forces certainly exist, because when a gas is cooled to a sufficiently low temperature, and the kinetic energy of the molecules thereby decreased, condensation to the *liquid state* occurs. This is what causes the *water vapour* in one's breath to condense on the car windshield in cold weather. It is also clear that the forces between molecules must become repulsive at shorter distances. Matter would otherwise go on condensing, or imploding, to ever-increasing densities. With the interatomic force thus being attractive at longer range and repulsive at shorter range, equilibrium must prevail at an intermediate distance at which the net force is zero. It turns out that the distance of closest approach between atoms in condensed matter is approximately 0.3 nm. If, for simplicity, atoms are regarded as spheres, the radius of a typical atom is therefore about 0.15 nm. It is encouraging to note that this is approximately equal to the radius of the electron probability clouds described in the previous chapter. This shows that although these are rather tenuous, they nevertheless play a decisive role in determining the forces between atoms. When two atoms or two molecules impinge upon one another, they do not interpenetrate to any marked degree unless their original kinetic energies are much higher than those encountered in normal *condensed matter*. This contrasts with the dramatic situation, so dear to the science fiction writer, when solar systems collide. During such a cataclysm, the two suns would hurtle towards each other, paying little heed to the orbiting planets.

The orbiting electrons determine the way in which one atom interacts chemically with another. In order to understand why the various elements display different chemical behaviour, we must pay particular attention to the systematic trends in the electronic arrangements described in the periodic table. The electrons arrange themselves among the different *quantum orbits*, some of these being filled while others are incomplete. The *inert gases* are special in that they have only completely filled orbits. They take their collective name from the fact that they do not readily enter into chemical interactions with other elements. They are also referred to as the *noble gases*; they display an aristocratic disdain for liaisons with each other and with other types of atom. Completely filled shells or orbits therefore produce low *chemical affinity*; they are primarily associated with repulsion. In general, when an atom has both completely and partially filled orbits, it is the latter that govern chemical bonding. This conclusion comes from two observations. All atoms in the same column of the periodic table display, by and large, the same chemical behaviour. For instance, in the first column we have the elements hydrogen, lithium, sodium, potassium, rubidium, and caesium. Lithium, sodium, potassium, and rubidium have long been known to be typical metals, and it has been found that hydrogen too becomes a metal at sufficiently high pressure. These elements have different numbers of completely filled inner orbits, but they all have a single electron in their outermost unfilled orbit, a clear indication that it is the latter that determines the chemical behaviour. If instead of a column we look at a row of the periodic table, a systematic trend in chemical behaviour is found. Consider,

The electron volt
Although not part of the Système International d'Unités, the electron volt (symbol eV) still enjoys widespread use in the science of materials because typical interatomic bond energies are comparable to this unit. The chemical process occurring in a battery or accumulator, for example, produces potential differences of a few volts. It is defined as the work done on an electron when passing through a potential rise of one volt (1 V). One electronvolt (1 eV) is equivalent to 1.6021×10^{-19} J (joules).

for example, the row that stretches from sodium to argon. We have just seen that sodium, a member of the alkali metal family, has a single electron in its outermost orbit while the inner electrons are arranged in completely filled orbits. The latter can be regarded as essentially the neon structure, neon being the element that immediately precedes sodium. Magnesium, with two electrons in its outermost orbit, belongs to the alkaline earth metal family together with beryllium, calcium, strontium, and barium. It displays a chemical behaviour quite distinct from that of sodium. Sodium chloride has the chemical formula NaCl, as we noted earlier, whereas magnesium chloride has twice as many chlorine atoms as magnesium atoms, and its formula is $MgCl_2$. Moving farther along the row, through aluminium, silicon, phosphorus, and sulphur, the number of electrons in the outer orbit increases from three to six, and further changes are seen in chemical affinity. In argon only completely filled orbits are present, and this element, like all the other members of its family, is chemically inert. It is thus the electrons in the outermost orbits that are responsible for the chemical behaviour of the elements. This is not surprising because it is just these electrons that are most loosely bound and therefore most likely to interact with electrons of surrounding atoms.

The formation of a two-atom molecule from two individual atoms is a *many-body problem*. The system includes the nuclei of the two atoms, as well as their electrons. But because the electrons in the inner filled shells are not chemically active, the problem can be simplified. We can consolidate the nucleus and the inner electron shells of each atom into a single working entity and, using an earlier definition, call it an *ion*; the total number of electrons in the inner shells is less than the number of protons in the nucleus. The ion is of the positive type, and its net charge equals the number of electrons in the outer unfilled shell. We may regard a sodium atom, for example, as a single negative electron rotating around an ion with a single positive charge, and magnesium as two electrons and a doubly charged positive ion. It is only the inert gases that cannot be treated in this way. They do not readily form ions, and this is directly related to their chemical quiescence.

Before considering the motion of electrons under the influence of not one but two positive ions, we can draw another contrast with a celestial situation. The Moon exploration programmes have made us familiar with the three-body problem involving the Earth, the Moon, and a space vehicle. Depending on its velocity, the latter can be either in an Earth orbit or a Moon orbit. And if its velocity is just right, it can describe an orbit that takes it in turn around the Earth and the Moon. In these explorations the orbits have frequently been of the figure-of-eight type, but a satellite could also move in a more distant orbit, with the Earth and Moon acting as a combined central body. The major gravitational interaction acts between the Earth and the Moon, and this keeps the latter in its terrestrial orbit. The satellite can orbit around both the larger bodies because it interacts with each of them. But these interactions are negligible compared with the Earth–Moon interaction, and they have no influence on the Moon's orbit around the Earth. In the two-atom molecule the situation is quite different. Again it is the smallest bodies, the electrons, which orbit around the larger objects, the ions. But the ion–electron interactions are not minor disturbances. They are the dominant factors in the overall stability of the molecule. Were it not for the orbiting electrons, the two ions, being both positively charged, would

Characteristics of some interatomic bonds

Because there are nearly a hundred different naturally occurring elements, the number of different possible atom–atom combinations is very large. The few bonds characterized in the following table have particular relevance for the organic and biological domains. The energies are in attojoules (symbol aJ) and the distances in nanometres (symbol nm).

Bond	Energy	Distance
H–H	0.72	0.074
H–C	0.69	0.109
H–N	0.65	0.100
H–O	0.77	0.096
C–C	0.58	0.154
C=C	1.04	0.133
C≡C	1.36	0.121
C–N	0.49	0.147
C=N	1.02	0.129
C≡N	1.48	0.115
C–O	0.58	0.143
C=O	1.23	0.122
N–N	0.26	0.148
N=N	0.69	0.124
N≡N	1.57	0.100
N–O	0.35	0.144
N=O	0.80	0.120
O–O	0.35	0.149
O=O	0.82	0.128

repel each other and a stable molecule would be impossible. The electrons act as a sort of glue, holding the two ions together and overcoming their repulsive tendency. When an electron is instantaneously closer to one of the ions, it still attracts the other.

This picture of a simple molecule is still rather classical, and we must inquire how the quantum constraints outlined in the previous chapter influence the situation. Once again Wolfgang Pauli's *exclusion principle* plays a major role. In a letter to Pauli, written in 1926, Paul Ehrenfest included the following passage: 'I have for a long time had the feeling that it is your prohibiting condition that above all prevents the atoms and through them the crystal from falling together. I suggested this some months ago in a popular lecture on "What makes solid bodies solid?" In the attempt rationally, at long last, to consider an ideal gas, to quantize the impenetrability of the molecules, I noticed for the first time the peculiar running about of this idea . . .'. The repulsion between completely filled electron shells arises because their mutual overlap would violate the Pauli principle. To avoid this, the offending electrons are forced into levels with higher *quantum numbers*, and this raises the energy. Max Born and James Mayer, in 1932, showed that this idea leads to reasonable values for the repulsive forces between atoms. The Pauli principle also applies to the interaction between incomplete shells. In 1918, Gilbert Lewis suggested that the important interatomic linkage known as a *covalent bond* consists of a shared pair of electrons. A decade later, Walter Heitler and Fritz London showed that such a bond is stable only if the spins of the two electrons are antiparallel (i.e. lie in opposite directions), as the Pauli rule would suggest. The composite paths by which electrons move around both ions in a molecule are known as *molecular orbitals*. The earliest quantum mechanical treatment of such an orbital, for the simplest case of one electron and two protons, was made in 1927 by Max Born and J. Robert Oppenheimer. Edward Condon's extension to the *hydrogen molecule*, with its extra electron, followed immediately, and much of the credit for the subsequent application to larger molecules is due to Linus Pauling, John Slater, and Robert Mulliken. But not all bonds are covalent; we must consider the electron–ion interaction in a broader context.

Although the arguments above are plausible enough, we have not yet established that atoms do unselfishly share some of the electrons under certain circumstances. After all, atoms prefer to be neutral; even the outermost electrons are sufficiently well bound to the nuclei that the neutral atom does not spontaneously break up into a positive ion and individual electrons. Why is it that the approach of a second neutral atom can induce some of the electrons to adopt a modified arrangement or configuration? There are actually two quantities that decide this issue, the *ionization energy* and the *electron affinity*, and they are both referred to the neutral atom. The ionization energy is the energy required to pull away the most loosely bound electron from the atom so as to leave behind a singly charged positive ion. This energy arises from the electric attraction between the negative charge on the electron that is extracted and the positive charge on the ion. Strictly speaking, this defines what is known as the first ionization energy, for it is possible to remove a second and further electrons, and the energies to pull these away will in general be different. Not surprisingly, the ionization energies of elements depend on their position in the periodic table. The lowest first ionization energies are displayed by the alkali metals. The reason for this lies in the way their electrons are arranged, with a single electron in the

Atomic Number → 26, Symbol → Fe, Atomic Mass → 55.85

Hydrogen and the Alkali Metals	Alkaline Earth Metals	1st Transition Metals					2nd Transition Metals			3rd Transition Metals		Boron and Carbon Groups		Nitrogen and Oxygen Groups		Halogens	Noble Gases
1 H 1.008																	2 He 4.003
3 Li 6.939	4 Be 9.012											5 B 10.81	6 C 12.01	7 N 14.01	8 O 16.00	9 F 19.00	10 Ne 20.18
11 Na 22.99	12 Mg 24.31											13 Al 26.98	14 Si 28.09	15 P 30.97	16 S 32.06	17 Cl 35.45	18 Ar 39.95
19 K 39.10	20 Ca 40.08	21 Sc 44.96	22 Ti 47.90	23 V 50.94	24 Cr 52.00	25 Mn 54.94	26 Fe 55.85	27 Co 58.93	28 Ni 58.71	29 Cu 63.54	30 Zn 65.37	31 Ga 69.72	32 Ge 72.59	33 As 74.92	34 Se 78.96	35 Br 79.91	36 Kr 83.80
37 Rb 85.47	38 Sr 87.62	39 Y 88.91	40 Zr 91.22	41 Nb 92.91	42 Mo 95.94	43 Tc 99.00	44 Ru 101.0	45 Rh 102.9	46 Pd 106.4	47 Ag 107.9	48 Cd 112.4	49 In 114.8	50 Sn 118.7	51 Sb 121.8	52 Te 127.6	53 I 126.9	54 Xe 131.3
55 Cs 132.9	56 Ba 137.4	57 La 138.9	72 Hf 178.5	73 Ta 181.0	74 W 183.9	75 Re 186.2	76 Os 190.2	77 Ir 192.2	78 Pt 195.1	79 Au 197.0	80 Hg 200.6	81 Tl 204.4	82 Pb 207.2	83 Bi 209.0	84 Po 210.0	85 At 210.0	86 Rn 222.0
87 Fr 223.0	88 Ra 226.0	89 Ac 227.0															

Rare Earths	58 Ce 140.1	59 Pr 140.9	60 Nd 144.2	61 Pm 147.0	62 Sm 150.4	63 Eu 152.0	64 Gd 157.3	65 Tb 158.9	66 Dy 162.5	67 Ho 164.9	68 Er 167.3	69 Tm 168.9	70 Yb 173.0	71 Lu 175.0
Actinides	90 Th 232.0	91 Pa 231.0	92 U 238.0	93 Np 237.0	94 Pu 242.0	95 Am 243.0	96 Cm 247.0	97 Bk 247.0	98 Cf 251.0	99 Es 254.0	100 Fm 253.0	101 Md 256.0	102 No 254.0	103 Lw 257.0

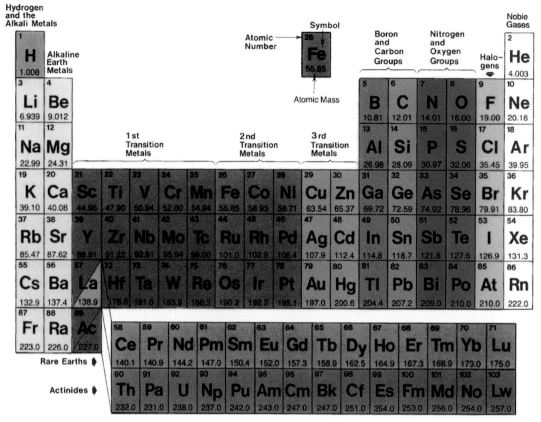

When two or more atoms combine chemically, the tendency of the individual participants to gain or lose electrons is determined by their electronegativities. The higher the value of this quantity, the greater is the propensity for gaining electrons, and the element is said to be *electronegative* (blue and orange). *Electropositive* elements (yellow and red) have low values and lose electrons relatively easily. The *inert gases* do not readily participate in bonding that involves electron redistribution (though metastable cases have been achieved). The electronegativities of some of the rare earths and actinides have not yet been determined. The standard abbreviations for the elements are used here; their full names can be found in the periodic table in Chapter 1.

outermost orbit and all other orbits filled. Orbiting electrons act as a sort of shield which partially cancels the positive charge on the nucleus. This effect is most complete if the electrons are arranged in closed orbits, and in the alkali metals the final electron is rather weakly bound to the positive ion. Proceeding along a row in the periodic table, it is found that the ionization energy systematically increases, the maximum value being found in the noble gases. This is consistent with the fact that closed electron shells are particularly stable.

The first ionization energy of sodium is 0.82 attojoules (aJ), while that of argon is more than three times this amount, namely 2.53 aJ. These elements are located at opposite ends of the third row of the periodic table. The increase of ionization energy within this row is not smooth because both s and p states are available to the electrons in the outermost orbit. The first ionization energy of magnesium is 1.22 aJ, and this corresponds to completion of the 3s orbit. For aluminium the value is only 0.96 aJ, because the third electron is in the 3p state. Such subtleties are of little concern here, and we turn to the other important quantity, the electron affinity. This is the energy gained when an electron is brought from infinity up to a neutral atom of the element in question. It is thus a measure of the strength of binding that stabilizes a negative ion. It is important not to confuse this with the ionization energy; one quantity refers to the attraction between an electron and a neutral atom, while the other concerns an electron and a positive ion. Electron affinities also show a systematic variation with position in the periodic table. Their values tend to be much lower than ionization energies,

Explosions

An explosion is a violent and rapid increase of pressure in a confined space, caused either by a chemical (or nuclear) reaction or by an external source of thermal energy, and characterized by copious production of gas. An explosive is a substance that undergoes rapid chemical change when heated or otherwise ignited, the volume of gas thereby produced far exceeding that of the original substance. The classic example is *gunpowder*, which is a mixture of potassium nitrate (KNO_3 – also known as *saltpetre*), powdered charcoal (carbon, oxygen and hydrogen) and sulphur. The first mention of this substance was in a letter from around 1267, written by Roger Bacon (1219–1294), although contemporary Arabian writings variously mention Chinese snow and Chinese salt, which reportedly could create a 'thunder that shakes the heavens'. Bacon recommended a mixture consisting of about 40% saltpetre, with the

balance comprising equal parts of charcoal and sulphur. The recipe was subsequently revised by Albertus Magnus (1206–1280), John Arderne of Newark (1266–1308), Berthold Schwartz (1318–1384) and others, all of whom suggested about two thirds saltpetre, two ninths charcoal and one ninth sulphur. The composition (by weight) used nowadays is 74.64% saltpetre, 13.51% charcoal and 11.85% sulphur. The chemical reaction that produces the explosion is surprisingly complex, namely:

$$\begin{aligned}74KNO_3 + 16C_6H_2O + 29S \Rightarrow\ & 56CO_2 + 14CO + 3CH_4 + 2H_2S + 4H_2 + 35N_2 + 19K_2CO_3 \\ & + 7K_2SO_4 + 2K_2S + 8K_2S_2O_3 + 2KSCN + (NH_4)_2CO_3 + C\end{aligned}$$

each kilogram of powder liberating 665 kcal of energy. The early use of gunpowder in China appears to have been confined to fireworks, whereas it quickly found military applications in Europe. Both the illustrations reproduced here appeared in artillery handbooks. The larger picture, from a seventeenth century German volume, shows a military artificer carrying out a test explosion, while the smaller picture, from a French handbook about a century later, shows the production of saltpetre. Gunpowder is still widely used in fireworks, but the explosions employed in mining, rock blasting and the like are achieved through use of *dynamite*. Invented in 1866 by Alfred Nobel (1833–1896), it consists of *nitroglycerine* absorbed in *kieselguhr* (hydrated silica, formed from the skeletons of minute plants known as diatoms).

a consequence of the preference that atoms have for existing in the neutral state.

The fate of the outer electrons when two atoms approach is determined both by the respective ionization energies and the electron affinities. Consider sodium chloride. The ionization energy of sodium is 0.82 aJ, while that of chlorine is 2.08 aJ. When neutral atoms of sodium and chlorine approach, it is thus far more likely that the sodium atom will lose an outer electron. Moreover, the electron affinity of chlorine is 0.58 aJ, whereas that of sodium is only 0.11 aJ. In this respect too a chlorine atom is better able to attract an extra electron. But we must remember that the above numbers

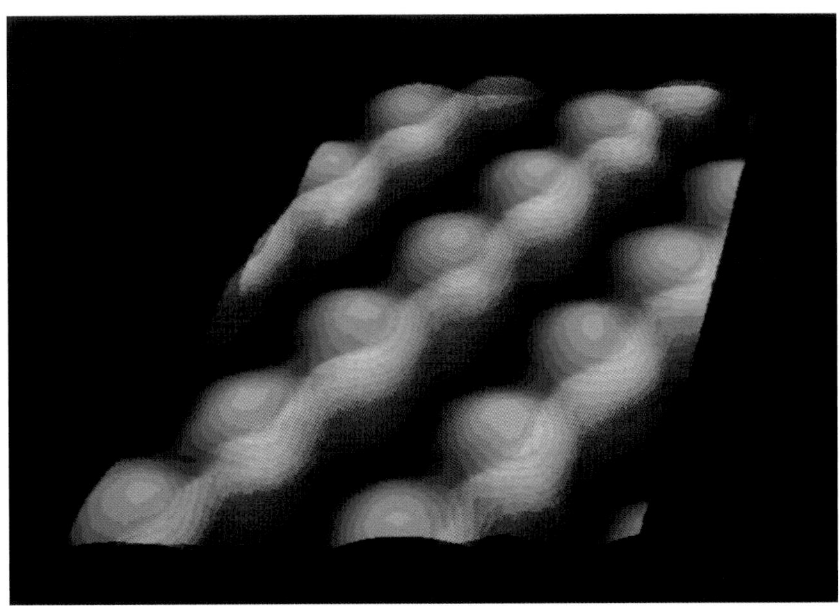

In the *scanning tunnelling microscope*, invented by Gerd Binnig and Heinrich Rohrer in 1979–81, an extremely sharp needle performs a raster over the specimen surface at a height of only a few atomic diameters. This permits the flow of a tunnelling current of electrons between them. The strength of the current (which is a quantum phenomenon) is critically dependent on the scanning height, and it is influenced by the surface's atomic-level contours. By allowing the tunnelling current to control the specimen–needle distance, the surface can be imaged with atomic resolution. The picture of a gallium arsenide surface shown here is actually a composite of two such images. The blue image is for tunnelling from needle to specimen, and it picks out the gallium atoms, while tunnelling from specimen to needle favours the arsenic atoms, which appear in red. Operated in the appropriate mode, the microscope is even able to reveal the interatomic bonds.

refer to ionization and capture processes defined in terms of removal of an electron to, or arrival of an electron from, a distance much greater than the atomic diameter. In the real situation the relevant distances are of atomic dimensions, so the numbers require correction by an amount that is not easily calculated. When the two neutral atoms approach, the sodium loses its single electron from the outermost orbit, and this becomes attached to the neutral chlorine, turning the latter into a negative ion. The neutral sodium atom becomes a positive ion, and the two ions attract each other. The uncorrected first ionization energy of sodium lies 0.24 aJ above the uncorrected electron affinity of chlorine, but the ion–ion force easily offsets this factor. Moreover, the chlorine atom has seven electrons in an orbit capable of holding eight, so the electron it gains from the sodium just fills that outermost orbit. Transfer of an electron from the sodium atom to the chlorine therefore satisfies the electronic ambitions of both the neutral atoms, and produces a situation in which two ions both having only completely filled shells of electrons attract each other electrically.

Let us replace the sodium by magnesium. The situation resembles the previous one, but the magnesium atom has two electrons in its outermost orbit rather than one. A molecule containing one magnesium and one chlorine atom is conceivable, although it would not be as stable as sodium chloride because the ionization energy of magnesium is higher than that of sodium. The chlorine atom requires just one electron to complete its outer shell, whereas the magnesium atom has two electrons to donate. A particularly stable situation will arise, therefore, if *two* chlorine atoms join with one magnesium atom. The end result will be a molecule in which a doubly-charged magnesium ion is joined to two singly-charged chlorine ions. This is the origin of that important chemical quantity, the *valence* of an element. It is a measure of the number of other atoms with which the atom of a given element tends to combine. Because binding derives primarily from the structure of the outermost orbit, this is usually referred to as the *valence orbit*. The normal valence of magnesium is two, and the chemical formula

Interatomic bonds and ideal mechanical strength

The energy required to break an object is primarily used in generating the new surfaces created at the site of fracture. Before these separate, there is a stretching of the interatomic bonds in the material. As the applied force builds up, the bonds ideally elongate until they are roughly 20% greater than their original equilibrium lengths, whereupon the material breaks. We will make a very crude estimate of the fracture stress by assuming the force to remain constant at its average value, during the bond stretching. For carbon the bond energy is 0.58 aJ, and the bond length is 0.154 nm. In an optimally oriented carbon fibre, closely packed planes of atoms lie along the tensile direction, and the planes are spaced about 0.35 nm apart. This gives an average bond density of approximately 2×10^{19} per square metre (m^2), and thus a fracture surface energy of 5.8 J/m^2, since each bond is shared between two atoms. The total surface energy, for the two surfaces created by the fracture, equals the product of the force and the distance of elongation, and this yields a stress value of 3.8×10^5 MN/m^2. The measured fracture stress of carbon fibres is lower than this by a factor of about 20. The discrepancy results from the slight imperfection inevitably present in real fibres, which can tolerate a strain of only about 1%.

for magnesium chloride is $MgCl_2$. On the other hand, the valences of both sodium and chlorine are one, and the chemical formula for sodium chloride is NaCl.

The valence of an element is a measure of its combining capacity, and it is possible to ascribe a consistent set of elemental valences from a knowledge of the atomic masses of the elements and their compounds. The familiar molecular formula of water, H_2O, indicates that two hydrogen atoms combine with one oxygen atom. Hydrogen is said to exhibit *minimal valence* because it always forms compounds having molecular formulas in which the subscript of the other element is one. An element showing this minimal behaviour has a valence of one, and in the case of hydrogen this is directly related to the single orbiting electron. In structural, as opposed to molecular, formulas, valence is denoted by one or more straight lines, which are usually referred to as bonds. Thus hydrogen is indicated by H–. The molecular formula for water then shows that the valence of oxygen must be two, and in structural formulas this is indicated by –O–. The chloride of hydrogen, which forms a well-known acid, has the molecular formula HCl. This shows that chlorine is monovalent, Cl–, while the chloride of magnesium is $MgCl_2$, so magnesium is divalent, –Mg–. We can now check the consistency of this system by considering the molecular formula for magnesium oxide. With both magnesium and oxygen being divalent, the structural formula can be written Mg=O with the two atoms joined by a *double bond*. The molecular formula of this compound should therefore be MgO, which is indeed found to be the case. Methane is CH_4, while ammonia is NH_3. This shows that the valences of those two important elements, carbon and nitrogen, are four and three, respectively.

Strictly speaking, these valences should carry a positive or negative sign to indicate whether an electron is lost or gained in the act of bonding. Thus the valence of hydrogen is +1, while that of oxygen is −2. Some elements exhibit two or more different valences. In *ferrous compounds*, for example, the iron ion is Fe^{2+}, the atom tending to lose two electrons. In *ferric compounds*, the neutral atom loses three electrons, giving the ion Fe^{3+}, and the element is trivalent. An atom of a particular element changes its valence when there is a redistribution of electrons in the available orbits. Thus in the ferrous form of iron the electron configuration is $1s^2 2s^2 2p^6 3s^2 3p^6 3d^6 4s^2$, while

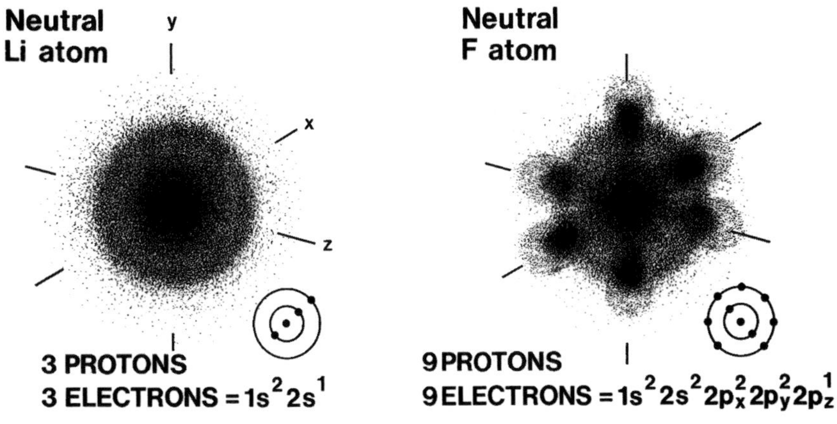

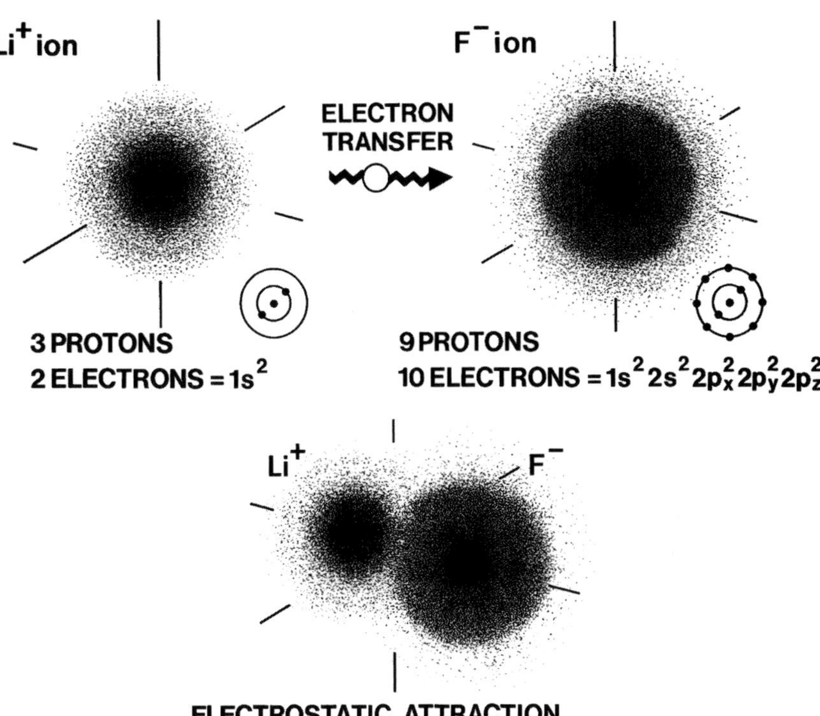

In *ionic bonding*, electropositive and electronegative atoms combine through the electric attraction between ions that are produced by *electron transfer*. Good examples are provided by the *alkali halides*, such as LiF. Each neutral (electropositive) lithium atom loses its 2s electron and thereby becomes a positive ion, somewhat resembling a positively-charged helium atom. Each neutral (electronegative) fluorine atom gains an extra 2p electron and becomes a negative ion, resembling a negatively-charged neon atom.

for ferric iron it is $1s^2 2s^2 2p^6 3s^2 3p^6 3d^5 4s^2 4p^1$. In the first case the atom has two electrons to donate, while it can give three electrons in the second.

The chemical behaviour of two elements combining to form a compound is determined by the respective ionization energies and electron affinities, but it is cumbersome to work with two quantities. Chemical practice commonly involves a single number which is a combined measure of both. This is *electronegativity*, and it is so defined that elements which tend to donate electrons have small values, while those preferring to acquire electrons have high values. The former are said to be electropositive, while the latter are electronegative. Linus Pauling has constructed an electronegativity scale in which the values vary from a low of 0.7, for caesium and francium, to a high of 4.0 for fluorine. Of the two elements involved in table salt, sodium

The type of bonding present in a particular piece of condensed matter is determined by the electronic structure of the participating atoms (shown here schematically, with the valence orbit denoted by a circle). If only electronegative elements are present, the atoms tend to share their *valence electrons* with surrounding neighbours, and the bonding is *covalent* as in the diamond form of carbon. When only electropositive elements are involved, as in the metals, the outermost electrons become detached, producing a lattice of positive ions in a sea of conducting electrons. *Ionic bonding* occurs when both electropositive and electronegative elements are present, electrons being transferred from the former to the latter so as to produce complementary ions which attract each other electrically. An example of ionic bonding is found in common table salt. In addition to these strong primary bonding types, weaker secondary bonds can occur both between closed electron shells, as in the noble (inert) gases such as argon, and between molecules which can share a hydrogen atom, as in water. The oxygen and hydrogen atoms in one water molecule are covalently bonded, but hydrogen bonding occurs between different water molecules.

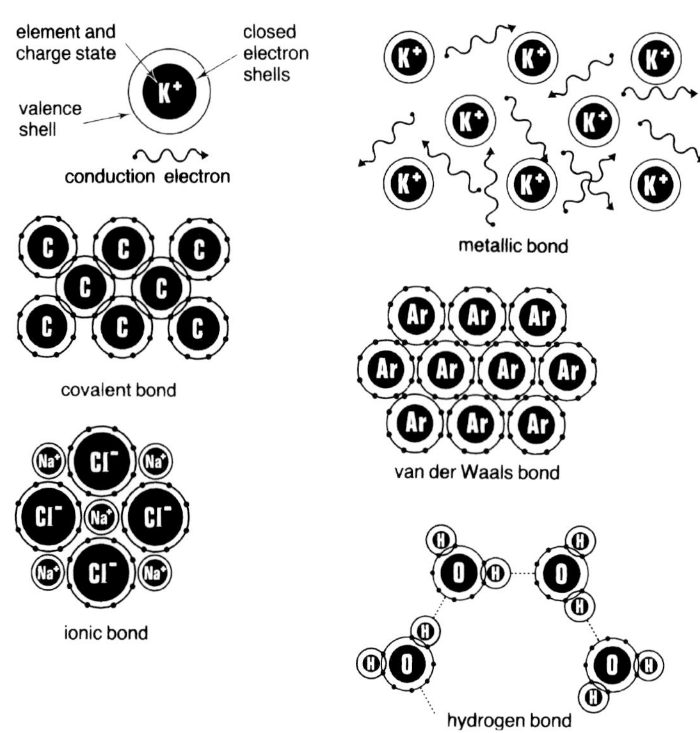

has an electronegativity of 0.9, and is thus electropositive, while chlorine has a value of 3.0, which puts it clearly in the electronegative class. When one is concerned not with the actual magnitude of the electronegativity, but merely with an atom's overall tendency, it is customary to work with what is known as the *polarity*. Electropositive atoms are said to have positive polarity because they tend to lose their outer electrons and become positive ions. Similarly, electronegative atoms have negative polarity. In the joining of two atoms to form a molecule, or many atoms to form a larger mass of solid or liquid, various combinations can arise. If the individual polarities are all negative, the bonding is *covalent*, a common alternative name for this type being *homopolar*. Exclusively electropositive atoms produces metallic bonding, while a mixture of electropositive and electronegative atoms gives *heteropolar bonding*.

In *metallic bonding* the electropositive atoms give up some of their outer electrons and become positive ions, while the electrons thereby released move relatively freely in the solid or liquid as a whole. The metallic bond finds its most perfect expression in the alkali metals, the atoms of which give up their single outer electrons and become spherical ions with a positive charge of one. In *covalent bonding* the electronegative atoms attempt to steal, as it were, one another's electrons. If the individual atoms have roughly the same electronegativity, they meet with equal success in this venture, and the net result is a compromise in which the electrons are shared. Shared electrons are subject to the same constraints that determine the orbital distribution in isolated atoms, and the covalent bond is directional. The classical example of covalent bonding is *diamond*, one of the crystal forms of pure carbon. In contrast to metallic bonding, covalent bonding is also found in small groups of atoms, as in many *organic compounds*. The covalent bond

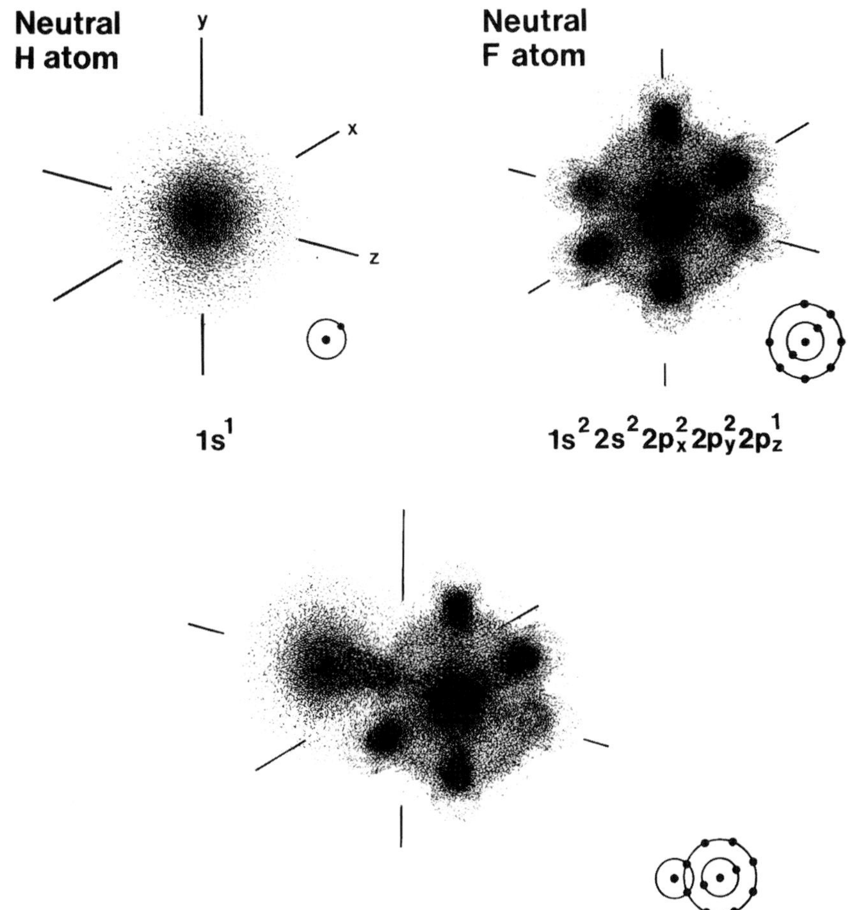

A simple example of *covalent bonding* occurs in the HF molecule. The incomplete orbitals, the 1s of hydrogen and the $2p_z$ of fluorine, overlap and the two electrons are shared in a *sigma bond*.

is also special because it is *saturated*. This follows from the limitation of the number of electrons in shared orbits imposed by the Pauli principle. If a third hydrogen atom closely approaches a hydrogen molecule, its electron is forced into a higher energy level. This makes the three-atom cluster unstable. Saturation is a vital factor in the structure and behaviour of organic compounds.

Heteropolar bonding occurs only for mixtures of electropositive and electronegative atoms, and therefore does not exist in pure elements. With one atom trying to capture electrons while the other is eager to shed them, mutual satisfaction is easily established. The final result is a pair of positive and negative ions which attract each other electrically, and this is the origin of the alternative term: *ionic bonding*. The best examples of such bonding are found in the alkali halides, in which the only extra electron that the alkali atom has becomes the only additional electron that the halide atom needs. The classic case of sodium chloride has already been discussed.

The three types discussed so far are known as *primary bonds*, and they are all relatively strong. Weaker types also exist, and these are called *secondary bonds*. We will consider the two most common: the *van der Waals bond* and the *hydrogen bond*. If the noble gases are cooled sufficiently, they form condensed phases, or condensed states, both liquid and solid. This shows that even though these atoms have only closed electron shells and zero electron

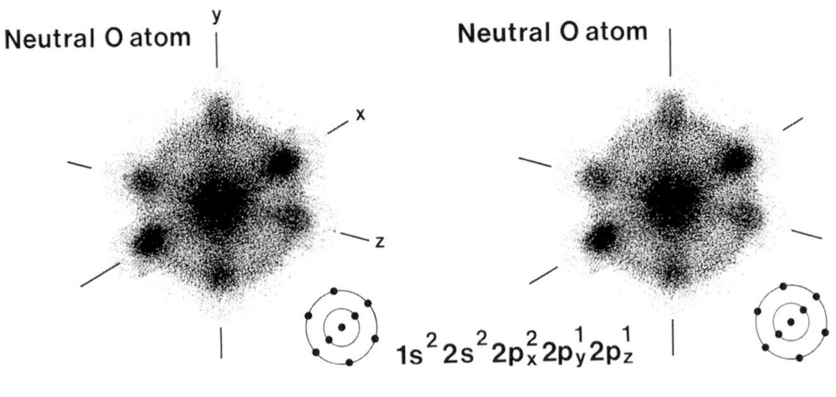

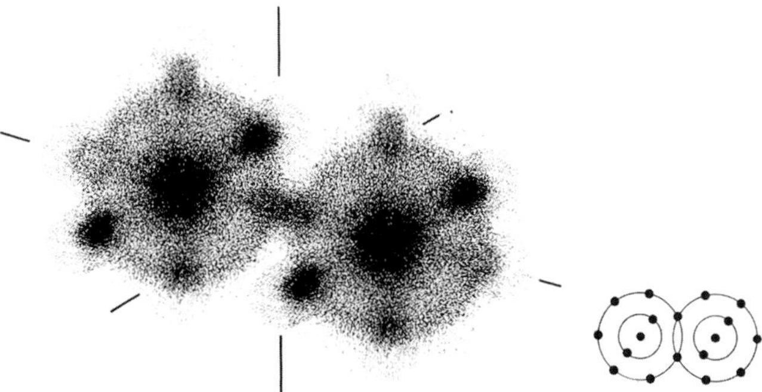

Electron sharing - σ bond of P_z orbitals

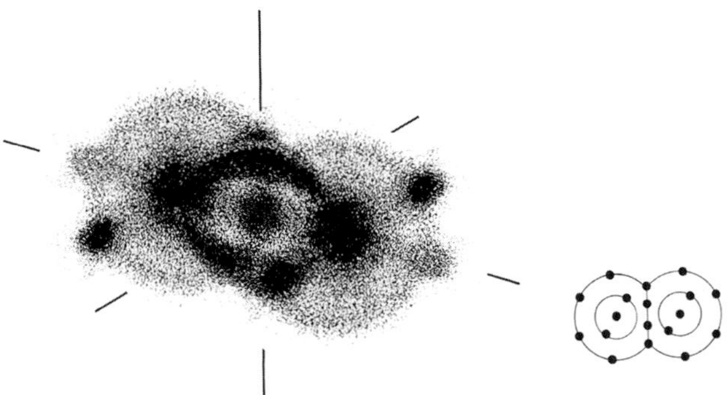

Further electron sharing - π bond of P_y orbitals

The covalent bonding between two oxygen atoms involves both a sigma bond, formed from the $2p_z$ orbitals, and a *pi bond*. The latter involves distortion and overlap of the $2p_y$ orbitals and confers axial rigidity on the molecule.

affinity, they are still able to attract one another weakly. Although the time-averaged electron distribution in such atoms is spherically symmetric, there are instantaneously more electrons on one side of the atom than the other. This produces what is known as an *electric dipole*: positive and negative charges separated by a small distance. Such dipoles are the electric equivalent of the magnetic dipole of the common bar magnet with its north and south poles separated by a distance approximately equal to the length of the bar. The individual electrons are constantly in motion in their orbits, however, so the dipole is perpetually changing both in orientation and strength.

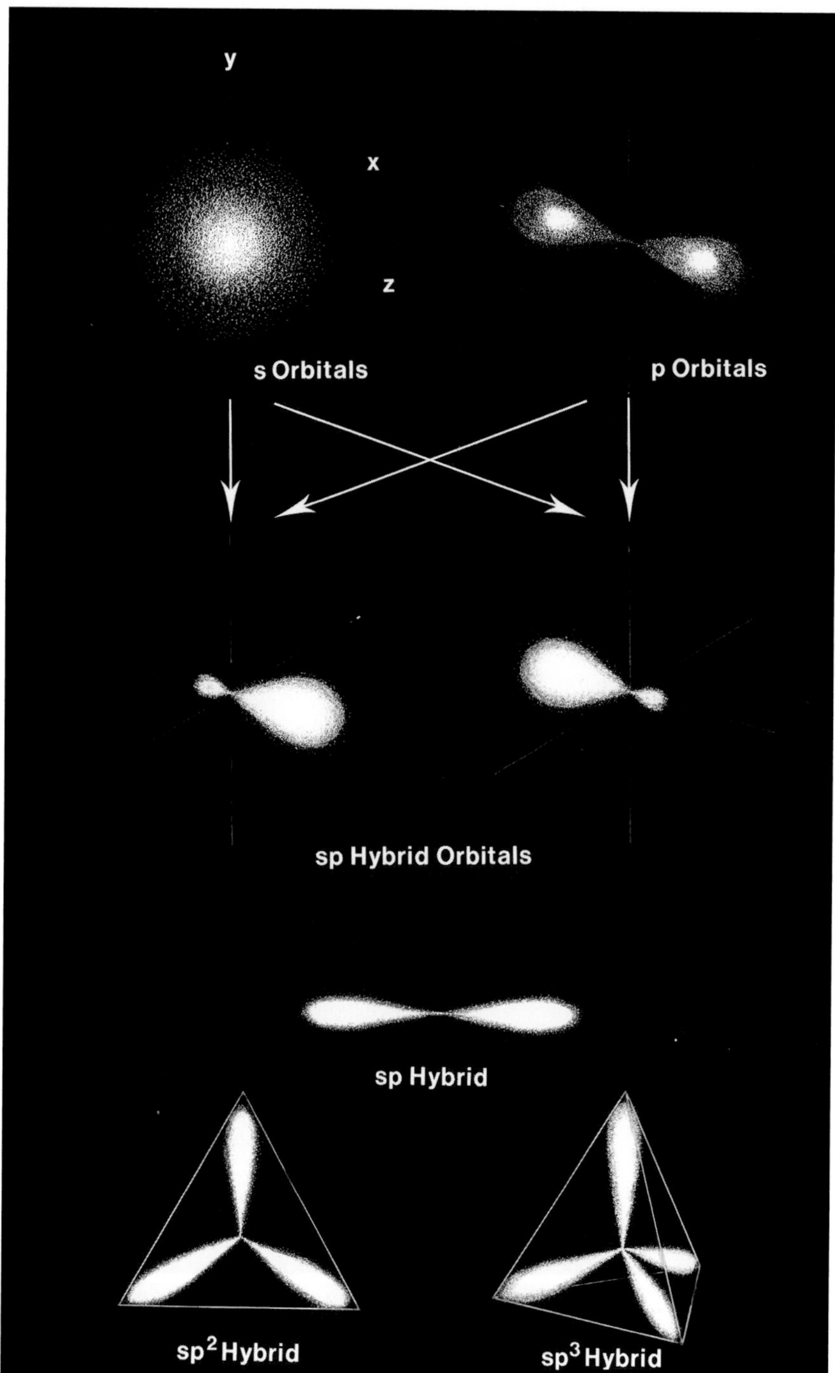

Any combination of *wave states* in an atom is itself a possible wave state and the result is known as a *hybrid orbital*. The number of the latter must equal the number of original orbitals. Thus an s orbital and a p orbital produce two different sp hybrids (top figure). The combination of the different hybrids produces a configuration whose geometry determines the spatial arrangement of interatomic bonds.

When two noble gas atoms approach each other, their individual fluctuating dipoles interact, a *resonance effect* enabling them to get into step with each other, and this causes an overall attraction. Fluctuating dipole bonds, which are non-directional, are also referred to as van der Waals bonds, after their discoverer Johannes van der Waals, and the forces associated with them are known as *dispersive attractions*. Permanent dipoles can also be induced when the field of one atom distorts the charge distribution of an adjacent atom,

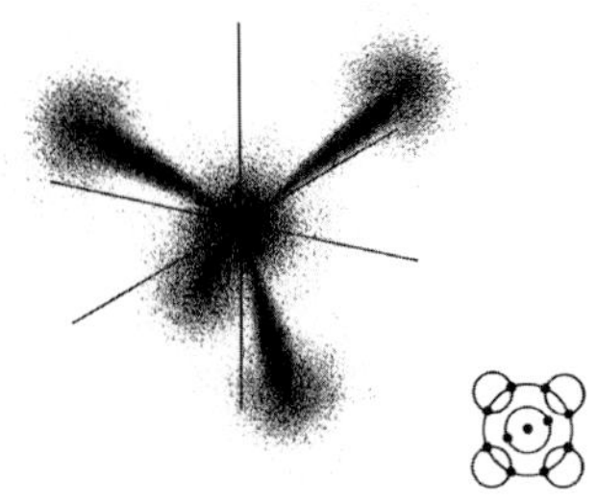

In the formation of a molecule of methane, CH_4, the hybridization of some of the carbon orbitals can be imagined as occurring in two steps. One of the 2s electrons is first promoted into the $2p_z$ state, and the remaining 2s state combines with the three 2p states to produce four sp^3 hybrids. It is the latter which produce the four sigma bonds with the 1s orbitals of the hydrogen atoms, giving the methane molecule its tetrahedral symmetry.

the susceptibility to such changes being determined by the *polarizability*. Unlike the van der Waals bond, the *permanent dipole bond* is directional.

A special type of permanent dipole is present in hydrogen bonding. When hydrogen donates a share of its single electron to directional bonding, with carbon or oxygen for example, its proton nucleus becomes incompletely screened. This leaves a net positive charge which can form an electric link with negative poles on surrounding molecules. The bond is always saturated because of the small size of the partially charged hydrogen ion. This sits like a dwarf between giants, and there is no room for a further ion to approach it. *Hydrogen bonding* is a prominent feature of water and many biological molecules. This is vital to life on our planet because the bond is weak enough to be overcome by thermal agitation at temperatures normally experienced at the Earth's surface. Hydrogen can also establish an ionic bond, in which it gains an electron from an electropositive partner such as lithium.

The characteristic directionality of the covalent bond is important to the chemistry of organic and biological structures. It has features not encountered in other types of bond, and they must now be considered. In Chapter 1 we saw that quantum mechanics requires electrons to be treated in terms of diffuse probability clouds. The individual electrons of isolated atoms are located in various orbits having different geometrical shapes and symmetries with respect to the nucleus. We saw that the s states have spherical symmetry, while p states have nodes at their centres and lobes stretching out along certain directions. Combinations of the different p orbitals can be formed, according to the rules specified by quantum mechanics, and these produce composite probability clouds with pairs of lobes stretching variously along the three Cartesian coordinate axes, x, y, and z. Because two electrons can be located in each orbital, one with positive spin and one negative, these composite orbitals can contain a total of six electrons. The Cartesian system is used only as a convenience in describing the relative orientations of the orbitals in two adjacent atoms. Providence has not actually inscribed such coordinates in space.

The lowest energy state of the hydrogen atom is the 1s, in which the probability cloud is a sphere. When two hydrogen atoms are brought together, there is overlap of those parts of the s orbitals that lie in the region between the two nuclei, and a bond is established if the spins are antiparallel. In the fluorine molecule, the p orbitals become important. The nine electrons in a fluorine atom are so distributed that five of them are in the 2p state. Of these, four give rise to two $2p_x$ and two $2p_y$ states, while the $2p_z$ state contains a single electron. There is nothing special about the z-axis, and the orbital pecking order of first x, and then y, and lastly z is an arbitrary but now traditional nomenclature. When two fluorine atoms approach each other, it is therefore their $2p_z$ orbitals that stretch their hands towards each other, so to speak. The two z-axes line up, and the two lobes of the $2p_z$ orbitals that lie between the nuclei overlap to establish the covalent bond. The remaining 2p electrons of the two atoms take no part in the bonding. They are still present, however, and give the lobes of the $2p_x$ and $2p_y$ orbitals sticking out in the two directions perpendicular to the common z-axis. Passing from the fluorine molecule to the chlorine molecule, we encounter outer electrons with a principal quantum number of 3 rather than 2. This means that the nuclei in the chlorine molecule have more electrons spherically distributed about them, but the bonding is still the responsibility of the outer p_z electrons. As a final example of this simple type of bonding, let us consider

The electron dot (Lewis) notation

There is a way of writing structural formulae which is a more informative alternative to the symbol-and-line method. This system, first proposed by Gilbert Lewis (1875–1946), symbolizes an electron by a dot. Only *valence orbit electrons* are shown, the idea being to indicate how they determine covalent bonding. Examples of the symbols for neutral atoms, using this system, are

H· :Ḟ̤· :Ċ̤l· ·Ċ̣· :Ṇ̇· :Ö̤ ·:N̤̈e:

In a two-dimensional representation, the positions of the dots have no relationship to the actual spatial arrangement, but they reveal which types of bonding are present. There are two rules: all bonding electrons must occur in pairs, and each atom is surrounded by eight electrons, except hydrogen and helium, which have only two. Thus two equivalent symbolic expressions of the hydrogen molecule are H:H and H–H, with their different ways of indicating the sigma bond. Other examples of molecules involving sigma bonds are

water

H:Ö̤:H

hydrogen fluoride

H:F̤̈:

ammonia

H:N̤̈:H
H

methane

H
H:C̤̈:H
H

Similarly, for a molecule involving one sigma bond and one pi bond, we would have

:Ö::Ö:

while for a molecule with one sigma bond and two pi bonds, this convention gives

:N:::N:

The system is also useful for indicating the electronic state of a *radical*. The latter is a group of atoms, present in a series of compounds, which maintains its identity during chemical changes affecting the rest of the molecule, but which is usually incapable of independent existence. The hydroxide radical OH–, for example, is indicated thus

:Ö̤:H$^-$

the minus sign being associated with the extra electron.

a molecule in which both s and p electrons are involved. The well-known acid, hydrogen chloride, is a case in point. The bond is established by overlap of electron probability density in the region between the two nuclei, on their axis of separation. From the above, we see that the $3p_z$ orbital of the chlorine atom is aligned along this same direction.

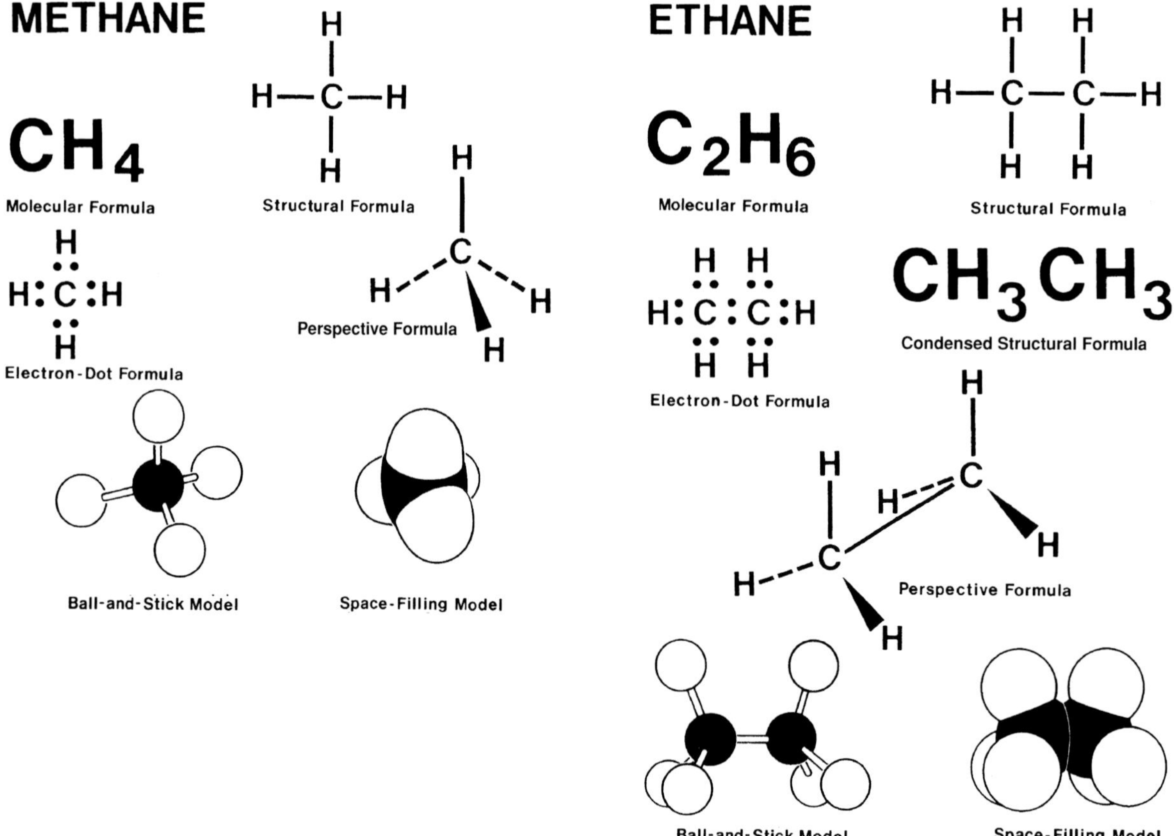

Although the structures of molecules are most accurately displayed by their electron probability distributions, simpler representations are usually adequate. The most common forms are illustrated here, for the smallest two members of the *alkane family*: methane and ethane.

In all our examples thus far the final electron distribution has been symmetrical with respect to the line joining the two nuclei. Such bonding is said to be the sigma (σ) type, and because of the symmetry one of the atoms could in principle be rotated with respect to the other without appreciably changing the energy. We now turn to a different type of bond in which this easy rotation is not possible. An example occurs in the oxygen molecule. Each oxygen atom contains eight electrons, the outer four of which are in the 2p state: two in a filled $2p_x$ orbital, and one each in the $2p_y$ and $2p_z$ orbitals, so that neither of these is filled. As in the fluorine molecule, there is bonding between the p_z orbitals of the two atoms, and this brings the respective incompletely filled p_y orbitals into close proximity. The latter provide further bonding, these orbitals bending towards each other and finally overlapping. The p_z bonding is of the sigma type, as before, but the p_y bonding is quite different. It produces an electron probability distribution that is asymmetric with respect to the line between the two nuclei. This is pi (π) bonding, and formation of a pi bond imposes resistance to mutual rotation of the atoms; such rotation could not be performed without elongation of the pi orbitals, and this would lead to an overall increase in energy. This is our first encounter with a major factor in the mechanical stability of covalent molecules.

After the fluorine molecule with its 18 electrons and the oxygen molecule with its 16 electrons, the logical next step is the nitrogen molecule. Each atom has only seven electrons, the outer three of which are in the 2p state, with one each in the p_x, p_y and p_z. In just the same way as for oxygen and fluorine, the unpaired p_z electrons in the two atoms overlap to give a *sigma bond*, and as in the oxygen molecule the p_y orbitals produce a *pi bond*. This leaves the unpaired electrons in the p_x orbitals, one on each of the atoms. These too are in close proximity, and they form a pi bond. The nitrogen molecule is even more rigid than the oxygen molecule because it has twice as many pi-bonding orbitals. These examples have involved different elements, but it must not be inferred that a given element can display only one type of bond multiplicity. Carbon–carbon linkages are a case in point. Single, double, and even triple bonds are encountered, and they have different rigidities, strengths, and geometries. The purely sigma type C–C bond is 0.154 nm in length, and it has an energy of 0.58 aJ. For C=C, with its additional pi bond, the values change to 0.133 nm and 1.04 aJ. The C≡C bond has a length of only 0.121 nm, because of the further inward pull of the second pi bond, and the energy rises to 1.36 aJ. The multiple bond characteristics of other covalent links, both homonuclear and heteronuclear, show similar trends.

We have so far considered only molecules containing two atoms. There are certain features of the bonding between atoms, however, that are best illustrated by larger structures. The common gas *methane* is the first member of the family of organic compounds known as the alkanes, or paraffins. Each of its molecules consists of one carbon atom and four hydrogen atoms. What would the above principles reveal about the structure of the methane molecule? How are the hydrogen atoms arranged in space about the carbon atom, and what is the distribution of electrons within the interatomic bonds? In the lowest energy state, or ground state, of carbon the outer two electrons are in unfilled $2p_x$ and $2p_y$ orbitals. Each of these has two lobes stretching out on either side of the nucleus, giving four lobes in all. A superficial glance at this arrangement might suggest that a hydrogen atom is located on each lobe, producing a molecule with the shape of an addition cross. Such a conjecture would be hopelessly incorrect. The shape of a p orbital can be misleading; each lobe seems capable of having another electron attached to it. But with the one electron already present, this would give a total of three, which is one in excess of the permitted number. Faced with this difficulty we could be excused for doubting that the compound CH_4 exists. Should not the hydride of carbon have the formula CH_2, with the two hydrogens subtending a right angle at the carbon atom?

The answer to this puzzle lies in an effect that we have not yet encountered: *electron promotion*. With the expenditure of a modest amount of energy, one of the 2s electrons can be transferred to the $2p_z$ orbital, producing a situation in which there are four unpaired electrons instead of two. The four hydrogen atoms, with their single electrons all in 1s states, can then be attached, and the net result is a molecule involving four bonds. The energy gained through establishing bonds with two extra hydrogen atoms more than offsets the energy expended in promoting the electron up to the $2p_z$ state, so that the end result is a stable molecule. This is analogous to the increased lifetime earning power that rewards a student for the modest investment of the relatively impecunious college period. But what about the distribution of the four bonds in space? Following the electron promotion, there are three p orbitals and one s orbital. By analogy with the hydrogen and hydrogen fluoride molecules, we might conclude that the methane molecule has three hydrogen atoms at fixed positions, pairs of them subtending right angles at the carbon atom, while the fourth is involved in a sigma bond with the 2s electron of the carbon atom. This is still not the correct answer, because we need just one more concept: *hybridization of electron orbitals*.

Although the behaviour of electrons in atoms can usually be accounted for by referring to electron wave intensity, which is the square of the wave amplitude, we must bear in mind that the amplitude is the more fundamental quantity. It can have both positive and negative values, a distinction that is lost when only the intensity is considered. The epoch-making experiment of Thomas Young, described earlier, can only be understood through reference to the amplitudes of the light waves. We have already seen that any combination of solutions arising from the Schrödinger wave treatment is itself a solution. The only restriction is that the number of composite orbitals must equal the total number of orbitals used in their construction. When two orbitals coalesce, or fuse together, the individual wave amplitudes are added. Let us consider such hybridization, as it is called, in the case of an s orbital and a p orbital. Although the two lobes of a p orbital have equal probability, the instantaneous situation corresponds to a negative amplitude on one side of the nucleus and a positive amplitude on the other. For the s orbital the amplitude is instantaneously positive everywhere or negative everywhere. Where the two orbitals overlap, the amplitudes will tend to reinforce each other on one side of the nucleus and cancel on the other, and this gives the hybrid orbital an asymmetry which skews the resulting wave intensity over to one side. Reversing the amplitudes of the initial p orbital gives the other hybrid, with its denser lobe lying on the other side of the nucleus. It was Linus Pauling, in 1931, who first appreciated the chemical possibilities inherent in these principles.

Let us now return to methane. After promotion of one of the 2s electrons into a $2p_z$ orbital, a total of four orbitals are available for bonding. To these must be added the four s orbitals of the hydrogen atoms. In a manner similar to the above example, the final configuration has four bonding orbitals that are hybrids between the s and p states of the carbon and the s states of the hydrogens, and four non-bonding orbitals. The four bonding orbitals are symmetrically arranged so as to locate the carbon atom at the centre of a tetrahedron whose corners are occupied by the hydrogen atoms. The geometry of the tetrahedron is such that the four lines that can be drawn from the four corners to the centre are of equal length, and pairs of them always subtend the same mutual angle of 109.5° This, then, is the structure of the methane molecule.

Promotion and hybridization are the key to the surprising discoveries of noble gas compounds in 1962. The absolute inertness of these elements had always been taken for granted until 1933, when Linus Pauling predicted that they might display chemical activity, under suitable conditions. The first such compound to be produced, by Neil Bartlett, was xenon hexafluoroplatinate, $XePtF_6$, and this was soon followed by the production of xenon tetrafluoride. In the latter, two of the electrons in the outermost p orbital of the xenon atom are promoted to the adjacent d orbital, and hybridization then produces the required four bonding lobes for the fluorine atoms. The compound is formed quite easily, by heating the two elements to 400 °C, and the resulting crystals are unexpectedly stable at room temperature. The ionization energies and electron affinities of the noble gases diminish with increasing atomic radius, so xenon and radon should be the most reactive members of this family. Helium and neon, and possibly also argon, are probably absolutely inert.

Given the variety of bonding types, it is not surprising that so many different classes of material are found on the Earth's surface. And their properties vary widely. The ionic bond is not saturated, and because the electric force is long-ranged each atom interacts with many of its neighbours. The positive and negative ions in ionic crystals tend to appear alternately along the atomic rows, and the crystals are strong and stable. Sodium chloride, for example, melts at 801 °C. In divalent examples, such as magnesium oxide, doubling the charge gives electrostatic forces that are four times stronger. This oxide melts at 2640 °C. The covalent bond is the strongest link of all. The hardness of *diamond*, the epitome of a covalent crystal, is well known. *Graphite*, which like diamond is pure carbon, is also partially covalent. It does not melt at atmospheric pressure, transforming instead directly to the gaseous or vapour phase around 4000 °C. Metallic bonds can be almost as strong. The covalent bond energy in diamond is 0.30 aJ, while the metallic bond energy in lithium is 0.27 aJ. Metals do tend to have lower melting temperatures, but this is probably attributable to their lack of directional bonding. The relative weakness of the hydrogen bonds between water molecules is manifested in the fact that this substance melts at 0 °C. The oxygen and hydrogen atoms in a single water molecule are held together by covalent bonds, and these are not ruptured in the melting transition. The van der Waals bond is weaker still, and argon, for example, melts at −189 °C.

As if this were not variety enough, further nuances of behaviour derive from the mixture of bonding types found in some substances. *Mica*, for example, is a layered structure in which weak van der Waals forces hold together the sheets of atoms within which the bonding is covalent. This gives such substances their characteristic flaky or soapy feel. And there is yet another source of diversity. Let us return to some simple examples of the covalent bond. In the hydrogen molecule it has an energy of 0.72 aJ, while for the fluorine molecule the energy is 0.25 aJ. We might expect the bond in a hydrogen fluoride molecule to have an energy intermediate to these two, but the value is found to be 0.94 aJ. Linus Pauling has shown that a new factor must be taken into account in such heteronuclear molecules, and it arises from the differing electronegativities of the two atoms. This means that the bonding, in spite of its basically covalent character, will derive a certain contribution from ionic effects. It is therefore wrong to regard the different bonding types as being mutually exclusive.

On our macroscopic level of consciousness, it is natural that interatomic bonds should be taken for granted. We become aware of their importance only when they expose their limitations, such as when an object subjected to excess stress breaks in two. We do not put the fracture surfaces together, hoping to make them adhere. It is not possible to establish sufficiently intimate contact, and anything less than this is inadequate because of the short-range nature of the interatomic bond. Even at a distance of a few atomic diameters, the positive charges on the nucleus and the negative charges of the orbiting electrons totally cancel each other's influence. The net force at this distance is thus zero even though the underlying electric force is long-ranged. When two objects are in contact, some of the atoms on one surface must of course be within range of atoms on the other. We see the effect of this limited bonding in the phenomenon of *friction*, and its extent can be increased by mutual deformation of the surfaces and lack of *lubrication*.

Summary

A chemical compound is stabilized by the short-range forces between the constituent atoms. The magnitude of the interatomic force is determined by the behaviour of the electrons of each atom when other atoms are in the vicinity. The two major factors are the ionization energy, which is the amount of energy required to pull an electron away from an atom, and the electron affinity, the strength of the attraction between a neutral atom and an extra electron. The two contributions are collectively measured by the electronegativity or, qualitatively, by the polarity. In strongly bound condensed matter, primary bonds exist between atoms of the same polarity (covalent bonds, metallic bonds) or opposite polarity (ionic bonds). The covalent bonds between atoms of electronegative elements involve electron sharing, and they are saturable and show stabilizing directionality. Double and triple bonds occur, and they influence the strength, geometry, and rigidity of the interatomic linkages. In the absence of primary bonding, molecules can be stabilized by weaker secondary forces of the van der Waals or hydrogen bond type.

3

All things began in order, so shall they end, and so shall they begin again; according to the ordainer of order and mystical mathematics of the city of heaven.

Thomas Browne (1605–1682)

The Garden of Cyrus

Atoms in concert: *states of matter*

Until now, our discussion has concerned isolated atoms and clusters of very few atoms. Considering that a single cubic centimetre of a typical piece of condensed matter, a block of stainless steel for example, contains approximately 10^{23} atoms, we see that there is still a long way to go in the description of actual substances. We must now inquire how large numbers of atoms are arranged with respect to one another, so as to produce the great variety of known materials. It will transpire that the observed arrangements are dictated both by the forces between atoms and the prevailing physical conditions. Before going into such details, however, it will be useful to examine the simple geometrical aspects of the question.

It is frequently adequate to regard atoms as being spherical. The two-dimensional equivalent of a sphere is a circle, so we can begin by considering the arrangement of circles on a piece of paper. The circles could be drawn at random positions, well spread out and with close encounters occurring only by occasional chance. This is a reasonable representation of the instantaneous situation in a simple monatomic *gas*. A variant of this drawing would replace the single circles by small groups of circles, and we would then have a molecular gas. Returning to the single circles, more of these could be added until ultimately no further circles could be placed without overlapping some of those already drawn. This would be equivalent to the situation in a very dense gas, while a picture of the *liquid state* could be obtained directly from it by compressing the assembly slightly so as to push the individual circles into direct contact with their neighbours. An example of the analogous situation in three dimensions is the packing of peas in a jar or apples in a barrel, the act of compression doing nothing to remove the randomness of the original arrangement. Now suppose that we set ourselves the goal of packing as many circles on the paper as possible. It would not require much contemplation to arrive at a solution in which they are systematically drawn in neat rows, with the rows stacked tidily against one another. The resulting picture would reveal various types of symmetry, depending on the direction of observation, and it would be a simple representation of a *crystal*.

This does not exhaust the possibilities. The circles could be strung together, so as to compose worm-like structures tortuously winding their way around the paper. An assembly of such strings would give a simple picture of a *polymeric substance*. The strings might not be independent. They could be joined together by smaller strings, as in a *cross-linked polymer*. Adding the third dimension makes little difference. In the crystal the lines of atoms become planes, stacked one upon another to form the

Although the idea of atoms, as the ultimately indivisible particles of matter, had already emerged in Ancient Greece, Plato (427–347 BC) and his colleagues emphasized the states of matter rather than chemical species in their classification of the elements. Their preoccupation with simple geometrical shapes is betrayed in the assignment of the cube, the least mobile polyhedron, to the element earth, while the pyramid, octahedron, and icosahedron were identified with fire, air, and water, respectively. The pyramid (tetrahedron) is the most pointed, whence its connection with heat, while the icosahedron has the roundest shape, which explained the soft feel of water. The intermediate nature of the dodecahedron, with its large number of faces, was vaguely associated with the cosmos.

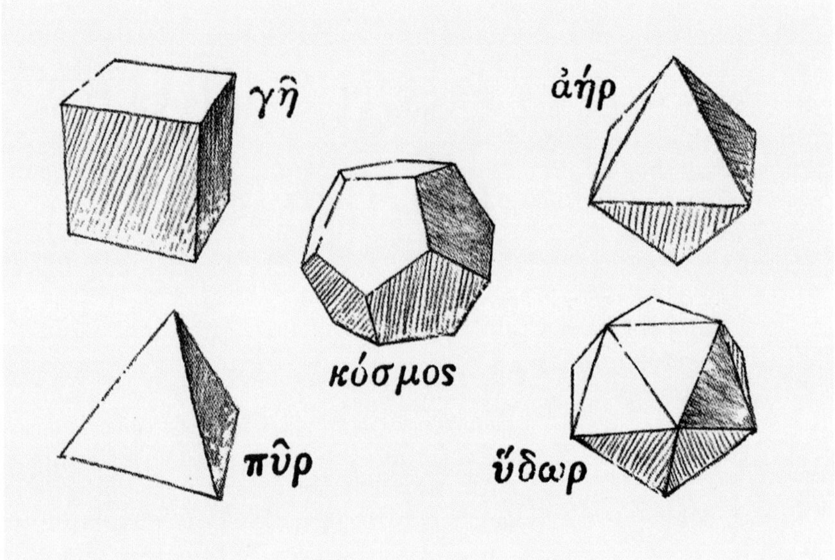

three-dimensional structure. In the case of a polymer the third dimension is topologically (i.e. spatially, in the sense of the manner in which the various regions are mutually connected) important because it allows the strings of atoms to avoid one another by passing over and under. It also enables them to adopt special conformations such as the helical arrangement that is of great importance in certain biological structures.

In drawing the circles we were given a free choice, and the patterns and structures were limited only by our powers of fantasy. In Nature, things are obviously not as arbitrary as this. The chief limitations are imposed by the *interatomic forces*, and the behaviour of a substance is determined by the interplay between these forces and the prevailing thermal conditions. A strong hint that this is so comes from our knowledge of simple *phase changes*. Ice, for example, melts if its temperature rises above 0 °C. The forces between water molecules are not appreciably temperature dependent, so the reason for this *melting* must be that the effect of the intermolecular forces is overcome as the temperature is raised. The ability of these forces to hold the molecules together in the crystalline state somehow becomes impaired. Similarly, when water is further heated until its temperature rises to 100 °C, at a pressure of one atmosphere, it *boils*, and the liquid is freely converted into *water vapour*. The *thermal agitation* has now become so violent that molecules on the surface are able to overcome the attractive pull of their neighbours and move off into the space above, at a rate exceeding that of the opposing arrival of vapour molecules at the surface. To appreciate the influence of temperature on the structure of materials, we must delve deeper into those important branches of science known as *thermodynamics* and *statistical mechanics*.

There have been many examples of scientific fields that began as a collection of abstract ideas, only later finding practical application. This was not the case with thermodynamics. It grew out of the needs of practical engineering, and it was a child of the Industrial Revolution. Engines were required both to drive the wheels of mass production and to pump the water from coal mines, the dwindling forests being unequal to the task of satisfying the

Entropy and free energy: a simple demonstration

The consequences of the first and second laws of thermodynamics were summarized in a famous statement made by Rudolf Clausius (1822–88): the energy of the universe is constant; the entropy is increasing towards a maximum. This indicates an inexorable trend towards molecular chaos, and an ultimate situation in which no further energy can be extracted from the environment. There is no cause for alarm, however; this *thermal death of the universe* is estimated to occur in several billion years (and some believe that it will be followed by a global shrinking, in what is referred to as the *Big Crunch*, the way then being clear for a new *Big Bang*). The Clausius statement refers only to the universe as a whole; there is nothing to prevent part of it from experiencing a decrease of entropy. If an elastic band, preferably a wide one, is rapidly stretched and placed against a moistened lip, it feels warm. In the unstretched state, the randomly twisted and kinked arrangement of the rubber molecules gives a high entropy. Stretching aligns them, and the entropy decreases. Because the change is sudden, the free energy rises, and this is subsequently dissipated as heat which causes the temperature to rise. If the stretched band is allowed to accommodate to the temperature of the lip, a sudden relaxation of the tension makes it feel cool. The process has been reversed, with an increase in entropy and a fall in free energy.

ever-increasing demand for fuel. There was naturally a commercial interest in making these engines, such as the steam engine perfected by James Watt in 1776, as efficient as possible. Through the studies of Sadi Carnot, Hermann von Helmholtz, James Maxwell, and others, certain fundamental laws were gradually established. The *first law of thermodynamics* states simply that energy is conserved: it can be neither created nor destroyed. We cannot get something for nothing. As a direct consequence of this, when a system is made to proceed from an initial state to a final state, the resulting change in its energy is independent of the way in which the change is brought about. The systems originally referred to by this law were engines, but all material things, whether they be inorganic crystals or living organisms, must obey the same rule. A *second law of thermodynamics* was also initially couched in very practical terms. It can be formulated in various, though equivalent, forms, the two most common of which are: in an isolated system heat cannot be made to flow from a colder region to a hotter region, and it is not possible for an engine to be 100% efficient. These laws exploded the pipe dream of perpetual motion. For our purposes here we must interpret the second law in atomic terms, and before this can be done, it is necessary to introduce a new concept: *entropy*.

Consider a container of compressed gas, such as a hairspray can, in a closed room. Depression of the valve releases some of the gas, enabling the system to be used as a primitive engine. As such it is capable of work: the spreading of a layer of liquid over the hair. If the valve is kept in the depressed position, all the compressed gas will ultimately escape, and the engine will have run down; it will be incapable of further work. Let us compare the initial and final states of this system. Considering only those gas molecules that started off in the can, we see that the initial state could be said to be associated with a higher degree of information, because the molecules were localized within a smaller volume. The smaller the can, compared with the room, the greater the relative organization of the gas molecules in the initial state. Information content and organization are indeed two aspects of the same concept. When the valve is depressed and the gas rushes out, the system becomes more disorganized. It achieves, on

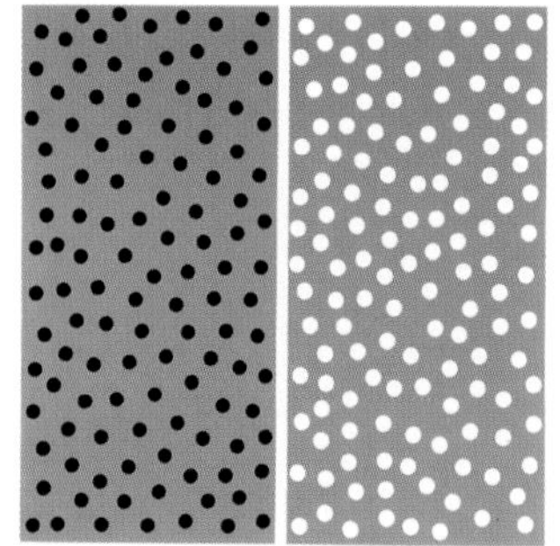

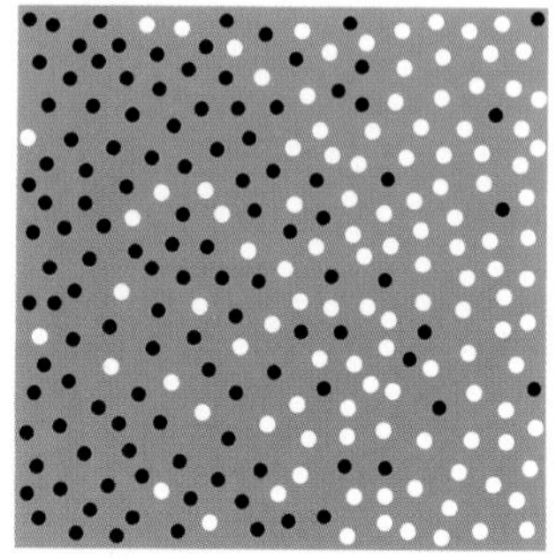

The interpretation of entropy given by Ludwig Boltzmann (1844–1906) is illustrated in this simple experiment in which two different gases are initially separated by a partition and are thus relatively confined. Removal of the barrier permits each gas to diffuse into the other and to gain access to the entire volume of the container. The entropy of the system, being inversely related to the degree of order, is clearly increased as a result of the mixing process. (It must be emphasized, however, that the term *degree of order* refers not only to the atoms' spatial positions but also to their velocities.)

the other hand, a more probable state. The gas molecules are in a constant state of motion within the room, and if the valve remains depressed, some of the molecules will actually re-enter the can. Experience shows us, however, that it is extremely unlikely that so many of the molecules will again be simultaneously located within the can that the original compressed state is re-established. Thus the natural tendency of the system is towards a state with higher probability, greater disorder, and lower information content. Ludwig Boltzmann, in 1896, interpreted entropy as being a measure of the probability of the state of a system. Entropy increases with increasing disorganization, that is with a lowering of the order, and decreasing information content. Thus entropy and information are inversely related, and the latter is referred to as negentropy in this context. We can now put the second law of thermodynamics on an atomic footing: in a closed system the change is always towards a situation having a higher entropy.

Before proceeding farther, we must have a clear idea of what is meant by the terms *heat* and *temperature*. The use of these two words as if they were interchangeable is incorrect. Temperature refers to the state of a substance, and it is independent of the amount of material involved; it is referred to as an *intensive quantity*. Heat, one form of energy, is an *extensive quantity*. At a given temperature the amount of heat associated with an object is proportional to its mass, and it is also determined by the *specific heat* of the substance concerned. Thus if a steel saucepan and a steel spoon have the same temperature, the saucepan, having the larger mass, will contain more heat. A landmark in the understanding of thermal effects was the observation, by Benjamin Thompson (Count Rumford) in 1798, that the temperature of the barrel of a cannon increases when it is bored. He concluded that this effect is caused by friction and that there must be a connection between mechanical work and heat energy. Their actual equivalence was subsequently demonstrated experimentally by James Joule. Because these two forms of energy had separate histories, they were naturally measured in different units: heat in calories and mechanical work in a unit named after Joule himself. He was able to establish the proportionality between the two, the recognized ratio being that one calorie equals 4.1868 J. At the microscopic level, the interpretation of heat in mechanical terms leads naturally to the idea that thermal effects arise from molecular motion. A piston acquires its pushing power from the countless collisions that the molecules of the working substance make with it. In the mechanics of motion, velocity is an intensive quantity, it being independent of mass, while kinetic energy is clearly extensive. Galileo's falling stone and feather had the same velocity, but the stone had the higher kinetic energy. As was realized by Lord Kelvin (William Thomson), the extensive quantities, heat and kinetic energy, are to be identified with each other, as are the intensive quantities temperature and velocity. The greater the molecular velocities, the higher is the temperature. There is a complication here. Maxwell and Boltzmann showed that in an assembly of moving molecules there must be a distribution of velocities; for a given temperature, at any instant, some molecules will be moving relatively quickly while others are more sluggish. This arises from the frequent collisions, the velocity of a molecule invariably being changed when it strikes another. The temperature of an assembly of molecules is therefore an averaged measure of all the motions, and it turns out that it is determined by the average squared velocity. Kelvin realized that the true

Calories and joules

Before Count Rumford (Benjamin Thompson 1753–1814) unified them in 1798, heat and mechanical work were regarded as being unrelated, and they were thus measured in different units. They are now recognized as different aspects of the same concept, energy, and 1 calorie is equivalent to 4.1868 joules. The calorie is the amount of heat energy required to raise the temperature of one gram of water through one degree on the Celsius scale. The dietician's unit for the energy value of foods is the capitalized Calorie, and this is 1000 calories. A 20-gram slice of bread gives about 50 Calories, and a 2-decilitre glass of milk about 140 Calories. We use up about 300 Calories per hour when walking quickly, and 800 Calories per hour climbing stairs. The rate is only 25 Calories per hour for reading, and about double that when we eat. An ideal adult diet provides approximately 2400 Calories per day. This is a power of 120 watts, equivalent to two medium-powered domestic light bulbs. Most of our intake is used in such vital functions as maintaining body temperature, circulation of the blood, transmission of nerve signals, and replacement of cells. Only a small fraction is actually used in breaking interatomic bonds, but it is instructive to compare our energy consumption with the total bonding energy in the body. Using the fact that one gram of carbon contains 5×10^{22} atoms, and assuming about three bonds per atom, we find that an adult human is held together by some 10^{28} bonds. Taking a typical bond energy as 0.6 aJ (i.e. 0.6×10^{-18} joules), we see that this is equivalent to a total bonding energy of 1.5×10^{6} Calories. This is roughly six hundred times the daily energy intake.

zero of temperature corresponds to the situation in which all molecular velocities are zero. This null point on the Kelvin scale lies at −273.16 °C.

We can now appreciate the significance of entropy and its inexorable increase during the changes of an isolated system. The loss of information, which is associated with the randomizing of molecular motions, is equivalent to a decrease in the amount of energy available for useful work, and this loss is greater the higher is the temperature. The essence of mechanical work is that it involves motion that is performed in a directed or coordinated manner. When it is converted to heat, the motions become those of the molecules, and these are *chaotic*; the energy has been *irreversibly degraded*. The heat can be used to provide further mechanical energy, but the overall loss is inevitably considerable. We would not attempt to obtain traction power from the heat generated by the brakes of a car, whereas the use of a brake-flywheel arrangement has proved profitable. Entropy-related energy can be called unavailable energy, to distinguish it from the energy available to drive a system. Employing slightly different approaches, Helmholtz and J. Willard Gibbs independently realized that the energy that can be used by a system is equal to the total energy minus the unavailable energy, and they called this net quantity the *free energy*. It is the existence of unavailable energy that makes an engine less than ideally efficient, and we will see that entropy is of paramount importance in determining the structure of condensed matter. Entropy is not always directly associated with motion. It can be present as a property of a molecular arrangement. The recording and erasure functions of a tape recorder are achieved with an electromagnet which, respectively, aligns and randomizes the magnetic domains contained in the recording tape. One can purchase both blank tape and tape with prerecorded music. The latter costs more even though the two products are identical in purely material content. The higher price of the tape with the music is of course related to the energy that has been expended in making the recording. Employing the terms used earlier, the tape with the music

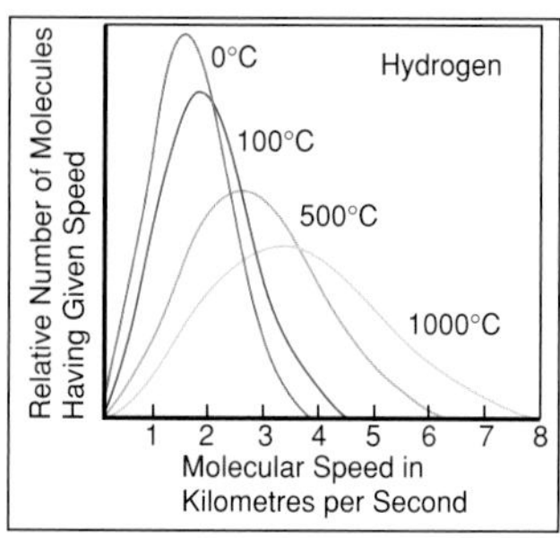

The apparent quiescence of a stationary gas belies a frenetic activity at the molecular level. The individual molecules move about at a rate which increases with rising temperature, colliding with each other and with any object that lies in their (straight) paths. At a given temperature there is a well-defined distribution of velocities. In hydrogen at room temperature, for example, many of the molecules are travelling at one and a half kilometres per second. This is about thirty times the speed of a hurricane, but we detect no disturbance because the motions are random. When a wind is blowing, the velocity distribution acquires a slight bias in one direction with the result that there is an overall drift of molecules down-wind.

contains more information than the blank tape, and this takes the form of alignment of the magnetic domains. But more information means lower entropy, and because the latter lies on the debit side as far as free energy is concerned, the tape with the pre-recorded music has a higher free energy.

The most important fact to emerge from the analyses of Helmholtz and Gibbs concerned the equilibrium state of a system. They were able to show that this condition prevails when the free energy is at a minimum. In other words, a system moves towards equilibrium by decreasing its total energy and increasing its entropy. It follows that if two possible states of a system are in competition at a given temperature, the more stable state will be that having the lower free energy. The beauty of this approach lies in its ability to make predictions regarding the macroscopic behaviour of a system, on the basis of the statistical properties of matter at the microscopic level. This sounds quite straightforward, but the problem lies in the fact that the free energy is generally a difficult quantity to calculate. In computing the energies of different possible microscopic states, one actually requires a knowledge of the positions of all the nuclei and electrons and of their interaction energies. Assuming that such information can be obtained, however, the analysis is then simply a routine evaluation of the relative probabilities of the different possible microscopic states for any given temperature. That many different states can correspond to the same temperature is a direct consequence of the range of velocities present in any assembly of molecules, and also of the inherent chaos of thermal motion.

The evolutionary trend of living things towards greater sophistication seems to defy the principle of entropy increase. The more complex an organism is, the greater is the amount of genetic information required to produce it. But plants and animals are not isolated systems, so they do not violate the laws of thermodynamics. They are built at the expense of their environments, and they survive by degrading energy. The assembly of a faithful copy of the genetic message from smaller molecules, which takes place before every division of a living cell, represents a local increase in free energy. This is more than compensated for by the decrease in free energy of the organism's surroundings, however. On a cosmic scale the entropy inevitably increases, and it seems that the Universe must gradually die away to a state of total chaos. This is probably not what T. S. Eliot had in mind, but a passage from *The Hollow Men* is appropriate: 'This is the way the world ends, not with a bang but a whimper'.

In the light of what has been discussed, let us consider what happens when a kettle of water is placed on a hot stove. Heat energy passes through the bottom of the kettle and into the water, and this gives the water molecules a higher kinetic energy. The average velocity, and therefore the temperature, increases. If the stove is sufficiently hot, the temperature of the water will ultimately reach 100 °C, which is its boiling point at a pressure of one atmosphere. The velocities of the molecules in the liquid are not equal, and there will always be some of them at the surface of the liquid with a velocity sufficiently high that they can overcome the attractive forces of their neighbouring molecules and escape into the space above. If the water was contained in a closed vessel, these molecules would exert the pressure that is known by the name *vapour pressure*. Pressure arises because of the impacts of the vapour molecules with the walls of the vessel. As the temperature of the water rises, the fraction of molecules that are able to escape from the surface increases, and the vapour pressure consequently rises. The boiling

Just below
melting temperature

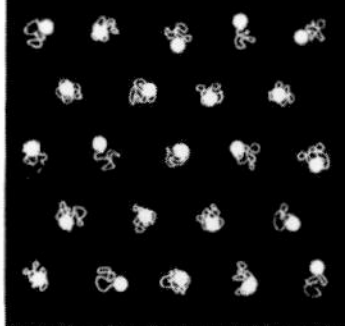

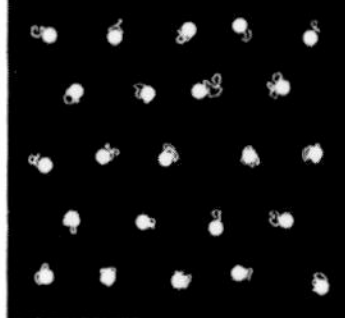

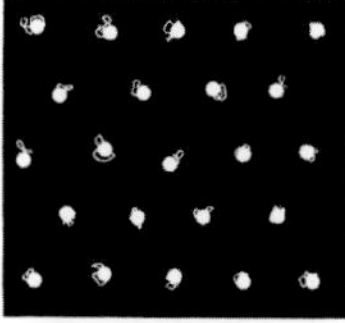

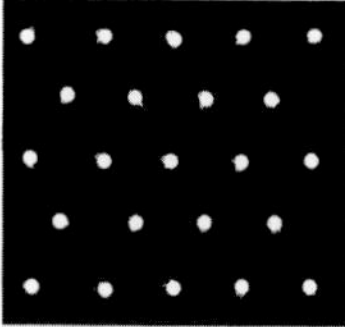

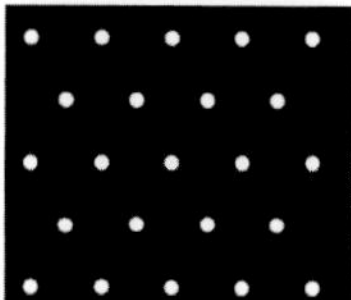

Absolute zero
temperature

The connection between *heat* and the *energy of motion* is revealed by this schematic illustration of the amplitude of thermal vibration of the atoms in a crystal. At the absolute zero of temperature, −273.16°C, all the heat energy is frozen out and the atoms lie motionless at their appointed positions in the lattice. (This neglects the small quantum effect known as *zero-point energy*.) As the temperature increases, the thermal motions become more vigorous until the crystal loses its thermodynamic competition with the liquid state and therefore melts. The atoms are indicated by the dots, and their paths by the irregular lines.

point corresponds to the situation in which the vapour pressure equals the external pressure. Atmospheric pressure decreases with increasing altitude, so the mountaineer's kettle boils at a lower temperature. The issue of interest here is the competition between the liquid and gaseous states of water, immediately above the boiling point. From our earlier discussion, it must be concluded that the free energy of steam is lower than that of liquid water for temperatures in excess of 100 °C, so we must consider the relative magnitudes of the components of that quantity for the two phases. The energy contribution that arises from the interactions between the molecules favours the liquid phase; in that state the separation distances between neighbouring molecules are closer to the distance corresponding to the minimum in the potential energy interaction. The entropy factor, conversely, favours the gaseous state because this is characterized by the greater degree of chaos. And as we saw earlier, this latter contribution increases with increasing temperature. At a particular pressure, there must be a temperature at which the difference in interaction energies of the two phases, referred to as the *latent heat*, is just offset by the difference in the entropy terms, and this is the *boiling temperature*. But the two phases in balance are physically distinct only if the temperature is below a critical value. For water at 100 °C and atmospheric pressure the specific volumes of the liquid and vapour phases differ by a factor of about 1600. This factor decreases as the temperature rises, and at the *critical point* (374.2 °C and 218.3 atm for water) the *liquid–vapour interface* becomes unstable; above this point the liquid state passes smoothly into the vapour as the volume increases. Such continuity is made possible by the structural similarity of the two states; they are both disordered.

An apparently more complicated phenomenon is observed at a lower temperature. If liquid water is cooled below 0 °C, at atmospheric pressure, it transforms to the solid form: ice. Again there must be competition between the free energies of the two phases, with exact equality obtaining at the transition temperature. Ice is crystalline, and the most significant characteristic of this form of matter is the orderly pattern in which the atoms or molecules are arranged. The liquid state of water, and indeed of any substance, is distinguished by its relative disorder, and this gives a relatively high entropy. The energies of interaction, on the other hand, are lower in the crystalline state. Thus again there are two factors which tend to offset each other, and there is balance at 0 °C.

The *melting transition* is still not fully understood. The reason for its comparative subtlety lies in the fact that both crystal and liquid are examples of what is known as *condensed matter*, and the volumes occupied by these two phases usually differ by no more than a few per cent at atmospheric pressure. In this transition the molecules are not able to escape from the influence of their neighbours, and yet something does happen to break down the regular atomic layout of the crystal into the disorder of the liquid. John Bernal, in 1962, proposed a simple model of a typical liquid which has been surprisingly successful in explaining many of the experimental observations. It was a development of earlier studies made by Peter Debye and H. Menke, and also Joel Hildebrand and W. E. Morrel. Using steel spheres which, if placed in a regular array, would represent a simple close-packed crystal, he simulated a liquid by packing these into an irregularly shaped container. A salient feature of such a model, apart from its lack of crystalline order, is that although the spheres are still in contact, they have, on average, fewer neighbour contacts than they would have in regular packing. This too is in

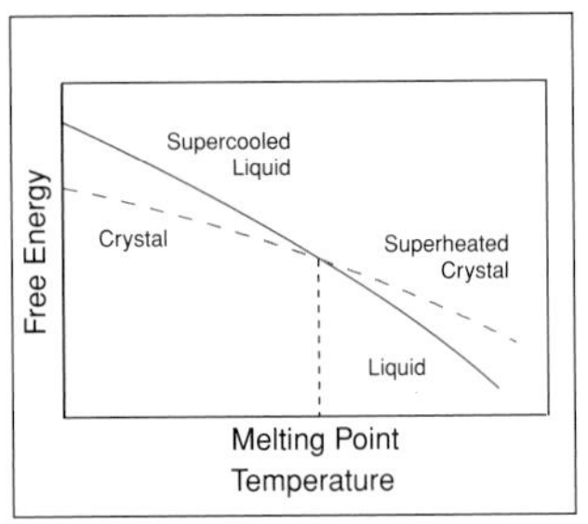

The most stable state for a given set of physical conditions is that for which the *free energy* is lowest. At a sufficiently low temperature the favoured form of all elements and compounds is the crystal. It has a lower entropy than the corresponding supercooled liquid, but this unfavourable factor is more than offset by the crystal's lower potential energy. As the temperature is raised, the free energy of the liquid state shows the more rapid rate of decrease, and at the *melting temperature* it becomes the thermodynamically stable form.

agreement with real liquids, as is the fact that randomization in the model leads to a volume expansion of several per cent. Water is anomalous in this respect because it actually contracts upon melting. The lower density of the crystalline phase is the reason why ice floats on liquid water, and it is also the cause of burst plumbing in winter. The inadequacy of the model is that it is static whereas the atoms in liquids, and in crystals near their melting points, are in rather vigorous motion. John Finney has shown that it is more applicable to a glass. Research into the nature of the melting transition continues, one idea being originally proposed by I. Frenkel, and later developed by Morrel Cohen, Gary Grest and David Turnbull: the fluidity of a liquid is conjectured as being caused by small regions of free volume which open up instantaneously at random locations, enabling nearby atoms to jump into them. Such free regions are presumably absent in the crystal. A different approach, adopted in 1986 by Rodney Cotterill and Jens Madsen, stresses the many-body nature of the problem, and requires consideration of what is known as the *configurational hyperspace*. This is a simultaneous description of the positions of all the atoms (or molecules), and there are fundamental differences between the hyperspace topologies of the crystalline and liquid states. No critical point for melting has been observed, and the essential incompatibility between crystalline order and liquid disorder suggests that such a temperature will never be found.

The physical states of pure elements and compounds in the terrestrial environment is dictated by the atmospheric temperature and pressure. Mercury is normally seen in its liquid state simply because its melting point lies well below room temperature, at −38.4 °C. For some elements, such as hydrogen, nitrogen, oxygen and chlorine, even the boiling temperature lies below room temperature, and these are commonly observed in the gaseous state. Michael Faraday was the first to observe that gases can be liquefied by compressing and cooling them to a sufficiently low temperature, demonstrating this with chlorine (boiling point at atmospheric pressure, −34.7 °C), in 1823. The need for compression in this process was explained fifty years later, by Johannes van der Waals, who showed that this brought the molecules within range of the attractive forces of their immediate neighbours. The cooling is necessary because the liquid phase can be produced only if the temperature is below the critical point. For the majority of elements the *melting temperature* lies well above room temperature, and they are normally experienced in the solid state.

The motions of atoms or molecules in condensed matter are opposed by energy barriers. These arise from the interatomic or intermolecular forces, and the situation is not unlike that of a person trying to squeeze through a crowd. A potential energy barrier can be traversed by a body if the latter has sufficient kinetic energy. Relative motion of atoms or molecules in condensed matter thus becomes easier with increasing temperature, and an increase of such motion produces greater fluidity, or lower viscosity. Many examples of the exploitation of this familiar fact are seen in the kitchen. Butter, cooking fats, and syrup all become more fluid as they are warmed. Atomic motions are required in the formation of a crystal when the temperature of a liquid falls below the *freezing temperature*. This follows from the different arrangement of the atoms in these two phases. Because the atoms become more lethargic in their motions as the temperature falls, this can present the crystallization process with a dilemma. If the liquid is cooled too rapidly past the freezing point, there will be insufficient time for the crystal

Boltzmann's constant
The *thermal energy* present in a material object increases with rising temperature. It is contained in both the *kinetic energy* of atomic motion and the *potential energy* of interatomic interaction. The average total energy of a vibrational mode (i.e. state) in a piece of material is equal to the product of the absolute temperature and a constant named after Ludwig Boltzmann (1844–1906). This constant, the symbol for which k_B or simply k, has the value 1.3806×10^{-23} joules per degree kelvin. Room temperature is about 300 kelvin, so the average vibrational mode energy of an atom in condensed matter at this temperature is approximately 4×10^{-21} joules, or 0.004 attojoules (aJ). A typical covalent bond energy is 0.6 aJ, while that of a hydrogen bond is about 0.03 aJ. Rupture of even the latter type of bond thus requires the cooperative effort of several vibrating atoms. However, the thermal energy at room temperature easily causes relative motion between the molecules in many liquids.

to be established, and what is known as a *supercooled liquid* is produced. Further cooling slows down the atomic motions, and a rigid but disordered solid is ultimately produced. This is a *glass*, an example of a *metastable state*: it has a higher free energy than the corresponding crystal, at the same temperature, but the two states are separated by a situation of even higher free energy. There is a free energy barrier against the *glass–crystal transition*, or *devitrification* as this process is usually called. In contrast to crystal melting, which involves a large and very sudden increase in fluidity, a glass melts gradually. This is a reflection of the fact that in a random assembly the energy barriers opposing atomic motion have a wide range of heights. The above-mentioned hyperspace analysis has also been applied to the glass–liquid transition, which is found to correspond to gradual changes in the hyperspace topology as the temperature rises. The glass-blower's art is made possible by the gradual nature of the glass–liquid transition.

The four states of matter can be classified in various ways. Gases and liquids are collectively referred to as *fluids*, while crystals and glasses are *solids*. If we wish to emphasize density, liquids, crystals and glasses are grouped together under the term *condensed matter*. Finally, we can use regularity of atomic arrangement as a means of distinction and thereby separate crystals from the *amorphous* forms of condensed matter, namely liquids and glasses. Sugar is a familiar example of a substance that can readily be obtained in several different physical states. The solid form, commonly used in the kitchen, is crystalline. If this is melted to produce liquid sugar and then rapidly cooled, by letting molten drops fall into water, a glass is formed which usually goes by the name of toffee. This well-known form of confectionery is metastable, and slowly devitrifies to become fudge. Because this reversion to the crystalline form is accompanied by shrinkage, and consequent cracking, the fudge is mechanically weaker than toffee and is thus easier to chew.

Although the surfaces of crystals and liquids are manifestly different, they both show a tendency towards minimization of surface area. This is because of the energy associated with such a boundary. An atom at the surface lacks neighbours on one side, and this implies a deficit of interatomic bonds. Because the zero-energy situation between two atoms corresponds to infinite separation, bond energies are negative, so the missing bonds at a surface correspond to a positive energy contribution. In the case of a crystal, *surface tension* arises directly from stretching of the interatomic bonds, whereas surface tension in a liquid should rather be looked upon in terms of the extra atoms that must be introduced into the surface when the latter increases its area. It is of course the surface tension that gives small drops of liquid, and also bubbles, their spherical shape. Interesting effects are observed at the interfaces between different substances. Whether or not two liquids are capable of being mixed to form a homogeneous, or uniform, liquid depends on the interatomic forces. An example of complete *miscibility*, important in many well-known beverages, is observed in the water–alcohol system. The behaviour at a crystal–liquid interface depends on the degree of bonding between the two phases. If the strength of the inter-phase bonding is competitive with that between the atoms in the liquid, the liquid is said to wet the solid. *Solders* and *brazes* function as they do because they readily wet metallic surfaces. Mercury does not wet glass, and it is for this reason that its surface, or *meniscus*, in a thermometer is concave towards the mercury side; it bulges outwards. Water wets glass because of the electric

The four common states of matter may be characterized by their degrees of order and the mobility of the individual atoms. In a *crystal*, the atoms are arranged in a regular pattern, so the entropy is low, and the dense packing makes for low atomic mobility. A *liquid* is somewhat less ordered, and the increased mobility enables it to flow and to adopt the shape of a container. The lowest order, and thus the highest entropy, is seen in the gaseous state. In *gases*, the atoms are so widely spaced that a confining surface does not form, and the substance completely fills a container. In a transition from one phase to another, which must occur at constant free energy, the internal energy must change by a certain amount (known as the *latent heat* of the transition) to offset the change in free energy associated with the change in entropy. This compensation can only be perfect at one specific temperature (i.e. the *melting point* or the *boiling point*). *Glasses* are not in thermal equilibrium, and their metastable state is produced by sufficiently rapid cooling of the corresponding liquid.

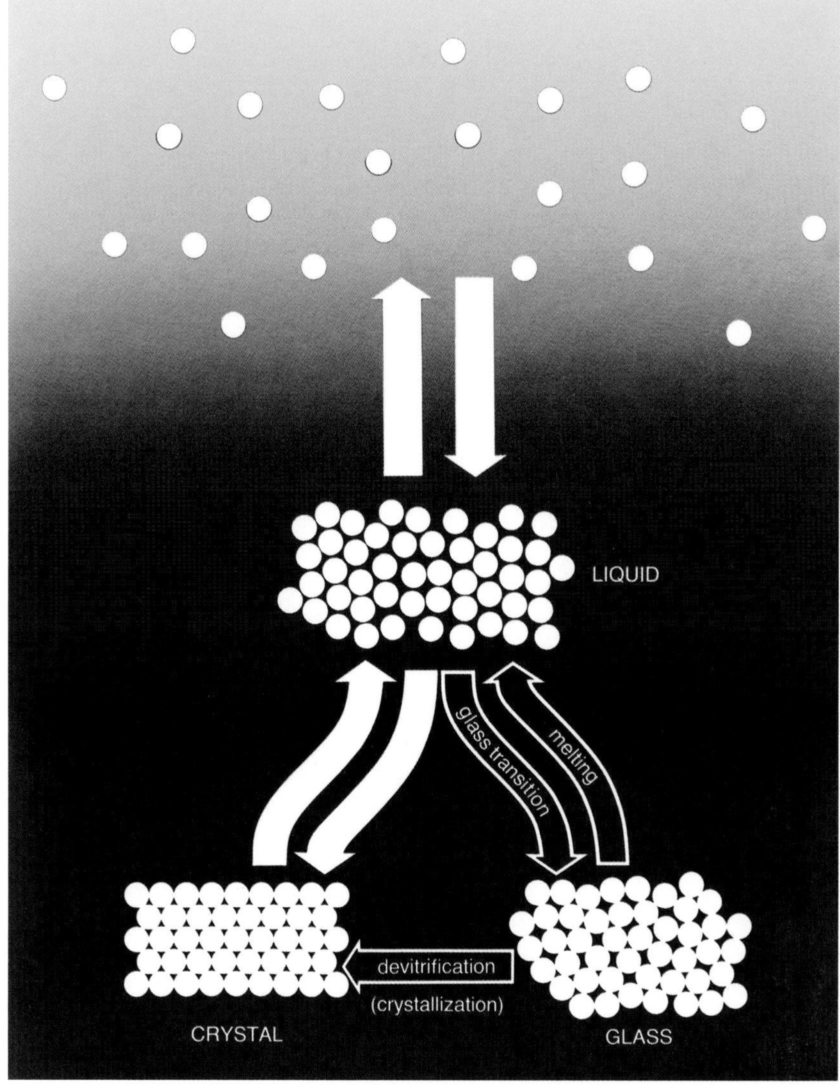

attractions between its molecules and those of the silica in the glass, and its meniscus is convex towards the water side; it bulges inwards. In a tube of sufficiently narrow bore, the inter-phase forces at the meniscus are capable of supporting the weight of a column of the liquid. The smaller the diameter, the longer is the column supported, the phenomenon being known as *capillarity*, from the Latin word *capillus*, meaning hair, and emphasizing the required fineness. Capillarity is extremely important in plant life, the very thin vascular systems being capable of drawing up threads of liquid through considerable distances.

Many important processes involve the intermixing of two substances to produce a third, composite substance. Such mixing must proceed by the passage of atoms or molecules of each species into the other, this process being known as *diffusion*. Of particular interest is the rate at which diffusion occurs, and the rules governing this were elucidated by Adolf Fick in 1855. Increasing temperature increases the diffusion rate. This is not surprising because atoms move faster at higher temperature, but the effectiveness of

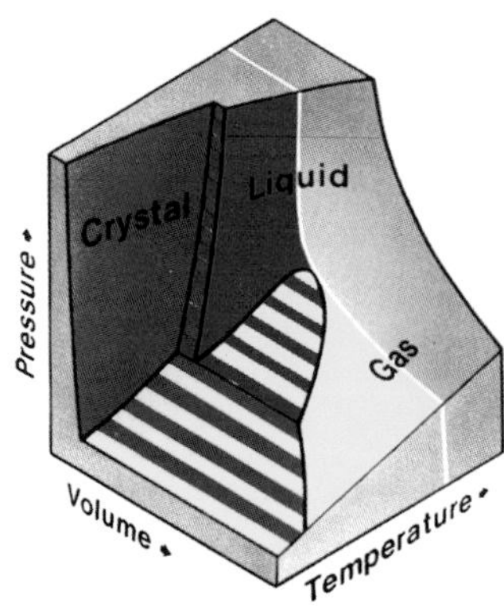

The physical state adopted by a substance depends on the temperature, the pressure, and the volume. If the temperature of a crystalline solid is raised sufficiently, at constant pressure, it melts to become a liquid. Further raising of the temperature ultimately brings the latter to the boiling point, at which there is a transition to the gaseous state. Similarly, sufficient compression of gas, at constant temperature, produces first a liquid and ultimately a crystalline solid. Indeed, if the temperature is low enough, the solid can be produced directly from the gas. The striped regions indicate coexistence of two phases (liquid and gas, for example) under the prevailing conditions. Above the *critical temperature*, indicated by the white line, the distinction between liquid and gas disappears, and compression produces the liquid state without a sudden decrease of volume.

thermal agitation depends on the heights of the energy barriers. These are determined by the configuration of the assembly of atoms through which the diffusing atoms are moving. If this is in the liquid state, diffusion is essentially governed by the exploitation of relatively open patches by atoms having sufficiently large kinetic energy. In a crystal the atomic packing is usually so tight that diffusion has to be mediated by the movement of defects in the regular pattern of atoms, known as the crystal lattice. These defects are known as *vacancies*. They are holes of atomic dimensions, and a diffusive jump involves the direct interchange of a vacancy and an atom of the diffusing species. In a typical metallic crystal close to its melting point the vacancy jump rate is about 10^9 per second. Being thermally activated, the rate falls away rapidly with decreasing temperature, however, and it is down to about 10^5 per second at half the absolute melting point. Other modes of diffusion are known, but they are less common. Although the individual diffusive jumps are made at random, the gradient of concentration gives the overall diffusion a clear directionality. A diffusing species tends to migrate from a position where its local concentration is high to one where the concentration is lower, and this trend continues until the distribution is uniform. The latter behaviour is often loosely referred to as *Fick's law*. Not surprisingly, this is consistent with a tendency towards increasing entropy, or decreasing order.

A redistribution in the spatial arrangement of atoms also occurs in chemical reactions, but in such cases it is frequently the free energy rather than diffusion that is the dominant factor. In the preceding chapter, using the analogy of a car moving through a series of hills and valleys, we saw that a system can be metastable against transition to a state of lower energy because its path to the latter is blocked by an intervening hill. The expenditure of energy in pushing the car to the crest of that hill is repaid with interest when the vehicle then rolls down into the lower valley on the far side. This simple example illustrates changes of potential energy, whereas a proper description of processes at the atomic level requires a full account of the changes in free energy. A system consisting of coal and oxygen is metastable at room temperature; the carbon in the fuel will not combine with the gaseous oxygen until it is heated to above a certain temperature, whereupon the resulting fire will then continue to burn until the supply of one of the ingredients fails. The heating supplies the necessary push to get the system over the barrier and into the lower free energy state in which the carbon and oxygen are chemically combined. A more dramatic example of the same principle is the triggering of a dynamite explosion by the application of a small amount of energy conveyed by an electrical conduit.

The rate at which a chemical reaction proceeds can frequently be increased appreciably by use of what is known as a *catalyst*. In this case the net reaction remains unchanged, but it is achieved by a different path in which an intermediate product is formed en route. Thus instead of the reaction INITIAL → FINAL we have INITIAL → INTERMEDIATE → FINAL, and the two energy barriers involved in the latter pathway are both lower than the barrier in the direct reaction. The net result will then be achieved with the investment of less energy or, at a given rate of energy input, more rapidly. Hydrogen peroxide decomposes only very slowly by itself, but in the presence of manganese dioxide the reaction is almost explosive. A mixture of alcohol vapour and oxygen reacts very sluggishly, but platinum speeds things up so much that the metal becomes red hot. A catalyst appears to work

All *chemical reactions* involve a change of free energy, this quantity being minimized in the case of thermal equilibrium. The chemical compound, glucose, contains more free energy than an equivalent amount of carbon dioxide and water, but it is metastable because a small amount of energy (an *activation energy*) must be expended in order to provoke break-up of a glucose molecule.

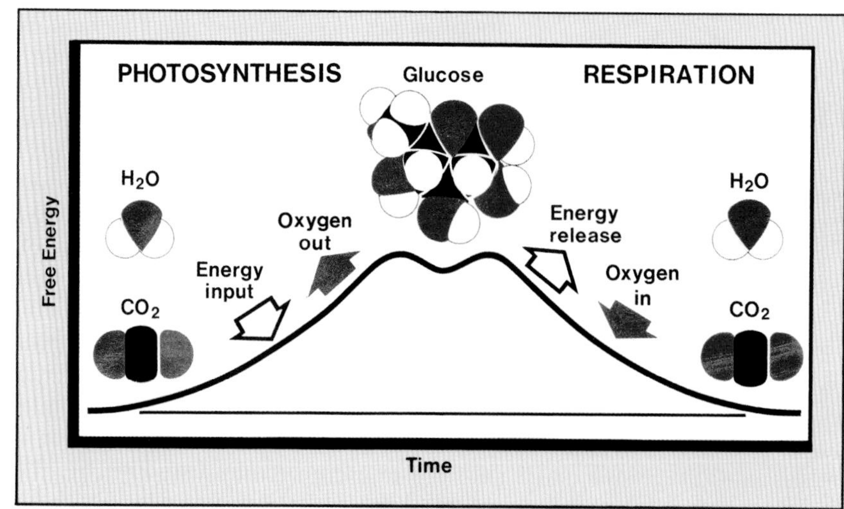

by holding the two reactants in close proximity with each other; it could be called a chemical matchmaker. A small quantity of catalyst is usually sufficient to achieve a vast increase in reaction rate, and the catalyst itself emerges unchanged after the reaction. The use of catalysts in the chemical industry is of course widespread. In biological systems catalysts are known as *enzymes*, and their presence is vital because speeding up of reactions by a rise in temperature is not a viable proposition. One molecule of the enzyme catalase can decompose 40 000 molecules of biologically poisonous hydrogen peroxide in one second at 0 °C. It is important to remember in this connection that atmospheric temperature is about 300 °C above the absolute zero temperature, so there is always activation energy available to be tapped, if only the barriers can be lowered sufficiently. Although they influence reaction rate, catalysts do not change the equilibrium of a chemical reaction. This is determined by the relative concentrations of the substances involved. Balance is achieved when the original substances are reacting to yield products at the same rate as these new chemicals are themselves interacting to produce the original substances. If the concentration of any component is altered, the equilibrium is shifted. If water vapour is passed over red-hot iron, iron oxide and hydrogen are formed. The passage of hydrogen over red-hot iron oxide produces the reverse process. When the reactions are made to occur in a confined space, equilibrium is ultimately reached, with all four substances present.

The great majority of the materials around us are not pure elements, and few of them are even pure compounds. Many indeed are not even single phases, but rather mixtures of different phases, such as solids in liquids, solids in gases, gases in liquids, and so on. Thomas Graham appears to have been the first to attempt a thorough study of such systems. In the period around 1860 he made observations on the penetrability of various substances through membranes such as pig's bladder or parchment. If such a membrane is used to divide a solution of salt, sugar or copper sulphate from pure water, it is found that some of the dissolved material passes through the membrane into the water. In a solution, the component which has the same physical state as the solution itself is known as the *solvent*, while the dissolved substance is referred to as the *solute*. It is found that the easy

Phases and phases

The word *phase* is commonly applied to two different concepts: a stage of development in a series of changes, and a separate part of a non-uniform, or heterogeneous, body or system. The first meaning of this important word is the more familiar, and in the scientific context it usually refers to the special case of regularly cyclic variations, as in the phases of the Moon. Points on the path of a wave are said to have the same phase if their displacements are equal and varying in the same manner. The phase of a single light wave, or an X-ray wave, cannot be determined photographically because the exposure time is invariably long compared with the time scale of the electromagnetic oscillations. But phase becomes permanently related to position in space when two coherent light beams mutually overlap. This was the basis of Young's interference experiment, described in Chapter 1, and it also underlies the holographic technique which will be discussed in Chapter 10. In its other meaning, the word phase defines any uniform state of matter, providing it extends over a volume considerably greater than that occupied by a single atom; the state is then said to be *homogeneous* and *macroscopic*. A mixture of ice and water is a good example of a *two-phase system*, while a solution of salt in water constitutes a single phase. All gases, no matter how complicated their compositions, are one-phase systems. Changes of state, such as the melting of ice to produce liquid water and the boiling of water to give steam, are formally referred to as *phase transitions*. More subtle examples are seen in solids when, because of a change in temperature or pressure, one crystal structure transforms to another.

passage does not apply to all solutes. Glue, albumen and starch, for example, remain on their original side of the membrane, which is therefore said to be semipermeable. Because of the differential action of the membrane, mixtures of substances such as salt and starch can be separated from one another, if they are first made up into a solution. Graham called the process *dialysis*, which means separation *(lysis)* by passing through *(dia)*. He gave the general name *colloid* to those substances which do not pass through the membrane, the word coming from the Greek *kolla*, which means glue. A solvent tends to flow through a *semipermeable membrane* in the direction of increasing concentration of the solute, the separated solutions thus tending towards equal strength. This effect is known as *osmosis*.

It is now clear that the ability of substances to pass through the minute holes in a membrane depends on the size of their molecules. If the solute particles are of approximately the same size as those of the solvent, separation is not possible with a semipermeable membrane. In a *colloidal suspension* the solute particles, which in this case are referred to as the *disperse phase*, have diameters in the range 1–100 nm. The question arises as to how particles of this size can remain buoyed up in the solvent, which for a colloidal system is referred to as the *dispersion medium*. The answer lies in an intriguing effect first observed in 1826 by Robert Brown: *Brownian motion*. By using an optical microscope, he noticed that pollen grains, and also small particles of the resinous pigment gamboge, performed erratic movements when suspended in water. It was Albert Einstein, in 1905, who first realized that these movements are the result of perpetual, but irregular, bombardment by the water molecules. These collisions are able to counteract the force of gravity if the particle is sufficiently small, the upper limit on the size being indeed about 100 nm.

In an earlier classification, a colloidal suspension in which the disperse phase is a solid and the dispersion medium a liquid was referred to as a *sol*, but this term is now generally applied to all colloidal systems. Sols are divided into two classes. If the disperse phase, having been separated from

Hydrogen and the Alkali Metals | Alkaline Earth Metals | 1st Transition Metals | 2nd Transition Metals | 3rd Transition Metals | Boron and Carbon Groups | Nitrogen and Oxygen Groups | Halogens | Noble Gases

Symbol → Fe; Boiling Point → 3000; Melting Point → 1536

H −252.7 −259.2																	He −268.9 −269.7
Li 1330 180.5	Be 2770 1277											B – (2030)	C 4830 3727g	N −195.8 −210	O −183 −218.8	F −188.2 −219.6	Ne −246 −248.6
Na 892 97.8	Mg 1107 650											Al 2450 660	Si 2680 1410	P 280w 44.2w	S 444.6 119.0	Cl −34.7 −101.0	Ar −185.8 −189.4
K 760 63.7	Ca 1440 838	Sc 2730 1539	Ti 3260 1668	V 3450 1900	Cr 2665 1875	Mn 2150 1245	Fe 3000 1536	Co 2900 1495	Ni 2730 1453	Cu 2595 1083	Zn 906 419.5	Ga 2237 29.8	Ge 2830 937.4	As 613• 817	Se 685 217	Br 58 −7.2	Kr −152 −157.3
Rb 688 38.9	Sr 1380 768	Y 2927 1509	Zr 3580 1852	Nb 3300 2468	Mo 5560 2610	Tc – 2140	Ru 4900 2500	Rh 4500 1966	Pd 3980 1552	Ag 2210 960.8	Cd 765 320.9	In 2000 156.2	Sn 2270 231.9	Sb 1380 630.5	Te 989.8 449.5	I 183 113.7	Xe −108.0 −111.9
Cs 690 28.7	Ba 1640 714	La 3470 920	Hf 5400 2222	Ta 5425 2996	W 5930 3410	Re 5900 3180	Os 5500 3000	Ir 5300 2454	Pt 4530 1769	Au 2970 1063	Hg 357 −38.4	Tl 1457 303	Pb 1725 327.4	Bi 1560 271.3	Po – 254	At – (302)	Rn (−61.8) (−71)
Fr – (27)	Ra – 700	Ac – 1050															

Rare Earths	Ce 3468 795	Pr 3127 935	Nd 3027 1024	Pm – (1027)	Sm 1900 1072	Eu 1439 826	Gd 3000 1312	Tb 2800 1356	Dy 2600 1407	Ho 2600 1461	Er 2900 1497	Tm 1727 1545	Yb 1427 824	Lu 3327 1652
Actinides	Th 3850 1750	Pa – (1230)	U 3818 1132	Np – 637	Pu 3235 640	Am – –	Cm – –	Bk – –	Cf – –	Es – –	Fm – –	Md – –	No – –	Lw – –

The strengths of the bonds between atoms are reflected in the magnitudes of an element's melting and boiling points, listed here in degrees Celsius. The metallic bonds in elements such as molybdenum and tungsten are obviously very strong, whereas the bonds between inert gas atoms are so weak that these elements condense to the liquid and crystalline forms well below room temperature.

the dispersion medium by evaporation or coagulation, can be returned to the colloidal state simply by addition of more dispersion medium, the sol is said to be *lyophilic*. A *lyophobic* sol is irreversible in that the disperse phase does not revert to the colloidal state by adding more dispersion medium. Both the dispersion medium and the disperse phase might be gas, liquid or solid, so we might expect to be able to classify all colloidal suspensions into one of nine groups. Gas-in-gas colloidal systems do not exist, however, since all gas mixtures are homogeneous. If the dispersion medium is a gas and the disperse phase either a solid or a liquid, the sol is referred to as an *aerosol*. This group is divided between *fogs* and *mists*, in which the disperse phase is a liquid, and *smokes* and *fumes*, when the disperse phase is a solid.

Turning to colloidal systems in which the dispersion medium is a liquid, we first have the case in which the disperse phase is a gas. Such a sol is known as a *foam*, and a common example is lather, or shaving cream. It is one of the peculiarities of gases that their solubilities in liquids decrease with rising temperature. If the disperse phase is a liquid, the liquid-in-liquid sol is known as an *emulsion*. Of the numerous examples of emulsions many are common in the culinary domain. The basis of mayonnaise, for example,

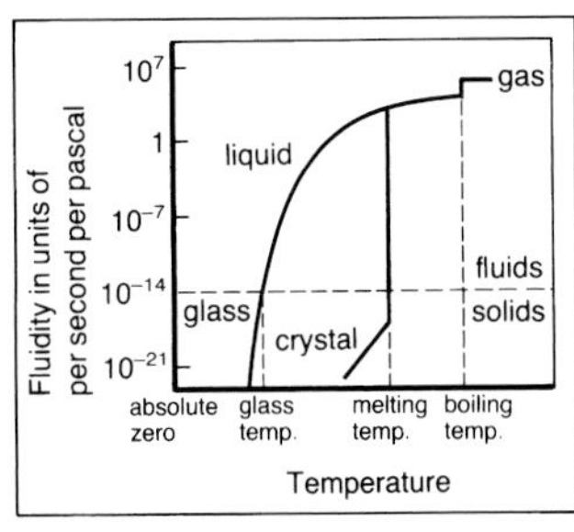

Few physical properties are as sensitive as *fluidity* to changes of state. This quantity abruptly increases by no less than seventeen decades (seventeen powers of ten) when a crystal melts, for example. The *glass–liquid transition* is far more gradual, a fact exploited by the glassblower.

is a colloidal mixture of vinegar with edible oil. It is an interesting substance that requires stabilization by a binder, or *emulsifying agent*, a role played by the phospholipid from egg yolk. The molecules of an emulsifying agent are *amphipathic*, in that they have one end that is compatible with one of the liquids while the other end is readily attached to the other liquid. Another example of such a binder is *soap*, which is used for dispersing greasy substances in water and thereby acts as a cleansing agent. A further example of an emulsion is the *slurry* that is a mixture of water and machine oil. This is used as a combined lubricant and coolant in metal working.

The most common type of colloidal suspension is that in which the dispersion medium is a liquid and the disperse phase a solid. Lyophilic examples include water-based suspensions, or *hydrosols*, of starch and gelatine, the latter being used in dessert jelly; animal resin in alcohol (shellac); rubber in benzine (rubber cement); and numerous oil-based paints. Particularly striking among the lyophobic examples are those in which very fine metallic particles are suspended in water. A gold sol, for example, can be made by adding a reducing agent to a dilute solution of gold chloride. Although the suspended particles are too small to be observed with an optical microscope, their presence can be detected by the *Tyndall effect*, named after John Tyndall, in which light passing through the medium is scattered to an extent that depends on the relationship between particle size and the wavelength of the light. The beautiful optical effects produced in this way were quite familiar to the alchemists and were no doubt exploited in order to impress prospective patrons. Lyophobic sols are usually stabilized by the electrical charges on the surfaces of the dispersed particles, and chemically induced changes in these charges can alter the state of aggregation. Michael Faraday showed that the red colour of a gold sol can be turned to blue by the addition of common salt. Particles scatter light most strongly when the wavelength of the radiation is comparable to their physical dimensions. Because the wavelength of blue light is shorter than for red, we must conclude that the addition of the salt causes a segregation into smaller particles. A particularly useful example of *coagulation* is the clotting of the negatively charged particles in blood by the addition of Al^{3+} ions. Yet another manifestation of the same principle is the aggregation, and consequent deposition, of the particles of clay minerals when the fresh river water in which they are suspended reaches the sea.

Liquid-based colloidal systems display an interesting effect upon partial evaporation of the dispersion medium. When the distance between the particles of the disperse phase is sufficiently small, they can interact with each other and the forces that are thereby brought into play can stabilize the structure. This produces what is known as a *gel*. Lyophilic examples of gels are dessert jellies and laundry starch; a stiffened collar becomes pliable when it gets wet. A very important example of a lyophobic gel is builder's cement when it is setting. If some lyophilic sols are not disturbed for a long time, the particles of the disperse phase can assemble themselves into a weakly stable network resembling a gel. When such substances are stirred, the network is gradually broken down, and the fluidity increases. Such behaviour is known as *thixotropy*, and it is frequently observed in paints.

Colloidal systems in which the dispersion medium is a solid are not so common, but examples do exist. When, the disperse phase is a gas, the colloidal suspension is known as a *solid foam*. Examples are meringue, when it has been cooked, and the fuel coke. Solid-in-solid colloidal suspensions

Catalysis

A catalyst is a substance which increases the rate of a chemical reaction, but which itself remains unchanged when the reaction has taken place. Typical examples are metals and metallic oxides, as well as enzymes in the biological domain. The chemical properties of one or both of the reactants are modified in the proximity of the catalyst's surface, and this lowers the energy barrier of the reaction. A particularly famous example is the catalytic mechanism discovered by Fritz Haber (1868–1934) and developed into an industrial process by Carl Bosch (1874–1940); indeed, the Haber–Bosch process has rightly been called one of the most significant discoveries of the twentieth century. It is used in the production of ammonia, for use in fertilizers, and it involves fixation of atmospheric nitrogen, the reaction being $N_2 + 3H_2 \Rightarrow 2NH_3$. The catalyst in the Haber–Bosch process is iron oxide. Nitrogen fixation processes used earlier in ammonia production included the one developed by Kristian Birkeland (1867–1917) and Sam Eyde (1866–1940), in which nitric oxide (NO) was formed from atmospheric nitrogen and oxygen by the action of an electric arc. The world's food supplies would be inadequate if it were not for the availability of industrial fertilizers, which replace the natural compounds that have become so depleted through intense soil cultivation and disposal of animal sewage in the sea. The picture shows typical catalyst particles used in the process; they measure a few millimetres across.

occur in some metallic alloy systems in which the minor constituent is precipitated out of solid solution as very fine aggregates of atoms.

When the heavier particles in a heterogeneous (i.e. non-uniform) system become too large to be supported by Brownian motion, they settle out under the force of gravity at a rate which depends on their size. A well-agitated mixture of sand and water rapidly separates when left, whereas the solid matter in orange juice remains suspended for a much longer period. Among the intermediate systems are *pastes* and *modelling clays*. In this type of substance, the behaviour is a hybrid between that of a *plastic solid* and a *viscous fluid*. Substances in which the rate of flow is proportional to the force are said to behave in a Newtonian manner. Isaac Newton was the first to make a quantitative analysis of such behaviour. Modelling clay performs differently in that it is almost rigid at low force levels, but becomes easily mouldable if the applied force is sufficiently high. This vital quality of clay is known as *Bingham flow*, after E. C. Bingham who made a study of the underlying physics.

In the powerful theoretical technique of *molecular dynamics*, which was pioneered by Berni Alder and Thomas Wainwright in 1957, the positions of the molecules in a small volume of matter, at different times, are calculated by computer. The underlying principles are those established by Isaac Newton (1642–1727), in the seventeenth century. In this picture of a two-dimensional liquid (taken from a computer simulation by the present author), the molecular positions at two different times, separated by an interval of a few picoseconds, are shown in yellow and blue. If the overall motion of a molecule during this time is zero, yellow and blue will overlap to give a green colour. The isolated colours thus indicate motion, and they show that molecules tend to move into regions where the molecular packing is locally more loose. This endorses an idea originally expressed by Yakov Frenkel (1894–1952) and subsequently developed by Morrel Cohen, Gary Grest and David Turnbull.

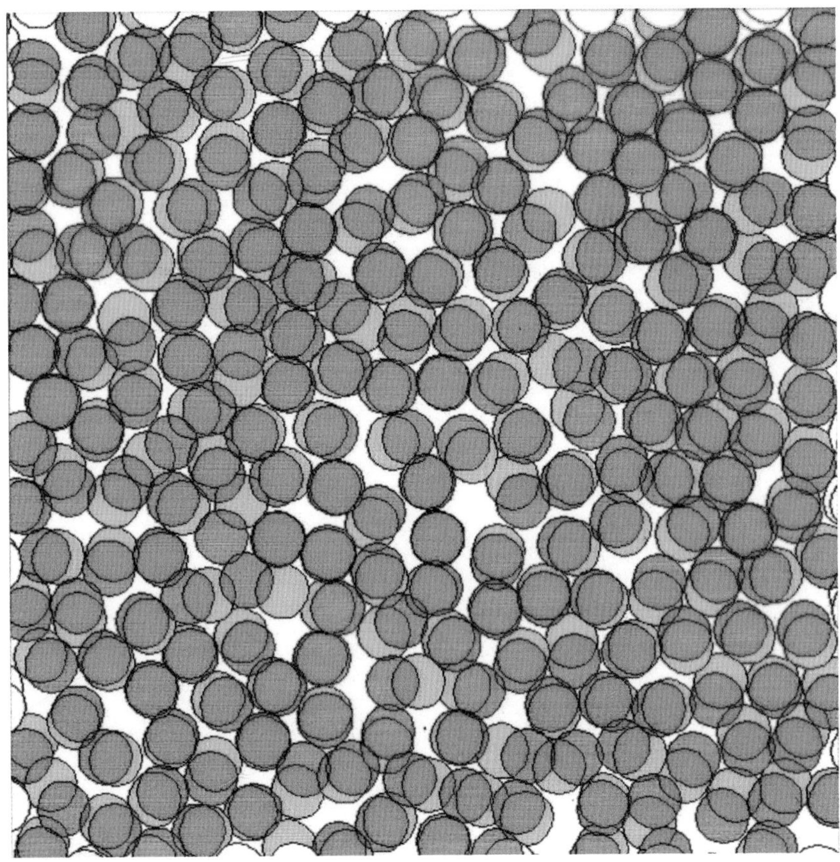

Apart from determining the *viscosity* of the various forms of matter, and their ability to mix, the atomic motions are also responsible for many other characteristic properties. The *conduction* of both heat energy and sound energy are examples of *transport properties*, and in a gas they are exclusively mediated by the motions of their atoms, or molecules in the case of a molecular gas such as air or steam. In *heat conduction*, atoms adjacent to the heat source receive kinetic energy from it, and subsequently pass this on to other atoms via atom–atom collisions. We can think in terms of an average free path between such collisions, and this is inversely related to the density. In air at atmospheric conditions the collision rate, for a single atom, is about 10^{10} per second, and the average free path is about 50 nm, which is approximately 100 atomic diameters. In the absence of the *convection currents* that are invariably set up when a gas (or a liquid) is subjected to a temperature gradient, heat conduction is slow. The atomic motions, and hence also the collisions, are chaotic. The energy is thus passed along in the haphazard manner referred to as a *random walk*. In the case of *sound transmission*, the atoms in the gas receive kinetic energy from the vibrating body, such as the diaphragm of a loudspeaker, and the disturbance is propagated as the succession of compressed and rarefied regions known as a *longitudinal wave*. In a gas carrying a *sound wave* the molecules are still colliding, of course, but the wave disturbance superimposes a directed motion in which momentum is conserved. In atmospheric air at sea level the *speed of sound*, about 330 m/s, is thus almost as great as the average speed of the molecules themselves.

Latent heats of melting and evaporation

Although the changes of state from crystal to liquid and liquid to gas occur at constant temperature, they involve a marked alteration of atomic arrangement. In the former transition, *melting*, the atomic configuration passes from the neat pattern of the crystal to the densely-packed disorder of the liquid. During the liquid–gas transition, *evaporation*, the degree of disorder becomes total and there is a pronounced decrease of density. Both of these *phase changes* involve absorption of energy from the surroundings: the *latent heat*. The energy changes caused by the phase transitions of terrestrial water are responsible for the weather, the atmosphere functioning as a vast thermodynamic engine. The figures below give the latent heats of melting and evaporation of a number of common substances, all entries being given in kilojoules per mole.

Element or compound		Latent heat of melting	Latent heat of evaporation
argon	(Ar)	1.18	6.52
chlorine	(Cl)	6.41	20.41
germanium	(Ge)	31.80	334.3
gold	(Au)	12.36	324.4
iron	(Fe)	15.36	351.04
mercury	(Hg)	2.30	59.15
sodium	(Na)	2.60	89.04
uranium	(U)	15.48	422.6
water	(H_2O)	5.98	47.23
salt	(NaCl)	30.18	764.9
benzene	(C_6H_6)	9.82	34.69

Conduction of heat and sound in a liquid differs from that in a gas in two important respects. Because of the denser packing in a liquid, each atom is always interacting with several neighbouring atoms; there is no time at which atoms are moving as independent entities. The passing on of kinetic energy involves groups of atoms moving simultaneously in a cooperative manner. It is for this reason that the velocity of sound in a liquid is higher than it would be in the corresponding gas. The other difference is that a liquid is capable of sustaining *transverse vibrations*: waves in which the disturbance is not directed along the direction of propagation. Such transverse waves in a liquid occur only for relatively high frequencies, however.

Because of the orderly arrangement of atoms in a crystal, the cooperativity is even more pronounced, and the range of frequencies over which transverse waves can be supported is not markedly different from that of the longitudinal waves. But wave motion is an inherent property of a regular lattice, and even heat is conducted in this manner. The waves cannot pass unhindered through the crystal, however, because they interfere with each other. The concept of an average free path is still useful, therefore, and its extent is about ten atomic diameters in a crystal such as sodium chloride at room temperature. In metals, heat is predominantly transported by the free electrons, which have average free paths about ten times greater. Electrons, moreover, travel about 100 times faster than lattice waves, and although only a small fraction of them contribute to conduction at room temperature, heat transport in a metallic crystal such as silver or copper is high. Sound, being associated with wavelengths very much longer than the average free path of lattice waves, is not subject to the limitations imposed on heat conduction. This is true even of *ultrasonic waves*,

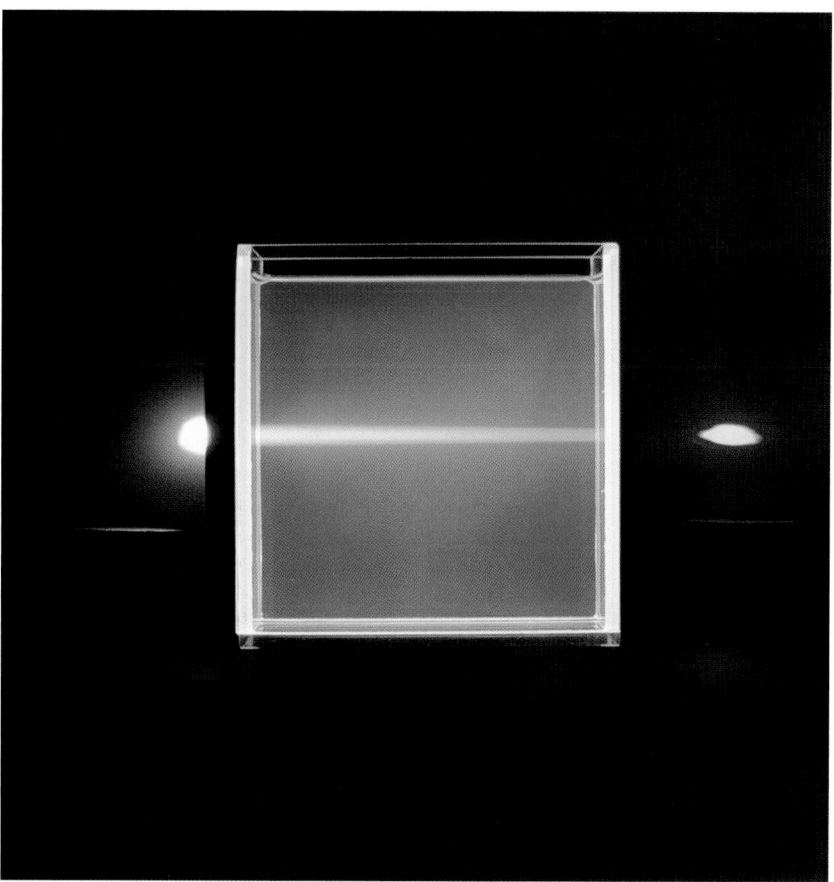

In a *colloidal suspension*, the particles are too small to be visible to the naked eye. They can nevertheless be detected by their ability to scatter light. The colour most strongly scattered depends on the particle size, large particles scattering long wavelengths and small particles giving strongest scattering at short wavelengths. The sulphur particles in the solution shown here have diameters comparable to the wavelength of blue light, about 200 nanometres, so they give strong scattering of this colour. A beam of white light, partly intercepted by the oblique white card at the right, thus becomes depleted of blue wavelengths, and emerges from the solution with a complementary red tinge. This is revealed by the second white card at the left. The same effect, named after its nineteenth-century discoverer, John Tyndall (1820–1893), is responsible for the visibility of smoke, the blue colour of the daytime sky, and the red glow at sunset.

the wavelengths of which are equivalent to millions of millions of atomic diameters.

Attenuation of wave motion, the gradual decrease of intensity with increasing distance from the source, occurs because of two factors. Sound pulses usually contain vibrations that stretch over a range of frequencies. If the transmitting medium shows *dispersion*, the velocity of propagation varying with frequency, the shape of a pulse will slowly change; the disturbance becomes more diffuse, although there is no actual loss of energy. As might be anticipated, because they involve more rapid variations, the higher frequencies are attenuated most strongly, and it is for this reason that the *echoes* returning from distant mountains sound mellow. Attenuation also occurs if the waves are scattered from suspended heterogeneities such as those present in mist, fog, and smoke. Similarly, there is decreased thermal conductivity in a crystal if this contains imperfections.

Interatomic forces are responsible for the elasticity of solid bodies. The elastic deformation takes the form of a *strain*, the definition of this quantity being that it is the ratio of the dimensional change to the original unstrained dimension. According to the type of deformation involved, the dimension can be of length, area or volume. Strain results as a response to *stress*, this being a force per unit area. The ratio of stress to strain defines an *elastic modulus*. For a given solid there are several different elastic moduli, corresponding to whether the deformation is *tensile*, a straight pull or push along a single direction; *shear*, involving a change of shape; or *bulk*, with a change of volume without altering the shape. The magnitude of each modulus depends on the strengths of the interatomic forces, and also on direction if the solid is a crystal. In general it is found that the strain is directly proportional to the stress producing it. This relationship was enunciated in the famous phrase of Robert Hooke, in 1666, '*Ut tensio sic vis*' ('As the strain so the force'). This is true, however, only for strains up to a certain value known as the *elastic limit*. It is not difficult to see why Hooke's law is obeyed for small strains, because the interatomic potential varies only rather gradually around its minimal value, the forces varying in direct proportion to distance. It is less obvious that the elastic limit should be encountered already with strains of just a few per cent, which is invariably found to be the case. The restoring force that holds the atoms together does not reach its maximum value until the distance between neighbouring atoms has been increased by about 15%. The reason for this discrepancy was discovered surprisingly recently, when we consider the rate of progress in related fields of scientific endeavour. The underlying cause is the presence of atomic-scale inhomogeneities. Below the elastic limit these inhomogeneities are immobile, and the solid returns to its original condition when the stress is removed. Beyond the limit, *plastic deformation* takes place, and the solid is found to have developed a permanent change in some of its dimensions when the stress is relieved. In the case of a crystal, the operative inhomogeneities are known as *dislocations*, which we will encounter in later chapters.

It is important to distinguish between *mechanical stability* and *thermodynamical stability*. A crystal structure might be mechanically stable and yet not exist at a given temperature, because there is an alternative crystal structure that has a lower free energy. There are many examples of elements and compounds in which different crystal structures are stable at different temperatures. A famous case occurs in iron, for which the stable crystal structure below 910 °C is of the body-centred cubic type, while above this temperature the most stable form is the face-centred cubic. Iron is particularly unusual because when the temperature is raised above 1390 °C, there is reversion to the body-centred cubic structure. It is interesting to note that the converse situation, thermodynamical stability without mechanical stability, is observed in a liquid. A complete analysis of the necessary and sufficient conditions for mechanical stability, in terms of the elastic moduli, was achieved by Max Born in 1939. A distinction between the elastic and plastic regimes is not always easy to make. Accommodation of applied stress is possible if the molecules are given the opportunity to readjust their positions. Thus the substance will flow if there is sufficient time, while for shorter times elastic behaviour is observed. Substances in which the relaxation time lies in the range from a fraction of a second to a few seconds behave in an apparently peculiar fashion. *Pitch* flows slowly, but shatters when struck sharply. Similarly, *silicone putty*, also known as bouncing silicone or silly putty, can be pulled out into long threads, just like chewing gum, but it bounces if thrown against a hard object such as a floor or wall. Similar intermediate relaxation times are the cause of the peculiar behaviour of *viscoelastic liquids*, which flow one way when stirred, but start running backwards when the stirring ceases. Another manifestation of the significance of time scale is seen when comparing the distances of penetration of bullets fired into ice and liquid water. Because of the high projectile velocity, the atomic motions in the liquid have insufficient time to permit relaxation, and the penetration distances in the two phases are comparable.

Summary

The equilibrium state of an assembly of atoms or molecules is determined both by the forces between the particles and the condition of the system regarding volume, temperature and pressure. A system seeks the situation corresponding to the lowest energy, but a distinction must be made between available (or free) energy and unavailable energy. The latter is related to the information content, and the tendency is towards loss of information and the development of a disorganized, but more probable, arrangement. This factor becomes more important with increasing temperature and causes the phase transitions from solid to liquid and liquid to gas. Atomic mobility also increases with temperature and causes the decrease of viscosity on heating. The reverse effect, increasing viscosity with decreasing temperature, is exploited by rapidly cooling a liquid to produce the metastable glassy state. The interplay of interatomic forces and thermal agitation determines the elastic and diffusion properties of condensed matter. These same factors govern the rate at which systems progress towards equilibrium. Many common materials are mixtures of phases, in which the dispersed groups of atoms are so small that molecular motion in the supporting medium overcomes the precipitating influence of gravity. These are colloids such as fogs, smokes, emulsions, metallic sols and cement gel.

4

Where order in variety we see,
and where, though all things differ, all agree.
Alexander Pope (1688–1744)
An Essay on Criticism

Patterns within patterns: *perfect crystals*

The development of the modern atomic theory of matter owes much to the more venerable science of *crystallography*. One of the major clues leading to the discovery of the periodic table was the recognition that the elements are divided into groups having a family likeness. This derived, in part, from the appearance of crystals containing the elements in question. Crystallography was established during a remarkably fruitful decade around the late 1660s. Studying the shapes of crystals of alum, which is potassium aluminium sulphate, and possibly comparing them with orderly piles of musket shot or cannon-balls, Robert Hooke concluded that a crystal must owe its regular shape to the systematic packing of minute spherical particles. A few years later, in 1669, Niels Steensen, also known by the name Nicolaus Steno, noticed that the angles between equivalent pairs of faces of quartz crystals are always the same. Subsequent studies revealed that the same is true of other crystals, even though their shapes are different from that of quartz. Around 1690, Christiaan Huygens succeeded in explaining the three-dimensional structure of crystals of Iceland spar, or calcite, on the basis of regular stacking of equal spheroids. These solid figures are generated by rotating an ellipse about one of its axes. The next significant advance came in 1781. While scrutinizing a calcite crystal, René Just Haüy accidentally dropped it and noticed that the planes of fracture were parallel to the outer faces of the crystal. This regular *cleavage*, exploited by the cutter of precious stones, suggested to Haüy that the elementary building blocks of a crystal, which he referred to as *molécules intégrantes*, must be stacked in regular and parallel layers. A similar idea, albeit less well defined, had occurred to Torbern Bergman in 1773.

In the decades following Steno, painstaking cataloguing of the crystal shapes of a great number of compounds established that all *crystals* can be placed in one of seven categories, known as *systems*, according to the sets of angles and relative edge lengths of the elementary units required to describe their overall shape. The simplest system is the cubic, in which all the angles are right angles and all the sides equal, examples being rock salt, or sodium chloride, and alum. At the other end of the scale of complexity is the *triclinic system*, in which all angles and all sides of the elementary building blocks are different. Examples of crystals falling in this system are copper sulphate, potassium dichromate, and the important clay mineral kaolinite. The calcite that was the object of both Huygens' contemplations and Haüy's fortunate mishap belongs to a system which has intermediate complexity: *trigonal*. Although none of its angles is 90°, they are at least all equal, and the sides of its building blocks are all equal in length.

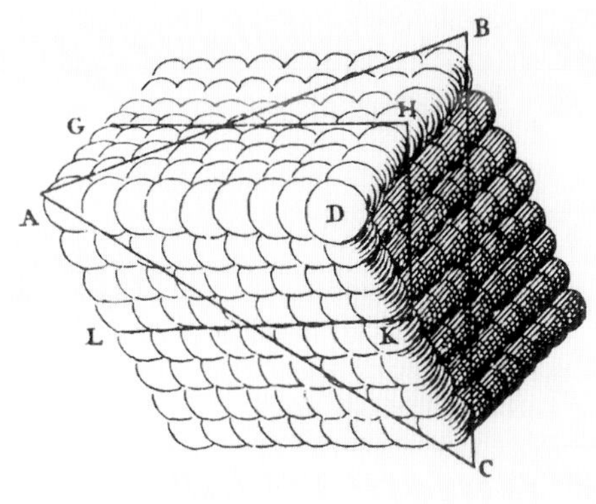

The science of crystallography can be said to date from the latter part of the seventeenth century and the emergence of several rudimentary theories which attempted to link the outer symmetry of crystals to a conjectured inner regularity. Prominent among these were speculations by Robert Hooke (1635–1703), Niels Steensen (1638–1686) and Nicolaas Hartsoeker (1656–1725). And a particularly famous example was the demonstration, around 1690, by Christiaan Huygens (1629–1695) that the shape of a crystal of calcite (see the photograph) could result from the regular stacking of equal spheroids (the diagram).

The crystals occurring in nature often have striking beauty, and the study of their geometry must have been aesthetically pleasing. Moreover, although it did not contribute directly to the study of atomic structure, crystallography included observations that gave rise to a great deal of stimulating conjecture. A particularly important example was the discovery that some elements can exist in more than one crystalline form. One type of tin, for example, is a *tetragonal* version, with all angles equalling 90° and two of the three side lengths equal. It has a white colour. Grey tin, on the other hand, is cubic. Such multiplicity of crystal structure is known as *allomorphism*. Probably the most striking example occurs with pure carbon, which is found in two very different forms, the transparent and hard diamond, which is covalently bonded, and the black and soft graphite, which is held together both by covalent bonds and van der Waals bonds. It is worth stressing that crystallography is an exact science, based on precise measurement, which now concerns itself more with microscopic aspects of the crystal interior than with external shape. Ice cubes are familiar objects, but they must not mislead us into thinking that household ice has a cubic crystal structure. We should rather be guided by the symmetries observed in snowflakes, which reveal that common ice belongs to the crystal class known as *hexagonal*. There is in fact a cubic form of ice, but this exists only at high pressures.

The great step that took crystallography from the macroscopic domain of crystal shape to the microscopic world of atomic arrangement occurred in the early part of the twentieth century. It had to await Röntgen's discovery of X-rays, and the later developments of quantum theory which showed that all rays are also waves. It was thus only a matter of time before attempts were being made to determine the *X-ray wavelength*. It soon became clear that this must be well below that of visible light, which has wavelengths around 500 nm. Seventeen years after Röntgen's discovery, Max von Laue hit upon the idea of using a crystal to get an approximate measure of this quantity. From data on the densities of crystals and atomic weights of the constituent elements, it had been estimated that atomic diameters must be about 0.3 nm, and von Laue saw this length as being conveniently small compared with wavelengths in the visible region. If the atoms were as regularly arranged as the crystallographers supposed, a crystal might produce *interference effects*, when illuminated with a beam of X-rays, analogous to those obtained with light by Thomas Young just over 100 years earlier. Von

Like several other elements, and many compounds, pure carbon can exist in different crystalline forms. At room temperature, and for pressures below about twenty thousand atmospheres, the more stable form is graphite (right). As originally demonstrated by John Bernal (1901–1971), the graphite structure consists of a parallel arrangement of layers, within which the atoms are joined by strong covalent bonds, while weaker van der Waals bonds hold the layers together. This is the origin of graphite's weakness against shearing forces, and its suitability as a lubricant. The hardness of diamond (left) derives from its non-planar distribution of exclusively covalent bonds.

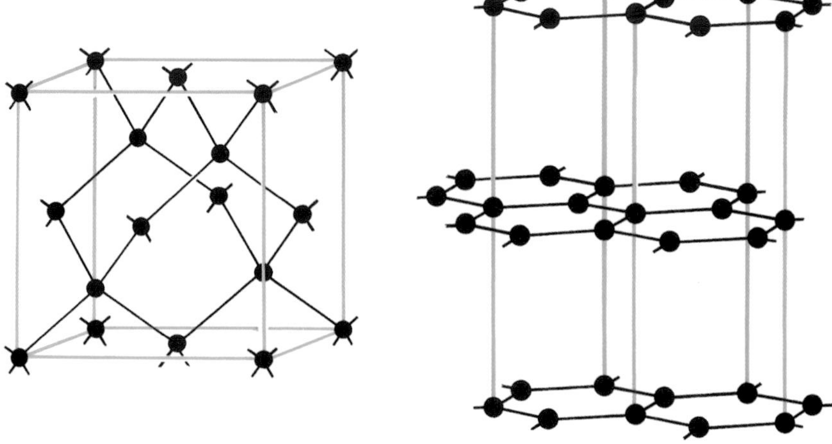

Laue's two colleagues, W. Friedrich and P. Knipping, aimed a thin beam of X-rays at a crystal of copper sulphate and, with the aid of a photographic plate, detected the anticipated interference (*diffraction*) pattern. From the geometry of the experiment, von Laue, Friedrich and Knipping concluded that the wavelength of the X-rays they were using was in fact about 0.1 nm. This experiment, which can be regarded as the most important ever undertaken in the study of condensed matter, was rapidly followed up by William Henry Bragg and his son, William Lawrence Bragg. They inverted the procedure by assuming the value for the wavelength, and they determined the structure of various crystals from the observed diffraction patterns.

It would be difficult to exaggerate the importance of the X-ray diffraction technique. It effectively enables us to see the arrangement of the atoms in a crystal. Surprises awaited the early practitioners of this new art. It was already reasonably clear that the atomic arrangement in substances such as calcite must be rather regular. But it was now discovered that the atoms in an irregularly shaped gold nugget, for example, are just as tidily arranged. And there were further surprises. It was found that in spite of the common usage of the term crystal glass, the atoms in a glass are *not* neatly arranged. One of the most astonishing facts to emerge from subsequent studies of more complicated substances was that even paper and hair are more crystalline than glass. The newly won access to the microscopic realm obviously put crystallography on a still firmer footing. One important result was the confirmation of a prediction by Auguste Bravais, in 1848, concerning the number of different types of arrangement of identical atoms that could be made to fill space. He found that there are just fourteen, and the *Bravais lattices*, as they are now called, represent a subdivision of the seven *crystal systems* referred to earlier. The development of atomic-level crystallography presented a most exciting prospect. Given the concurrent developments in the quantum theory of the electronic structure of atoms, the great challenge now was to see how this related to the arrangement of atoms in the crystalline state. Are the symmetries of atomic pattern a reflection of the symmetries of the electron probability clouds? The answer to this fundamental question is, in principle, yes. We have already seen how the directionality of covalent bonding dictates the relative positions of atoms in molecules, and

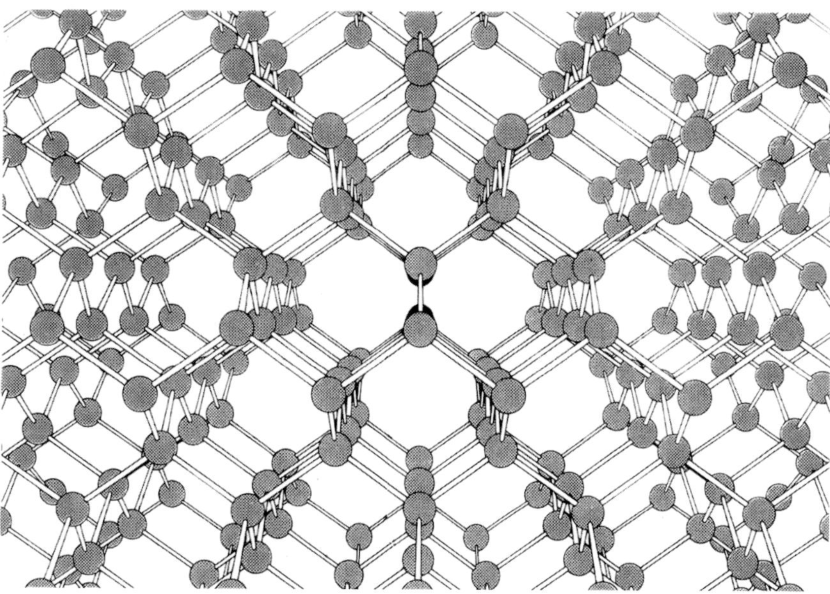

The study of crystal structure by *X-ray diffraction*, which was pioneered by Max von Laue (1879–1960), William Henry Bragg (1862–1942), and William Lawrence Bragg (1890–1971), owes much to the original demonstration by Thomas Young (1773–1829) of optical interference effects. The regular arrangement of atoms in a crystal permits this to coherently scatter incident X-ray radiation along certain directions, producing a pattern of spots that is analogous to the line pattern observed by Young. In this illustration, the hexagonal arrangement of atoms (right) in a silicon crystal (top left) is reflected in the prominent six-fold symmetry of the corresponding diffraction pattern (below left).

the same applies to covalent crystals. But in metallic crystals the situation is more complicated, as we shall see later.

To appreciate the contributions of crystallography, we should consider crystals in the broader context of the different phases of matter. Of the three common states, solid, liquid and gas, the solid is the easiest to describe in terms of the positions of its constituent atoms. Only in this form do the atoms stay put for a reasonably long time. In a liquid the atoms move around so rapidly that they are constantly changing their neighbourhood, and we must be content with describing the surroundings of any atom in terms of an average environment. In a gas, chaos reigns and each atom is essentially on its own almost all the time. Although the description of structure is easiest for a solid, the problem is not a trivial one. Indeed, for the two subclasses, *crystalline solids* and *amorphous solids* (i.e. glasses), the first have been reasonably amenable to structure determination ever since those pioneering efforts of von Laue and the Braggs, whereas the latter still obstinately defy an unambiguous treatment. To begin with, we need a working description of what is meant by a crystal. This is simply a structure built up of the regular repetition of identical units. It is implicit that the units be single atoms or groups of atoms, because this definition might otherwise include macroscopic patterned objects like tiled floors and wallpaper. The contemplation of such things can nevertheless help us to appreciate the intrinsic beauty of crystallography, and they serve to teach some of the basic rules of crystal symmetry. The word itself comes from the Greek for ice: *krystallos*. It became associated with quartz, which was believed to be ice that had been petrified by extreme cold. Only in the late Middle Ages did it acquire its present more general meaning. Use of the crystal to symbolize purity does not conflict with the scientific connotation; perfect crystals are by definition pure.

It would not be pedantic to wonder how the finiteness of real crystals affects crystallinity. The free surfaces constitute an atypical environment, and the outermost atoms are subjected to forces that must be quite different from those experienced by their brethren deep in the interior. Since

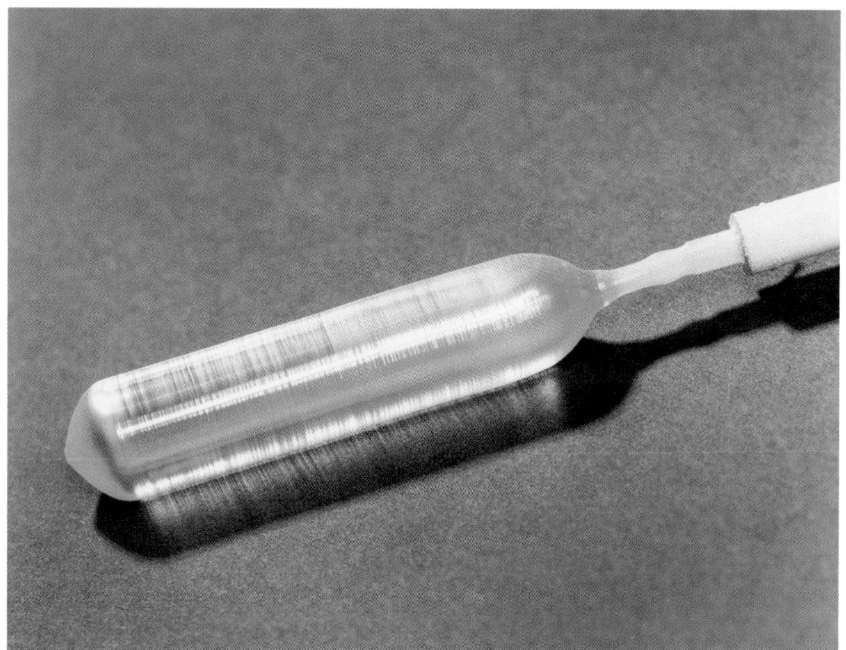

In the as-grown state, crystals fabricated for industrial purposes usually do not have shapes that reflect their internal crystallographic symmetry. This lead molybdate crystal was produced for use in acousto-optical devices.

it is the forces that determine structure, this implies distortion and hence disturbance of the regular pattern. But typical crystals of a few cubic centimetres contain about 10^{24} atoms, and the number of repeated units along any straight line within such a crystal is 10–100 000 000. The relatively few distorted units at the surface are unimportant. But if we consider smaller and smaller assemblies of units, a point is reached at which the rules can be violated. It is possible to arrange a finite, but relatively small, number of identical units in a stable configuration that is inconsistent with crystallinity. Further units cannot be added indefinitely to such a structure so as to build up a regularly repeating pattern. Alan Mackay has suggested that these be referred to as *crystalloids*. Examples are found in abundance in the viruses, where the repeating units are protein molecules, and occasionally in small clusters of atoms in the inorganic state. A crystalloid form of gold has been observed which contains 147 atoms in an icosahedral array. The structure is onion-like with four shells containing 1, 12, 42 and 92 atoms, and the outer surface has a five-sided symmetry that cannot be added to so as to make a crystal.

Let us turn now to the actual arrangements of atoms in crystals and seek the reason for the complexity hinted at earlier. The wallpaper analogy serves to illustrate the essentials of the problem. The pattern might be so unimaginatively regular, like the black–white–black–white of a chess board, that it can be grasped at first glance. But a designer has all sorts of devices for introducing subtlety: obtuse or acute angles between the axes, for instance, and colours other than black and white. The repeated sequence could also have its own internal structure. Simple geometrical figures can be replaced by objects from the living world: vegetables and fruits for the kitchen walls, flowers for the lounge, fish and shells for the bathroom, and so on. And further complexity can be introduced by mixing the figures; by giving them a variety of orientations; by periodically omitting a figure, and so on. We see that the number of different possible patterns is truly infinite, but there are

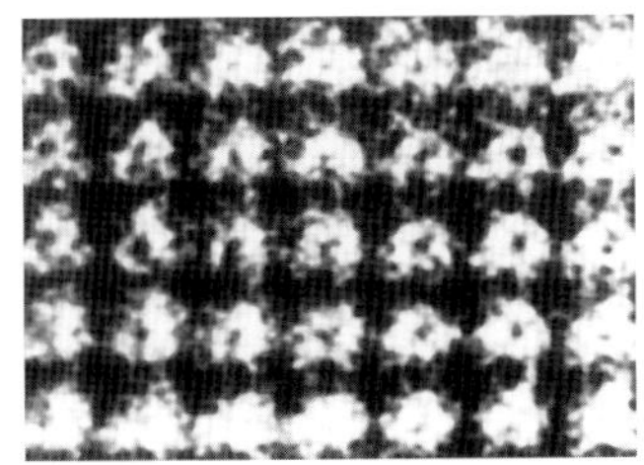

Although *X-ray diffraction* yields information from which a crystal's structure may be determined, X-ray lenses are not a practical proposition so the technique is incapable of producing direct images. This is possible if the X-rays are replaced by electrons. And the resolving power of the transmission electron microscope has been so improved in recent years that one can now distinguish individual atoms, under ideal circumstances. This electron photomicrograph shows the orderly arrangement of atoms in a gold crystal, the picture being about two nanometres wide.

nevertheless inherent limitations at the fundamental level. At this point, as we begin to discuss the underlying principles, it will be necessary to introduce a few definitions. A *space lattice* is a collection of points, in the mathematical sense, so arranged that each of them is surrounded by precisely the same configuration; if such a lattice is looked at from one of the points, the view always appears the same, independent of which point is chosen as the observation site. A space lattice is generated by the points of intersection of three families of parallel planes, all planes in each family being uniformly spaced. Such a lattice divides space into parallelepipeds (i.e. polyhedra that are bounded by flat faces, opposite pairs of which are parallel), and these are known as *primitive cells*. In two dimensions the planes are replaced by lines, and the primitive cells become parallelograms. The situation is thus analogous to that observed in a commercial orchard. Irrespective of the dimensionality, it follows that each lattice point lies at a *centre of symmetry*; for every lattice point lying on one side of the chosen point there is another that is identically situated on the opposite side.

For a three-dimensional array of points, there are just fourteen different types of primitive cell that can be stacked together to fill space, as Bravais had proved. In two dimensions the number of possibilities drops to five. Turning again to wallpaper, we see that in spite of the great variety of available designs, each can be placed in one of just five classes. Individuality derives from the manner in which specific items are worked into the primitive cell. The nature and arrangement of these items can be arbitrarily complex, and it is in this sense that there are an infinity of possibilities. In crystals, similarly, there is no limitation on the number of atoms in the primitive cell. This number can be one, but it might be several thousand in the case of a crystal composed of protein molecules. The arrangement of atoms within the primitive cell of a crystal is known as the *basis*, and the prescription for constructing the crystal can be concisely stated thus:

$$\text{Bravais lattice} + \text{basis} = \text{crystal structure.}$$

The primitive cell shapes range from the simple cubic, with all sides equal and all angles 90°, to the triclinic with the three non-equivalent sides all of different length, and the three non-equivalent angles all different, none of these being 90°. Between these extremes lie units with some sides or some angles equal. From the basic set of fourteen Bravais lattices we could design an infinite number of different crystals simply by replacing each mathematical point with an atom or a group of atoms. A good example of a rather complicated crystal is the regular packing of haemoglobin molecules that can be achieved under appropriate chemical conditions. The overall pattern of the structure is basically simple, but each unit contains over 10 000 atoms, the haemoglobin molecule being almost devoid of symmetry.

Before proceeding farther with the discussion of actual crystals, it would be advisable to define just what is meant by *symmetry*. The word itself has a Greek root, *metron*, which means measure, while the prefix *sym* means together. The compound word is used in the sense of having the same measure, and equality of certain dimensions lies at the heart of the symmetry property. Let us first ignore the spatial extent that any lattice must have and concentrate on symmetry in the environment of a single point. Such regularity is identified by what are known as *symmetry operations*, and the principles can be illustrated by imagining the placement of a number of copies of an unambiguous figure with respect to a single point. Let this figure be the print

The seven crystal systems and fourteen space lattices

In 1848, Auguste Bravais (1811–1863) showed that there can be only fourteen mutually distinguishable space lattices. Each of the fourteen Bravais lattices is associated with one of the seven crystal systems, so some crystal systems have more than one associated Bravais lattice. There are for instance three cubic Bravais lattices. The following table lists the seven crystal systems, their associated space lattices, and examples of crystalline substances from the systems.

System	Defining Characteristics	Space lattices	Examples
Cubic	Three axes at right angles, all equal length	Simple Body-centred Face-centred	Caesium Chloride Sodium Copper
Hexagonal	Two equal axes subtend 120° angle, each at right angles to third axis of different length.	Simple	Zinc
Tetragonal	Three axes at right angles, two of equal length.	Simple Body-centred	Barium titanium oxide Indium
Trigonal (rhombohedral)	Three equally inclined axes, not at right angles, all equal in length.	Simple	Calcite
Orthorhombic	Three axes at right angles, all of different lengths.	Simple Base-centred Body-centred Face-centred	Lithium formate monohydrate Uranium Sodium nitrate Sodium sulphate
Monoclinic	Three axes, one pair not at right angles, all of different lengths.	Simple Base-centred	Lithium sulphate Tin fluoride
Triclinic	Three axes, all at different angles, none of which is a right angle, all of different lengths.	Simple	Potassium dichromate

The unit cells of the fourteen Bravais lattices are shown below with the hierarchy of crystal systems (top right) and the diamond structure (next page bottom right), which is not a lattice. The hierarchy of crystal systems expresses the fact that a cubic lattice is a special example of a tetragonal lattice as well as of a trigonal lattice, whereas a tetragonal lattice is a special example of an orthorombic lattice; the other lines in the hierarchy represent similar relations.

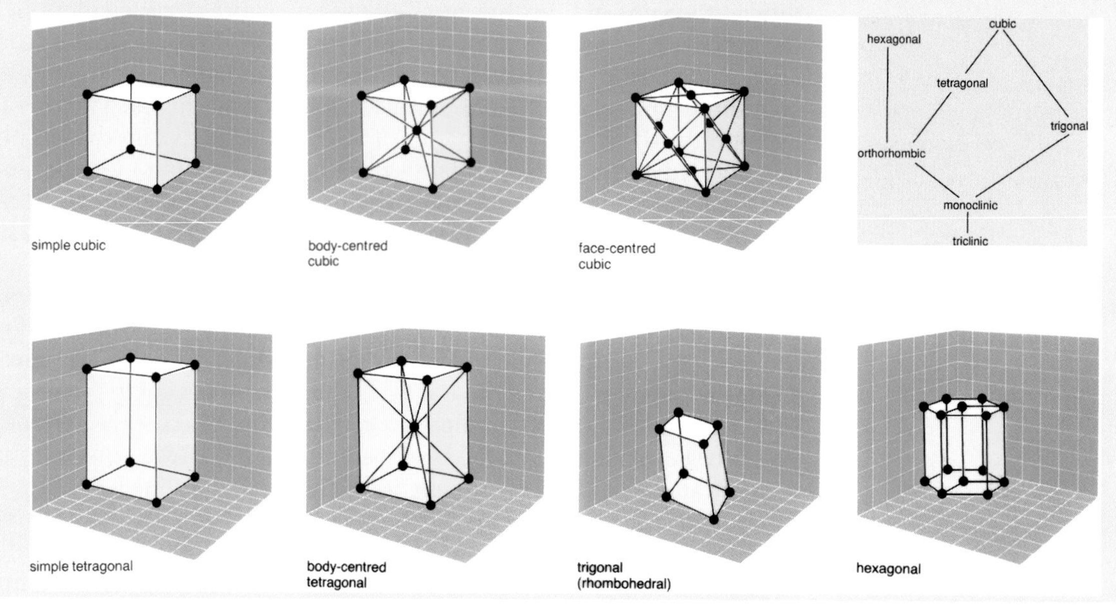

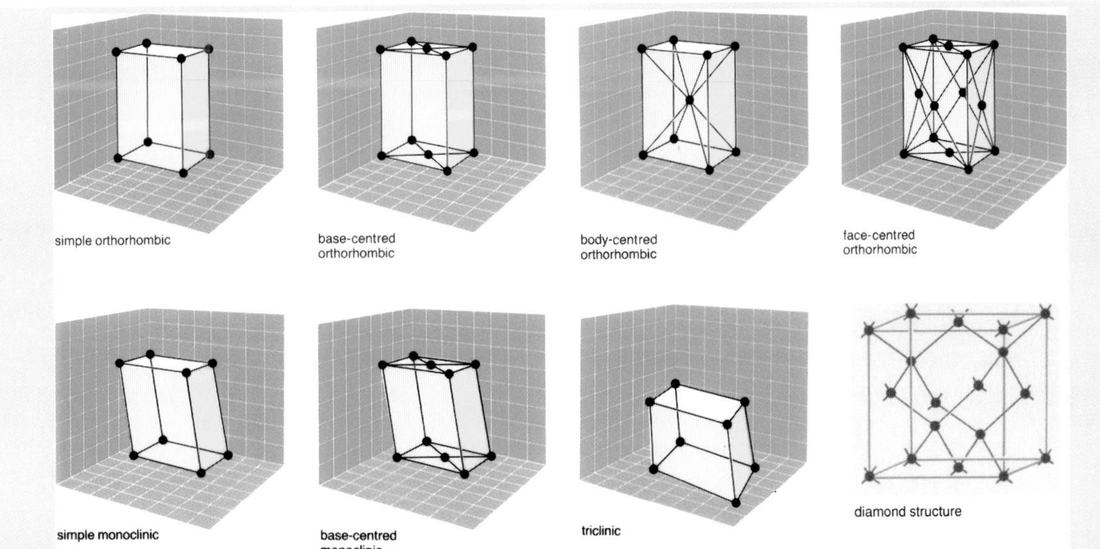

The unit cells of the Bravais lattices are not all primitive cells containing just one lattice point. For instance the unit cell of the face-centred cubic lattice contains four points, because each of the eight corner points is shared between eight adjacent unit cells, and each of the six points on the faces of the cube is shared between two adjacent unit cells. Likewise the body-centred cubic unit cell contains two points, one corner point and the centre point. The primitive cell for the body-centred cubic lattice is spanned by sides joining a corner point with each of three adjacent centre points, and it can be considered to be the special case of a trigonal lattice with a top angle of 109° 28′. Each of the fourteen unit cells of the Bravais lattices fills space when repeated along the directions of its three independent sides.

Physical crystals can be described by means of a Bravais lattice and a basis, the specific arrangement of atoms in the unit cell of the lattice. For some elements the basis comprises just one atom. Notable examples are copper with a face-centred cubic structure, sodium with a body-centred cubic structure and indium with a simple tetragonal structure. Some Bravais lattices are not known to be associated with any physical crystals with only one atom in the basis. The triclinic lattice is an example. In most cases there is more than one atom in the basis. This is obviously the case for compounds, but diamond has the face-centred cubic structure with two carbon atoms per lattice point, the second atom being displaced along the diagonal of the unit cell by one quarter of its length.

Some elements and compounds, said to be polymorphic, display a change in crystal structure under certain conditions of temperature and pressure. Tin, for example, is known in both the diamond structure, which is stable below 13 degrees Celsius, and the body-centred tetragonal structure. Crystals of the former are grey and the latter white. Carbon is also polymorphic, being either diamond or graphite.

The face-centred cubic structure is the basis for a close packing of identical spheres. It can be seen as a stacking of close-packed planes, each of which has a triangular arrangement of lattice points. The close-packed planes comprise each of three corner points, which are adjacent to one corner point in the face-centred cubic unit cell. Two of these planes are seen nearly edge on in the picture above. When stacking the close-packed planes, the projected position of the points in the next plane can fall in either of two equivalent sets of points, which are designated B and C in the diagram below, A representing the position of the original points. If the second plane is positioned in the B position, then the third plane could be in either the A or C positions. In the case of the stacking sequence ABCABC . . . , the face-centred cubic lattice appears. But the stacking sequence ABABAB . . . is also possible, and the result is the hexagonal close-packed structure. This is not a lattice, because an extension of the line from an A position in the first plane to a B position in the second plane does not intersect the third plane in a lattice point. It is in fact a hexagonal structure with two points in the basis, which would look like an off-set body-centred hexagonal unit cell (note that the hexagonal unit cell depicted above is not primitive, it actually comprises three primitive cells, of which one is indicated in the foreground with sides in solid lines). Nevertheless it is an important structure, which is for instance found in crystalline Ti, Co, and Zn.

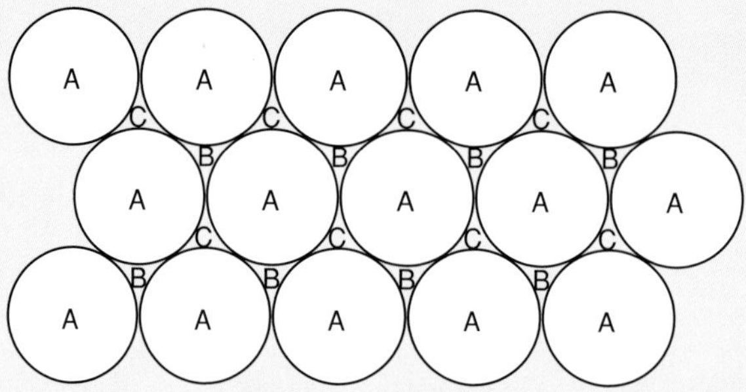

In 1611, it was conjectured by Johannes Kepler (1571–1630) that it is impossible to pack identical spheres more closely than by the arrangements represented by the face-centred cubic or the hexagonal close-packed structures or by any structure represented by other stacking sequences such as ABACABAC . . . , but a mathematical proof has been hard to find. A proof based on computer analysis was found by Thomas Hale in 1998, but it remains controversial as to whether it finally settles the issue. The controversy is formal rather than substantial – it is not thought that the Kepler conjecture could be false.

of a right hand, and let the problem be confined to two dimensions. Two fundamental symmetry operations are possible. A rotation moves the figure around a circular path with the point as centre. If half a complete rotation brings one figure into precise coincidence with another, the pattern is said to have two-fold *rotational symmetry*. Coincidence achieved by every quarter turn gives four-fold rotational symmetry, and so on. There is no limit to such symmetry; seventeen-fold rotational symmetry is just as acceptable as six-fold. If the half rotation had caused the hand to lie over another with the thumb pointing in the wrong direction, we would conclude that two-fold rotational symmetry was lacking. But coincidence could be achieved in this situation by *mirror reflection* in a line perpendicular to and bisecting the line joining the two hands. There is a third type of *point symmetry* operation, the *inversion*, but this is not fundamental since it can always be achieved by a combination of rotation and mirror operations. The *point group* is defined as the set of all symmetry operations that can be applied to the arrangement of items around a point so as to leave the pattern unchanged. Such a group is sufficient to preclude all ambiguities, and we have already seen that there is no limitation to its complexity. The point group is stated concisely according to a standard convention, an example being $2m$. This terse recipe states that the pattern has two-fold rotational symmetry about a certain point, or axis in three dimensions, and that there is also a mirror line, or plane in three dimensions.

In considering spatially repeating patterns, we require an additional symmetry operation: translation. A pattern is said to possess *translational symmetry* if a rigid push in some direction, and through a particular distance, returns the pattern to precise registry. The *space group* is defined as the set of all point group operations and translation operations that can be applied to the spatial arrangement of items so as to leave the pattern unchanged. As with the point group, the space group admits of secondary operations.

The body-centred cubic structure served as inspiration for the Belgian exhibition building called Atomium, at the 1958 World Fair in Brussels.

Translation plus rotation gives rise to *screw symmetry*, for example, while translation plus mirror reflection gives what is known as the *glide operation*. As for the point groups, space groups are usually denoted by a succinct set of alphabetic and numeric symbols. It can be shown that the total number of possible space groups is 230, this being first demonstrated by Evgraph Fedorov and Artur Schoenflies, independently, in 1890.

In real crystals, the requirement that these can be composed of repeating units puts a severe constraint on the point groups. There are 32 different point groups having symmetry consistent with that of a space lattice. This was proved by Johann Hessel in 1830 and, more elegantly, by Axel Gadolin in 1867. It is found that the only rotational operations compatible with translational symmetry are the two-fold, three-fold, four-fold and six-fold. As early as 1619, Johannes Kepler had demonstrated that a seven-fold axis of symmetry cannot exist in a lattice, and he reported this fact in his *Harmonice Mundi*. If crystals are categorized by the type of rotational axes that they possess, it is found that there are just seven different systems. It was these that had already been identified by crystallographers in the pre-microscopic era. The systems can be arranged in a hierarchy, in which each is a special case of the ones lying below it. If the categorization takes both point group and translation operations into account, there are fourteen possibilities in all, and these are the lattices that were discovered by Bravais.

Until now, we have ignored the basis of the crystal. There is no obvious requirement that the point groups of the basis and the Bravais lattice be identical. Indeed the basis could have a seven-fold axis, which the lattice certainly cannot. In wallpaper, the complexity of the basis is limited only by the fantasy of the designer, and there is no essential relationship between the arrangement of the objects in the primitive cell and the symmetry of the lattice. In real crystals, however, the atoms feel each other's influence because of the interatomic forces. This is always the case, irrespective of whether the basis is simple or complicated. We should therefore expect the point group of the basis to influence the choice of Bravais lattice actually adopted by the crystal. Let us consider extreme examples. The lattice showing the highest degree of symmetry is the cubic, the senior member of the hierarchy referred to earlier, while the lowest is found in the triclinic. We could draw a triclinic lattice of spherical atoms, such as those of the noble gas argon, and a cubic lattice composed of molecules of a large and asymmetric enzyme. But contemplation of the interatomic forces involved would lead us to expect that the reverse situation, cubic argon and triclinic enzyme, is closer to the truth. And there are further complications. The basis itself cannot be perfectly rigid; it will be deformed by the overall crystal, according to the latter's lattice symmetry. With basis and lattice exerting an influence on each other, the result will be a compromise, the exact outcome of which usually cannot be predicted. There are, nevertheless, certain rules which provide a useful guide. The point group of the basis, for example, must form part of the point group of the crystal lattice. The Dutch artist, Maurits Escher (1898–1972), closed the gap between the unrestricted domain of wallpaper and the constrained world of real crystals, by constructing pictures in which the repeating units are made to fill up all the available space. The resulting requirement that the pieces interlock precisely reveals the essential interdependence of the symmetries of the basis and the lattice.

The goal of relating the symmetries of electron orbits around nuclei and atomic arrangements within crystals has now been achieved, but the

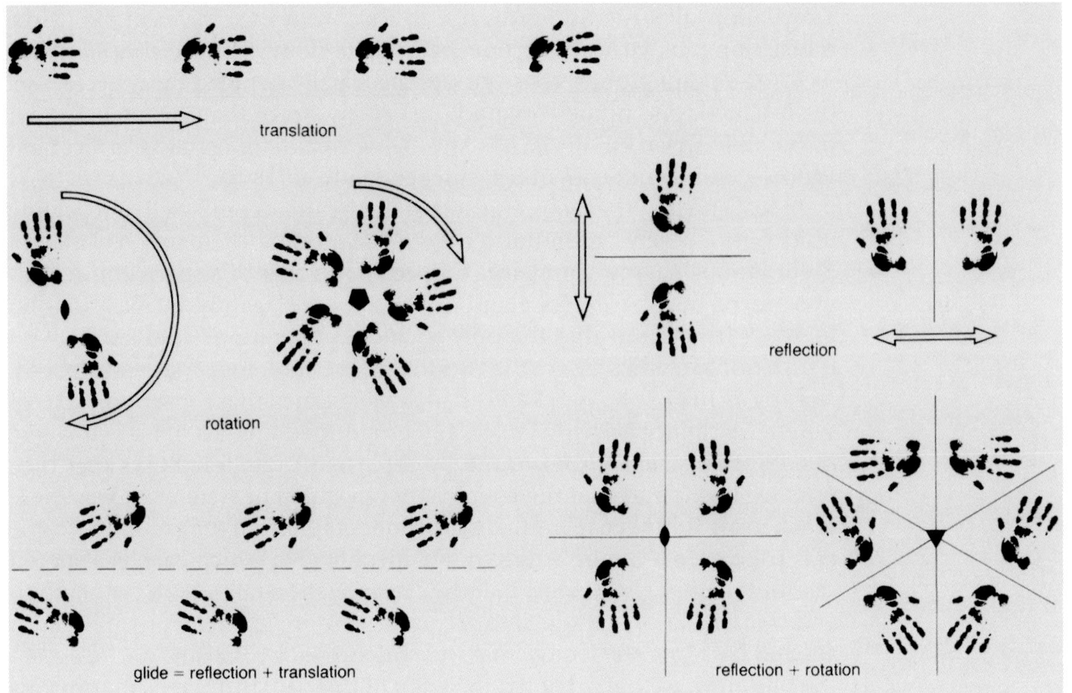

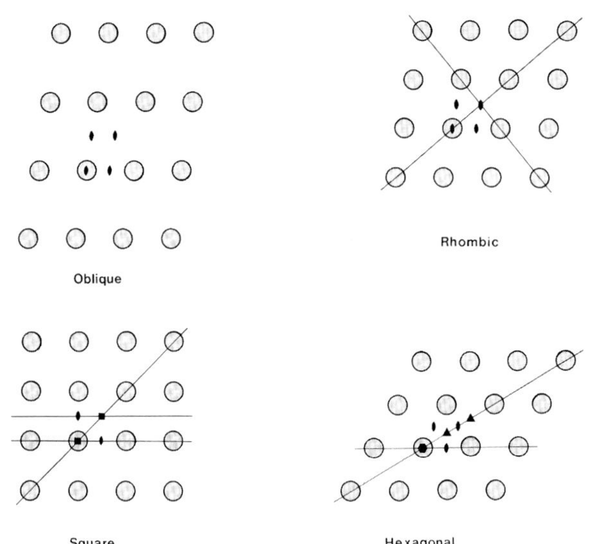

Three of the symmetry operations (top figure) are fundamental: *translation*, the only operation which must be possessed by all lattices; *rotation*, for which two-fold and five-fold examples are shown; and *reflection*. *Glide* is equivalent to a combination of reflection and translation. The *screw operation* (not shown) combines rotation and translation. The oblique two-dimensional lattice has four non-equivalent two-fold rotation points. In the rhombic net, two of the latter become equivalent and two reflection lines appear. The square net has two non-equivalent four-fold rotations, one independent two-fold rotation, and three independent reflection lines. Finally, the hexagonal net has two reflection lines and just three independent rotation points; one six-fold, one three-fold, and one two-fold.

The surface of this cadmium sulphide crystal, etched with a weak hydrochloric acid solution, has six-sided pyramid-shaped pits which reflect the hexagonal nature of the underlying atomic lattice. Viewed by optical microscopy, the surface produces *interference effects* between the incident and reflected light, the coloured concentric bands providing a contour map of the depressions. The deepest *etch pit*, which therefore has the most bands, measures about fifty thousand atomic diameters across and is about five thousand atomic diameters deep.

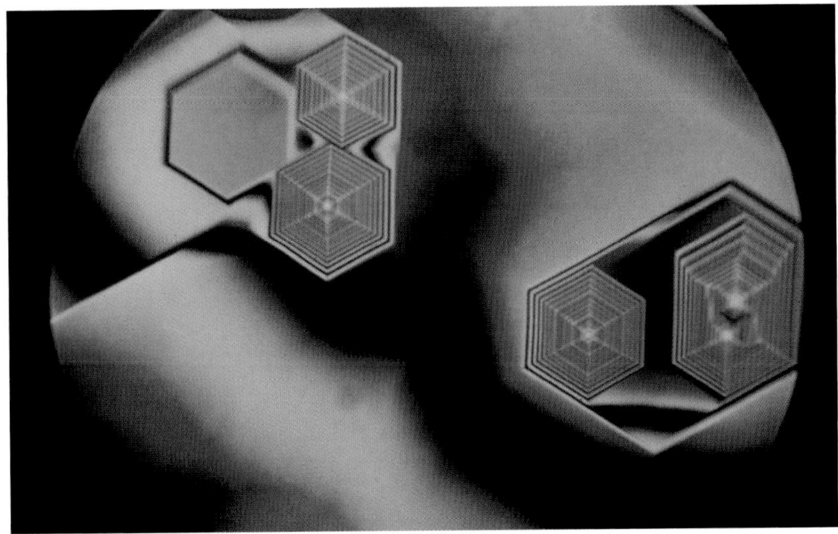

relationship is not quite so straightforward as might have been anticipated. To begin with, it is only the outer electrons of an atom which determine how this is arranged with respect to its neighbours. Then again, the electronegativity of some atoms is so low that they do not succeed in holding on to these electrons when they participate in bonding. Thus in a metal, with its most loosely bound electrons joining the mutual stake of the collective electron sea, which provides both electrical conductivity and a sort of cohesive glue, the packing of the positive ions is governed primarily by simple geometry. This indeed is very close to what had been anticipated by Hooke, Huygens and Haüy. The *coordination number*, which is the number of other atoms that are the immediate neighbours of a given atom, is consequently rather high. In many metals this number has its highest possible value, namely twelve. This is the case for copper, silver, gold, aluminium, zinc, cadmium and several others, the tight packing being related to their *face-centred cubic* (fcc) or *hexagonal close-packed* (hcp) crystal structures. In some metals each atom is even surrounded by fourteen other atoms, but in this *body-centred cubic* (bcc) case, eight of the neighbouring atoms are slightly closer than the other six. The bcc structure is, in fact, slightly less densely packed than the fcc and hcp arrangements, by about 8%. Examples of elements that adopt the bcc configuration at room temperature are iron, chromium, tungsten and sodium.

The converse situation occurs in covalently bonded crystals, where the tendency for each atom to hold on to its outer electrons imposes restrictions on the coordination number. Whence the relatively open lattice structure adopted by silicon, germanium, and carbon in its diamond form. A similar saturation effect is responsible for the openness of the hydrogen-bonded lattice of ice. Intermediate to these extreme types of behaviour, we have the alkali halides, such as common salt, in which the two species of ion, one electrically positive and the other negative, can be regarded as essentially spherical. An ion of one species is surrounded by as many ions of the other species as can be fitted in geometrically. Again the result is a rather densely packed structure. This rather gross distinction between tightly and loosely packed crystal structures is frequently inadequate. When space must be found in a lattice for the insertion of small numbers of atoms of a foreign

Quasicrystals

One charm of the scientific enterprise is its ability to produce occasional surprises. Just when one of its established branches seems to have been explored in its entirety, a new facet is discovered that gives the subject a fresh lease of life. In crystallography, it was well established that a *sufficient* condition for *long-range order* is that the elementary building blocks obey certain rules of symmetry. Two-dimensional crystal lattices cannot possess five-fold symmetry points, for example, because it is not possible to fill a plane completely with a network of regular pentagons (see the red and white diagram). Similarly, the icosahedron is inconsistent with three-dimensional crystallinity, although regular icosahedral arrangements of a limited number of atoms (here coloured yellow) are possible, and are believed to be a feature of some glasses. During the 1970s, Alan Mackay pointed out that building blocks possessing the well established crystalline symmetry elements might not be *necessary* for long-range order, which could result from hierarchic packing or self-similarity, rather than from simple lattice repetition, and he began exploring the possibility of a more generalized science of crystallography. He referred to atomic arrangements such as the icosahedron by the term *crystalloid*, and he called the not-quite-periodic superstructures that can be built up from them mock-lattices, later renaming them *quasi-lattices*. He was intrigued to discover that such lattices are related to what is known as the *golden section*. If a straight line is divided into two segments, the greater with length A and the smaller with length B, such that $A/B = (A + B)/A$, that division is said to have conformed with the golden section. Working through the algebra, one then finds that $A/B = 1.618$, to three decimal places. This number had been known from antiquity, and it was much in evidence in such buildings as the Great Pyramid and the Parthenon. One even sees the same proportion used in many a famous old painting. Mackay's ideas elicited little response, despite the fact that they harmonized with a classic mathematical study of nearly periodic functions carried out over forty years earlier by Harald Bohr (1887–1951). Then in 1984, Daniel Shechtman was surprised to discover ten-fold symmetry in an electron diffraction pattern he took of a rapidly cooled aluminium-manganese alloy. His result initially met with much scepticism, but such

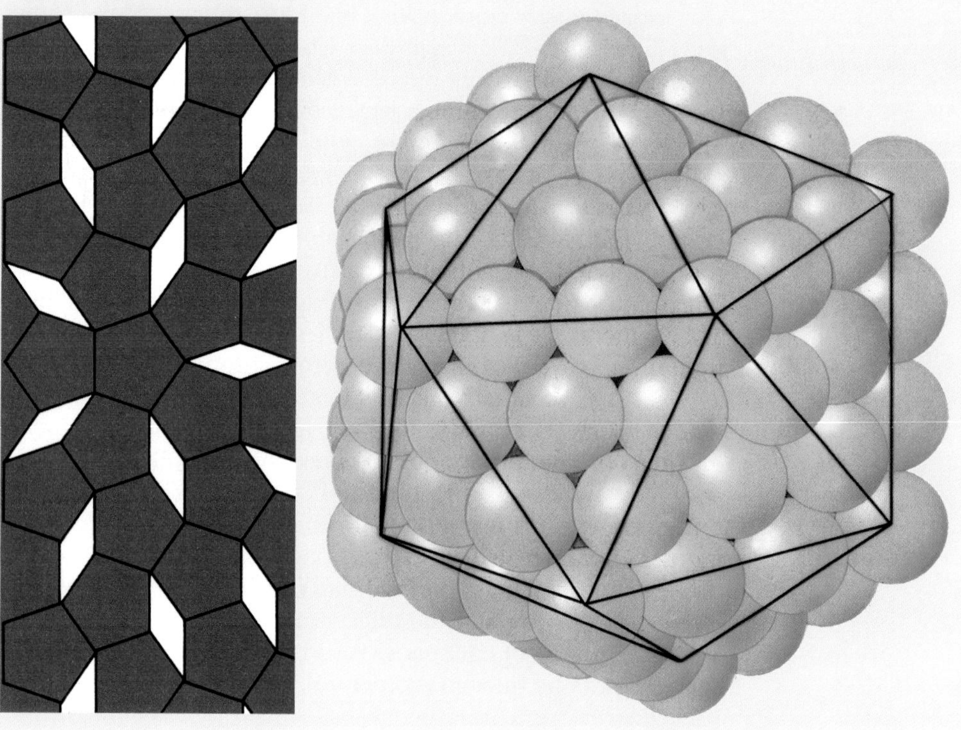

'forbidden' symmetry has now been observed in the diffraction patterns of numerous other alloys. Dov Levine and Paul Steinhardt suggested that the translational order in such materials is quasiperiodic, rather than periodic, and they coined the term *quasicrystal*. Ten-fold symmetry is clearly visible in the diffraction pattern reproduced here, for example; it was taken by Conradin Beeli, from an aluminium-manganese-palladium alloy. Meanwhile, it transpired that Roger Penrose had been independently studying the non-crystalline filling of two-dimensional space – the technical term is *tessellation* – by polygons such as the rhombuses shown in the blue and green diagram. In this example, the two types of rhombus have acute angles of 72° and 36°, and both the lack of crystallinity and the ten-fold symmetry are apparent upon closer scrutiny. The aesthetically pleasing *Penrose tilings* now enjoy considerable prominence. It is interesting that here too one finds the golden section emerging; the ratio of the numbers of blue and green rhombuses tends toward the limiting value of 1.618 as the pattern is extended to sufficiently large radii. It is indeed intriguing that precisely the same limiting ratio crops up in an infinite integer series that had been studied by Leonardo of Pisa (1175–1250). (Guillaume Libri referred to him as Fibonacci, in the nineteenth century, and the name has stuck ever since.) That series runs 1,1,2,3,5,8,13,21,34, . . ., the rule being that each number is simply the sum of the two preceding numbers. The *Fibonacci series* is frequently encountered in nature. It accounts for the arrangement of seeds in the head of a sunflower, for example, and for the pattern of units seen in a fir cone. Quasicrystals now routinely provide the starting point for materials with novel properties. Rónán McGrath, Renee Diehl and their colleagues deposited a thin layer of copper on the five-fold surface of an aluminium-manganese-palladium quasicrystal, in order to produce a conductor with unusual electrical properties. They discovered that the copper atoms arranged themselves – the technical term is self-organized – into a pattern that had two distinct spacings which followed the Fibonacci sequence. Uwe Grimm, Elena Vedmedenko and Roland Wiesendanger have been investigating the influence of quasicrystallinity on the arrangement of atomic spins, this holding the promise of magnetic properties different from those found in bulk material.

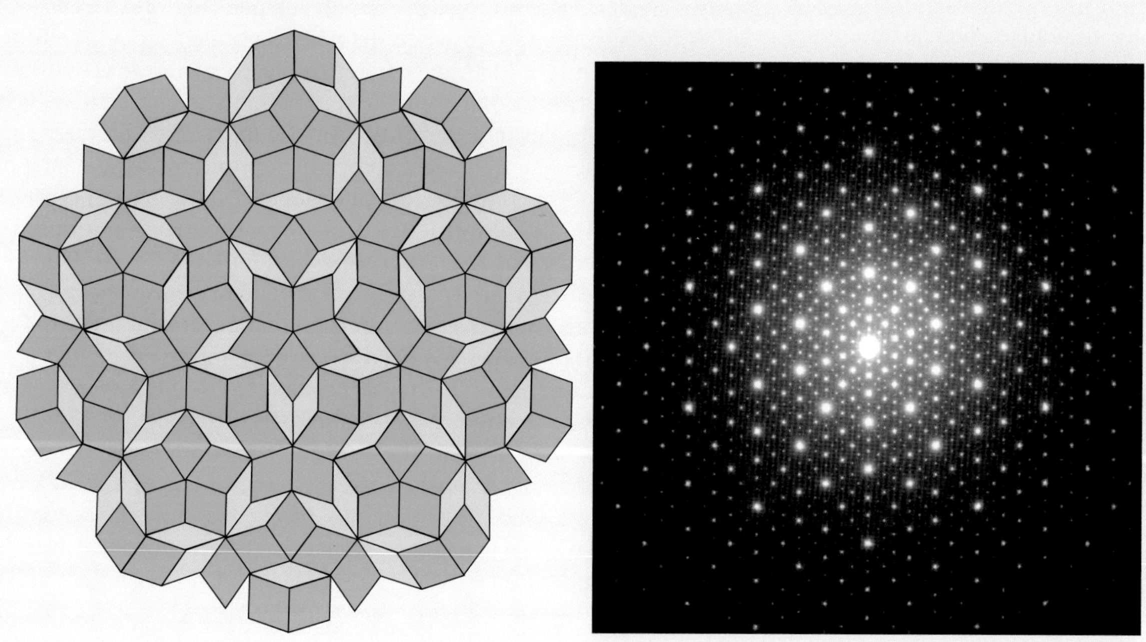

Although not inspired by crystallography, the periodic patterns in many of the drawings of the Dutch artist Maurits Escher (1898–1975) illustrate the principles of *lattice symmetry*. The colour picture involves mirror lines and three-fold rotation points, while the ingenious interlocking dog pattern has glide lines.

element, it is often necessary to differentiate between two crystal structures in which the packing densities differ by only a few per cent. These subtle differences are responsible for the abundance of different crystal structures observed in metallic alloys.

In engineering practice two of the most important attributes of a material are its *thermal conductivity* and its *heat capacity*. Depending on whether we seek to minimize heat transfer, as when insulating a boiler or a refrigerator, or encourage it, in a household radiator for example, we can choose between materials having widely differing abilities to transmit heat energy. The heat capacity is a rigorously defined quantity, but the term is also familiar in colloquial usage. Some materials seem to be able to hold more heat energy than others. The discovery of the beautiful order and symmetry in crystals raised the question as to what influence these regular patterns might have on the traditional theories of thermal properties. In 1819, Pierre Dulong and Alexis Petit had discovered an empirical rule concerning the specific heat capacity of the various forms of condensed matter, this quantity being defined as the amount of energy required to raise the temperature of 1 g of the substance in question by 1 °C. They demonstrated that the *specific heat* of a solid is roughly twice as large as that of the corresponding gas. Building on the earlier ideas relating heat to the energy of motion, they suggested that particles in a gas have only *kinetic energy* whereas those in a solid have equal amounts of kinetic energy and *potential energy*. Just how the potential energy came into the picture was not really clear, and in those pre-atomic days this was understandable.

The first attempt at a modern theory was made by Albert Einstein in 1907. By using the quantum ideas of Max Planck, he suggested that the kinetic and potential energies of the atoms are related to their vibrations about appointed positions in the solid, and because of the underlying symmetry in a crystal he assumed that each atom has the same vibrational frequency. The Planck theory had postulated that frequency is proportional to energy. Noting that the rules of *statistical mechanics* dictate the number of oscillations

The crystal structure of sodium chloride (table salt) has a face-centred cubic *Bravais lattice* and a *basis* consisting of one sodium ion, Na^+ (shown in grey), and one chlorine ion, Cl^- (shown in green). It can thus be regarded as two interlocking face-centred cubic sublattices, one of sodium and one of chlorine, mutually displaced by a distance equal to half the edge length of the unit cube.

simultaneously occurring at any temperature, Einstein saw that the specific heat must decrease with decreasing temperature. He had, however, made the simplifying assumption that the vibrating atoms act independently of one another. In view of the obvious importance of the interatomic forces, which are responsible for holding the atoms in their appointed places in the crystal, this was rather naive. These interactions were properly taken into account by Peter Debye in 1912, and shortly thereafter in more complete analyses by Max Born and Theodore von Karman, and by Moses Blackman. They showed that the forces between neighbouring atoms cause these to vibrate in unison so as to set up a series of waves that are constantly criss-crossing throughout the crystal. They too found a decrease in specific heat with decreasing temperature, but, in agreement with experiment, the falling away as zero temperature is approached is not as rapid as in the Einstein theory. In retrospect, it is interesting to note that Lord Kelvin had already worked out the types of wave motion that can be supported by a regular lattice, as early as 1881.

It was Debye himself, in 1914, who realized that *wave motion* also provides the key to understanding the thermal conductivity of a crystal. A travelling wave carries energy, playing the same role in a crystal as do the moving molecules in a gas. That a wave can accomplish in one phase what a particle does in another comes as no surprise, in view of the wave–particle duality that is so fundamental to quantum theory. The name *phonon*, used for a quantum of crystal wave energy, emphasizes the particle aspect. It is reminiscent of the word *photon* introduced earlier to describe the wave–particle quantum of electromagnetic radiation in the optical region of the spectrum. The natural *vibration frequencies* in crystals have a maximum that lies around 10^{12} Hz, and this is in the infrared region. The emission of heat radiation by crystals is related to this fact, and a wide variety of agencies can stimulate phonon vibrations: irradiation with particles, such as gamma rays; simple mechanical agitation, such as banging with a hammer; or thermal agitation by heating.

At any instant, the vibrations within a crystal produce a complex situation. There are waves of various amplitudes, wavelengths, and *phases* (i.e. stages of development) travelling in all possible directions. A given wave cannot be expected to travel indefinitely in an undisturbed fashion. Sooner or later it must impinge upon a region in which the crystal is locally distorted. This disturbance might be caused by another phonon, and thus have a transient nature, but it could also be of the permanent type associated with a *crystal defect*. We will consider such departures from perfection in the next chapter. If the average free path between phonon-distortion collisions is comparable to the width of the crystal, waves can transport energy across the latter in an unhindered manner, more characteristic of radiation than conduction. Such a state of affairs does indeed exist in defect-free crystals at low temperatures because the number of waves present at any time is small, and their amplitudes are small, both of which lower the phonon–phonon collision probability. Frederick Lindemann showed in 1912 that the vibrational amplitude always reaches the same relative value at *melting*, about 10% of the *nearest-neighbour distance*, so a temperature must be judged as being high or low depending on its relationship to the melting point. It is for this reason that diamonds feel cold to the touch. These gemstones are stable up to about 1200 °C, so room temperature is relatively low, and the corresponding average free path for phonons is large.

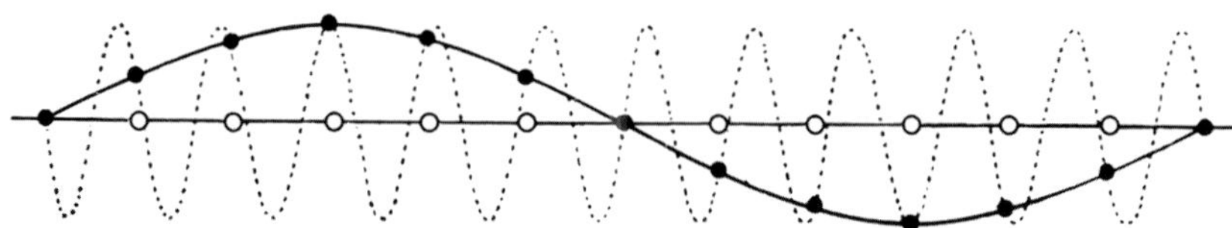

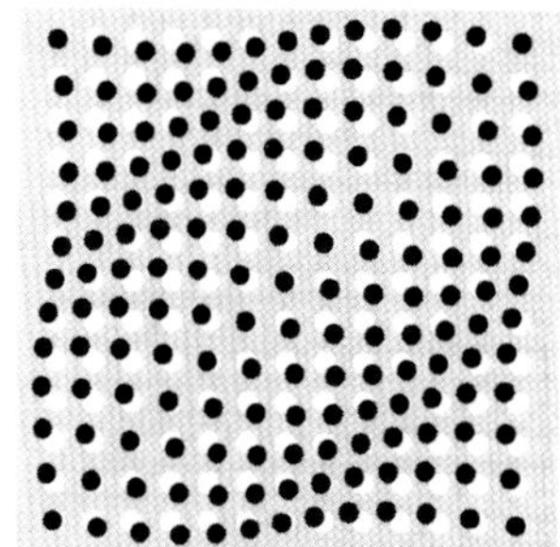

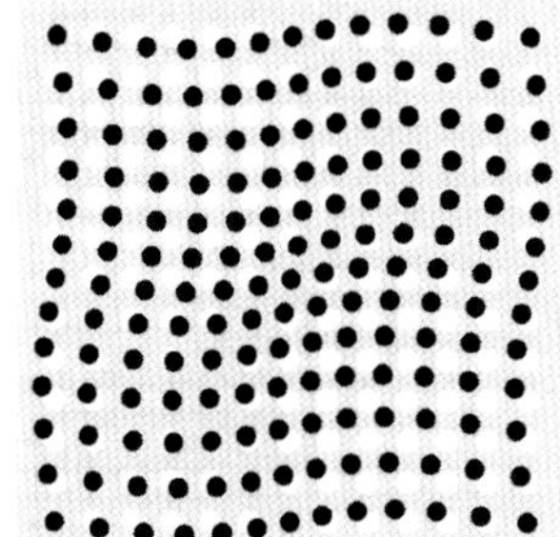

As was first explained by Peter Debye (1884–1966), energy is conducted through crystals by lattice waves. Two basic types are possible: longitudinal, in which the periodic displacements are along the direction in which the wave travels, and transverse, with displacements at right angles to this propagation. These are illustrated with simple linear chains (upper diagrams), the open circles indicating the equilibrium (undisplaced) lattice positions. A wave (dotted line) with wavelength below the fundamental limit, which is equal to twice the distance between adjacent lattice points, will always give displacements indistinguishable from a wave of wavelength above this limit. Two different instantaneous situations in a two-dimensional square lattice are shown in the lower diagrams. In one, a single longitudinal wave travels in a direction lying 45° to the horizontal (left). In the other (right), there is a combination of four waves, one transverse and one longitudinal in both the horizontal and vertical directions.

The collision between two phonons raises an interesting point. When a moving body hits another travelling more slowly in the same direction, kinetic energy is transferred from one to the other. Similarly, we should expect many collisions to produce from two impinging phonons a new phonon having a higher energy than either. This is certainly the case, but, as Rudolf Peierls realized in 1929, such a consolidation process can have peculiar consequences. Higher energy means lower wavelength, and the discrete nature of the crystal lattice imposes a natural lower limit to phonon wavelength. In the minimum wavelength situation, each atom is always moving in the opposite direction to its immediate neighbours in the direction along which the disturbance travels. A *phonon–phonon collision* that would otherwise produce a new phonon with wavelength lower than the allowed minimum is found to produce one of greater wavelength travelling in the opposite direction. A rather loose analogy to this is the peculiar effect seen when observing the spokes of a rotating wagon wheel. At low speeds the motions can readily be followed by eye, but the persistence of vision sets a limit to the speed that the eye can cope with, and above this limit the spokes suddenly appear to be moving slowly in the wrong direction. Peierls called this type of collision an *umklapp process*, the word coming from the German for flopover.

Two final points should be made regarding phonons in crystals. One concerns the familiar *thermal expansion* with increasing temperature. As was first pointed out by E. Grüneisen in 1912, phonon energies are lowered when a crystal expands because the distances between the atoms, and hence also the wavelengths, are increased. A longer wavelength gives a lower frequency and the latter is directly related to energy, following Planck's theory. This is

Sound, heat and elasticity

The *phonon* waves that travel along the various directions in a crystal transport energy. In an *insulator*, in which there are no free electrons to contribute to the process, these waves are the sole means of thermal conduction. Their velocity is determined by the rate at which displacement is supplanted from one atom to another along the direction of propagation, and this is related to the magnitude of the interatomic forces. The latter are reflected in the elastic stiffness of the crystal lattice. The rate of transmission of *sound energy* is also determined by a crystal's elastic properties. In all crystals, except helium below about 0.9 degrees kelvin, collisions of the phonon waves with such static defects as impurities and dislocations, as well as with other phonons, lead to thermal resistive processes, and *heat conduction* is an essentially diffusive process. The propagation is nevertheless governed by the elastic properties of the crystal lattice. The relationship between the various quantities are illustrated for three non-metals in the following table, in which the longitudinal *elastic modulus* is given in giganewtons per square metre, the *thermal conductivity* at room temperature in watts per metre per degree kelvin, and the *speed of sound* in metres per second.

Material	Elastic modulus	Thermal conductivity	Speed of sound
Diamond	1 076.0	1 800	17 530
Silicon	165.7	168	8 430
Germanium	128.9	67	4 920

partially compensated by the increase in potential energy due to stretching of the interatomic bonds. But these are small effects compared with the expansion caused by the asymmetry of the interatomic potential, a feature which was illustrated in Chapter 2. The repulsive forces increase rapidly with decreasing interatomic distance, whereas the increase in the attractive forces with increasing interatomic distance is less pronounced. As the thermal energy rises, the atoms thus prefer to lie farther apart. The other point refers to the situation when long-range crystallinity is absent, as in a glass. Cooperative motion of long rows of atoms is not possible in such a structure, and the vibrations are localized to relatively small groups of atoms. As we might expect, the thermal conductivities of glasses are generally lower than those of crystals, and these substances find frequent use in heat insulation.

Summary

The atoms in a crystal are neatly arranged in rows, columns and planes, like bricks in a wall, and this is the origin of the regular geometrical shapes of minerals such as calcite and rock salt. Apparently amorphous solids such as gold nuggets and hammered iron are found to be crystalline, when studied by X-ray diffraction, which probes atomic arrangement. Rather surprisingly, glass is less crystalline than paper, and even hair. The atomic pattern is formally described by the point group, which specifies the rotation and mirror reflection operations that are possible for a particular crystal, and the space group, which includes both point and translation operations. Just fourteen different basic patterns are possible in three dimensions: the Bravais lattices. The elementary unit of the pattern in a real crystal, known as the basis, consists of a single atom or a group of atoms, and because the basis can be arbitrarily complicated, crystal variety is truly unlimited. Vibrational energy in a crystal manifests itself in transverse and longitudinal waves, or phonons, which propagate along the various directions. They are responsible for heat conduction.

5

All nature is but art unknown to thee,
All chance, direction which thou canst not see;
All discord, harmony not understood;
All partial evil, universal good;
And, spite of pride, in erring reason's spite,
One truth is clear,
Whatever is, is right.

Alexander Pope (1688–1744)
An Essay on Man

The inevitable flaw: *imperfect crystals*

The bonding between atoms is governed by the structure of the atoms themselves, and because all atoms of a given element are identical, we might expect the atomic arrangement in a crystal to be regular. Crystals, and indeed solids in general, were believed to be perfect for many years after the atomic structure of matter had become generally accepted. There are now so many common phenomena whose very existence is the result of a lack of perfection that it seems surprising that the concept of *crystal defects* developed only recently. The idea emerged in several different guises, and there was an important precursor which, although not specifically related to crystals, played a major role in heralding the new era.

The first person to delve into the realm of imperfection was Alan Griffith, in 1920, following studies of the strengths of glass rods and fibres. He found that when the diameter decreases to about 10 μm, the strength of a fibre becomes markedly higher than that of a relatively thick rod. Assuming that such thin fibres must be free of the flaws that plague thicker specimens, Griffith analyzed the energy balance at idealized cracks and its dependence on applied stress. The crux of the matter lies in the fact that when a *crack* grows, there is an increase in energy due to the new surfaces generated, but also a decrease caused by the release of stress. The first of these factors is proportional to the crack area, but the stress-relief contribution is proportional to crack volume. Now volume increases at a faster rate than area; doubling the crack diameter causes a four-fold increase in crack area but an eight-fold increase in crack volume. If there is sufficient mechanical energy stored in the material around the crack, therefore, the total energy can be lowered by growth of the crack. This ultimately leads to rupture of the specimen. The density of mechanical energy is related to the applied stress, so there is a critical stress level above which a crack of given size will spontaneously propagate. The smaller the crack, the higher the danger limit, and Griffith's very fine fibres were strong simply because any cracks in them would necessarily have been very small. Today's flourishing *fibre glass* industry is a fitting monument to his pioneering efforts.

The deformation of crystals proved to be a surprisingly difficult problem, which baffled scientists for three decades following some intriguing observations reported by James Ewing and Walter Rosenhain in 1900. Examining specimens before and after slight deformation, they discovered that metals develop a characteristic surface marking when strained: a series of fine parallel lines which, from X-ray studies, were subsequently shown to coincide with the directions of certain crystal planes. If the deformation causes these planes to slide past one another, this must produce steps on the surface,

Until 1920, crystals were assumed to be perfect, each atom permanently residing in its appointed place in the lattice. In that year, Alan Griffith (1893–1963) suggested that solids are inherently weak because they contain minute cracks, like those in the iron specimen shown here. Such defects are relatively gross, but since Griffith's introduction of the concept, imperfection has been recognized as occurring right down to the atomic level. The width of the picture is about a millimetre.

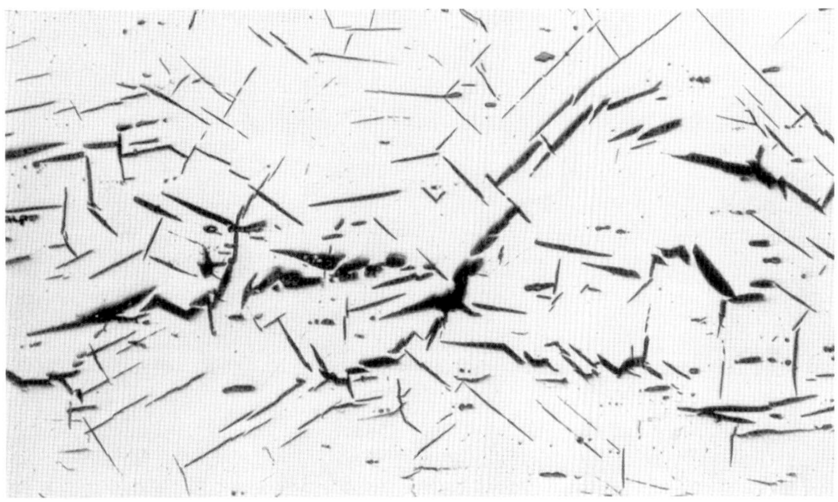

like the echelon arrangement seen when a deck of playing cards is pushed. Because the surface steps were clearly associated with a permanent change of specimen shape, they promised to shed light on the nature of the deformation process. Robert Hooke's investigations had shown that when a metal wire is subjected to a tensile stress, a weight suspended from one of its ends for example, the elongation is completely reversed when the load is removed provided the length of the wire has not been increased by more than 1% or 2%. If this is exceeded, removal of the stress leaves the wire permanently stretched. The critical elongation is known as the *elastic limit* beyond which the deformation is said to be *plastic*. The observation of *slip lines* revealed the significance of this limit. Up to this point the application of stress causes the distances between the atoms to be slightly increased, but no irreversible relative motion occurs. For stresses above the critical value, sliding of adjacent atomic planes occurs, and a mechanism similar to that of a ratchet prevents them from returning to their original positions when the stress is removed.

We have only to contemplate the orderly arrangement of atoms in a plane to find the origin of the ratchet mechanism. Looked at from any direction, each plane is seen to form a repeating pattern of hills and valleys over which the next plane must slide. If the sliding atoms have traversed one hill and are on their way up the second when the stress is removed, they will fall back to the valley between the first and second hills, and the crystal will acquire a permanent *deformation*. Slip occurs preferentially on planes in which the hills are low and the valleys shallow, and this is the case for those in which the atoms are most densely packed. Nearly all the common metals have crystal structures belonging to one of three types: the face-centred cubic (fcc), body-centred cubic (bcc), and hexagonal close-packed (hcp). Of these, the first and the last have particularly densely packed planes, and metals having these structures are easiest to deform. Aluminium (fcc) and magnesium (hcp), for example, are quite *ductile* if pure. The more open nature of the bcc structure means higher hills and deeper valleys, and metals such as iron which have this structure are more rigid and *brittle*.

This approach to deformation is quite straightforward, but it turns out to be utterly inadequate. If one plane is assumed to pass rigidly over its neighbour, with all atoms moving simultaneously, the force calculated to

Babel revisited

The desire of the biblical people of Shinar to build, at Babel, a tower so tall that it would reach into heaven, displeased God. He confounded their efforts by replacing the single universal language with a host of disparate tongues, thus denying them the coordination necessary for such an ambitious undertaking. In fact, the enterprise was already doomed to fail because it would have been thwarted by the inherent limitations of *interatomic bonds*. We calculate the maximum possible height of a tower, assuming it to be built of the strongest material available at that time, namely granite. Like all polycrystalline and amorphous substances, this rock fails by shear along a plane lying 45° to the direction of compression, when it is subjected to a sufficiently high stress. The ultimate failure strength of granite is 20 meganewtons per square metre, and its density is 2.7×10^3 kilograms per cubic metre. A stress of 1 kilogram per square metre is approximately equal to 10 newtons per square metre. Because the 45° angle introduces an extra factor of two, we find that the danger point would have been passed if the tower exceeded a height of 1445 metres, well short of the target. In 1956, Frank Lloyd Wright (1867–1959) designed a 528-floor skyscraper, to be built in Chicago, Illinois. The projected height was one mile, or about 1609 metres. Although the building would no doubt have been reinforced with steel, it is probably fortunate that it got no farther than the drawing board. As Victor Weisskopf (1908–2002) emphasized, even the elimination of shear failure does not permit structures to reach unlimited heights; the pressure on the material cannot exceed that required to cause melting. (This applies only to substances which, like ice, contract upon melting.) And because the weight of a given mass is determined by the gravitational attraction, the limiting height of a mountain decreases as the planetary size increases; the larger the planet, the smoother must be its surface.

bring about this motion is higher than the observed level by a factor of about a thousand. The slipping process occurs not simultaneously but via a consecutive motion which passes sequentially from one atom to the next. At any instant most of the slipping plane is stationary, the disturbance being limited to a narrow *line defect* known as a *dislocation*. Dislocations made their appearance in theoretical studies by Egon Orowan, Michael Polanyl and Geoffrey Taylor, in 1934. One form of the defect can be looked upon as an unpaired, or dangling, bond. The sequential shifting of this bond can be achieved with surprisingly low stress, and when dislocation motion is invoked, theory and experiment fall nicely into agreement. This was demonstrated by Rudolf Peierls and Frank Nabarro.

The very low stress required to move a dislocation is the result of a situation that is essentially the reverse of a tug-of-war, at the atomic level. Atoms lying immediately ahead of the defect resist its approach, because it will displace them from their positions of lowest energy, namely their lattice sites. Those positioned immediately behind the dislocation, which have already been displaced by the defect, try to push it forward, out of the way, so that they can drop into their new stable positions. Caught between the opposing pushes, the dislocation is subjected to a net force which is quite small, and the resistance to its movement is almost negligible. The motion of a dislocation is not unlike that of a caterpillar. This creature ambles around by constantly passing a disturbance down the length of its body; at any time it has just one or two pairs of feet raised. That the dislocation is a natural consequence of bending a crystal could have been anticipated from a study of curvature in the builders arch. Each successive layer contains an extra brick. Indeed, a hint of such an arrangement in deformed crystals was already present in a suggestion by Ulrich Dehlinger, in 1929, and his *verhakung* is what would now be called a dislocation.

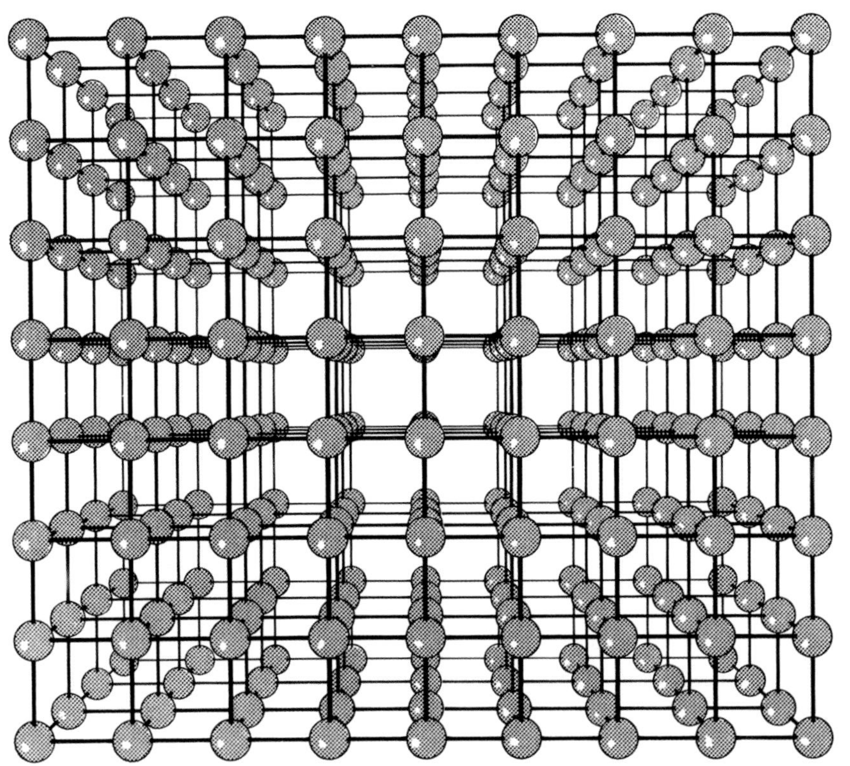

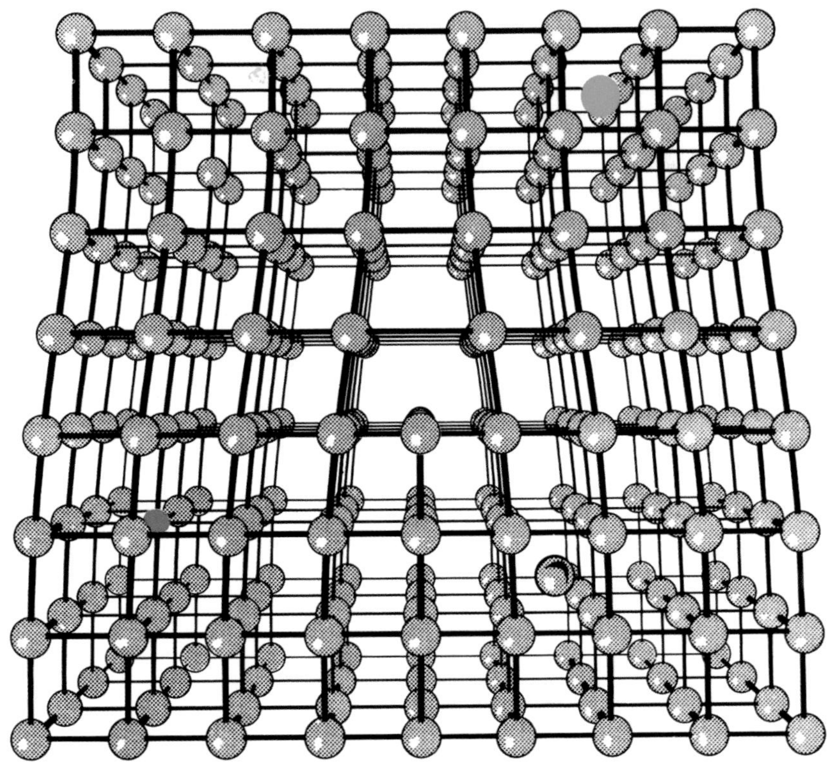

The simple cubic lattice shown above is perfect, a condition which never prevails in nature, if the crystal is sufficiently large. The structure below contains five common defects: an *interstitial* (lower right); a *vacancy* (upper left); interstitial and substitutional *impurities* (upper right and lower left, respectively); and a *dislocation* in the edge orientation (centre).

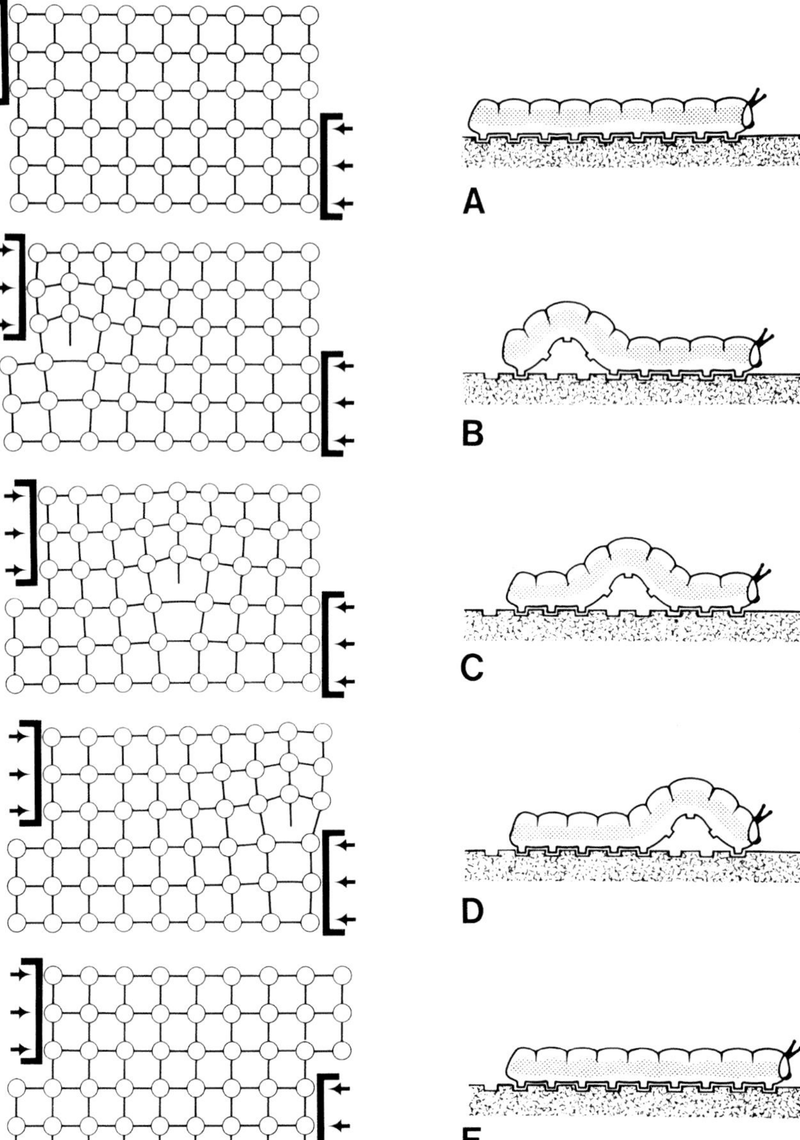

The low stress level required to provoke *dislocation motion* is a consequence of the fact that the disturbance is spread out over many atomic diameters, the strain being appreciable only at the central region, known as the core, in a sort of reverse tug-of-war, the atoms ahead of the core oppose the motion while those behind promote it. The consecutive movement of the core, from site to site along the *slip plane*, is analogous to the crawling of a caterpillar, which eases the burden of movement by having only a few legs off the ground at any instant.

The almost exact balance between the inherent forces on a dislocation, the pushing influence from the atoms behind it and the restraining tendency of those lying ahead, is conditional on there being a relatively large number of atoms involved; the transition from the un-displaced region to the displaced region must be spatially extended over many interatomic distances. The amount of stress that must be applied externally to destroy this balance, and make the dislocation move, is then small, and the crystal can be deformed easily. This is generally the case in metals because their free-electron bonding imposes no directional restriction. The face-centred cubic and hexagonal examples such as copper and zinc, respectively, have wide dislocations, and these metals are quite ductile. Body-centred cubic metals, being less densely packed, have somewhat narrower and consequently less mobile dislocations. The relative stiffness of iron and its structural cousins

The characterization of line defects

Like all defects, *dislocations* and *disclinations* cause discontinuity in the pattern of atoms or molecules in condensed matter. They both belong to the class known as *line defects*. The dislocation disrupts translational symmetry, and it is a common defect of crystals, in which it mediates *plastic deformation*. The disclination disturbs rotational symmetry, and because this would cause excessively large energies, disclinations are not present in crystals. They are a common feature of *liquid crystals*, however, and it has been suggested that they may be found in glass. The strength and direction of the disturbance associated with a dislocation is measured by describing the circuit recommended by Johannes Burgers (1895–1981). In the perfect crystal this starts (green point) and finishes (red point) at the same place (top left), whereas it fails to close (middle left) if the circuit encloses a dislocation. The closure failure is known as the Burgers vector (blue arrow). This vector is not necessarily a translation vector of the lattice; it can be smaller, in which case one refers to a *partial dislocation*, or larger, as in the superdislocation that is commonly implicated in fracture. When a positive and negative dislocation of equal strength are close together, a Burgers circuit enclosing both defects must close itself. This dipole configuration, as it is called, is shown below left. A disclination can be similarly characterized by keeping track of direction while describing a circuit around the defect, as suggested by Frank Nabarro. The two figures at the right show a crystalline pattern based on hexagons, and the arrows indicate direction. The central site of the disclination has five-fold symmetry and the rotational angle is $+60^\circ$. A disclination centred on a point of seven-fold symmetry would have a -60° rotational angle. In 1965, Doris Kuhlmann-Wilsdorf suggested that *melting* is associated with the spontaneous generation of *dislocation dipoles*, and support for this conjecture was obtained in the early 1970s in computer simulation studies of melting carried out by Laust Børsting Pedersen, Ernst Jørgen Jensen, Willy Damgaard Kristensen and the present author. The studies demonstrated the role of dislocation proliferation in the melting process, although the liquid state is not appropriately described as a crystal saturated with dislocations.

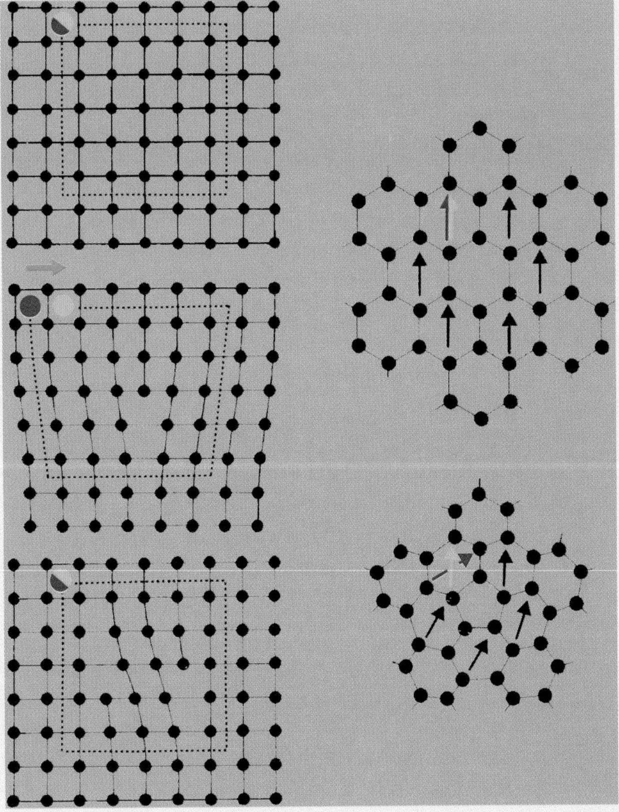

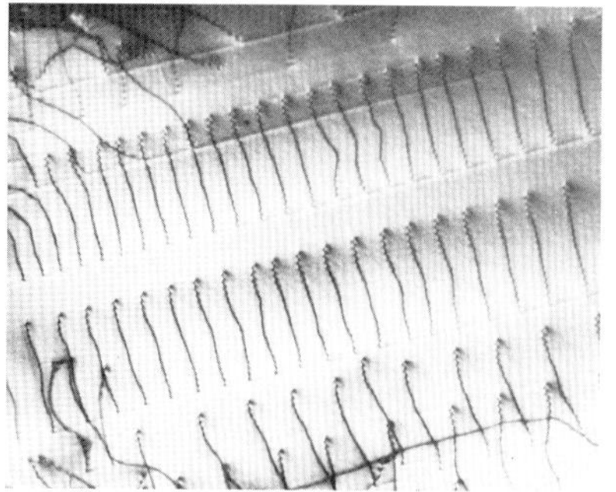

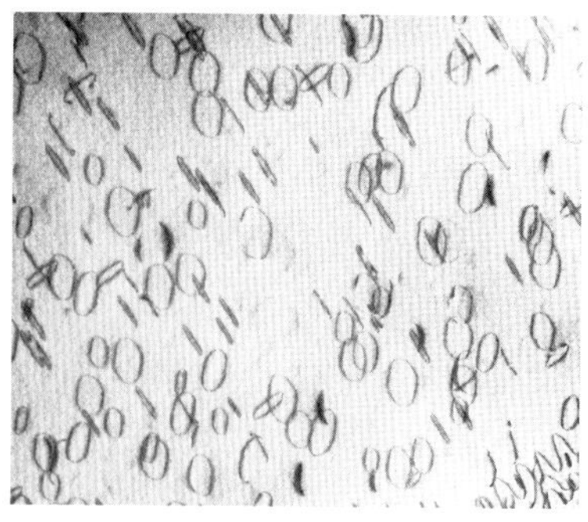

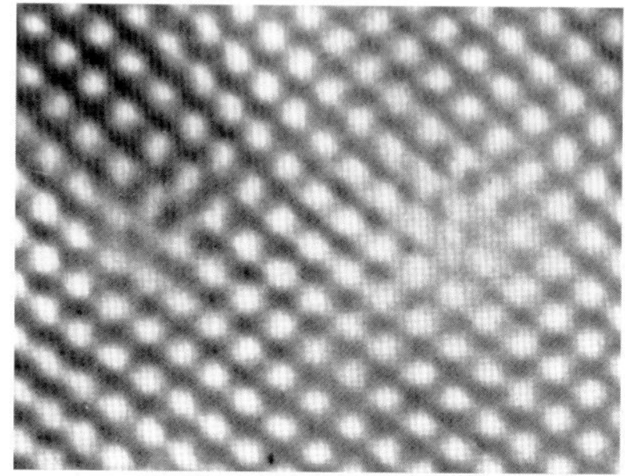

Transmission electron microscopy enjoys the advantage of revealing imperfections which lie within crystals that are optically opaque. And because the electron wavelengths are considerably less than an atomic diameter, disturbances that occur at the atomic level can easily be seen. The roughly straight lines on the left-hand picture, the height of which is about two micrometres, are *dislocations*. The photomicrograph is simultaneously imaging the entire thickness of the stainless steel specimen, so the extremities of the dislocations are the points at which they intersect the upper and lower surfaces. This interpretation is inescapable because dislocations cannot end abruptly within a crystal. They can, however, form closed loops, as seen in the right-hand picture of an aluminium −4% copper specimen, the height of which is about five micrometres. The photomicrograph below was taken at such a high magnification that the individual atoms of germanium can be distinguished. Two dislocations can be detected by viewing along the atomic rows, from a low angle.

can be compensated for by raising the temperature, and this is why they are rolled red hot. Covalent bonding, with its highly directional nature, represents the other extreme, and crystals such as silicon and diamond have very narrow dislocations and are very hard. One theory of glass structure compares this with a crystal saturated with dislocations having widths comparable to the interatomic distance. This approach still lacks experimental justification, but it does at least explain the intrinsic hardness of the glassy state.

The dislocation is the line of demarcation between two regions of a crystal, one of which has slipped relative to the other. From this follows an important topological property of the defect; it cannot end abruptly inside the crystal, but must terminate at the surface or on another dislocation. A closed loop of dislocation satisfies this requirement, and *dislocation loops* are indeed common in crystals. A less obvious attribute of this defect is that the amount of slip associated with it is exactly the same, in both magnitude and direction, along the entire length of the dislocation line. Because the properties of dislocations are so well defined, we have a complete description of the crystal interior if we know where the dislocations are located, and which displacements are associated with them. We are therefore relieved of the far more laborious task of keeping track of the vast number of individual atoms. During the deformation process, the dislocations move around in the crystal like pieces of imaginary string. Indeed, they can form complicated tangles, and this impedes their motion. But unlike pieces of real string, dislocations can also cut through one another, although this does involve the expenditure of energy. The tangling process is one of the underlying causes of the familiar phenomenon observed when a crystal is worked in the cold state. A beaten or rolled piece of metal, for example, becomes progressively harder because the increase in the number of dislocations caused by the deformation raises the probability of making these inhibiting tangles. If the metal is heated to near its melting point, the *work hardening* can be

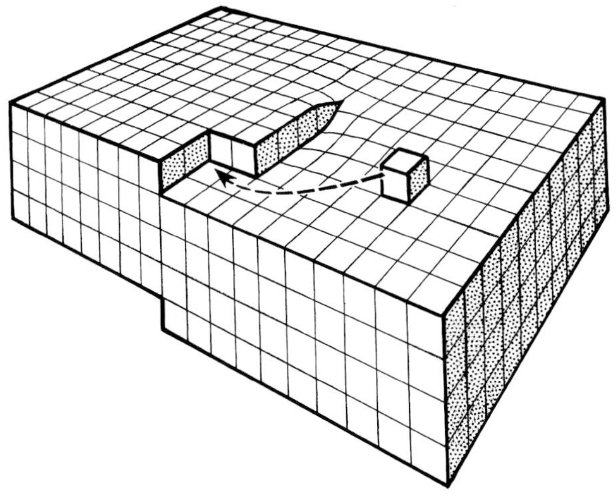

A *screw-oriented dislocation* converts a set of parallel crystal planes into a single continuous helical sheet. Where this helix intercepts the surface, a step is formed (see diagram) which cannot be removed by adding further atoms or molecules; the crystal grows as a never-ending spiral. The photograph shows a *growth spiral* on the surface of a polypropylene crystal.

removed, as every blacksmith knows. The curative ability of the heat works through the agency of *diffusion*, the rate at which the atoms migrate being higher at elevated temperatures. This atomic motion repairs the damage to crystalline perfection caused by the dislocations, by shuffling these defects out of the crystal.

It requires no special act of faith to believe in dislocations. Direct observation of these defects is now a routine matter. Initially, it required considerable ingenuity to render them visible, and exploitation of the presence of another crystal imperfection; the *impurity atom*. Alien atoms usually have a size that is inconsistent with the interatomic spacings in the host crystal and are thus associated with a certain *strain energy*. This can be relieved if they migrate to the vicinity of a dislocation, where the lattice is distorted, and a dislocation in a transparent crystal can become visible if it has a sufficiently dense collection of impurity atoms around it. This was first demonstrated by J. M. Hedges and Jack Mitchell, in 1953, using silver halide crystals. They employed an *optical microscope*. It later transpired that dislocations are visible even when not thus decorated, if the operating wavelength of the microscope is comparable to the smallest interatomic distances in the crystal. This is so for an *electron microscope* operating at 50 000 V or more, and *transmission electron microscopy* of dislocations, and other defects, has become widespread since the pioneering efforts of Walter Bollmann, Peter Hirsch, Robert Horne and Michael Whelan, in 1956. The technique has two advantages; the crystal need not be optically transparent, and the dislocations can be observed even when in motion.

Because the displacement associated with a given dislocation is constant, both regarding magnitude and direction, whereas the line defect itself is reasonably flexible, various situations can arise. The presence of a dislocation can be detected, and its displacement characterized, by a simple prescription recommended by J. M. Burgers. Like an ant crawling along the grooves between paving stones, one traces out a path which, in the perfect crystal, would make a closed circuit, starting and finishing at the same point. If a dislocation is now introduced, the same path will not form a closed circuit if, but only if, it encircles the dislocation line. The closure failure of this circuit defines the displacement associated with the dislocation: the *Burgers vector*. In general, the Burgers vector and the dislocation line make an arbitrary

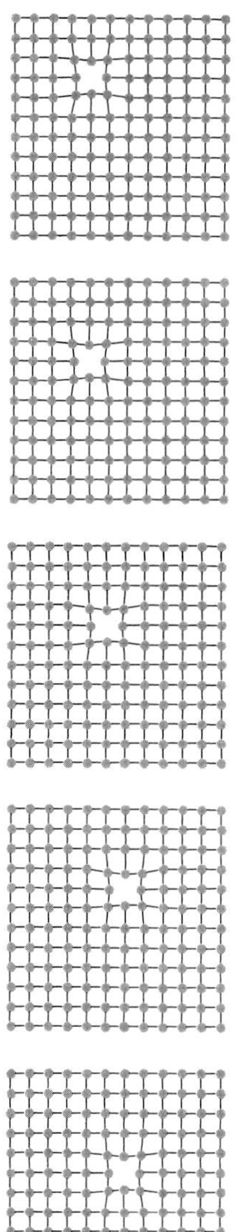

At any temperature, all the atoms in a crystal are vibrating about their appointed positions in the lattice, and the latter contains a certain proportion of vacant sites. As the temperature rises, both the equilibrium concentration of these *vacancies* and the amplitudes of the *atomic vibrations* increase. A vacancy can move around in the crystal because neighbouring atoms can jump into the hole, filling it up, but leaving a new vacancy at the site thereby vacated.

angle with each other, while the extremes, vector and line either parallel or mutually at right angles, give what are known as the screw and edge orientations, respectively. The *screw-oriented dislocation* is an intriguing defect which converts a set of independent and parallel planes into one continuous helical plane that is reminiscent of a spiral staircase. Any dislocation must make a step where it emerges at a crystal surface, and Charles Frank showed that the step associated with the screw orientation would lead to a special type of crystal growth: helical and never-terminating. The *edge-oriented dislocation* is equivalent to a semi-infinite plane of atoms, terminating at the dislocation line. It must be emphasized that dislocations are generally flexible, and that their screw or edge character can readily be modified by changes of position provoked by external stress. Only rarely is a dislocation locked into a particular orientation. Crystallographic symmetry imposes severe limitations on the types of Burgers vector that can occur in a given structure. When two dislocations meet, and if it is energetically favourable for them to do so, they combine to form one new dislocation with Burgers vector equal to the sum of the original individual vectors. In the special case where the latter have equal magnitude but exactly opposite direction, the combined Burgers vector has zero magnitude, and the two dislocations mutually annihilate.

It was speculation about the atomic motions involved in diffusion that led to the discovery of several other types of crystal defect: types which are simpler than the dislocation. A wallpaper analogy serves to illustrate the salient points. The dislocation is the defect that occurs when a careless paper-hanger fails to line up one strip in exact registry with the next; trying to correct the fault by sliding the paper, he finds that some of the paste has already dried. The repair is only partially successful and the offending strip of paper develops tell-tale wrinkles: the dislocations. But the paper itself might have errors. A blossom might be missing in a floral design, or an extra one pushed in between the neat rows. These *point defects* are the *vacancy* and the *interstitial*, respectively. The existence of vacancies was first postulated by W. Schottky in 1935, and the interstitial by I. Frenkel in 1926.

There was no lack of candidates for the responsibility of *diffusion*. The problem lay, rather, in determining which defect actually plays this role. The most obvious way for an atom to move is to leave its appointed position in the lattice and squeeze through the gaps between other atoms. Such gaps are narrow, however, and the associated *energy barriers* are prohibitively high. They can be avoided if several atoms move simultaneously in a coordinated manner, a particularly attractive possibility being the cooperative motion of a ring of atoms. A third possibility is vacancy-mediated motion. This defect provides a convenient hole into which the diffusing atom can jump. If the next jump takes the atom back to its original position, no net diffusion will have been achieved. But if a constant supply of moving vacancies is presented to the diffusing atom, it will be able to travel in the direction required by *Fick's law*, namely towards a region where the concentration of its species is lower. Yet another type of diffusive motion involves the interstitial. An atom moving away from its normal position might hit an interstitial and push it into another lattice site, the atom thereby displaced becoming a new interstitial. A succession of such events would cause diffusion. The relative probabilities of these different types of defect motion were calculated by Hillard Huntington and Frederick Seitz, in 1941, using reliable representations of the interactions between atoms. They found that

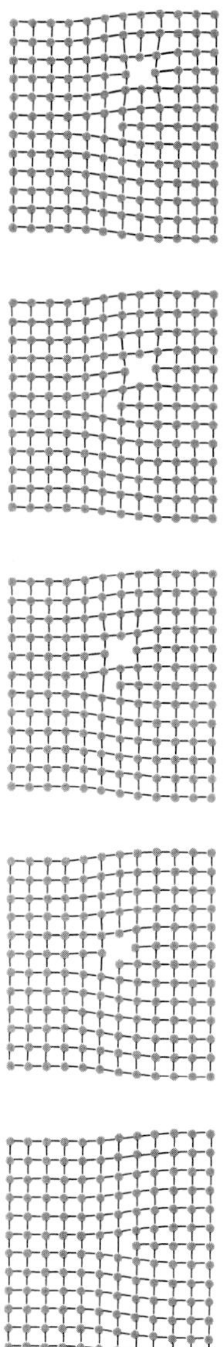

A *dislocation* is normally constrained to move along its own *slip plane*; this motion is said to be conservative, and it is referred to as *glide*. But movement out of the slip plane can occur if the dislocation absorbs or emits a *point defect*, such as the *vacancy* shown here. *Interstitial* absorption would make the dislocation move in the opposite direction. In either case, the motion is said to be non-conservative, and it is known as *climb*.

for crystals in which the atomic packing is reasonably dense, as is the case in metals, the vacancy mechanism is by far the most likely. The formation and migration of defects provides a good illustration of the application of *thermodynamics* to condensed matter. The probability of a defect appearing in a crystal is related to the *free energy* associated with its formation; the lower the free energy, the greater the probability. The higher the temperature, the greater is the *thermal activation* available to generate a defect, so its concentration must increase. In a typical metal close to its melting temperature, the vacancy concentration is about one for every thousand atomic sites.

Having identified the modes of diffusion in a crystal, we can clarify the thermally activated removal of dislocations. These defects unavoidably accompany the relative slipping of adjacent parts of a crystal. There is a topological requirement that the plane in which a dislocation moves contains both the dislocation line and the Burgers vector, and the resistance to such slip, or *glide*, depends on the strength of the interatomic interactions. If the resistance is low, the dislocation moves easily and is said to be *glissile*. High resistance gives lower mobility, and in the stationary extreme the defect is said to be *sessile*. The position of a sessile dislocation can nevertheless be shifted by inducing motion out of its slip plane, but this requires the diffusive motion of individual atoms, which must be either added to or taken from the edges of the incomplete atomic planes associated with the defect. Glissile dislocations can undergo the same process. This type of motion is referred to as *climb*. Vacancies can be either captured by or ejected from a dislocation, and these processes cause climb in opposite directions. For interstitial capture or ejection, the direction of climb is reversed. A climbing dislocation must ultimately reach the crystal surface and thus be eliminated. The velocity of climb is governed by the rate at which point defects can migrate to the dislocation. Since that rate is governed by diffusion, which is temperature dependent, climb proceeds more rapidly with increasing temperature. Thus the *annealing* of a dislocation-laden crystal, to remove these defects, is best carried out at a temperature near the melting point. This is the origin of the curative effect, by which heating is able to return a hardened metallic bar to a softer condition. We should note the special circumstances of the screw-oriented dislocation, with respect to the glide and climb processes. Because its Burgers vector lies along the dislocation line, there are in principle an infinite number of possible slip planes. The crystallographic constraints reduce these to a limited set, but the dislocation is nevertheless provided with the opportunity of changing its plane of travel. This is useful, because it can thereby avoid obstacles by moving around them, a process known as *cross-slip*. But the screw-oriented dislocation wins this advantage by giving up the possibility of climb; it cannot accomplish such motion because there are no incomplete atomic planes associated with its particular configuration.

The atoms surrounding a vacancy tend to relax inwards towards the vacant site, so vacancy motion is not quite as easy as we might expect. Nevertheless, a jumping rate of 10^9 per second is quite common in a metal near its melting point. In a *covalently bonded crystal*, such as silicon or diamond, the highly directional bonds permit the vacancy to maintain a rather open configuration, but they also tend to hinder the motion of the atoms, even when these are adjacent to a vacant site. In *ionic crystals* such as the alkali halides, the situation is complicated by the presence of two different ionic species that usually have quite different sizes. In sodium chloride, for example,

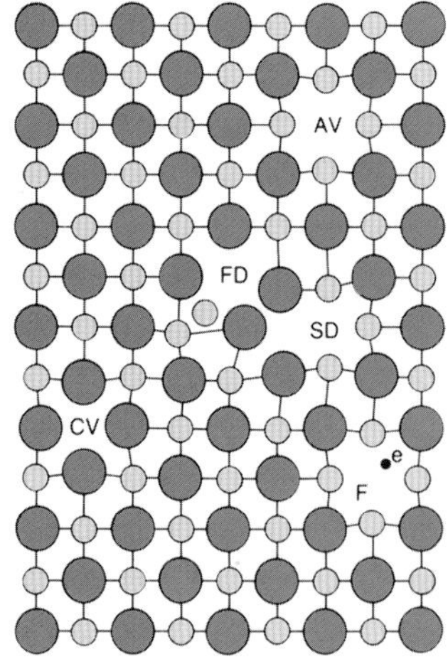

The variety of defects in *ionic crystals* is particularly rich because of the presence of two different types of electrically charged *ion*. But this also imposes constraints, the crystal attempting to maintain electrical neutrality. The *anion vacancy* (AV) is positively charged, because there is a local deficit of a negative ion, while the *cation vacancy* (CV) is negatively charged. Electrical neutrality is preserved by the *Schottky defect* (SD), which is equivalent to an anion vacancy and a cation vacancy in adjacent sites, and by the *Frenkel defect* (FD) in which a displaced cation becomes an interstitial but remains close to its original site. The positive charge of an anion vacancy can attract an electron (denoted by the letter e), and the latter can occupy any of a series of bound energy levels similar to those of a typical atom. The transitions between some of these levels correspond to wavelengths in the visible region, and the compound defect is usually known by the German name *farbzentren* (F), i.e. *colour centre*. Ionic defects are important in the *photographic process*.

a small sodium atom, known as a *cation*, would be expected to be more mobile than a larger chlorine atom, which is termed an *anion*. There are, however, restrictions imposed by the requirement of overall electrical neutrality. A *cation vacancy*, being equivalent to a missing positive charge, has an effective negative charge, while an *anion vacancy* is positively charged. There are two ways in which point defects can exist in an ionic crystal and avoid electrical imbalance. If an ion is displaced from its normal site but not removed from the crystal, all the original charges will still be present. The displaced atom will leave behind a vacancy, and it will come to rest in one of the interstices, becoming an interstitial. There will be a residual electrical attraction between the two defects thus produced, and there will be a tendency for them to form a bound pair, known collectively as a *Frenkel defect*. Such a defect is more likely to involve the smaller ion because of the relatively favourable energies involved. The other possibility is for two unlike vacancies, one cation vacancy and one anion vacancy, to form an electrically neutral pair. This is the *Schottky defect*. Beautiful optical effects, first studied by Robert Pohl in the 1930s, can be produced at positively charged vacancies in transparent ionic crystals due to the capture of electrons. The latter arrange themselves into atom-like quantum orbits, the transitions between some of which correspond to frequencies in the optical region. These defects are known as *colour centres*, and they will be discussed in Chapter 10.

Vacancies and interstitials are not the only point defects. The other important member of this group is the foreign atom or *impurity*. It can either occupy a regular lattice site, as if filling a vacancy, when it is referred to as a *substitutional impurity*, or it can be an *interstitial impurity*. This is not to say that both configurations are equally likely for a given alien element. The most probable location is governed by the thermodynamics of the situation, and this in turn is related to the interactions between the atoms of the impurity and those of the host crystal. Just what is meant by the terms pure and impure depends critically on the actual elements involved and the properties of interest. A crystal of silicon or germanium is useless to transistor technology if it contains as much as 1 part in 10 000 000 of impurities, whereas 1% of impurity in many metal components can be regarded as negligible. In *alloys*, atoms of a foreign element, or elements, are added to the main constituent deliberately, the relative proportion of these often reaching as much as 50%. The actual distribution of impurity or alloying atoms depends on the thermodynamics of the situation and, therefore, on the temperature. The tendency for the entropy to have its maximum value favours their dispersion throughout the host crystal in as random a fashion as possible, but this might be opposed by an inherent preference of the impurities to establish interatomic bonds with their own species. It is such conflicting factors that are responsible for the rich variations of microstructure exhibited by many alloy systems. At elevated temperatures a minor element is usually distributed as single solute atoms in solid solution. In some cases, as with gold–silver alloys at any relative composition, this situation persists even down to room temperature and below. In other systems, lowering the temperature induces the solutes to cluster together, there then being insufficient thermal activation to break the favoured impurity–impurity bonds. Important examples occur in metals, and the clusters of atoms of the minor element are known as *precipitates*. An understanding of the influence of impurity atoms on the electrical, optical, and magnetic properties of crystals has led to the development of novel materials of technological utility.

Viewed along any particular direction, a perfect crystal always appears to be composed of a series of atomic planes stacked in a regular sequence. In the *face-centred cubic* structure, for example, this sequence runs ABCABCABC if the crystal is observed along a direction that is equally inclined to all three of the principal axes of the unit cell (upper two diagrams). A *stacking fault* can be introduced over a limited area in various ways: by locally removing the atoms lying in one of the planes so as to give the sequence ABCACABC, for example; by adding atoms so as to locally produce a new plane and turn the sequence into, say, ABCABACABC; or by sliding a sheet of the atoms within their plane until they reach a metastable configuration. An example of the latter would produce the sequence ABCABABCA. The bottom left diagram shows an A layer, say, on top of which are stacked atoms in a C layer, to the left, and in a B layer, to the right. The picture shows an electron micrograph of a stacking fault in silicon, the width of the picture being about fourteen micrometres.

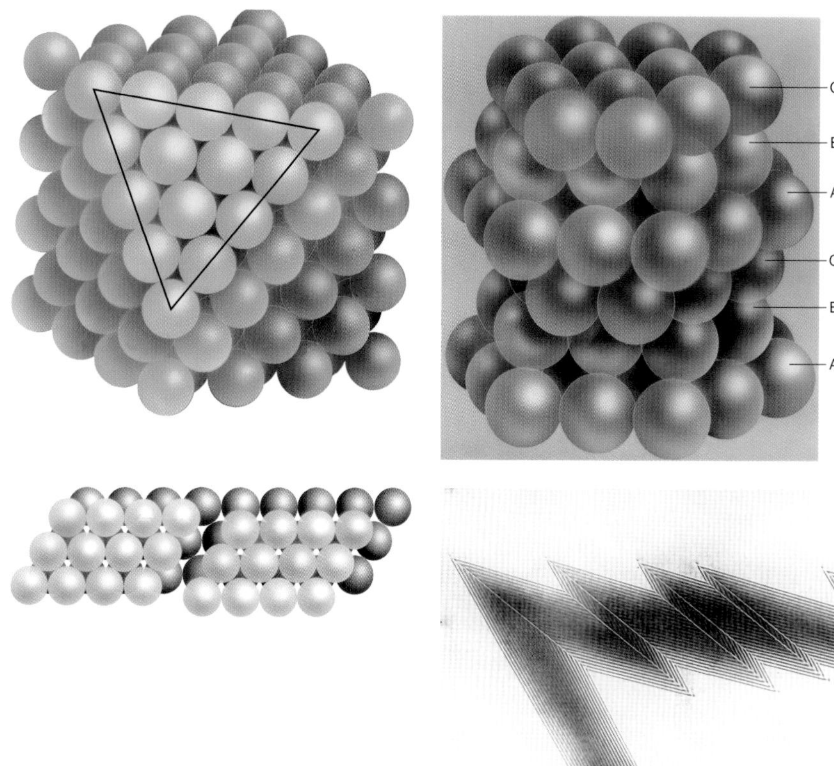

Among the mechanical effects associated with impurities, the most important is their inhibition of dislocation motion. Deployed individually, they provide a natural hindrance by dint of their distortion of the lattice, while in the form of a precipitate they can become an impenetrable barrier.

Point defects, which are essentially imperfections of occupation, are clearly more localized disturbances than the line defect represented by the dislocation. They involve a single atom, or atomic site, plus small displacements of a few surrounding atoms. The dislocation is an imperfection of pattern; the distortion is subtle but long-ranged, and even atoms lying far from the centre of the defect can feel its influence. There is a third class of defects which cause a different type of disruption of the crystal lattice. These are *interfacial defects*, the prominent members of this family being the stacking fault, the twin boundary, the grain boundary, and the free surface. These too are interruptions of pattern, but the disturbance is abrupt and rather localized. Once again, a wallpaper analogy proves useful.

The paper is retailed in rolls of standard length, whereas walls vary in height, and this presents the decorator with considerable scope for error. Imagine that a design for a bathroom consists of rows of fish (F), sea horses (S), and crabs (C) in regular repetition: FSCFSCFSC. If one roll ends with crabs while the next starts with sea horses, a decorator who is not alert might introduce a defect and discover that his pattern contains the local sequence FSCFSCSCFSCFSC. This is an example of a *stacking fault*. If a crystal consists of atoms of three different elements, each arranged in neat layers, a stacking fault of this type might well occur. Stacking faults can be formed, however, even in crystals of a pure element, because different atoms might

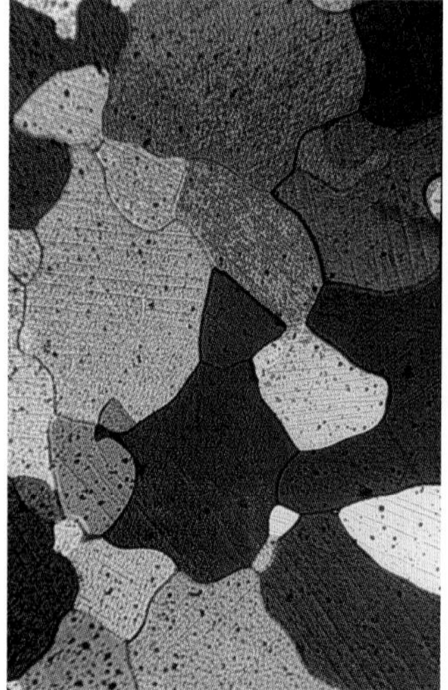
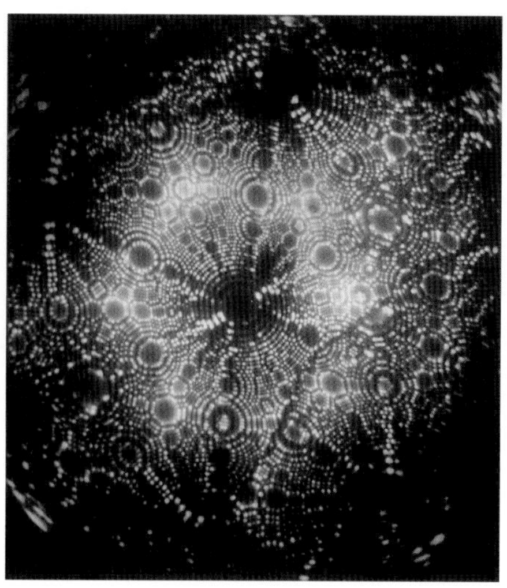

Most crystalline materials are composed of aggregates of small crystallites, or *grains*, which are mutually misoriented (left). The interfaces between these particles are *grain boundaries*. Viewed with an *optical microscope*, using polarized light, the various grains preferentially reflect different wavelengths, and the polycrystalline surface of this piece of copper looks like a patchwork quilt (centre). The actual atomic arrangement around a grain boundary is revealed by *field-ion microscopy*. In this picture of the surface of a tungsten specimen (right), a grain boundary can be seen running from the lower middle to the middle right.

have different relative positions in the lattice pattern. This is just as effective a way of labelling atoms as is giving them a different atomic number. Stacking faults are frequently associated with dislocations; the introduction of one of these defects at the extremities of an accidentally slipped region might be such as to leave a stacking fault in an otherwise registered pattern.

With pictures of such things as sea horses in his wallpaper, the decorator is unlikely to hang a strip upside-down, but with a more subtle design such an accident could occur. Unless the symmetry of the pattern is trivially simple, the error will introduce the defect known as a *twin boundary*. If the correct sequence involves the three non-equivalent layers, or positions, XYZ, the twin boundary will have the form XYZXYZYXZYX. The stacking sequence is inverted into the mirror image of itself by the defect. By comparison, the influence of a stacking fault is limited to a few atomic planes, and this defect is detectable only by devices which are capable of resolving detail at the near-atomic level. *Twinning faults* were well known to the early crystallographers, who could observe the boundary separating two twin crystals with the naked eye. Stacking faults appeared on the scene only in the 1940s, the pioneering effort being due to William Shockley and Robert Heidenreich. The chief significance of both the stacking fault and the twin boundary is that they impede the motion of dislocations and thus make a crystal more difficult to deform.

If the wallpaper equivalents of the stacking fault and the twin boundary reflect carelessness on the part of the decorator, the *grain boundary* implies gross negligence. This third interfacial defect arises when two crystalline regions are mutually misoriented. In the two-dimensional world of wallpaper the scope for such error is limited to a single angle of rotation plus horizontal and vertical displacements, and the offending panel will be visible because of the mismatch around its perimeter. In real crystals there are more *degrees of freedom*; grain boundaries need not be, and usually are not, planar. A piece of crystalline material that contains no grain boundaries or twin boundaries is known as a *single crystal*, and naturally occurring examples are often prized for their aesthetic qualities. Most crystalline substances exist as conglomerates of small crystals and are referred to as *polycrystals*. The small crystals themselves are known as *grains*, hence the term grain

The surface of a crystal is the location of its interaction with the environment. The microscopic structure of the outer layer depends upon the treatment to which a crystal is subjected. The picture above left shows the dramatic sculpturing of an aluminium crystal caused by a solution of sodium hydroxide, while the upper right picture shows the surface of a copper crystal that has been bombarded with forty-kilovolt argon ions. The pictures were taken by *scanning electron microscopy*, and their widths are about a hundred micrometres. In both cases, the surfaces appeared bright to the naked eye. The optical photograph (right) of a five-millimetre diameter single crystal copper sphere, which was heated in oxygen at 250 °C for 30 minutes, exposes the dependence of oxidation rate on the local crystallographic direction. The resulting oxide film has a thickness that varies from ten to over two thousand nanometres, and the various *symmetry elements* are clearly visible. The flower-like appearance of the copper-plated nickel whiskers (second row, left) is an example of (in this case unwanted) *dendritic growth*. The five photographs taken by time-lapse optical microscopy show a remarkable event during the growth of perovskite crystals from a saturated solution. The four faces of the rectangular crystal and also the four larger faces of the hexagonal crystal are all of the same crystallographic type, whereas the remaining faces of the hexagonal crystal have a lower energy per unit area. The two crystals touch, and atoms diffuse across a bridge, from the unfavourably oriented rectangular crystal to the hexagonal crystal, which grows upwards in the shape of a pyramid.

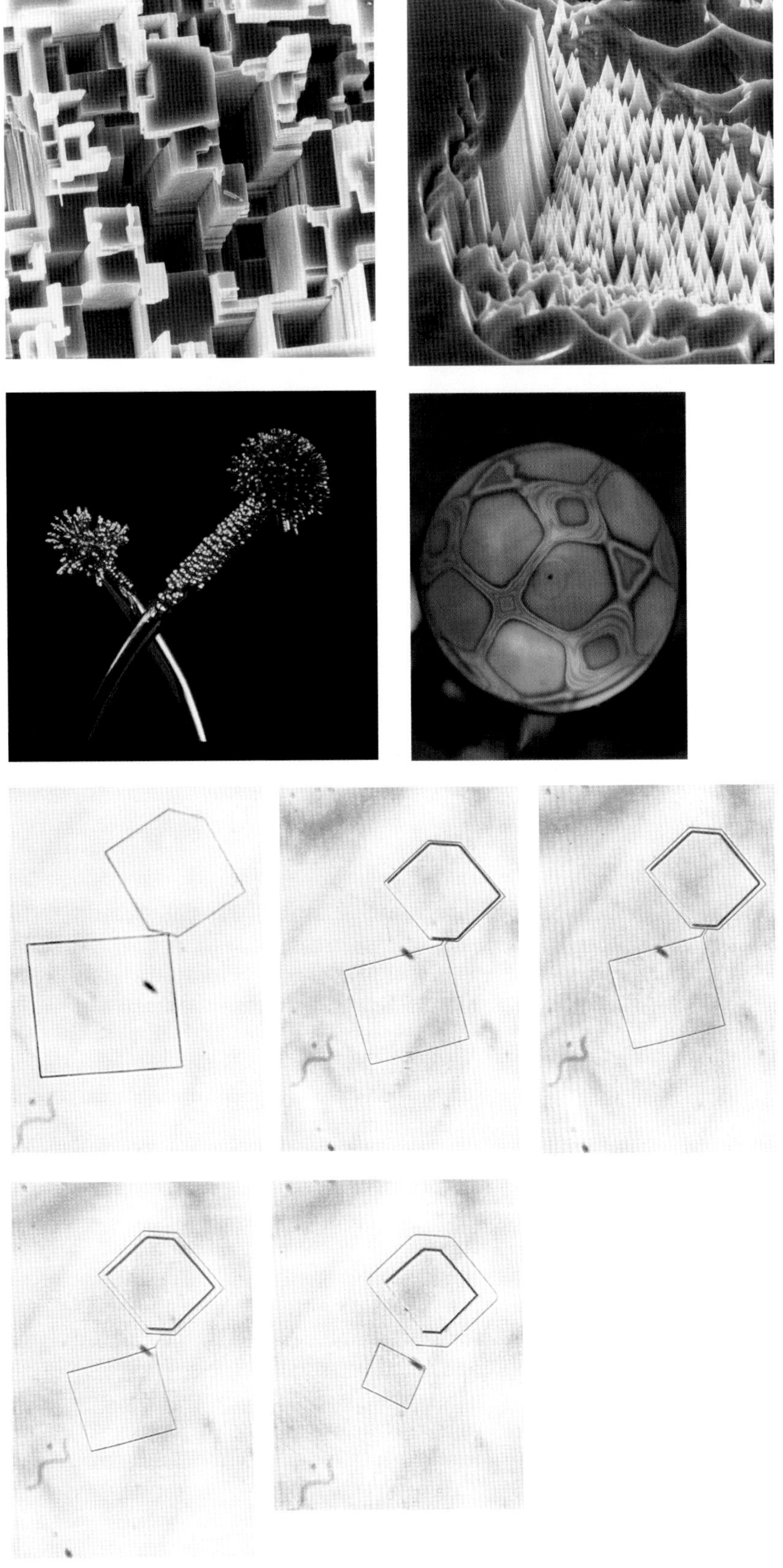

Disclinations are not a common feature of crystals; the associated strain energies would be prohibitively large. These defects are seen in small two-dimensional arrays, however. The left photograph shows $+60^\circ$ and -60° wedge disclinations in small rafts of soap bubbles. Disclinations are quite common in *liquid crystals*, in which the restraining forces are much weaker. The right picture shows the lithograph *Prententententoonstelling*, by Maurits Escher (1898–1975). It is based on a centrally-located defect known as a *distraction* which depends upon a type of symmetry that crystals cannot have: *similarity symmetry*; among the buildings overlooking a water-front is a print gallery in which hangs the same picture of buildings overlooking a water-front, and so on *ad infinitum*.

boundary for the interfacial surface over which two grains are joined. The reason why the crystalline nature of metals was discovered so much later than that of substances such as calcite and rock salt lies in the fact that the former are usually polycrystalline, with rather small grain diameters. Hence their irregular shapes.

The bonding between atoms lying at a grain boundary is not as complete as it is deep in the interior of a grain. Some atoms at such an interface make close encounters with almost their full complement of neighbouring atoms, but most have environments that are affected by the mismatch between the contacting grains. We must regard a grain boundary as a region of inherent weakness. Indeed, this is exploited in the *metallographic technique* introduced by Henry Sorby in the 1860s. The surface of a polycrystal is first made smooth by polishing it with successively finer abrasives, and then subjected to the mild chemical attack known as *etching*. Because of their slightly more open structure the grain boundaries are less resistant than the interiors of the grain, and the grooves which develop along them are visible by optical microscopy, and sometimes even to the naked eye. The grains that intersect the surface of a polycrystal can also be distinguished by examining the latter in polarized light, the various crystallographic orientations giving different degrees of reflection. Oddly enough, in view of its vulnerability to chemical attack, the grain boundary promotes hardening. The lack of lattice continuity across the interface represents a barrier to dislocation movement. We must therefore add the grain boundary to the list of defects which, by inhibiting dislocation-mediated deformation, give a crystal *mechanical strength*. At high temperature, however, this benevolent

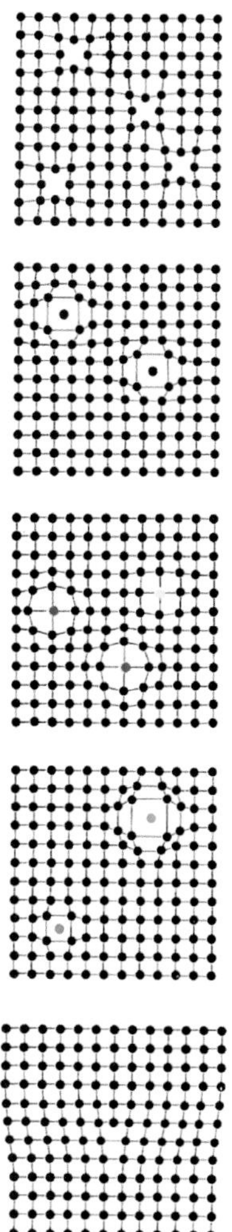

All *defects* distort the crystal lattice, the nature and degree of the disturbance depending upon the type of imperfection. The atoms adjacent to a *vacancy* (top) usually relax inwards, and the distortion is quite localized. *Interstitials* involve more marked violations of the equilibrium distances (upper middle) and the distortions extend over greater distances. The degree of disturbance caused by substitutional (middle) and interstitial (lower middle) *impurities* depends on the atomic numbers of both host and alien atoms. The *dislocation* (bottom) causes a more long-ranged disruption of the lattice.

effect is counteracted by the onset of thermally activated sliding of grains past one another.

The average grain size in a polycrystal provides a crude measure of the ease with which the crystalline phase is nucleated. When the liquid material is cooled below the freezing temperature, the free energy of the crystalline form suddenly becomes lower than that of the liquid. Whether or not the crystalline phase can actually establish itself, however, depends on the probability of a sufficient number of atoms simultaneously arranging themselves in the crystalline configuration to produce a nucleus. The size of such a nucleus is determined by the competition between two factors. The difference between the free energies of the liquid and crystalline phases is proportional to the volume of transformed material, whereas the energy of misfit between a crystalline region and the surrounding melt is proportional to the surface area of the interface. Since volume increases at a faster rate than area, the volume-related factor must predominate if the crystalline region is greater than a certain critical size. Because the difference in the free energies of crystal and liquid increases with decreasing temperature, the critical nucleus size diminishes as the degree of supercooling increases. For a typical metal cooled to 20% below its absolute freezing temperature, the critical nucleus contains a few hundred atoms. In a situation where such *homogeneous nucleation* proves to be difficult, formation of the crystalline phase can be promoted by *seeding*, that is by the introduction of small crystals into the melt. Easy nucleation makes for many small grains, while difficult nucleation produces a polycrystal with fewer larger grains or, in the extreme case, a single crystal.

An interesting situation can arise at the crystal–liquid interface if the liberation of the *latent heat of crystallization* is large enough to locally raise the crystal temperature above that of the liquid. If the interface is not quite flat, a bulging region will push out into colder liquid, where the effective degree of supercooling is greater. This will produce a higher growth rate, and the bulge will be self-reinforcing. The result is a highly non-uniform interface consisting of a series of spikes, out of which may grow secondary spikes, tertiary spikes, and so on, such structures being known as *dendrites*. *Dendritic growth* is also observed for the transition between the vapour and crystal phases, the classic example being the *snowflake*.

The most obvious interfacial defect, the *free surface*, is also the most commonly overlooked. It is the boundary between the crystal and the surrounding environment, which might be a vacuum, or a gas such as air. The surface is an abrupt termination of the regular crystalline pattern, and this produces the imbalance of forces that gives rise to *surface tension*. The quiescent appearance of a typical crystal surface belies the frenetic activity that is actually occurring at the atomic level. Even at room temperature, atoms lying at the outermost level move about the surface by making periodic jumps over the local energy barriers, and this activity increases as the crystal is heated. Such migrations make possible the phenomenon of *faceting*. Surfaces of various crystallographic orientations have, in general, different free energies. The orientation corresponding to the lowest free energy is favoured on purely thermodynamic grounds, but this factor can be partly offset by geometrical considerations. If a crystal initially having the shape of a perfect sphere is heated to near its melting point, atomic migrations lead to a faceting that permits low-energy surfaces to grow at the expense of high-energy orientations. The shape gradually changes to that of a polyhedron, but this must

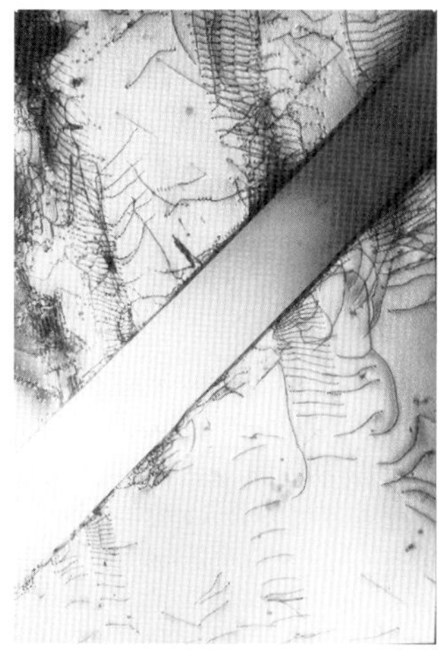

The defect structures in typical engineering components are usually quite complex. Dislocations interact with other dislocations to produce complicated networks and tangles. These decrease the dislocation mobility and consequently cause *hardening*. The complexity is further compounded by the presence of other defects, such as the *twin boundaries* that are acting as a barrier to dislocation motion in this specimen of stainless steel. The visibility of defects, by *electron microscopy*, is critically dependent on crystal orientation, and no dislocations are seen within the twinned region which runs diagonally across the picture. The height of the electron photomicrograph is about five micrometres.

have a larger surface area and further development is ultimately prohibited when the total free energy has reached its minimum value.

When some of the atoms at the surface originate in the environment rather than the crystal interior, as with *oxidation* in the atmosphere, the problem is more complex. Bonding between atoms of different elements must occur, and both the barriers to atomic migration and the free energies of the various surfaces will reflect the heterogeneity of the situation. There are other processes that reveal the dependence of surface energy on crystallographic orientation. The rate of chemical attack, for instance, is faster in some directions than it is along others. Whence the differential rates of erosion of different grains which, in metallography, augments the grain boundary etching technique. Finally, there is the development of regular, and often dramatic, features during *sputtering*. This term refers to the removal of surface atoms by bombardment with ions or other subatomic particles.

There are yet other imperfections which are not readily classified under the three standard groups, *point defects, line defects* and *interfacial defects*, already enumerated. A case in point is the *void*, which is essentially a hole of near-atomic dimensions in the crystal interior. It is equivalent to a cluster of vacancies, but it is also a spherical or polyhedral surface. When such a void is so large that opposite faces cannot feel each other's influence, the surface will be no different from those that define the outer extremities of a crystal. The overall situation will still be unique, however, because the void is an isolated region in which foreign atoms can become trapped. Under some circumstances the latter can be regarded as a captive bubble of gas.

In recent years, a type of line defect other than the dislocation has become the subject of much attention: the *disclination*. Dislocations are generated when two parts of a crystal suffer a mutual translational displacement. In a disclination the displacement is one of rotation. These *orientational defects* are not observed in crystals, but they are frequently found in *liquid crystals*, which will be discussed in Chapter 13. They are also an important feature in the coats of viruses. Disclinations crop up in the familiar problem of the groomed coat of the domestic pet. Any attempt to consistently brush the animal's fur in the same direction is thwarted by the overall continuity of this covering, and the inevitability of boundaries between conflicting brushing directions. The term itself was coined by Charles Frank in 1948, and the story goes that his original choice of the longer term disinclination was deprecated by a professor of English, who felt disinclined to grant a new meaning to that word. A circuit analogous to that of Burgers, described earlier in connection with dislocations, has been prescribed by Frank Nabarro for detection and characterization of the disclination. The closure failure of the Nabarro circuit is one of orientation rather than displacement. Two basic forms of the disclination exist: the *wedge disclination*, in which the net rotation is made around the line of the defect, and the *twist disclination*, in which the rotations occur in a plane containing the line of the defect. An example of a twist disclination is the well known *Moebius strip*, named after A. F. Moebius. A rectangular ribbon-shaped strip is joined end to end after it has been twisted through 180°, the result being a continuous surface bounded by a continuous line. Disclinations are a common, and indeed unavoidable, feature of *human fingerprints*. Disclinations of the 180° wedge type are known to the dermatologist as *loops* and *triradii*, while the 360° wedge variety are called *whorls*.

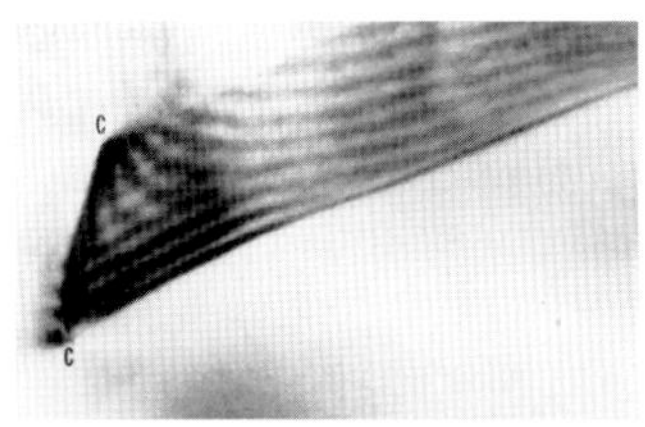

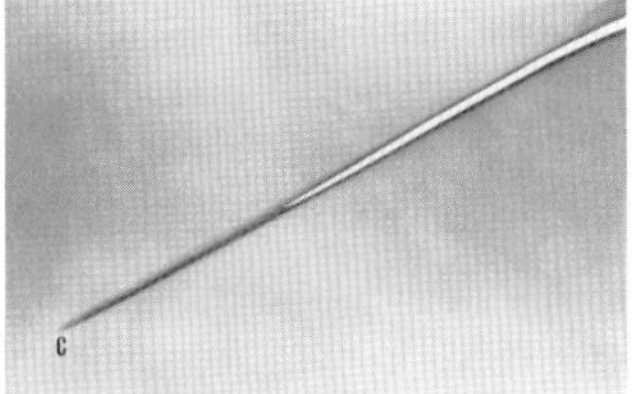

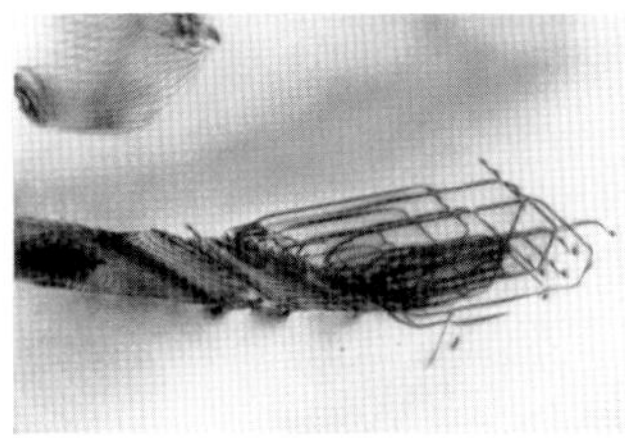

The main difference between *brittle fracture* and *ductile rupture* is seen in the microstructure of the region surrounding the tip of an advancing *crack*. If this tip locally deforms the crystal, producing dislocations which can relax the stress, the rupture is ductile. In the brittle regime such dislocations are not generated, and the crack spreads through the crystal with no effect other than the breaking of interatomic bonds. Silicon is brittle below about 500 °C and the upper and middle electron photomicrographs show a dislocation-free crack tip in this material, in plan and side views respectively (the letter c denotes the tip). The lower picture exposes the presence of dislocations around the tip of a crack in silicon above 500 °C, when the material is in the ductile condition. The widths of the pictures are about two micrometres.

To complete this review of the fundamental defects, mention should be made of the *dispiration* and the *distemperation*. The former, discovered by William Harris in 1979, differs from the dislocation and the disclination in that it cannot exist in an elastic continuum, that is in a medium lacking atomicity. It occurs only under conditions such as those that obtain in a crystal, where the symmetry and discrete nature of the lattice permit unambiguous definition of position and orientation. The dispiration is a defect in the screw symmetry of a crystal, and it is equivalent to a simultaneous translation and rotation. Dispirations are believed to be important in the deformation of crystalline polymers. Harris has suggested that dislocations, disclinations and dispirations should be grouped together under the general designation *distraction*, the word traction referring to displacement. Some other neologism might have been preferable; our professor of English could find this new meaning of an old word too distracting. The distemperation is a dislocation in time rather than space. The most famous example is the *International Date Line*, at which there is a temporal discontinuity of exactly one day. In the crystalline domain, distemperations can arise in any phenomenon in which time plays a role. In a *spin wave*, the vector that characterizes the strength and polarization of the elementary atomic magnet precesses (i.e. varies systematically) from site to site. The net precession around a closed circuit should ideally be zero, but the phase can be shifted by 360°, 720°, and so on, by introducing one or more distemperations.

The technological performance of a piece of crystalline material depends on its *microstructure*, and this is determined by the concentrations and arrangement of various types of defect. But although the scene has been set and the characters introduced, we do not know which are to play the leading roles and which the supporting parts. We must turn to *thermodynamics* for guidance and consider the changes in *energy* and *entropy* associated with each defect. The energy is related to the amount of distortion caused by the defect. For point defects such as the vacancy, and interfacial defects like the grain boundary, this energy is small, whereas it is much higher for dislocations. The entropy is determined by the number of possible arrangements of the defect in the crystal and is thus related to the defect's topology. Point defects, being independent entities, can occupy any position, and their entropy is high. Conversely, the dislocation is a line that is forced to trace a continuous path through the crystal, cutting down the possible choices and hence decreasing the entropy. The same is true, to an even greater extent, of the grain boundary.

Taking the two contributions together and remembering that the entropy factor becomes more important with increasing temperature, we see which defects are favoured by dint of their low free energy. The vacancy has low energy and high entropy, both of which encourage its presence. Vacancies are therefore common, the thermal equilibrium concentration rising to about one for each thousand atomic sites at the melting point, for a typical metal. The interstitial concentration is lower because the energy is higher; the extra atom finds it a tight squeeze sitting in the confined space between atoms on normal lattice sites. For impurity atoms, the energy depends on which elements are involved, and we must differentiate between solvent–solvent bonds, solvent–solute bonds, and solute–solute bonds. It is this multiplicity of factors which can give rise to richly structured *phase diagrams*, such as that of the iron–carbon system discussed in Chapter 9. Grain boundaries, having low free energies, are tolerated, as witnessed by the widespread existence of *polycrystals* and the difficulty of producing single crystals. For the dislocation, both energy and entropy are unfavourable, and the thermal equilibrium value is negligibly small. Dislocations can be introduced by mechanical deformation, but they are trespassers; the crystal is always trying to rid itself of them.

The characters have been identified, but just how the drama is to be played out depends on the situation in question. Different processes call for different scenarios, and it is not always easy to decide whether a particular defect is to be looked upon as hero or villain. One of the most important qualities of a crystal is its ability to withstand mechanical burden. Engineering components usually must not buckle or crack. Service stresses can be of a variety of types: high, but of short duration, as with *impact*; low, but long-lasting, in which there might be the risk of *creep*; cyclic, with the possibility of ultimate *fatigue fracture*; and so on. The failure to account for such time-dependent phenomena was a particularly serious indictment of the defect-free picture of a crystal. But in view of the amount of knowledge of defects that has accumulated, it comes as a disappointment to find that there are still no universally accepted theories of *mechanical deformation*. The reason lies in the intrinsic complexity of the microstructure in a stressed crystal. Several types of defect are present simultaneously, and there are large numbers of each. If the plot is difficult to follow, it is because the stage is crowded; too many characters are getting into the act.

A case in point is the hardening of a crystal when it is *cold-worked*. Very few dislocations are free to move when a stress is applied. The easy motion of a dislocation through a crystal is possible because of the translational symmetry of the lattice. Anything that upsets the regular arrangement of atoms is a potential impedance and will thus be a possible hardening agency. This is true of precipitates, and even of individual impurity atoms in some cases. It is also true of grain boundaries, and indeed of other dislocations. A dislocation line can be thought of as having a certain tension, as would a string, and the bowing out of the region of line lying between two points, where it is pinned down, must increase its length. This involves an increase in energy, and the stress level needed to cause bowing to a critical radius, beyond which further length increase is spontaneous, is inversely proportional to the distance separating the pinning points. In 1950, Charles Frank and Thornton Read proposed an intriguing process that could operate in deformed crystals, producing a dramatic increase in the number of moving dislocations. In this mechanism, dislocations continue to be emitted from the region between the two pinning points, and they pass out over the slip plane as a series of concentric loops. Experimental evidence of such *dislocation multiplication* was reported by Heinz Wilsdorf and Doris Kuhlmann-Wilsdorf in 1954.

The rate of deformation of a crystal is determined by the speed with which adjacent planes of atoms slip relative to each other, and this in turn is governed by the number of slipping dislocations and their velocities. Initially, therefore, the operation of *Frank–Read dislocation sources* makes for easy deformation. But there is a limit to how much deformation can be achieved in this manner, because there are soon so many dislocations moving in the crystal that they begin to interact. Although it is true that these defects can cut right through one another, this requires expenditure of energy, and the ultimate result is a tangle of dislocations in which each of them is denied further movement by the collective influence of the others. The situation is often thought of in terms of individual dislocations trying to cut their way through dislocation forests. It is at this stage that the familiar deformation-induced hardening effect sets in. The metallurgist refers to it as *work hardening*. A crystal that has been softened by heat treatment, or annealed, is relatively easy to deform, but it gets tougher as the deformation proceeds.

A grain boundary can be shown to be equivalent to a dense array of static dislocations, and it presents a strong barrier to individual moving dislocations. The smaller the size of the grains in a polycrystal, the shorter will be the distance that a dislocation can travel, on average, before coming up against such a boundary. This suggests that the strength of polycrystals should be inversely proportional to the average grain diameter, a fact that was established by E. O. Hall and N. J. Petch in the early 1950s. The hardening caused by dislocations and grain boundaries is operative irrespective of whether the crystal is pure or impure. In the latter case, additional hardening derives from the heterogeneity of the situation. A clump of atoms of the minor element represents a particulate obstacle which can be introduced either during crystal growth from the melt or *in situ*, after solidification, as a precipitate. Even individual minority atoms can cause an increase in strength. If such atoms are too big for the lattice, they can relieve some of the misfit strain by moving into the distorted region around the centre of a dislocation. This will tend to function as a brake on the defect; if the dislocation moved away, the impurity atoms would again find themselves creating local strains in the lattice. As was first pointed out by Alan Cottrell, each dislocation attracts a cloud of impurity atoms which then anchor it in position.

When crystalline materials must withstand moderate loads for long periods of time, *creep* can occur. This is a gradual change of shape, and the rate depends on both the stress level and the temperature. It is not constant, there being an initial transient period of relatively slow deformation before the creep builds up to its steady and faster value. This was demonstrated in 1910 by E. N. da C. Andrade, who found that it applied to a wide range of materials. An increase of temperature raises the creep rate, and in practice the rate of deformation is sufficient to cause concern only if the temperature exceeds half the absolute melting point. This refers to the Kelvin temperature scale in which 273.16 degrees are to be added to the Celsius value. The element lead melts at 600.6 K (327.4 °C), so half the melting point corresponds to the temperature 300.3 K (27.1 °C), which is typical of the domestic environment. We must conclude, therefore, that those parts of the lead plumbing that are under stress are creeping at a significant rate, and this is indeed the case. Iron and its common alloys, on the other hand, have much higher melting points, and we can safely assume that a car chassis, for example, maintains its dimensional integrity.

Various defects may be operative in the creep process, depending on both the temperature and the stress level. High-temperature creep, caused by the migration of vacancies, was analyzed by Conyers Herring and Frank Nabarro in 1950. The application of stress dynamically disturbs the otherwise uniform vacancy distribution, so as to give a continuous flow of these defects from regions of high stress to places where the stress is lower. When a vacancy moves one way, an atom must move in the opposite direction, so this directed migration gives a net transport of atoms and hence a change of shape. *Herring–Nabarro creep* proceeds at significant rates for surprisingly low stress levels. At lower temperatures, but somewhat higher stresses, the situation is dominated by dislocation movement. The underlying principles are essentially those discussed earlier in connection with hardening, but here the interplay between the various defects is quite different. The important processes are: *dislocation glide*, caused by the applied stress; *dislocation climb*, which is controlled by vacancy diffusion and is thus temperature dependent; and *dislocation annihilation*. In the steady state of this technologically important regime, dislocation loss by mutual annihilation of two of these defects is compensated for by the generation of new dislocations at other locations, by Frank–Read sources, and a steady rate of deformation is maintained.

Of the many processes controlled by defects, it might seem that fracture is one of the simplest. This is true insofar as the *Griffith criterion*, as described at the beginning of this chapter, always applies, but processes other than *cleavage crack propagation* might also be operative, making the rupture more complicated at the microscopic level. The opening of a crack involves the sequential breaking of interatomic bonds over a particular plane of weakness. An obvious competitor to this process is the deformation of the surrounding material by the movement of dislocations. If dislocations can relieve the applied stress at a rate comparable to that due to the opening of the crack, the plastic deformation represents an additional factor in the Griffith energy balance, as originally suggested by Egon Orowan, and by G. R. Irwin. Other defect-related processes are now known to contribute to the local changes of internal shape that ultimately lead to rupture. If impurities are present, for example, and show deficient adherence to the host material, fissures can develop around them at sufficiently high stress. The atomic transport associated with diffusion eventually causes these voids to coalesce, and failure ensues when one of the voids reaches the critical size. Similarly, voids can open up along grain boundaries, leading to *intergranular fracture*. Cohesion failure at grain boundaries also occurs when a second element is present which wets the boundary, because it has a low interfacial energy with the host material. Such *grain boundary embrittlement* is a familiar headache of the metallurgist.

Two extremes of fracture failure are identified: brittle and ductile. *Brittle fracture* is sudden, and it occurs without detectable plastic deformation of the surrounding material. It is the type colloquially associated with such words as snapping, cracking and crumbling, and it was this mode that was the object of Griffith's original study. In *ductile fracture*, which more correctly should be referred to as *ductile rupture*, because its final stage nevertheless involves brittle fracture at the atomic level, the stressed body suffers a change of shape before the ultimate failure. It is a gradual tearing rather than a sudden snapping. The term ductile comes from the Latin word *ducere*, which means to lead. If the applied stress is of the simple tensile type, the deformation takes the form of the familiar local thinning, known as *necking*. Very pure specimens will neck down to a fine point or chisel edge, whereas impurities give rise to voids. The plastic deformation occurring in ductile rupture involves the movement of one or more types of defect, and this implies a temperature dependence. Indeed, most materials are capable of displaying both brittle and ductile behaviour depending on the temperature, the brittle state corresponding to the colder region. If a metallic mug is dropped on the floor, it is dented, but a wine glass shatters. In the mug,

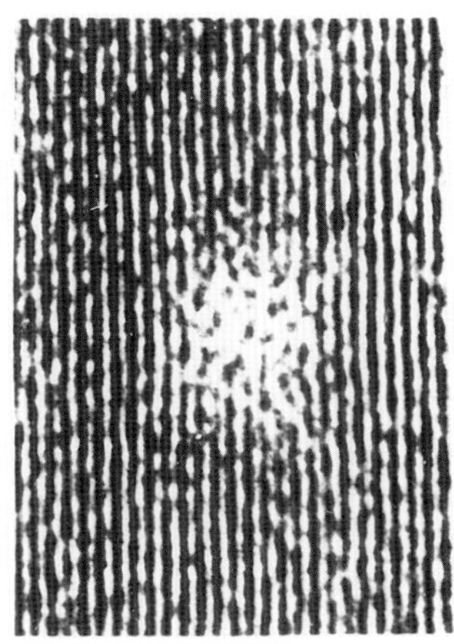

When a sufficiently energetic *subatomic particle* such as a neutron, alpha particle, or fission fragment, is travelling in a crystal, there is a calculable probability that it will hit an atom and displace it from its normal lattice site. If the collision is violent enough, the displaced atom can dislodge others, producing a *collision cascade* which ultimately leads to the formation of a damaged zone. The centre of the latter is equivalent to a cluster of *vacancies*, while the displaced atoms come to rest as *interstitials*. The entire process takes just a few picoseconds. This electron photomicrograph, the height of which is about fifty nanometres, shows a damage zone in platinum phthalocyanine caused by a *fission fragment*.

dislocation movement relieves the stress, preventing it from reaching the level required for fracture. The glass has no such safety valve, and the stress builds up until it passes Griffith'sdanger limit. In many crystalline materials, under certain conditions, fracture is preceded by the gradual dimensional change that characterizes creep or the progressive change in microstructure that foreshadows *fatigue failure*. Because fatigue is particularly troublesome in metals, we will defer discussion of this phenomenon to Chapter 9.

Defects can be produced when a crystal is irradiated by sufficiently energetic particles. Interest in such effects was generated primarily through the development of nuclear power. In terms of economic impact, the major manifestation of the phenomenon is seen in the construction materials of nuclear reactors. But *irradiation damage* is also provoked by other sources of radiation, including X-rays and natural radioactivity, and it takes a variety of forms. Firstly there is the transmutation of atoms of one element into different elements. Uranium and plutonium nuclei can be split into smaller fragment nuclei, for example, and the noble gases krypton and xenon, which are common products of such fission, become impurities of the host crystal. Then there is the *ionization* of atoms due to the removal of some of their electrons by the impinging particles. In organic materials, including living tissue, this can cause severance of covalent bonds and ultimately break-up of molecules. Ionic solids are also susceptible to this type of damage, and the resultant change in conductivity can be used for measuring purposes. The most spectacular exploitation of these effects is in the *photographic process*, the damaging radiation being light photons. Metals are immune from ionization damage because their bonding electrons are not localized.

The third aspect of irradiation damage is the actual displacement of atoms in the target material. With a probability that depends both on the nature of the radiation and on the elements present in the irradiated material, an atom in the latter can suffer a direct hit by an incoming particle. If the imparted energy is sufficient, the struck atom can be displaced from its normal lattice position. Unlike the incident particle, which can move past many atoms before scoring a hit, this moving atom, known as the *primary knock-on*, will interact strongly with its neighbours and cause further displacements. These, in turn, can provoke still more displacements, and the final extent of the *collision cascade* depends on the actual energy of the primary knock-on. In a crystal, the distribution of damage also depends on the direction in which the primary knock-on moves. The cascade terminates when none of the moving atoms has sufficient kinetic energy either to displace other atoms or to surmount the energy barriers against atomic motion through the lattice. Where the moving atoms come to rest, they must become interstitials, except in those rare cases where an atom drops into a vacancy. Conversely, the sites that they leave behind are vacancies, and for each primary knock-on the net result of the damaging event is a vacancy-rich core surrounded by an interstitial-rich region. It was originally believed that such a *Seeger zone*, named after Alfred Seeger who proposed it in 1956, would be roughly spherical and that the core would be a void. This picture has now been replaced by one which has small groups of vacancies spread out along the path of the primary knock-on, with regions of perfect crystal in between, and interstitials forming a more distant cloud.

Because the perfect crystal is the state of lowest free energy, displacement of atoms to produce vacancies and interstitials must increase this quantity; irradiation damage leads to a storage of energy in crystals. Eugene Wigner, in the mid 1940s, foresaw that this could lead to trouble in nuclear reactors. In order to produce the desired chain reaction, the fast neutrons released from the fission of uranium nuclei must be slowed down to speeds at which they are better able to provoke further fission. The slowing down was originally achieved by allowing the fast neutrons to hit and displace the atoms in graphite crystals. But the damage to the latter, known as a *moderator*, caused a build up of *stored energy*. Wigner realized that when the density of damage is sufficiently high, the stored energy might be spontaneously released, leading to dangerous overheating of the reactor core. This must happen when the density of Seeger zones is so high that they overlap to a marked degree, with interstitials of one zone falling into vacancies of another, thereby liberating the formation energy of both these defects. An accident of this type occurred at the *Windscale reactor* in England in 1957. Trouble is obviously not encountered if a liquid moderator is employed, and even for the solid type periodic deliberate heating removes the damage before it can build up to a dangerous level. The various structural materials in a reactor are also subject to irradiation, and this can lead to impaired mechanical performance and dimensional changes. Both are caused by changes in microstructure that involve vacancies, interstitials and dislocations. Voids and grain boundaries also play a role under some circumstances. Any departure from perfect crystallinity increases the volume, and swelling is particularly pronounced if voids are formed; the material displaced from these pores can be regarded as being added to the outer surface. Nucleation of voids appears to depend on a subtle difference between vacancies and interstitials. The latter are more mobile and their strain energy is higher, both of which promote capture by dislocations, which thereby climb. This leaves a surfeit of vacancies in the regions between the dislocations, and they aggregate to form the voids.

Summary

Crystals invariably contain imperfections that influence, or even dictate, their macroscopic physical properties. The types of imperfection include point defects, line defects and interfacial defects. The point defects are the vacancy, a lattice site from which an atom is missing; the interstitial, an extra atom lying between the normal sites; and the impurity atom. The vacancy commonly mediates diffusion, and both vacancies and interstitials are produced when crystals are subjected to sufficiently energetic irradiation. The line defects are the dislocation, a disruption of translation; the disclination, which disturbs rotational symmetry; and the dispiration, a defect of screw symmetry. The latter two are common in liquid crystals and polymers, respectively, while dislocations are important in the mechanical deformation of crystals, particularly of the metallic variety. The grain boundary, an interfacial defect, functions both as a barrier to dislocation motion and as a location of inherent susceptibility to fracture. Strengthening effects are produced by certain impurities, grain boundaries, and, at sufficient high strain, by the formation of dislocation tangles. Creep results from defect motion: vacancies at high temperature, dislocations at high stress.

6

Keep up your bright swords, for the dew will rust them.
William Shakespeare (1564–1616) ***Othello***

The great mediator: *water*

Water is the most common substance on Earth, and it no doubt has the best known of all chemical formulas: H_2O. It is one of only two liquids that occur naturally in appreciable quantities, the other example being petroleum. More than three quarters of the Earth's surface is covered with water, and if it were not for this compound, there would be no life. About 60% of the weight of the human body is water, it being present in the interior of every cell. It also accounts for the bulk of such specialized aqueous media as *blood* and *mucous*. The central role played by water was of course clear from the earliest times, and Aristotle regarded it as one of the four elements, together with air, fire and earth. Its chemical structure became clear in the late eighteenth century, when Henry Cavendish and Antoine Lavoisier established that it is a combination of hydrogen (discovered by Cavendish in 1766) and oxygen (discovered by Joseph Priestley in 1774). Cavendish burned measured quantities of these two gases together and found that they produced water, no change in total weight having occurred during the reaction. The reverse process, decomposition of water into hydrogen and oxygen, was first achieved by Lavoisier. In spite of its abundance and apparent simplicity, water is an unusual compound, and compared with other liquids we could call it exotic. It has special qualities which enable it to play its vital role in both the biological and inorganic domains. It is an almost universal solvent, and when it provides the medium for acid–base reactions, it is not just a bystander but an active participant.

Although the name *water* is colloquially associated only with the liquid phase, this term more correctly applies to the chemical compound formed by the combination of two hydrogen atoms and a single oxygen atom. Those other two common substances, the solid form, *ice*, and the vapour form, *steam*, are thus both water. If we are to understand the structure and properties of this ubiquitous compound, we must take a closer look at the structure of the *water molecule*. From what has been said in connection with other apparently simple molecules, it is not surprising to find that there is more to know than its simple formula might suggest. *Spectroscopic experiments*, of the type which played such a crucial role in unravelling the nature of atoms, have been used on the water molecule. The technique has now been extended to include the range of natural frequencies of atom–atom vibrations, and the data can be interpreted to produce remarkably accurate values for molecular dimensions. Thus the geometrical description of the water molecule can be given to five-figure accuracy. Each of the oxygen–hydrogen bond distances is 0.095 718 nm, while the angle between the two oxygen–hydrogen bonds is 104.52°. We should not be surprised at the value of the bond distance,

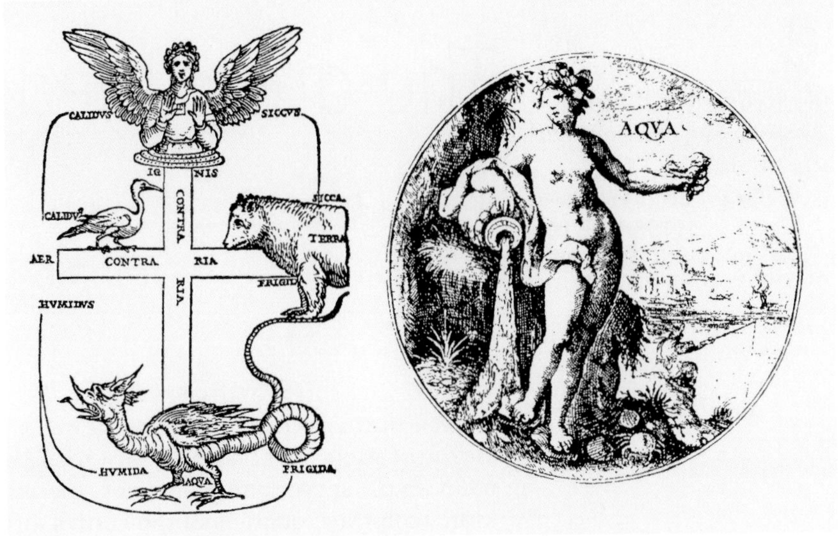

Artistic impressions of water, as one of the four elements of antiquity, varied widely. In two extreme interpretations of this member of the quartet envisaged by Aristotle (384–322 BC), we see water symbolized by a dragon and by the gentle flow from Mother Nature's pitcher. The dragon appeared in the *Pretiosa Margarita Novella*, an encyclopedia of alchemy published in 1546, while the softer view was expressed in an engraving by the Flemish artist Crispijn van der Passe fifty years later.

because the radii of the oxygen and hydrogen atoms, based on a scheme of consistent values for all atoms derived by Linus Pauling, are 0.066 nm and 0.030 nm, respectively. The peculiar bond angle, on the other hand, might well cause some speculation.

In trying to construct a picture of the arrangement of the three atoms in a water molecule, it is natural to turn to the electron orbitals in the individual atoms. The hydrogen atoms have a single electron in the 1s state. The distribution of electron probability in this state is spherical, with almost all of it occurring within a radius that is roughly equal to the *Pauling radius*. Oxygen has eight electrons: two in the 1s state, two in the 2s, two in the $2p_x$, and one each in the $2p_y$ and $2p_z$ states. If the oxygen–hydrogen bonds simply consisted of the overlapping of the 1s states of the hydrogens with the $2p_y$ and $2p_z$ states of the oxygen, we should expect the H–O–H angle to be exactly 90°. The fact that this differs by approximately 15° from the observed value shows that an important factor is being overlooked. An obvious candidate is the repulsion between the two hydrogen atoms, but we find that this could increase the angle by not more than 5°. It turns out that the dominant factor is the *hybridization* of the 2s, $2p_y$ and $2p_z$ orbitals of the oxygen atom, so as to form four new orbitals. We encountered this interesting and important effect in connection with the discussion of methane and other molecules. In fact, the final hybridized arrangement comes close to exhibiting perfect tetrahedral symmetry.

In a tetrahedron, the angle subtended by any two of the four corners at the centre is 109.5°, a mere 5° greater than the H–O–H angle in the water molecule. The 2s–2p hybridization produces four lobes of electron probability density extending outwards from the centre of the slightly distorted tetrahedron. Two of these reach two of the corners and overlap the 1s clouds of the hydrogen atoms located there, while the other two lobes extend only part of the way toward the remaining two corners. The major lobes provide the basis of the oxygen–hydrogen bonds. The two minor lobes, which are referred to as *lone-pair hybrids*, are small concentrations of negative electric charge. The final configuration of the water molecule thus has the oxygen nucleus at the centre of a slightly distorted tetrahedron, positive charges

The dominant factor determining the shape of the *water molecule* is the hybridization of the 2s and 2p orbitals of the oxygen atom. The four hybrid orbitals have lobes which project out from the oxygen nucleus in the direction of the four corners of a tetrahedron. Two of these orbitals provide bonds for the two hydrogen atoms, while the remaining two become *lone-pair orbitals*. The latter are negatively charged, and attract the hydrogen atoms on neighbouring water molecules. The colours on this electron wave picture, and on the smaller space-filling model, have no physical significance, but they conform with the standard convention for oxygen and hydrogen.

on two of the four tetrahedral corners, and two negative charges in the direction of other corners, but not quite reaching those points. The nearly perfect tetrahedral symmetry of the water molecule is the major factor in determining the arrangement of atoms in the most common form of ice crystal.

It has proved difficult to develop a universally accepted picture of the *liquid state*. This is true even for relatively simple liquids such as those of the closed-shell elements like argon and krypton. Given the more complicated water molecule, it is clear that the development of a reliable theory of liquid water is a formidable task. In the 1930s, the most popular model regarded it as a mixture of small aggregates of water molecules. There were various versions of this approach, the most prominent being that of H. M. Chadwell. It was based on a mixture of H_2O, $(H_2O)_2$ and $(H_2O)_3$, the latter two complexes being known as *dihydrol* and *trihydrol*, respectively. Data from several different types of experiment have now shown that this early idea was incorrect. *Vibrational spectroscopy*, for example, reveals that the proposed aggregates could remain bound to one another for no longer than about 10 ps. Another serious deficiency of the small aggregate approach is that it does not explain the marked angular correlation between the water molecules, which electrical measurements on the liquid show must be present. These measurements, probing what is known as the *dielectric constant*, detect any tendency of the molecules to become oriented in the direction of an applied electric field.

The current picture of the liquid state, albeit a rather incomplete one, can be traced back to the pioneering work of John Bernal and Ralph Fowler in 1933. They emphasized the need for adopting what might be called the global view by considering the three-dimensional arrangement of the water molecules. Many different types of experiment have shown that one of the most important factors determining this arrangement is the hydrogen bonding between adjacent molecules. Because a *hydrogen bond* can be formed only when one of the two positive poles of one water molecule is directed towards one of the two negative poles on an adjacent water molecule, this imposes certain restrictions on the possible arrangement of the molecules in space. Indeed, in the common crystal form of water it is the arrangement of hydrogen bonds which actually gives the ice its observed lattice structure. In the liquid form the temperature is higher, and this is presumed to permit so much distortion of the hydrogen bonds that the arrangement in the liquid can be looked upon as an irregular but nevertheless four-coordinated arrangement of the molecules. The *glassy state* is often treated in a similar manner. Indeed, the term *random network* is now frequently used when describing both the instantaneous structure of (liquid) water and (solid) glass.

In order to appreciate the difficulty in constructing an acceptable picture of the collective behaviour of the molecules in liquid water, we must consider what might happen to one molecule during a brief period. Experiments at around room temperature indicate that a water molecule oscillates within a surrounding cage of other molecules for about 4 ps, before jumping out of this position and into an adjacent cage. During this brief period the molecule would be able to execute approximately 40 oscillations. There is nothing unusual about this behaviour. Even though other liquids have different characteristic times, the motion of an individual molecule is qualitatively similar. Complexity arises from the fact that water molecules can

The links between water molecules, in both the liquid and solid states, are primarily provided by *hydrogen bonds*. The positive electrical charge on a hydrogen atom of one molecule is attracted towards one of the negatively-charged lone-pair orbitals on another molecule. The underlying tetrahedral symmetry of a single water molecule thus provides the basis for the similar symmetry found in ice crystals.

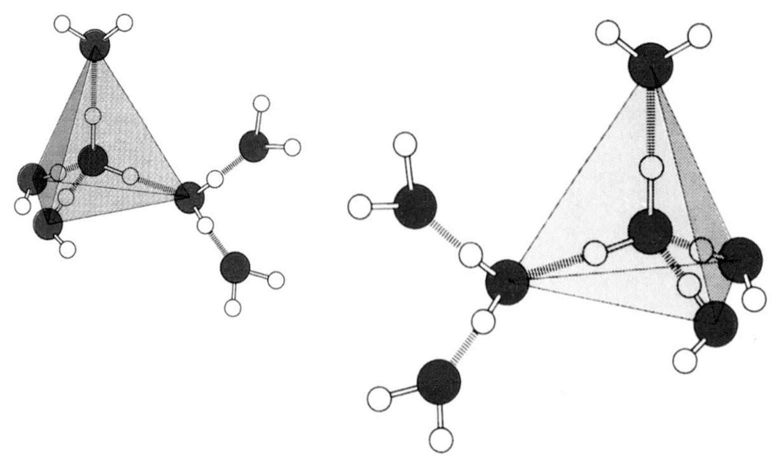

also spontaneously break up, so as to form hydrogen (H^+) and hydroxyl (OH^-) ions. The rate at which this dissociation occurs has been found to be 2.5×10^{-5} per second, which means that each molecule in liquid water dissociates on average once in 11 h. At any instant there are approximately 6×10^{16} *hydrogen ions* present in one litre of water, and of course the same number of *hydroxyl ions*. The number of undissociated water molecules in a litre is approximately 3×10^{25}, so the hydrogen ions are well spaced; about 800 water molecules could lie along a line running from one ion to its nearest neighbour. The same is true of the hydroxyl ions, so the average spacing between adjacent hydrogen and hydroxyl ions is about 400 water molecules. There is strong evidence that both of these ions are strongly associated with neutral water molecules, the hydrogen ions forming *hydronium*, H_3O^+. But the ions are also known to be rather mobile. A hydrogen ion, for example, stays associated with a given water molecule for only about a picosecond before moving on. This means that any given water molecule will catch a hydrogen ion once every half a millisecond or so, and hold on to it for about a picosecond. It is interesting to note that the mobility of hydrogen and hydroxyl ions in water is noticeably higher than the mobilities of other small ions such as Na^+ and K^+. The reason for this stems from the fact that in the former case the solvent and the diffusing species are composed of the same two elements. This makes it possible for a hydrogen ion to migrate as a sort of disturbance that moves more rapidly than the individual atoms. An analogy to this type of motion is what is commonly observed during the shunting of railway rolling stock. If a row of trucks is struck at one end by an additional truck, the disturbance rapidly passes down the row and ultimately leads to the ejection of a truck at the other end. The same thing occurs in water, the disturbance being passed along rows of molecules by the shifting of hydrogen bonds.

The question of dissociation of the water molecule is an important one. In principle, any interatomic bond can be broken if sufficient energy is available. In the case of pure water at room temperature, the available thermal energy is sufficient to break up some of the water molecules into hydrogen ions, which are simply protons, and hydroxyl ions. In the equilibrium state, which must prevail in quiescent pure water, there will be as many recombination events bringing hydrogen and hydroxyl ions together to produce neutral water molecules. The number of free hydrogen ions existing at any

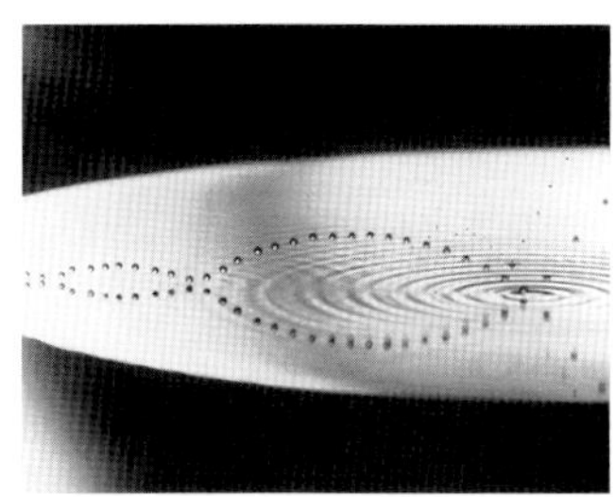

Anyone who has made a flat stone bounce on the surface of water will be aware of the need to achieve a near horizontal angle of incidence. When the projectiles are themselves water, there is the added complication of possible coalescence, which depends upon the *surface tension*. Whether bouncing or coalescence occurs depends upon droplet size, and impact velocity and angle. The droplets in the picture have a radius of 150 micrometres.

instant must be the same as the number of free hydroxyl ions, since there must be overall neutrality. Under the above conditions, the concentration of either species of ion is 10^{-7} moles, or about 6×10^{16} ions/l (ions per litre). In an aqueous solution, there can be other sources of hydrogen and hydroxyl ions than the dissociation of neutral water molecules. But the presence of these other *ion donors* will not affect the rate at which water molecules break up. This means that when a solution is in equilibrium, the rate at which hydrogen and hydroxyl ions recombine to form neutral water must be the same as it is in the case of pure water. That rate is related to the probability that hydrogen and hydroxyl ions collide with one another, and this is simply given by the product of their concentrations. In pure water the concentration of each type of ion is 10^{-7} mol/l (moles per litre), so this constant product must equal 10^{-14}. This means, for example, that if the concentration of hydrogen ions rises to 10^{-3}, the concentration of hydroxyl ions must drop to 10^{-11}. A convenient shorthand way of denoting hydrogen ion concentration was suggested by Søren Sørensen in 1909, and it uses the symbol pH, this being simply the negative power of ten in the *hydrogen ion concentration*. Thus a pH of 1 corresponds to a concentration of 10^{-1}, while a concentration of 10^{-14} gives a value of 14. The ammonia used in smelling-salts has a pH of about 12, while orange juice and vinegar have the approximate values of 4.5 and 3 respectively. The acid in a fully charged accumulator, or car battery, has a pH of roughly 1.

Measurement of the very short lifetimes of ions in water is an impressive achievement, and it is worthwhile to digress briefly and consider how this was accomplished. To begin with, we recall that the breaking up of a compound into its constituents involves the overcoming of *interatomic forces*; there is a *free energy barrier* to be traversed. The source of the required energy is the thermal (kinetic) energy of the molecules themselves, and this increases with temperature. The effect of a change in temperature is to shift the balance between the concentrations of the reacting species. Because the establishment of *equilibrium* in a reaction involves movement of the participants, it cannot be achieved infinitely quickly; a *relaxation time* is inevitably involved. We see an analogous effect, but on a quite different time scale, in the relatively slow accommodation of the sea's temperature to the changing seasons. In order to probe such relaxation times, and hence obtain a measure of reaction rates, it is necessary to change the temperature, or the pressure, within a period that is short compared with the relaxation times. The idea of effecting rapid temperature changes by discharging an electrical condenser through the solution being studied occurred independently to Manfred Eigen, and to George Porter and Ronald Norrish in 1954, the technique being known as *flash photolysis*.

According to a suggestion by Svante Arrhenius in 1887, an *acid* is a chemical compound that dissociates in solution to yield hydrogen ions, while a *base* is a compound yielding hydroxyl ions. The pH of neutral water is 7, and the presence of an acid increases the concentration of hydrogen ions and thus lowers the pH value. It will, at the same time, decrease the concentration of hydroxyl ions. For the same reason, a basic compound gives an increased pH value in aqueous solution. J. N. Brønsted and T. M. Lowry, independently in 1922, broadened the definitions of acid and base by proposing that the former is any compound that can lose a proton, while a base is any compound that can accept a proton. Finally, in the following year, Gilbert Lewis defined an acid as a substance that can accept a share in a pair of

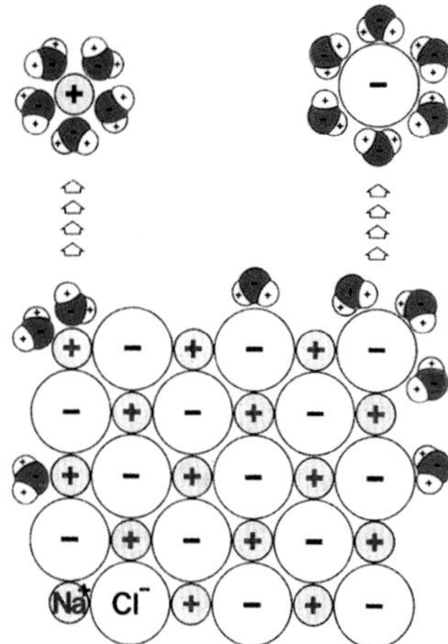

The effectiveness of water as a *solvent* for ionic compounds stems from the dual charge on its own molecules; irrespective of whether an ion is positively or negatively charged, a water molecule can always rotate so as to confront the ion with a complementary charge.

electrons held by a base so as to form a *covalent bond* with the latter. This is reminiscent of one of the meanings of *oxidation*: the process of removal of an electron from a compound. The other meaning of the word oxidation is the obvious one of addition of oxygen to a material to make an oxide. An *oxidizing agent* is one that tends to accept electrons. Indeed, the name oxygen actually means acid generating.

The importance of acids and bases in both the biological and inorganic realms could hardly be exaggerated. The rearrangement of molecular fragments in *chemical reactions* occurring in aqueous solution is determined by the relative affinities that they have for each other and for the water molecule. The greater the tendency for an acid or a base to dissociate and interact with water, the stronger it is said to be. Acids have a sour taste, like that of lemon, and they are detected chemically by their action on the organic dye, *litmus*, which they turn from blue to red. If an acid is sufficiently strong, it will react with metals and release hydrogen. Bases, also commonly referred to as *alkalis*, have a bitter taste, and they turn litmus from red to blue. They have a characteristic soapy feeling, owing to their interaction with the oily substances in the skin. Another definition of acids and bases couples them together in that an acid neutralizes a base to produce a *salt* and water.

The sour taste of an acid is the result of the presence of hydronium ions, these having been formed by the interaction of water with hydrogen ions that follows from the Arrhenius definition of an acid. The bitter taste of a base is caused by the hydroxyl ions. Some examples serve to illustrate typical dissociation reactions. Hydrochloric acid is one of the strongest acids, and it interacts with water in the following manner.

$$HCl + H_2O \rightarrow H^+ + Cl^- + H_2O \rightarrow H_3O^+ + Cl^-$$

This set of reactions runs from left to right, as written, because there is always a tendency to produce weaker acids and bases from stronger reactants. HCl is a stronger acid than H_3O^+, and H_2O is a stronger base than Cl^-. We can call H_3O^+ an acid because it fulfills the Arrhenius definition in that it can dissociate and produce an H^+ ion.

$$H_3O^+ \rightarrow H_2O + H^+$$

Other examples of *strong acids* are perchloric, $HClO_4$; sulphuric, H_2SO_4; and nitric, HNO_3. The hydronium ion is an intermediate acid, as is phosphoric acid, H_3PO_4. Organic and biological acids tend to be fairly weak. We will discuss amino acids, nucleic acids, and fatty acids later. Other *weak acids* are carbonic acid, H_2CO_3, and acetic acid, CH_3COOH. In water, the latter dissociates to produce a negatively charged ion and a proton, the latter immediately joining with water to produce hydronium. The reaction is, using structural formulas for greater clarity:

```
     H   O                      H   O
     |  //                      |  //
  H—C—C          →         H—C—C          + H+
     |   \                      |   \
     H   O—H                    H   O−
```

We see that the activity of acetic acid takes place at the COOH end of the molecule, and this is referred to as the carboxyl group. It is very important in biological acids, and we will encounter several examples in later chapters. One of the strongest bases is sodium hydroxide, NaOH, which is also known as *caustic soda*. It dissociates in water to produce a positively charged metallic ion, Na^+, and a negatively charged hydroxyl ion. (Because the latter comprises more than one atom, it should strictly be referred to as a *radical*.) The liberation of a hydroxyl ion in this reaction is consistent with the Arrhenius criterion.

The initial stage of *rusting* involves the building up of a continuous *oxide layer* on the surface of the iron. While this layer is still quite thin, as shown here, it produces optical interference colours like those seen when water is covered with a thin film of oil. The different shades of colour seen in this optical photomicrograph result from slight variations in the thickness of the oxide film, while the rectangular pattern reveals the underlying cubic symmetry of the iron crystal.

Two of the above reactions illustrate concepts already discussed. The dissociation of hydrochloric acid

$$HCl + H_2O \rightarrow H_3O^+ + Cl^-$$

increases the effective concentration of hydrogen ions and, by definition, lowers the pH. This is the hallmark of an acid. If sodium hydroxide is now added, this too dissociates

$$NaOH \rightarrow Na^+ + OH^-$$

and the relatively large number of effective hydrogen ions can be decreased by the reaction

$$H_3O^+ + OH^- \rightarrow H_2O + H_2O$$

thereby increasing the pH value. This is the *acid–base neutralization* referred to earlier, and it has the net effect of putting positive sodium ions and negative chlorine ions into solution. If the water is gradually evaporated, the concentration of these ions in solution increases, and ultimately the solution will become saturated. Beyond this point sodium chloride, NaCl, will be precipitated out of solution, and it will appear as crystals at the bottom of the container. The choice of sodium chloride as an example of a salt might be confusing in view of the common name of that compound. It should therefore be mentioned that the compound rubidium nitrate, $RbNO_3$, produced by the reaction of RbOH and HNO_3 in solution, is just as good an example of a salt.

If we wish to use a solution of common salt, in the kitchen for instance, we do not employ the above neutralization reaction between hydrochloric acid and caustic soda. It is far easier, and certainly less dangerous, simply to dissolve some of the solid material. The technical name for this process is *hydrolysis*, meaning splitting (*lysis*) by water (*hydro*). When the compound is in solution, it is broken up into sodium and chlorine ions, and it is interesting to see just why water should be so efficient at bringing about the disruption of a salt crystal. The reason lies in the electrical structure of the water molecule, with its two positive and two negative electrical charges jutting out from the central oxygen atom. This means that irrespective of the charge on an ion, the water molecule will always be able to arrange itself so as to bring an opposite charge adjacent to it. In solution, this electric attraction persists, and the ions move around carrying a small retinue of bound water molecules known as the *hydration shell*. It causes the mobility of the ion to be lower than would otherwise be expected. We should note that the easy solution of salt in water is an example of the fact that substances with similar bonding, in this case ionic, tend to mix readily. Common salt is not easily dissolved in benzene, whereas that liquid is an excellent solvent for most fats.

Sodium chloride is the salt of a strong acid and a strong base, and it dissolves in water to give an approximately neutral solution. This is certainly not the case for all salts. A solution of sodium acetate, for example, is basic, its pH being greater than 7, while a solution of ferric chloride is acidic, with a pH less than 7. Sodium acetate, CH_3COONa, is the salt of a weak acid, acetic acid, and a strong base, sodium hydroxide. Like virtually

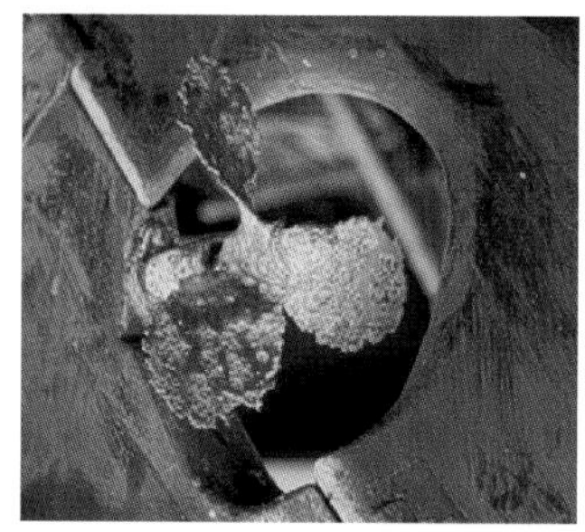

The rate of *aqueous corrosion* is higher in sea water than in fresh water because the sodium and chlorine ions increase the electrical conductivity. The eating away of metallic components below a ship's water-line can be prevented by use of a *sacrificial anode*. This is a piece of metal, fastened to the hull near the part to be protected, and made of an element higher in the *electrochemical series* than the vulnerable item. It is therefore preferentially attacked. Zinc anodes are frequently used in this situation.

all salts, it is almost completely ionized in water, the reaction being as follows.

$$CH_3COONa \rightarrow CH_3COO^- + Na^+$$

Because the words strong and weak in this context refer to ionizing tendency, we see that of the two ions produced in this reaction, it is the acetate that shows the stronger inclination to be neutralized by the water. This occurs by combination with hydrogen ions, which it can take from the hydronium, the reaction being as follows.

$$CH_3COO^- + H^+ \rightarrow CH_3COOH$$

But the removal of hydrogen ions from the water leaves behind unpaired hydroxyl ions. The solution is thus alkaline. By the same type of reasoning, we see why a solution of ferric chloride, the salt of a strong acid and a weak base, should be acidic.

At this point it would be natural to speculate as to what determines the pecking order which dictates that some radicals are strong in solution while others are weak. Since ions are involved, electric effects must be in evidence, and the question would seem to be one of strength of affinity; something like that involved in determining *electronegativity*. If a zinc rod is placed in a solution of copper sulphate, it acquires a layer of metallic copper, while some of the zinc goes into solution, as zinc ions, Zn^{2+}. If a copper rod is placed in a zinc sulphate solution, the copper does not go into solution, and no metallic zinc deposit is formed. By repeating this experiment with pairs of metals in turn, it is possible to build up a scale showing the relative stability in solution of all the metals. Indeed, the list can be extended to cover all possible types of ion in solution. This list is the *electrochemical (or electromotive) series*, the relative strengths (potentials) of the various ions usually being given in volts. Metals such as copper and gold, occurring low in this series, are more resistant to corrosion than those appearing higher in the list, such as zinc and sodium.

Electrochemical potentials are typically a few volts, a fact which is evocative of the common *battery*. When a piece of metal is surrounded by an aqueous solution, there is a tendency for ions of the metal to pass into solution. An investigation of this phenomenon by Hermann von Helmholtz showed that the loss of positive ions leaves the surface of the metal negatively charged, and this charge then attracts the ions that have passed into solution. The latter thus tend to remain in the vicinity of the metal surface, and a double layer of charge, positive and negative, is thereby established. J. F. Daniell, in the early nineteenth century, exploited the differential tendency for solution of different metals to form a practical battery. He set up a liquid receptacle containing two compartments, separated by a barrier that was porous to ions but which would prevent convective mixing of the aqueous solutions on the two sides. One of the compartments contained a zinc rod submerged in a zinc sulphate solution, while the other had a copper rod in a copper sulphate solution. Zinc is higher up in the electrochemical series than copper, it having a potential of −0.76 V, relative to the value 0.0 V for the hydrogen ion. The corresponding potential for copper is +0.34 V. Because the electrochemical potential is a measure of the tendency to ionize, we see that the potential on the zinc rod will be more negative than that of the copper rod. Hence a current of (negatively charged) electrons will flow from the

The electrochemical series

When two dissimilar metals are separated by an electrically conducting compound which, in solution or in the molten state, is simultaneously decomposed, a voltage difference is established. The separating medium, known as an *electrolyte*, conducts electricity, not by the flow of electrons but through the motion of ions. If the circuit is closed by an external conductor connecting the two metal *electrodes*, an electric current flows. This effect is the basis of *electric batteries*, and it is the underlying cause of *galvanic corrosion*. The following table lists some of the more important *electrode potentials*, relative to the standard hydrogen electrode at 25 °C, and for one mole (i.e. 6×10^{23}) of ions per kilogram

Element	Metal electrode	Ion electrode	Potential in volts	Element	Metal electrode	Ion electrode	Potential in volts
lithium	Li	Li^{+}	–3.05	nickel	Ni	Ni^{2+}	–0.23
potassium	K	K^{+}	–2.92	tin	Sn	Sn^{2+}	–0.14
calcium	Ca	Ca^{2+}	–2.87	iron	Fe	Fe^{3+}	–0.045
sodium	Na	Na^{+}	–2.71	hydrogen	H_2	H^{+}	0.000
magnesium	Mg	Mg^{2+}	–2.36	copper	Cu	Cu^{2+}	0.345
aluminium	Al	Al^{3+}	–1.67	copper	Cu	Cu^{+}	0.522
zinc	Zn	Zn^{2+}	–0.762	silver	Ag	Ag^{+}	0.797
chromium	Cr	Cr^{2+}	–0.71	mercury	Hg	Hg^{2+}	0.799
chromium	Cr	Cr^{3+}	–0.50	platinum	Pt	Pt^{2+}	1.20
iron	Fe	Fe^{2+}	–0.44	gold	Au	Au^{3+}	1.50
cadmium	Cd	Cd^{2+}	–0.42				

Relative to the neutral hydrogen, metals with a negative voltage are anodic, and are thus corroded, while those with a positive voltage are cathodic, and are protected.

zinc to the copper, if they are connected externally by a wire. When such a current flows, the zinc rod will tend to lose its negative charge and thereby disturb the equilibrium of the Helmholtz double layer. At the end of the nineteenth century, Henri-Louis Le Chatelier expounded a principle (the *Le Chatelier principle*) which states that if a system that is in equilibrium is subjected to a changing tendency, the system will react in such a way as to oppose the influence of that change. Applied to the case in point, this indicates that the flow of electrons from the zinc rod will be followed by the passing of more zinc ions into solution. Conversely, the copper rod receives electrons, and these can combine with copper ions in the adjacent aqueous layer to produce neutral copper atoms, which are deposited on the copper rod. Thus we see that equilibrium can be maintained in spite of the passage of electrons along the connecting wire. The *Daniell cell* is therefore a working battery, with a voltage of 1.1 V (i.e. 0.76 + 0.34 V). The two metal rods are referred to as *electrodes*, and in the example just discussed the zinc rod is the *anode*, or *electron donor* (supplier), while the copper rod is the *cathode*, or *electron acceptor*. The reaction occurring at the anode, the dissociation into positive ions and negative electrons, is known as *oxidation*, while the cathodic reaction, recombination of positive ions and negative electrons to produce neutral atoms, is called *reduction*. The loss of material from the anode is known as *galvanic corrosion*, after its discoverer, Luigi Galvani.

The hydrolysis of a salt and the development of electrode potentials are two facets of a larger story. The study of these two phenomena proceeded concurrently with investigations of two related effects: *electrolysis* and *aqueous conduction*. Indeed, experiments on several of these effects were frequently carried out by the same investigator. The first person to observe that electrical energy, and also heat energy, can be obtained from a chemical reaction was Alessandro Volta, in 1792. He showed that electricity is produced when two unlike metals are separated by a moist material such as leather saturated with salt water. Volta, after whom the unit of *potential difference* is named, also hit upon the idea of connecting several of these primitive batteries in series so as to give a larger voltage. A different, but no less relevant, type of observation had been made a few years earlier by Henry Cavendish, who noticed that the electrical conductivity of water is greatly enhanced if it contains dissolved salt. Liquids displaying an electricity-carrying capacity came to be known as *electrolytes*, and it was soon realized that substances such as oil were to be classified as *non-electrolytes*. Then in 1802 came the first dissociation of a compound by the passage of electricity, with the electrolysis of water by W. Nicholson and A. Carlisle. The most thorough investigation of electrolysis was due to Michael Faraday, in the period around 1830. It was he who realized that the current in the electrolytic process must be carried by charged particles in solution, and he called these *ions*, after the Greek word *ienai* which means to go. It was clear to Faraday that both positive and negative ions must be present in an electrolyte. Those ions which proceeded to the anode were called *anions*. Faraday showed that these ions were the charged radicals related to the substances, such as oxygen and chlorine, observed to be liberated at the anode, and these are what we have earlier referred to as *electronegative elements*. Conversely, the ions which move towards the cathode are *cations*, and these are radicals of *electropositive elements* such as the metals and hydrogen. An important embellishment to the story was made in 1887 by Arrhenius, who showed that the degree of ionization of a salt, when it is used to produce an electrolyte, is quite large and can indeed approach 100%.

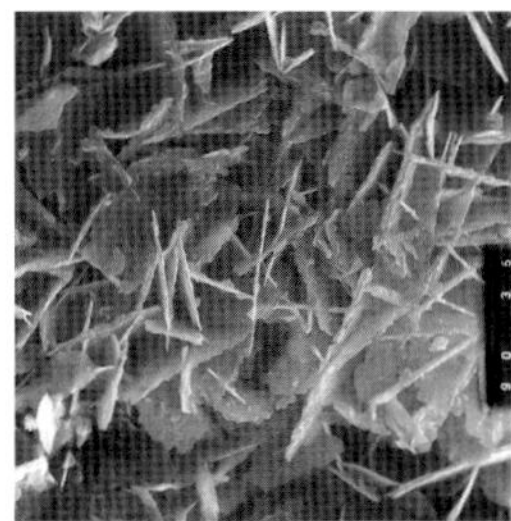

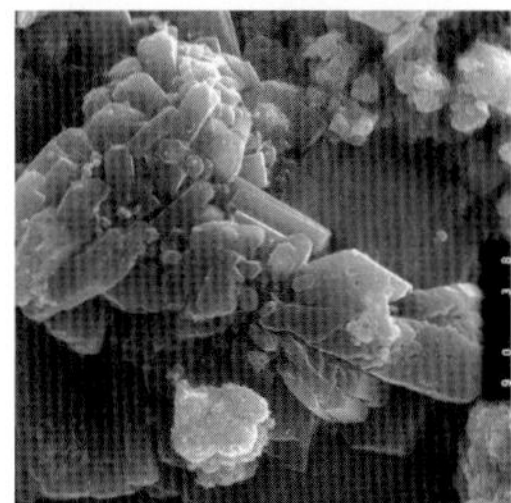

During the initial stage of *rusting*, iron is covered with a continuous film of the hydroxide $Fe(OH)_3$, which is rapidly changed to FeO. The latter usually contains a proportion of Fe^{2+} ions which give it a green colour. As the reaction progresses, rust crystals of various compositions are formed with diameters of up to several micrometers. In addition to the green variety, there are also crystals of black rust, Fe_3O_4, and brown rust Fe_2O_3. The angular crystals seen in the scanning electron photomicrograph at the top are green rust while the column-shaped and rounded crystals seen in the middle picture are black rust and brown rust respectively. Because this type of microscope does not use

Electrolysis is defined as chemical decomposition by an electric current passed through the substance, when it is either in the molten or dissolved state. As a typical example of the process, let us consider a receptacle containing molten sodium chloride and equipped with two carbon rod electrodes that dip into the salt, and which can be connected to a battery. The molten salt consists of dissociated sodium and chlorine ions, which can move relatively easily through the melt. The sodium ions, being positively charged, are attracted towards that electrode which has been connected to the negative terminal of the battery: the cathode. Because electrons are available at this electrode, the following cathodic reaction can occur

$$Na^+ + e^- \rightarrow Na$$

the symbol e^- denoting an *electron*. The concurrent anodic reaction must involve the giving up of electrons, in order to maintain the electron current being drawn from the cathode. It thus takes the form

$$2Cl^- \rightarrow Cl_2 + 2e^-$$

the chlorine molecules being liberated in the gaseous form, as bubbles at the anode. The metallic sodium liberated at the cathode cannot escape in this manner, and it is deposited instead as a layer on the cathode. If instead of molten sodium chloride an aqueous solution of that compound

Paleoclimatology

Because water expands upon freezing, the solid form – ice – is less dense than the liquid, upon which it therefore floats. According to a principle first propounded by Henri Le Chatelier (1850–1936), a system in equilibrium will, when subjected to a stress, respond so as to relieve that stress. So when ice is subjected to pressure, which tries to compress it, the result will be melting if the pressure is sufficiently large. This phenomenon is exploited by the ice skater, the pressure due to the blades being enough to create a thin layer of lubricating liquid water. The pressure can in principle be supplied by a column of ice itself, and this will cause melting if the column is sufficiently thick. It transpires that the required thickness is about 3 km, and this is the case in both the arctic and the antarctic, where the permanently sub-zero temperature permits the gradual build up of a sufficiently thick layer, because of water precipitation from the atmosphere. In the arctic, the average rate of precipitation is such that the 3 km of deposited ice corresponds to about 120 000 years, while the lower rate of precipitation in the antarctic means that about a million years of precipitation remains stored in the ice-cap. During the 1950s, it occurred to Willi Dansgaard that something else is preserved in the arctic and antarctic ice-caps, namely a record of the variation of atmospheric temperature during those geological periods. His reasoning was as follows. The two most abundant isotopes of oxygen are ^{16}O and the slightly heavier ^{18}O, and their differing masses must be reflected in slightly differing rates of precipitation, the water molecules containing the heavier isotope naturally falling faster. But, as Dansgaard realized, this differential rate of precipitation would also be temperature dependent, so determination of the relative abundances of the two isotopes in a specimen of ice would provide a measure of the atmospheric temperature at the time the precipitation took place. There is one additional factor, however, this stemming from the fact that rain falling immediately prior to passage of a warm front has been formed at a greater altitude than rain that falls when the front actually passes. And the temperature at the higher altitude is considerably lower. The upshot is that the earlier precipitation is relatively depleted of ^{18}O, so measurement of the relative abundances effectively serves as a thermometer for the conditions that prevailed during various eras in the past. Dansgaard and his colleagues systematically obtained samples – known as *ice cores* – from different depths in the Greenland ice-cap, from which the abundance ratio could be determined. They were thereby able to determine how the local mean atmospheric temperature has varied during the above-mentioned 120 000 years. It was found, for example, that the most recent glacial period ended about 10 000 years ago. Claus Hammer, a colleague of Dansgaard, has extended the ice-core technique to determine the times of major volcanic activity.

optical wavelengths, the different rust colours are not actually seen in such pictures. The lower picture shows iron oxide crystals that have formed on the surface of a 12% chromium steel superheater tube from a power plant, during exposure to 500°C steam for 11 000 hours. The width of all three pictures is about 25 μm.

is electrolyzed, hydrogen is liberated at the cathode and both oxygen and chlorine at the anode. The proportion of oxygen increases as the solution becomes more dilute. The reason for the difference can best be seen by first considering the electrolysis of pure water. The cathodic reaction is

$$2e^- + 2H_2O \rightarrow H_2 + 2OH^-$$

while at the anode we have

$$2H_2O \rightarrow O_2 + 4H^+ + 4e^-$$

When sodium chloride is present in dilute solution, the sodium ions proceed to the cathode, where their positive charges tend to neutralize the hydroxyl ions. The chloride ions, simultaneously, migrate to the anode and neutralize the hydrogen ions. Recalling the principle of Le Chatelier, we see that these neutralization processes promote further electrolytic dissociation of the water.

Viewed from this angle, the crystal structure of the most common form of *ice* clearly has the hexagonal symmetry that is reflected in the six-fold patterns seen in *snowflakes*. This computer-generated picture also indicates the choice of position available to the hydrogen atoms; a large number of different arrangements are possible, all satisfying the dual conditions that two hydrogens are attached to each oxygen while only one hydrogen is located along each oxygen–oxygen line (closer to either one or the other of the oxygen atoms).

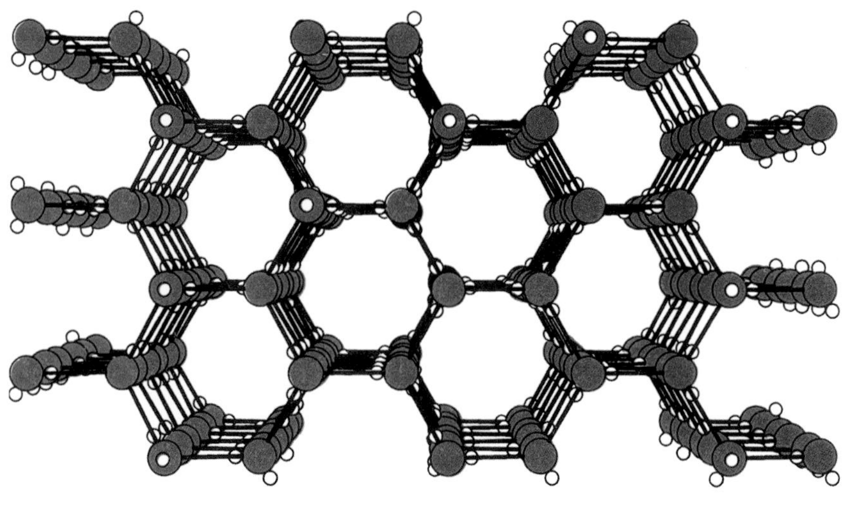

The outcome of Faraday's systematic studies of electrolysis was the emergence of two fundamental laws. The first of these states that the weight of a substance produced at the appropriate electrode is directly proportional to the amount of electricity (i.e. current multiplied by time) that has passed through the electrolytic cell. The second law states that the weight of a substance deposited is proportional to its *equivalent weight*, the latter being the atomic weight divided by the valence. In retrospect, we can see how close his studies brought Faraday to the nature of the atom; he linked atomic weight and valence with the occurrence of positive and negative charges. The lead was followed up by G. Johnstone Stoney who, in 1874, concluded that the proper interpretation of Faraday's observations lay in the existence of a certain standard quantity of electricity that traversed the electrolyte each time a chemical bond was ruptured. In 1891 he proposed that this natural unit of electricity be given the name *electron*. Thus, the unit electric charge is associated with the fundamental particle, bearing the same name, which was discovered by Joseph Thomson just six years later. Another important consequence of the pioneering work on electrolysis was that it rapidly became the accepted view that chemical compounds are held together by the attraction between positive and negative electric charges.

Probably the most important manifestation of electrolytic effects is *aqueous corrosion*. Let us consider a typical situation. An item that has been fabricated from cast iron is coated with a thin layer of tin to prevent oxidation, iron being more readily oxidized than tin. The desired protection is achieved until the item sustains a scratch that is deep enough to cut right through the tin, exposing the underlying iron. In the presence of moisture, the system becomes an *electrolytic cell*, since two different metals are present. Iron, being higher up in the electrochemical series than tin, becomes the anode, and it is therefore subject to dissociation into electrons and positive ions. In the absence of soluble impurities, the water that is present will be approximately neutral, and because there will be some dissolved oxygen available the cathodic reaction will be

$$O_2 + 2H_2O + 4e^- \rightarrow 4OH^-$$

In 1611, Johannes Kepler (1571–1630) gave his benefactor, a Regensburg city counsellor by the name of Wackher, an intriguing New Year's present. It was an essay entitled *On the Six-Cornered Snowflake*, and in it Kepler attempted to explain why these beautiful objects have hexagonal symmetry. His ideas were based on the packing of equal spheres, and have long since been superseded. The modern theory of these structures recognizes the importance of *dendritic growth*, and relates the overall pattern to the underlying six-fold symmetry of ice crystals.

The surface area of the cathode is much larger than that of the exposed area of anode, so the electric current will be concentrated around the scratch. At the anode, the dissociation reaction is

$$Fe \rightarrow Fe^{3+} + 3e^-$$

and the ferric ions can combine with the hydroxyl ions

$$Fe^{3+} + 3OH^- \rightarrow Fe(OH)_3$$

to produce ferric hydroxide, commonly known as *rust*. The factor that limits the chemical reaction is the availability of oxygen, the reaction slowing down when the supply is used up. This important role played by oxygen is clearly demonstrated in marine installations that are subject to tidal variations in sea level. It is observed that the most rusty parts on a pier, for example, are those that lie between the high and low water marks. Prevention of aqueous corrosion can be achieved by introducing an expendable second metal that lies higher up the electrochemical series than the metal to be protected. Hence the zinc plating, also known as *galvanizing*, of iron articles, in which the zinc is the anode. If such an article is scratched and subsequently becomes moist, a cell is formed in which zinc ions pass into solution, ultimately to be redeposited on the iron. In this type of arrangement, the expendable metal is known as a *sacrificial anode*.

It has long been realized that corrosion is a major commercial factor. In 1975, a United Nations committee assessed its impact on the economies of the industrially developed nations, finding it to represent a deficit that is typically a few per cent of a country's gross national product. Even before the industrial revolution led to an explosive increase in the use of fabricated metallic components, corrosion was influencing technology. A good example is seen in the hull design of ships. The increasing call for vessels to spend protracted periods at sea increased the depredations of barnacles, worms, and seaweed. In an effort to forestall such ravages, lead sheathing of hull bottoms was tried, in the 1600s, and this metal gave way to copper in the following century. Oblivious of the principles of electrochemistry, marine architects continued to specify iron bolts for the hull joints, and these provoked electrolytic action with the copper cladding. The repercussions were sometimes tragic. The entire bottom of the naval vessel *Royal George* fell out while she was anchored in Portsmouth harbour, and she sank immediately with terrible loss of life. Shortly thereafter bronze bolts became standard in this application, and the problem disappeared. *Corrosion resistance* has become an important branch of metallurgy, and one of its famous mileposts occurred in 1913 when Harry Brearley, investigating steel alloys with various compositions for the manufacture of gun barrels, discovered that a 14% chromium content gives a particularly durable product: *stainless steel*.

The *hydrogen bond* is weak compared with the typical *covalent bond*, involving approximately one-tenth the energy of the latter. It is significant only in reactions which occur with small changes of energy. In *biochemistry*, the making and breaking of hydrogen bonds is of paramount importance. Biologically active molecules are frequently held together partly by such bonds, and all biological reactions take place in an aqueous environment, in which hydrogen bonds play a vital role. Bearing in mind the inadequacy of two-dimensional representations of three-dimensional objects, we may depict the hydrogen bond thus

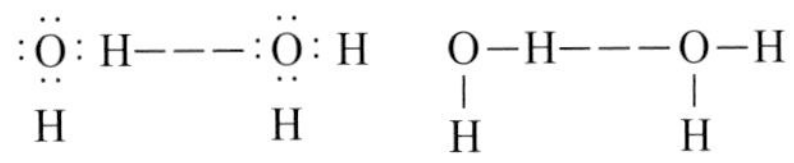

```
 ..          ..
:O: H---:O: H     O-H---O-H
 ..          ..      |       |
 H           H       H       H
```

the figure to the left employing the *Lewis notation*, while the one on the right uses the structural formulation. The oxygen atoms exert a strong pull on the *lone electrons* supplied by the hydrogen atoms; to the oxygen atom of a nearby water molecule any of the hydrogen atoms appears almost totally devoid of screening electrons. This effect is directional because of the small size of the hydrogen ion compared with that of the two oxygen ions; it is like a dwarf between two giants. For the same reason, the bond is saturated; there is no room for any further ions to approach the hydrogen ion and establish electric bonding. The hydrogen bond is usually almost straight.

Although the solid form of water is less important than the liquid, it is nevertheless interesting. The first study of the crystal structure of *ice* was carried out in 1922 by William Henry Bragg, who succeeded in elucidating the arrangement of the oxygen atoms. The *X-ray diffraction technique*, which he employed, cannot locate the hydrogen atoms, because their associated electron density is so low. This difficulty can be overcome by replacing the X-rays with a beam of low-energy neutrons. John Bernal and Ralph Fowler, in 1933, and Linus Pauling two years later, noted the similarity in many physical properties of ice and liquid water and concluded that the H_2O molecules in ice are intact. Knowing the chemical formula for water and the arrangement of the oxygen atoms in the crystal, we might expect the assignment of the hydrogen locations to be a straightforward matter, but this is not the case. Pauling realized that there must be two conditions to which the hydrogen positions are subject. The first is that each water molecule is oriented so that its two oxygen–hydrogen bonds point towards two of its four nearest neighbouring oxygen atoms. The second demands that the orientations of adjacent water molecules be such that only one hydrogen atom lies on the axis joining adjacent oxygen atoms. These conditions are seen to guarantee collectively a maximum of hydrogen bonding, but one finds that the hydrogen positions are not uniquely specified; there is still a certain degree of choice. Pauling himself realized that there is this *residual entropy* in ice, and his elegant calculation of its magnitude was subsequently verified experimentally.

The *hydrogen bonding* in the common form of ice produces a rather open crystal structure; this is the reason for that peculiar property of water: its solid form is less dense than the liquid. Solid iron sinks to the bottom of the smelter's crucible, but the iceberg floats. If the expansion upon freezing is opposed by application of pressure, the ice melts again. This is exploited by the ice skater, who glides on a thin film of liquid water. It turns out that there are several other crystalline forms of ice, most of which are in fact more dense than the liquid. The first of these exotic types of ice, which exist at higher pressures and lower temperatures, was discovered in 1900 by G. Tammann, and several others were found by Percy Bridgman during the period 1912–1937. Like other crystals, ice contains such defects as vacancies, dislocations and grain boundaries, the actual concentration of these imperfections depending on the thermodynamic and mechanical history of the material. Their presence has relevance to the question of *glacial flow*, which is another example of *creep*. There are other defects which are peculiar to ice. They involve errors of water-molecule orientation and were discovered by Niels Bjerrum in 1951.

Summary

The peculiar properties of water such as its ability to dissolve a wide range of substances derive from the unusual physical structure of the H_2O molecule. It has two positive and two negative electric poles and can thus attach itself both to anions such as Cl^- and cations such as Na^+. In liquid water, some of the molecules are dissociated into hydrogen ions, which join with neutral water molecules to form hydronium, and hydroxyl ions. The effective hydrogen ion concentration determines whether a solution is acidic or basic (i.e. alkaline). The tendency of an element to dissociate into electrons and positive ions is measured by its position in the electrochemical series. When two different metals are placed in an electrolyte, which is a liquid containing both positive and negative ions, their differing electrochemical forces produce a voltage difference, and this can be exploited to produce a battery. Such a set-up also leads to dissolution of that metal which is higher in the series, the anode, and this is the cause of aqueous corrosion. Ice crystals are held together by hydrogen bonds, the directionality of which gives an open structure. Solid water is thus less dense than liquid water, upon which it floats.

7

Deep in unfathomable mines
Of never failing skill
He treasures up his bright designs
And works his sovereign will

William Cowper (1731–1800)
Olney Hymns

From mine, quarry and well: *minerals*

The way in which the world came into being, and the time of this important event, have of course been the subject of frequent speculation down through the ages. The issue reveals an interesting schism between oriental and occidental philosophies. The Hindus, for instance, have long maintained that the Earth has existed for thousands of millions of years. In sharp contrast to this, we have the views of Archbishop Ussher and Dr Lightfoot, who, in the middle of the seventeenth century, made a systematic scrutiny of the Bible. They found that a consistent chronological sequence could be established and were able to announce that creation commenced promptly at 9 am on Sunday, 23 October, in the year 4004 BC. The remarkable fact is that this result continued to be taken seriously until not much more than a hundred years ago. And there are many who *still* believe it.

The various scientific methods that have been brought to bear on the question during the past century have successively increased the estimates of the age of the Earth and its neighbouring planets. One can now state, with reasonable confidence, that the solar system is about 6000 million years old. This makes it between a third and a half the age of the Universe itself. Recent studies, using *radioactivity techniques*, indicate that the Earth was formed either at the same time as the solar system or shortly thereafter, because the oldest rocks on Earth are about 3800 million years old. The oldest rocks on the Moon have been dated at around 4700 million years, but we must remember that the absence of weathering effects on the Moon makes for easier extraction of geological information than is the case here on Earth.

Reliable data are now available both for the relative abundances of the elements in the Earth's crust and in the Universe as a whole, and if they are compared, one is immediately struck by the dramatic difference; a planet like our own is clearly an atypical corner of creation. The two elements that are by far the most common in the Universe are hydrogen and helium, atoms of the former outnumbering those of the latter by about four to one. (Because the atomic mass of helium is about four times that of hydrogen, however, the total masses of helium and hydrogen in the Universe are roughly equal.) The third most common element is oxygen, but there is only one oxygen atom to about a thousand hydrogen atoms, and the occurrence of the other elements is roughly inversely related to their atomic weight. By contrast, almost half of the *Earth's crust* is oxygen, and over a quarter of it is silicon. Most importantly, for it is this which has enabled us to develop our technology, we find relatively large quantities of aluminium, iron,

Mining aroused little interest until the middle of the sixteenth century, when the increasing demands of technology caused an acute shortage of wood, and coal rapidly became the chief object of the subterranean activity. This illustration from *De re metallica*, written by Georgius Agricola (1494–1555), shows the state of the art five hundred years ago.

calcium, sodium, potassium and magnesium. Lesser, but nevertheless useful, amounts of copper, lead, tin, zinc, mercury and silver are also found.

What cosmic series of events took place to provide such riches? The currently accepted view is that the formation of a solar system comes about through a sequence of *thermonuclear reactions* which use hydrogen as the basic building material and, through a succession of nucleus–nucleus fusions, produce successively heavier elements. From what was described earlier regarding the structure of the hydrogen atom, it is not immediately

The Universe and the Earth: composition

The Earth is by no means a typical region of the Universe, as can be seen from the two tabulations of *elemental abundance* given below. The various entries are listed as percentages of the whole by weight, in each case.

The Universe

H	He	O	Fe	N	C	Si
50.3	47.0	1.07	0.61	0.46	0.14	0.12
Mg	**Ni**	**S**	**Ca**	**Al**	**Na**	**Cl**
0.088	0.051	0.040	0.011	0.010	0.0044	0.0037

The Earth

O	Si	Al	Fe	Ca	Na	K
46.5	28.0	8.1	5.1	3.5	2.8	2.5
Mg	**Ti**	**H**	**C**	**Cl**	**P**	**S**
2.0	0.58	0.20	0.20	0.19	0.11	0.10

obvious how this consolidation into heavier elements occurs. The nucleus of each hydrogen atom is a single proton, and this bears a single positive charge. When two hydrogen atoms approach each other, therefore, the two positive charges repel, and this inhibits the formation of a compound atom with a nucleus containing both protons. However, it turns out that the intrinsic electric repulsion between the protons can be overcome if they are pushed together with sufficient force. Then, when they are separated by a distance comparable to their diameters – about a femtometre – the strongly attractive but short-ranged *nuclear forces* take over and stability is achieved. The heavier isotope of hydrogen, known by the name *deuterium*, has one proton and one neutron in its nucleus. When two deuterium nuclei come together, therefore, only two of the four nucleons repel electrically.

Stars such as the Sun are probably formed in the following manner. The individual hydrogen atoms in a tenuous interstellar cloud begin to feel the gravitational attraction of one another, and this slowly leads to condensation. As the cloud shrinks, the concentration of mass at its centre increases, and this in turn leads to ever-increasing velocities of the hydrogen atoms towards the centre. Since velocity and temperature are equivalent concepts, we can think of this process as a conversion of gravitational energy into heat energy. The velocities attained by protons ultimately exceed that required to overcome the long-range electric repulsions, and fusion occurs. The first products of this process are helium atoms, but as the effective temperature is increased, ever larger atomic nuclei are produced. The temperature for helium production is slightly in excess of ten million degrees Celsius, while a temperature of over a thousand million degrees is required for the production of an iron nucleus. For all elements up to and including iron, the fusion process leads to a net release of energy, but for heavier elements energy is absorbed, and it is not immediately obvious that they can be produced in the collapsing process. One possibility is that *supernovae*, which are essentially stellar explosions, eject nuclear fragments at high velocity, and these subsequently unite with other nuclei to form the heavy elements. Ninety-one of the known elements occur naturally on Earth, and over seventy of these have already been shown to be present also in the Sun.

Native (elemental) *copper* deposits usually occur in association with basic *igneous rocks*, the metal having been formed by the chemical reaction of copper-bearing solutions with iron minerals. This type of deposit is found in the Keweenaw area of northern Michigan, USA, the source of the specimen shown here. Small amounts of native *silver* are sometimes found in such copper, underlining the chemical similarity between the two elements. The lighter patch on the surface of this specimen reveals the presence of silver.

As the fiery ball that was eventually to become the Earth cooled, it lost some of its original components. Some of these escaped by dint of their volatility, while others wandered away because they were so light that the Earth's gravitational attraction was insufficient to hold them. It is interesting to note in this connection that the planet Jupiter, which is far larger than the Earth, has managed to hold on to a much larger proportion of its original hydrogen. As the Earth cooled, by radiating energy out into space, the point was reached at which the outside began to solidify. This would at first have occurred in isolated patches around the surface, but ultimately a solid crust completely enclosed the molten mass. At this time, various processes were taking place on a vast scale, which have led to the rich variety of geological structures that are now found on the terrestrial surface. The various elements differ in such physical properties as density, freezing point, viscosity, and thermal conductivity, and all these factors contributed to the separation process. In addition, there was the factor of *chemical affinity*, which caused some elements to form compounds with one another while shunning liaisons with other elements.

Knowledge of the present structure of the Earth's interior comes primarily from *seismological data. Shock waves*, including those produced by earthquakes, travel at a speed that is related to the density of the material through which the wave moves. Studies of such waves have shown that the Earth consists of various layers, somewhat reminiscent of the concentric shells of an onion. Starting from the centre, there is a *solid core*, its existence having been established by Inge Lehmann in 1936. It has a radius of about 1400 km and density of up to 17 g/cm^3. It consists primarily of a mixture of iron and nickel. Surrounding this is a liquid region, about 2000 km thick, with a density of between 10 and 15 g/cm^3. The liquid is known by the general name *magma*. Outside this lies the *mantle*, with a density of 3.4 g/cm^3 and a thickness of roughly 2900 km. This layer is thought to consist of a variety of different igneous rocks, but their composition is still a matter of speculation. After this comes the *Moho discontinuity*, named after Andrija Mohorovičić, and finally the outer crust, which is not one continuous mass but rather a number of adjoining patches. It consists of three types of rock: igneous, metamorphic and sedimentary. The latter type occurs exclusively at or near the outer surface, while the others are distributed in a more complex fashion.

Down to about 15 km the mean density is 2.7 g/cm^3, and the rock, known by the collective name *sial*, consists primarily of silica and alumina. Below this the density increases to 3.9 g/cm^3 and the predominant constituent is *sima rock*, which is a mixture of silica and magnesia. The separation processes referred to earlier have produced a distribution in which the greater part of the heavier elements such as iron have found their way to the centre of the Earth, while the lighter elements such as calcium, silicon and aluminium have stayed near the surface. It should be noted, however, that this is not simply a question of elemental density, but rather that of the density of the compounds in which the different elements participate. In this respect we are fortunate that iron combines fairly readily with oxygen; iron's oxides are relatively light and are thus abundant in the Earth's outer layer. The temperature at the very centre of the Earth has been estimated to be around 4000 °C. Since this lies above every known elemental melting point, it seems surprising that the central core is nevertheless solid. The answer lies in the

Uraninite, UO_2, commonly called *pitchblende*, is both an important ore of uranium and a source of radium. Uranium is radioactive, and nuclear disintegrations through a succession of elements occur continuously in the ore. Lead, the stable end product of these reactions, is thus always present as an impurity. The radioactivity of the specimen shown here was demonstrated by placing its polished surface in contact with a photographic plate. The result of a three-day exposure was the picture, known as an *autoradiograph*, shown on the right.

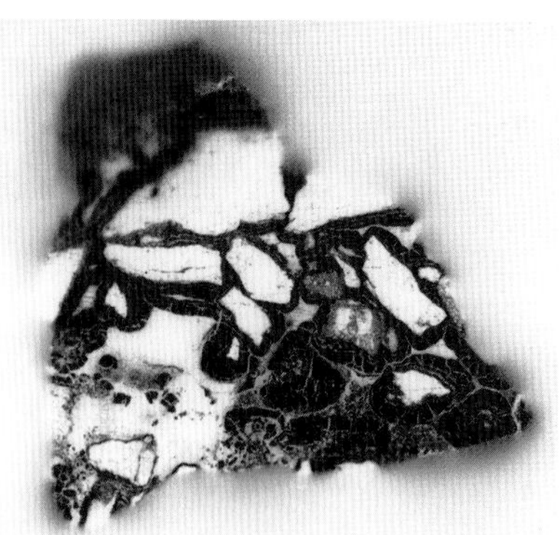

fact that most melting points increase with pressure, and the centre of the Earth is indeed under enormous pressure.

The natural separation processes pushed the lighter elements such as hydrogen, carbon, oxygen and nitrogen to the surface, where they existed both in the elemental form and in gaseous compounds like water vapour, methane, carbon dioxide and hydrogen sulphide. These compounds were continually being formed and disrupted, but competing factors led to the gradual diminution of some of the components. The breaking down of water by the Sun's energy, for example, would sometimes be followed by combination of the resultant oxygen with a surface metal to form an oxide, and escape of the hydrogen atoms to the upper regions of the atmosphere. As the inhospitable atmosphere cooled, condensation set in, producing precipitation of monumental dimensions which added its contribution to the sculpturing of the Earth's surface that the shrinkage-induced cracking had already begun.

It is believed that the first signs of life appeared on Earth about 3500 million years ago. The condensation referred to above must have been accompanied by vast electrical storms, and these might have supplied the energy to combine some of the light elements of the early atmosphere into such biochemical building blocks as amino acids and short lengths of organic polymer. The build up of these compounds in the primaeval oceans, and the passage of great periods of time, ultimately produced matter that was capable of primitive reproduction and *metabolism*, which is the extraction of energy from organic materials in the environment. One of the most significant evolutionary mileposts was the development of cells capable of using solar light energy to combine water and carbon dioxide into such hydrocarbons as sugar, cellulose and starch, by the process known as *photosynthesis*. Oxygen is liberated as a byproduct of this process, but when plant cells die, they decay by oxidation and thus reabsorb oxygen. With the massive erosion that was occurring concurrently, large amounts of plant matter were buried before they could decay, and this produced a net increase in the oxygen content of the atmosphere. The latter thereby slowly developed the benevolent composition that it has at the present time.

Large-scale relative movements of the Earth's crust continued to occur, and as recently as 200 million years ago there was a breaking up of a large land mass, a sort of supercontinent known as *Pangaea*, into the familiar continents that we see depicted in today's atlases. Francis Bacon and Alfred

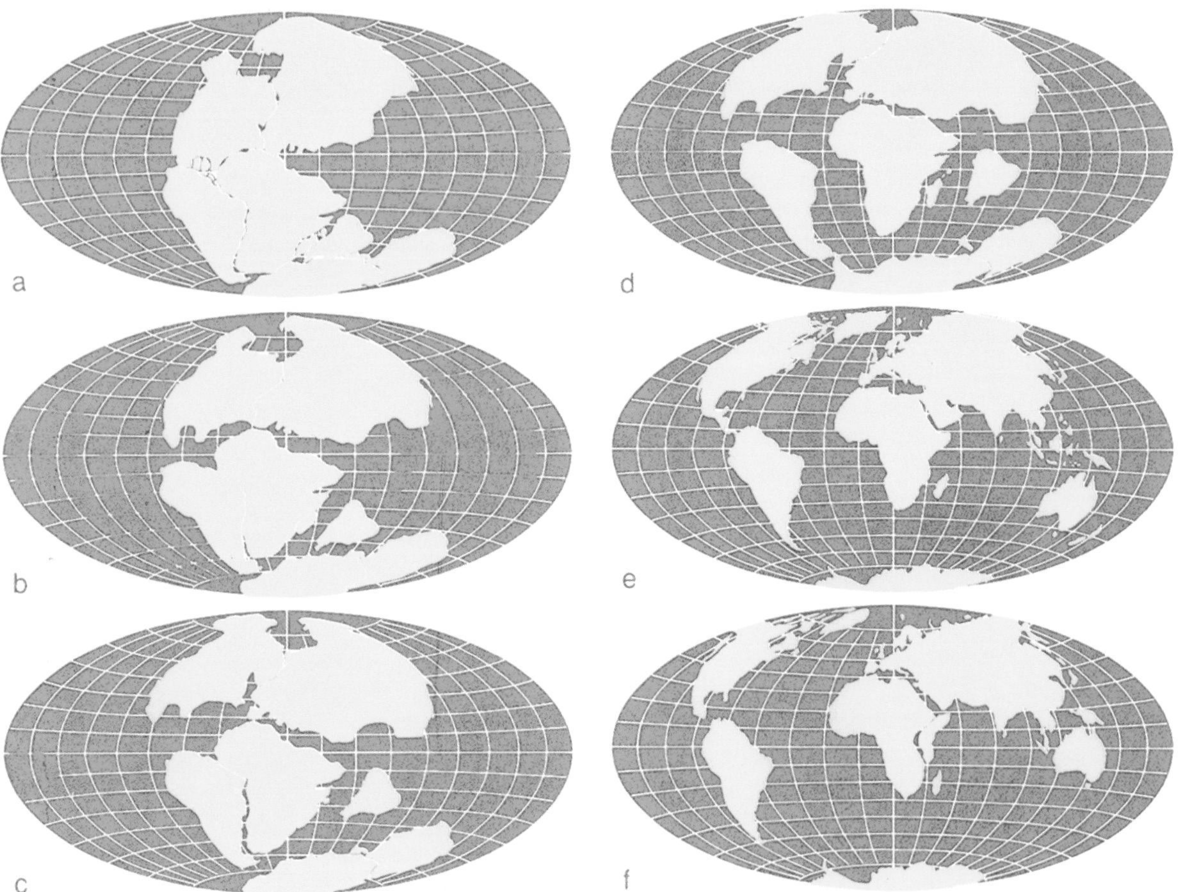

In 1620, Francis Bacon (1561–1626) noted that the east coast of the Americas closely resembles the west coast of Europe and Africa. This was impressive, given the primitive state of cartography at that time. Georges-Louis LeClerc (1707–1788) suggested in 1749 that the land mass separating those continents – *Atlantis* – sank, thereby producing the Atlantic Ocean. A rationalization of Bacon's observation was offered by Antonio Snider-Pellegrini, in 1858, when he suggested that material from the Earth's interior presses itself against the surface crust, causing it to crack, pieces of the crust then being forced apart. The hypothesis of *continental drift*, put forward by Alfred Wegener (1880–1930) in 1915, initially met with considerable opposition from the geophysical community. The modern theory of *plate tectonics* has not only vindicated Wegener's idea, it also charts the break-up of the original land mass, *Pangaea*, into the present continents. After about 20 million years of drift, this supercontinent had split into northern and southern portions known as *Laurasia* and *Gondwana*, respectively. This situation, about 180 million years ago, is shown in the second figure. The following two figures show the distribution of land masses after further periods of 45 million years and 115 million years, respectively, and the fifth their present positions. The final figure shows what the world might look like 50 million years from now, if present trends are maintained. It has been established that the continental plates move around on a softer layer lying at a depth of between 100 km and 250 km. *Earthquakes* are generated when the edges of two adjacent plates move with respect to each other. Such edges are far from smooth, so there is large-scale friction, which creates shock waves. When the mutually moving plate edges lie under the sea, the resulting earthquakes can provoke a *tsunami*, or *tidal wave*.

Fossils, Paleontology and Geology

Geology deals with the origin, history and structure of the Earth, while *paleontology* is concerned with prehistoric forms of life. The study of *fossils* bridged these two scientific disciplines, the 1660s being the pivotal decade. Robert Hooke (1635–1703), a native of the Isle of Wight, a few miles south of the English coast, explored the cliffs near his birthplace in Freshwater (seen in the background of this photograph of the adjacent fossil-rich Compton Bay). He noticed shells such as those of oysters, limpets and periwinkles, as well as plantlike structures, buried in the layers of clay, sand, gravel and loam, and concluded that they had been deposited there long before, and subsequently became petrified. About the same time, in 1666, some French fishermen caught an enormous shark, weighing almost 1800 kg, and its head was studied by Niels Steensen (1638–1686, also known as Nicolaus Steno). He was surprised to find thirteen rows of teeth in the lower jaw, the innermost being clearly less mature than those at the front, and he surmised that there must a constant turnover of teeth throughout the creature's life. (The copper plate of the head, reproduced here, appeared in *Metallotheca Vaticana* by Michaelis Mercati.) Even more significantly, he noticed that the teeth closely resembled the mysterious triangular 'tongue stones' (also known as *glossopetrae*) he had found embedded in the steep cliffs around the Etruscan town of Volterra, deep inland and far removed from the Italian coast. He concluded that these stones were really petrified shark teeth, and that the site of the town must thus have lain under water during an earlier period. This brilliant piece of scientific induction was celebrated by the postage stamp reproduced here. But it transpired that both Hooke and Steensen had been scooped by Leonardo da Vinci (1452–1519), who had come to similar conclusions almost two centuries earlier, following his own speculations about sea shells found deep inland. It is intriguing that the sciences of paleontology and geology were pioneered by three of history's greatest polymaths. Mary Anning (1799–1847) has been called the finest fossil collector the world has ever known. She lived in Lyme Regis, on England's south coast, which boasts some of the world's richest fossil deposits, and she made some of the most spectacular discoveries of dinosaur fossils from the Jurassic period, including the first ichthyosaur and the first plesiosaur, examples of which are shown here. Fossils now provide the best way of dating sedimentary rocks.

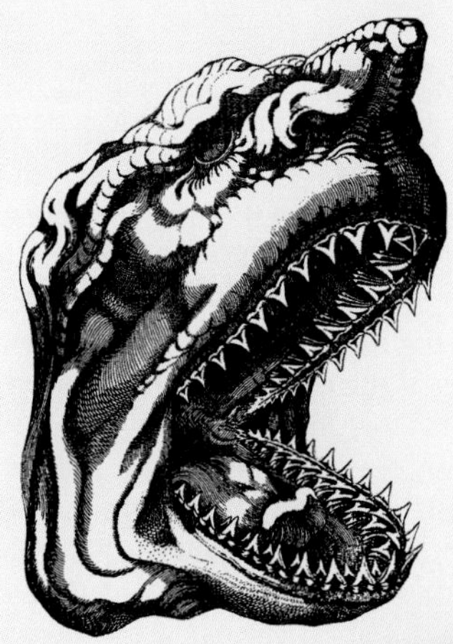

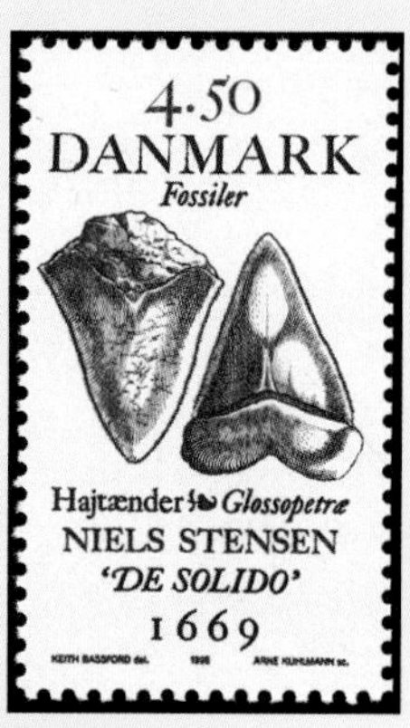

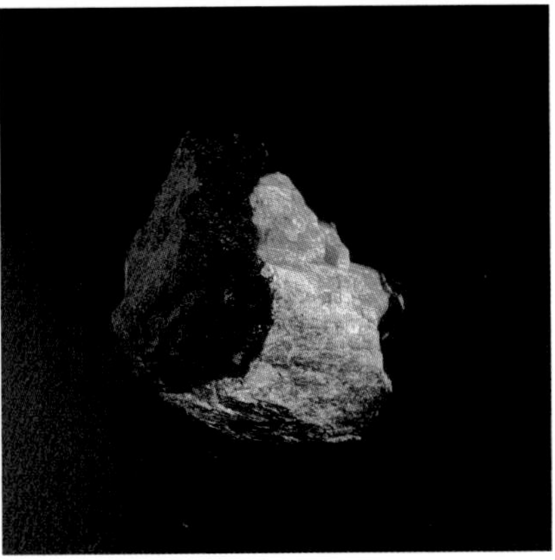

The mineral *franklinite* has only ever been found in the Precambrian limestone rocks at Franklin and Sterling Hill, New Jersey, USA. It is predominantly $ZnFe_2O_4$ but manganese impurities cause reddish-brown streaks in the black crystals. As is often the case, the mineral is found in close association with others. Its common partners, calcite and willemite, can readily be detected under ultraviolet light, because they both show pronounced fluorescence (right-hand picture). The *calcite* appears red and the *willemite*, a zinc silicate, green.

Wegener were the first to appreciate the significance of the fact that there is a remarkable resemblance between the east coast of South America and the west coast of Africa, and they individually suggested that these must originally have been adjacent parts of the same continent. It is now known that this was indeed the case and that they, together with India and Australia, once formed a region of Pangaea known as *Gondwana*. Recent geological investigation of movements of the sea-bed have led to elaboration of Wegener's *continental drift* hypothesis. The Earth's crust is seen as comprising seven major pieces, or plates. They have a rigid exterior, the *lithosphere*, which is some 100 km thick, and a more plastic region, the *asthenosphere*, lying immediately below this. *Plate tectonics*, the science of the relative movement of these crust fragments, gives a quite detailed picture of the most recent 200 million years. Where the plates drift apart, new *basalt* rock is added by the welling up and solidification of magma. These cracks are the *mid-ocean rifts*, which are still separating at the rate of a few centimetres per year. Elsewhere the plate margins must move towards one another, and the scraping at the interface causes *earthquakes*. A well-known example is the San Andreas fault in California. In such regions one plate edge is pushed down while the other rides up to produce new mountains. The location of *volcanoes* also shows a marked correlation with the positions of the interplate boundaries.

The various types of surface activity produced (and of course still do produce) three distinct classes of rock. *Igneous rock* was produced by the simple solidification of the molten material, the magma referred to earlier. *Sedimentary rock* was formed by the deposition of material released during the erosion process. Finally, there are the *metamorphic rocks* which were produced by the action of heat and pressure on rocks belonging to the other two classes. It should be noted that all three types of rock were being produced concurrently. New igneous rock could be produced in a region that had lain quiescent for some time, for example, by the sudden outbreak of volcanic activity. Such surface eruption continues in the present era, a fact for which the Roman city of Pompeii provides grim testimony. A great amount of movement has occurred in the Earth's crust over the ages, and the three types of rock have become thoroughly mixed together to produce today's heterogeneous outer surface. The sedimentary rocks are particularly important to the geologist because they contain the fossilized remains of the various creatures that have inhabited this planet down through the

The geological time scale

Era	Period	Epoch	Million years before present	Geological events	Sea life	Land life
	Quaternary	Holocene		Glaciers recede. Sea level rises. Climate becomes more equable.	As now.	Forests flourish again. Humans acquire agriculture and technology.
	Pleistocene		0.01	Widespread glaciers melt periodically causing seas to rise and fall.	As now.	Many plant forms perish. Small mammals abundant. Primitive humans established.
			2.0			
	Tertiary	Pliocene	5.1	Continents and oceans adopting their present form. Present climatic distribution established.	Giant sharks extinct. Many fish varieties.	Some plants and mammals die out. Primates flourish.
		Miocene	24.6	Seas recede further. European and Asian land masses join. Heavy rain causes massive erosion.	Bony fish common. Giant sharks.	Grasses widespread. Grazing mammals become common.
Cenozoic era		Oligocene	38.0	Seas recede. Extensive movements of Earth's crust produce new mountains.	Crabs, mussels, and snails evolve.	Forests diminish. Grasses appear. Pachyderms, canines and felines develop.
		Eocene	54.9	Mountain formation continues. Glaciers common in high mountain ranges.	Whales adapt to sea.	Large tropical jungles. Primitive forms of modern mammals established.
		Paleocene		Widespread subsidence of land. Seas advance again. Considerable volcanic activity.	Many reptiles become extinct.	Flowering plants widespread. First primates. Giant reptiles extinct.
			65			
	Cretaceous	Late Early	97.5	Swamps widespread. Massive alluvial deposition. Continuing limestone formation.	Turtles, rays, and now-common fish appear.	Flowering plants established. Dinosaurs become extinct.
			144			
Mesozoic era	Jurassic	Malm Dogger Lias	163 188	Seas advance. Much river formation. High mountains eroded. Limestone formation.	Reptiles dominant.	Early flowers. Dinosaurs dominant. Mammals still primitive. First birds.
			213			
	Triassic	Late Middle Early	231 243	Desert conditions widespread. Hot climate gradually becomes warm and wet.	Icthyosaurs, flying fish, and crustaceans appear.	Ferns and conifers thrive. First mammals, dinosaurs, and flies.
			248			

Era	Period	Epoch	Million years before present	Geological events	Sea life	Land life
	Permian	Late Early	258 286	Some sea areas cut off to form lakes. Earth movements form mountains.	Some shelled fish become extinct.	Deciduous plants. Reptiles dominant. Many insect varieties.
	Carboniferous	Pennsylvanian Mississippian	320 360	Sea-beds rise to form new land areas. Enormous swamps. Partly-rotted vegetation forms coal.	Amphibians and sharks abundant.	Extensive evergreen forests. Reptiles breed on land. Some insects develop wings.
	Devonian	Late Middle Early	374 387 408	Mountain formation continues. Seas deeper but narrower. Climatic zones forming.	Fish abundant. Primitive sharks. First amphibians.	Leafy plants. Some invertebrates adapt to land. First insects.
Paleozoic era	Silurian	Pridoli Ludlow Wenlock Llandovery	414 421 428 438	New mountain ranges form. Sea level varies periodically.	Large vertebrates. Coral reefs develop.	First leafless land plants.
	Ordovician	Ashgill Caradoc Llandeilo Llanvirn Arenig Tremadoc	448 458 468 478 488 505	Shore lines still quite variable. Increasing sedimentation.	First vertebrates.	None.
	Cambrian	Merioneth St David's Caerfai	525 540 590	Earth's crust forms. Surface cools. Vast rainfall produces seas. Mountains formed. Much volcanic activity.	Shelled invertebrates.	None.
Sinian era		Vendian		Shallow seas advance and retreat over land areas. Atmosphere uniformly warm.	Seaweed. Algae and invertebrates.	None.

geological eras. The systematic study of the various strata has led to the dividing up of the most recent 600 million years into various eras. The *Paleozoic* (meaning ancient life) era stretched from about 590 to 248 million years ago. This was followed by the *Mesozoic* (middle life) era, while the most recent era is the *Cenozoic* (recent life), which began around 65 million years ago. These great eras have been subdivided, the first 85 million years of the Paleozoic, for example, being known as the *Cambrian Period*. The burial of large amounts of vegetable material, which produced today's coal fields, occurred in another period of the Paleozoic era known as the *Carboniferous*, which lasted from around 360 to 286 million years ago.

Sulphur is one of the few pure elements to occur in amounts large enough to have economic importance. It is frequently found in regions of recent volcanic activity, being deposited around the gaseous outlets known as *fumaroles*, but it is more commonly found in association with gypsum and limestone, in *sedimentary rocks* of Tertiary age. Substantial quantities of sulphur occur in the cap-rock of some of the salt domes in the Gulf coastal regions of Texas and Louisiana. The specimen shown here is in the form of a *stalactite*, produced by deposition from the dripping solution.

Before considering the impressive variety of substances found on the Earth's surface, and within the accessible depths beneath it, it is necessary to define some common terms. The word *rock* is used in a rather loose sense and is generally taken to mean the macroscopic masses of material of which the Earth's crust is composed. A rock can consist of a single mineral, but most rocks are formed from an aggregate of minerals. Some rocks are even composites of mineral and non-mineral substances. The word *mineral*, on the other hand, has a more precise definition though it too is commonly used in a more general sense. We often think of all natural substances as being divided into three groups or kingdoms: animal, vegetable and mineral. According to this classification, the mineral kingdom comprises all non-living matter and as such it is a vast domain. The mineralogist uses the term in a much more restricted sense. Of the large number of inorganic substances, only relatively few are properly qualified to bear the name mineral, and until now only about 2000 different types have been discovered. A common dictionary definition states that a mineral is something obtained from a mine, but this is misleading because we might conclude that all substances found in the outer layer of this planet are thereby included. The strict definition of a mineral is that it is *a homogeneous inorganic crystalline solid occurring in the Earth's crust*. It is interesting to note what is thus excluded. The solidified remains of plant and animal secretions cannot be minerals, so we must rule out such interesting things as *bitumen*, *pitch* and petrified resins like *amber*. Then again, the rock-like products formed by the deposition of animal or plant remains must not be classified as minerals, and this excludes things like *coral* and *petrified wood*. In the purest definition, minerals must be crystalline, so naturally occurring compounds that prefer to adopt a glassy form, such as *obsidian*, cannot be included.

Even the most cursory glance at a list of the chemical compositions of different minerals reveals a striking fact: a large number of electropositive elements are represented, but only few electronegative elements. The classification of minerals groups them according to the anion, or anions, present. The first class consists of pure elements. It contains such things as gold, silver, copper, sulphur, and carbon in its two common forms: graphite and diamond. Then there are the separate classes in which a single anion is present such as the sulphides, oxides and halides. Finally, there are those minerals that have more than one anion, examples being the carbonates, which contain both carbon and oxygen, phosphates (phosphorus and oxygen), sulphates (sulphur and oxygen), and silicates (silicon and oxygen). Minerals containing more than a single anion can have complicated chemical compositions, and quite elaborate crystal structures. It would be outside the scope of this book to attempt a more detailed classification. Instead,

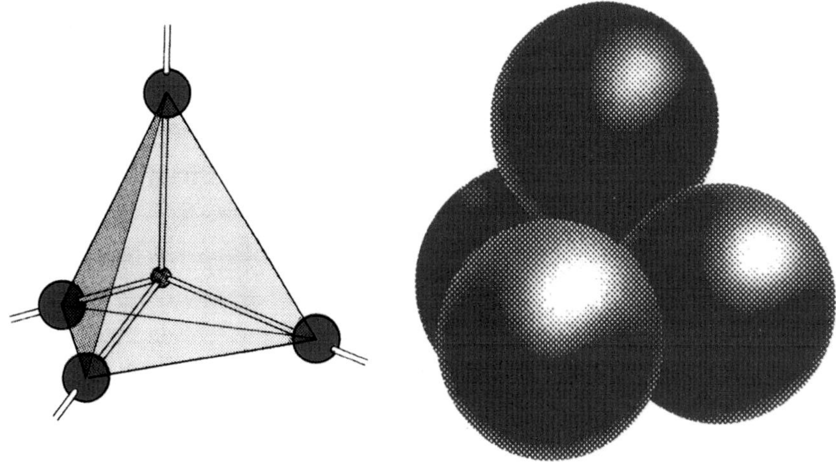

The most common coordination number of *silicon*, when it associates with oxygen, is four, and it lies at the centre of a tetrahedron that has oxygen atoms at its corners. This arrangement leaves each oxygen atom with an extra valence bond which can be part of an adjacent tetrahedron. This is clearly shown in a ball-and-stick model on the left. The space-filling model shown on the right is physically more realistic, but it completely obscures the small silicon ion.

a description of certain minerals will be given, as an illustration of typical features encountered by the mineralogist. There is one fact that should be emphasized at this point: of all the Earth's minerals, well over 90% are silicates.

Because of their purity and crystallinity, minerals can be readily identified. They have distinct shapes that can be recognized with the naked eye, or with a lens of modest power. Other distinguishing qualities are colour, ability to be cleaved, and hardness. If subjected to sufficient stress, all materials can be broken, but only crystals can be cleaved. Transparent minerals can also be identified by their optical characteristics, and a few of them display *optical activity*, a remarkable effect in which a single beam of light entering the crystal is split up into two separate beams of different polarizations. Apart from simple physical appearance, hardness is probably the simplest test that can be used to distinguish between different minerals. The standard *Mohs scale* runs from one to ten and is based on common examples with increasing order of hardness.

Rocks generally consist of aggregates of several minerals, but there are some minerals that tend to occur by themselves. These are known as *vein minerals*, and their occurrence is important. In fact, most of the economically significant minerals are vein formers, and this has facilitated their extraction from the Earth's crust. Let us take a closer look at some of the common types of mineral that fall within each group, starting with the rock formers. These are nearly all silicates, with minor amounts of oxides, although one common type, which forms the sedimentary rock known as *limestone*, is a carbonate. The different rock-forming minerals are characterized by the arrangement of silicon and oxygen atoms. The most stable configuration is an arrangement in which an oxygen atom is located at each of the four corners of a tetrahedron, while a single silicon atom lies at the tetrahedron's centre. The binding between the silicon and oxygen atoms is covalent, but as we shall see in the following chapter it is also partly ionic.

These silicon–oxygen tetrahedra can be arranged in various ways. In the *olivine* group, the minerals of which are magnesium iron silicates, the tetrahedra occur as separate entities. This produces a structure that is equally strong in all directions. This mineral does not cleave, there being no plane of particular weakness. Olivine is a common constituent of basic, as opposed to acidic, igneous rocks such as *basalt*. (The terms acidic and basic as used

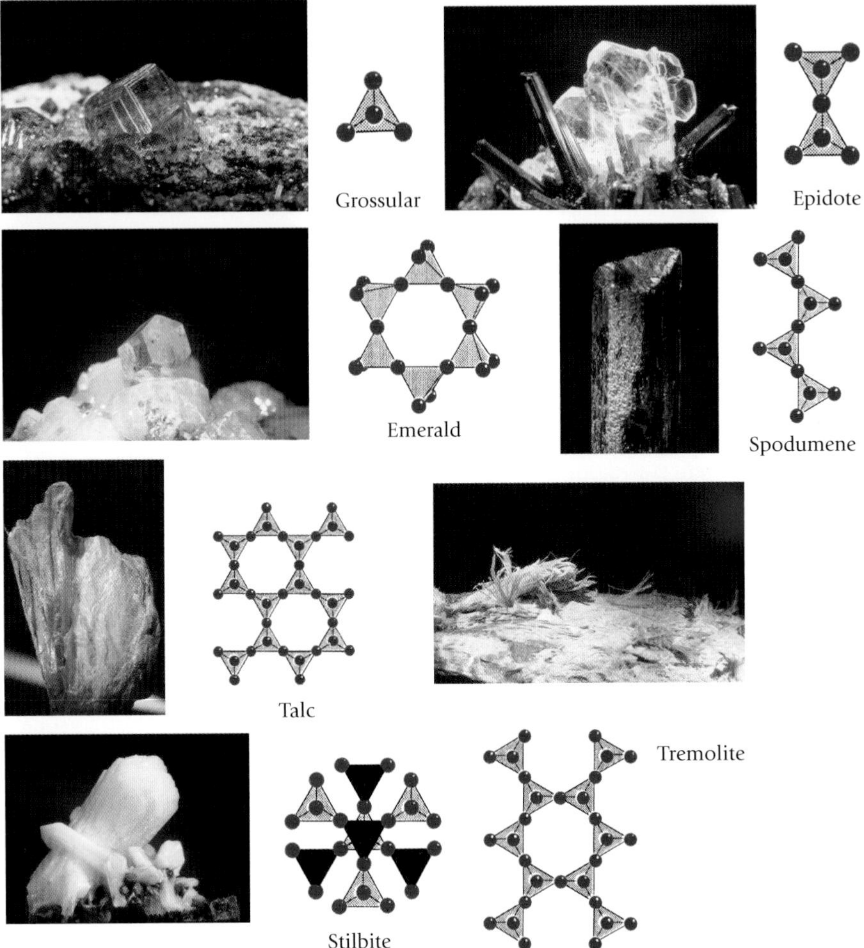

The *silicates* are the most abundant minerals in the Earth's crust. They all comprise silicon–oxygen tetrahedra, but because these can be arranged in various ways the silicates display a great variety of crystal structures and physical properties. These minerals are consequently divided into a number of subclasses, which are here illustrated with typical examples and their corresponding tetrahedral dispositions. In the usually dense and hard nesosilicates, represented by grossular, a type of garnet, the silicate groups are isolated. (The specimen was found at the Jeffrey Mine, Asbestos, Canada.) Pairs of tetrahedra sharing an oxygen atom are the hallmark of the sorosilicates, a typical member of this subclass is the epidote (from Untersulzbachtal, Austria) shown here. Cyclosilicates, as their name implies, contain rings of tetrahedra. A good example is the gem emerald (Bogotá, Colombia). The rings, which usually have between three and six tetrahedra, stack on each other, producing elongated crystals that are moderately dense and very hard. The inosilicates have chains of tetrahedra. The chains can be single, each tetrahedron is linked to the next by a common oxygen atom, or double, with cross-linking through alternate tetrahedra. These two types are represented by the pyroxenes and the amphiboles, respectively, and the examples shown here are spodumene (Minas Gerais, Brazil) and tremolite (Kragerø, Norway). The characteristic feature of the phyllosilicates is tetrahedra that share three oxygen atoms, forming the sheets that give these minerals their flaky quality. The talc (New Hampshire, USA) shown here is a good example. Finally there are the tektosilicates, in which the tetrahedra form three-dimensional arrangements because all their oxygen atoms are involved in intertetrahedral links. This produces rather open crystal structures, and these silicates have relatively low densities and only moderate hardness. The tekto subclass includes some of the most important rock-forming minerals. The illustration shows stilbite (Berufjord, Iceland), a typical example.

The Mohs scale of mineral hardness

Friedrich Mohs (1773–1839) introduced a simple definition of hardness, according to which one mineral is said to be harder than another if the former scratches the latter. The Mohs scale is based on a series of common minerals, arranged in order of increasing hardness. It spans the observed range of that property, and runs from 1 to 10. On this scale, a polymer like polyethylene has a hardness of about 1, a finger nail 2.5, a copper coin 5, window glass 5.5, and the blade of a pen-knife 6.5. Tool steel has a hardness of about 7. and thus readily cuts glass. Most gemstones have a hardness exceeding that of sand and silica, and are therefore not susceptible to scratching by such materials when the latter are present in airborne dust. The complete Mohs scale is as follows:

Mineral	Composition	Hardness
talc	$Mg_3Si_4O_{10}(OH)_2$	1
gypsum	$CaSO_4.2H_2O$	2
calcite	$CaCO_3$	3
fluorite	CaF_2	4
apatite	$Ca_5(PO_4)_3F$	5
orthoclase	$KAlSi_3O_8$	6
quartz	SiO_2	7
topaz	$Al_2SiO_4F_2$	8
corundum	Al_2O_3	9
diamond	C	10

by geologists are discussed later in this chapter.) Its colour ranges from light green through various shades of brown to almost black, if the iron content is high. It has a moderate density, 3.5 g/cm^3, and its Mohs hardness number is 6.5. In the *pyroxene* group the silicon–oxygen tetrahedra are linked together in a chain arrangement, and the chains are packed parallel to one another. Minerals belonging to this family can be cleaved on planes that do not intersect the chains. A typical example of a pyroxene is augite, which contains calcium, and its crystal structure is monoclinic. The pyroxenes occur in both igneous and metamorphic rocks. They are less dense and less hard than the olivines. In the *amphibole* group, the silicon–oxygen tetrahedra form a double-chain arrangement, which again gives crystals that are readily cleaved on planes that do not intersect the chain axes. The density and hardness of the amphibole minerals are comparable with, but slightly lower than, those of the pyroxenes. A particularly common amphibole, which occurs in both igneous and metamorphic rocks, is hornblende. Crystals of this mineral are monoclinic, and they have a black and almost metallic appearance. At a still lower density there is a group that is particularly rich in different varieties of mineral: the *feldspars*. Their silicon–oxygen tetrahedra are arranged in a three-dimensional network which cleaves readily on two planes that are roughly at right angles to each other. The hardness is comparable to that of minerals in the olivine and pyroxene groups. The feldspars are conveniently divided into two classes. The *plagioclase feldspars* are triclinic, and they contain sodium or calcium, or both sodium and calcium, while the *alkali feldspars* are similar, but have potassium instead of calcium. They have various crystal structures. Alkali feldspars frequently occur in such acidic igneous rocks as *granite*. A quite common example is orthoclase, which has a monoclinic crystal structure. Plagioclase feldspars occur in intermediate and basic igneous rocks such as diorite, andesite, gabbro, dolerite, and *basalt*. Good examples of the plagioclase class are: albite,

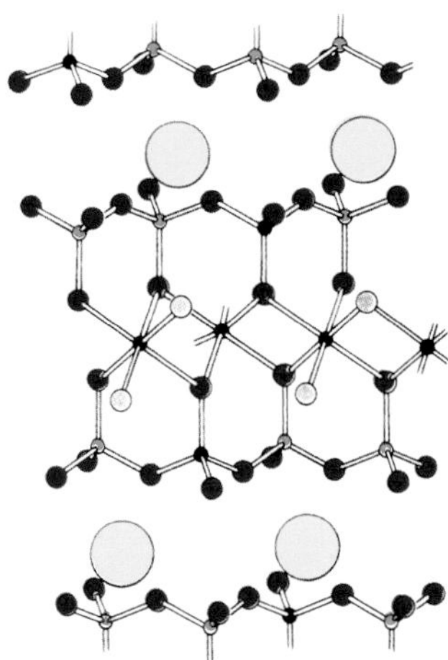

Silicate mineral crystals are not simply arrangements of silicon–oxygen tetrahedra. Atoms of other elements are usually present, and they determine how the tetrahedra are positioned in the crystal structure. A good example is muscovite mica ($KAl_3Si_5O_{10}(OH)_2$). Each of its sheets comprises two layers of tetrahedra, with an octahedral layer sandwiched between. One out of every four silicon atoms (coloured blue) is replaced by an aluminium atom (black), in the tetrahedral layers, while all the octahedra are based on combinations of aluminium with oxygen (red) and hydroxyl ions (pink). The sheets are linked together with potassium ions (yellow). The structure has several features in common with that of kaolinite, which is illustrated in Chapter 8. Muscovite commonly occurs in the rocks known as schists, gneisses, and granites. Very large crystals have been found in Custer, South Dakota, USA.

which contains sodium, anorthite, which contains calcium, and labradorite, which contains both of these elements.

The least dense mineral groups based on silicon–oxygen tetrahedra are mica and quartz. Although their densities are comparable, mica is very soft while quartz is quite hard. This difference can again be traced to the arrangement of the tetrahedra. In *mica*, they are arranged in sheets. The bonding within the sheets is covalent, while adjacent sheets are held together by weak secondary bonds. Because of this, the sheets can readily be cleaved from one another, and it is this which gives mica its familiar flaky appearance. In *quartz* the tetrahedra form a tightly packed three-dimensional structure in which each of the four corners of every tetrahedron is linked to adjacent tetrahedra. This produces hard crystals that cannot be cleaved. Mica is common in acidic and intermediate igneous rocks such as granite, trachyte and syenite, while the appearance of quartz in igneous rocks is restricted to the most acidic varieties such as rhyolite. Mica has for many years done Trojan service in applications requiring a material which can withstand heat and provide electrical insulation. The hardness of quartz makes it suitable for use in bearings, like those found in watches, and accurately cut crystals are used as frequency standards in radio transmitters and receivers. The rich variety of different quartz gems stems from the many different impurities commonly associated with this mineral. Among the better known coloured forms of quartz are jasper (red), flint (black), amethyst (mauve), bloodstone (dark green with red spots), and agate (alternating white and coloured bands, the most common colours being red, pink, brown and yellow). Although there are over 2000 known minerals, a mere handful of them make up the bulk of the Earth's crust. Feldspars alone account for 59.5% of the total, by weight, while the amphiboles and pyroxenes together contribute 16.8%. Quartz represents about 12% of the whole.

Our brief review of silicate minerals would be incomplete without mention of a few other igneous and metamorphic examples. Topaz and tourmaline are both associated with acidic igneous rocks such as granite, and they are both quite dense and rather hard. *Topaz* is a mixed aluminium silicate and fluoride, and its orthorhombic crystals may be either colourless or tinted with blue, yellow, orange or light red. High quality crystals of topaz are used as gems. *Tourmaline* is basically a silicate of one of several metals, but it also contains boron. Its trigonal crystals are based on a ring structure consisting of six silicon–oxygen tetrahedra with the most open spaces in between containing triangular units of one boron and three oxygen atoms. The arrangement of subunits in this mineral gives its crystals several interesting physical properties. The absorption of light in tourmaline is markedly dependent on direction of incidence. Its crystals are *pyro-electric*; a voltage difference appears between opposite faces of a tourmaline crystal when it is heated. Like quartz, tourmaline is also *piezo-electric* in that a voltage can be induced by applied pressure. Tourmaline crystals have been found in more different colours than is the case for any other mineral. Indeed, virtually every imaginable colour has been observed. *Garnet* is a metamorphic mineral, often found in conjunction with diamond, and its cubic crystals are mixtures of either calcium or aluminium silicates with another metallic silicate. They are very dense, in excess of 4 g/cm^3, and their relatively high hardness makes them suitable as bearings. Because garnet crystals are often quite large, frequently being several centimetres across, and of

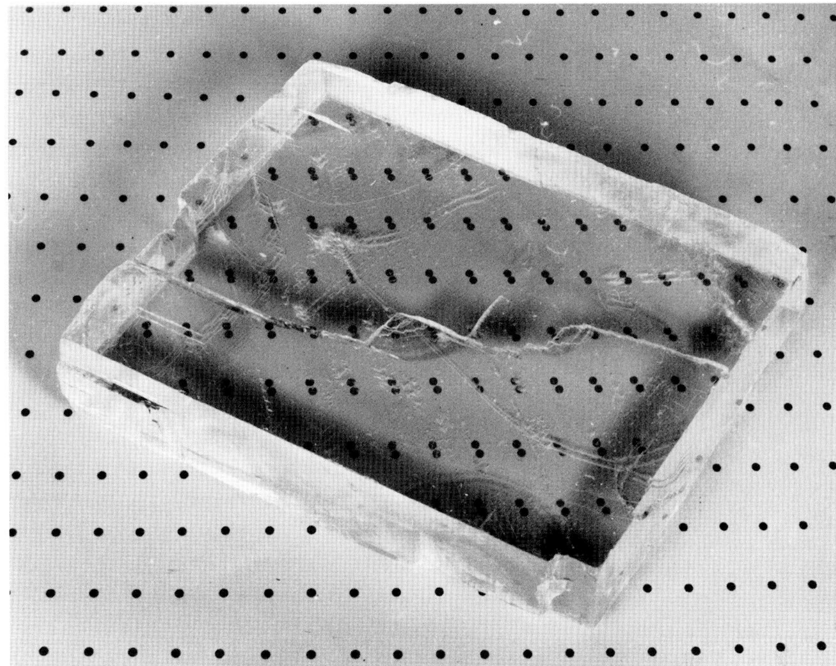

In crystals belonging to systems other than the cubic, an incident light ray gives rise to two refracted rays rather than one. For most such crystals the speeds of the two refracted rays are almost equal, but there are some materials that form two well separated refracted rays because the velocity of light in them depends markedly on crystallographic direction. The classic example of *double refraction*, as this effect is called, is provided by calcite ($CaCO_3$). Its ability to split one ray into two is here demonstrated by the doubling of the spots in that part of the underlying pattern covered by the crystal. The effect can also be used to determine which surface irregularities lie on the bottom of the crystal and which on the top.

beautiful colouring, particularly red, green and black, they are popular as semiprecious stones.

Carbonates are a prominent exception to the rule that rock-forming minerals are usually silicates. The most familiar of these are *calcite*, which is a calcium carbonate, and dolomite, a calcium magnesium carbonate. They are both slightly more dense than quartz, but their hardness is much lower. Both of these minerals have a rhombohedral crystal structure, and calcite is a *double refractor*; objects observed through a calcite crystal appear double. The *stalactites* and *stalagmites* found in some caves are basically calcite.

Vein minerals, in contrast to the rock formers, are usually non-silicates. They are, nevertheless, frequently found in association with silicate minerals. Most vein minerals, in common with the rock formers, are compounds, but a few of them are fairly pure elements. These are the metals gold, silver, platinum and copper, and the non-metals sulphur and carbon. The relative rarity of veins of gold, silver and platinum explains why these are referred to as *precious metals*, and much adventure and romance was associated with the hunt for the mother lode, a particularly rich vein. Much has been written about the occurrence of gold nuggets in certain streams in California, the discovery of which, in 1849, led to the famous *Gold Rush*. Our rationality, concerning this particular element, hardly bears close scrutiny. Almost all the gold ever mined has been re-deposited below the Earth's surface, in storage vaults. Quite concentrated seams of elemental copper are not uncommon, particularly good examples being found around Lake Superior in North America. Although veins of the other precious metals have been found, the prospector has usually had to be content with the relics of eroded seams at the Earth's surface. Native carbon is found in two different allomorphic forms: graphite and diamond. These distinct crystal structures of the same element could not be more different in their appearance and

Magnetite, Fe_3O_4, is one of the most widespread oxide minerals, and it is a major source of industrial iron. It usually occurs as a coarse granular mass of octahedral crystals, major deposits being located in Utah in the USA, Kiruna in Sweden, and the Ural Mountains of the former USSR. Its name may derive from the Magnesia area of Greece, and the name is of course related to the word magnetism. The magnetic property of the mineral is readily revealed by the simple demonstration illustrated here. Some specimens actually display polarity, and are then known by the name *lodestone*.

physical properties. *Graphite* is very soft, opaque with a jet-black colouring, and it is able to conduct electricity. *Diamond*, on the other hand, is the hardest naturally occurring material. It is transparent, and it is an electrical insulator. Graphite finds one of its most common applications in the humble pencil, while diamonds are amongst the most expensive of jewels.

Several of the vein-forming oxide minerals are among the most economically significant substances in the Earth's crust. There are, for example, three common ores of iron that are all oxides. These are magnetite, haematite and limonite. Bauxite, the most important aluminium ore, is also an oxide, in a hydrated state. Finally, there is cassiterite, which is the chief ore of tin. It too is an oxide. The black cubic crystals of *magnetite* are magnetic, and this mineral was actually the first magnetic substance to be discovered, it being known earlier by the name *lodestone*. The trigonal crystals of *haematite* usually have a reddish colour, rather reminiscent of blood. Indeed the red colour of blood and that of haematite both stem from the same valence state of the iron atom and the word haematite is based on the Greek *haima* for blood. *Limonite* is a less pure form of iron oxide, and it has neither a well-defined crystal structure nor definite colour. The mineral usually has a dull red or brown appearance, and it is in fact the most common colouring agent found in soil.

There are several important vein minerals that are sulphides. The most prominent are: pyrite, which is an iron sulphide; galena, a lead sulphide; chalcopyrite, a copper iron sulphide; and blende, also known as sphalerite, a zinc sulphide. The cubic crystals of *pyrite* are fairly dense and reasonably hard. Pyrite is a less important source of iron than the three oxides named earlier. It is, however, a valuable source of sulphur, and it is used in the production of sulphuric acid. The tetragonal crystals of *chalcopyrite* are less dense than pyrite and markedly softer. This enables it to be distinguished from iron sulphide, because the form and appearance of the two minerals are otherwise very similar. They both have a golden colouring, and both are known by the nickname *fool's gold*. The dark grey cubic crystals of *galena* are rather dense and quite soft. This mineral, which is frequently found associated with calcite and quartz, is the main industrial source of lead. *Blende* is an important source of zinc, and it is often associated with calcite, quartz and fluorite (which also goes by the name *fluorspar*). Its cubic crystals have a dark brown colour, with medium density and hardness.

Among other significant vein minerals are certain phosphates, sulphates, carbonates and halides. The only really common phosphate is that of calcium, which is known by the name apatite. Its economic importance lies mostly in its use as a fertilizer. The hexagonal crystals of *apatite* have medium density and hardness, and they occur in several different colours, green being the most usual. The two most prominent sulphates are gypsum, which is a hydrated calcium sulphate, and barytes, a sulphate of barium. The monoclinic crystals of *gypsum* are light, transparent, and rather soft. This white mineral is used in *Plaster of Paris*, and it is also a vital constituent of modern *cement*. The orthorhombic crystals of *barytes* occur in a variety of pastel shades, which are sometimes so faint that the crystals have a transparent or milky appearance. They occur in cleaved flakes that often have surface markings somewhat reminiscent of lace-work. Of the vein-forming hydrated carbonates, mention should be made of the green-coloured *malachite* and the blue-coloured *azurite*, both of which contain copper, and the light brown *siderite*, which is an iron carbonate. Finally, there are the halides, the most

Lapis lazuli, a rare mineral valued for its beautiful blue colour and its mysterious twinkle, takes its name from the Latin word for stone (lapis) and the Arabian word for blue (azul). It was among the first stones used in jewellery, and was the object of a busy trade in the city of Ur, on the Euphrates river, almost seven thousand years ago. The sole source in those days were deposits in Sar-e-Sang, a remote mountain valley in the Badakhshan district of Afghanistan, but today it is also mined in Siberia and Chile. The mineral is actually a composite, the main constituent being *lazurite* – $(Na,Ca)_8(AlSiO_4)_6(SO_4,S,Cl)_2$ – the sulphur content giving the blue colour. The attractive twinkle stems from the presence of *pyrite*, FeS_2, and the other components are calcite, haüynite, sodalite and nosean. It is translucent and the hardness is about 5.5. Pulverized lapis lazuli was once the only available ultramarine pigment, and it was thus of great importance in the art world; its permanence, compared with other pigments, is the reason for the prominence of the blues in Old Master paintings. The specimen shown here is unusually large; its length is about 10 cm.

familiar of which are *fluorite*, which is calcium fluoride, and the common *rock salt*, also known by the name *halite*, which is of course sodium chloride. Specimens of fluorite have been found with almost every conceivable colour, although the shades are usually rather light. Fluorite, like gypsum, also known by the name *alabaster*, has been used down through the centuries in the manufacture of ornaments.

Having briefly considered some of the more significant minerals, it is appropriate that we turn to the way in which they are formed and look in turn at the three different types of rock mentioned earlier. Just which minerals will be formed when a magma cools to produce igneous rock depends on the original composition of the liquid. Several important physical factors are determined by the relative amounts of the different elements present, a particularly strong influence deriving from the silicon–oxygen content. The strong covalent bonding between the silicon and oxygen atoms permits the silicon–oxygen tetrahedra to exist essentially unaltered even in the liquid state. We can therefore refer to the silica content of a magma, and this usually lies between 40% and 70% of the total mass. One measure of acidity is related to the oxygen content. Low-silica (and hence low-oxygen) magmas are therefore called basic, while high-silica magmas are acidic. Acidic magmas have crystallization temperatures that lie as much as 200 °C below those of the basic variety. Viscosity increases as temperature decreases. The viscosity of a basic magma, at any temperature, is greater by a factor of several thousand than that of the acidic type. In addition to the acidity, the important property of a magma is its content of different elements, for it is these that determine which minerals can be formed during cooling. These minerals have different melting points, and their densities also show considerable variation. Each mineral is precipitated out of solution when the magma temperature falls below the corresponding freezing point, and it sinks to the bottom of the molten mass at a rate that is determined by the difference between the mineral density and the average magma density, as well as by the viscosity of the magma. The texture of the final rock depends not only on the mineral content, but also on the size of the various crystals, or grains. This in turn is determined by the rate at which the magma cools. Slow cooling permits the growth of some grains at the expense of others, and a coarse structure evolves. Rapid cooling, on the other hand, produces a texture consisting of many small grains. Indeed, if the magma is cooled sufficiently rapidly, no crystallization occurs and a glass is formed. Just how

Every mineral has its own specific *solidification temperature*. The different components in a cooling *magma* therefore appear at different times, and being generally more dense than the liquid they tend to sink upon crystallization. The composition of the magma is thus continuously changing, and this *fractionation*, as it is called, accounts for much of the heterogeneity of rock structure. The sequential crystallization is referred to as a *reaction series* and the example shown here is typical of a basic magma. The approximate crystallization temperatures are indicated for olivine (the specimen shown having been found in St John Island, Egypt); augite (a pyroxene – Mt Kenya, Kenya); hornblende (Arendal, Norway); labradorite (a plagioclase feldspar – Labrador, Canada); microcline (an orthoclase feldspar – Hilbersdorf, Germany); muscovite (an acidic mica – Custer, South Dakota, USA); and quartz (Kongsberg, Norway). The plagioclase feldspar family actually displays a wide range of crystallization temperatures, calcium-rich members such as anorthite appearing at around 1000 °C whereas those with a high concentration of sodium, such as albite, do not crystallize until the temperature of the magma has fallen to around 800 °C. The crystals shown all have lengths of a few centimetres.

1100 °C

Olivine

1000 °C

Augite

Hornblende

900 °C

Labradorite

800 °C

Microcline

700 °C

Muscovite

600 °C

Quartz

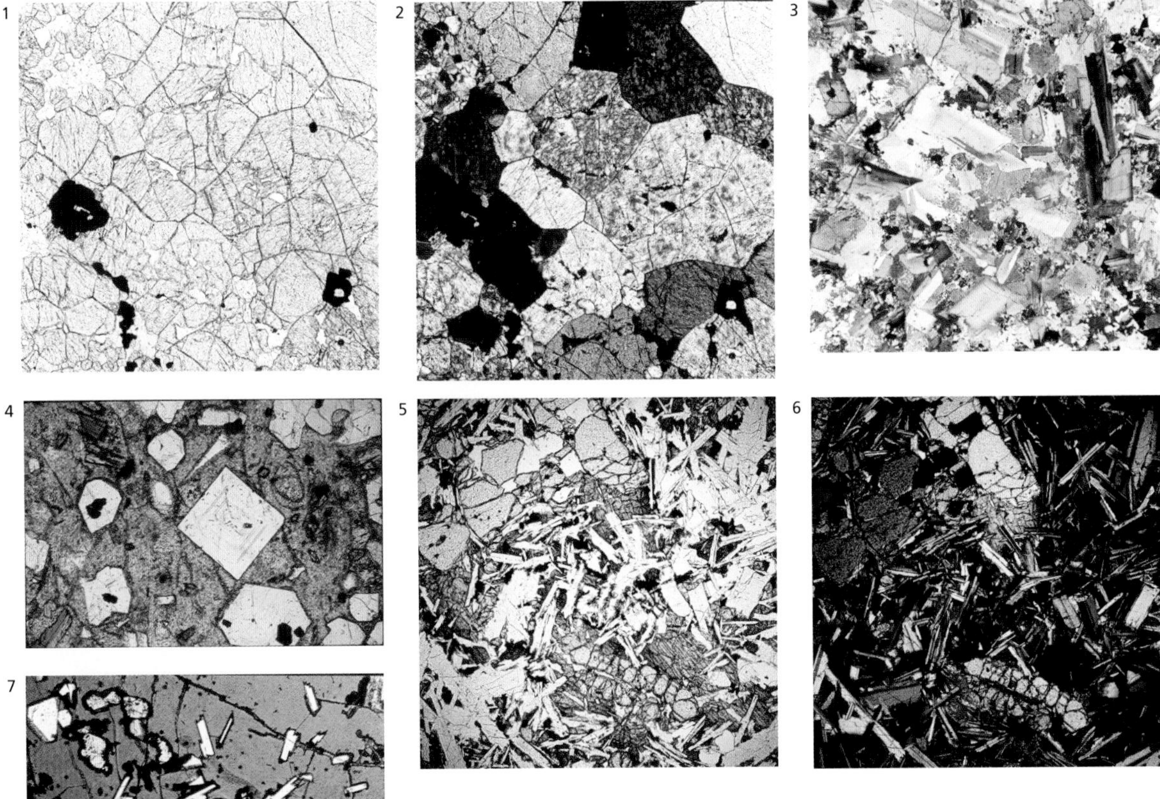

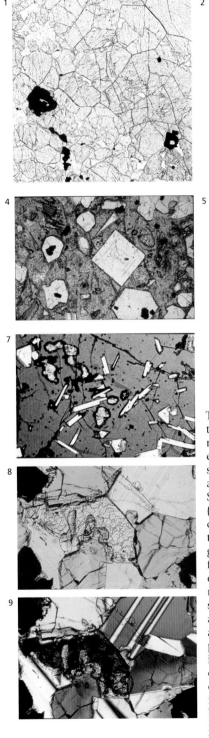

The variety of *minerals* in a single type of rock is revealed by examination of a thin-section in an optical microscope. The different minerals can often be recognized by their characteristic crystal shapes, and the task is made even easier if the observations are made through *crossed polarizers*. The examples shown here illustrate some of the igneous types. Peridotite is strongly basic, and pictures (1) and (2) show a thin specimen from the Isle of Skye, Scotland, viewed both in direct light (1) and through crossed polarizers (2). The coloured regions in the latter picture are all olivine, by far the major constituent of this rock. Pictures (3) and (4) represent the other extreme; they are both acidic, and both specimens are from northwest Iceland. In the granophyre (3) the various green shades which dominate are quartz and feldspars, the orange parts are pyroxene and the small black patches are iron oxides. Granophyre is somewhat less acidic than granite. The regularly-shaped crystals in the rhyolite (4) are plagioclase feldspar and the smaller pieces are quartz. The green fragments are hornblende, while an amorphous acidic glass makes up the background bulk, which is referred to as the matrix. The remaining rocks are all basalt, which is not so basic as peridotite. Pictures (5) and (6) show a specimen from Surtsey, the volcanic island southwest of Iceland that was formed during the period 1963–7. It contains olivine, which appears brown in normal light but black between crossed polarizers, augite, a pyroxene which produces the pretty colours, and narrow crystals of plagioclase feldspar, some of which are seen to be crossed by twin boundaries (the thin black lines). Picture (7) shows a specimen that is remarkable for its large amount of non-crystalline material, (the brown background); the crystals are plagioclase (long and thin), olivine (roughly triangular), and pyroxene (irregular). Pictures (8) and (9) show a specimen of *lunar basalt* recovered during the Apollo 17 mission. The central grain, which appears almost black between crossed polarizers, is crystobalite, a form of pure silica. The bright colours are pyroxene, while the striped regions are plagioclase.

Amber, also known as *succinite*, is the fossilized resin of a now-extinct species of pine tree, and is thus actually organic in origin. This is the reason for its surprising softness and low density. The trees in question grew in Northern Europe during the Tertiary period and the resin became incorporated in the so-called blue Earth layer during the Oligocene epoch. It was subsequently dispersed by glacial action during the Ice Ages, thereby becoming concentrated in waterways. Particularly rich deposits are found in the Eastern Baltic, in the area once known as East Prussia, and specimens are often found washed up on the shore, following stormy weather. Minor sources have been found in Transylvania, in the Mediterranean, in North America and in such Western European rivers as the Elbe and the Oder, but these are regarded as curiosities. The translucent (and often transparent) yellowish-brown solid contains succinic acid, and the occasional body of a trapped insect. The specimen shown here is unusually large; its measures 13 × 11 × 5 cm and weighs 420 g.

rapidly a magma must be cooled in order to produce this non-crystalline form depends on the actual composition of the magma. For some substances such as obsidian the glassy form is the natural state, and this is produced even if the cooling rate is very low.

A typical sequence of events in a cooling magma would run as follows. When the temperature has fallen to slightly below 1100 °C, olivine and anorthite, a calcium plagioclase feldspar, are precipitated and slowly fall to the bottom of the melt. When the temperature falls below 1000 °C, augite and labradorite, a sodium–calcium plagioclase feldspar, come out of solution and settle above the olivine and anorthite crystals. Then, when the temperature is slightly above 900 °C, hornblende crystals are formed, and these are followed, slightly above 800 °C, by biotite mica and albite, a sodium plagioclase feldspar. When the temperature has fallen to around 700 °C, orthoclase feldspar appears, and this is followed by muscovite mica, around 650 °C, and quartz, around 550 °C. It should be remembered that, with the precipitation of each mineral, the composition of the remaining magma will be changed. Because the minerals occurring early in this precipitation sequence are low in silica content, the remaining magma becomes increasingly enriched in silica. The final mineral to come out of solution, namely quartz, has a higher silica content that any of its predecessors. Just which proportions of the minerals in this sequence, known as a *Bowen's reaction series*, are present in the final rock will be determined both by the acidity and the original elemental content.

Igneous rocks are characterized by the acidity and coarseness of the final texture. Acidic rocks contain relatively large amounts of quartz and orthoclase feldspar, and smaller quantities of the plagioclase feldspars, mica, and hornblende. The coarse variety is known as *granite*, the medium-grained type

Meteors and Meteorites

Meteors are solid bodies that impinge upon the Earth from outer space, and are thus of great age. Roughly a million of them reach our planet every day. They become incandescent upon entering the atmosphere, because of friction with air molecules, and are then visible as *shooting stars*, smaller examples simply burning up in the process. Those large enough to reach the Earth's surface are known as *meteorites*, and they are usually mainly composed either of stone or of iron. It is likely that these different types are merely the remnant fragments of larger composite bodies having iron cores and stone exteriors, just as in the case of the conjectured former planet *Theia* discussed in the Prologue. The twenty-ton iron meteorite shown here in the courtyard of Copenhagen's geological museum with its discoverer, Vagn Buchwald, is the fifth largest iron meteorite ever found. It was the second-largest member of the multi-meteor shower that landed in Cape York, Northern Greenland, in prehistoric time, and it was located in 1963 after Buchwald's extensive search. He named it *Agpalilik*, which in Eskimo language means *auk mountain*, the place where it was found. It is the largest meteorite ever to be sectioned. One of the cut surfaces is shown in the right picture, and it reveals the presence of troilite (FeS) inclusions. The meteorite is composed of 90% Fe, 8% Ni, 1.3% S, 0.5% Co plus traces of other elements such as carbon and phosphorus, and the main body consists of an iron-nickel single crystal. Such large-scale perfection is hardly surprising, given that the original meteor was annealed over millions of years. If the current melting trend of the Greenland ice-cap continues, other meteorites from the Cape York shower may be revealed.

as *granophyre*, while *rhyolite* consists of fine grains. Intermediate rocks contain smaller amounts of quartz and orthoclase feldspar, quite large proportions of plagioclase feldspars, and smaller quantities of mica, hornblende, and augite. The coarse-grained varieties are *diorite* and *syenite*, while those of medium grain go by the names microdiorite and microsyenite. The fine-grained types of intermediate igneous rocks are *andesite* and *trachyte*. Basic igneous rocks are up to 50% plagioclase feldspar, while the remaining half consists of roughly equal proportions of hornblende, augite and olivine. The coarse, medium and fine varieties go by the name *gabbro*, *dolerite*, and *basalt*, respectively. Finally there are the ultrabasic igneous rocks which contain up to 90% of *olivine*, with the balance being made up of *augite* and plagioclase feldspar. The viscosity of the magma that produces such rocks is usually so low that the texture cannot avoid being rather coarse-grained, and the resulting rock is known as *peridotite*. We now see why the recovery of substances from rocks, as opposed to veins, is usually not an economic proposition; rocks contain too many different elements.

As mentioned earlier, *sedimentary rock* is formed by the deposition either of material released during the erosion of pre-existing rocks or material from plant and animal remains, or both, and it could be looked upon as a sort of combined refuse dump and graveyard. The surface of the Earth is always in a state of change and a number of different sedimentary processes occur simultaneously. Meteorological, plant and animal influences are perpetually at work breaking down the exposed surfaces and liberating minute particles, which the surface moisture washes away and deposits elsewhere. Water is an anomalous substance in that it expands upon freezing. When the water that has seeped down between the microscopic crevices freezes, therefore, it tends to widen these cracks and fracture the surface into small particles. In regions where the latter are not rapidly removed, wind hurls the resulting dust against the surface of the rock, and this leads to further erosion. In some latitudes the daily swings of temperature can be so large that a considerable thermal fatigue results, and this too can have a devastating effect on the surface layers. The innocent-looking roots of plants also play a role in breaking down rocks as they burrow downwards in their search for water. Yet another factor is the alimentary activity of microscopic organisms. Ingestion, digestion and elimination by these creatures cause a finer division and homogenization of various substances at the Earth's surface. Finally, there is the deposition of vegetable and animal remains. Sedimentary rocks are conveniently divided into three groups. The products of surface fragmentation are referred to as *clastic*, or fragmental, rocks. Those that are formed from the remains of plants and animals are known as *organic rocks*. Finally, there is the *chemical group*, rocks of which are formed by precipitation from sea water of particles that have been so finely divided that they are essentially in solution.

The clastic group is subdivided into three classes according to the size of particle. The coarsest variety goes by the name *rudaceous* and ranges from pebbles, with particle diameters of a few millimetres, through cobbles, with diameters of a few centimetres, to boulders. If the particles have sharp edges, the rock is known as *breccia*, while *conglomerate* is composed of smooth stones. The *arenaceous* group has particle sizes in the range 0.05–2.0 mm, and rocks in this group are commonly known as *sandstone*. The most common sand former is quartz, but some sands contain mica and feldspar. Finally, for smaller particle sizes there is the *argillaceous* group, which includes clay, shale, marl and mudstone. It should be noted that names such as these have a rather general connotation in that they usually contain a mixture of several different minerals. Marl, for example, is predominantly calcium carbonate, but not exclusively so. Some clastic rocks have a mixture of particles of different sizes. A good example of this is a greywacke sandstone, which is a composite in which arenaceous sand particles are cemented together by an argillaceous matrix.

Originating from material that is in solution, members of the chemical group are the most homogeneous of the sedimentary rocks. The precipitation is caused by a change in the chemical composition of the solution. This can be brought about by a simple increase in concentration, so that the solution becomes supersaturated. A common cause is the evaporation of water. This process is what produced the Dead Sea and the Great Salt

Lake in Utah, USA. Another example of change of composition occurs at the mouth of a river, where fresh water meets sea water. The chemical group is subdivided according to the type of anion present. By far the most common carbonate is the calcium variety which precipitates in the sea water near some coastlines to produce the familiar *limestone* or *chalk*. Also quite common is the precipitated silica rock, known by the name *flint*, that is frequently found in conjunction with chalk. Other examples in this group are *laterite*, an iron hydroxide; *bauxite*, a hydrated aluminium oxide; and *rock salt*, which is of course sodium chloride.

Two very common examples serve to illustrate the organic group of sedimentary rocks. The decay of plant material can be prevented if it does not have access to oxygen. This will be the case if it becomes buried relatively rapidly, and decay is effectively inhibited if bacteria are unable to thrive under the prevailing conditions. Such trapped plant remains first form *peat*, and this is slowly transformed by pressure and heat to *lignite*, which is also known as *brown coal*. This, in turn, gradually transforms to *bituminous coal*, often called *black coal* or simply coal. A further stage of development is possible, and it produces *anthracite*, which is almost pure carbon. The classic example of an organic sedimentary rock that derives from animal life is the *limestone* deposited on the sea bed. It consists of the skeletons of myriad sea creatures. It is interesting to note that two common features of the coastline, the chalk cliff and the coral reef, are composed of the same chemical compound even though their origins are quite distinct.

The occurrence of fossilized plants in coal seams shows the accepted explanation of the formation of coal to be substantially correct. Because they are not in the solid state, oil and natural gas are unable to provide such tell-tale evidence, and the origin of these important fossil fuels is still a matter of speculation. These substances are usually grouped together under the names *petroleum* or *mineral oil*. Such terms actually cover a mixture of hydrocarbons and other organic compounds, and the composition of petroleum varies considerably from one source to another. Thus petroleum extracted in North America is relatively rich in paraffins, while the Asian variety contains a higher proportion of cyclic hydrocarbons such as benzene, which will be discussed in Chapter 12. The word petroleum itself comes from the Greek word *petra* meaning rock, and the Latin word *oleum*, meaning oil. This 'rock oil' appears to have been familiar to the peoples of Mesopotamia and the surrounding areas, as early as 3000 BC. The primitive technology of those days would not have permitted the boring of wells, and the oil in use at that time was found oozing out of the surfaces of certain types of rock. Such seepages probably varied in viscosity and composition from the water-like consistency of *gasoline* to the almost solid bitumens, *tar* and *asphalt*, and *pitch*. These substances were originally used as waterproofing for boats and household items, as lubricants, in illumination, and the heavier varieties were mixed with sand to produce a surface covering for roads. Petroleum products were also valued in medical practice, both for their laxative properties and as liniments. Drilling for oil is a relatively recent development, the first recorded well being that used by E. L. Drake in Pennsylvania, USA, in 1859. This modest effort, with a depth of just over 20 m, foreshadowed an industry which really started to flourish with the advent of the internal combustion engine and the arrival of the car.

Diamonds

The main source of *diamonds* is the ultrabasic igneous rock *kimberlite*, in which they are found as scattered crystals. *Kimberlite pipes* are the remains of *volcanic plugs* that have been forced up to the Earth's surface, and examples have been located with diameters up to a kilometre. The pipe shown here is the famous *Big Hole*, the result of opencast mining at Kimberley in South Africa. The largest gem quality diamond ever found was the 3106-carat (621-gram) crystal named after Thomas Cullinan (1862–1936). In the group photograph, taken shortly after its discovery in 1905, the priceless find is held by the general manager of the Premier mine, M. W. Hardy, who is flanked by surface manager Frederick Wells, who actually found the diamond, and Cullinan himself, the mine's owner, on the left. The *Cullinan diamond* was presented to King Edward VII for his 66th birthday, and it was cut into 9 major gems and 96 smaller brilliants by Joseph Asscher. The largest of the gems, the pear-shaped 530.2-carat Cullinan I (shown here), also known as the Star of Africa, is the world's largest cut diamond; it is now in the head of the royal scepter of the British crown jewels. The second-largest is the cushion-shaped 317.4-carat Cullinan II; it is set in the centre-front of the British imperial state crown, fashioned for the coronation of Queen Elizabeth II in 1953. The final picture shows a diamond in its parent rock. It is of the yellow variety, and it displays the characteristic octahedral shape. The finest yellow specimen yet discovered was probably the 51-gram crystal found in 1964, and named after Ernest Oppenheimer (1880–1957), who played a major role in organizing the diamond mining industry.

Although there is no universally accepted explanation of the origin of *petroleum*, a consensus does appear to be evolving. Particularly significant among the various pieces of evidence is the fact that petroleum is invariably associated with sedimentary rocks, especially those that have been formed in a marine environment. There are, on the other hand, no examples of petroleum occurring in conjunction with igneous or metamorphic rocks. Irrespective of the source, petroleum frequently contains compounds, known as *porphyrins*, that are formed from the colouring matter of living organisms. The latter might be the green chlorophyll of plants or the red haem of blood. Finally there is the evidence provided by the abundance ratio of the two carbon isotopes, ^{13}C and ^{12}C. This ratio can be used as a sort of fingerprint. It is found that the carbon isotope ratio of petroleum is strikingly similar to that of the lipid, or fat, molecules commonly found in living organisms. Taken collectively, this evidence suggests that petroleum was derived from organic materials associated with ancient sea life. There was, in fact, an abundance of primitive aquatic life, plankton, algae and the like, dating from before the Paleozoic era. Such primitive organisms could have been the source of petroleum that has been found associated with sedimentary rocks dating from the early Paleozoic era. Petroleum, which is presumably still being formed, is not actually found in the strata in which the marine organisms were originally deposited. Once it had been transformed to its present state, it migrated under pressure and came to rest in what are termed *reservoir rocks*. These are porous seams which lie beneath other layers of rock that are impermeable to oil, and even to natural gas. A typical example of a petroleum-trapping geological structure is the *salt dome*. Such a dome forms because the salt is less dense than the surrounding material, so it rises, causing the layers lying above it to bulge upwards. Salt is impermeable to petroleum, and hence its ability to act as a trap. It is for this reason that the oil prospector takes the appearance of salt, in his borings, as a promising sign.

The outer surface of our planet enjoys a fairly constant environment. The atmosphere protects it from the large variations of temperature that are experienced by the surface of the Moon, for example. The atmospheric pressure too stays roughly constant. Below the surface, however, high temperatures and pressures act upon rocks both of the igneous and sedimentary types and can bring about changes in their structure. This produces *metamorphic rocks*. Another factor in these changes is the viscous flow of the subterranean rock, which permits adjacent layers of atoms to slide past one another, enabling them to find arrangements of lower free energy. The classification of metamorphic rocks is somewhat more complicated than is the case for other rock types. The divisions are made partly according to chemical composition, but texture and particle size are also seen as useful means of differentiation.

A fairly obvious distinction can be made between metamorphic products of rocks having sedimentary or igneous origins. Metamorphic rocks that were originally argillaceous sediments tend to have a layered structure that cleaves readily into flakes. Examples of this kind are *slate* and *mica schist*. If the sediment was of the arenaceous variety, on the other hand, a monolithic rock would be produced upon metamorphosis, an example of this being *quartzite*. One of the most familiar metamorphic rocks is *marble*, which was formed from sedimentary limestone. Rocks formed by the metamorphosis of igneous material are classified according to whether the latter was acidic or basic. An example of the former type is *gneiss*, which before transformation comprised quartz, feldspar and mica. A beautiful example of a metamorphic rock that derived from basic igneous material is *eclogite*, in which crystals of red garnet are surrounded by a matrix of green pyroxene. To get an idea of the way in which the properties of the end product depend on the original material, we might compare mica schist with *hornfels*. The flaky texture of the former is related to the sequential manner in which the original sedimentary rock was deposited. Hornfels, which is metamorphosed shale, also contains flakes of mica, but these are distributed with random orientations, and the final metamorphic rock cannot be cleaved.

Diamond is a good example of the metamorphic formation of a rock crystal. To begin with, let us look at the relationships of the two naturally occurring forms of pure carbon, namely graphite and diamond. If the latter is heated to about 1200 °C, it slowly transforms to graphite. The rate of transformation increases with temperature. If this is raised to about 2000 °C, the transformation is very rapid, and the crystal shatters into a heap of graphite powder. It has been found that diamond can survive these high temperatures if it is subjected to a pressure of several thousand atmospheres. Conversely, diamond can be produced by heating graphite to about 2500 °C under a pressure in excess of 100 000 atm. It is this latter process that is responsible for the production of diamonds in nature. The events which led to the present understanding of the natural formation of diamond, and indeed to a proper appreciation of metamorphic processes in general, have a distinctly romantic flavour. Before the middle of the nineteenth century, virtually all diamonds had been found in or near surface streams, and most of them were discovered either in India or Brazil. In the year 1866, the children of a Boer farmer, Daniel Jakobs, were playing with pebbles on a bank of the Orange River at Hopetown near Kimberley in South Africa. Their attention was excited by one particular pebble which sparkled in the sunlight as it was turned. It transpired that the object was a diamond weighing 21.25 carats, which is slightly more than 4 g. Subsequent exploration in the area showed that the diamond-bearing deposit was a roughly circular region a few hundred metres in diameter. It was soon realized that such a circle is the intersection with the Earth's surface of a cylindrical 'pipe', which has been pushed upwards from a region lying deep down in the Earth's crust. Such a geological feature is now known as a *kimberlite pipe*, after the location of this decisive discovery. It is believed that these pipes are in fact the relics of fossilized plugs of ancient volcanoes, which originate at depths of perhaps more than 150 km. At such great depths, both the pressure and temperature necessary for the transformation of graphite to diamond are known to be attained.

Summary

The distribution of elements in the Earth's crust is not typical of the Universe as a whole, there being disproportionately large quantities of the heavier members. Minerals are sometimes fairly pure single elements such as carbon, sulphur and gold, but they are usually compounds. The most common of these are based on combinations of silicon and oxygen. Rocks invariably contain fragments of several different minerals, while the economically important vein minerals occur as isolated concentrations of a single species. Both of these arrangements can be produced in one of three ways. Igneous rocks, such as granite and basalt, are formed by the solidification of the molten material that now lies beneath the Earth's surface. Sedimentary rocks are the deposited remains either of the products of erosion, as with clay, or of dead plants (e.g. coal) or animals (e.g. limestone). The action of heat and pressure on either of these types produces metamorphic rock, three examples of which are slate, marble and diamond. The origin of oil deposits is still uncertain.

8

All that is, at all,
Lasts ever, past recall,
Earth changes,
But thy soul and God stand sure,
Time's wheel runs back or stops:
Potter and clay endure.

Robert Browning (1812–1889)
Rabbi ben Ezra

Almost forever: *ceramics*

Of all the different types of material, ceramics might be the most difficult to define. Many would regard them as falling in a small and rather restricted group, and the only examples that come readily to mind would probably be bathroom fixtures, tiles, and the insulators in spark plugs and on telephone poles. The term *ceramic* actually covers a large variety of natural and artificial substances that share the desirable qualities of hardness and resistance to heat, electricity and corrosion. Just how large and how important the ceramic domain is can be gauged by some of its members: stone, brick, concrete, sand, diamond, glass, clay and quartz. If there has been a lack of understanding of ceramics, it is excusable because even dictionary definitions tend to be rather narrow. We find them restricted to either pottery or porcelain in most cases, and even the better efforts usually go no farther than 'products of industries involving the use of clay or other silicates'. The word ceramic actually comes from the Greek word *keramos*, which means burnt stuff. This is too broad a term to be useful here. The best working definition uses a combination of chemical and physical criteria. A ceramic is a solid composed of a mixture of metallic, or semi-metallic, and non-metallic elements in such proportions as to give the properties described above of hardness, durability, and resistance. The non-metallic element is most frequently oxygen, but this is not a strict requirement. The extremely hard compound tungsten carbide is a ceramic, for example, and so is silicon carbide, commonly known by the name *carborundum*. There are even nitrogen-based ceramics such as the strong *silicon nitride* used for turbine blades, and ceramics which contain one of the halogens.

The choice of a chemical and physical basis for the definition of a ceramic is in keeping with the philosophy of this book; if we know how the atoms are arranged and how they are interacting with one another, we should be able to understand why the material behaves as it does.

One of the most common components found in naturally occurring ceramics is the silicate unit, a combination of silicon and oxygen. It would be wise to begin, therefore, by asking how the atoms of these two elements are arranged with respect to each other, and how they enter into interatomic bonding. In oxygen, the outer electron shell is incomplete to the extent of two electrons. For silicon, there are four electrons in an outer shell that is capable of holding eight. The question arises as to what happens when the two atomic species compete for each other's electrons. The qualitative similarity in the outer shells of the two elements might lead us to expect that this competition will end in a stand-off. This is not the case, however, because the issue is decided by the electronegativities. For oxygen this is 3.5,

Brightly coloured stones were highly valued in ancient civilizations as a source of ornamentation. They were sometimes ground into a powder and applied to a surface with a suitable adhesive, but decoration more frequently involved mosaics like those used in this serpent mask representing the Aztec god-king Quetzalcoatl Ce Acatl. It was among the treasures sent home by Hernando Cortés (1485–1547) to the emperor Charles V (1500–1558), during the Spanish conquest of Mexico in the period 1519–1526. The mosaics are of turquoise, which is a basic aluminium phosphate. The blue and green colours are caused by traces of copper.

while for silicon it is only 1.8. The result is not a stalemate, therefore, and it is the oxygen atoms that end up with the lion's share of the electrons.

The resulting redistribution of electrons has an important effect on the sizes of the two atomic species. The normal radius of a silicon atom is 0.132 nm, but if it loses all four of its outer electrons the radius drops to 0.041 nm, so it becomes a rather tiny ion. The oxygen atom, on the other hand, grows to a radius of 0.140 nm when it gains the two electrons required to fill its outer shell. The question of arranging the silicon and oxygen ions thus becomes one of placing dwarves among giants. As mentioned briefly in the previous chapter, a common solution is the arrangement of the oxygen atoms in a tetrahedral configuration with the silicon atom occupying the small hole at the centre, equidistant from the four oxygen atoms. It is interesting to note that we might have arrived at this structure by simple geometrical reasoning. If three hard spheres are arranged in an equilateral triangle with each sphere touching the other two, and a fourth sphere is placed on top so as to touch the other three, as in the stacking of

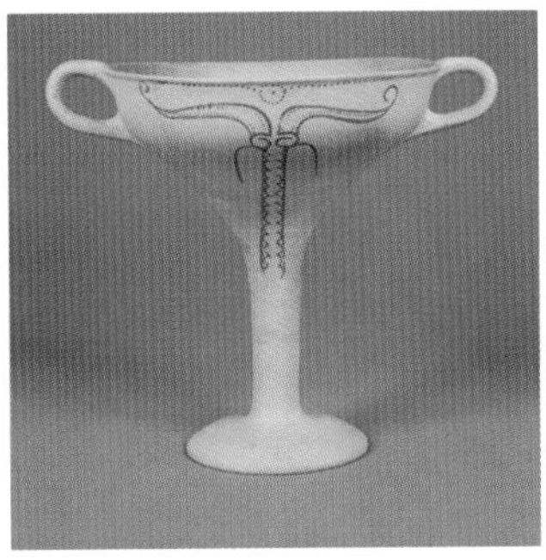

Being essentially fully oxidized, ceramics are remarkably resistant to corrosion. They are notoriously brittle, however, and large intact antique specimens are relatively rare. This Mycenaean drinking cup dates from about 1250 BC. It was found in Kalymnos, Greece, and has a height of nineteen centimetres.

cannon-balls, for example, the largest sphere that could be contained in the space at the centre would have a radius only 22.5% of that of the original spheres. Allowing for the fact that real atoms interpenetrate one another to a certain extent, it is not surprising that the silicon ion manages to squeeze its way into the gaps between the close-packed oxygen ions.

A situation in which the electrons given up by atoms of one element are captured by the atoms of another element produces *ionic bonding*, the best examples being seen in the alkali halides. Although identification of the different types of bonding involves a perfectly valid distinction between various extremes, it would be wrong to conclude that all interatomic bonds have to fall entirely within a single class. An alkali halide such as sodium fluoride happens to be a particularly good example of ionic bonding because the one electron in the outer shell of sodium fits nicely into the single vacancy in the outer shell of fluorine. Moreover, the electronegativities show a greater disparity, 4.0 for fluorine and 0.9 for sodium, than is the case for silicon and oxygen. Linus Pauling has stressed that many examples of interatomic bonds do not belong to a single class, but rather represent a mixture of different types. He derived a scale which links the electronegativity difference between two elements to the percentage ionic character of bonds between them. The interaction between silicon and oxygen is a case in point: the bonds between these atoms are about 50% ionic and 50% covalent. This can be compared with the case of sodium fluoride, in which the degree of ionic character is slightly in excess of 90%.

In tetrahedral *silicate units*, the silicon atom is surrounded by four oxygen atoms, and it gives up its four valence electrons to them, one to each. This means that each oxygen atom receives one of the two electrons that it would like to acquire. It can obtain a second electron if it is also close to a second silicon atom, and this is the key to the silicate structure. This tendency of the oxygen atoms can be satisfied by arranging the tetrahedra so that they share corners. With each oxygen atom thus simultaneously belonging to two tetrahedra, the chemical formula for silica is SiO_2 and not SiO_4.

There are many different ways in which an extensive array of tetrahedra can be built up, while still satisfying the corner-to-corner requirement. They can be arranged in one-dimensional rows, as in a *linear polymer*, and this is what happens in the case of *asbestos*, a calcium magnesium silicate. Alternatively, they can form a regular two-dimensional pattern, as in a mica such as *muscovite*, a potassium aluminium silicate. Then again they might be stacked in a three-dimensional crystalline array, as in *cristobalite*, which is one form of quartz, and composed of pure silica. Even this does not exhaust the possibilities, because the tetrahedra can be disposed in a fairly random fashion, with only negligible violation of the corner-to-corner rule. This is what occurs in the case of an amorphous structure, and common *window glass* is indeed such a random distribution of silica molecules. The development of the *X-ray diffraction technique*, by William Henry Bragg and William Lawrence Bragg, was vital to modern ceramic science. These pioneers, their colleagues John Bernal and William Taylor, and subsequent generations of crystallographers, have systematically determined the structures of a large number of these materials.

Of the several virtuous properties displayed by ceramics none is more significant than their resistance to chemical attack. In their ability to withstand the dual insults of high temperatures and corrosive environments they have the field to themselves. The basis of their remarkable performance lies

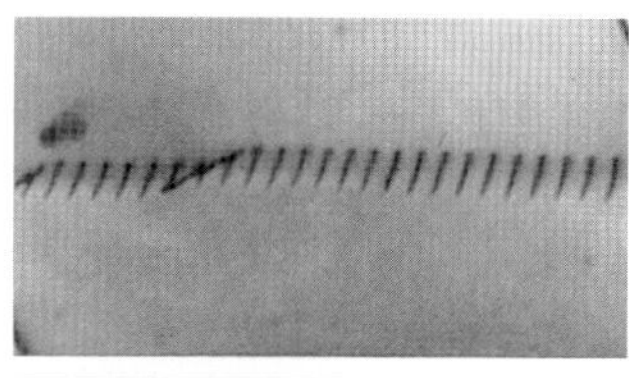

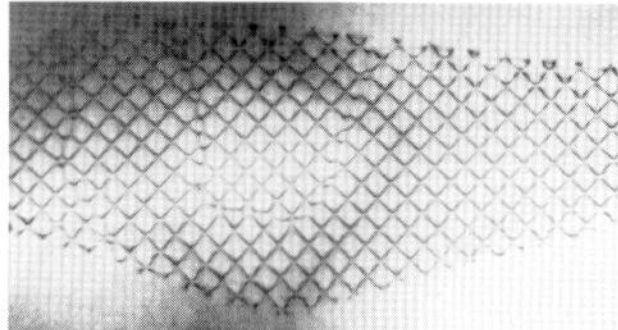

Dislocations do not move easily in a typical ceramic, because the interatomic forces are directional. Such forces resist the strains associated with line defects. There is a tendency for dislocations to be confined to groups in which the individual defects partially cancel out each other's distorting influence. *Grain boundaries* having particularly small misfit angles are equivalent to regular arrays of dislocations, and relatively strong forces resist the escape of individual dislocations from such a structure. Two examples are shown here: a *low-angle boundary* in alumina, Al_2O_3 (above) and a two-dimensional network, known as a *twist boundary*, in nickel oxide. The width of the picture is about one micrometre, in each case.

in their composition and structure. The economical use of space, with the small cations fitting into the spaces between the anions, leads to dense packing, and this in turn has two physical consequences. Dense packing implies a high concentration of bonds, so the thermal energy required to shake the atoms from one another is unusually large. The melting temperature is therefore high. Similarly, the close placement of the atoms provides a bulwark against chemical attack. In the case of the many ceramics that contain a large proportion of oxygen, there is another important factor contributing to their chemical inertness. Materials tend to deteriorate in the presence of some chemicals because they become oxidized, but the oxygen-based ceramics can be looked upon as being already highly oxidized, and it is this which gives them their impressive resistance. The technological impact of ceramics in the metal processing industry could hardly be exaggerated. One can trace the links in their historical developments from the primitive crucibles of early times to the massive kilns of the modern steelworks.

The hardness of ceramics stems from the nature of their interatomic bonding. We can best appreciate this by comparing them with metals. The malleability of a metal is a consequence of non-directional interactions; the positive and roughly spherical metallic ions are held together by the collective sea of conduction electrons, and the packing of the ions is essentially governed by geometry. Deformation occurs by the sliding of one plane of ions over another, in a sequential manner by dislocation movement. Slipping is easy because the associated distortion to the bond angles has little influence on the metallic crystal's energy. The interatomic bonds in a ceramic are partially covalent, however, and the most significant characteristic of such bonds is their directionality. Thus the slipping motion of one plane over another is unfavourable, and a ceramic cannot be readily deformed. The sluggish motion of dislocations in these materials was established by W. G. Johnston and John Gilman, who measured the velocity of the defects under a given stress, in 1956. One can hammer a lump of iron into shape but not a piece of ceramic. Although they are hard, ceramics are also quite brittle, and this too is related to the bonding. Materials ultimately fail under a tensile load when their interatomic bonds are broken, the critical stress level being given by the Griffith crack criterion. If this occurs with little or no prior deformation, the substance is *brittle*, while measurable deformation before final rupture is the hallmark of the *ductile* solid. The gradual change of shape, which gives ductile rupture its tearing quality, involves atomic motion, most commonly by dislocation movement. We should expect brittle behaviour, therefore, when dislocation motion is difficult, as in the case of ceramics. Their performance under compression is exemplary, but they are notoriously unreliable under tension. In industrial, as opposed to laboratory, ceramics there is another source of weakness. These are usually produced from powdered raw materials, and the final product invariably contains inter-particle flaws, which function as *Griffith cracks*.

Mention should be made of the interesting electrical effects observed in ceramics, some of which have been investigated with a view to exploitation for energy storage. A crystal always contains a certain concentration of vacant atomic sites, which are important agencies of diffusion. They play this role in ceramics and permit the important structural changes during the firing process that are a crucial aspect of production. By allowing the migration of ions, which jump by exchanging places with them, the *vacancies* also cause a movement of charge, and can thus produce an electrical

The chemistry of early cement making

The chemical reactions that were used by ancient builders are closely related to those employed in present cement technology. Let us consider the processes exploited by the Etruscans. When limestone is heated, carbon dioxide is liberated, and the residue is lime. The reaction is

$$CaCO_3 \rightarrow CaO + CO_2$$

Addition of water gives the reaction

$$CaO + H_2O \rightarrow Ca(OH)_2$$

and the resulting slaked lime can be moulded to the desired shape until hardening takes place through the reaction with atmospheric carbon dioxide. This latter process is described by

$$Ca(OH)_2 + CO_2 \rightarrow CaCO_3 + H_2O$$

The original material has come full circle, the net result being a change in shape of the calcium carbonate from the arbitrary lumps of limestone rock to the desired form of the final hardened paste.

current. This is easily detected in an experiment in which opposite faces of a ceramic crystal are connected to a battery and a current meter. It must be emphasized that different species of ion, and thus different types of vacancy, are present in a ceramic and their mobilities will not usually be equal. Most commonly, the current is carried by the (positive) cations, which are usually smaller than the anions and therefore migrate more readily through the crystal. There are many exceptions to this rule, however. Thus in barium fluoride it is the fluorine that moves, while replacement of barium by another element of the same family, calcium, gives a crystal with a mobile cation. In certain ceramics, calcium fluoride being a case in point, the crystal structure contains rows of relatively open sites along which a sufficiently small cation can move with negligible hindrance. The same is true of *zeolites*. The differential permeability to positive ions in this type of ceramic, usually referred to as a *superionic conductor*, promises to be useful in novel batteries.

Apart from these intrinsic diffusion effects, a variety of additional migration phenomena can result from deviations of *stoichiometry*. The latter refers to the ideal composition, as indicated by a compound's chemical formula. Stoichiometric calcium fluoride, for example, contains equal numbers of calcium and fluorine atoms. In ionic and covalent materials, where charge neutrality and electron sharing effects give the interatomic bonds their directional nature, considerable departures from stoichiometry can be tolerated. The crystal sometimes accommodates the imbalance by having extra vacancies in the minority sub-lattice, that is to say vacancies in addition to those already in thermal equilibrium with the crystal. These extra defects confer an enhanced mobility on the sub-lattice element. A proper description of non-metals requires inclusion of *charge localization* effects. In some cases an extra electron is associated with a particular defect, while in others there is a local electron deficit, which can be regarded as a positive charge known as an *electron hole*. The motion of such local charges must also be taken into account when considering diffusion effects. With the abundance of factors arising from ionic size, charge, crystal structure and stoichiometry, each case is best treated individually and generalizations regarded with suspicion. The important compound *uranium dioxide* is interesting in that the oxygen is mobile both in the stoichiometric and non-stoichiometric varieties. But vacancies do not mediate atomic mobility in this case. Diffusion occurs by an interstitialcy mechanism, in which an interstitial oxygen ion moves into a regular oxygen sub-lattice site, bumping the lattice ion into a neighbouring interstitial position. Another important ceramic, because of its relevance to corrosion, is iron oxide. Of the several varieties, FeO and Fe_2O_3 are the most significant. ($Fe_2O_3.H_2O$ is *red-brown rust*, one of the common rust forms.) In FeO, vacancies are most common in the cation sub-lattice, while for Fe_2O_3 it is energetically more favourable for these defects to appear in the oxygen anion sub-lattice. Because of this defect structure, the corrosion proceeds by the movement of oxygen through the oxide film, which therefore grows downward into the metal. This produces compressive stresses, which can be relieved by rupture of the rust film. It flakes off, exposing a new metallic surface, and the entire corrosion process is repeated cyclicly. The film provides no protection in this case. The opposite extreme is seen with *aluminium*. Its oxide resists the passage of both ions and electrons, providing a protective and passivating layer.

There are two ways in which heat can be conducted through a crystal. The energy can be transported as kinetic energy by those electrons which are able to move throughout the crystal: the *conduction electrons*. Heat can also be conducted along the atomic rows by the waves referred to earlier as *phonons*. Because ceramics are electrical insulators, the first of these transport properties is not possible, so all heat conduction must be accomplished by phonons. If these waves are able to travel undisturbed through the crystal, the thermal conductivity is high. This is the case in most ceramics at relatively low temperatures because the available thermal energy is sufficient to generate only a few long-wavelength phonons. Just as an opaque object scatters a light wave, so will any imperfection scatter phonons and hence diminish the thermal conductivity. A common imperfection is the grain boundary, and the heat resistance of a piece of *firebrick*, for example, is caused largely by its polycrystalline structure. In fact, even a complicated crystal structure is sufficient to cause decreased phonon transport. Finally, as the number of phonons per unit volume increases with increasing temperature, these waves interact and effectively scatter each other, to their mutual detriment. This leads to a fundamental decrease in thermal conductivity at high temperature, the latter being defined with respect to the melting point. Gem-quality diamonds feel cool because they conduct heat so well. This in turn is attributable to their simple crystal structure, their lack of imperfections, and to the fact that room temperature is a relatively low temperature for this material. It is interesting to note the difference between *sapphire*, a good conductor, and a piece of *refractory brick*, which is prized as an insulator. Both are essentially alumina, also known as *corundum*, but the former is one large crystal whereas the brick is a conglomerate of many small crystals, with a high density of grain boundaries and voids, which obstruct phonon transport.

The ceramics industry has roots firmly planted in antiquity, and the bulk of its products are based on techniques pioneered by artisans of earlier millennia. The ceramic art is far from stagnant, however, and the current trend is towards specialized materials with unique properties or outstanding performance. Among the exciting newcomers are various nitrides and carbides with impressive abrasive qualities, *magnetic ceramics* having sharply defined threshold levels, which make them ideal for switching units in computers, and high-temperature examples capable of withstanding the extreme

Two microscopic techniques emphasize different aspects of the structure of *cement clinker.* The scanning electron photomicrograph on the left shows the granular appearance of a fracture surface, the largest grains being about fifty micrometres in diameter. The green colour is produced by the phosphorescent screen of the microscope, and has no physical significance. A special etching technique was used to differentially colour the various clinker components seen in the optical photomicrograph on the right. *Alite* appears brown, *belite* blue, *ferrite* white, and *tricalcium aluminate* grey.

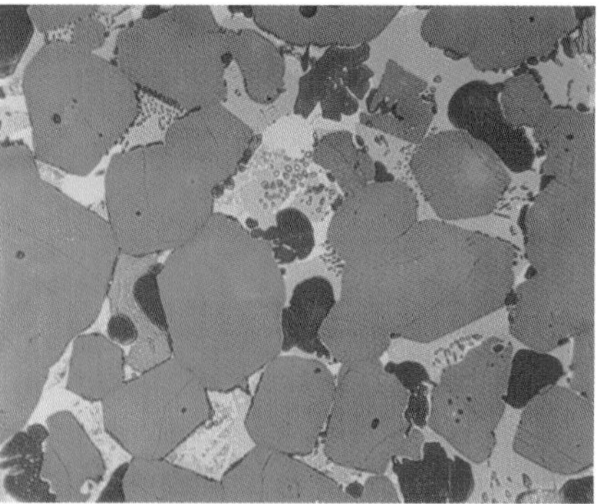

conditions experienced by space-vehicle nose cones and rocket housings. The nuclear fuel *uranium dioxide* is unique in its ability to maintain structural integrity over prolonged irradiation exposures. One new group, known as *cermets,* are metal–ceramic composites. They distinguish themselves in combining high strength with high-temperature stability. *Pyroceram* (*Pyrex*) is the name of a new type of material developed at Corning Glass Works, New York, in 1957. The initial fabrication produces a uniform glass, and nucleation agents such as titanium dioxide promote controlled crystallization during subsequent thermal treatment. The outcome is a polycrystal of very fine grain size, with great strength and high-temperature durability. Finally, there are the *molecular sieves,* which are used to separate molecules of different sizes. This differential effect is achieved through the unusually large interatomic spacings in certain ceramics, sufficiently small molecules being able to pass through the lattice. The industrially important zeolites have already been mentioned. Ceramics can be usefully divided into three groups: naturally occurring ceramics; ceramics produced by chemical reaction, and ceramics produced by heat treatment. The amorphous varieties of these, known commonly by the name glass, will be considered separately in a later chapter. Naturally occurring ceramics, such as marble and quartz, have already been discussed in connection with rocks and minerals. Particularly prominent examples of the other two groups are cement, brick and pottery, and we will now discuss these in considerable detail.

Two important types of ceramic are represented in this photograph in which Joseph Aspdin (1778–1855), who invented *Portland cement* in 1824, poses in front of one of his early brick-built kilns.

The exploitation of the chemical reaction between calcium oxide, which is also known by the names lime, *quicklime* and *burnt lime,* and water to provide a binder in building construction dates at least from the Etruscans. They added water to lime to form calcium hydroxide, or *slaked lime,* which they used to cement together sand and stone aggregate into what would now be called a primitive *concrete.* Quite how the Romans discovered that burning a mixture of limestone and silica improves the process remains a mystery, but it is known that they mixed volcanic ash with calcium carbonate, which is found as limestone and chalk, and burned it to produce a *cement* that was so durable that many examples of it have lasted to the present time. The modern era of this technology could be said to date from 1824 and the award of a patent to Joseph Aspdin for what he called *Portland cement;* its colour reminded him of the Jurassic building stone that is so common in the Portland area of Dorset on the south coast of England. The superiority of this new product was immediately recognized, and within a year it was being used in major engineering projects. Marc Isambard Brunel, for example, used it in his Thames Tunnel, construction of which started in 1825. Aspdin's

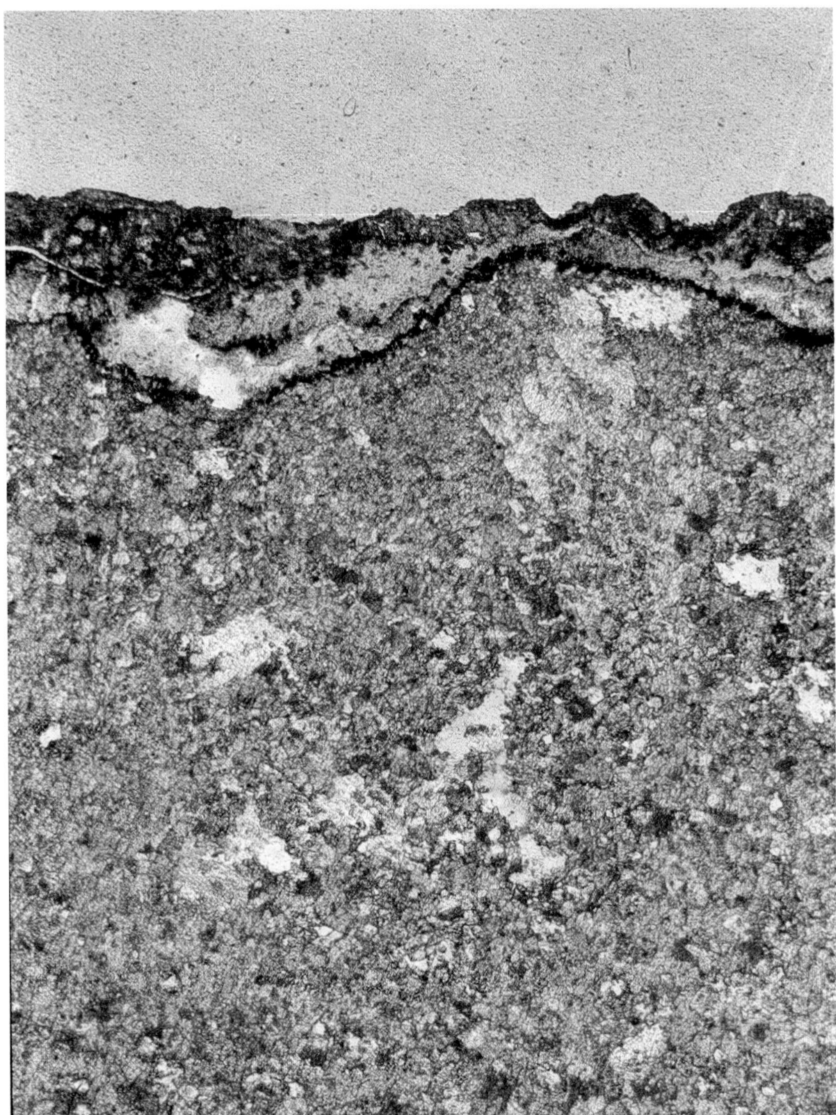

Several of the components present in modern cement powder are also found in the natural clinker *larnite*, which is *beta*-dicalcium silicate. As can be seen in this optical photomicrograph of a thin section, weathering effects at the surface of the mineral produce a primitive form of *concrete*.

original specification called for the mixing of burnt limestone, clay and water into a slurry, which was then heated so as to drive off the water and ultimately led to a partial melting of the mixture. The latter was then cooled down, ground into a powder, and was ready to be mixed with a suitable amount of water and aggregate to form concrete. This method is essentially the one that is still used today. An important distinction between the early lime-based material and Portland cement is that the latter will set and harden even under water.

Cement paste, which is simply a mixture of cement powder and water, can be used either together with sand, when it is referred to as *mortar*, or with stones as well so as to produce *concrete*. (The term *stone* is used if the mean particle diameter exceeds 2 mm. Particles under this size are referred to as *sand*, and a mixture of stones and sand is called *aggregate*.) When used in the foundations of buildings, concrete is called upon to withstand only compressive loading, a purpose for which it is admirably suited. Its tensile

The *hardening process* in cement involves the growth and interlocking of several different types of crystal, including plate-shaped monosulphate crystals ($3CaO.Al_2O_3.CaSO_4.12H_2O$) and needle-like crystals of both calcium silicate hydrate and ettringite ($3CaO.Al_2O_3.3CaSO_4.32H_2O$). All these types are visible in this scanning electron microscope picture. The width of the area shown is about thirty micrometres.

strength is only one-tenth of its compressive strength, however, so if there is any risk that the concrete will be subjected to either tension or shear, it is nowadays common to reinforce it with steel rods. In 1845, William Fairbairn built an eight-storey refinery in Manchester, England, the floors of which were wrought-iron I-sections over which concrete had been poured and allowed to set. Four years later, Joseph Monier was producing concrete tubs for orange trees in which a mesh of iron rods had been embedded, and at the Paris Exposition in 1855, J. L. Lambot exhibited a concrete boat that had been moulded on iron plates. Thaddeus Hyatt published the first engineering theory of *reinforced concrete* in 1877, and the first buildings to make major use of this new material began to spring up about 20 years later. François Hennebique completed a mill in 1895, the load-bearing components of which were reinforced concrete, and Tony Garnier incorporated it in the scheme for his *Cité Industrielle* a few year later. The first building to be based entirely on reinforced concrete was Frank Lloyd Wright's Unitarian Church in Oak Park, Illinois, USA, completed in 1906. These buildings were the forerunners of what came to be known as the International Style of architecture. Cement itself is often the weakest component of concrete, and the mechanical properties of the composite depend to a great extent on its performance.

In modern cement production, the starting point is a mixture of approximately 80% limestone and about 20% clays, which are essentially aluminosilicates with many other minor components. This mixture is ground with water to a slurry and then slowly passed down a kiln, which is usually in the form of rotating cylinders lying a few degrees from the horizontal. As the mixture is heated, it first loses its water, and then *calcination*, or carbon dioxide liberation, occurs. Finally the remaining material, which amounts to some 35% or 40% of the initial solid mass, reaches the burning zone of the kiln, where the temperature is 1200 °C to 1500 °C. During this latter stage a combination of melting and sintering causes it to form clinker. (*Sintering* is an adhesion process of adjacent particles by an actual exchange and inter-penetration of their atoms.) The *clinker* is cooled down and ground into a fine powder. In most modern cements a small amount of *gypsum* is ground together with the clinker. The surface-to-volume ratio of the particles in the final cement powder determines the speed of reaction when water is added to it, and the different common grades of cement are usually given designations that indicate how rapidly the final paste becomes rigid and gains strength.

The production of a stone-like solid, from the wet paste obtained when water is mixed with cement powder, depends on the chemistry of the interaction of water with the chemical compounds present in the powdered clinker and gypsum. It has been found that the clinker contains no less than four major compounds. The most abundant of these are alite and belite. There are also two less abundant compounds, neither of which represents more than 10% of the whole. *Alite* is a silicate phase, its full chemical name being tricalcium silicate. *Belite* too is a silicate phase: dicalcium silicate. The two less abundant compounds are tricalcium aluminate, commonly referred to as the *aluminate phase*, and tetracalcium aluminoferrite, which is known as the *ferrite phase*. Modern Portland cement contains between 40% and 65% of alite, from 10% to 25% belite, up to 10% each of aluminate and ferrite, and between 2% and 5% gypsum. There are invariably also minor amounts – a fraction of a per cent usually – of free lime, magnesium oxide,

Modern cement chemistry

The *Portland cement* now in common use has five main constituents and a number of minor components, none of which represents more than a fraction of a per cent of the total. The chief compounds are listed in the following table

Compound	chemical composition	% of total
alite	$3CaO.SiO_2$	40–65
belite	$2CaO.SiO_2$	10–25
aluminate	$3CaO.Al_2O_3$	up to 10
ferrite	$4CaO.Al_2O_3.Fe_2O_3$	up to 10
gypsum	$CaSO_4.2H_2O$	2–5

Extensive studies have been made of water's interaction with some of these compounds. The alite–water and aluminate–water reactions are

$$3CaO.SiO_2 + H_2O \rightarrow Ca(OH)_2 + (CaO)_x{\cdot}(SiO_2)_y{\cdot}(aq)$$
$$3CaO.Al_2O_3 + 6H_2O \rightarrow 3CaO.Al_2O_3.6H_2O$$

The value of x lies in the range 1.8–2.2, while y is close to one. The letters (aq) indicate an aqueous suspension. These reactions rapidly lead to formation of a *colloidal gel* of silicate and aluminate particles. The calcium hydroxide, which stays in suspension, is also known as *portlandite*. Belite's reaction with water is similar to that of alite, but it is much slower; it gives the same product. When water is added to cement powder, a number of reactions proceed simultaneously. Particularly important is the growth of the needle-shaped crystals of *ettringite*, which appear at the surfaces of the aluminate particles. The reaction is

$$3CaO.Al_2O + 3CaSO_4.2H_2O + 30H_2O \rightarrow 3CaO.Al_2O_3.3CaSO_4.32H_2O$$

This slows down the aluminate-water reaction, which then gradually builds up a network of calcium-silicatehydrate fibrils. The interaction between aluminate, gypsum, and water also produces the beautiful plate-shaped monosulphate crystals, which have the composition $3CaO.Al_2O_3.CaSO_4.12H_2O$. These penetrate the fibril network, and growth continues until the crystals interlock to give the paste its mechanical strength.

sodium sulphate and potassium sulphate. These *trace compounds* can influence the final product to a much greater degree than their abundances in the cement powder might suggest. At their worst they represent the Achilles heel of the final solid.

The central question in cement science concerns what happens when water combines with the various compounds, but it is clear that this is a complicated issue. The rates of reaction of some of the compounds with water are influenced by the presence of the other compounds. Then again, the rate of heat liberation due to the hydration process is different for the different compounds. Finally, hydration of a compound changes both its thermal conductivity and diffusion characteristics, so the overall problem is complex and challenging. As we might expect, one strategy of the cement scientist involves separate study of the different compounds in cement, and acquisition of information about their individual characteristics. What follows is a brief summary of the more pertinent results.

The addition of water to alite leads to the formation of calcium silicate hydrate and calcium hydroxide, which is also known by the name *portlandite*. The hardened paste has high strength when finally set, and because alite is also the most abundant compound in cement, it makes the dominant

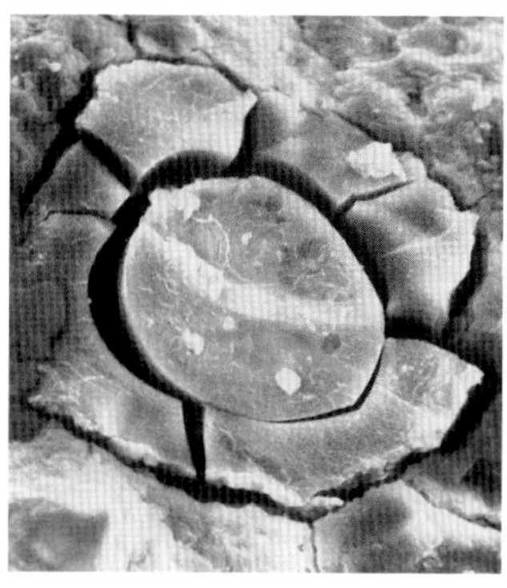

Stones are deliberately added to cement in the production of *concrete*, but certain minority minerals are detrimental. This optical photomicrograph shows a particle of *opal* which has been transformed into a gel by reaction with the alkaline components in the surrounding cement. The gel expands, due to the absorption of water, and this causes cracks in the concrete.

mechanical contribution to the final hardened product. The hydration proceeds at an appreciable rate a few hours after the addition of the water, and lasts up to about 20 days. It is accompanied by the moderate release of heat, about 500 J/g, and it involves acquisition of about 40% by weight of water, of which a little more than half is chemically bound, while the rest is physically absorbed. The hydration of alite is accelerated in the presence of aluminate, and gypsum has a similar influence.

Belite's reaction with water is rather similar to that of alite, and it yields the same end product. But the reaction is much slower. The hardening process does not really get underway until about the second day, and it is not completed until more than a year has passed. Its ultimate strength, however, is rather similar to that of hydrated alite. The amount of water taken up by the hydration process is about the same as for alite, on a weight basis, while the amount of heat liberated is only about 250 J/g. Belite produces less calcium hydroxide than does alite, because it contains less calcium in the original state. It is interesting to note that Nature has given us a geological example of the hydration process for the dicalcium phase. The naturally occurring mineral *larnite*, which is actually *beta*-dicalcium silicate, has been able to interact with water over a period of many million years. Inspection of the free surface of larnite specimens frequently reveals the presence of the hydrated product, which, because it has been developing over a period that is measured in geological time, represents a sort of primaeval cement paste.

In almost every respect the aluminate phase is the most spectacular of all the components of cement. The two most abundant compounds in cement interact with water over a period of days and months. Relatively speaking, if belite is to be looked upon as the marathon runner of the team, aluminate plays the role of a 100 m sprinter. Its reaction with water is essentially over within the first few hours, but during this time the material that is actually transformed can incorporate in its final product up to twice its own weight of that liquid. It is also the prime source of heat during the hydration process of cement powder, because it gives off no less than 900 J/g. But in spite of these heroic performances, it turns out to be something of a let-down when it comes to mechanical integrity. In fact, its contribution to the ultimate strength of the cement paste is rather limited, and, as if to add insult to injury, it is susceptible to attack by dissolved sulphates. These interact with it and cause expansion, which can have a dangerously weakening influence on the final product.

Compared with the firework display of aluminate hydration, the ferrite phase gives a most subdued performance. In fact, its key word is moderation: moderation in heat evolution, about 300 J/g; moderate use of water, between about 35% and 70% by weight; and its contribution to the ultimate strength of the paste is modest. Like alite and belite, however, it is not susceptible to attack by dissolved sulphates. Moreover, ferrite does play a cosmetic role in that the amount of this compound present determines the final colour of the hardened paste.

Although it represents only between 2% and 5% of the total weight, gypsum can be regarded as the critical component in modern cement. Its great importance lies in its regulation of the activity of the aluminate phase. It does this by forming crystals of *ettringite*, a calcium sulphoaluminohydrate, on the surface of the aluminate grains. This has a passivating influence on the aluminate, and the reaction of the latter with the water thereafter

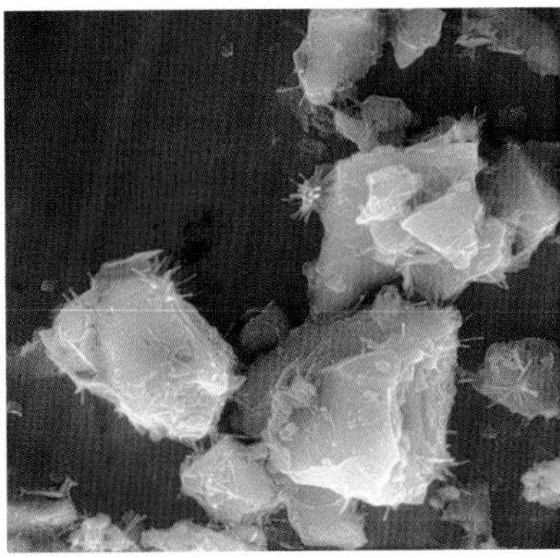

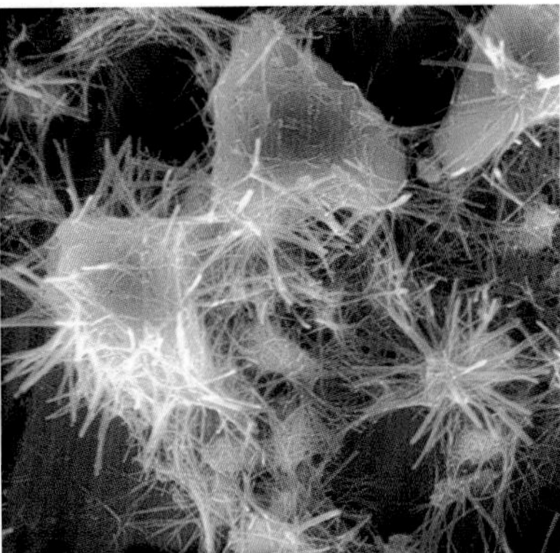

The rate of a cement's reaction with water depends upon the surface area of the powder per unit of weight. *Rapid cement* has a grain diameter that is typically ten micrometres, and this gives so much reactive surface that *gypsum* is required to slow down the initial phase, and limit the heat liberation. These scanning electron photomicrographs record three stages during the hydration. The upper left picture shows unreacted grains, while the upper right picture reveals the presence of needle-formed *ettringite* crystals after five minutes of hydration. They are produced by the reaction of gypsum with tricalcium aluminate. The lower picture shows the extensive needle-shaped crystals of calcium-silicate hydrate that are typical of the final microstructure. They are formed by the reaction of water with the calcium silicates, and it is their mutual interlocking which gives mechanical stability.

proceeds at a much slower rate. The wet paste is thus kept in mouldable form long enough to be worked into the desired shape, before it begins to set and harden. The effect of the ettringite is also to slow down the heat liberation and hence the temperature rise in the wet paste. Particularly important is the fact that the influence of gypsum on the aluminate phase reduces the amount of volume change under conditions of varying humidity, although exactly how it achieves this desirable result is not yet fully understood.

The influence of the other minor compounds is not always advantageous, and modern cement standards specify a maximum content of these agents. The free oxides of calcium and magnesium both react with water to produce hydroxides, and these are accompanied by an increase in volume. Any expansion that occurs when the solidification processes of the other compounds are well underway can open up small cracks, which can seriously weaken the final solid product. The other two common minor compounds in cement, sodium sulphate and potassium sulphate, behave in a rather different fashion. They can be looked upon as having an effect which is opposite to that of the gypsum. They accelerate the hydration reaction and promote rapid setting of the wet paste. Such early congealing is inconsistent with high ultimate strength. If the cement is used to produce concrete, and alkali-reactive matter such as flint is present in the aggregate, alkali silicate compounds can be formed which will expand under humid conditions and give cracking. One particularly welcome development in modern cement technology has been the emergence of a product containing rather little of these alkalis.

Cement paste prepared with the optimal amount of water is rather viscous, and use of more than the recommended proportion of liquid makes for easier handling. But the final hardened cement is consequently more porous and cracks readily. The result can resemble the inferior piece of workmanship shown here.

So much for the individual compounds, but what happens to them collectively when water is added? It would be an exaggeration to claim that the process is understood right down to the minutest detail, but certain things are now reasonably clear. Before going into these, some general observations should be made. The familiar presence of the cement mixer on any building site could easily lead us to underestimate the subtleties of the hardening process. It is worth remembering that freshly mixed cement paste would be useless if it solidified much more slowly or rapidly than it actually does. The utility of the material stems from the fact that it quickly achieves a mechanical stability, while still quite wet and mouldable, and at a time when the hardening process is far from complete. The modern view of the setting and hardening of cement paste is actually traceable to the middle of the eighteenth century. As early as 1765, Antoine Lavoisier had showed that the setting of *Plaster of Paris* is attributable to 'a rapid and irregular crystallization, the small crystals becoming so entangled that a very hard mass results'. Henry Le Chatelier borrowed this idea, in 1887, supposing the cement to dissolve in water, and subsequently be precipitated when the solution becomes supersaturated. A new aspect of the story emerged in 1893, when W. Michaëlis suggested that the hardening involved the formation of a *colloidal gel*. John Bernal established that particles in the gel were crystalline, and the vital synthesis of the two approaches was made by J. Baykoff, in 1926. He suggested that the setting of cement paste is caused by the formation of the gel, while the hardening is achieved by the ensuing crystal growth.

The advent of the *scanning electron microscope*, with its superior depth of focus compared to the optical instrument, has been important to cement science. In the 1970s, it played a prominent role in the discoveries by John Bailey, Derek Birchall and Anthony Howard, and also by David Double and Angus Hellawell, regarding the hydration reactions. When water is added to the mixture of cement powder and aggregate, there being up to six times as much of the latter, the initial hydration lasts a few minutes and involves the alite–water and aluminate–water reactions. This rapidly leads to formation of a colloidal gel of aluminate and silicate particles. To begin with, we can ignore the complications caused by the gypsum and consider the physical nature of the products of these chemical reactions. The colloidal gel is produced at the interface between the water and the cement grains, and it quickly forms a sort of enveloping sack around the latter. The doubly charged calcium ions diffuse quite rapidly out of the gel and into the surrounding water, where the calcium concentration is maintained at an equilibrium value by the precipitation of the angular crystals of calcium hydroxide. The calcium-depleted gel is essentially silicic acid. Separation of the constituents of the originally homogeneous gel leads to a build up of *osmotic pressure*, and this periodically causes rupture of the water–gel interface. Where this occurs, the calcium hydroxide and silica components are brought into contact, and a *precipitate* of calcium silicate hydrate is formed. This is deposited in a configuration that is reminiscent both of a volcanic crater and the ring-shaped mound of a hot spring. Further rupturing of the water–gel interface takes place preferentially at these craters, and this leads to the formation of hollow needle-shaped projections known as *fibrils*. These stick out from the cement grains like the quills of a hedgehog. Through continued fibril growth the cement particles begin to interact, and they ultimately become locked together like a clump of thistles.

The important clay mineral *kaolinite* consists of parallel sheets, each of which comprises five atomic layers. These alternate between anions and cations, the latter being silicon and aluminium ions. Each silicon ion is tetrahedrally coordinated to four oxygens, while the octahedrally-bounded aluminium ions have two oxygens and four hydroxyl ions in their immediate environment. This figure shows how the twenty-eight electrical charges of either sign (positive and negative) are arranged within the unit cell, to give overall electrical neutrality. The two outer layers of each composite sheet are both negatively charged, so adjacent sheets repel each other electrically. This accounts for their relatively easy shearing past one another, which gives *clay* its plastic quality.

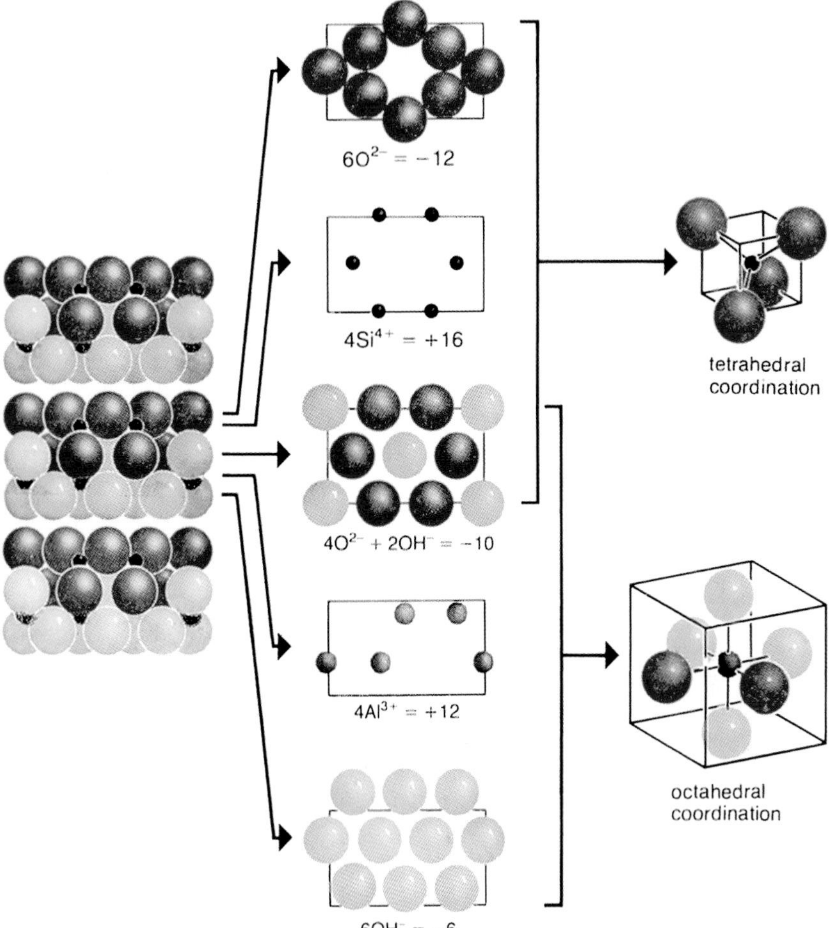

In the practical situation, gypsum is the important extra factor. Within a few minutes of the addition of water, needle-shaped crystals of ettringite appear at the surfaces of the aluminate particles, owing to their interaction with water and gypsum. The aluminate–water reaction is thereby slowed down, while the alite–water interaction continues to proceed at its rather slower rate, producing calcium–silicatehydrate fibrils. With the help of the sulphate from the gypsum, the beautiful plate-shaped monosulphate crystals are also formed. The belite–water interaction yields similar products but at a slower rate. Exactly what is produced by the ferrite–water interaction remains unknown, but the compounds are expected to be even more complex in composition. After about five hours the cement paste is said to be set. At this stage, it has a heterogeneous structure, consisting of an open three-dimensional network filled with colloidal gel particles. The strength is quite low, but the paste is mechanically stable. The hardening process then sets in, and this lasts up to about a month. It involves an increase in the amount of hydration products, with the fibrils, which are either amorphous or fine-grain crystalline, increasing in length. The calcium hydroxide crystals show a concurrent multiplication and growth. As these reactions proceed, more and more of the available volume becomes filled, and ultimately there is a considerable amount of interaction and bonding between

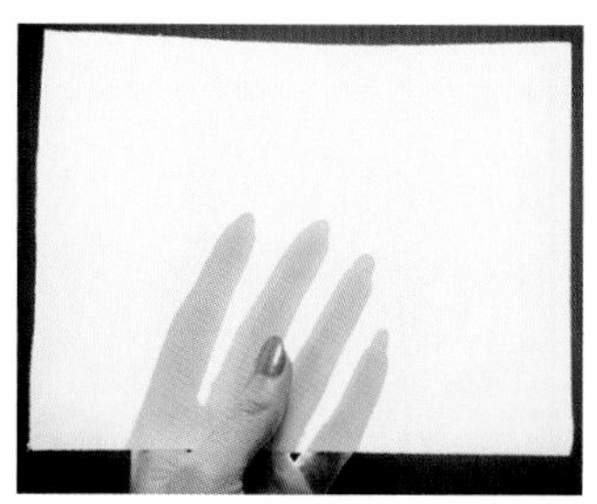

The gaps between the grains in cement are in the micrometre range, and they set a limit on the tensile strength of the material. An important development has been the addition of *silica powder* to the cement, prior to mixing with water. These fine particles fill up the inter-grain spaces, and prevent them from acting as crack-nucleating sites. The sheet of this new material shown here is thin enough to be translucent, and it is almost as pliable as a piece of paper.

the individual structures. It must be this which gives the final composite its mechanical strength.

As to possible flaws in the final product, we have already noted the detrimental role played by the minor oxides and sulphates during the setting of cement. Trouble can also be encountered if the hardened cement is to be subjected to sea water, as in harbour constructions. Dissolved sulphates interact with the aluminate phase to produce further ettringite, and the associated increase in volume can lead to cracking. Nowadays, it is common to use low-aluminate cement for such purposes. Trouble sometimes arises on building sites owing to incorrect mixing of cement. The wet paste is easier to handle if it contains more than the minimum amount of water required for reaction with the individual components, and one can appreciate the temptation this holds for the less scrupulous builder. The excess water is not used up in the chemical reactions, and this leads to a rather porous solid with a strength below the ideal. As can be concluded from the amount of water required to hydrate the various components, the best results are obtained for a water-to-cement ratio of about 0.4 to 0.5, the strength being approximately 50 MN/m^2 (meganewtons per square metre). As the water-to-cement ratio rises, performance steadily deteriorates, and when the ratio is 0.7, the strength is down to a mere 10 MN/m^2. A promising recent innovation has been the addition of *silica powder* to cement. It has a particle size of about 0.1 μm, which is smaller than that of cement by a factor of around ten to a hundred. Such small particles can fill up the gaps between the cement grains, preventing them from acting as *Griffith cracks*. This gives the hardened paste a compressive strength up to three times that of the ordinary product, and the increased density gives improved durability.

The third and last group of ceramics are produced by heat treatment, the technical term for which is *firing*. This type includes pottery, earthenware, china, glass and brick, and the essential features of production are the same for all these materials. The origins of the art are particularly well documented. Because *pottery* is generally fragile and thus vulnerable to breakage, domestic demand has always required a large and steady production. The relics of ancient civilizations are invariably rich in fragments of kitchen vessels and the like, and it has been easy to establish that fired pottery was already in use by 6500 BC in western Asia. These early items would have been produced in open bonfires, in which the temperature can easily reach 800 °C – quite adequate for firing. Glazed examples appeared in Egypt before 4000 BC, but just how the glass-producing technique was acquired is not yet clear. It might have come through observation of the accidental glazing effects produced in copper smelting, but it could also have been a byproduct of attempts to reproduce the beautiful colours of such natural ceramics as the blue-green *turquoise*, an aluminium phosphate with traces of copper, and the blue *lapis lazuli*, a sodium aluminium silicate with traces of sulphur.

Production of pottery and bricks involves fewer processes than is the case for concrete, so it is not surprising that this art was learned well before limestone was being burned to produce cement. The widespread availability of *clay*, the basic material from which these products are made, suggests that they were among the first substances to be utilized by primitive civilization. There are obvious physical differences between a bone china cup and a builder's brick, but the basic structure of these items is very similar. We can therefore confine our attention to one representative example: the *brick*. The brick-making process was probably discovered in one of the warmer

The chemistry of bricks

There are obvious physical differences between bone china and a builder's brick, but their underlying chemistry is quite similar. The starting material is basically clay, which is a secondary mineral. It is formed from minute particles of *feldspars*, *micas*, and ferromagnesium minerals. In a sufficiently acidic aqueous environment, potassium feldspar, for example, undergoes *ion exchange*:

$$KAlSi_3O_8 + H^+ \rightarrow HAlSi_3O_8 + K^+$$

The product of this reaction is an aluminium derivative of silicic acid, which combines with hydroxyl ions to yield *kaolinite*, a white clay:

$$4HAlSi_3O_8 + 2OH^- + 2H_2O \rightarrow Al_4Si_4O_{10}(OH)_8 + 8SiO_2 + O_2 + 2H^+$$

The particles of this secondary mineral are only about a hundred nanometres in diameter and are thus small enough to be *colloidally suspended* in water. Their precipitation, due to a change in the acidity of the water, ultimately produces clay. When the latter is heated to between 700 °C and 900 °C, chemically bound water is liberated:

$$Al_4Si_4O_{10}(OH)_8 \rightarrow 2Al_2O_3.4SiO_2 + 4H_2O$$

This is followed by adhesion of the particles into a rigid solid, the process being known as *sintering*.

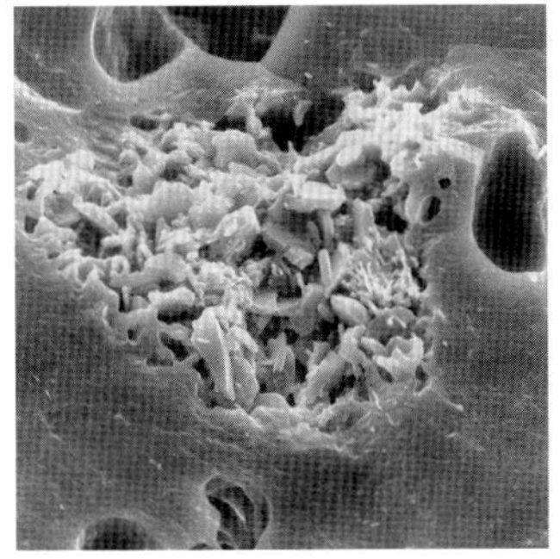

The heat treatment of clay, known as *firing*, to produce *brick* or *porcelain*, involves atomic diffusion on the crystal surfaces in the raw material. The individual particles are thereby consolidated into larger regions and the end result of this *sintering*, as it is called, is a single polycrystalline piece of solid. This scanning electron photomicrograph shows an area of clay specimen in which the temperature was only barely high enough for sintering. The central region contains incompletely sintered clay particles. The width of the picture corresponds to about a fifth of a millimetre.

regions of the Earth, where the Sun's heat is sufficient to dry out a mouldable mixture of clay and water. In the several thousand years that have passed since this early achievement, the production of bricks has developed from a haphazard to a rather exact technology. The underlying chemical processes, on the other hand, have remained essentially unchanged.

Clay is a secondary mineral in that it is produced by the weather-induced disintegration of certain rocks. Incessant pounding by rain, over many thousands of years, has caused small particles of these rocks to be broken loose and washed downstream to regions of rivers at which the rate of flow is insufficient to keep them suspended. Descending to the river bed, they have become *alluvial deposits* of clay. The process has been accelerated in those places where the periodic freezing and melting of water in tiny crevices has produced microscopic fractures of the rock surfaces. Clay formation is not a purely mechanical process, however. There are also chemical changes due to the interaction of water with the microscopic erosion products, and it is these that must be considered in some detail.

Many different rock minerals act as primary sources of the material ultimately to become clay, and just how much of the final product the individual minerals represent depends on their natural abundance. The composition of clay therefore varies from one geographical location to another. Broadly speaking, the primary sources are feldspars, micas and certain ferromagnesium minerals. *Feldspars* are aluminosilicates of the alkali metals, potassium, sodium and calcium. The *micas* involved in clay production are also based on potassium aluminosilicate, but in the hydrated form, *muscovite mica* being a typical example. In the case of *biotite mica*, some of the aluminium atoms are substituted by iron and magnesium. The *ferromagnesium minerals* do not involve aluminium at all, and they can be looked upon as complicated silicates containing magnesium, iron and, in some cases, calcium. The important members of this group are *olivine*, *pyroxene* and *hornblende*. Particles of these primary minerals, eroded away from the rocks and washed downstream, interact chemically with water in a process known as *hydroxylation*,

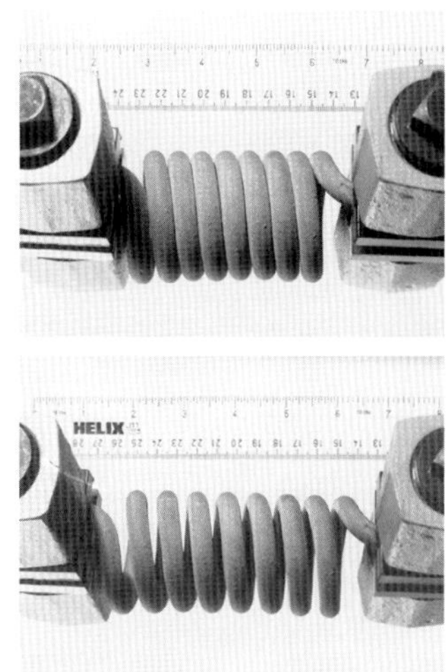

Cement performs well under compression, but its achievements in situations involving shear or tension are at best mediocre. The traditional way of compensating for this inadequacy has been to reinforce the material with steel rods, but the introduction of small silica particles, to fill up potential cracks, also produces a marked improvement. The *silica-cement spring* shown here (upper picture) is capable of withstanding an extension of about thirty per cent (lower picture).

which is simply the addition of hydroxyl ions. Close to their sources, the hydroxylation process involves water and single primary minerals, but farther downstream, where the particles from different minerals are in close contact with one another, it leads to *chemical exchange*. The net result is that three secondary minerals are formed, and it is these which together make up what is known as clay. They bear the names *kaolinite, montmorrilonite* and *illite*. All three are *aluminosilicates*, but in the case of illite there is also some potassium present. Kaolinite is perhaps the best known because of its use in the manufacture of porcelain. This fine white substance takes its name from the mountainous district in north China where it was first discovered: Kaoling, which means high hill.

The chemical reactions by which the secondary clay minerals are produced are fairly complex. An impression is provided by the case of *potassium feldspar*. Potassium and hydrogen ions are first exchanged, this reaction requiring that many free hydrogen ions are available. The aqueous condition must therefore be acidic. The silicic acid thus formed combines with hydroxyl ions to produce the kaolinite. The particle diameters of the secondary minerals are of the order of 100 nm, and these are small enough to stay colloidally suspended in the water. Hydrogen ions are released as the kaolinite is formed. This suggests that a high concentration of such ions, as occurs in an acidic environment, will be unfavourable to the stability of the kaolinite. Under acidic conditions, therefore, the particles of kaolinite cluster together so as to minimize the total area presented to the aqueous solution, a process known as *flocculation*. The larger particles thus formed then tend to be deposited out of suspension, producing sedimentary clay. A sudden increase in acidity occurs naturally at the mouths of the rivers, where fresh water meets sea water.

Clay's mouldability stems from the interaction of its constituent minerals with water and arises from certain features of their crystal structure. They are all composed of layers, the forces between which are of the weak van der Waals type. In this respect, the crystal structures are reminiscent of graphite. In the case of the clay minerals, the layers are alternately composed of silica and alumina, the former lying in the particularly strong octahedral configuration. Because of their layered nature, the clay minerals cleave readily by the breaking of van der Waals bonds, and each time this occurs, two new faces are formed which expose oxygen and hydroxyl groups. The exposed oxygen and hydroxyl ions cause the mineral particles to attract water molecules. The water actually becomes attached to the particles, through hydrogen bonding. This bound water, referred to as *mechanical water*, to distinguish it from the chemically combined water already present in the crystals, can be looked upon as acting like a lubricant between the flakes. It is this that gives the clay its important mouldability.

Let us turn now to the actual manufacture of bricks. The raw material, as excavated, usually consists of clay mineral with particles of mica, feldspar, and quartz suspended in it. Other minerals such as limestone and ferric oxide are usually also present. Siliceous material, of various types, including shale and slate, are added, and all the ingredients are ground together. The optimal plasticity and strength are obtained for a clay mineral content of approximately 20%. The mechanical water content is then brought up to about 20% of the weight of the clay. The amount of water that actually has to be added to the mixture to achieve this composition depends on the relative abundances of the three clay minerals in the excavated material.

The first building to be constructed entirely of reinforced concrete was the Unitarian Church in Oak Park, Illinois. Designed by Frank Lloyd Wright (1867–1959), it was completed in 1906.

Montmorrilonite, for example, contains roughly 15% of mechanical water, whereas for illite and kaolinite the corresponding amounts are about 3% and 1%, respectively. At the correct composition the mixture is moulded into the desired shape, either by simply pressing or by extrusion and cutting, and it is then ready for the heating process.

Moulded material that is ready for firing is referred to as *green brick*. During the firing process, the temperature is gradually raised to between 950 °C and 1150 °C. The mechanical water begins to evaporate as soon as the temperature is above 100 °C, but the chemically combined water is far more tenacious. It is not all removed until the composite material is glowing white hot. At these elevated temperatures, between 700 °C and 900 °C, the limestone is broken down into calcium oxide and carbon dioxide, while the various iron compounds are oxidized. The final stage of the firing cycle is referred to as *vitrification*, a term that could be misleading because the adjective vitreous is commonly used for the glassy state. The final brick material is not a glass, although the grain size is admittedly down in the micrometre range. Two important processes occur during vitrification, which sets in at about 900 °C, the first being a sintering together of what is left of the clay mineral particles. The second process involves melting of the oxides of the two alkalis, potassium and sodium, and the two alkaline earths, calcium and magnesium. These molten oxides are referred to as *fluxes*, and they spread themselves in a thin liquid layer around the sintered clay particles. The actual rates at which these microscopic processes take place is of obvious interest to the brick technologist. Much of the relevant experimental information, for the important component materials, was collected in the 1950s. J. E. Burke, R. L. Coble and W. D. Kingery were particularly prominent in analyzing these data in terms of atomic-scale mechanisms.

The strength of the final brick derives partly from the sintering and partly from the cohesive effect of the solidified fluxes. The colour of the cooled brick is determined largely by its iron content. If there is very little iron, the brick has a light brown colour. If iron is present to an appreciable degree and vitrification occurs under oxidizing conditions, the ferric form of iron,

Fe^{3+}, is produced, and the bricks are red. If, on the other hand, vitrification occurs under reducing conditions, the ferrous form, Fe^{2+}, gives the bricks a blue shade. It is interesting to note that one also encounters the colours red and blue in the different forms of blood, arterial and venous, and here too it is the iron atom that is the underlying cause.

Several of the stages in the manufacture of bricks are reminiscent of the way in which cement clinker is produced; both processes involve the heating of a mixture of water and soil minerals. They are both ancient arts, and for both the new era can be said to have started in the middle of the nineteenth century. As we saw earlier, Aspdin's Portland cement process dates from that time, and the *Hoffmann kiln*, named after Friedrich Hoffmann, in which all stages of brick production are carried out continuously under a single roof, made its appearance in 1858.

Summary

The hardness and insulation properties of ceramics derive from their combination of covalent and ionic bonding. Many ceramics are based on silicon and oxygen, the atoms of these elements being arranged in a tetrahedron. The tetrahedra may form chains, sheets or a three-dimensional network, depending on which other elements are present in a particular ceramic. Being already saturated with oxygen, many ceramics cannot be further oxidized, and this makes them highly resistant to corrosion. The directionality of the bonding makes dislocation movement difficult, and this is why they are hard but brittle. Apart from those that occur naturally, this class of materials can be divided into ceramics that are formed by heating, such as brick and porcelain, and those produced by chemical reaction, such as cement. Concrete, which is a composite of cement, sand and the small stones known as gravel, is strong in compression but weak in tension and shear. This shortcoming can be compensated for by the incorporation of steel rods, the resulting composite being referred to as reinforced concrete.

9

Gold is for the mistress –
silver for the maid –
Copper for the
craftsman cunning at
his trade.
'Good!' said the Baron,
sitting in his hall,
'But Iron – Cold Iron – is
master of them all.'

Rudyard Kipling (1865–1936)
Cold Iron

Monarchs of the cave: *metals*

It is no exaggeration to say that of all the materials in common use none have had a greater influence on our technological development than *metals*. Their unique combination of ductility and high electrical and thermal conductivity, together with their ready alloying, qualifies them to play a role for which there is no stand-in. Although they were not the first substances to be pressed into service, their use does go back at least 8000 years. The order in which the various metallic elements were discovered was inversely related to the ease with which they form compounds with the non-metals, particularly oxygen. Thus, with the possible exception of meteoric iron, the first metals known to man were probably gold, silver and copper, these being the only three common examples that actually occur as the metallic element. Conversely, there are metals which so tenaciously hold on to oxygen that they were not extracted from their ores until quite recently. One can imagine early man being attracted by the glint of gold nuggets, and he must have been intrigued by their ease of deformation compared with the brittleness of flint and other stones. The major use of gold has been in ornamentation. Down through the ages it has been employed for little else than jewellery, coinage and, more recently, dental fillings and electronic circuitry. Indeed, the bulk of all of this metal that has been mined has never been gainfully employed; it languishes in storage vaults. Gold has been the stuff of romance and intrigue, but it did not lend its name to an epoch. The Stone Age gave way not to a gold age but to the Bronze Age.

Although *bronze* consists primarily of copper, its manufacture had to await an important discovery. The alloy contains about 10% of *tin*, which can be obtained only by smelting one of its ores. The first metal to be extracted by *smelting* (which takes its name from the Germanic word for melting) was probably *copper* itself. Just how early man stumbled upon the process is not known, but it was probably through the chance use of copper-bearing rock in a primitive hearth. Since the fuel must have been wood, charcoal was present and ready to reduce the metal with the aid of the fire's heat. When the fire had burned itself out, the melted and solidified copper would have been visible in the depressions at the base of the fireplace, and this ultimately led to the idea of casting. The discovery of bronze, man's first alloy and his first really useful metal, was probably also the product of luck. It required the availability of copper and tin ores at the same location, and there is considerable evidence that the vital advance was made in the region of Sumer at the northern tip of the Persian Gulf, around 3000 BC.

A rudimentary metallurgy could be said to have established itself at about this time. It was already appreciated that metals become hardened when

Because it resists oxidation, *gold* occurs in its elemental state, and it was one of the first metals used by ancient artisans. It is usually found in a quite pure form, and this enhances the natural *ductility* which permitted the early craftsmen to produce exquisitely detailed ornamentation. The gold model of a chariot shown here, produced in the Bactrian area of western Asia around 500 BC, is a good example of Achaemenian (Persian) art. It was found in 1877, in the dry bed of the Oxus River, which now divides Afghanistan from Turkmenistan.

they are *cold-worked*, and it must have been immediately obvious that cast bronze is harder than cast copper. Then again, the basic idea of *corrosion* had probably made its mark, because scratching the dull surface of copper would have revealed the brightness of the underlying metal. It is interesting to reflect on the way in which Nature seems to lend most of its metals, ultimately reclaiming them through corrosion. Looking at a rusty nail or a green copper roof, we see a metal which has already started on its return journey. The chemical side of metallurgical practice is concerned not only with the coaxing of metals out of their ores but also with persuading them to remain in the metallic form as long as possible.

Metallurgy is, then, one of our most ancient arts, but it is often referred to as one of the youngest sciences. It has come of age only in recent decades, and its rapid progress has been made possible only by developments in physics and chemistry. It is, moreover, a rather large domain. The metallurgist is called upon to provide pure metals and *alloys* for components with a wide variety of sizes and shapes, and these are required to perform under a broad range of conditions. That the design criteria for many of the metallurgist's products become increasingly more stringent can be appreciated by contemplating some of the newest applications: metals to withstand irradiation in nuclear reactors; metals to tolerate the high re-entry temperatures experienced by space vehicles; metals to perform correctly at temperatures as low as −270 °C in superconducting magnets; metals that can absorb large quantities of hydrogen and thereby serve as fuel reservoirs; and metals capable of being fabricated in the Lilliputian dimensions of miniaturized electronic circuits. In the face of such demands there is no room for the trial-and-error approach of earlier times. Today's metallurgist must know why

The chemistry of early iron making

The chemistry underlying modern *steel production* has much in common with that of the processes used by the Iron-Age artisans. When the iron ore *haematite*, Fe_2O_3, is heated together with a source of carbon, such as *charcoal*, the following reaction occurs:

$$3Fe_2O_3 + 11C \rightarrow 2Fe_3C + 9CO$$

The oxygen is removed from the ore by making it combine with carbon to form carbon monoxide gas. An oxide has been exchanged for a carbide. The latter is known as *cementite*, and it contains a little under 7% of carbon by weight. This is not the only possible reaction; in practice there is also

$$Fe_2O_3 + 3C \rightarrow 2Fe + 3CO$$

in which pure iron is actually produced. Iron and cementite are mutually soluble, and the mixture of these two components present in the early hearths, known as *bloomeries*, must have contained about 5% of carbon. If only iron and carbon had been present, this composition would have corresponded to a melting temperature slightly above 1100 °C, but with the various impurities present in the ore and charcoal, the melting point was low enough to fall within the capability of an open fire. Iron is brittle if it contains too much carbon, and the heating, folding, and beating procedure used by the early blacksmiths unwittingly removed some of the carbon, and thereby made the product more *malleable*. The key process was the formation of the oxide, FeO, on the outer surface of the metal, when it is heated in air. The iron oxide appears in the form of a black scale, and the term *blacksmith* arose from the ubiquitous presence of this substance in the early forges. The repeated folding ensured that a sufficient amount of the oxide came into contact with the carbide so that the reaction

$$Fe_3C + FeO \rightarrow 4Fe + CO$$

could occur, converting the cementite into the pure metal.

metals behave as they do, and this requires an understanding of them at the atomic level. He or she must know the ideal structure of their material and also which imperfections are present, because it is the latter which determine the microstructure, and through this the macroscopic properties. But before the rolling mills, hydraulic presses, lathes, drills, soldering irons and welding torches can convert the raw ingot into the finished product, the elementary metals must be extracted from their ores. It is therefore appropriate that we begin with this part of the story.

As is only to be expected, the degree to which different metals are used is determined largely by economic considerations. Contributory factors are the availability of suitable ores and the amount of difficulty encountered in recovering the metal from them. Progress in extractive methods has frequently produced dramatic changes in the overall picture. In the middle of the nineteenth century, the cost of producing aluminium was so prohibitive that it was used only to about the same extent as gold, and in similar applications. Then again, the present prominence of steel owes much to the fact that its price was reduced by more than a factor of ten during the second half of the nineteenth century. Another important factor favouring *iron* and *steel* is the richness of several of the iron ores. *Haematite*, Fe_2O_3, occurs in ores containing as much as 50% iron by weight, while for *magnetite*, Fe_3O_4, there are ores with up to 70% of the metal. In the ores of copper, lead and nickel the yields are usually 4% at best. At present, the global output of steel is about ten times larger than that of all the other common metals put together. More than half of the elements are metals, and there is not room

The industrial extraction of lead

Few ores are as rich in their respective metals as are those of iron. The extraction of the desired metal invariably requires several steps, including a final refinement to separate out impurity metals. The process typically starts with a concentration operation such as roasting, to change the chemical association of the metal, or the impurities, or both. Concentration is also achieved physically by crushing and gravity separation, and chemically by *leaching*. The latter involves solution in an inorganic liquid, such as dilute sulphuric acid, and subsequent selective *precipitation*. An example of the *roasting process* is the extraction of *lead* from *galena*, PbS, which is not readily reduced by carbon. Roasting in a current of air produces the reaction

$$2PbS + 3O_2 \rightarrow 2PbO + 2SO_2$$

The air is then cut off, more of the ore is added, and the temperature is raised. This gives the molten metal, via the reaction

$$2PbO + PbS \rightarrow 3Pb + SO_2$$

The validity of such reaction expressions depends on compatibility with the free energy balance under the prevailing conditions. This is determined by the changes of internal energy and entropy during the reaction, but it is also influenced by temperature. The entropy contribution to the free energy increases with rising temperature. It is for this reason that the second of the above processes requires a somewhat higher temperature.

here to discuss them all. Because of its leading position in the field, iron is our best choice as an illustrative example.

Somewhat offsetting the many advantages enjoyed by iron is its rather high melting point, 1536 °C, which is almost 500 K above that of copper. It was not possible to reach such a high temperature during the *Iron Age*, which started around 1400 BC, and the only really metallic iron available to our early forebears was that obtainable from *meteorites*. We can be reasonably sure that this was the case because examples of iron surviving from the period contain unusually large amounts of nickel, a meteoric characteristic. Moreover, religious writings of the time frequently refer to iron as being sent from Heaven. All iron ores are essentially oxides, and the extraction of the metal is a question of overcoming the natural affinity between iron and oxygen. Fortunately, carbon has an even stronger attraction for oxygen, so it can be used to remove the latter from the ore. The smelting of an iron ore, in the presence of carbon, produces an impure type of iron if the temperature reaches about 800 °C. This is a good example of the lowering of the melting point by the addition of a second element. The primitive hearths were capable of reaching this temperature, and the product was loose granular aggregates that are referred to as *blooms*. This type of iron, known as *sponge iron*, contained not only the metal but also bits of charcoal, iron carbides, and various other impurities. As was already appreciated during the Iron Age, most of these undesirable components can be removed by reheating the crude ingot to about 750 °C and hammering it. This produces *wrought iron*, the purest form of iron normally available, and nearly free from carbon. It is very tough and has a fibrous structure. The technique must have developed empirically, and the systematic beating and folding carried out by the early forgers of swords and other weapons assumed the form of a religious ritual. The oldest surviving example of the blacksmith's art is a

Coalbrookdale, in Shropshire, England, became a leading centre of the iron industry during the eighteenth century, due to its proximity to coal deposits and the technical advances achieved there by Abraham Darby (1678–1717) and others. This impression of the region was painted by Philip de Loutherbourg in 1801. It is entitled 'Coalbrookdale by Night', although the actual scene is now believed to depict the nearby Bedlam Furnace.

dagger blade dating from about 1350 BC, recovered from the tomb of the Egyptian monarch Tutankhamen.

The fact that blowing on a piece of charcoal raises its temperature must have been obvious from the earliest times, and there is ample archaeological evidence that the early smelters were quick to hit upon the idea of using bellows. The steady improvement of these devices eventually led to the passing of a significant milepost. The temperature attainable reached the magic figure of 1130 °C, at which a mixture of iron and about 4% carbon melts. Material of this composition would gradually develop through the diffusion of carbon into the iron, and when the dual requirements of temperature and composition were satisfied, the alloy became a liquid and ran out the bottom of the hearth. The emerging liquid metal was made to flow into a series of long narrow channels in a sand bed, where it solidified. The depressions were in the form of a comb, and a fanciful resemblance to a sow and its litter suggested the name *pig iron*, a term that has survived even though pig beds are no longer used. With this successful production of pig iron, the *blast furnace* had come of age, and most of the subsequent developments in the industry centred around making improvements to its mode of operation. A key development was the introduction of *limestone*, which combined with the unwanted earthy minerals associated with the ore. This produced a *slag* which, being lighter than iron, floated to the top of the mixture, from where it could be readily removed.

One of the early uses of pig iron was in the manufacture of cannon-balls, and the consumption of these by marine artillery underlines an essential conflict of interests. The wood used to produce charcoal required in the smelting of iron was also needed for the production of naval vessels. Attempts to use coal instead of charcoal met with little success because it contains too much sulphur and phosphorus, and these elements make the iron unacceptably brittle. At *forging* temperatures, sulphur combines with iron to produce a sulphide, which weakens the grain boundaries and causes crumbling. This is known as *red shortness*, or *hot shortness*. Similarly, phosphorus and iron produce iron phosphide, which gives the embrittling of

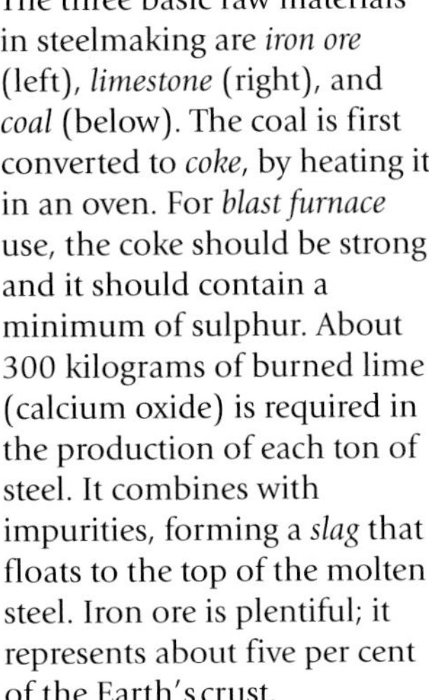

The three basic raw materials in steelmaking are *iron ore* (left), *limestone* (right), and *coal* (below). The coal is first converted to *coke*, by heating it in an oven. For *blast furnace* use, the coke should be strong and it should contain a minimum of sulphur. About 300 kilograms of burned lime (calcium oxide) is required in the production of each ton of steel. It combines with impurities, forming a *slag* that floats to the top of the molten steel. Iron ore is plentiful; it represents about five per cent of the Earth's crust.

iron at atmospheric temperature referred to as *cold shortness.* A most welcome development, therefore, was the use of *coke*, notably by Abraham Darby, who patented the process in 1709. Coke has much to recommend it. It is strong, yet light and porous, and it contains much less sulphur than does coal. Coke is thus able to support the ore, to prevent it from becoming too compact, and to allow the upward flow of hot air and the downward flow of molten iron. It was first employed at Coalbrookdale near the River Severn in Shropshire, England, and this area became one of the cradles of the *Industrial Revolution.* The switch in one of the basic raw materials caused a relocation from areas with large forests to places endowed with rich coal deposits.

Pig iron contains too much carbon to make it a practical material, and if it is used directly for cast-iron objects, these are likely to suffer from serious deficiencies. A classic example of this shortcoming was the unreliability of cast-iron cannons against fracture. Since this would invariably occur during actual use of the weapon, the consequences were most dire, and in 1784 the time was ripe for another major advance. In that year, Henry Cort patented his process for *puddled wrought iron.* The chemistry was not new, since it embodied something that had been exploited earlier, namely reduction of the carbon content by making it interact with the oxide. Cort's contribution was to introduce the oxide *in situ.* He designed what is now known as the *reverberatory furnace*, the chimney of which contained a short horizontal section with a bowl-shaped depression. In the latter, a quantity of pig iron

The electrolytic production of aluminium
The best example of the commercial production of a metal by an electrolytic process is that of *aluminium*. It was pioneered by Charles Hall (1863–1914) and Paul Héroult (1863–1914), independently in 1886, and it was responsible for the sudden drop in manufacturing cost that assured aluminium its present industrial prominence. The input material is *alumina*, Al_2O_3, which is extracted from the most economic ore, bauxite, $Al_2O_3.2H_2O$, by the *Bayer process*. The ore contains iron oxide and silica impurities, which are removed by solution in hot concentrated caustic soda (i.e. sodium hydroxide) to produce sodium aluminate. The reaction is

$$Al_2O_3.2H_2O + 2NaOH \rightarrow 2NaAlO_2 + 3H_2O$$

and the impurities remain undissolved. Aluminium hydroxide crystals are added to the filtered solution, to cause *seeding* of the aluminium hydroxide precipitate. The latter is then converted to almost pure alumina by roasting at 1100 °C. The *Hall–Héroult process* itself hinges on the discovery that alumina dissolves in molten *cryolite*, Na_3AlF_6. Metallic aluminium is extracted from this electrolyte in a cell operated at 6 volts, with expendable carbon anodes and an optimal temperature of 960 °C. The carbon anodes are burned to carbon dioxide, and the resultant energy liberation is used to heat the cell. The overall reaction is

$$2Al_2O_3 + 3C \rightarrow 4Al + 3CO_2$$

and the approximately 99.6% pure aluminium appears as a solid upper layer floating on the electrolyte.

was melted so as to form a puddle, into which was stirred an appropriate quantity of iron oxide, the latter usually being scale recovered from the rolling mill. Stirring was performed with a *rabble*, an implement reminiscent of a garden hoe, and gradual removal of carbon caused the melting point of what was left in the puddle to increase steadily. A point was reached at which the material became solid, and it was then removed from the furnace. The final product could be readily worked, because of its low carbon content, and it was in fact *wrought iron*.

The term *steel* refers to a mixture of iron and carbon when the content of the latter is under about 2%. Pig iron contains as much as 4% carbon, and this is clearly more than is desirable for most uses, because the material is brittle. Wrought iron, on the other hand, is too soft because it contains virtually no carbon. A decisive advance in the making of steel was secured by the efforts of Henry Bessemer, who unveiled his radically new converter in 1856. It was based on his own discovery that the heat given out during oxidation of carbon and silicon impurities in iron is sufficient to raise the temperature into the range 1600–1650 °C, which is actually in excess of the melting point of pure iron. His original aim was to remove the carbon and silicon from the iron by oxidizing them, but the fact that his process is exothermic means that the *Bessemer converter* also requires no fuel. The vessel has the shape of a large open-mouthed bottle which can be rotated so as to lie in either a horizontal or vertical position. It has openings at the bottom, called *tuyères*, through which a blast of air can be forced. With the bottle horizontal, liquid pig iron is admitted at the mouth, and the device is then rotated into the vertical position. Air is bubbled up through the melt for about 20 min, and the bottle is lowered again to the horizontal position so that the steel can be poured into a convenient solidification vessel. The reaction is quite violent, and a shower of flame and sparks gushes forth from the mouth of the bottle during the 'blow'. The process also has the economic merit of speed, in which it is clearly superior to the puddling method.

The first all-metal bridge was this thirty-metre long structure that crosses the River Severn at *Ironbridge* in Coalbrookdale, Shropshire (top, with detail to right). It was made of cast iron, and was opened to the public in 1779. (The present author's wife and father are just visible at the centre of the bridge.) The thousand-metre long *Forth Bridge*, made of *open-hearth steel*, was completed just 110 years later.

The main disadvantage of the original Bessemer design, namely that the output material contained too much oxygen, was removed by Robert Mushet, in 1857. He showed that the steel could be deoxidized by the addition of a small amount of manganese. This method of removing oxygen, which is carried out before the final casting, is referred to in the trade as killing. Bessemer's goal was the large-scale, rapid, and cheap production of wrought iron, but the metal recovered from his converter differed from this quite significantly. It was what is now known as *mild steel*, and it is chemically rather like wrought iron. But the two are physically quite distinct. Wrought iron retains a thread-like residue of slag, which confers a certain degree of corrosion resistance, and acts as a *flux* (i.e. a substance which assists fusion) when the metal is welded. Mild steel contains no slag, and it is inferior both regarding corrosion resistance and weldability. It is, on the other hand, quite superior in its mechanical properties. Nowadays, mild steel can easily be *welded* electrically or by an oxygen-gas torch.

The original Bessemer process was not entirely trouble free. It did not remove phosphorus, a common constituent of iron ore, so the cooled metal

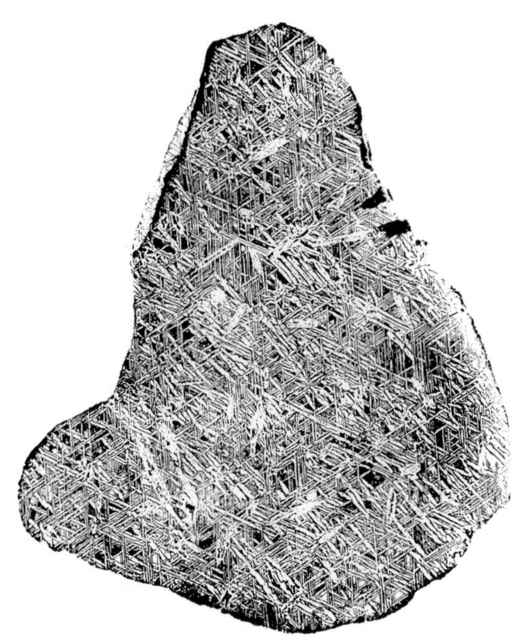

Until the microscopic studies of the surfaces of meteorites, carried out by Alois Von Widmanstätten (1754–1849), it was believed that the crystalline state of matter is almost exclusively confined to minerals. In the early nineteenth century, Widmanstätten polished a cross-section of an *iron–nickel meteorite*, and then *etched* it in acid. His surprising observation of a network of straight lines, reproduced here, revealed that metals also consist of planes of atoms, arranged in a regular manner.

tended to suffer from cold shortness. This difficulty was resolved by S. G. Thomas, in 1879, who realized that both phosphorus and the lining of the converter were chemically acidic and would thus show no mutual affinity. Use of a basic liner should, he argued, lead to removal of the phosphorus by chemical combination. Calcined dolomite, MgO.CaO, was tried, and it worked perfectly. There was even a bonus: the resulting slag was rich in calcium phosphate and could be ground to a powder and sold as fertilizer. The Bessemer method has now been largely superseded by the *open-hearth process*, which permits a greater degree of control over such production variables as fluidity, carbon content and the removal of phosphorus. Introduced in 1863, by C. W. Siemens, the open-hearth furnace exploits a regenerative principle to recover waste heat, and thereby reduce fuel consumption. It differs from the earlier reverberatory furnace in that it is fired, by coal gas, from either end alternately, the heat absorbed from the waste gases during one half cycle of operation being transferred to the fuel and air at the next reversal. A still later development, dating from the early 1950s, has been the emergence of *Basic Oxygen Steel*, produced by blowing oxygen at high pressure, through a water-cooled nozzle, onto the surface of the molten iron. The heating for much of today's steel production is supplied by the *electric arc*, another Siemens innovation. A dramatic impression of the rapid advance of the industry can be gained by comparing the small cast-iron bridge that spans the 30 m width of the River Severn at Coalbrookdale, and the mighty railway bridge, of open-hearth steel, that makes the 1000 m leap across the River Forth in Scotland. The former was completed in 1779, while the latter was opened just 110 years later.

The motivation for the industrial enterprise involved in extraction of metals lies in the unique qualities of the metallic state. Its distinguishing feature is the delocalizing of the outer electrons. This is a consequence of the typically low values of electron affinity and ionization energy displayed by metals, as collectively expressed by their marked electropositivity. The electrons in the outermost, non-filled orbits become detached and wander

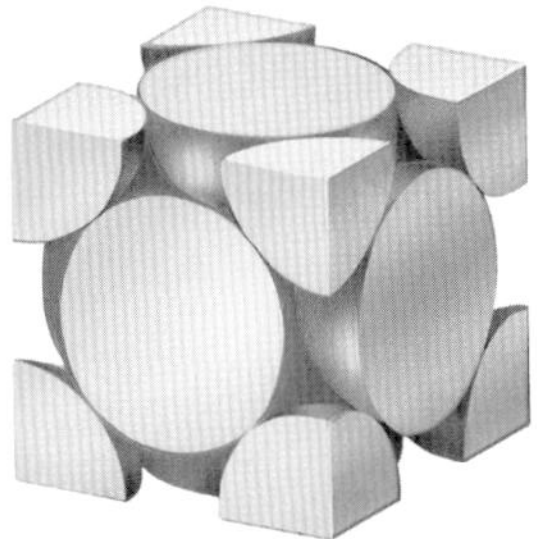

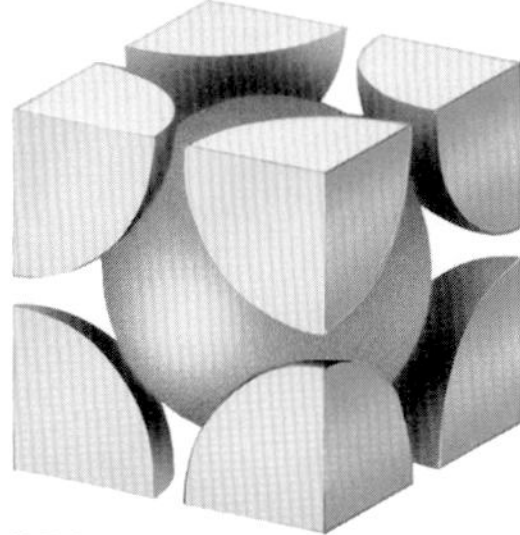

Depending upon the temperature, pure iron can adopt either of two different crystal structures: *face-centred cubic* (fcc) and *body-centred cubic* (bcc). The density of atomic packing is higher in the fcc case, but it nevertheless contains *interstitial holes* that are larger than any found in the bcc form. The open space in the bcc structure is more extensive, but it is also more uniformly distributed. The positions of the largest holes are clearly visible in these cut-away models of the two unit cells. A carbon atom is readily accommodated in the fcc structure but it causes considerable distortion in a bcc lattice.

throughout the crystal, leaving behind positive ions. The atoms thereby donate a share of their electron complement to a collective sea of electrons, which both guarantees overall cohesion and supplies the means of electrical conduction. Most of the defining characteristics of metals are related to the peculiarities of this bonding. The non-directional nature of the latter shows that the packing of the ions is determined largely by considerations of geometry. Robert Hooke had already stumbled on the basic idea in 1665, when he demonstrated the connection between the macroscopic shapes of crystals and the different possible arrangements of regularly stacked musket balls. We now see that his models were particularly appropriate to metals, since each ball can be imagined as representing a metallic ion, whereas his simulation has dubious validity for covalent crystals. With their ions neatly packed like apples in a fruit store's display, it is not surprising that metals are the densest of substances, and it is easy to see why they so readily lend themselves to *alloying*; the occasional substitution of an orange for an apple will not upset the stacking. Easy alloying provides the basis for *soldering* and *welding*. The non-directional and indiscriminate nature of the metallic bond is the very antithesis of what occurs in the organic and biological domains, in which specificity is one of the vital factors. With the ions in a metal neatly arranged in a crystalline array, and in the absence of the restraining influence of directional bonds, we might guess that one plane of atoms can easily slide over an adjacent plane. Such slipping is indeed a common phenomenon in metals, and it is the underlying cause of their easy deformation.

Much of the progress in understanding the properties of materials has come about through advances in the examination of their microstructure. Such observations have been particularly important for metals because their outer surfaces give few clues as to what is going on inside. Asked to compare the faceted exterior of a calcite crystal and the smooth curves of a gold nugget, we would probably not guess that both of these substances are crystalline. The first demonstration of an underlying symmetry was achieved by Alois von Widmanstätten, during his studies of an iron–nickel meteorite. He cut a section through this object, polished one of the surfaces, and then attacked it chemically with an acid solution. The latter process, referred to as *etching*, revealed a well-defined geometric pattern of minute crystals. This early observation was made almost 60 years before the event usually accepted as the birth of *metallography*. The modern era was ushered in when Henry Sorby made his first studies of metallic surfaces with the aid of a *reflected-light microscope*, in 1864. When an etched surface is thus examined, it is seen to be divided up into areas that are bounded by fairly straight sides. Observations such as this have now become a common part of metallurgical practice, and they are fully understood. The different polygons are the intersections with the specimen surface of three-dimensional polyhedra, each polyhedron being a small crystal referred to as a *crystallite* or *grain*. The grain diameter usually lies in the range 0.01–1.00 mm. The lines of demarcation seen in the microscope are simply the borders between the crystallites: the *grain boundaries*. In many cases these are particularly susceptible to chemical attack, and this is why the etching process often renders them visible. Modern metallography makes considerable use of *polarized-light microscopy*. Inspected with this technique, a polycrystalline surface takes on an appearance reminiscent of a patchwork quilt. The individual grains have different colours, which change kaleidoscopically as the

The iron–carbon phase diagram

The thermodynamically stable forms of alloys, for various compositions and temperatures, are indicated on what is known as a *phase diagram*. This example shows the iron-rich end of the *iron–carbon system*, and it covers the range of *common steels*. The diagram has been coloured according to a convention in which pure colours represent single phases and intermediate colours represent two-phase regions. Below 910 °C, the stable form of pure iron has the bcc structure; interstitial solid solutions of carbon in this structure (pure blue) are known as alpha (α)-iron, or *ferrite*, the maximum carbon solubility being 0.025% at 723 °C. Between 910 °C and 1390 °C, pure iron adopts the fcc form, and interstitial solid solutions of carbon in this structure (pure red), up to a maximum of just under 2% at 1130 °C, are called gamma (γ)-iron or *austenite*. Above 1390 °C, pure iron reverts to the bcc modification, so the pure blue colour is again appropriate for the interstitial solid solutions, which are called delta (δ)-iron and which show a maximum solubility of about 0.1% at 1493 °C. Liquid iron–carbon is indicated by the pure yellow colour. The other pure state does not actually appear as such on the diagram, but its implied colour code is white. This is the orthorhombic intermetallic compound iron carbide, Fe_3C, known as *cementite*. Two important *eutectic points* are indicated. (The word eutectic is derived from the Greek word for 'easily melted', and it denotes that composition which yields the lowest melting point for a given phase.) The lowest melting point for an iron–carbon alloy occurs at 1130 °C, for 4.3% carbon. Cooling this eutectic liquid produces the approximately one-to-one mixture of austenite and cementite known as *ledeburite*. The slow cooling of austenite corresponding to the eutectic at 723 °C and 0.8% carbon gives the characteristically patterned mixture of ferrite and cementite known as *pearlite*, the proportions being about seven-to-one. Rapid cooling of austenite with this same composition produces the hard metastable form known as *martensite*.

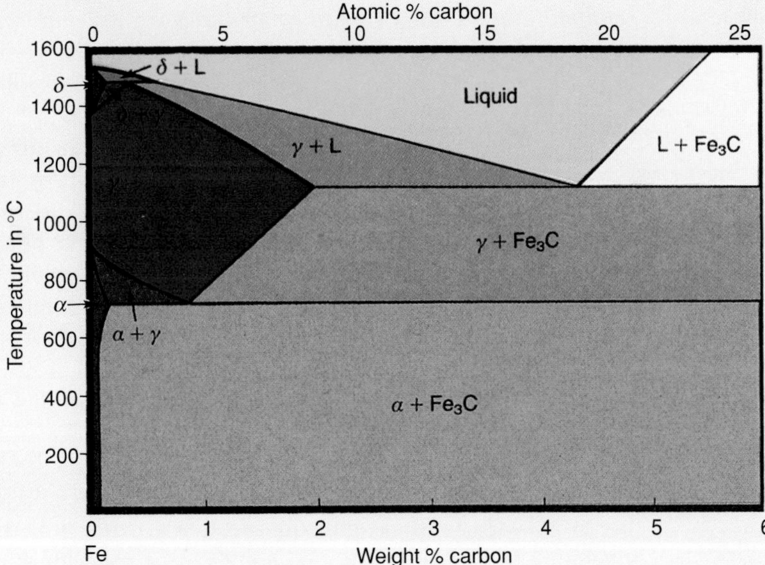

surface is rotated. This beautiful effect is produced because the degree to which light waves of the various wavelengths are reflected depends upon the orientation of the crystal planes.

In the decades following the pioneering work of Sorby, several other types of *crystal defect* were discovered, most of them more subtle than the grain boundary and, as a consequence, more difficult to observe. These were introduced in Chapter 5, the list including dislocations, stacking faults, twins, vacancies, interstitials and impurity atoms. All have now been made visible through the development of the *electron microscope*, invented by Ernst

The arrangement of grains in a *polycrystal* is determined during *casting*, which is solidification by cooling from the molten state. The pattern of solidification follows that of the temperature gradients in the mould, high gradients giving the long grains known as *columnar dendrites*. This *zinc* polycrystal was rapidly cooled from 800 °C in a roughly cylindrical mould. This cross-sectional view, which covers an actual diameter of about five centimetres, reveals the columnar dendrites. Their long dimensions lie along the direction of maximum temperature gradient, which is approximately at right angles to the surface of the mould. Directional solidification is usually undesirable and it can be minimized by maintaining a fairly uniform temperature throughout the cooling. This is often achieved by casting in sand, which is a good thermal insulator.

Ruska in the 1930s, the *field-ion microscope*, and latterly the *scanning tunnelling microscope*. The latter two instruments, the former invented by Erwin Müller in the 1950s and the latter by Gerd Binnig and Heinrich Röhrer in the 1980s, actually permit observation of metals at the atomic level. The increasing sophistication afforded by these instruments has not caused the *optical microscope* to be pensioned off. On the contrary, the mechanical behaviour to be expected of a metallic component can often be gauged by those features of the microstructure that are observable at the optical level. This applies not only to the grain size but also to the nature and distribution of *precipitates*. Metallographic examination of ancient objects reveals the microscopic basis for the processes that were developed empirically. The repetitive heat, fold and beat procedure used by the early blacksmiths was effective because it gave a small grain size. The two faces brought into contact at each fold would not in general match up at the atomic level and would therefore form a grain boundary. By repeating the process many times, a large number of boundaries were introduced, increasing the mechanical strength of the metal considerably.

By dint of the quantity of iron and steel used industrially, the most important example of strengthening by impurities is that seen in the iron–carbon system. The reason why some iron-rich alloys of iron and carbon are ductile and others brittle lies in differences in crystal configuration. To begin with there is pure iron which, below 910 °C, has the body-centred cubic (bcc) structure and is known as *alpha-iron*, or *ferrite*. Heated to above this temperature, it transforms to *gamma-iron*, or *austenite*, which has the face-centred cubic (fcc) structure. Iron is rather unusual in that it actually returns to the bcc structure between 1390 °C and the melting point at 1536 °C. This high temperature phase, *delta-iron*, will not concern us here. Inquiring how much carbon can be in solution in these different forms of iron, we discover something puzzling. The maximum amount of carbon that can be dissolved in the bcc phase is 0.25% by weight, and this composition is achieved at 723 °C. At the same temperature, the fcc phase can dissolve up to 0.8% by weight. The fcc structure is more densely packed, and yet it can accommodate more carbon. The reason for this apparent paradox lies in the way in which the open space is distributed in these two structures. The bcc lattice contains many small holes whereas the fcc arrangement has fewer larger holes. In fcc iron, the centres of the largest holes are surrounded by six iron atoms, and the available space has a radius of 0.106 nm. This is more than enough for the carbon atom, with its 0.092 nm radius. In the bcc structure, the largest hole has a radius of 0.085 nm, and the carbon atom is presented with a rather tight squeeze. It is worth noting that similar considerations apply to the choice of metals to be used for storage of hydrogen. This prime candidate as a vehicular propellant of the future will probably be transported in vanadium or magnesium.

The maximum amount of carbon that can be dissolved in austenite is just under 2% by weight, and this occurs at 1130 °C. We must note that the atomic weight of carbon is much less than that of iron, so the traditional use of weight in the description of alloy composition gives a rather misleading picture in the case of the *iron–carbon system*. It is worth emphasizing, therefore, that 2% of carbon by weight corresponds to 8.7% of carbon by atomic composition. Any composition below about 2% carbon can always be brought into a situation of complete solid solution, simply by adjusting the temperature, and such an alloy is a *steel*. For compositions above 2%

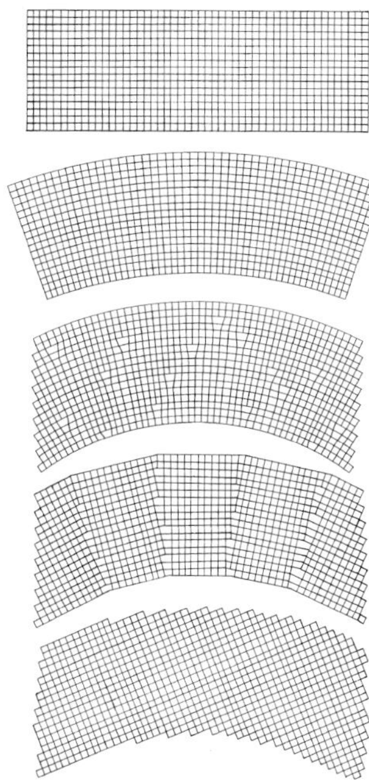

Five idealized states of a crystal are identified here. The perfect crystal (top) can be deformed elastically (upper middle) up to a few per cent strain, and still recover its original condition when the stress is removed. If the *elastic limit* is exceeded, the resulting *plastic deformation* introduces dislocations which permanently change the shape of the crystal (middle). The strain energy can be reduced by *polygonization* (lower middle), during which a polycrystalline structure is established; the *dislocations* rearrange themselves, primarily by *glide*, into groups which constitute the *grain boundaries*. If the temperature is sufficiently high (bottom), the boundaries are removed by the *climb process* described in Chapter 5. The crystal has returned to its perfect state, but it has adopted a new shape.

carbon, no heat treatment will be capable of preventing the formation of either the carbide, Fe_3C, referred to as *cementite*, or graphite, or both these heterogeneities. Alloys containing more than 2% of carbon belong to the class known by the general name *cast iron*, of which there are several forms. Steels are divided, for convenience into three groups: *mild steels*, containing less than 0.3% by weight of carbon; *medium steels*, with between 0.3% and 0.7% carbon; and *high-carbon steels*, having from 0.7% to 2% carbon. Mild steel has only moderate strength, but it has the great advantage of being easily worked. The higher carbon steels must be heat treated after fabrication, so as to fully develop their strength potential. About 90% of all steel used for structural purposes is of the mild variety.

Anyone who has watched a blacksmith at work will have noticed that hot iron or steel components are occasionally plunged into water, as part of the fabrication process. This is clearly not just for convenience of handling, because the object continues to be held in a pair of tongs. It is exploitation of the fact that rapid cooling, or *quenching*, produces a *metastable phase*. The process is best appreciated by considering first what would happen if the alloy was allowed to cool slowly. Let us assume that the material is an austenitic alloy containing 0.5% carbon by weight, and that its temperature is 1006 °C. If the temperature is lowered to below 723 °C, the alloy leaves the austenitic region, with its stable solid solution of carbon, and passes into a region in which ferrite and cementite can coexist in a stable mixture. The separation into these two phases occurs through diffusion of carbon atoms, and the rearrangement is from a situation in which they were uniformly distributed throughout the material to one in which there is a heterogeneous division between regions with a small amount of carbon in solution and regions in which every fourth atom is a carbon atom. Under the optical microscope this slowly cooled material has a characteristic scaled appearance, and it goes by the name *pearlite*. If, instead of cooling slowly, we quench this same alloy, there will not be sufficient time for the diffusion process to redistribute the carbon atoms, and they will be caught out of place. The carbon atoms start out at the high temperature sitting snugly in the large holes in the fcc structure, but these holes shrink in size as the arrangement of the iron atoms goes over to the bcc configuration. In the final situation the atoms around each carbon atom actually adopt a tetragonal configuration, so as to better accommodate it. The structure is known as *martensite*. It is essentially ferrite with a uniform distribution of tetragonal distortions, and the latter act as barriers to dislocation motion. Martensite is in fact the hardest component in quenched steel. The austenite–martensite transition involves some interesting physics, because it is not limited by the usual diffusion process. It can therefore proceed at impressive velocities, and small crystals of martensite are formed in less than a millionth of a second.

Judged according to a dual criterion of strength and cheapness, cast iron is second only to structural steel. Some elements, when present in iron–carbon alloys, promote the formation of graphitic clusters of carbon atoms. This is particularly true of silicon. In *white cast iron*, the silicon content is kept below about 1.3%, and all the carbon exists in the form of cementite. This alloy, which takes its name from the bright appearance of its fractured surfaces, is brittle and too hard to be machined. Usually ground into shape, it finds use in brake shoes and in the edges of large cutting tools such as plowshares. *Grey cast iron* contains between 2% and 5% of silicon, and its

Metals can be broadly classified into two groups, depending upon the way in which their dislocations move. The main constraint is imposed by the atomic arrangement around the defect; if the energy required to produce a given area of *stacking fault* is low, the dislocations adopt what is known as the extended form, in which two *partial dislocations* are located at either end of a faulted region. A typical example of this type is shown in the *stainless steel* specimen on the left, and the extended configuration tends to confine the dislocation to a single atomic plane. An extended dislocation is visible at the centre (A), while one of the partial dislocations of a similar defect lying to the left (B) has advanced into the *grain boundary* which runs along the top of the picture. The width of the picture is about two hundred nanometres. At the other extreme, exemplified here by an aluminium–7% magnesium alloy, the extension of the dislocations is negligibly small and they can easily *cross slip* from one plane to another, producing *dislocation tangles* and *dislocation sources*. The width of the picture is about five micrometres.

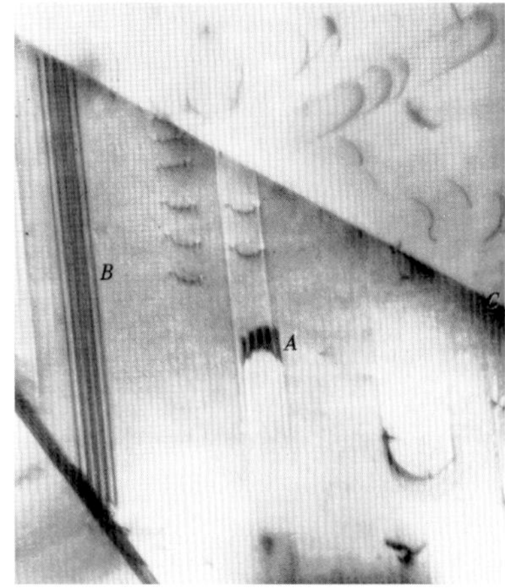

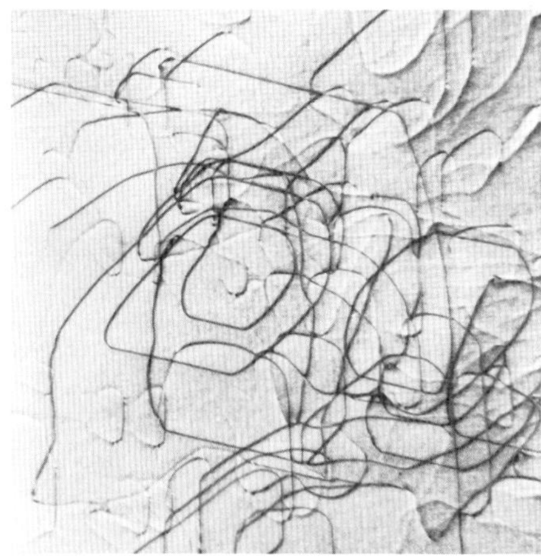

carbon is predominantly distributed as graphite flakes in a mixture of ferrite and pearlite. It owes its widespread use to ease of casting and machining, and is seen in drain pipes, lamp posts, domestic ovens, railings, engine cylinder blocks and a host of other items. The main disadvantage of grey cast iron stems from the tendency for cracks to develop at the edges of the graphite flakes. This is avoided in *black cast iron*, which is produced from the white variety by prolonged heating at 900 °C, to rearrange the cementite into rosette-shaped *graphite nodules*. The lack of sharp-edged inclusions gives a material that is a compromise between good machining and excellent mechanical strength, while retaining the advantage of being cast. Its use is standard in agricultural and engineering machinery.

Knowledge of the iron–carbon system permits us to appreciate several significant technological advances. An early hardening treatment involved heating wrought iron in contact with a source of carbon such as charcoal dust or scrap leather. The carbon diffused into the iron, but the layer thereby converted to steel was extremely thin, and the improved edge of a cutting tool, for example, would be removed after a few sharpenings. This *case-hardening* later gave way to the *cementation process*, in which a deeper penetration of carbon was achieved by packing the iron with charcoal, sealing both in an airtight clay box, and heating for several days. The product was known as *blister steel*. In a further improvement, achieved by Benjamin Huntsman in 1740, blister steel was melted, so as to even out the carbon distribution, and then cooled slowly. The carbon content of this *crucible steel* could be readily adjusted and matched to the intended use: 0.9% carbon for cutlery steel, for example.

Alloy steels designed for specific purposes emerged in the latter part of the nineteenth century. The first of these, developed by Robert Mushet in 1868, contained a substantial amount of tungsten and was intended for machine tools. Mushet's alloy hardened itself when heated and allowed to cool naturally, and its microstructure was then so stable that the hardness was retained even if the tool subsequently became very hot. The vital additive of an alloy steel introduced by R. A. Hadfield in 1887 was manganese, and the product was tough and wear-resistant without being brittle. It was suitable for crushing and rolling equipment. Finally, mention should be made of the *austenitic stainless steels*, developed by Harry Brearley in 1913, which contained appreciable amounts of chromium and nickel. The best

The change of shape provoked by mechanical stress on a piece of metal is caused by the movement of such crystal defects as vacancies, dislocations, and grain boundaries. In a typical situation, all of these contribute to some extent, their respective roles depending upon the temperature relative to the absolute melting point. Lead has a low melting point (601 K) so atmospheric temperature (about 300 K) is relatively high for this metal. As seen in the picture of old lead plumbing, even the mechanical stress due to its own weight is sufficient to cause considerable *creep*. The melting point of steel is much higher (around 1700 K) and this metal is best rolled when red hot, even though the underlying mechanisms are similar for both materials.

known version has 18% and 8% of these elements, respectively, and this is the origin of the designation 18/8 sometimes stamped on stainless ware. Interest in the development of iron-based alloys remains undiminished after more than three thousand years. In 604 BC, Nebuchadnezzar is said to have carried off into captivity a thousand blacksmiths from Damascus. Today's foundry engineer is a no less valued member of the technological fraternity.

The foundry is, in fact, the starting point for all metal technology. Irrespective of how they are subsequently processed, metals must first be cast. This term is now usually reserved for those objects, such as the engine block of a car, that are given what is essentially their final shape in this manner. Even for metals that are to be rolled, forged, drawn or extruded, however, the nature of the casting is of prime importance. It is this which determines the size and arrangement of the grains, and hence the mechanical behaviour. *Casting* is the solidification of a liquid metal into a particular shape. This is done in a mould, and the freezing produces a polycrystal. The process variables are the rate of cooling and the thermal gradient, the latter being determined by the temperature difference between the cooling liquid and the mould. A large gradient often gives a cast piece of metal having an unacceptable non-uniformity of mechanical properties. The gradient can be reduced by casting in sand, which is a good thermal insulator.

The advancing interface between the cooling liquid and the growing crystal is flat only in the rare cases of specially controlled solidification and high purity. The normal situation produces the tree-like crystalline growths known as *dendrites*, the word coming from the Greek for tree. It is this type of multiply branched growth that gives the limb of a snowflake its characteristic appearance. A *grain boundary* is formed when the surfaces of two independent dendrites make contact, and this region can become the depository of impurities expelled by the growing crystalline regions. The dendrites thus determine the shape of the grains. If the temperature gradient is high, the growth perpendicular to the mould wall is rapid, giving what are known as *columnar dendrites* and consequently elongated grains. This is the origin of the undesirable unevenness. A low thermal gradient gives dendrites with no marked elongation in any direction. It is found that the best mechanical performance is obtained if the arms of the dendrites are many and finely spaced, and such a structure is achieved if the cooling rate is high. As T. Z. Kattanis has shown, the reason for this lies in the fact that not all dendrite arms continue to grow after they have been formed. Some simply dissolve again, giving a coarsening of the arm spacing that is more pronounced for slow cooling.

About 80% of all metal products undergo mechanical processing after casting, and this is usually carried out at elevated temperature because the

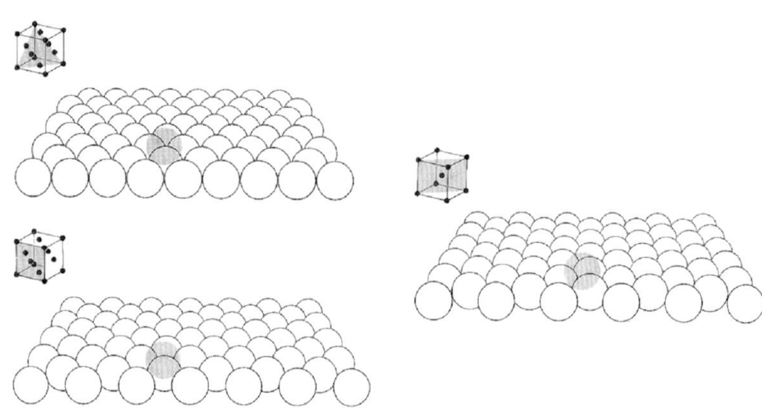

The mechanical deformation of a crystal proceeds through the consecutive motion associated with *dislocations*, rather than the simultaneous sliding of one atomic plane over another. The stress required to provoke dislocation-mediated *slip* is nevertheless related to the frictional force between adjacent planes, and this in turn is determined by the mutual penetration of the atoms across the interface. These three diagrams show all the atoms in one plane, and (shaded) the position of a single atom in an adjacent plane when this first makes physical contact with one of the underlying atoms during sliding. The upper left picture reveals why *face-centred cubic crystals* like gold and copper are easily deformed on the close-packed cube diagonal planes; the interpenetration is rather small. The cube face planes have greater interpenetration, and they are less favourable for slip. The *body-centred structure* (right) shows a more pronounced interpenetration, even for the most favourable type of plane, and *plastic deformation* therefore requires higher stress levels. This is the basic cause of iron's brittleness compared with copper, say. But these diagrams are strictly applicable only to zero temperature. At working temperatures, a reliable picture is obtained only if the thermal vibrations of the atoms are taken into account.

workability is greater when the material is hot. In considering this important property of a piece of metal we must bear in mind that it will usually be a polycrystal. Every grain changes its shape as the material is worked, but its deformation is subject to a severe constraint. The total volume of metal in each grain remains unchanged, and it can be shown that deformation will cause gaps to appear along the grain boundaries unless five independent *glide systems* are operative in each grain; there must be five different sets of crystallographic planes on which dislocations can slip. *Dislocation motion* is easier at high temperature, and the above constraint is not restrictive in hot-working. In colder crystals with little symmetry, however, the five-system condition is often not fulfilled, and rolling causes *brittle crumbling*. This is true of such metals as zinc, magnesium, titanium and uranium.

Even in *hot-working* the rate of deformation must not be too rapid. *Forging* is often carried out in several stages, and rolling speeds are kept below certain limits. This gives the metal time to dynamically accommodate to the deformation, the vital changes occurring in the distribution of dislocations. The basic mechanisms have much in common with those of *creep*, but they operate at a very much higher rate. Dislocations are generated, and they multiply, climb and annihilate. Two types of process are identified: recovery and recrystallization. *Recovery* is the decrease of dislocation concentration by mutual annihilation, while in *recrystallization* dislocations are removed by advancing grain boundaries. In each case the polycrystal lowers its energy by decreasing the defect content. Recovery becomes operative at a lower temperature than does recrystallization.

Peter Hirsch, Archibald Howie, Michael Whelan, and their colleagues have found that metals can be divided into two broad classes, according to the distribution of dislocations produced by deformation. Their difference lies in the intimate connection between dislocations and stacking faults. At the highest magnification, a *dislocation* is seen to be not really a line in the mathematical sense. It is extended along its *slip plane*, and the region thereby encompassed is a *stacking fault*. If the energy associated with the latter is low, the width of the extension is large, and vice versa. One of the above classes, exemplified by iron and aluminium, is associated with high stacking-fault energy. Deformation of metals in this group produces a cellular arrangement of *dislocation tangles*. As the amount of strain increases, new dislocations are generated at the walls of these cells, and they can traverse several cells before being stopped at another wall. Chance encounters initiate new walls, and the cell size is gradually decreased. It ultimately reaches a minimum

Metal fatigue

It is rare for a metallic component to experience only constant mechanical loading throughout its service life. Most items are subjected to the fluctuating stresses associated with such variables as non-uniform road surfaces, changing winds, wave motion, temperature and pressure cycles, mechanical vibration and periodic power surges. Whereas the single application of a static stress seldom causes mechanical breakdown, repeated loading at a lower and apparently safe level frequently leads to the catastrophic effect known as *fatigue failure*. The first systematic study of the problem was reported in 1829 by Wilhelm Albert (1787–1846), who examined the effects of fluctuating stress on iron chains. August Wöhler (1819–1914) later investigated the phenomenon in the particularly vulnerable axles of rolling stock; their frequent failure plagued the early days of railway. The term *fatigue* appears to have first been associated with failure under cyclic loading by Jean-Victor Poncelet (1788–1867) in 1839. During the following century, studies of fatigue explored such things as the relationship between stress level and average number of cycles to rupture, the influence of environmental factors, and the gradual changes in the surface of the periodically loaded component. A particularly prominent feature of the latter is the development of the surface cracks caused by intensely localized glide known as persistent slip, as illustrated in the atomic force micrograph of a copper surface reproduced here in the colour photograph. Only since the mid 1950s has it been possible to examine the internal structure of fatigued metal specimens, one of the first detailed studies by *transmission electron microscopy* being reported by Peter Hirsch, Robert Segall and Peter Partridge. This revealed the gradual development of a cellular structure in which walls composed of tightly packed dislocations are separated by regions almost free of these defects. Such a structure is now referred to as a *persistent slip band*, the first example actually being discovered by optical microscopy in 1903, by James Ewing and J.C.W. Humfrey. An example in fatigued copper is shown in the other picture, the width of this transmission electron micrograph being about 50 μm. The way in which this pattern gradually builds up, and the manner in which it is related to the surface cracks, is currently being explained, thanks to the efforts Mick Brown, Hael Mughrabi and Ole Bøcker Pedersen, and their respective colleagues. Fatigue failure caused serious set-backs to the British aircraft industry in the early 1950s, and not the least to the de Havilland company. The prototype de Havilland 110 (later in service as the Sea Vixen) disintegrated while flying at the 1952 Farnborough Air Show, killing test pilot John Derry, his flight observer Anthony Richards, and 29 spectators who were hit by one of the two jet engines. (The present author was a terrified eye-witness.) A series of fatigue-provoked disasters to de Havilland Comet airliners during the following two years led to the withdrawal of these pioneering machines, and delayed the full-scale introduction of jet passenger travel by almost a decade.

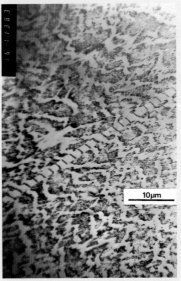

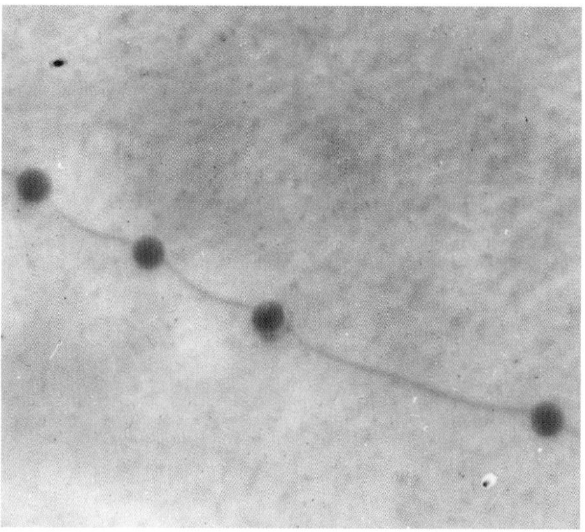

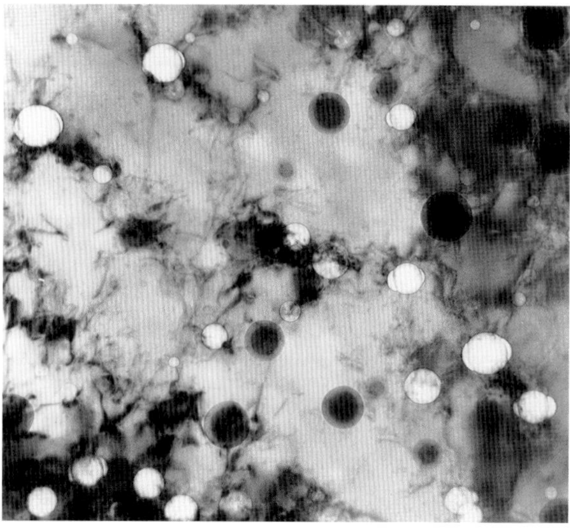

The distribution of atoms of a minority element in a crystal is determined by thermodynamics, and depends upon the temperature and the strengths of the various interatomic forces. Under certain circumstances, the alien atoms show a marked preference for bonding with their own species, and this produces the clusters of atoms known as *precipitates*. These can have a pronounced influence on mechanical properties, as do the oxide precipitates shown here. In the nickel specimen (left), they are acting as barriers to dislocation movement, and are thus causing hardening, while the silica particles in the copper–0.3% silicon alloy (right) are providing sites for embryonic *cracks* because the surrounding metal is failing to adhere to the precipitates. Under increasing stress, such cracks grow and coalesce, and ultimately cause *ductile rupture*.

value of about one micrometre, after which further dislocation movement is very difficult. The development of this cellular structure depends on the ability of the dislocations to *cross slip*, and this is easy if these defects are narrow. In the other group, which comprises metals such as brass, bronze and stainless steel, the dislocations are wide and cross slip is difficult. A cell structure is not developed in these metals, the dislocations remaining on the slip planes in which they are generated. With increasing strain the concentration of dislocations rises, and the defects push against obstacles such as grain boundaries like cars backed up at a traffic jam.

As the temperature rises, the curative process known as *annealing* becomes increasingly effective. Just as in creep, but at a much higher rate, some of the dislocations climb and annihilate, and those that remain adjust into configurations of lower energy. The distances involved are not great; in a typical metal rolled to a 90% thickness reduction, the spacing between adjacent dislocations is about a hundred atomic diameters. This is the tidying up process of *recovery*, and it appears to be limited to the cell-forming metals. Depending on the temperature and rate of deformation, these metals can in fact recover dynamically while being worked. The dislocation rearrangements convert the cells to grains of small mutual misorientation, known as *subgrains*, and these are constantly being formed and broken up. At a higher temperature, metals of both types recrystallize, and the removal of dislocations by the advancing grain boundaries makes the material softer. P. Blain, D. Hardwick, C. Rossard, C. Sellers and W. McGregor Tegart have found dynamic recrystallization in metals of the non-cell variety. This phenomenon is useful in hot-deformation because the boundary migration prevents the opening up of *intergranular cracks*, which could lead to a *fracture*. Recrystallization is a mixed blessing, however, because a large grain size means low strength in the final product. Control can be achieved by adding small amounts of selected impurities, such as niobium to low-carbon steel, to retard recrystallization.

Having discussed the way in which metallic objects are produced, we turn to their performance under service conditions. We have already seen how this can be gradually impaired by creep, or improved by the various hardening processes. We must now consider the types of failure by

In 1751, the Pennsylvania Assembly commissioned production of a bell to hang in the new State House in Philadelphia. There was no local expertise, so the task was given to the Whitechapel Foundry of London. The roughly 4000-kilogram bell arrived on the *Hibernia* about a year later, but it cracked on the very first ring. John Stow and John Pass undertook to melt and recast the bell, and they added about a hundred kilograms of copper to decrease its brittleness. Unfortunately, the new bell had a poor tone, and a second recasting was carried out in 1753. Although the tone was still disappointing, this third version remained in service until it developed a crack in 1835, while tolling the death of John Marshall. It was damaged beyond repair a year later while pealing to commemorate George Washington's birthday. (The *Liberty Bell* is perhaps trying to show us how fragile liberty really is.) A recent analysis has shown the bell to contain about 70% copper, and between 25% and 30% tin depending on position. Such a composition is well known to give a brittle product, although it should produce the best ring. Artillery commanders of the past, when short of metal for cannon, often melted down church bells. The makeshift weapons were dangerously brittle, unless a sufficient amount of copper was added, and they frequently burst during use, injuring the gun crews. Such disasters were usually attributed to divine revenge. The alloy known as *gun metal* has only 9% tin, while 30% tin gives the metal *speculum* which was used in ancient mirrors.

which the performance is abruptly terminated. The problem of *fatigue* is always a threat when components are subjected to fluctuating stress, which is unavoidable in machinery with moving parts such as engines, springs, drills, cutting equipment, and the like. Other seemingly static structures that are also liable to fatigue include aircraft wings and bridges. The *fatigue life*, or endurance, depends on the maximum level of stress reached during each fluctuation cycle. For high stress, failure might come after a few reversals, while at low stresses a component can survive for thousands of millions of cycles. Impending rupture can be detected in the form of a *fatigue crack*, which ultimately reaches the critical size given by the *Griffith criterion*. When such a crack is well developed, it is visible to the naked eye as a sharp wedge-shaped fissure, but surface microscopy reveals that an incipient crack is already present after rather few cycles, and that this actually grows during each reversal. It has a characteristic structure consisting of a crater, or *intrusion*, and an adjacent elevated region known as an *extrusion*. Crystallographic examination of these structures shows them to lie parallel to the planes on which the dislocations can glide, and the natural conclusion is that they result from dislocation motion. Why this should be so is not immediately clear. Dislocations that move through a certain distance during the first half of a cycle might have been expected to move back to their original positions when the stress is reversed, restoring the crystal to its original shape. Observation of the internal microstructure of fatigued specimens, however, has exposed an interesting heterogeneity: the dislocation distribution is not uniform. Instead, these defects gradually become arranged in quite well-defined groups, which are remarkably straight and in which the dislocation density is very high. Between these groups lie regions which are almost totally devoid of dislocations, and the overall structure resembles candy striping. The movements which ultimately give rise to the intrusions and extrusions appear to occur in some of the gaps between the dislocation-rich areas. These gaps develop a ladder-like arrangement of dislocations, and are known as *persistent slip bands*. The word slip applies to the inferred dislocation motion, but the exact nature of the process remains stubbornly obscure.

Metallic components frequently become unserviceable simply by becoming deformed out of their original shape, but they fail most dramatically when they *fracture*. As was discussed in Chapter 5, there are two extremes of this bothersome phenomenon: a gradual tearing apart, properly known as *ductile rupture*, and a sudden brittle snapping, with very little deformation of the material surrounding the crack. In engineering practice it is usually desirable that a metal be in the ductile condition, and the brittle situation often spells disaster. As a simple example, consider the bumper of a car. It is designed to deform gradually, thereby absorbing a considerable amount

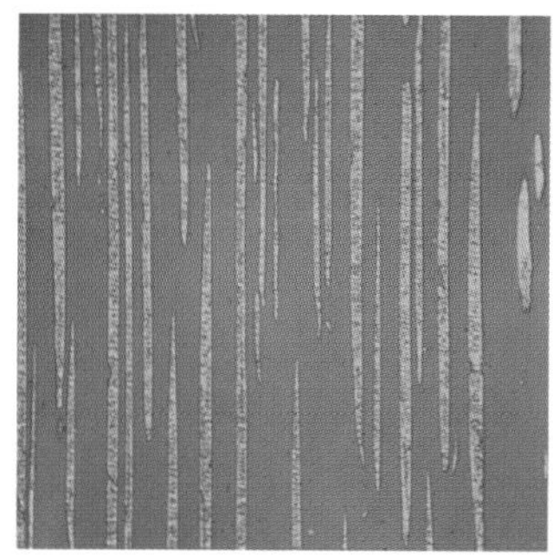

Most *fibre composites*, such as fibre glass, are fabricated in stages; the fibres are made first, and these are then imbedded in what is referred to as the *matrix*. The composite shown here is special in that it was produced by a single process: the cooling of a *eutectic alloy* of niobium and carbon. This yields an aligned array of niobium carbide fibres, known as *whiskers*, in a niobium matrix. The requirements of an ideal matrix are threefold: that it does not scratch the fibres, and produce *Griffith cracks* on their surfaces; that it flows readily under typical service deformations, thus putting virtually the entire load on the fibres themselves; and finally that it adheres well to the surfaces of the fibres, preventing them from simply being pulled out of the composite. When these requirements are fulfilled, the mechanical performance of the composite far exceeds that of either of the individual components.

of energy, which might otherwise be transmitted to more vulnerable parts of the vehicle, not to mention the driver and passengers. A bumper would be quite useless if it shattered immediately upon being struck. Inspection of the surfaces of a ductile rupture shows them to be pitted with small holes, indicating that the final failure occurs because of the formation of microscopic *voids*, or *pores*. These voids are usually associated with *precipitates* or other small particles, the cavities opening up when the stress overcomes the adhesion between the latter and the host metal. In copper specimens, for example, the void-generating particles are usually of copper oxide. If great care is taken to eliminate these, the stressed specimen can actually be made to neck right down to a point. The shape change associated both with the initial uniform deformation and with the ultimate necking of a ductile specimen occurs through the large-scale movement of dislocations. Metals having the face-centred cubic or hexagonal close-packed structure, such as copper and zinc, respectively, are always ductile because the dislocations are able to move at all temperatures. Dislocation movement is not so easy in the body-centred cubic structure, and metals such as ferritic iron and tungsten usually become brittle if the temperature falls below a certain value.

The *ductile–brittle transition* point for structural steel usually lies below atmospheric temperature, but brittle failures nevertheless do occur, often with disastrous consequences. The culprit is frequently an inhomogeneity produced during a fabrication process such as welding, and the brittle crack propagates at a supersonic velocity of about 2000 m/s. When service conditions require metallic components to withstand both chemical and mechanical burdens, *stress-corrosion cracking* may become a problem. This was first recognized in the nineteenth century with the failure of brass cartridge cases exposed to atmospheres containing traces of ammonia. It continues to cause trouble in boilers, where high stress is combined with large hydroxyl ion concentrations. The corrosion is usually quite localized, and the cracking has been attributed to the anodic dissolution of the metal at the points of maximum bond distortion. In the case of brass, this is believed to be exacerbated by preferential dezincification at the grain boundaries, along which the cracking is observed to propagate.

Although about half of the elements are metals, the engineer, faced with the task of choosing one for a particular job, is not confronted with an embarrassment of riches. Seeking a balance between effectiveness and price, he invariably finds his selection quite limited. The intended application usually makes possession of certain properties mandatory, but access to these must not compromise economic viability. The cost of putting a metallic component into service includes that of mining the raw materials, extraction from the ores, purification, production of a suitable alloy, and the various casting and machining operations. There is also the question of durability and the cost of maintenance. With the growing problem of diminishing resources, even the scrap value is becoming a major consideration for several metals. The growing interest in what used to be regarded as esoteric elements is simply a reflection of increasing specialization, and the trend can be expected to continue. These minor metals are sometimes used alone, but they are more frequently employed as *alloying agents*. Some, such as molybdenum, tungsten, tantalum and vanadium, have gained prominence because of their high melting points. Steel continues to be the industry's leader, and for structural applications where the emphasis is on strength at low cost, it is without serious competition. Its disadvantage of poor corrosion resistance can be decisive, but it is often cheaper to use steel and bear the added expense of painting, tinning, or galvanizing. Ships, for instance, will continue to be built from steel, and their owners will continue to do battle with the ravages of sea water. Copper is the first choice where one seeks high thermal conductivity with relatively modest investment, and its ease of soldering is a great advantage. The good conductivity is exploited in car radiators, engine components, and heat exchangers in chemical and power plants. Copper finds widespread use in alloys such as brass and bronze. It was also the standard conductor of electricity, but has recently been challenged by aluminium. Indeed aluminium is now displacing other metals in applications that range from aircraft fuselages and boat superstructures to cooking utensils and beverage cans. It is light, easy to work, and remarkably corrosion resistant, and it is not expensive when evaluated on the basis of volume rather than weight. Metals are eminently useful, but they also possess a natural beauty, and their recent re-emergence, in the hands of the architectural community, is particularly gratifying. Copper roofs have long been in evidence, of course, but our cities are now becoming metallurgical show-cases for anodized aluminium, stainless steel, bronze, and iron deliberately rusted.

Summary

Most metals exist naturally in compounds, and their industrial exploitation begins with extraction from a suitable mineral. The prominence of iron is related to the richness and abundance of its ores, and to the variety of properties available through alloying it with carbon. The useful qualities of metals, such as ductility, high electrical conductivity, and ease of alloying, stem from their underlying atomic arrangement: a closely packed assembly of positive ions held together by a sea of mobile conduction electrons. Dislocations in face-centred cubic metals move under low stress at all temperatures, permitting ready deformation. In body-centred cubic metals this is true only above the brittle–ductile transition temperature. All metals are initially cast, and some are given their final form by this process. The favoured uniformity of grain size is achieved with low temperature gradients in the mould, and rapid cooling gives the greatest strength. In forging, rolling, extrusion and drawing, the vital property is the workability. Dislocations must be able to slip on five crystallographically independent systems of planes in a deforming polycrystal, to prevent cracking at the grain boundaries. The behaviour of a metal during hot-working is determined by the degree of recovery and recrystallization, the former resembling rapid creep. The occurrence of the surface cracks that ultimately cause fatigue failure is related to the development of bands in which dislocation slip is particularly pronounced.

10

There was no 'One, two, three, and away', but they began running when they liked and left off when they liked, so that it was not easy to know when the race was over.

Lewis Carroll (1832–1898)
Alice in Wonderland

The busy electron: *conductors and insulators*

Of all technological accomplishments, none so epitomizes human skill and ingenuity as the harnessing of electricity. Because so much of modern existence is based on it, electrical energy is inevitably taken for granted. Only occasional power failures are capable of reminding us just how dependent we are on this invisible commodity. The practical use of electricity is a surprisingly recent development. It is hard to believe that many of the scientists who precipitated this revolution, made their discoveries as recently as in the nineteenth century, and frequently wrote up their laboratory notes by candle-light. The great advances in atomic physics, early in the twentieth century, paved the way for a second revolution: *solid state electronics*. Its impact on society has been enormous, giving rise to new industries, new professions, new communication techniques and new organizations. Both revolutions depended, to an exceptional degree, on progress in fundamental science. The people responsible for this breathtaking series of innovations were primarily concerned with abstract ideas, but their efforts have given society radio, television, the telephone, the computer, new forms of heating, lighting, power and transport, and a host of other developments.

Many electrical phenomena are transient in nature, whereas *magnetism* has the advantage of relative permanence. It is not surprising, therefore, that some of the earliest observations in this area were made on magnets. Petrus Peregrinus de Maricourt plotted the field distribution around a piece of *lodestone*, in the thirteenth century, using a magnetic compass needle, and called the two points from which field lines were found to emanate *magnetic poles*. William Gilbert, in his monumental volume *De Magnete*, published in 1600, described both the Earth's magnetism and the generation of electric charge by friction. He also discovered the concepts of electric field and electric charge and demonstrated that the forces between charges could be both attractive and repulsive. Credit for the discovery of *electrical current* must be given to Stephen Gray, who, in 1729, touched one end of a damp string with an electrified glass rod and found that it could transmit charge over distances measurable in tens of metres. Benjamin Franklin will always be remembered for the episode with his kite, establishing the electrical nature of *lightning*, but he also recognized the existence of positive and negative electricity. He was the first to enunciate the law of *charge conservation*. The discovery of *metallic conduction* was reported in 1791, by Luigi Galvani, who had observed the twitching of a frog's muscle when connected by a wire to a source of *static electricity*. Development of the practical aspects of electricity soon followed. By 1792, Alessandro Volta had constructed the first *battery*, based on copper and zinc electrodes, and a moist pasteboard. This

The emergence of controllable electrical phenomena in the early eighteenth century caught the imagination of biologists, and Luigi Galvani (1737–1787) showed that a frog's leg could be made to twitch by connecting it to an electrical generator. In 1786, he noticed spontaneous convulsions in a leg that was not in contact with such a device, and concluded that limb movement is driven by animal electricity rather than by animal spirits. But the leg had been suspended from a hook arrangement involving two different metals, and Alessandro Volta (1745–1827) soon established that this produces electrical effects having an inorganic origin. Galvani's subsequent demonstration of spasms in a frog's leg muscle that was brought into contact with the frog's bared spinal cord proved that there really are sources of electricity inside the animal. This illustration appeared in his report published posthumously in 1791. A generator is positioned on the left side of the work bench, and two legs are seen hanging from a hook on the left-hand wall.

was 40 years in advance of Michael Faraday's elucidation of the nature of *electrolysis*, and the Daniell and Leclanché cells followed in 1836 and 1866, respectively. In the *Daniell cell*, a negative pole of amalgamated zinc stands in a pot of dilute sulphuric acid, which in turn is placed in a container of copper sulphate solution, with a copper plate as the positive terminal. The corresponding components of the *Leclanché cell* are zinc, ammonium chloride, manganese dioxide and carbon respectively. The *lead–acid accumulator* emerged in 1881, and still survives, essentially unchanged. The missing link between electricity and magnetism was found in 1820, by Hans Christian Ørsted, who noticed that a magnetic compass needle is influenced by current in a nearby conductor. His discovery literally set the wheels of modern industry in motion.

With the discovery of the *electron*, by Joseph (J. J.) Thomson and John Townsend in 1896, it became clear that the electrical conductivity of a material must be related to the behaviour of its electrons. But this represented a considerable challenge. The range of conductivities is enormous, the best examples such as copper and silver being 10^{32} times as conductive as insulators like polythene and porcelain. We recall, from Chapter 1, that electrons, in their motion around the nucleus of an isolated atom, are constrained to describe orbits which correspond to discrete energy levels. The level having the lowest energy is the *ground state*, and above this lie the first *excited state*, the second excited state, and so on. When the energy of an orbital electron is increased to the extent that it exceeds the *ionization energy*, it can no longer be bound to the *nucleus*, and it breaks away from the atom. Like an artificial satellite moving beyond the grasp of the Earth's gravitational attraction, it will have acquired a velocity in excess of the escape velocity. We must ask what happens to the orbiting electrons when a number of atoms are brought together.

The outer electrons of an atom will be the first to feel the influence of the electrons and positive nuclei of surrounding atoms. One of the

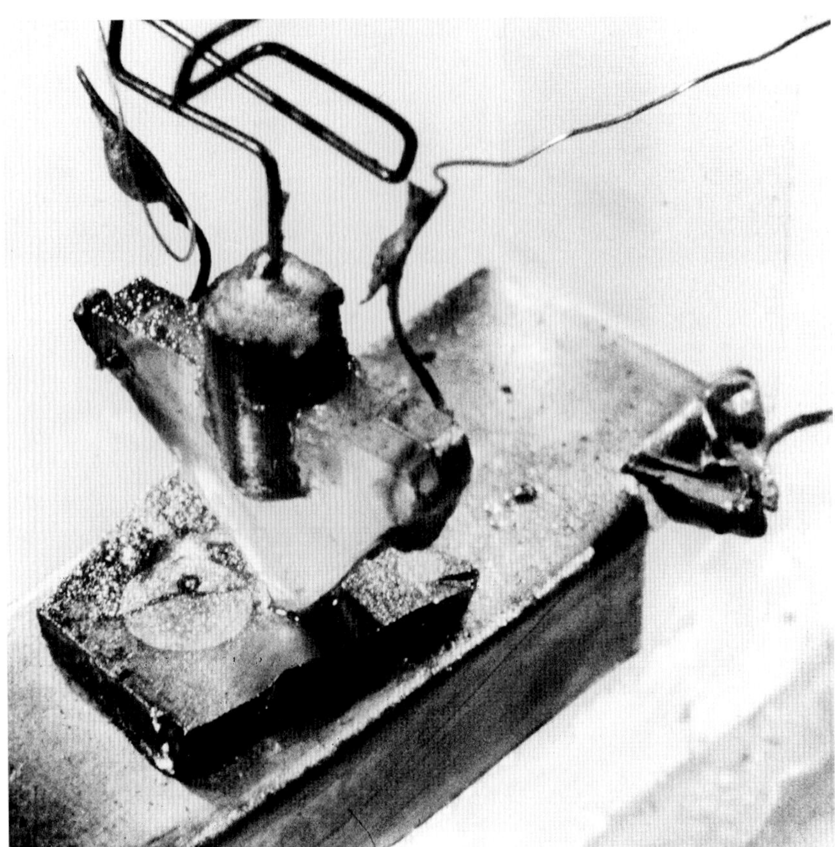

The first *transistors* assembled by their inventors, John Bardeen (1908–1991), Walter Brattain (1902–1987), and William Shockley (1910–1989), at Bell Laboratories, were primitive by today's standards, but they revolutionized the electronics industry. This example, the original point-contact version, made its debut on 23 December 1947. It amplified electrical signals by passing them through a solid semiconductor material, basically the same operation performed by present junction transistors.

defining characteristics of a metal is that it is *electropositive*, to the extent that the outermost electrons become detached from their own atoms and move throughout the entire crystal. The suggestion that some of the charged particles contained in a metal are free to migrate was made by Paul Drude in 1900. During the next five years, Hendrik Lorentz refined Drude's model, assuming that the mobile negative charge carriers are a single species of electron, and that their velocities are distributed according to a rule established by James Clerk Maxwell and Ludwig Boltzmann. (This rule depends on applicability of Isaac Newton's laws governing the motion of what are now called classical bodies, a designation that includes atoms in an ideal gas, billiard balls and planets.) The *Drude–Lorentz theory* thereby provided a qualitative explanation of electrical and thermal conduction. In 1913 E. Grüneisen explained why the electrical resistance of most metals decreases under pressure. Following Lorentz, he related this phenomenon to the electrons' average free path between successive collisions with the metallic ions. With increasing pressure, the latter are squeezed together and thus vibrate with a smaller amplitude, decreasing the collision probability and hence also the resistance. The Drude–Lorentz approach also accounted for the reflection which gives a metal its characteristic lustre. *Maxwell's theory* of *electromagnetic waves* had revealed the existence of an oscillatory electric field associated with a beam of light, and this was presumed to interact with the *free electrons*, causing them to vibrate in sympathy with the incident radiation and re-emit a large fraction of it.

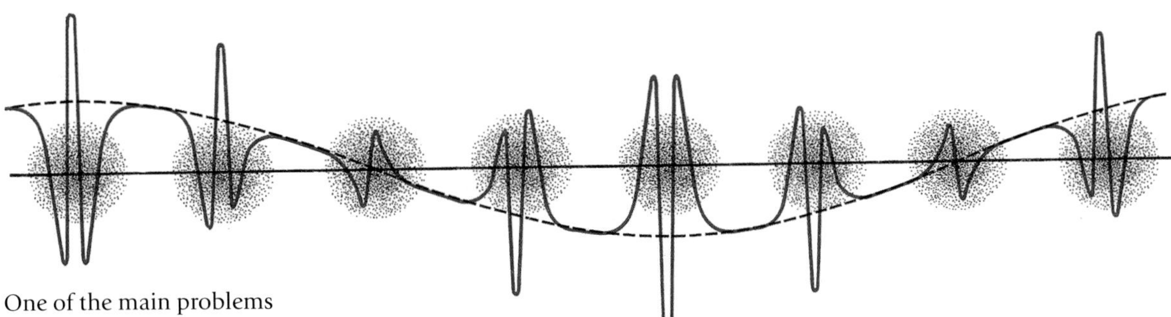

One of the main problems that bedeviled theoretical physicists early in the twentieth century concerned the apparent easy motion of electrons through metal crystals. It was not clear how an *electron wave* could move through a lattice of ions, here indicated by the grey spheres, without suffering frequent collisions with them. Then in 1927, Felix Bloch (1905–1983) showed that the ionic lattice has a negligible influence on an electron wave so long as the amplitude of the latter, shown by the red line, is suitably modified when it is in the vicinity of each ion. The violent oscillations of the wave amplitude every time it is near an ion, imply that the kinetic energy of the travelling electron is higher in these locations. The wave modifies its form in the vicinity of the ions, in order to accommodate to their strongly perturbing influence, and this permits it to move past them without having its amplitude diminished.

There were things left unexplained by this neat picture, however. K. Baedeker had discovered that substances lacking metallic bonding are nevertheless capable of sustaining a current of electrons. Moreover, some of these display an anomalous *Hall effect*. The latter, discovered by E. H. Hall in 1879, concerns the flow of current under the combined influence of electric and magnetic fields. Just as in the experiments which led to Thomson's discovery of the electron, these fields are mutually perpendicular, with the magnetic force imparting an additional drift, the direction of which is determined by the sign, positive or negative, of the moving charges. Metals showing a negative Hall effect were deemed normal in that this indicated negative charges, but some were anomalous, seeming to have positive charges. The Hall effect also provides a measure of charge concentration and had indeed shown that for many metals this amounts to about one free electron per atom. In 1911, Niels Bohr demonstrated that although this appeared to validate the Drude–Lorentz approach, the theory was incapable of accounting for *paramagnetism* and *diamagnetism*. We will return to these phenomena later in the chapter. The major difficulty with the classical model, as opposed to one based on *quantum theory*, was its failure to explain why the *specific heat* of a metal was orders of magnitude lower than the predicted value. This quantity refers to the amount of energy required to raise the temperature of a substance through a standard interval. Why should a metal's electrons, apparently so numerous and mobile, contribute so parsimoniously to the thermal energy?

The resolution of this puzzle came in 1928, with Arnold Sommerfeld's semiclassical model, the way having been paved two years earlier, simultaneously and independently, by Enrico Fermi and Paul Dirac. They had investigated the statistical behaviour of particles which, unlike the classical ideal bodies considered by Maxwell and Boltzmann, are subject to Pauli's exclusion principle. Treating the electrons as if they had no interaction with each other or with the positive ions, Sommerfeld envisaged them as occupying a hierarchy of quantized energy states, up to a maximum level that is now referred to as the *Fermi energy*. We have already compared the discrete energy levels of Bohr's model atom with the apartments of a high-rise residential building, with the occupants desiring to live as close to the ground as possible. This analogy is again useful, but the entire piece of metal, with about 10^{23} free electrons per cubic centimetre, must now be treated as a single entity. The levels are so crowded together that the inevitable smearing out which is a consequence of *Heisenberg's uncertainty principle* turns them into a single continuous band. The Fermi energy is analogous to the uppermost occupied apartment, with all floors above empty. Acquisition of thermal

The energy of an electron in a crystal can be increased by acceleration in an electric field, such as that imposed when a metallic conductor is connected to a battery, or by absorption of a light photon or a heat quantum. But the electron can accept the extra energy only if an appropriate higher energy state is available. In a piece of material, the discrete energy levels of the individual atoms are replaced by a series of allowed *energy bands* (coloured green), which are separated by forbidden regions, or *band gaps* (coloured red). Because of the Pauli exclusion principle, the electrons sequentially fill the available states until all electrons are accounted for, the uppermost occupied state being the *Fermi level*. In a typical conductor the Fermi level lies in the middle of an allowed band, the *conduction band*, while the hallmark of an *insulator* is coincidence of the Fermi level with the top of the *valence band*. An *intrinsic semiconductor* is characterized by an unusually narrow gap between a full (valence) band and an empty (conduction) band. *Extrinsic semiconductors* have impurities which give rise to localized levels lying within the gap, either just below the conduction band (n-type) or just above the valence band (p-type).

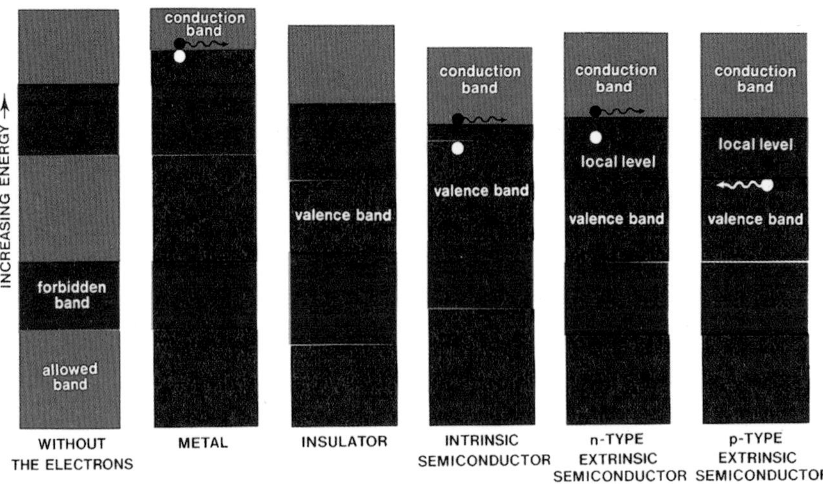

energy by electrons is similar to the building's inhabitants being provoked to move to apartments lying a few floors higher. Unless all moved at precisely the same moment, a highly unlikely event, the only change would be that due to the people at the top taking advantage of the empty apartments above them. Sommerfeld's Fermi–Dirac hierarchy had an analogously limited scope for energy change, and could thus contribute only modestly to the thermal energy. This was the reason for a metal's surprisingly low specific heat. Oddly enough, Ralph Fowler had used the same reasoning two years earlier to resolve the astrophysical riddle of what happens when a *White Dwarf* becomes a *Black Dwarf*, which is the lowest macroscopic quantum state of a *star*.

Sommerfeld's theory, attractive as it was, appeared to be based on untenable assumptions. It neglected interactions which in an isolated atom are of vital importance to the stability. As Felix Bloch realized, in 1927, the main problem was to explain how the electrons could sneak past all the ions in a metal, so as to avoid having an average free path (between collisions) comparable to the distance between adjacent atoms. He tackled Erwin Schrödinger's wave equation for an electron moving along a line of atoms, the potential energy therefore varying in a periodic manner, and found to his delight that the acceptable wave solution differed from that of a totally free electron only by a periodic modulation. By adjusting its form close to the ions, so as to accommodate their strong disturbing influence, the wave is able to move past them without being attenuated. The situation is reminiscent of what Friedrich Engels said in a quite different context: freedom is the recognition of necessity. The next advance, by Rudolf Peierls in 1929, was the explanation of the anomalous positive Hall effect seen in some elements. He demonstrated that, for certain ranges of wavelength, the electrons would move counter to the direction suggested by the applied field, as if they bore a positive charge. To appreciate the Peierls idea, it is necessary to consider certain implications of the Bloch theory, and it will prove convenient to anticipate a subsequent elaboration by Alan Wilson. For a *free electron wave*, there is no limitation on the energy; it can take any value. Most energies are not accessible in an isolated atom, only certain discrete values being permitted. When an electron moves through a periodic lattice, there are also forbidden energy values, but they arise in a different

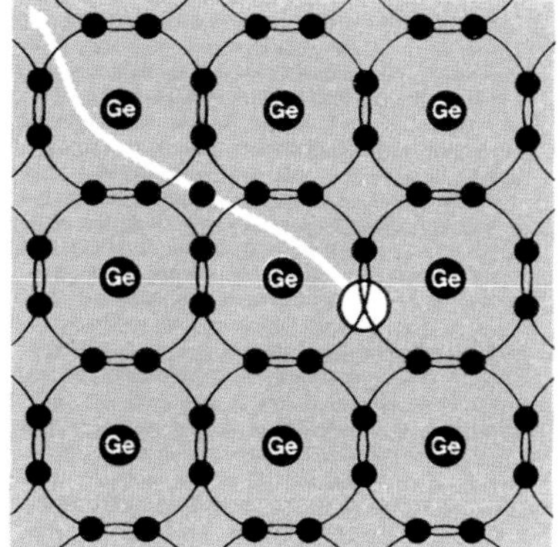

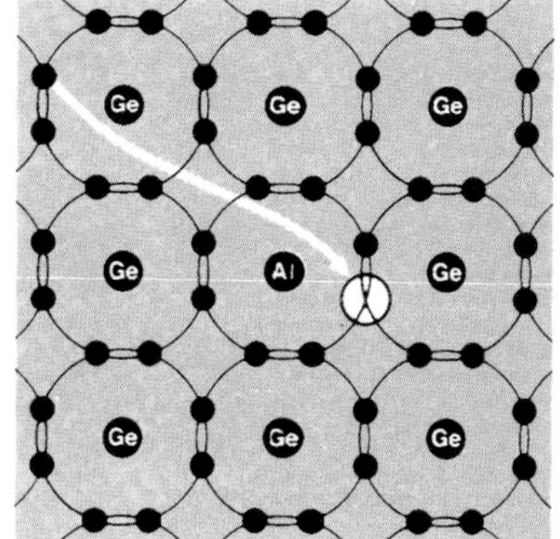

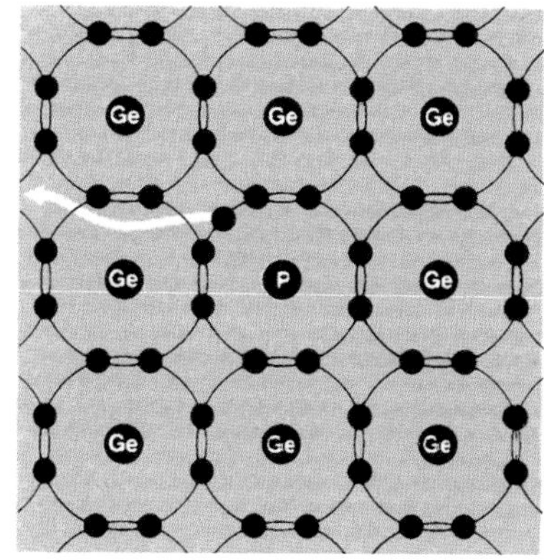

In an *intrinsic semiconductor* (left) an electron can be transferred from the *valence band* to the *conduction band* only if there is sufficient energy to raise it across the *band gap*. When it is in the valence band, an electron contributes to the *covalent bonding*. Promotion of an electron across the gap partially disrupts such bonding, but it provides the crystal with a carrier of electrical current. *Extrinsic semiconductors* have lower conduction threshold energies because impurities provide local energy levels within the gap. An aluminium atom, for example, has one less electron than the surrounding atoms and this deficiency can be made up for by capture of an electron from a germanium atom, which thereby acquires a hole that can move through the lattice as if it were a positively-charged electron (middle). A phosphorus impurity atom has one extra electron, and this can detach and move through the lattice (right). The aluminium impurity is referred to as an *acceptor*, and it produces a p-type semiconductor, while the *donor* phosphorus impurity gives an n-type.

way. The ion lattice can diffract the electron waves in much the same way as it does X-ray waves when the latter are used to determine crystal structures by Bragg diffraction. When the electron wavelength matches the period of the lattice, or a sub-multiple of it, the perturbing influence of the ions can be so large that the electron wave becomes strongly attenuated; it cannot be transmitted as one of Bloch's modulated waves. Even when there is not a perfect match, the disturbance of the wave by the ions can be so large that its progress is very laboured, as if the electron had become sluggishly overweight. Remembering that the wavelength of the electron wave determines its energy, this means that for certain ranges of energy a propagating electron wave is not possible. These are referred to as *forbidden bands*, and they appear as gaps in an otherwise continuous energy spectrum. The electrons are distributed in the available bands according to the *Pauli exclusion principle*, filling successively higher levels until the Fermi energy is reached. When a *band gap* is encountered, the next electrons are forced to have an energy higher than their lower-level brethren by at least the gap energy. We can imagine this filling prescription as proceeding up through the allowed bands and jumping over the forbidden bands, until the crystal has its full complement of electrons.

A crystal can be characterized by the position of its Fermi energy level with respect to the limits of the electron band in which it lies. In a typical simple metal it is approximately in the middle of an allowed band. Therefore, of the available states in this band, referred to as the *conduction band*, only some are occupied. Electrons near the Fermi level can be given extra kinetic energy because there are unfilled states available immediately above. The extra energy can be imparted to the electrons in a number of ways. They can be subjected to an electric field, by connecting opposite ends of the metallic crystal to the poles of a battery. They can also gain energy from thermal motion, if the crystal is heated, and a third possibility is to irradiate the crystal with light energy. Because there is a continuum of available energy states, with the Fermi level lying within the conduction band, it is possible to excite the uppermost electrons with vanishingly small extra kinetic energy. Given sufficiently sensitive instruments, the metal can be subjected to the minutest voltage, and a current will still be detected.

Wilson saw that Bloch's theory had in fact proved too much. Before the latter's breakthrough it had been difficult to understand metals, whereas afterwards it was the existence of *insulators* that required explanation. This is where the Wilson picture comes into its own. A special situation prevails if

Electrical resistivity

The resistance to electric current displayed by a material depends upon the nature of the interatomic bonds in the latter. For some types of bonds, each atom's electrons remain localized around the nucleus, while in others they are permitted a degree of freedom that ranges from the occasional hop, as seen in semiconductors like silicon, to the almost unhindered flow in the typical metal. The standard measure is the *resistivity*, values of which are tabulated below in units of ohm metres, at 0 °C. As is immediately apparent, this quantity shows great variation from one class of material to another.

Material	Class	Resistivity
copper	metal	1.7×10^{-8}
zinc	metal	5.9×10^{-8}
iron	metal	9.7×10^{-8}
uranium	metal	2.9×10^{-7}
graphite	semiconductor	5×10^{-5}
germanium	semiconductor	$10^{-3} - 10^{-1}$
silicon	semiconductor	$10^{-1} - 10^{1}$
water	ionic	$10^{2} - 10^{5}$
sodium chloride	ionic	$10^{3} - 10^{5}$
glass	ceramic	$10^{9} - 10^{11}$
diamond	ceramic	$10^{10} - 10^{11}$
porcelain	ceramic	$10^{10} - 10^{12}$
mica	ceramic	$10^{11} - 10^{15}$
rubber	polymer	$10^{13} - 10^{15}$
paraffin wax	polymer	$10^{13} - 10^{17}$
polyethylene	polymer	$10^{14} - 10^{18}$
fused quartz	ceramic	$10^{16} - 10^{18}$

the Fermi level just coincides with the top of a band. In this case none of the electrons can be excited unless they are given enough energy to jump across the gap that separates the top of the completely filled band, referred to as the *valence band*, and the bottom of the totally empty conduction band. This is the situation in a typical insulator such as the diamond form of carbon, and the lack of any detectable electric current results from the fact that every electron moving in one direction is counterbalanced by one moving in the opposite direction. The gap is so large that the crystal would suffer breakdown before the electrons could be given sufficient energy to jump across it. In some materials, known as *intrinsic semiconductors*, the situation is qualitatively similar, but the gap between the filled valence band and the empty conduction band is rather small; less than a couple of volts or so. In such a case, the electrons can overcome the gap with easily achievable excitation energies. Typical examples of intrinsic semiconductors are silicon and germanium, with band gaps of 1.1 V (0.18 aJ) and 0.7 V (0.11 aJ), respectively.

Experimental endorsement of the *band theory* was provided by Herbert Skinner's studies of emission and absorption of X-rays by metals. When a solid is bombarded by electrons with sufficient energy, one or more of the orbital electrons may be ejected, leaving behind an incomplete shell. This situation persists until an electron in a higher orbit falls into the vacancy, the transition being accompanied by the emission of an energy quantum. For the inner levels, such quanta have energies lying in the X-ray region, the energy corresponding to the difference between the two levels involved.

Superconductors: toward a technological revolution

At temperatures below a certain threshold, a number of pure metals and alloys offer no resistance whatsoever to the passage of electrical current. If a current is set up in a closed loop of such a material, it will in principle continue to flow forever. The threshold temperature is a characteristic of the material in question, and for many years after the discovery of *superconductivity*, by Heike Kamerlingh-Onnes (1853–1926) in 1911, all the known superconducting materials had thresholds around a few degrees Kelvin. Thresholds closer to room temperature would bring superconducting phenomena within range of relatively easy technology, and presage radical changes in computing, transportation and power transmission. J. Georg Bednorz and K. Alex Müller produced a material with a threshold of 35 K, in 1986, and Paul Chu and his colleagues had pushed the record up to 92 K by early 1987. Space has been left in the chronological table given below, so that the reader can fill in the latest developments in this burgeoning field of scientific endeavour.

Year	Material	Threshold
1911	Hg	4.2 K
1913	Pb	7.2 K
1930	Nb	9.2 K
1941	Nb–N	16.0 K
1952	V–Si	17.1 K
1956	Nb–Sn	18.5 K
1973	Nb–Ge	23.7 K
1986	Ba–La–Cu–O	35.0 K
1987	Ba–Y–Cu–O	92.0 K
1988	Bi-Sr-Ca-Cu-O	112 K
1988	Tl-Ba-Ca-Cu-O	130 K
1993	Hg-Ba-Ca-Cu-O	164 K
1996	Hg-Tl-Ba-Ca-Cu-O	168 K

As was realized by Skinner, Nevill Mott and Harry Jones, in 1934, the intensity of the X-ray emission, at any energy, provides a measure of the density of electron states at the level from which an electron jumps. Skinner's results, for magnesium, showed a clear indication of an energy gap.

In real crystals, the regular spacing of ion positions extends in all three dimensions, and the repeating pattern is also periodic in directions other than the three fundamental directions. The *band structure* can, and usually does, vary with direction in a quite complex fashion. Hans Bethe realized that the Fermi level is actually a *Fermi surface*, and Lev Landau subsequently showed that this theoretical concept has physical reality, susceptible to experimental measurement. The first Fermi surface was determined by Brian Pippard in 1957. The metal was copper, and the surface had the shape of a spiked ball, rather like one of the early sea mines. Other metals were investigated, especially by the Cambridge school of Pippard, David Shoenberg, and their colleagues, and a catalogue of Fermi surfaces established. Many of these were found to have quite complicated shapes, some even suggestive of the organic domain. Harry Jones was the first to build a theory of alloy structures on the interplay between the Fermi surface, band structure and crystallographic symmetry.

There is a second type of semiconductor, which has become particularly important in the electronics industry. It is the *extrinsic semiconductor*, and since its arrival on the commercial scene, in the early 1950s, it has revolutionized wireless, television, computing and telecommunications. This development was based on experimental phenomena observed a number of decades earlier. Wilhelm Röntgen was the first to find that irradiation of rock salt with X-rays induces conductivity in that material. Robert Pohl, in

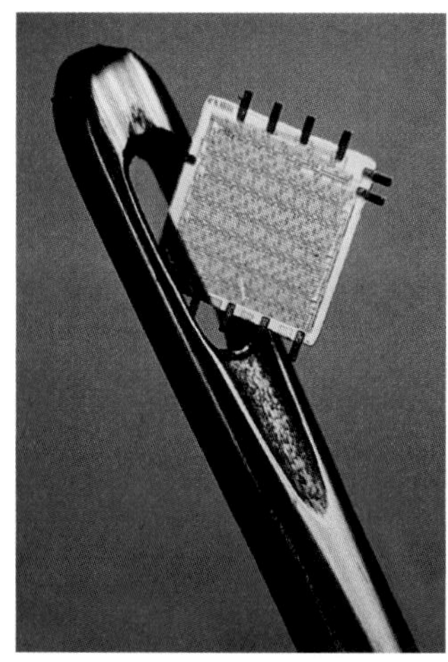

The drive towards *miniaturization* received a significant boost with the emergence of *integrated circuit chips*, in which complete electrical circuits are accommodated on a single piece of semiconductor crystal. The early example shown here is about one millimetre square and a hundred micrometres thick; small enough to pass through the eye of a needle. The degree of miniaturization achieved subsequently has produced printed circuits with components that lie beyond the resolution of an optical microscope!

the mid 1920s, showed that this effect is associated with vacancy defects in the sodium chloride lattice, and it later transpired that the (positive) anion vacancy can actually capture an electron, forming a system with its own discrete energy levels. The importance of this work for subsequent developments with extrinsic semiconductors lay in the realization that individual defects, of atomic dimensions, could give rise to *localized electron states*. This led Pohl to predict replacement of vacuum tubes in radios by crystals, a prophecy spectacularly vindicated: by the early 1980s semiconductor production had outstripped the entire world's shipbuilding industry in terms of market value. Extrinsic semiconductors form the basis of the transistor, which we will discuss shortly. They are fabricated by introducing to a host crystal *impurity atoms* with a different number of electrons in their outermost orbits, a technique known by the term *doping*. Consider, for example, silicon doped with phosphorus. Silicon has two 3p electrons, while phosphorus has three. The two 3s electrons and two of the 3p electrons of phosphorus become hybridized, just as do the corresponding electrons in the silicon atoms, so the phosphorus impurity atoms are readily incorporated into the tetrahedral bonding pattern of the host silicon lattice. The outcome is that the phosphorus atoms masquerade as silicon atoms, each with an extra electron localized at its position in the crystal. This gives rise to an extra feature in the band structure: a *localized energy level*. If such a composite crystal is subjected to a sufficiently strong electric field, the excess electrons become detached from the phosphorus atoms, while the bonding electrons are unaffected. Because of this, the localized energy level corresponding to the phosphorus lies above the upper limit of the valence band, and it is closer to the conduction band. This reduces the energy gap. The energy required to excite an electron into the conduction band, from this localized level, is therefore less than the gap energy in pure silicon. The localized state is called a *donor level* because the phosphorus atoms donate conduction electrons to the crystal as a whole. This type of extrinsic semiconductor is known as *n-type* because the current carriers are electrons, the initial letter denoting negative.

The opposite situation occurs if the impurity atoms have less outer electrons than do atoms of the host. An example is the addition of aluminium atoms to a silicon crystal. Aluminium has only one electron in the 3p state, so its atoms lack an electron compared with silicon. They can be incorporated into the silicon lattice, and they accept electrons from it, if these are given enough excitation energy to promote them into the localized energy level. Just as in the previous case, this level lies between the top of the valence band and the bottom of the conduction band; it is easier for electrons to jump from a silicon atom to an adjacent aluminium atom than to break away from a silicon atom and move through the crystal. There is an important difference between this type of extrinsic semiconductor and the n-type described earlier. In the n-type, electrons are excited away from localized states into the crystal as a whole, where they provide the means of electrical conduction. In this second case, on the other hand, the electrons are excited to a localized state, and they cannot directly mediate conduction. But the excitation of electrons out of the valence band leaves the latter incompletely filled and therefore capable of permitting charge movement. The current is still caused by electron motion, but it is carried by the *hole* in the valence band, which, being the absence of a negative charge, is effectively a positive entity. This extrinsic semiconductor is known as a *p-type* because the charge carrier is positive. The concept of electrical conduction through motion of electrons is reasonably straightforward, whereas the idea of an electrical hole is more subtle. An analogy is useful. Let us imagine a number of firms all specializing in the same enterprise, employing similarly qualified personnel. If all positions in all firms are occupied, the situation is totally static, permitting no movement of employees from one firm to another. When an employee at one firm is promoted to a different type of position, a vacancy is left, and this can produce a series of personnel movements. The person moving into the vacant position creates a new vacancy, into which another can move, and so on. When a person moves from one position to another, a vacant position moves in the reverse direction. This is analogous to what happens in a p-type semiconductor, the moving employees being equivalent to electrons and the vacancy to a hole. Recalling the work of Peierls, we see that it is the movement of holes which gives an *anomalous Hall effect*.

A junction between n-type and p-type semiconductors has a useful property known as *rectification*. With no applied voltage, there is a weak current flow across the junction owing to thermally excited motion of electrons on the n-side and holes on the p-side. When electrons and holes meet at the interface, they mutually annihilate. This produces a current, because there is a net electron flow towards the junction on one side and a net electron flow away from the junction on the other. The current flows in the direction from n to p. If a voltage is applied to the junction, such that the n-side is negative and the p-side positive, the current will increase because the electric field drives both electrons and holes towards the junction. This so-called forward bias therefore gives high electrical conduction, and a heavy current is observed. A switch in polarity of the applied field to reverse bias has the opposite effect, and very little current flows; both the electrons and the holes will now be driven away from the junction. A device that favours current flow in one direction but prohibits it in the other is said to be a *rectifier*, and this simple *p–n junction* is an example.

Semiconductor rectification is exploited in the *transistor*, which is a back-to-back arrangement of two p–n junctions. Two configurations are possible, and both are used: the p–n–p and n–p–n types. Let us consider the former, in a circuit with the central n-type crystal earthed and therefore neutral, one of the p-type regions being at a positive voltage and the other negative. The p-type region with the positive voltage is called the *emitter*, while the negative side is the *collector*. The n-type region is referred to as the *base*, and its junction with the emitter is biased in the forward direction. Conversely, the junction between base and collector is biased in the reverse direction. If the base region in such a composite arrangement is quite wide, the distance between the forward and reverse biased junctions being of the order of millimetres, nothing useful is gained. When the thickness of the base crystal is sufficiently reduced, however, it becomes possible for some of the holes injected from the emitter into the base to survive long enough to reach the reverse-biased junction between base and collector. Because there is no barrier for holes at that junction, those reaching it from the base greatly increase the reverse current. Moreover, because an increase in voltage difference between emitter and base increases the supply of these excess holes arriving at the junction between base and collector, this will cause a corresponding rise in the current at the latter junction. The circuit can therefore function as an *amplifier*, the gain of which can be adjusted by varying the voltage between base and collector. The n–p–n transistor works in a similar manner. The first transistor was produced in 1947 by John Bardeen and Walter Brattain. It was rather primitive, consisting of two closely spaced points in contact with the surface of a semiconductor crystal. Such a device is now referred to as a *point-contact transistor*. The idea of rectification by point contact is reminiscent of the *'cat's whisker'* that was a familiar feature of the early days of radio reception. Their discovery was quickly followed by the *junction transistor*. This variant, first produced by William Shockley, consists of three semiconductor crystals of the appropriate type in close contact with one another.

In 1938, J. H. De Boer and E. J. W. Verwey pointed out that nickel oxide presents band theory with a dilemma. The 3d band of this compound, capable of holding ten electrons, contains only eight, so it should be a conductor of electricity like a metal. But it is in fact a transparent insulator. The answer to this puzzle was supplied by Nevill Mott in 1949. Speculating on how charge mobility depends on the interatomic spacing, he considered the transfer of an electron from one atom to another in condensed matter. The removal of a single electron leaves behind a positive hole, and the two entities attract each other electrically. As more electrons are removed from the same atom, a new factor begins to make itself felt. These electrons form a cloud which tends to screen the positive charge, diminishing the attraction which is trying to restore electrical neutrality. Mott demonstrated that a point is reached, rather suddenly, at which electron transfer is no longer possible. For all atomic densities lower than a critical value the material is an insulator, and this is apparently the case for nickel oxide. Mott's theory is evocative of a prediction by Eugene Wigner and Hillard Huntington, in 1935, that hydrogen might become metallic (and hence a conductor) at sufficiently high density.

The element having the lowest gas–liquid transition temperature is *helium*. This element remains in its gaseous form down to a temperature of −269 °C or, on the absolute scale of temperature, 4.2 K. After his successful first production of the liquid form of helium in 1908, Heike Kamerlingh Onnes undertook systematic studies of the properties of materials in this very low temperature range. In 1911, he made the dramatic observation that the electrical resistance of mercury, which at these low temperatures is in its solid metallic form, suddenly falls to zero at about 4 K. The frictional drag that normally acts upon electrical current had completely disappeared. Kammerlingh Onnes and his colleagues demonstrated this by an elegant experiment in which current was set flowing through a conductor in the form of a closed ring. This ring was then isolated from all magnetic and electrical disturbances. A year later it was found that the current was still flowing at an undiminished strength, and its discoverer referred to this new effect as *superconductivity*. Superficially, it looked as if he had stumbled upon the elusive perpetual motion, but only if one ignored the fact that power was required to keep the refrigerator running.

The explanation of the frictionless motion of electrons was not found until 1957. The *BCS theory* of superconductivity, which takes its name from the initial letters of its three discoverers, John Bardeen, Leon Cooper and J. Robert Schrieffer, is based on a peculiarity in the interaction between two electrons in a crystal. Since all electrons bear a negative charge, we would normally expect repulsion between any two of these elementary particles. At low temperatures, however, this repulsion is more than compensated for

The lack of resistance in a *superconductor* permits heavy currents to be carried without the usual power losses. The cross-sectional view (upper picture) shows the individual filaments of a niobium-tin *superconducting cable* for use in the coils of a large magnet. The small permanent magnet (lower picture) floats in thin air above the superconductor surface because of a physical phenomenon known as the *Meissner effect* (named for Walther Meissner, 1882–1974), whereby the superconducting material excludes a magnetic field. The large-scale equivalent of this *levitation* would revolutionize transport, providing frictionless travel at much higher speeds than those currently employed.

by an attractive force that arises in the following manner. One electron interacts with the lattice of positive metallic ions by attracting it, thereby causing a local small deformation in the crystal. This deformation produces a small region with net positive charge, which can attract a second electron. The two electrons therefore become bound together through the mediating agency of the positive ion lattice, and their motion is constrained as if they were tied together by an elastic string. Such bound electrons are referred to as *Cooper pairs*, the formation of which conforms with the *Pauli exclusion principle*. The resistivity of a metal in its normal state arises from the scattering of the electron waves by static *imperfections* such as dislocations, vacancies, grain boundaries and impurity atoms, and the dynamic imperfections referred to as *phonons*. If one electron in a Cooper pair encounters an imperfection, any disturbance to its motion is quickly compensated for by the restoring pull of the other electron. The imaginary elastic string allows the pair to exhibit perfect recoil. This remarkable situation persists only if the amount of available thermal energy is insufficient to overcome the binding between the two electrons. Because the latter, equivalent to an energy gap, is about a ten thousandth of an attojoule (0.0001 aJ), even in the best superconductors, this limits the effect to very low temperatures. It is instructive to see how the low binding energy is reconciled with the wave nature of the particles involved. We have earlier seen that rapid spatial variation of the wave function is associated with high energy. The wave function of a Cooper pair is therefore spread out over a characteristic *coherence length* which is typically about 1 μm, as was demonstrated by Brian Pippard. But superconductivity is not merely a dance for two electrons; the ballroom is filled with couples. Each pair has zero net velocity in the absence of an electric field, and even when the crystal is carrying a large *supercurrent*, all the pairs have the same small velocity. They therefore constitute a single coherent state whose wavelength is very much larger than atomic dimensions. The scattering by individual atoms is therefore negligible. It is not surprising that a great deal of effort has been put into the search for materials that are still in this remarkable state near room temperature; such superconductors would save the expense of refrigeration. A bold step in that direction was taken in 1986, when Alex Müller and Georg Bednorz discovered that superconductivity in certain copper compounds persists for temperatures as high as 80 K, which is above the temperature of (relatively cheap) liquid nitrogen. But an explanation of these *high* T_c *superconductors*, as they are called, has presented theoreticians with a dificult challenge.

Electrons can be excited to a higher energy state if the crystal containing them absorbs light photons. The fate of light energy falling upon condensed matter depends on whether the illuminated substance is transparent or opaque. The relative positions of the Fermi level and the bands dictate the electronic behaviour, and they thus determine the optical properties of a substance. Let us consider first a typical transparent material: a thin crystal of common rock salt. White light falling on one face passes through the crystal and emerges, essentially unchanged, from the other side. It is not at all obvious that we should expect this null result; half the atoms in the crystal are sodium, and some of the spacings between electron energy levels in that element correspond to light in the visible region. Indeed, the brilliant yellow light that is characteristic of burning sodium was used as an example of the emission that occurs when electrons move from one energy level to another. It seems reasonable to expect part of the yellow component of white light to be absorbed in the salt crystal, and the emerging light to have a bluish tinge (that is to say, white light depleted of the yellow colour). Before considering this point, it will be useful to discuss the question of *absorption* and *emission* in general terms.

When an atom absorbs energy, an electron moves from one level to another lying at a higher energy. This can happen only if the level to which the electron moves does not already contain the full complement as permitted by the rules of quantum mechanics. Conversely, energy is emitted when an electron falls from one level to another having lower energy. When sodium is burned in a flame, thermal stimulation of its atoms causes such inter-level transitions, and the emission of light having the familiar yellow colour. Let us consider a beam of the same yellow light falling upon gaseous sodium. Because the incident light has just the right energy, it can be absorbed by the atoms, causing electrons to be excited into a higher energy level. When these fall back to their original level, a process referred to as decay, the sodium atoms emit light of the same yellow colour. In the steady state, there must be equal numbers of electrons undergoing decay and excitation, so we might expect the incident beam to emerge from the sodium gas with undiminished strength. This would be incorrect, however, because the radiation is emitted in all directions. The process of absorption and re-emission decreases the strength of the beam in the forward direction, converting part of it into radiation in all other directions.

In the salt crystal, neither the sodium atoms nor the chlorine atoms are in the isolated state. The atoms interact, and the *discrete energy levels* are replaced by *energy bands*. In sodium chloride, the energy gap between the *valence band* and the *conduction band* is 1.36 aJ. The salt crystal is transparent to the entire spectrum of white light because this covers the range of wavelength from approximately 800 nm to 400 nm, the corresponding energy range being 0.24–0.50 aJ. Thus there are no photons of light in the visible spectrum having energies sufficiently great to excite an electron from the valence band across the gap into the conduction band. The band gaps in intrinsic semiconductors are in the range 0.1–0.2 aJ, so these materials are opaque in the visible region of the electromagnetic spectrum, because the light photons have sufficient energy to provoke trans-gap electron jumps. The photons of infrared radiation have energies below 0.24 aJ, and those of sufficiently low energy falling on an intrinsic semiconductor with a broader gap will pass through; the semiconductor will be transparent to such radiation. A naturally transparent crystal will become coloured if gaps are introduced with energies below those of the visible range. This is the case for crystals containing impurities that give rise to localized energy levels. For this to occur the impurity atoms must not be so compatible with the host crystal that they can essentially accommodate to its band structure. Many beautiful examples of induced colour are seen in the natural crystals known as *gems*. The green colour of *emerald* is caused by the chromium impurities in a crystal of beryl, a beryllium aluminosilicate, while the same impurity element in corundum, an aluminium oxide, produces a precious stone with a gorgeous red hue: *ruby*. If, instead, the corundum contains iron and titanium impurities, it develops a rich blue colour and becomes *sapphire*.

Localized energy level states associated with interesting electrical and optical phenomena in insulators are not always associated with impurity atoms. They frequently occur at *crystal defects*, and the pretty effects seen in ionic crystals under some conditions are a case in point. Vacant lattice

Among the many uses of laser beams is the amplification of other laser beams. In this experimental set-up, the green beam from a copper-vapour laser (which is not shown) enters a chamber containing dye solution and produces the non-equilibrium condition of *population inversion* in that material, by what is referred to as *optical pumping*. One of the frequencies of the transitions by which the excited states decay lies at the red end of the spectrum. A beam of this frequency entering the pumped medium can thereby be amplified.

sites are inevitably present in any crystal, the thermodynamically dictated equilibrium concentration increasing with temperature. In a pure metallic element there is only one type of vacancy, a missing positive ion. In sodium chloride two different types are possible: a missing sodium atom and a missing chlorine atom. According to the nomenclature introduced in connection with electrolysis, the former is a *cation vacancy*, while the missing chlorine defect is an *anion vacancy*. The anion vacancy is a region in which there is locally more positive than negative charge, because of the loss of a negative ion. An electron in the vicinity of the anion vacancy feels the influence of this positive charge. As was first realized by Ronald Gurney, the attraction is so strong that the electron is in a *localized bound state*. The situation is somewhat reminiscent of the hydrogen atom with its single electron in orbit around a central positive charge. The localized state in sodium chloride has various energy levels, the electron normally residing in the lowest. Light energy with the correct wavelength can cause excitation to a higher energy level, and in sodium chloride and several other alkali halides some of the transitions correspond to wavelengths in the visible spectrum. When white light falls upon a crystal of potassium chloride, for example, those anion vacancies having a bound electron cause absorption in the yellow part of the spectrum and the crystal takes on a deep blue colour. This phenomenon was discovered by Robert Pohl in 1925, and he referred to vacancy-plus-electron defects as *Farbzentren*, a German term meaning *colour centres*. They are commonly known as *F-centres*. In the *photo-voltaic effect*, first reported by E. Becquerel in 1939, photo-generated minority charge carriers, holes or electrons, are separated from the majority carriers of the opposite species by a built-in electric field. The resultant charge flow and recombination in an external circuit can be used as a source of power and holds promise for conversion of solar energy.

The local energy levels associated with certain impurity atoms in optically transparent materials can be exploited to produce light beams of very high energy density, coherence and spectral purity. The device that produces this special form of light is known as a *laser*, the name being acronymic for **l**ight **a**mplification by **s**timulated **e**mission of **r**adiation. To understand how it works, we must return to the question of absorption and emission of radiation by atoms. Thus far we have considered only two processes: excitation and decay. When an electron is in an excited state, decay to the lower energy level occurs in a spontaneous and statistically random fashion, it being possible to both calculate and measure the mean lifetime of the excited state. In 1916, Albert Einstein realized that another type of decay transition is possible. It occurs when radiation having the energy corresponding to the difference between the two levels passes sufficiently close to an excited atom. When this happens, the electron can be induced to fall to the lower energy level, giving the energy of its transition to the passing radiation wave. It is found that the photon, or wave, thus emitted by stimulation is exactly in step, or in phase, with the stimulating photon. This resonance effect is known as *stimulated emission*, and the process can occur at any time after the excitation process. An excited atom can thus decay by stimulated emission after a time that is much shorter than the natural lifetime of the excited state.

Let us now consider a large number of atoms at a particular instant, some being in the ground state and some excited. If such a system is irradiated with light having the correct energy, all three types of process will occur.

Electron tunnelling and the Josephson effect

A central assumption of *quantum mechanics* is that particles are not completely localized in space. One important consequence is that an electron has a certain probability of passing through a potential energy barrier even though classical mechanics would indicate that the electron's energy is insufficient to accomplish such penetration. This is the *tunnel effect*, and it is observed when two metals or semi conductors are separated by an insulator having a thickness of a few atomic diameters. In 1960, Ivar Giaever showed that tunnelling can even take place between two superconductors. According to the Bardeen–Cooper–Schrieffer theory (named after John Bardeen 1908–1991, Leon Cooper and J. Robert Schrieffer), the electrons in a superconductor exist in pairs, and the waves of all the pairs are locally in phase (i.e. in step with one another). It was originally believed that the correlated tunnelling of a pair of electrons would be a most rare event, but Brian Josephson showed theoretically, in 1962, that *pair tunnelling* is just as likely as single electron tunnelling. He also made the surprising prediction that a tunnelling electric current can flow between two super conductors even when these are at the same voltage, the actual magnitude of the current depending on the difference in phase across the insulating gap. The latter is now known as a *Josephson junction*, and the predicted current across such a device was first detected by Philip Anderson and John Rowell. The phase difference is conveniently expressed as an angle, and when the waves are exactly in step this is 0°, the junction current then being zero. The maximum current is typically one milliampere and corresponds to a 90° phase difference. For non-zero voltages across the junction, up to a value related to the energy required to disrupt the electron pairs, the supercurrent oscillates at a frequency that is proportional to the voltage, the ratio being 484 terahertz per volt. In 1933, Walther Meissner (1882–1974) and Robert Ochsenfeld (1901–1993) observed that the *magnetic flux* is excluded from the interior of a superconductor; it is confined to a surface layer about 0.1 micrometres thick, known as the *penetration depth*. This implies that the magnetic flux passing through the middle of a loop-shaped superconductor will be trapped. In 1950, Fritz London (1900–1954) realized that there is a severe constraint on the *electron waves* in such a loop; they must not get out of step, so the phase difference for one complete trip around the circuit can only be 0°, 360°, 720°, and so on. This is reminiscent of de Broglie's wave interpretation of the stable Bohr orbits in an atom, and it means that the trapped flux can only take certain values (i.e. it is quantized), a fact verified experimentally by B. S. Deaver, R. Doll, W. M. Fairbank, and M. Näbauer in 1961. If the loop is interrupted by two Josephson junctions, phase differences other than those stipulated above can be developed, and will depend upon the magnetic flux threading the loop. By connecting the two superconducting segments to an external detection system, an apparatus is produced which can be used to measure the strength of a magnetic field. This is the basis of the *superconducting quantum interference device*, or *SQUID*, invented by Robert Jaklevic, John Lambe, James Mercereau, and Arnold Silver in 1964. Like other equipment based on Josephson's discovery, it is exquisitely sensitive; changes of magnetic flux density as small as 10^{-14} tesla can be detected, and the device has been used to map out the magnetic fields surrounding the human head. Integrated circuits incorporating SQUIDs, pioneered by Hans Zappe and having operation times as small as 10 picoseconds, may provide the basis for ultrafast electronic computers. The picture below shows a device made by Michael Cromar and Pasquale Carelli; a SQUID lies at the centre, surrounded by external coupling coils.

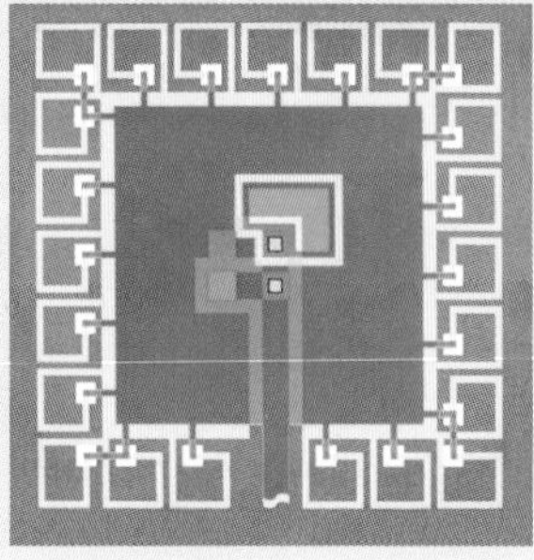

Electrons and the photographic process

Electron *band theory* and the physics of *crystal defects* find an interesting joint application in the analysis of the *photographic process*. Photographic film consists of a cellulose acetate sheet coated with a thin emulsion of gelatin in which are suspended fine crystalline grains of silver bromide. Silver halides are sensitive to light, which induces *photochemical decomposition*. Brief exposure to light slightly decomposes some grains, and small clusters of silver atoms are produced. During subsequent development in an alkaline solution of an organic reducing agent such as hydroquinone, these sensitized grains are reduced to metallic silver, whereas the unsensitized grains remain unchanged. The latter are removed, during the fixing process, by making them combine with thiosulphate ions, $S_2O_3^{2-}$ from sodium thiosulphate, $Na_2S_2O_3.5H_2O$ which is commonly referred to as hypo. The reaction is

$$AgBr + 2S_2O_3^{2-} \rightarrow Ag(S_2O_3)_2^{3-} + Br^-$$

and the silver thiosulphate is soluble.

The net result is the formation of a film with silver deposits where the light has been incident: the photographic negative. In 1936, James Webb suggested that the latent image is produced by trapped photoelectrons. Two years later, Ronald Gurney and Nevill Mott (1905–1996) explained the production of the silver particles by motion of electrons and silver cations in the silver bromide crystal lattice. Absorption of a photon raises an electron into the conduction band of the silver bromide, and this enables it to diffuse to a cluster of silver atoms, where it becomes trapped. The resultant negatively charged silver attracts the positive silver ions, and those of the latter which are in interstitial sites, as members of Frenkel pairs, are able to diffuse and attach themselves to the metallic silver cluster. The nucleation of a cluster was attributed to the formation of an embryonic particle of silver sulphide, sulphur being present in the gelatin. Gurney and Mott suggested that the conduction level of silver sulphide lies lower than the corresponding level in silver bromide, and that this localizes the electrons. Jack Mitchell and John Hedges subsequently accounted for the experimental observation of silver particle formation in the interiors of grains. Until their work, in 1953, only surface cluster formation was understandable. They demonstrated the presence of dislocations within the grains. Their contribution was doubly impressive: it revealed the nature of the internal traps, and it provided the first direct evidence of *dislocations*. The photon-induced ejection of an electron from an atom, and the subsequent movement of both the electron and a silver cation, are indicated in the diagram below (left). Also visible is the growing cluster of silver atoms which will ultimately produce the metallic silver particle that remains after development. Several such particles, or specks, are visible in the photograph (below right) of an undeveloped emulsion.

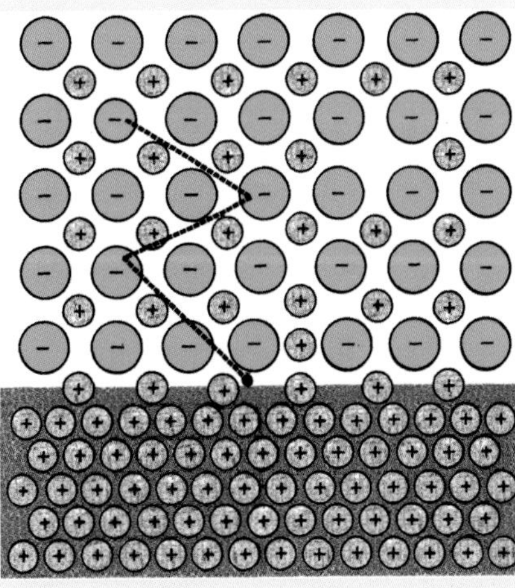

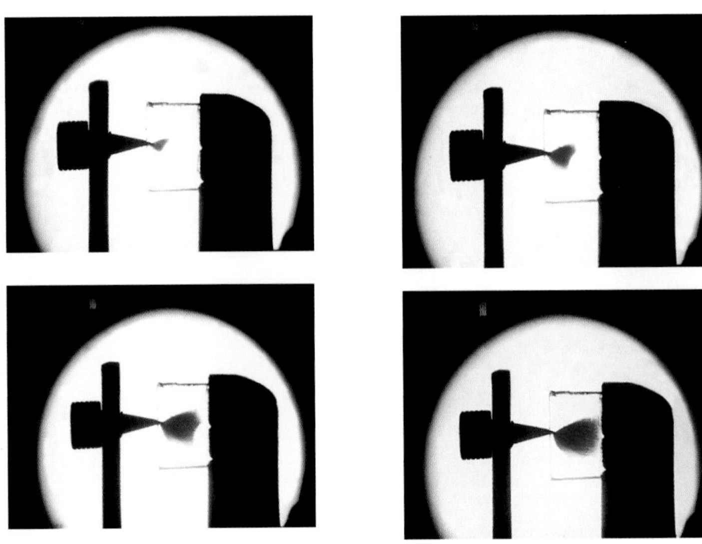

Anion vacancies in an ionic crystal are equivalent to positive charges, which are able to trap electrons. The latter become bound, and are able to occupy any of a series of *energy levels*. Some of the transitions between these levels correspond to frequencies in the visible region, and the otherwise transparent crystals which contain these centres appear coloured. The diffusion of *colour centres* is demonstrated in this series of pictures which were taken at one-second intervals. The approximately one centimetre square potassium chloride crystal, at a temperature of about 550 °C, is subjected to an electrical potential of approximately 300 volts applied between the metal holder and a sharp stainless-steel point. The latter acts rather like a lightning conductor, injecting electrons into the crystal and producing a steadily increasing concentration of colour centres.

Energy will be absorbed by atoms in the ground state, transferring them to the excited state. Some excited atoms will decay by *spontaneous emission* of radiation, while others will be induced to emit radiation by the stimulation process. The normal state of affairs is such that far fewer atoms are in the excited than in the ground state, simply because the lifetimes of excited states are usually very brief. If the intensity of the incident radiation is sufficiently high, however, it is possible to obtain a situation in which there is actually a majority of atoms in the excited state, a condition known as *population inversion*. If the irradiating light source is suddenly removed, the first excited atom to spontaneously decay thereafter will emit radiation. This radiation will encounter more atoms in the excited state than in the ground state. It is thus more likely to cause stimulated emission than excitation; the most likely outcome is that the wave will be reinforced by the stimulation process. It turns out that whereas spontaneous emission is isotropic, with all directions of photon propagation being equally probable, the stimulation process is highly directional. The stimulated photon travels in the same direction as the photon which stimulated it. As long as the state of population inversion lasts, this process rapidly reinforces itself, and there is a sort of avalanche effect, which produces a beam of very high intensity, coherence and directionality.

In the absence of a more specified geometry, the direction of the laser pulse would be essentially random. It is determined by the line between the original atom that decays by spontaneous emission and the first atom that it provokes into stimulated emission. In practice the direction of the emergent beam is controlled by placing parallel mirrors at either end of the lasing medium. A light beam falling at right angles on either of these is re-directed so as to pass time and time again through the assembly of excited atoms. This makes the length of lasing material in the direction perpendicular to the two mirrors effectively infinite. The final lasing pulse is extracted from the apparatus by having one of the two mirrors only half silvered, so that it allows part of the energy falling on it to be transmitted rather than reflected. The idea for a device of this type was put forward by Arthur Schawlow and Charles Townes in 1956, and the first working laser, which used a rod of ruby as the lasing material, was built by Theodore Maiman in 1960.

The direct and reflected light beams from this Scandinavian wall fitting have different colours. The reasonably white spectrum emitted from the candle is modified by the polished brass surface, because the metal does not reflect all wavelengths with equal efficiency; the reflecting power for a given wavelength is intimately connected with the way in which the electrons in the brass alloy are distributed among the available energy levels.

Metals are opaque to optical radiation, and their *lustre* is associated with a characteristically high *reflectivity*. Energy can be absorbed from incident electromagnetic radiation because there are empty energy states available immediately above the Fermi level. As John Bernal has remarked: 'all that glisters may not be gold, but at least it contains free electrons'. There are complications, however. The optical properties of a metallic surface depend on the frequency of the incident radiation. If this is high, the electrons can oscillate in phase with the light wave and transmit it through the metal, which is therefore transparent. For the alkali metals this onset of transparency occurs in the ultraviolet range, as discovered by R. W. Wood in 1933. At lower frequencies the stimulating electric fields and the oscillating electrons get out of phase, and the reaction of the electrons to the incident radiation is influenced by electron–ion collisions. This retarding factor causes the stimulation and response to get out of phase, and the wave is rapidly attenuated, within a characteristic distance that is actually less than a single wavelength. Even at radio frequencies this *skin depth*, as it is called, is only about 0.1 mm. The wave is thereby absorbed and part of it is re-emitted in the reverse direction. This *reflection* is so strong in a metal that relatively little energy is dissipated as heat in the crystal. Most metals have a silvery appearance, the notable exceptions being the red of copper and the yellow of gold. As Nevill Mott and Harry Jones first suggested in 1936, this appears to be caused by the overlapping of the bands in these metals. The bands are not as tidily distinct as we have made them out to be, and in both copper and gold electrons in the filled d band can be transferred to vacant s and p levels. These transitions give absorption and re-emission in the observed colour regions.

One of the most mysterious manifestations of the motion of electrons in crystals is that familiar phenomenon, *magnetism*. It is probably the most complicated of the electron-dependent properties, and there are no less than five different forms of it. To begin with, however, we need only consider the most common type: *ferromagnetism*. This occurs naturally in the mineral *magnetite*, a magnetic oxide of iron, mention of which appeared in Greek literature as early as 800 BC. The mineral takes its name from the Greek province of Magnesia where it was first mined. Magnetite was also known by the name *lodestone*, which means leading stone, this following from its first application in the navigator's compass. As we might anticipate from the name itself, the first observations of ferromagnetism in a metal were made on iron. William Gilbert reported in his *De Magnete* that the freshly smelted material becomes magnetic if it is drawn into the shape of a bar, stroked with a permanent magnet, or even tapped lightly while lying in a north–south direction.

The atomic-scale processes that give rise to magnetism are not difficult to understand because they are the counterparts of familiar phenomena occurring on a much larger scale. Electric motors are based on the fact that a wire carrying an electric current produces magnetic effects. This had been discovered in the early part of the nineteenth century by Hans Christian Ørsted, and independently by Gianni Romagnosi, an Italian lawyer. A flow of electricity also occurs at the atomic level, because charged electrons move in their orbits, and these elementary currents must similarly give rise to a magnetic effect. But the electrons in atoms are subject to the laws of *quantum mechanics*, and it cannot be taken for granted that all the tiny magnetic effects associated with the individual electrons will add up to give

Holography

The *laser* finds one of its more spectacular applications in the lensless photographic technique known as *holography*. In this process, invented in 1947 by Denis Gabor (1900–1979), one records not the image of an object but rather the light waves reflected from the latter. The word *hologram* literally means whole picture, and this implies that one is able to overcome the familiar inadequacy of a photographic emulsion, namely that it can record the amplitude of incident light but not its phase. The trick is to convert the transient temporal variations of the light wave into permanent spatial variations of photographic density, and this is achieved by making the wave reflected from the object interfere with a second reference wave, the two waves being mutually coherent. Just as in Thomas Young's original interference experiment (see Chapter 1), the recorded interference pattern, the hologram, contains a series of points with increased density where the two waves arrived in phase and reinforced each other, and diminished density where they were out of phase. The hologram bears no resemblance to the original object; it consists of a confusion of blobs and whorls. It nevertheless contains all the encoded optical information about the object, and that information can be retrieved by what is known as the reconstruction process. With the object removed, the hologram is illuminated by the reference beam alone and an image can then be seen at the location originally occupied by the object. It is as if the waves originally scattered by the object were frozen into the hologram, subsequently to be released during the reconstruction process. The entire image of the original object can be reproduced from any portion of the hologram, however small, but the photographic resolution diminishes as the size of the holographic fragment decreases. The diagram shows the recording and reconstruction arrangements. The pictures show two reconstructed images of the same object, and demonstrate that the hologram even contains information regarding parallax, which is the relative displacement of different parts of an object when viewed from different directions. Gabor's original work preceded the invention of the laser, and the holographic technique became practical only after the introduction of that device. The first holograms recorded by laser light were obtained in the early 1960s by Emmett Leith and Juris Upatnieks.

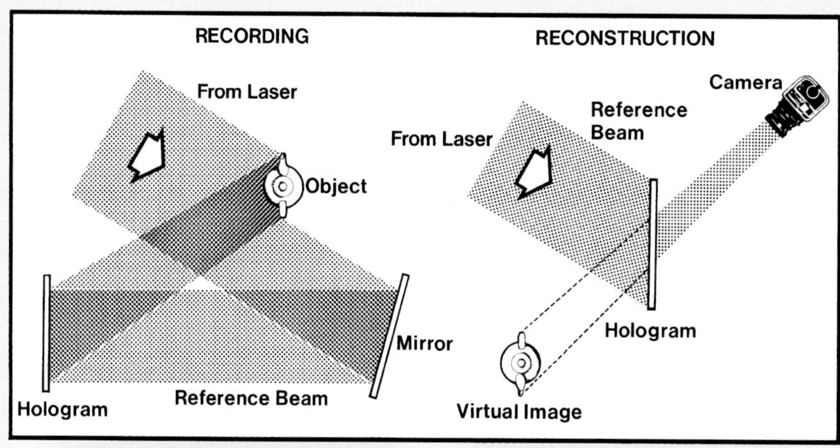

a measurable macroscopic effect. It has already been emphasized that electrons tend to arrange themselves in pairs, permitting them to cancel out each other's magnetic effect. Such cancellation could not be perfect for atoms containing an odd number of electrons, so we might expect half the elements in the periodic table to be magnetic. This is certainly not the case, however, because the unpaired electrons in atoms of odd-numbered elements are usually in the outermost orbits, where they can pair off with electrons on adjacent atoms. This suggests that *no* elements should show magnetism. But there are elements in which the unpaired electrons occupy some of the inner orbits. These are the *transition elements*. Among the prominent members of this group are iron, nickel and cobalt, all of which do indeed display ferromagnetism. A second group of elements in which this inner-orbit unpairing occurs is the *rare earths*, and these also show ferromagnetism.

The idea that bulk magnetism is ultimately the result of the elementary magnetic effects associated with electron motion is due to André-Marie Ampère. In order to appreciate how the arrangement of these atomic-scale magnets determines the total magnetism of a piece of material, we must recall some historical developments. Ampère and François Arago showed that steel needles exhibit stronger magnetism when placed inside a coil that is carrying an electric current. This observation was generalized by Michael Faraday, who showed that all substances become magnetic to a greater or lesser degree when under the influence of a magnetic field. Given the existence of the atomic-scale magnets, these early observations do not appear surprising, especially in view of the alignment of compass needles by the Earth's magnetic field. But there remains the mystery of why some materials are permanently magnetic. Wilhelm Weber realized that a substance must lose its magnetism if the atomic magnets are randomly arranged, but further progress was not possible until Pierre Curie had carried out his classical experiments at the turn of the century. He found that for those materials that are magnetic even in the absence of an external field, the ferromagnets, the magnetism increases when they are placed in a magnetic field. Above a certain temperature that is characteristic of the material, the *Curie temperature*, the strong ferromagnetism disappears and is replaced by a much weaker type of magnetism. This latter is known as *paramagnetism*, and it decreases in strength with rising temperature. Paramagnetism too is proportional to the strength of an applied field and is aligned in the same direction. It was explained in 1926 by Wolfgang Pauli who, like Arnold Sommerfeld in his successful theory of specific heat, appreciated the significance of the *Fermi–Dirac statistical distribution*. A third type of magnetism, *diamagnetism*, is actually present in all materials. It is quite weak and its direction is such as to oppose that of an external field.

Major theoretical advances were provoked by Curie's work. Paul Langevin explained the temperature variation of paramagnetism by the misaligning effect of thermal agitation on the atomic magnets, but the real breakthrough was achieved by Pierre Weiss who was the first to perceive the true nature of ferromagnetism. Weiss proposed that a ferromagnet consists of a collection of domains, in each of which all the atomic magnets are lined up parallel to one another. Because the direction of magnetism varies from one domain to another, the different domains are in partial, or even total, opposition, and the net magnetism is less than the sum of the individual domains. An applied magnetic field aligns the domains and thereby increases the

The *magnetic domain walls* in this iron crystal have been rendered visible under optical microscopy by coating the surface with magnetized iron oxide. The domains themselves are formed by the spontaneous subdivision of the originally uniformly magnetized material into a pattern of mutually opposing regions, here indicated by superimposed arrows showing the local direction of magnetization. The subdivision lowers the external field energy of the material, but this advantage is partially offset by the energy required to form the boundaries, known as *Bloch walls* (after Felix Bloch, 1905–1983), between the domains. The degree of spontaneous subdivision is dictated by a compromise between these opposing factors. When the material is placed in a magnetic field, the positions of the walls are changed, increasing the size of the domains whose magnetization is parallel to the applied field and diminishing those whose magnetization opposes that field. Crystal defects such as grain boundaries hinder such wall movements and this is a source of energy loss in components which must undergo periodic magnetic reversal.

material's magnetism. The existence of domains was later demonstrated by Francis Bitter, who deposited small iron oxide particles on the surface of a bar magnet and observed them with a microscope. Such particles tend to collect at the boundaries between the domains, rendering them visible.

The origin of the parallel arrangement of the atomic magnets within a given domain was an obvious theoretical challenge. A full quantum mechanical treatment was achieved through the collective effort of Paul Dirac, I. Frenkel, Werner Heisenberg, Edmund Stoner and John van Vleck. Ferromagnetic materials contain electrons whose spins are not compensated by electrons with the opposite spin, and thus constitute small atomic magnets. Because the electrons obey *Pauli's exclusion principle,* they interact with a so-called *exchange force* which tends to align neighbouring magnetic moments parallel to each other. The discovery by Heisenberg and Dirac of the exchange force, which is much stronger than the classical force between magnetic moments, was one of the first triumphant applications of *quantum mechanics* to the understanding of the properties of solids. The final result may be a collective alignment of all the atomic magnets within a domain, but this is contingent on the temperature being low enough that the ordered state is not disturbed by thermal fluctuations.

The interactions between adjacent atomic magnets does not always lead to a situation in which they are all parallel. In manganese fluoride, for example, they adopt a head-to-tail configuration. Materials showing such antiparallel alignments are referred to as *antiferromagnets,* and they were discovered by Louis Néel. The *Néel temperature,* above which the antiferromagnet becomes a paramagnet, is analogous to the Curie temperature in a ferromagnet. Still other arrangements are possible. In magnetite, two thirds of the atomic magnets point in one direction, while one third point in the opposite direction. This leads to a situation in which the magnetism is about one third of the value that would be obtained if the material had been a ferromagnet. Néel discovered this type of material in 1948 and referred to it as a *ferrimagnet.*

It must be emphasized that ferromagnetism, antiferromagnetism and ferrimagnetism are observed only when a material has what have been referred to as atomic magnets, these in turn arising only if the atoms have unpaired electrons in an inner orbit. The much weaker types of magnetism are manifestations of processes that are far more subtle. It turns out that two electrons that have formed a pair cancel out each other's magnetic effect only in the absence of external stimulation. When a field is applied, one of two things can happen. A slight imbalance between the orbits of the two electrons can occur, giving rise to *diamagnetism,* or a small lack of compensation is established between the two spins, producing *paramagnetism.*

The boundaries between the domains in a ferromagnet are referred to as *Bloch walls,* in recognition of Felix Bloch, who made an intense study of them in the 1930s. A Bloch wall is the plane of demarcation between two domains in which the atomic magnets are pointing in mutually opposite directions. Formation of such a wall requires expenditure of energy since the most stable situation at the microscopic level would be obtained when all the atomic magnets were lined up. On the other hand, the head-to-tail arrangement of the domains is favoured macroscopically because it decreases the strength of the magnetic field surrounding the material. This latter factor is related to the fact that two bar magnets attract each other most strongly when they are placed with the north pole of one bar and the south pole of the other adjacent to each other, and vice versa. Lev Landau and E. Lifshitz were the first to appreciate, in 1935, that these conflicting microscopic and macroscopic factors would lead to a compromise situation in a ferromagnet. The microscopic factor favours large domains and few walls whereas the macroscopic consideration favours small domains and many walls. For a typical ferromagnetic material the two factors are in balance when the domain diameter is between about 0.01 cm and 0.1 cm.

The most significant property of Bloch walls is that they can be readily moved by an external magnetic field. This can be demonstrated by the *Bitter technique*, and it takes only relatively weak fields to cause wall motion. The movements nicely confirm the original predictions of Weiss, because they are such as to increase the size of the domains in which the magnetization is parallel to the applied magnetic field. At the same time the size of the domains that do not lie in this direction is decreased, so the overall effect is to strengthen the magnetization of the material. If the direction of the external field is then changed, the Bloch walls are found to move accordingly, and the direction of strongest magnetization in the material now lies parallel to the new direction of the field. There is a certain degree of resistance to the motion of Bloch walls, this arising from crystal defects and impurities. When the resistance is particularly high, the material is said to be *magnetically hard*, while low resistance gives *magnetic softness*. Both these extremes find ready application in technology. Hard magnetic materials are good *permanent magnets*, while the soft variety are important when the magnetization in a component must constantly be changed, as for example in electric motors, dynamos and transformers. In these latter applications, resistance to Bloch wall motion causes the magnetization to lag behind the applied field, causing a loss of efficiency. The effect is known as *hysteresis*, from the Greek expression for coming late.

Although the applications of magnetism do not show wide variety, there is much magnetic material in use. Motors, generators and transformers account for the major part by weight. Recent years have seen a proliferation of devices in which relatively small amounts of magnetic material play a central role. Small ferromagnets are used in *computer memories*. The two possible directions of magnetization correspond to 0 and 1 on the binary system, which is commonly used in such devices. Magnetic storage is also exploited in *recording tape*, which consists of a suspension of fine iron particles in a thin strip of polymer material. The iron particles are at random in a new or erased tape, a magnetic head supplying the biasing influence on the Bloch walls during recording. This technology is now being superseded, of course, by optical recording on compact discs, but electronic recording is already making a comeback in the form of the highly convenient memory stick.

Summary

The electrical, optical and magnetic properties of materials are determined by the distribution of the outer electrons of the constituent atoms among the available energy bands. A partially full upper band permits electron mobility, the origin of electrical conduction and optical reflection in a metal, while a completely full conduction band produces an insulator. Narrow gaps between full and empty bands are the hallmark of semiconductors:

the basic components of transistors. A pairing of electrons, which survives only at sufficiently low temperatures, produces superconductivity. Insulator crystals are generally transparent because optical photons have insufficient energy to promote electrons across the gap between bands. Point defects are responsible both for colour centres and, at sufficiently intense stimulation, the lasing of certain insulators. Strong forms of magnetism, such as ferromagnetism, arise from the imbalance caused by unpaired electrons in incompletely filled quantum orbits. The weaker effects, paramagnetism and diamagnetism, are related to subtler perturbations of electron orbits by applied fields.

11

Like harmony in music; there is a dark inscrutable workmanship that reconciles discordant elements, makes them cling together in one society.

William Wordsworth (1770–1850) ***The Prelude***

A mysterious harmony: *glass*

The products of the glass maker are often aesthetically pleasing, but it would be unfair to claim that *glass* is intrinsically more attractive than any other type of material. Rather, by its presence in the family of materials, glass extends the range of useful properties and artistic qualities. As has transpired only recently, glass also extends the range of atomic structures in condensed matter, through its unique lack of order. Indeed, it is this atomic-level attribute, rather than the traditional ones of transparency and brittleness, which is now acknowledged as the defining characteristic of this type of material. *Oxide glasses* such as the common silicate varieties have been in use for at least 4000 years. Glass beads, dating from about 2500 BC, have been found in Egypt and other parts of the Near East, although exploitation of the plasticity of hot glass is more recent. These oxide materials so dominated the scene that the existence of the glassy state was believed to be intimately connected with the presence of *covalent bonds*. Glass has long been known to be a *supercooled liquid*, and covalent bonding was regarded as a prerequisite if crystallization was to be avoided. These attitudes underwent radical revision in the 1960s. With the production of metallic examples, it was suddenly realized that a glass need not be transparent, brittle, or insulating.

It is not surprising that the most familiar examples of the glassy state belong to the covalent family. The chief prerequisite of a *glass former* is high viscosity near the freezing temperature. The *viscosity* of a substance is a measure of the rate at which the atoms move about, a high viscosity being associated with low atomic mobility. To get from the disordered arrangement of the liquid state to the regular configuration of the crystal, the individual atoms must migrate, so it is clear that the crystallization process can be bypassed if the viscosity is sufficiently high. The two factors that determine the rate of atomic movement in condensed matter are the strengths of the interatomic bonds and the temperature. The bonds between the atoms in a typical covalent substance are not only directional but also link the atoms together into a network. At a given temperature, relative to the freezing temperature, this makes the material far more viscous than a liquid metal. As the temperature decreases, there is less *thermal energy* available to enable the atoms to jump over the *energy barriers*, so the viscosity increases. Therefore, if it were possible to lower the freezing temperature, the glass-forming tendency would be increased, because the atoms would be more sluggish when seeking to establish order. *Freezing points* can indeed be lowered by using mixtures rather than pure substances. Thus the common types of glass are mixtures of inorganic oxides.

Although glass was already in widespread use more than four thousand years ago, transparent examples are more recent. Early glasses were opaque, and frequently coloured. The Romans appear to have produced the first transparent variety. This cup was found near Colchester in England, and dates from the first century AD. The transparency of some glasses was first explained in the mid 1970s, not the least through the efforts of Philip Anderson and Nevill Mott (1905–1996).

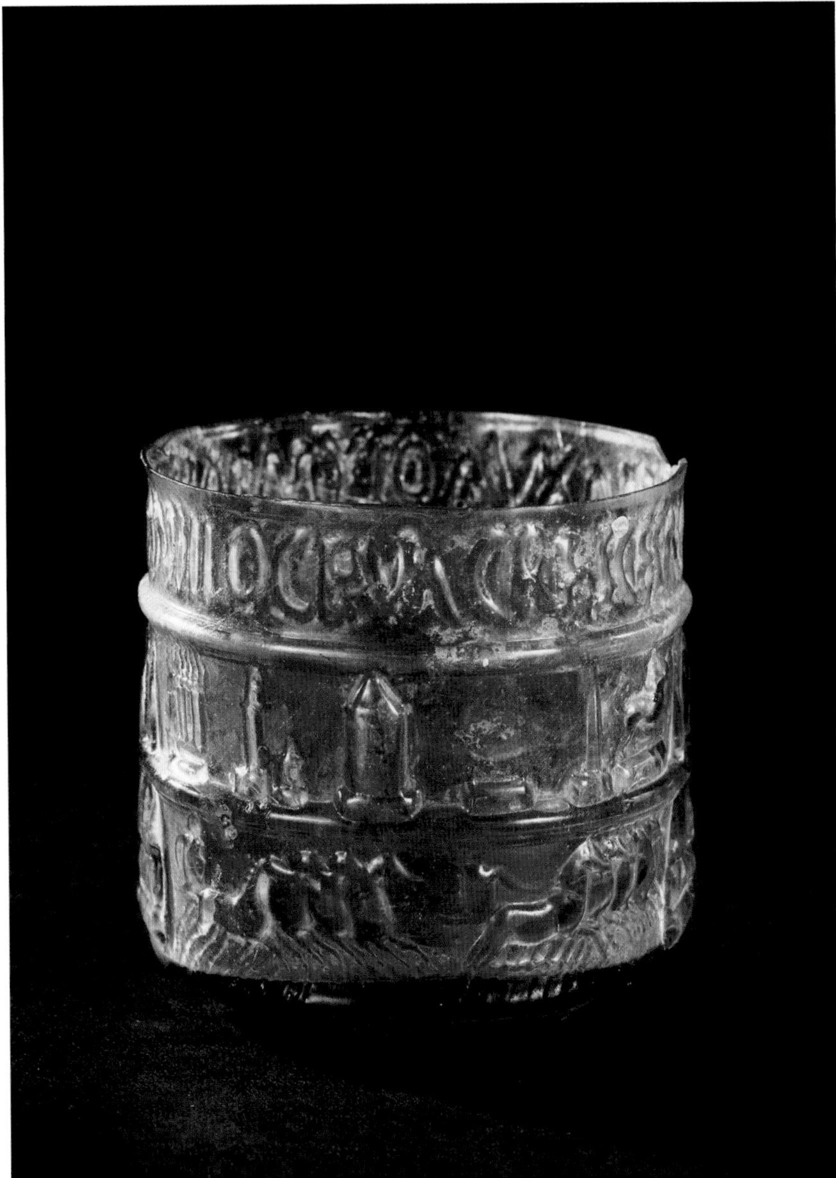

An understanding of the glassy state requires knowledge of the actual atomic arrangement. The standard way of probing the atomic structure of materials is by *X-ray diffraction*. The pattern produced when the diffracting specimen is crystalline is a characteristically regular array of spots, while the corresponding result for a liquid is a set of concentric and rather diffuse halos. Diffraction of X-rays from a typical glass produces a picture that is essentially indistinguishable from that of a liquid. The term *crystal glass* is therefore a misnomer, and it is interesting to note that substances such as wool, and even paper, are actually more crystalline than glass. X-ray studies of the non-crystalline, or *amorphous*, state were pioneered by B. E. Warren in the mid 1930s.

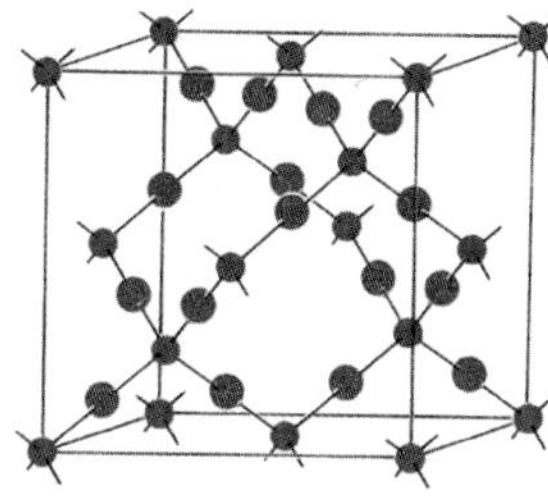

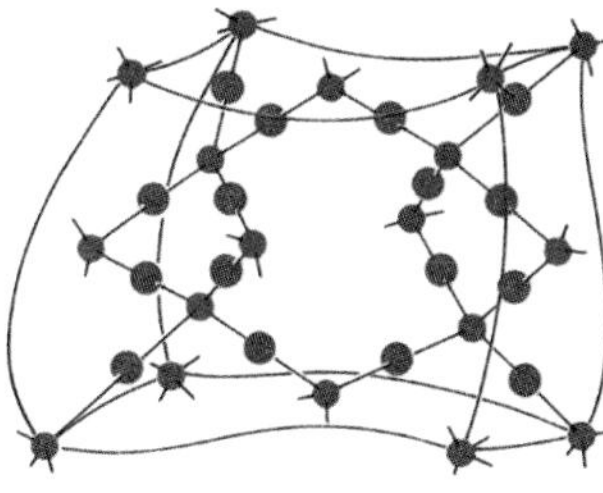

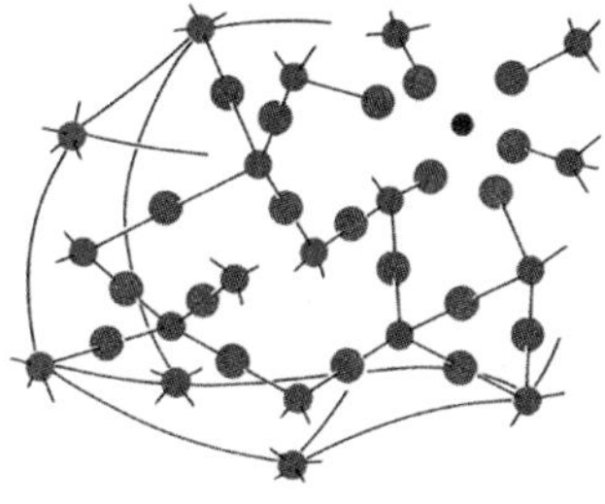

Silica, SiO_2, can exist in both crystalline and glassy forms. One of its crystalline modifications is *cristobalite*, a high-temperature form of *quartz*. Its crystal structure (top) is reminiscent of diamond, with the carbon atoms of the latter replaced by silicons. But there is an additional oxygen atom between each neighbouring pair of silicon atoms. The glassy form (centre) can be regarded as a distorted version of the crystal, both with respect to the positions of the atoms and the way in which they are linked together. Addition of what are referred to as *fluxing ions*, such as the small positively-charged sodium ion indicated on the lower figure, can be looked upon as totally disrupting the cristobalite structure. The ion attracts the oxygen atoms, causing them to sever some of their bonds with the silicons.

The theoretical picture of the glass structure that has gained the widest support was proposed by Willie Zachariasen in 1932. It is known as the *random network model*. The glassy form of pure silica is not a particularly convenient substance to work with, because it has such a high *softening temperature*, but it serves as a useful illustration. There are three different crystal forms of *silica*. These are *quartz, tridymite* and *cristobalite*, which are stable in the ranges below 870 °C, 870–1470 °C and 1470–1713 °C, respectively. In these crystal structures, as discussed in Chapters 7 and 8, each silicon atom resides at the centre of a tetrahedron of oxygen atoms, and the tetrahedra are linked to one another at their corners. Zachariasen showed that a material is likely to be a good glass former if it obeys four rules. These are that each oxygen atom is linked to not more than two cations; that the *coordination number* of the cations is fairly small; that the *oxygen polyhedra* share corners with each other, but not edges or faces; and finally that the polyhedra are linked in a three-dimensional network. If one builds a model according to these rules, with each tetrahedron sharing not less than three of its corners, it transpires that the structure lacks the repetitive regularity of the crystalline state, and it has no *long-range order*. It is precisely this latter deficiency that makes the X-ray diffraction pattern of a glass so diffuse and ambiguous. This brings us to the limit of present knowledge, because it is still not known just how irregular the structure of a glass actually is. One can build these random network models, but there is unfortunately no unique connection between such a model and the observed diffraction pattern.

Zachariasen's ideas make it easier to appreciate what happens when melted silica is cooled from its liquid state. Well above the freezing temperature the bonds between the silicon and oxygen atoms are constantly being broken, rearranged and reformed, in such a way that there are nevertheless many tetrahedra in evidence. As the liquid is cooled, its volume continuously decreases, at a fairly constant rate with respect to temperature, until the glass temperature is reached. The rate of decrease of volume with temperature is then markedly diminished. It becomes, in fact, similar to that observed in the crystalline state. If the material had been able to crystallize, its volume would have shown an abrupt decrease at the freezing temperature. The *glass temperature* always lies below the *freezing temperature* for the same material. When the glass temperature is reached, during the cooling process, the atomic configuration becomes frozen in, and this situation persists more or less unchanged as the temperature is further lowered. If the temperature is maintained constant for a sufficiently long time just below the glass temperature, the structure becomes stabilized in what is believed to be a *metastable state*. This is characterized by the linking together of the tetrahedra in the random network. The metastable state is presumably separated from the crystalline form by an energy barrier that cannot be overcome at the glass temperature, because of the very high viscosity. Below the glass temperature, the rate of change is so slow that the situation is essentially static. It is important to note that the glass temperature is not a fundamental quantity, like the melting and boiling points. If the rate of cooling is decreased, the glass temperature is lowered; a lower cooling rate gives the atomic configuration more time to rearrange itself, in an effort to establish crystallinity. It still falls short in this quest, but it does achieve a denser packing than would have been produced at a faster cooling rate. A measure of just how far a given configuration is from equilibrium is provided by the *fictive temperature*. This is defined as the point at which the glass

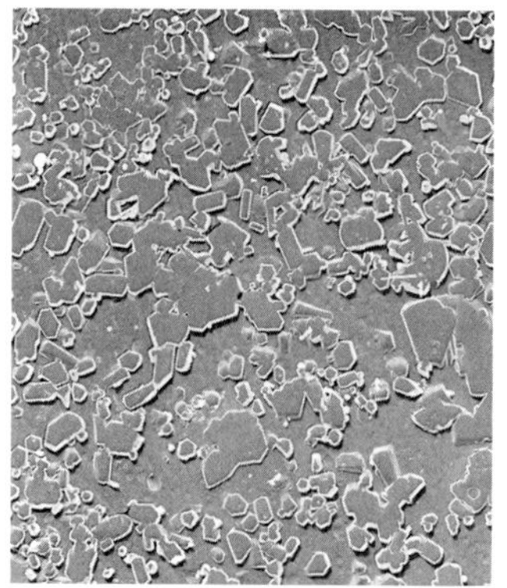

Among the newer forms of glass are those which are actually partly crystalline. They are referred to as *glass-ceramics*, and a typical example is seen in this photomicrograph of an oxynitride variety. It was produced by adding 46% by weight of silicon nitride to a common glass material containing silica and alumina. The result is a composite of glass and crystal which is stable up to a temperature of about 1800 °C. In this picture, the width of which is about ten micrometres, the crystalline regions appear as islands in the amorphous background.

would be in equilibrium if suddenly brought there in its current state. If the actual temperature coincides with the fictive temperature, the configuration must already be in equilibrium.

In connection with the various fabrication processes encountered in the glass industry, it is desirable to define several other characteristic temperatures. The highest temperature from which a glass can be rapidly cooled without introducing dangerously high internal stresses is known as the *strain point*. Somewhat higher than this lies the *annealing point*, at which such internal stresses can be relieved within a few minutes. At a still higher temperature comes the *softening point*, at which a fibre of specified size elongates under its own weight. The standard fibre used in this definition is 24 cm long, 0.7 mm in diameter, and the elongation rate is 1 mm per minute. Finally, there is the highest of these defined temperatures, namely the *working point*. In practice, the various characteristic temperatures are defined through standard viscosities. For the soda-lime–silica glasses the approximate values are 470 °C for the strain point, 510 °C for the annealing point, 700 °C for the softening point, and 1000 °C for the working point. The corresponding temperatures are about 200 K higher in the aluminosilicate glasses.

The *soda-lime–silica glasses* that are used in vast quantities for windows, bottles, and jars consist of 70%–75% silica (SiO_2) 13%–17% sodium oxide (Na_2O), 5%–10% lime (CaO), 1%–5% magnesia (MgO), and minor quantities of some other oxides. Such a glass is produced commercially by melting together silica in the form of sand, limestone ($CaCO_3$), and soda (Na_2CO_3). *Borosilicate glasses* have high resistance to heat and expand only slightly with rising temperature. Hence their use in ovenware of the *Pyrex* type. These glasses are also durable against chemical attack and find extensive use in laboratory hardware. Their composition is 60%–80% silica, 10%–25% boron oxide, 2%–10% sodium oxide, and 1%–4% alumina. In the high-temperature range, the *aluminosilicate glasses* perform even better. A product with the approximate composition 55% silica, 23% alumina, 8% magnesia, 7% boron oxide, and 5% lime does not show any noticeable softening until the temperature is above 800 °C, well into the red-hot range. If the alumina content is decreased to 15% and the lime component raised to about 17%, the resulting material has high durability that is useful in the *fibre glass* and *glass wool* industries. Finally, among the common types, there are the *lead-silicate glasses* which contain 20%–65% silica, 20%–80% lead oxide, and 5%–20% of sodium oxide or potassium oxide. In the lower range of lead oxide content, these glasses are used in lamps and various electronic devices which exploit their high electrical resistivity. As the lead oxide content increases, so does the *refractive index*, and this produces a range of glasses for optical purposes and also for the so-called crystal glasses that are used in tableware. The highest lead oxide contents are employed in glass radiation shields, the lead providing strong absorption because its high atomic number implies a high density of electrons with which the radiation photons can collide.

Little has been said yet about the structural chemistry of the different common types of glass. To begin with, let us consider the addition of monovalent ions, as occurs when a small amount of sodium oxide is mixed with silica. Unlike silicon, which holds on to its electrons and forms bonds which are at least 50% covalent, the sodium atom gives up its outer electron to become a positive ion. The sodium oxide thus tends to dissociate and contribute oxygen to the random network, upsetting the balance between

Although window glass provides adequate protection against rain, its interaction with water produces considerable weathering at the microscopic level. Terry Michalske and Stephen Freiman have suggested that a water molecule can force its way into a silicon–oxygen–silicon link, producing the new sequence silicon–oxygen–hydrogen–oxygen–silicon, with an extra hydrogen atom attached to one of the silicons. The final stage completely severs the link, leaving two silicon atoms with terminal hydroxyl groups. This scanning electron photomicrograph, which covers a width of about twenty micrometres, clearly shows surface rupture in a glass specimen exposed to 70 °C water for two weeks. J. E. Shelby and J. Vitko Jr., who produced the picture, demonstrated that there is indeed an unusually high concentration of hydrogen at the fracture surfaces.

oxygen and silicon ions. The number of oxygen ions is more than twice that of the silicon ions, so not every oxygen is able to bond with two silicon ions. Every silicon ion, on the other hand, still finds it easy to be tetrahedrally bonded to four oxygen ions. The outcome is a situation in which some oxygen ions are bound to only one silicon, and these are said to be *non-bridging*. On an atomic scale, fairly large holes are thereby opened up in the network and the sodium ions move into them. By so doing they tend to preserve local electrical neutrality, because the non-bridging oxygen ions are negatively charged. The sodium ions are in fact surrounded by six oxygen ions on the average. The sodium ions are referred to as *fluxing ions*, and they lower the viscosity of the material. They break some of the strong and directional silicon–oxygen links and replace them with ionic bonds that are both weaker and non-directional.

In general, we can divide the various components in the common type of glass according to the role they play in the final structure. First there are the *network formers*, the classical example of which is silica. Other common network formers are phosphorus pentoxide and arsenic pentoxide, both of which form tetrahedral units, and boron trioxide which forms triangles. The sodium oxide just discussed is an example of a *network modifier*: a disrupter of network continuity. Other examples of network modifiers are potassium oxide, lime, and magnesia. Finally there are what are referred to as the *intermediates*, which can either fit into the backbone of the network or occupy the holes in the structure. Examples of this latter group are such oxides as alumina, beryllium oxide and titanium dioxide. The compositions of the various glasses have been arrived at empirically over the centuries during which man has practised the art. Although it would be an exaggeration to say that all aspects of *glass making* are fully understood, it is now a reasonably established science in which success is synonymous with control; fine control of both composition and temperature.

The various steps in the commercial production of glass are quite straightforward, and only a few details need be added to what is fairly common knowledge. It should be remarked, however, that pure silica glass is a twentieth-century development because it was not previously possible to reach the high melting point of quartz. The melting of the raw material takes place either in large ceramic pots, which often hold up to several hundred kilograms, or in large tanks operated in a continuous fashion, with the components fed in at one end and the melt emerging from the other. As the temperature is raised, the constituents having the lowest melting points melt first, and the others then dissolve in them. In practice, the input material usually contains up to 30% of scrap glass that has become surplus in the post-melt processing. Such recycled material is known as *cullet*, and because it has the correct composition and therefore the lowest melting point, it serves the useful purpose of promoting the melting of the entire batch. One of the more difficult stages of the process is removal of the small bubbles formed in high concentrations in the melt. They are usually brought to the surface by addition of certain chemicals which decompose with the liberation of large bubbles. These sweep the smaller bubbles with them as they rise to the surface.

Alternatively, the bubbles are removed by the application of a mild vacuum. In the modern manufacture of *plate glass*, the melt is passed directly from the output end of the tank furnace into a *float bath* which contains a large dish of molten tin. The liquid glass is less dense than the tin, so it floats

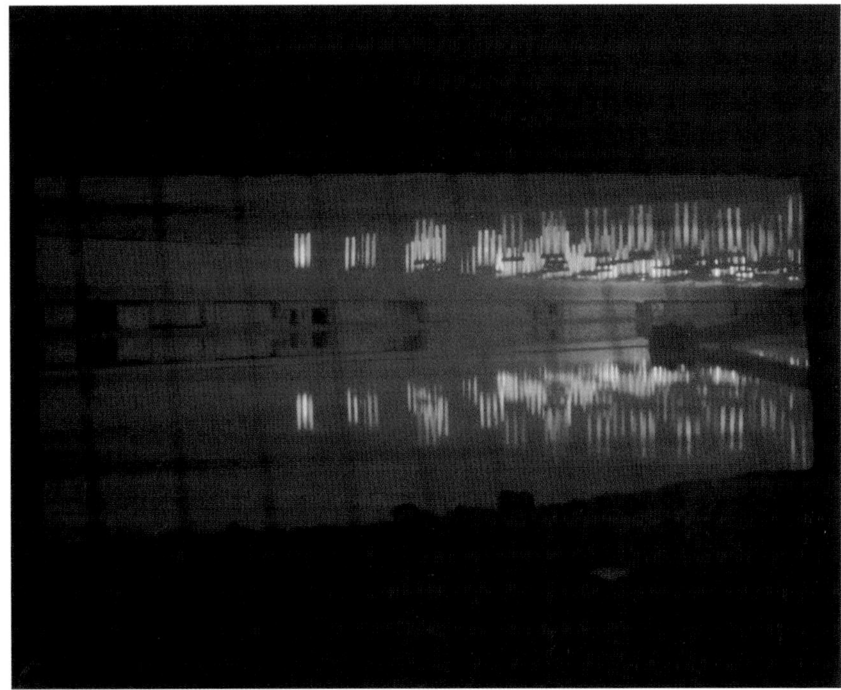

The perfectly flat sides of plate glass are produced by cooling a layer of the fused silica from the molten state. This would normally guarantee the flatness of only one of the surfaces, but the second surface will also be flat if the molten glass lies on a liquid metal such as tin. This view inside an industrial *float bath* shows the overhead heaters which control the temperature.

on the surface. Moreover, because both materials are in the molten state, the upper and lower surfaces of the glass are perfectly flat. Further along the float bath the temperature is lowered to a point which is under the glass temperature but still above the melting point of tin. The glass becomes rigid in this region, and it emerges as solid sheet. The process by which *glass wool* is produced is similar to that used in the making of *candy floss*, a familiar sight in amusement parks. The molten glass is poured continuously into a *spinner*, a container riddled with fine holes that is rapidly rotated on its axis. Fine threads are *extruded* from these holes and are removed in a convenient direction by application of a gentle draught.

The most significant property of silicate glass is its *transparency*. This is so familiar that we might take it for granted. The ease with which light passes through belies the complexity of the atomic processes involved in this transmittance of energy. Looking deeper into the question, we find that there is more to it than meets the eye; the phenomenon is less transparent than the material itself. Indications of underlying subtlety include the fact that a glass prism splits white light into the colours of the rainbow, while common glasses absorb radiation in both the ultraviolet and infrared regions of the *electromagnetic spectrum*. Glass is useful because it happens to display low absorption for those wavelengths to which the eye responds most strongly. The splitting of white light into its constituent colours was explained by Isaac Newton at the end of the seventeenth century. He showed that the change in direction of a beam of light as it passes from one medium to another, from air to water or from air to glass for example, is caused by the fact that the *velocity of light* depends on the substance in which it is travelling. This bending of a light beam is known as *refraction*, and the splitting of white light into colours, also called *dispersion*, occurs because different light frequencies have different velocities. The degree of bending is determined by a material's *refractive index*, which is defined as the ratio of the speed of light in vacuum to the speed in the substance in question. As we might expect, light travels faster in vacuum than through any collection of atoms, so the refractive index of all substances is greater than unity.

The electrical and optical properties of amorphous materials such as glass represented a major dilemma for theoretical physics, to which we must now turn. The combined efforts of Felix Bloch, Rudolf Peierls and Alan Wilson, in the early 1930s, provided a satisfactory explanation of the distinction between metals, insulators and semiconductors. An electron in a perfect crystalline lattice was shown to move as if it were free, without being deflected by the ions. This situation was assumed to prevail unless the electron's *de Broglie wavelength* is such as to give *Bragg diffraction* from the lattice, the result then being a forbidden *energy gap*. *Pauli's exclusion principle* permits only two electrons with opposite spins in each energy state, and the allowed states are successively filled up to the *Fermi level*. When the latter lies within an energy gap, an *insulator* results, the number of electrons moving in one direction being exactly balanced by the number moving in the opposite direction. The transparency of a *crystalline insulator* was shown to be a consequence of the gap, photon energies in the visible range being insufficient to excite an electron across the forbidden energy region. But glass too is transparent and must therefore also have an energy gap. In this case the gap can be traced back to the behaviour of the electrons in the constituent silica molecules, which do not absorb visible light. When randomly packed together as glass, they disturb each other sufficiently little that light can still be transmitted through them, and we can use this wonderful material to let in the sunlight while keeping the wind out.

An important contribution to the understanding of electron motion in non-crystalline matter was presented by John Ziman, in 1961. Noting that the experimentally determined *average free path* for electrons in liquid metals is considerably larger than the interatomic separation distance, he concluded that the scattering by each atom is quite small. Although the resulting distortion of the electron waves is far from negligible, the Ziman theory works surprisingly well, even for a free path comparable to the spacing between adjacent atoms. Mercury appeared anomalous, however, its resistivity in the liquid state differing from that predicted by the Ziman theory. It was apparently not amenable to this approach, and Nevill Mott realized that *electron traps* must be a natural attribute of the non-crystalline state. At a sufficiently high degree of disorder, electrons in the conduction band cannot move freely as *Bloch waves* because all states in the band have been turned into traps, with *quantized energy levels* and *localized wave states*. When this is the case, an electron can move from one wave state to another only with the help of thermal motion of the atoms. Such *thermally activated hopping*, as it is called, was first described by A. Miller and E. Abrahams in 1960.

Experimental information on the nature of traps was obtained by B. T. Kolomiets and his colleagues in the early 1960s. They stumbled on the surprising fact that certain *amorphous semiconductors* cannot be *doped*. A case in point is arsenic telluride, which is transparent in the near infrared. In both the crystalline and glassy states of this material each tellurium atom has two arsenic neighbours, while each arsenic is adjacent to three tellurium ions. If a silicon atom is added to the crystal, it replaces an arsenic. Three of its four electrons form bonds with neighbouring tellurium atoms, while the fourth remains weakly bound. The silicon thus serves as a *donor*. But in the glass, with its lack of perfect order, a silicon atom can find a location in which all four of its electrons are involved in bonding to neighbouring tellurium ions, none being surplus to act as a source of conduction. As Nevill Mott emphasized, such a structure would be likely to have the lowest *free energy*, and X-ray evidence suggests that this is the situation for all glasses produced by supercooling the corresponding liquid.

Mott went on to predict that there will be a condition of *minimum metallic conduction*, beyond which the disorder is such as to produce traps, and Morrel Cohen, Hellmut Fritzsche and Stanford Ovshinsky referred to the energy separating localized from non-localized states as the *mobility edge*. The fundamental origin of localization had actually been elucidated by Philip Anderson in 1958, well before much of the decisive experimental evidence had been accumulated. Anderson studied the diffusion of a wave state through a random array of sites, and its dependence on the degree of randomness, and found that when the latter exceeds a certain value, the states become localized. Recognizing that a continuum in energy does not necessitate a continuum in space, and realizing the need to consider distributions rather than averages, Anderson had divined the true nature of *trapping*. The gist of the matter is that there are two opposing factors: the farther an electron is prepared to hop, the greater will be its chance of finding a new site at an energy conveniently close to its own; but the larger the jump length, the greater is the attenuation because the wave amplitude dies away with increasing distance. Putting it more colloquially, we could say that an electron, unable to surf-ride on a propagating wave, as it would do in a crystal, has to find its way around in a glass by hopping from one atom to another as if using them as stepping stones. As the arrangement of these becomes more random, a point is reached at which the electrons are left stranded at particular atoms, and cannot conduct electricity.

An exciting discovery was made in 1975, by Walter Spear and his colleagues. Some *amorphous semiconductors* actually *can* be doped. These materials are produced by condensation from the vapour phase rather than by cooling from the liquid state. *Amorphous silicon* is deposited on a cooled surface, and doping with phosphorus is achieved by introducing a small amount of phosphine into the silane gas. Although much of the phosphorus is still triply coordinated, with no loosely bound electrons, some phosphorus ions occupy sites with four-fold coordination and become donors, just as they would in a silicon crystal. Such condensed semiconductor films seem to contain few defects, and hence fewer three-fold sites to trap and nullify the phosphorus atoms. Another important factor is the incorporation of hydrogen, which attaches itself to the unsatisfied orbitals that would otherwise be occupied by phosphorus. This technological breakthrough holds promise for commercial exploitation, particularly regarding *solar cells*. Amorphous materials are considerably cheaper than their crystalline counterparts, and the energy conversion efficiency for solar radiation falling on amorphous silicon is double that for large single crystals.

Amorphous, or glassy, selenium is widely used in the *photocopying process* known as *xerography*. For this element too, single crystal production is prohibitively expensive, and thin film deposition of an amorphous layer provides large uniform areas at low cost. During the copying process the upper surface of the photoconductive selenium layer is first given a positive charge by *corona discharge* from a wire. The latter is held at a high electric potential, and it is moved across the surface. The document to be copied is then imaged on the layer, electron and hole pairs being created by the photons that are reflected preferentially from the light areas of the original. The lower surface of the amorphous selenium film is in intimate contact with an earthed metal plate. The resultant field across the film causes the holes to drift toward the plate while the electrons move to the upper surface and locally neutralize the positive charge. The *drift mobility* during this vital part of the process appears to be determined by traps of the *point defect* variety. At this stage the surface of the film has positive charges in those areas corresponding to the dark regions of the document, and all that remains is to convert this latent image to a printed copy. This is accomplished by depositing negatively charged *carbon black* on the surface and then transferring it to paper with the aid of another corona discharge. The print is made permanent by a final heating of short duration.

Transparency was not the only mystery surrounding the glassy state. We still have to explain why a given glass absorbs radiation of some frequencies while transmitting others, and there is also the question of why different frequencies travel at different speeds. Light has two complementary natures: it simultaneously exists as discrete particles and as waves. Long before this *wave–particle duality* was proposed, James Maxwell had suggested, in 1867, that light consists of regularly oscillating electric and magnetic fields. These act at right angles to each other, and both are at right angles to the beam direction. Since all substances consist of positive and negative charges, we must expect them to interact with such fields. The charges themselves are moving, and they have their own characteristic frequencies, related to the motion of electrons around atomic nuclei and to the natural vibrations between atoms. If the frequency of the incident radiation is such as to promote electron *inter-band transitions*, energy is transferred from the beam to the material through which it is travelling. Radiation energy is converted to the kinetic energy of electron motion through the conduction band, and subsequent electron–atom collisions lead to a further conversion to heat energy. If the transfer is sufficiently large, the material will be opaque to radiation of the frequency in question. In the absence of inter-band transitions, the material is transparent, the atoms both absorbing and re-emitting light energy. The emission process is analogous to the transmission of radio waves from an antenna. In the latter it is the alternating flow of electrons along a metal rod several centimetres in length that is responsible, whereas in the optical process it is the oscillation of electrons over atomic dimensions. The *speed of light* of a given frequency in a transparent medium depends on how rapidly the electrons can respond to the electromagnetic waves in the incident beam, and high-frequency blue light travels more slowly than low-frequency red light simply because the response to the former is most influenced by the inertia of the individual electrons. Two of the best known optical materials are *crown glass*, a soda-lime–silica compound, and *flint glass*, which has a lead–potash–silica composition. Flint glass has a higher *refractive index* and also a higher *dispersion* than the crown variety. For flint the refractive index is about 1.65 at the blue end of the spectrum, and about 1.60 at the red end. For crown glass the corresponding range is 1.52–1.50. These two types of glass are often combined in a back-to-back arrangement so that their dispersions mutually compensate.

Coloration of glass is caused by the presence of impurity atoms. The amount of these required to give a perceptible tint is usually rather small, and the production of *colourless glass* actually presents a problem of purification. Because of this, the earliest glasses were probably so strongly coloured as to

The first vessels made entirely of glass appeared during the 18th Dynasty of Ancient Egypt, and survival of a large number of these early pieces suggests that a well-organized industry existed. But the artisans of that period were apparently unaware of the workability of the material when hot, and items such as this cosmetic container were moulded in sections. The container, which is about twelve centimetres high, dates from approximately 1400 BC. The blue and yellow colours are produced by ions of cobalt and chromium, respectively.

be essentially opaque. In today's cheap glass containers, such as those used for bottling beer, no effort is made to remove the Fe^{3+} ions that give the green colour. The investigation of which impurities produce what colours, in the various common types of glass, has proceeded as an empirical art over the centuries, the most active practitioners being those who produced *church windows*. The colour that a particular element imparts to glass is frequently the same as that observed in the corresponding crystal, and this serves as a very rough guide. Thus the colour of copper sulphate is royal blue, and the same hue is produced when a small amount of copper is added to soda-lime–silica glass. When this same impurity is added to borosilicate glass, on the other hand, it produces a green colour that is similar in tone to that of the copper mineral, malachite.

Perception of colour depends on the way the eye responds to electromagnetic radiation. It is sensitive to a limited range of wavelengths, stretching from 400 to 800 nm. If all wavelengths in the visible spectrum are present, and with roughly equal intensities, the light appears white. When some wavelengths are absent, the eye detects colour. Thus the removal of blue wavelengths from a white beam causes the eye to sense yellow. The question of colour is therefore intimately connected with the process of absorption. An element produces different colours in different glasses because its *absorption bands* depend on the local atomic environment. Cobalt gives a pink colour to borosilicate glass, a purple colour to potash–silica glass, and it colours soda-lime–silica glass blue. Then again, the colouring induced by an impurity depends on the electrical charges on its ions. For instance, Mn^{3+} gives a deep purple shade to soda-lime–silica glass, whereas Mn^{2+} promotes a faint pink. Similarly, Cr^{3+} colours the same glass green, whereas Cr^{6+} gives it a yellow tint. Finally, the colour given to the glass depends on the quantity of the impurity present. If 0.1% of cobalt oxide is added to soda-lime–silica glass, this takes on a blue colour, but a black glass is produced when ten times this amount is present. In all of the above examples, the coloration occurs because the ions in question have associated with them broad adsorption bands covering the complementary colours. The cobalt ion, for example, gives absorption of red light so an incident white beam is transmitted with a bluish colour.

The most common sources of colour in glass are the electronic processes associated with *transition elements*. These have inner electrons which are not involved in covalent bonding, and which are therefore available for absorbing and scattering light radiation. Moreover, many of their transitions fall within the visible range of the spectrum. In the periodic table, the row of transition elements that starts with scandium are the first ones encountered with electrons in d orbitals. These are special in that their electron probability density distributions have a multiplicity of lobes extending out along various directions. This characteristic, which distinguishes them from the more symmetric lower energy states, leads to an enhanced sensitivity to the local atomic environment. In an isolated atom of a transition element, the d-type energy levels are *degenerate*; they all have the same energy. In condensed matter, on the other hand, the relatively large spatial excursions made by the d electrons periodically brings them into close proximity with surrounding ions. This causes the various levels to adopt different energies, an effect which in the case of crystalline matter is known as *crystal field splitting*. The splitting also occurs when the transition metal atom is in a glass, but the lower degree of order makes for a more complicated arrangement of levels. Moreover, transitions involving these levels are only possible if the d orbitals are partially filled.

Another important way of colouring both glass and opaque ceramics is to produce in them small aggregates composed of impurity atoms, having a diameter comparable to the wavelength of light. Lord Rayleigh was the first to study the scattering from such small particles, and he showed that the intensity of the scattered light increases in proportion to the square of the particle volume, but decreases as the fourth power of the wavelength. Hence short wavelengths are scattered far more strongly than long wavelengths. The classic success of the Rayleigh analysis was its explanation of the *blue colour of the sky* and the *red appearance at sunset.* When we look up into the sky, in a direction away from the Sun, our eyes receive light scattered from the gas particles in the atmosphere, and we predominantly perceive blue because low wavelengths are scattered most strongly. At the close of day, when looking towards the Sun, we see those wavelengths which are scattered least, at the red end of the spectrum. These same principles apply to colloidal particles of gold, about 40 nm in diameter, when present in lead–silicate glass. The result is a beautiful red colour. Precipitation of particles with the correct size is achieved by careful heat treatment after the usual melting and cooling stages of the glass-producing process. This technique is known as *striking the colour*. Reversible precipitation of *colloidal metal particles* can also be achieved *in situ* when certain impurities in glass receive radiation of the appropriate wavelengths. As the precipitation proceeds, the colour of the glass changes. Such *photochromic glasses* are now occasionally used in spectacles and window panes.

It is not often that branches of science and technology as venerable as metallurgy and glass fabrication witness a genuine revolution. When both these fields simultaneously experienced such an upheaval, in 1960, it is not surprising that the repercussions were widespread. Until this event, glass had always been characterized as being transparent, brittle, and insulating, while metals were taken to be crystalline. In producing the first appreciable amounts of *metallic glass*, Pol Duwez, William Klement and Ronald Willens effectively rewrote the definitions of both classes of material. Until their breakthrough, the distinction between these forms of matter had been reflected in a marked difference in their chemical compositions. The work horses of metallurgical practice are iron, copper, zinc, aluminium, lead, nickel and cobalt. Glasses, on the other hand, are commonly formed from compounds, and we have seen that silicon, oxygen and sodium are particularly prominent. Even the way in which their atoms are joined together divided the traditional glasses and metals into tidily distinct groups. Most of the properties of glasses derive from their strong and directional covalent links, while metals owe their ductility, lustre and high conductivity to a unique collective and non-directional bonding. The free energy barrier separating the non-crystalline and crystalline states of metals is rather small, and slow cooling of a metallic melt does not yield a glass. The main technique employed in this advance has its roots in metallurgical tradition. It had long been known that metals are capable of exhibiting properties other than those of their natural state. The simple trick is to cool them so rapidly that a *non-equilibrium state* is frozen in, this process being known as *quenching*. Cooling rates of tens of millions of degrees per second are required, and these can be achieved in a variety of ways. The original method involved propelling molten globules against a cold flat surface, a process known as

Metallic glass can be produced at rates up to two kilometres per minute by squirting the molten metal against the surface of a rapidly-rotating metal cylinder. The jet is thereby instantly solidified and the glass is spun off the cylinder in the form of a ribbon a few micrometres thick.

splat quenching. More recent methods include pouring the liquid between rapidly counter-rotating drums and squirting a thin jet of the liquid against the cooled surface of a revolving cylinder. The last of these produces a continuous ribbon, at rates up to 2 km/min. The quite different technique referred to earlier in connection with amorphous semiconductors, *vapour condensation*, also works for metals, and there were reports in the 1930s, notably by J. Brill and J. Kramer, of metal glass formation by this method. Credit for the revolution must also be accorded to these pioneers.

These congealed metallic melts are frequently referred to as *amorphous alloys*, emphasizing the fact that they are mixtures rather than pure metals. They have invariably belonged to one of two groups: a transition, or a noble, metal with a smaller metalloid; or a mixture of two transition metals. Examples of the first type are $Au_{75}Si_{25}$ (which was in fact the first combination studied by Duwez, Klement and Willens), $Pd_{80}Si_{20}$ and $Fe_{80}B_{20}$. The other group includes $Ni_{60}Nb_{40}$ and $Cu_{66}Zr_{34}$. Many of the more recently developed metallic glasses have more complicated compositions, such as $Fe_{75}P_{16}B_6Al_3$ and $Pd_{16}Au_{66}Si_{18}$. Production cost is relatively low because by producing the glassy alloys directly from the molten metal, it is possible to bypass expensive and energy-consuming stages such as casting, rolling and drawing, which have been necessary when working with the crystalline state of the material. One technological limitation arises from the fact that a metallic glass reverts to the crystalline form if the temperature is raised to roughly half the melting point on the absolute Kelvin scale.

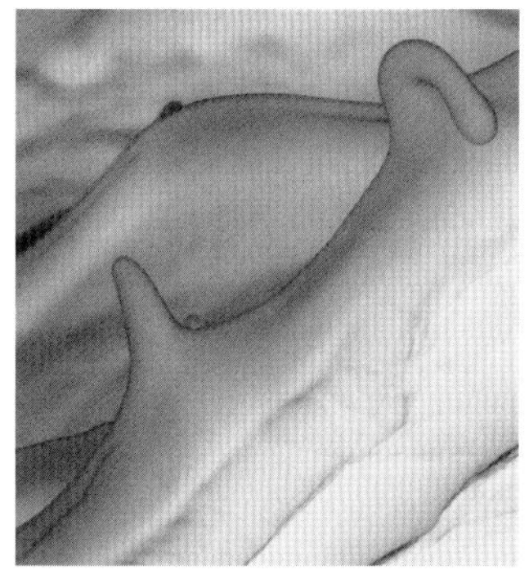

Observed at low magnification, the *fracture surfaces* of a *metallic glass* bear characteristic markings which resemble a river and its tributaries. A closer look at the junctions in this pattern reveals evidence of local ductility, as seen in this scanning electron photomicrograph, which covers a width of about five micrometres.

One of the best known models of the liquid state is that usually accredited to John Bernal, although variants of the same approach had been discussed by Peter Debye and H. Menke, and also by Joel Hildebrand and W. E. Morrell, both in the 1930s. Bernal's version, dating from 1959, involved pouring spheres into an irregularly walled container to produce what is known as *random close packing*. As John Finney subsequently demonstrated, the model is actually more applicable to metallic glasses, particularly when the structure is produced with two different sizes of sphere, with appropriate diameters. It must be emphasized that such models do not exhibit the total randomness of an ideal gas, because the dense packing imposes quite

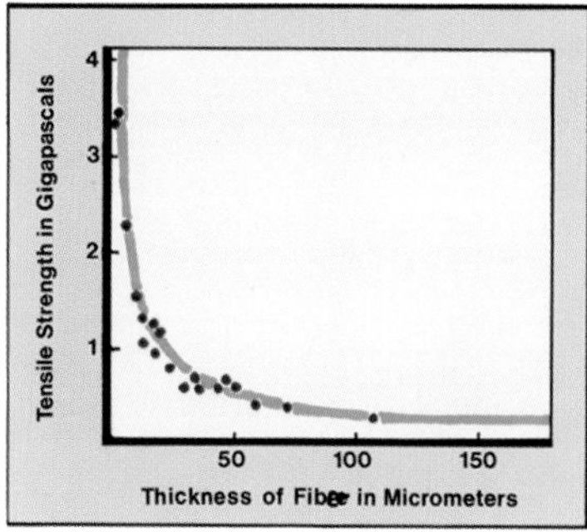

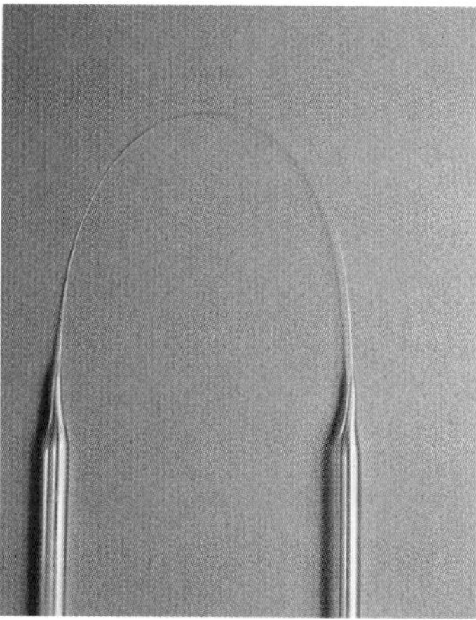

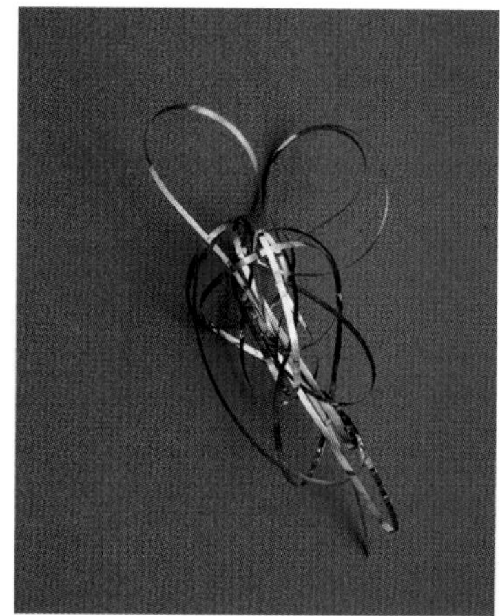

The high strengths of *silicate glass* and *metallic glass* fibres might have the same origins. In 1920, Alan Griffith (1893–1963) discovered that silicate glass is very strong when in the form of thin threads a few micrometres in diameter, and correctly concluded that this form is free from weakening cracks. The graph is a plot of Griffith's original experimental results while the left photograph shows an eighty-micrometre diameter *glass fibre* being strained to about 0.15%. The breaking strain of this relatively thick fibre was only approximately twice this value, consistent with Griffith's findings. Metallic glass ribbon, an example ($Pd_{80}Si_{20}$) of which is shown in the other photograph, has an extremely hard surface and it is very resistant to corrosion. Surface cracks are therefore uncommon in such materials, and they can usually survive strains of up to 50%.

The strengths of metallic glasses

Crystalline metals deform by the movement of *dislocations*, which cause relative slipping between adjacent planes of atoms. This source of weakness is present in all metal crystals except the special forms known as *whiskers*, which are consequently very strong. The traditional glasses, such as those based on silicates, do not possess well-defined atomic planes, and the dislocation mechanism is not operative: stress levels build up until the material snaps. In *metallic glasses*, a third mode of deformation is observed. These materials also lack regularity of microstructure, but they nevertheless fail by shearing along a fairly narrow plane lying at 45° to the direction of strain. At the time when metallic glasses could be fabricated only as thin ribbons, their strengths were compared with similarly thin examples of other materials. In the following table all entries are in meganewtons per square metre.

Material	Ultimate tensile strength
silicate glass fibre	1 400
crystalline steel wire	2 000
$Pd_{80}Si_{20}$ glass	1 360
$Cu_{60}Zr_{40}$ glass	2 000
$Ni_{75}Si_{8}B_{17}$ glass	2 700
$Co_{75}Si_{15}B_{10}$ glass	3 000
$Fe_{80}B_{20}$ glass	3 200
$Fe_{78}Si_{10}B_{12}$ glass	3 400

severe geometrical constraints. A certain degree of *short-range order* exists, even though *long-range order* is completely lacking. This has led some to conclude that metallic glasses are actually *polycrystals* with a grain size of near atomic dimensions. Indeed, it has been suggested that a glass might be essentially a crystal that is saturated with dislocations, but it is doubtful that one can identify such line defects when there is no regular lattice left to

The earliest metallic glasses could be produced only as thin ribbons in the micrometre range, and this threatened to limit their application to cables and the like. That drawback has now been overcome by rapidly cooling liquid metals to below their liquid–glass transition temperatures, and by using compositions comprising several different elements. Avoidance of crystallization is also favoured by ensuring that the relative abundances of the various-sized atoms are incompatible with crystallinity. The head of a golf club shown here was fabricated by injecting the high-temperature melt into a mould held below the liquid–glass transition temperature. The composition of this Vitreloy material is indeed quite complex: a combination of the early transition metals Zr and Ti with the late transition metals Cu, Ni and Nb, and with various minor amounts of Al, Sn, B, Si and Be. Just a few years ago, anyone predicting that such a piece of sporting equipment would one day be fabricated from glass would not have been taken seriously.

act as a reference structure. The dislocation concentration would have to be so high that the concept might lose its normal significance. A more promising approach was put forward by Colin Goodman, and independently by Rong Wang and M. D. Merz, in 1975. Noting that compounds which form glasses have as a major constituent a material that exists in several polymorphic crystalline forms, differing only slightly in free energy, they suggested that *polymorphism* might actually be a prerequisite for glass formation. With roughly the same free energy, and therefore approximately the same probability of occurrence, the various polymorphs would vie with one another as the liquid cools towards the glass temperature. Because of the competition, none would dominate, and the result would be a mixture of clusters of different polymorphs. There would inevitably be mismatch at the cluster interfaces. Initially, the assembly of clusters would be in a dynamic state, individual atoms constantly being shuffled from one polymorphic cluster to another. At the glass temperature, the situation would become static, with the clusters firmly bonded at their contact surfaces.

The random close-packed structure has won virtually universal acceptance as a reliable, though approximate, model of a metallic glass. When presenting it, Bernal emphasized the primitive state of our understanding of random geometry, and he suggested that the structure should be analyzed in terms of the shapes and sizes of the *interstitial holes* it contains. Somewhat surprisingly, rather few different types of hole account for nearly all the space between the atoms. The most common type is the tetrahedron, about 85% of all holes having this form, while no other type accounts for more than 6%. The other common types are the octahedron, the trigonal prism, the Archimedean square anti-prism, and the tetragonal dodecahedron. In 1972, D. E. Polk suggested that the spheres in a random close-packed assembly can be compared with the larger atomic species in a glassy alloy, and that the smaller atoms fit into the largest holes in the structure. The tetrahedral hole is too small to be useful in this respect, but the octahedral and trigonal prismatic holes appear to be viable candidates, especially if the smaller atoms are ionized. It was subsequently found that the metalloid elements are indeed present as ions when they are constituents of metallic glasses.

The remarkable resistance to corrosion exhibited by many glassy metals might imply absence of well-defined crystal boundaries and other defects. In crystalline materials, defects are notoriously vulnerable to attack by corrosive agents, because they are slightly more open regions of the structure and thus susceptible to penetration. The lack of such inhomogeneities in these glasses would be a strong factor in reducing their tendency to oxidize. If this interpretation is valid, it is evidence against the polycrystal and dislocation models of the amorphous state.

Much of the interest in metallic glasses can be traced to a prediction, by I. A. Gubanov, that some of them would display *ferromagnetism*. In view of the interactions between atoms underlying the ferromagnetic condition, and the reinforcing factor arising from crystal symmetry, Gubanov's conjecture sounded improbable. *Ferromagnetic glasses* were nevertheless discovered, and hold great technological promise, particularly for use in transformers. Energy dissipation in traditional polycrystalline transformer cores represents a loss which, in 1980, was estimated to amount to $200 million in the USA alone. Glassy $Fe_{86}B_7C_7$ has a *saturation magnetization* that is almost as high as the best crystalline examples, but with the advantage of a low *coercivity*; it is easily magnetized and demagnetized, and the *hysteresis losses* are thus greatly reduced. It is believed that this behaviour too is a result of the absence of grain boundaries. These and other defects are known to constitute barriers to the motion of *magnetic domain walls* (Bloch walls), and their absence probably produces *magnetic softness*. This property also suggests applications to magnetic memory devices, like those in computers, and tape recording heads, because it would facilitate rapid recording and erasure of information. Unfortunately metallic glasses fall somewhat short of the ideal because inhomogeneities in the form of slight variations of composition are difficult to eliminate completely. Some *ferromagnetic glasses* have extremely high *permeability*, making them potentially useful for miniature microphones and transformers. Others display unusually low *magnetostriction*, that is change in geometrical dimensions when placed in a magnetic field, which suggests use in transducers. Similarly, their low attenuation of sound and ultrasonic waves might lead to use in novel acoustic devices.

One of the most important and remarkable properties of metallic glasses is their unusual combination of strength and plasticity. The strengths of some examples exceed three times that of stainless steel. Covalent glass, such as the common window variety, even when in the favourable form of thin filaments, cannot be strained in excess of 1%. A metallic glass specimen having the same shape can withstand a local plastic shear strain well in excess of 50%. One obvious application that would capitalize on these qualities is their use as fibres for reinforcement, for example in car tyres. Not surprisingly, the mode of deformation exhibited by glassy metals is quite different from that observed in crystalline metals. The inherent weakness of a crystal lies in its ability to be readily dislocated. The virtually unavoidable presence of *dislocations* is responsible for the fact that the actual strength of a crystal is seldom more than one-thousandth of the ideal value. It takes surprisingly little force to move a dislocation, and stresses are thus unable to build up to potentially dire levels. The lack of translational symmetry in the glass precludes a deformation mechanism based on normal dislocation movement, and this is why covalent glasses are always brittle. Remarkably, metallic glasses nevertheless appear to fail by *ductile rupture*, and the mechanism by which this occurs has understandably aroused interest. In 1972,

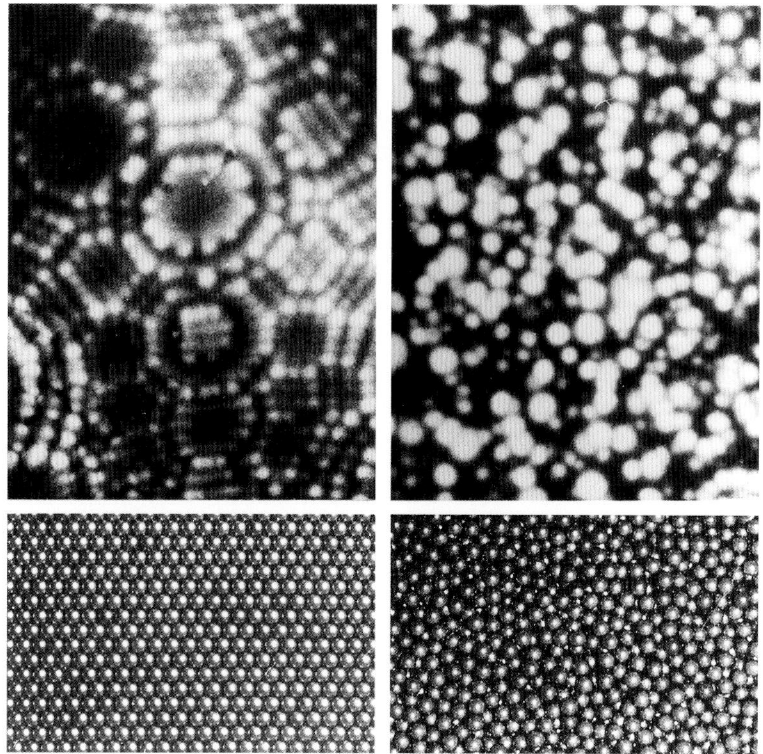

The *field-ion microscope* reveals the atomic arrangement on the surface of an electrically conducting specimen. The arrival of metallic glasses therefore permitted the first atomic-level views of the *glassy state*. These field-ion photomicrographs compare the structures of $Pd_{80}Si_{20}$ glass (upper right) with crystalline tungsten (upper left), and the figure also compares what such microscopy reveals with the corresponding hard sphere models of a crystal and a glass. The field-ion micrograph of a glass, taken by the present author in 1974, was the first atomic-scale picture of the glassy state ever produced.

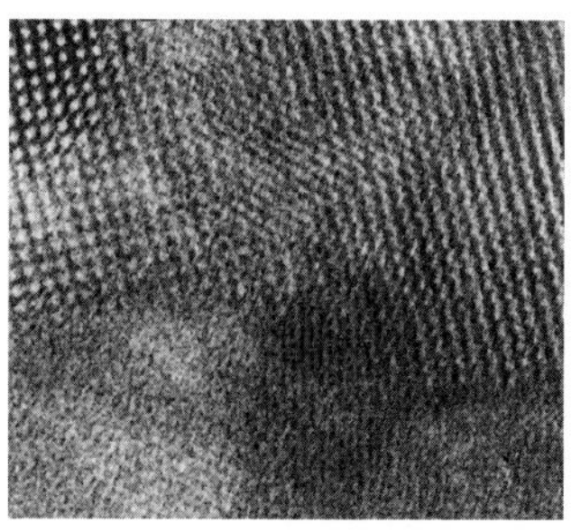

The narrowness of the *crystal–glass interface* is demonstrated in this transmission electron photomicrograph of synthetic *zeolite*, a sodium aluminium silicate. The lower part of the picture, which covers a width of about thirty nanometres, shows a region which has become amorphous under the damaging influence of the electron beam. The atomic rows are clearly visible in the upper area, and the interface extends over a distance of no more than a couple of atomic diameters.

H. S. Chen, H. J. Leamy and T. T. Wang observed surface lines, or *striations*, on a deformed specimen, suggesting some sort of *shear band* arrangement. P. Donovan went on to show that if the surface of such a specimen is polished and then etched, the striations reappear, indicating that the sheared zone has been modified by the shearing process. D. E. Polk and David Turnbull explained this on the basis of alteration of the short-range compositional bias, caused by the shear movements. Fractured surfaces of glassy metals display a characteristic vein or ridge pattern, this first being reported by Turnbull and Franz Spaepen, and independently by C. A. Pampillo and R. C. Reimschuessel. S. Takayama and R. Maddin compared such markings with those formed when two smooth surfaces held together with grease are separated. As one can readily demonstrate with two table knives and a small amount of butter, this produces a pattern of primary and secondary lines reminiscent of an aerial view of rivers and tributaries.

Summary

Glass is the rigid meta-stable solid produced by cooling the liquid form of the same substance rapidly enough to prevent crystallization, the stiffening occurring predominantly at the glass temperature. It is characterized by an arrangement of atoms or molecules which is irregular, and which thus contrasts with crystalline order. Silicate glasses are frequently transparent, with a colouring that is determined by impurities, and they are hard, brittle and electrically insulating. Metallic glasses, the production of which requires very high cooling rates, are opaque, corrosion resistant, electrically conducting, very strong and quite flexible. Several examples also have potentially useful magnetic properties.

12

'tis, we musicians know,
the C Major of this life.
Robert Browning
(1812–1889)
Abt Vogler

To the organic world: *carbon*

Paintings and museum reconstructions depicting the activities of medieval *alchemists* can easily give the wrong impression of the chemical knowledge of that period. Inanimate objects such as crystals are seen sharing shelves with the preserved bodies of small creatures, a common feature of the early laboratory. This could be taken to suggest that the scientists of those days had grasped the unity of chemistry: that in spite of great differences in their external appearance, Nature makes no distinction between chemical combinations in living and dead substances. Nothing could be farther from the truth. The universally held belief was that the matter contained in living organisms possessed an essential extra ingredient, a *vital force*, the mysterious origin of which was attributable to divine powers. This attitude was epitomized in the succinct classification to be found in the book *Cours de Chyme*, published by Nicholas Lemery in 1685. His division is still to be found in the standard first question of a popular parlour game: 'animal, vegetable or mineral?'.

By the end of the eighteenth century, the vital force theory was in rapid decline. Antoine Lavoisier, analyzing typical organic compounds, found them to contain inorganic substances such as hydrogen, oxygen, nitrogen, carbon, sulphur and phosphorus. The basic principles that govern chemical change, such as *conservation of total mass*, were being established during this period, and in 1844 Jöns Berzelius showed that they apply to all matter irrespective of whether it is inorganic or organic. The final demise of vitalism can be said to date from 1828, and an accidental discovery by Friedrich Wöhler. Attempting to prepare *ammonium cyanate*, by heating a mixture of ammonium chloride and potassium cyanate, both of which are inorganic salts, he was astonished to discover that he had produced crystals of *urea*. This compound was well known at the time, but until then it had always been associated with the process of life; it is present in the urine of mammals, for example. Through this stroke of good fortune, Wöhler had shown the barriers between the organic and inorganic domains to be illusory. He had also chanced upon something else that is particularly common in the organic realm. Ammonium cyanate had indeed been formed as an intermediate product of his reaction, but the atoms in the molecules of that substance had been rearranged, by the applied heat, to form molecules of urea. Two compounds whose molecules have the same atomic composition but different arrangements are known as *isomers*, after the Greek words *iso*, meaning same, and *meros*, meaning parts.

Although the chemistry of living and non-living things had thus been shown to be one and the same, *organic chemistry* survived as a separate field

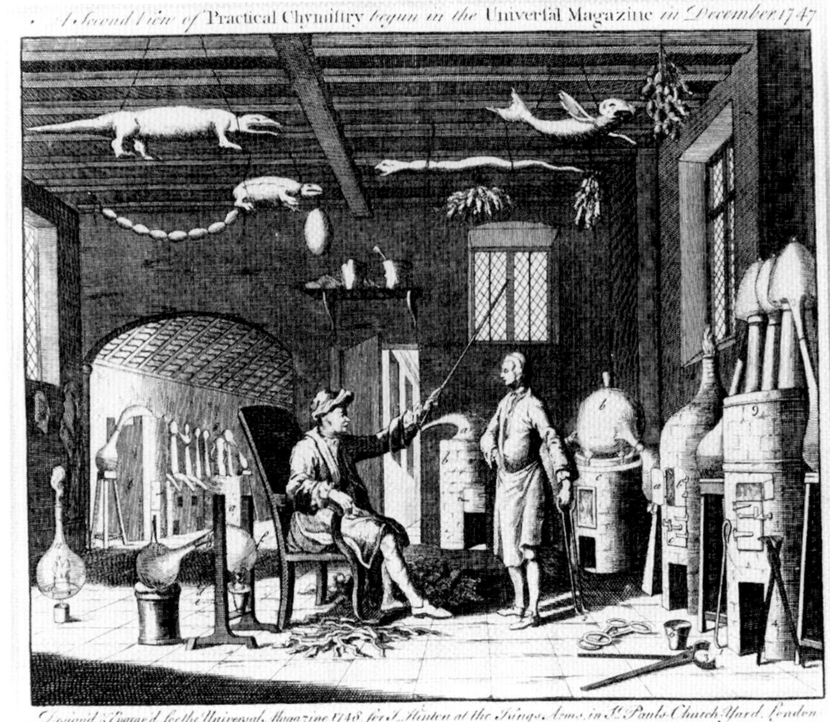

The accidental production, by Friedrich Wöhler (1800–1882) in 1828, of the organic compound *urea*, from the inorganic compounds ammonium chloride and potassium cyanate, revealed the universality of chemical principles. Before this pivotal discovery, organic substances had been regarded as possessing a mystical extra ingredient which conferred vitality. Pictures illustrating chemical activity in the pre-Wöhler era, such as this example from 1747, should not be interpreted too literally. They no doubt reflect the influences at work on the artist, such as his market, the climate of ideas, and patronage. But the lofty positions accorded the zoological specimens in this view of an alchemist's laboratory do seem to betray a certain reverence for animate objects.

of scientific endeavour. The reason for this can be traced to the further efforts of Berzelius, whose systematic study of numerous compounds having animal or vegetable origins led him to the conclusion that the essential element in all of them is *carbon*. This characteristic was formalized in Johann Gmelin's *Handbook of Chemistry* (1846), which recommended that organic chemistry be regarded as a science of the compounds of carbon. The subsequent observation of many characteristics peculiar to the compounds of that element has tended to reinforce this attitude, and the division between inorganic and organic chemistry has survived to this day. The distinction is certainly useful. The number of different organic compounds discovered or synthesized to date is well in excess of a million. The list includes fuel, foodstuffs, fabrics, medicines, explosives, cosmetics, disinfectants, pesticides and cleansing agents. There are numerous physical and chemical differences between the inorganic and organic classes. Organic compounds tend to have low melting and boiling points and, unlike many inorganic compounds, they often exist as liquids and gases at room temperature. They are usually inflammable, whereas the majority of inorganic compounds are difficult to ignite, and another point of difference is their low *solubility* in water. This latter property is related to the fact that their solutions do not conduct electricity, again unlike most inorganic compounds. Their slowly reacting, covalently bonded, molecules frequently exhibit isomerism, while the fast reacting, and often ionically bonded, inorganic molecules only rarely show this effect. By convention, such common carbon-containing compounds as carbon monoxide, carbon dioxide, carbon disulphide, and carbonates, bicarbonates, cyanides and carbides do not rate inclusion in the organic class of substances.

The alkanes

Although more than two million different organic compounds have been produced, keeping track of them is not as difficult as one might believe. They are arranged in families known as *homologous series*, members of which are chemically uniform. Within a given series there is a regular gradation of such physical properties of molecular weight, melting temperature, and boiling temperature. An example is the *alkane*, or *paraffin*, series: a family of saturated hydrocarbons in which adjacent members, or *homologues*, differ by a CH_2 unit. The structural formulae of the first eight members are given below, together with their melting (T_M) and boiling (T_B) temperatures at atmospheric pressure.

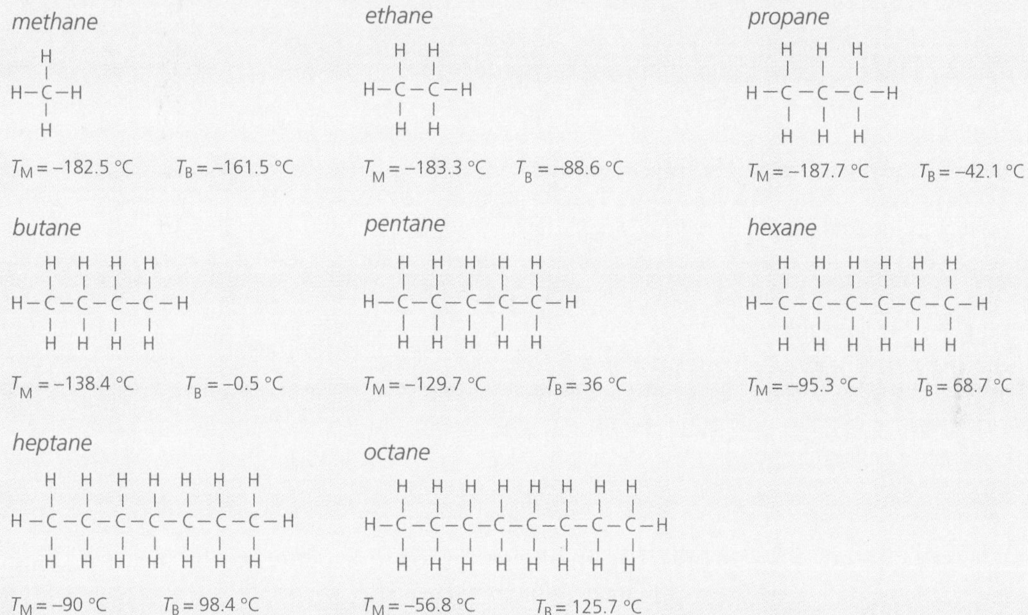

The alkane series is quite straightforward up to and including propane, but complications are encountered for all higher homologues; there is a choice of non-equivalent forms sharing the same molecular, as opposed to structural, formulae. These are referred to as *isomers*. The two isomers of butane are the one given above, which should strictly be called normal butane or *n*-butane, and *iso*-butane. The structure formula of the latter is

iso-butane

```
      H
      |
    H-C-H
      |
  H   |   H
  |   |   |
H-C - C - C-H
  |   |   |
  H   H   H
```

$T_M = -159.6$ °C $T_B = -11.7$ °C

All carbon–carbon bonds in this compound have the same length, the long bond in the diagram merely reflecting the inadequacy of two-dimensional representations. It is interesting to note that the two isomers have slightly different melting and boiling temperatures. For higher members of the series the number of different isomeric forms increases rapidly; octane has eighteen.

Asphalt, pitch and tar

Road surfacing is one of the most conspicuous materials in developed countries, but it is not always clear how it should be described; is that black stuff asphalt, tarmac or what? The etymologically senior term appears to be *tar*, which derived from the Old Teutonic for a substance obtained through destructive distillation of wood, coal or other organic matter. Its strongly antiseptic qualities were recognised in the earliest times, and tar soaps can still be purchased today. The term also refers to mixtures of certain hydrocarbons with resins and alcohols, widely used in preserving wood and rope. Applied to canvas, it produced the waterproof material *tarpaulin* much prized by the mariner, who thereby acquired the nickname 'tar' in 1647. Then there is *bitumen*, adopted in 1460 from the Latin, which actually subsumes some of the more familiar terms. It refers to various solid or highly viscous mixtures of hydrocarbons occurring naturally or produced during the distillation of petroleum. The word *pitch* comes from the Old English for a solid substance obtained by boiling tar. It is jet black and it has of course become an eponym for darkness. It found widespread use in the caulking of ships' hulls, when these were constructed from timber planks. *Asphalt* is the trickiest member of the group because the term is used in different contexts. In strictly chemical terms, it is the very viscous liquid produced by oxidation of high molecular-weight hydrocarbons in naturally occurring oil deposits. It was thus present at the surfaces of the major oil seepages discovered in ancient times at Baku, in what is now Azerbaijan. The word also refers to the mixture of the viscous liquid with sand and stones used for paving roads. Finally, there is the word *tarmac*®, the superscript icon indicating that it is in fact a registered trade mark. The word is short for tar macadam, and honours John Macadam (1756–1836), who pioneered the switch in road building from smooth stones to crushed stones, the latter being less prone to mutual slipping because of their irregular shapes. The addition of tar gave an even more stable surface, with the added advantage that it prevented dust from forming during use.

In order to gain a better appreciation of the peculiarities of organic chemistry, we might ask what makes carbon so special. There are several contributory factors, the main one being the way in which carbon links up with atoms of other elements. The bonding is of the covalent type, the combined effects of *electron promotion* and *orbital hybridization* producing electron clouds that project from a carbon atom in certain fixed directions, like outstretched arms. This makes for more specific, and limited, bonding than is observed, for example, in a metal. On the other hand, carbon lies in the middle of a row in the periodic table, and this gives it the relatively high *valence* of four. Its combining ability is consequently greater than the other elements with which it forms organic compounds. It thus enjoys a position of compromise between choice and specificity and this usually produces chain-like structures rather than three-dimensional networks, making for a high degree of mechanical flexibility. The classic exception to this rule occurs when all four bonds associated with each carbon atom provide links with other carbon atoms. This produces *diamond*, in which the covalent bonds lie along the tetrahedral directions, pointing symmetrically to opposite corners of the cube. Diamond thus belongs to the cubic crystal system. It could be argued that silicon, by dint of similar placement in its own row of the periodic table, should display a chemistry having much in common with that of carbon. This is true, to a certain extent, but carbon has the advantage of its light weight, which makes its compounds more susceptible to thermal agitation. Under the conditions prevailing at the Earth's surface, many of these compounds exist close to their limit of stability, and they are able to participate in chemical reactions. The heavier silicon compounds require the high subterranean temperatures to achieve a similar degree of reaction,

Unsaturated hydrocarbons

If two carbon atoms are linked by a *multiple bond*, the compound of which they are a part contains less than the maximum possible number of hydrogen atoms, or other monovalent appendages. Such a compound is said to be *unsaturated*. The lowest member of the *olefine series*, also known as the *alkenes*, is *ethylene*. Its alternative name is ethene, and its structural formula is

ethylene

```
H  H
|  |
C=C
|  |
H  H
```

The double bond is an example of *sigma–pi bonding*. At room temperature ethylene is a colourless inflammable gas with a sweet smell. Members of the *acetylene series*, also known as the *alkines* or alkynes, are even less saturated. The smallest member, *acetylene*, also called ethyne, has a triple bond

acetylene

$$H{-}C{\equiv}C{-}H$$

It is produced by the action of water on calcium carbide, CaC_2. The flame produced when acetylene is burned in oxygen can reach 4000 °C, which makes it useful for *welding*.

and at the Earth's surface they appear quite inert. These factors conspire to produce the difference between two members of the same chemical group. Carbon gives life, while silicon gives sand.

The *ball-and-stick models* frequently used as teaching and investigative aids are particularly relevant to covalent compounds. The interconnecting sticks are reasonably valid representations of shared electron bonds and they underline the importance of relative direction. Innumerable variations of structure can be achieved with such models, and the organic chemist builds compounds in much the same way, sequentially adding more atoms to produce molecules of increasing size and complexity. The reactions involved in such syntheses are often speeded up by the use of *catalysts* which provide a surface template that guides the participating components into the correct relative positions, and lowers the relevant energy barriers. The simplest organic compound is *methane*, the molecules of which consist of one carbon atom and four hydrogen atoms. Hydrogen is monovalent, and its molecules can thus participate in only one bond. Except in the special case of *hydrogen bonding*, the hydrogen atom always represents a terminus. In a methane molecule, therefore, each of the four covalent links terminates at a hydrogen, and the tetrahedral structure has no further bonds available for extra atoms. Chemical variety arises through replacement of one or more of the hydrogens with a polyvalent atom such as oxygen, nitrogen, sulphur, phosphorus, or indeed another carbon. Organic molecules frequently have a shape that is elongated in one direction. This produces the cigar shape of the *liquid crystal mesogen* and the long thread-like molecules of the *polymer*, classes that are discussed in the following three chapters.

In spite of the vast number of individual compounds found in the organic realm, their classification into groups presents no special problem. This is because similar geometrical arrangements and terminal units tend to produce similar, and readily recognizable, physical and chemical properties. The various organic compounds are found to belong to one or other family,

Oxygen in aliphatic compounds

The introduction of even a single atom of a third element, such as *oxygen*, to a hydrocarbon permits considerable variety of chemical properties. We will consider five types of *aliphatic compound* which involve oxygen: alcohol, ether, aldehyde, ketone, and organic acid. The two simplest alcohols are *methyl alcohol* and *ethyl alcohol*

```
methyl alcohol      ethyl alcohol

    H                  H   H
    |                  |   |
H — C — O — H      H — C — C — O — H
    |                  |   |
    H                  H   H
```

When the latter is produced from a sugar such as glucose, by *fermentation*, carbon dioxide is liberated. The reaction is

$$C_6H_{12}O_6 \rightarrow 2C_2H_5OH + 2CO_2$$

and in the production of *champagne* some of the gas is of course retained. The divalence of oxygen is exploited in the *ethers*, an atom of that element acting as a bridge between two alkyl groups. An example is *di-ethyl ether*

```
di-ethyl ether

    H   H       H   H
    |   |       |   |
H — C — C — O — C — C — H
    |   |       |   |
    H   H       H   H
```

which is a common solvent and *anaesthetic*. It can be prepared by heating an excess of ethyl alcohol with concentrated sulphuric acid

$$C_2H_5OH + H_2SO_4 \rightarrow C_2H_5HSO_4 + H_2O$$

the sulphovinic acid thereby produced reacting with more alcohol

$$C_2H_5HSO_4 + C_2H_5OH \rightarrow C_2H_5OC_2H_5 + H_2SO_4$$

to produce the ether, and sulphuric acid which is recycled. Compounds such as methyl alcohol and ethyl alcohol should strictly be referred to as *primary alcohols*, in that only one of the three links from the terminal carbon is not to a hydrogen atom. When such a compound is oxidized, an *aldehyde* is produced. An example is

```
methyl alcohol              formaldehyde

    H                           H
    |                           |
H — C — O — H + O   →       H — C = O + H2O
    |
    H
```

and we note that two terminal hydrogens have been traded off against the formation of a double bond. Traces of *formaldehyde* in food have a preserving effect, and are produced when meat and fish are mildly smoked. A *ketone* is produced by oxidation of a *secondary alcohol*.

```
acetone

    H   O   H
    |   ||  |
H — C — C — C — H
    |       |
    H       H
```

An example is *acetone*, which is a colourless inflammable liquid, cold to the touch, and used in the plastics industry. Finally, we have the *organic acids*. Unlike the ketones, which cannot be oxidized, the aldehydes always have at least one terminal hydrogen that is susceptible to attack; its replacement by a hydroxyl group produces an organic acid, as in the reaction

```
formaldehyde          formic acid
                        H
    H                   |
    |                   O
H–C=O + O  →            |
                      H–C=O
```

Organic acids are characterized by the COOH terminal, known as the *carboxyl group*; it comprises the carbonyl CO and hydroxyl OH units. The acidity stems from the tendency of the latter unit to lose its hydrogen atom.

formally referred to as a *homologous series*, having the same general skeleton of carbon atoms. The members of such a series share a general molecular formula and, when they are listed in terms of increasing molecular weight, display a regular gradation in physical properties. The major division of organic compounds divides them into two groups, each containing numerous homologous series, depending on whether the carbon atoms appear in rings or in open chains. Members of the former group are said to be *cyclic*, while those of the latter are referred to as aliphatic, which comes from the Greek word for fat. This is a reflection of the fact that *fats* are among the most characteristic members of the open-chain group. Because of their relative simplicity, *aliphatic compounds* will be considered first, examples being drawn from several of the most important homologous series.

Several hydrocarbon families provide an ideal starting point, both because of their simplicity and their technical importance. They contain only carbon and hydrogen. The simplest of these aliphatic hydrocarbons are the *alkanes*, also commonly known as *paraffins*. The latter name derives from the Latin *parum affinis*, meaning little affinity, and this series is indeed characterized by its stable and unreactive compounds. The first few members are methane, ethane, propane and butane, which comprise one, two, three and four carbon atoms respectively. *Ethane* can be imagined as being formed by removing one hydrogen from each of two methane molecules and then joining the two broken bonds so as to produce a molecule having six hydrogen atoms. Insertion of more units, each comprising one carbon and two hydrogens, into the backbone produces further members of the series. This systematic building up of ever larger members enables us to predict that a compound having 60 carbon atoms and 122 hydrogen atoms must belong to the alkane series, and such a substance is indeed known, it being a white waxy solid. At room temperature the first four members of the alkane series are gases, while the next eleven members are all liquids. They are the chief constituents of *paraffin oil*, or *kerosene*. All higher members are solids at room temperature, and they in turn are the main constituents of *paraffin wax*. *Methane* is the essential component of natural gas. In some regions it seeps directly out of the Earth's crust and, if it becomes ignited, appears as a persistent flame. The 'holy fire' of Baku, near the Caspian Sea, which so amazed the peoples of antiquity, was a famous example. Methane is produced during the bacterial decomposition of vegetable matter, in the absence of air, a condition which invariably obtains in stagnant water. Hence

A labour-saving convention

The compounds considered thus far in this chapter have reasonably small molecules, but many organic substances are much more complex, and drawing their structural formulae can be time-consuming. This task is eased by the convention in which the chemical symbols of all obvious carbon atoms, but not their related bonds, are omitted. Because hydrogen atoms are monovalent, and therefore always located at structural dead ends, neither these nor their associated bonds are drawn if their implied presence is unambiguous. This does not apply to *hydrogen bonds*, of course, and the number of implied hydrogen atoms can be deduced from the fact that the valence of carbon is four. According to this convention

n-butane and *trans-2-butene*

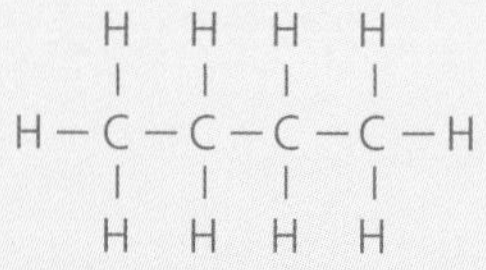

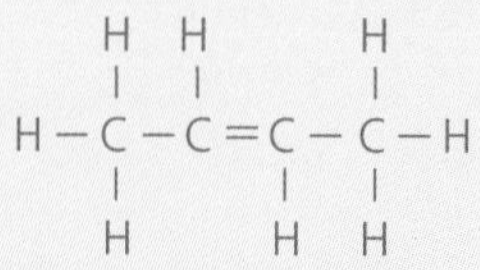

```
    H   H   H   H            H   H       H
    |   |   |   |            |   |       |
H - C - C - C - C - H    H - C - C = C - C - H
    |   |   |   |            |       |   |
    H   H   H   H            H       H   H
```

are simply drawn as

and

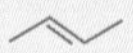

while the *cis* version of 2-butene becomes

Such diagrams have the added merit that they give a better indication of the mutual angles between the bonds, but a two-dimensional diagram can never do full justice to a three-dimensional molecule. Using the same rules, the ring-formed version of the compound C_6H_{12}, cyclohexane, is simply drawn thus

cyclohexane

its other common name, *marsh gas*, and it is frequently formed in conjunction with *phosphine*, a phosphorus–hydrogen compound which causes it to spontaneously ignite. This produces the ghostly flame known as *Will-o'-the Wisp*, or *ignis fatuus*, occasionally seen in marshes. The gas is particularly dangerous when it occurs in coal mines, because it forms explosive mixtures with air. Whence another of its aliases: fire-damp. *Gasoline* is mainly a mixture of hexane (six carbons in chain) and heptane (seven carbons), with lesser amounts of pentane (five carbons). The higher homologues are also inflammable, but fuels become cruder with increasing molecular weight of the constituents, because of the higher boiling points and the resultant increasing difficulty of ignition.

There are no double bonds in the alkane molecules, so all the carbon–carbon linkages are *saturated*. These are the *sigma bonds* that we encountered earlier, and their freedom of rotation, together with the absence of mutual interference between the small hydrogen atoms, makes for flexibility. However, such interference is common when groups of atoms rather than single atoms are appended to the carbon backbone, this effect being known as *steric hindrance*. Thus far, the alkane series is quite straightforward because methane, ethane and propane have no isomers. Complications first begin to arise in the case of *butane*, because the extra carbon, together with its attendant hydrogen atoms, can be added either to one of the terminal carbons or to the carbon at the centre. The resulting two molecules are clearly different since one is a straight chain while the other is branched. The linear chain version of this alkane is known as normal butane, or *n-butane*, the ordinary type, while the branched isomer is *iso-butane*. The use of condensed structural formulae is very convenient when dealing with different isomers, and the two examples just discussed are denoted respectively by $CH_3{\cdot}CH_2{\cdot}CH_2{\cdot}CH_3$, for the *normal* form and $(CH_3)_2CH{\cdot}CH_3$ for the *iso* form. As we proceed to higher members of the series, the number of different possible isomers increases rapidly. Octane, for instance, has no less than 18 different forms. It will not be necessary, here, to go into the niceties of nomenclature required to maintain control over the host of possibilities.

Just as in inorganic chemistry, a group of atoms that can be transferred as a whole from one compound to another is referred to as a *radical. Organic radicals* resemble their inorganic cousins in that they too confer characteristic properties upon the compounds containing them. The radical of methane is formed by the removal of one if its four hydrogen atoms, and the radical thus formed is referred to as the *methyl group*. Since it has one of its valences unsatisfied, it cannot exist in isolation as a stable species. The general name for a radical of an alkane is *alkyl*, and the individual terms are logical modifications of the corresponding names for the alkanes: butyl, heptyl, etc. One exception to this rule is the *amyl radical*, thus named because of its connection with starch, the Latin name for which is *amylum*. It has five carbon atoms, and it might otherwise have been known as the pentyl radical. The alkanes can be made to react with various elements or radicals to produce a great variety of compounds. In methane, for example, each of the hydrogen atoms can be replaced by another monovalent element or radical. The family of compounds between methane and chlorine is a case in point. Replacement of one of the hydrogen atoms with a chlorine produces *methyl chloride*, also known by the name *monochloromethane*. Addition of a second chlorine atom gives *dichloromethane*, which is also called *methylene chloride*, while the still more chlorinated member is *trichloromethane*, more commonly known as the anaesthetic *chloroform*. In the ultimate member of this series, all four of the hydrogens have been replaced by chlorine atoms, and the compound is *tetrachloromethane*, which is more familiar as the cleansing agent *carbon tetrachloride*.

One of the more obvious physical properties of the alkanes is a reticence to mix with water, a behaviour which stems from their lack of electric polarity. In this respect they are quite unlike ionic compounds in general, and indeed water in particular. It was emphasized earlier that some of the most conspicuous properties of water are related to the fact that the water molecule has both positive and negative surface charges. The ever-changing constellations of molecules in liquid water involve the matching of oppositely charged poles. The presence of an uncharged alkane molecule would interrupt such patterns, and it would locally diminish the number of different possible orientations available to a water molecule. *Free energy* increases with decreasing flexibility of arrangement, because of the lower *entropy*, and this is what happens upon introduction of an alkane molecule. The situation is thus unfavourable, and the water can be said to shun the alien molecule. The alkanes, and many similar families of organic molecules, are therefore said to be *hydrophobic*, or water hating. The existence of hydrophobic and *hydrophilic* interactions are of immense importance in the biological world.

A second important family of aliphatic compounds are the *alkenes*, and they differ from the alkanes in that they all involve *double bonds* between one or more of the backbone carbon atoms. Because this requires a molecule

Theoretical chemistry and the techniques of *chemical synthesis* are now so advanced that new *organic compounds* can be designed and constructed almost on demand. Control of the structures actually produced in the test tube relies heavily on observations of the way in which they interact with various forms of radiation. The liquid shown here is a solution of the five-ringed compound *perylene*, $C_{20}H_{12}$, in toluene. A beam of white light enters from the left, and a part of the spectrum provokes excitation of the perylene molecules. This absorption occurs in the ultraviolet–blue region, so the emerging beam has a yellow-green colour, as seen at the right. The excitations decay by emission of blue light. In contrast to the case of the *Tyndall effect*, which was described in Chapter 3, the observed colours are not exactly complementary because the *internal vibrations* of the perylene molecules are involved in the absorption and emission processes. The emission process is known as *fluorescence*; it differs from *phosphorescence* in that it ceases almost as soon as the source of light is removed.

containing at least two carbon atoms, the alkene series starts with a compound that is related to ethane rather than methane. This is *ethylene*, and it is not flexible like its alkane cousin, because of the *pi bond*. Another consequence of the latter is that some of the combining capacity of the two carbon atoms is not utilized. Comparing ethylene with ethane, we see that the former has two less hydrogen atoms, and in this respect it is said to be *unsaturated*. It is a common feature of unsaturated molecules that they display a marked tendency for *chemical activity*. This is because of the relative instability of the double bond, as first explained by Adolf von Baeyer. Assuming that the tetrahedral arrangement of the valence bonds in methane represents the lowest free energy state, he argued that the formation of a double bond would have to involve displacement of some of the valence arms away from their normal positions, and the resultant strain must cause weakening. Conversely, splitting of one of the bonds in a double bond permits restoration to the tetrahedral form with a consequent lowering of energy. The reactivity of ethylene is clearly demonstrated by the ease with which it combines with the halogens, to form compounds such as *ethylene dichloride*. This heavy, colourless and oily liquid was discovered in 1795 by four Dutch chemists, and it is known as *Dutch oil*. They went on to discover ethylene itself, and called it *olefiant gas*, meaning oil-forming gas, from which the alternative name for the alkenes derives, namely the *olefins*. The names of the higher members of the series follow logically: propylene, butylene, etc. Isomerism in this series is first observed with butylene, there being three isomeric forms, $CH_3{\cdot}CH{:}CH{\cdot}CH_3$, $(CH_3)_2C{:}CH_2$ and $CH_3{\cdot}CH_2{\cdot}CH{:}CH_2$, where the colon denotes a double bond as usual.

A third homologous series of aliphatic hydrocarbons, the *alkines* or *alkynes*, starts with *acetylene*, which is a gas at room temperature. The most interesting feature of molecules of this group is the presence of one or more *triple bonds*. In general, members of the alkyne series are even more unsaturated than the alkenes containing the same number of carbon atoms. According to the *von Baeyer strain theory*, acetylene should be even more reactive than ethylene, and this is certainly the case. Liquid acetylene, which boils at $-84\,^{\circ}C$, is in fact highly explosive, and the controlled burning of acetylene with oxygen produces temperatures in the region of $4000\,^{\circ}C$. Whence its common usage in the cutting and welding of steel components. Acetylene can be obtained through the action of water on calcium carbide. The reaction is quite straightforward, and calcium hydroxide appears as the other reaction product. The controlled production of the sweet-smelling gas was earlier used as a source of illumination, particularly in bicycle lamps, by burning it in air.

It is not possible to produce a linear hydrocarbon involving a quadruple bond between two adjacent carbon atoms, but this does not mean that all the possibilities have been exhausted. Indeed, insofar as every carbon–carbon bond in an alkyl chain can become unsaturated, the number of possible variations on this theme is truly unlimited. Discussion of such molecules is best made in connection with the questions of *polymers* and *fatty acids*, which will appear later. Neither have all hydrocarbons been covered, because there are important compounds involving *rings of carbon atoms*. Because of the natural chemical relationship between *linear molecules*, it will be logical to continue with these, and this is the point at which it is necessary to introduce a third element. It is oxygen, and there are no less than five important groups of compounds to discuss.

An *alcohol* is an organic substance containing one or more *hydroxyl groups*, which are directly attached to a hydrocarbon. If there is only one such group, the alcohol is said to be monohydric, while two and three hydroxyl groups give a dihydric and trihydric alcohol, and so on. The monohydric variety is clearly the simplest, molecules of this type being essentially hydrocarbons in which a hydrogen atom has been replaced by an oxygen and a hydrogen. Thus the compounds CH_3OH and C_2H_5OH are methyl alcohol and ethyl alcohol, respectively. They are the most familiar members of this series, and they are both liquids with boiling points somewhat lower than that of water. *Methyl alcohol* can be made by heating carbon monoxide and hydrogen to 450 °C at a pressure of 200 atm, but is also produced by the destructive distillation of wood, and hence its alternative name: *wood alcohol*. In smell, it resembles ethyl alcohol, but it is a most dangerous poison that, even when consumed in modest quantities, produces blindness or even death. It finds widespread industrial use, including the production of *dyes*, *perfumes*, *polishes* and *varnishes*, and it is still used in refrigeration. *Ethyl alcohol*, which is the compound commonly known as alcohol, is of course the basis of a vast industry whose intoxicating beverages include beer, wine and spirits. The *fermentation process* by which ethyl alcohol is produced has remained essentially unchanged since ancient times. It is achieved by the enzymes *zymase* and *invertase*, produced by *yeast*, which is the common name for the fungus *Sacchoromyces*. These two proteins catalyze the reaction in which sugar is broken down, with the release of energy. Yeast does this in order to obtain the energy that, unlike higher plants, it cannot obtain by oxidation, using the oxygen in the atmosphere. If the starting point is *cane sugar*, also called *sucrose*, the invertase breaks the molecules of this compound into two smaller sugar products: the two isomers *grape sugar*, also known as *glucose* or *dextrose*, and *fruit sugar*, the alternative names of which are *fructose* or *levulose*. The zymase then takes over and catalyzes a reaction in which one sugar fragment is made to oxidize another. In the production of sparkling wines and *champagne*, the second reaction product, *carbon dioxide*, is also retained. This is achieved by bottling the product when the fermentation process is not quite complete. In sufficiently large quantities, ethyl alcohol is just as poisonous as the methyl variety, but in small doses it acts as a stimulant. It mixes with water in all proportions and, when in the human body, tends to increase thirst rather than quench it, despite the claims of many beer producers.

Several other important types of organic compound can be produced from alcohol. In the ethers, the two valence bonds of oxygen enable an atom of that element to act as a bridge between two alkyl groups. An example is *diethyl ether*, the substance usually referred to simply as *ether*. It can be prepared by heating an excess of ethyl alcohol with concentrated sulphuric acid. Sulphovinic acid is produced as an intermediate, and this then reacts with more alcohol to produce the ether, and release sulphuric acid for repetition of the two-stage process. The pleasantly smelling product is a very volatile liquid, and this gives it a characteristic cold feeling; the liquid evaporates by extracting heat energy from the fingers. It is chemically inactive and useful both as a *solvent* and as an *anaesthetic*, although its inflammability makes caution imperative. The two compounds ethyl alcohol, CH_3CH_2OH, and dimethyl ether, CH_3OCH_3, are a classic example of *isomerism*. Their molecular compositions are identical, but their physical, chemical and physiological properties are quite different.

Stearic acid and *palmitic acid* are common *fatty acids* which are both used in the *soap industry*. Their melting points are 69 °C and 64 °C respectively. When the hydrogen atom at the carboxyl end of one of these molecules is replaced by a sodium atom, the result is a molecule whose non-fatty end shows an even greater affinity for water, and this is the origin of its efficacy as a cleansing agent. This space-filling diagram shows the lighter of these two varieties of soap: *sodium palmitate*.

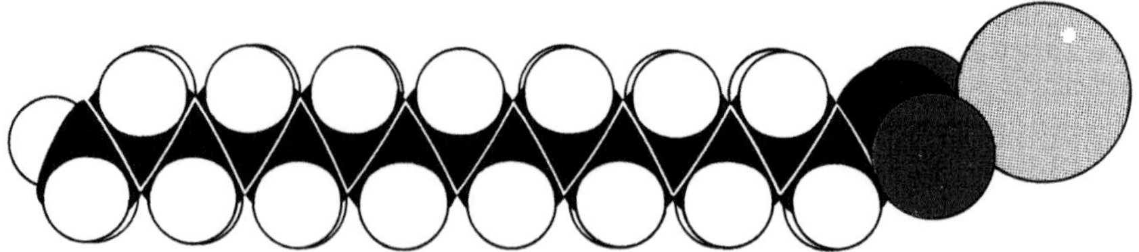

Oxidation of what is known as a *primary alcohol*, a term which will be enlarged upon shortly, produces a compound belonging to the aldehyde group. Once again, a simple looking reaction produces a profoundly different type of substance. The *aldehydes* find frequent use as *disinfectants* and as *preserving media* in the medical sciences. The simplest member of the series is *formaldehyde*, and it is produced from methyl alcohol by mixing the vapour of that substance with air, and heating it over a copper surface, the latter acting as a catalyst. The reaction involves the loss of two terminal hydrogens and the formation of a double bond between a carbon and an oxygen atom. The latter is a divalent configuration known as a *carbonyl group*. Formaldehyde is a gas at room temperature, and it is frequently used as a solution in water, when it goes by the name *formalin*. Traces of formaldehyde are formed when many organic substances are incompletely burned. Because of the disinfecting action of this chemical, its production is exploited in the preservation of meats and fish that are subjected to a mild smoking. The next member in the series, *acetaldehyde*, is a liquid and like formaldehyde it finds widespread industrial use in the production of *plastics, textiles* and *dyes*. The monohydric compounds methyl alcohol, ethyl alcohol, propyl alcohol, and so on, can be represented by the general formula RCH_2OH, and they are called *primary alcohols*. The R denotes an *alkyl radical* (or in the simplest of a series, a single hydrogen atom), and this formulation clearly reveals the connection with the corresponding *aldehydes*, the formulae for which are of the form RCHO. Thus far only the primary alcohols have been considered, but other types exist. There are valence bonds available on the central carbon for two, and even three, radicals to be attached. Thus a *secondary alcohol* can be written $(R)(R')CHOH$, while a tertiary alcohol would be $(R)(R')(R'')COH$. In these general formulae R, R′ and R″ denote radicals that might or might not be identical.

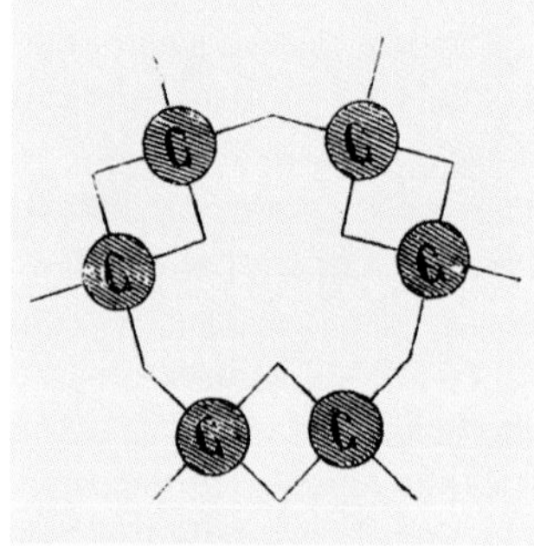

Before the middle of the nineteenth century, the concept of *molecular structure* was limited to rather vague interpretations of the compositions described by chemical formulae. Kekulé's theoretical work on the benzene ring was the first bold step towards the belief that such formulae might also represent the actual spatial arrangements of atoms in molecules. This figure is taken from his *Lehrbuch der Organische Chemie*, published in 1862, and shows his original model of the benzene molecule.

We have already seen that oxidation of a primary alcohol produces an aldehyde. Oxidation of a secondary alcohol produces a ketone, while oxidation of a tertiary alcohol causes such a molecule to split up into smaller products. The *ketones*, the general formula of which is $(R)(R')CO$, are among the most frequently used organic solvents, being particularly conspicuous in the *synthetic textile industry*. The lighter members of the series are liquids at room temperature. The simplest ketone, and the best known, is *acetone*, in which both R and R′ are CH_3 methyl groups. Like common ether, it evaporates readily and is cold to the touch. The major difference between aldehydes and ketones is that the latter cannot be oxidized. An aldehyde always has at least one terminal hydrogen that can be attacked by oxygen, and the replacement of this hydrogen by a hydroxyl group transforms the aldehyde into an *organic acid*. The simplest example is the transformation of formaldehyde into *formic acid*. The defining characteristic of an organic acid is a terminal group of four atoms in which a carbon is bonded both to a hydroxyl group and a carbonyl oxygen. This unit is known as a *carboxyl group*,

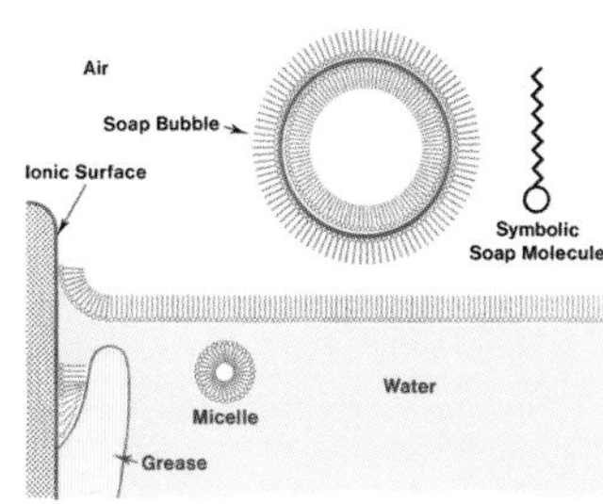

A *soap molecule* is said to be *amphiphilic* in that its two ends display opposing behaviour with respect to water. The polar end is attracted to that liquid while the hydrocarbon tail is shunned. These different tendencies dictate the arrangement of soap molecules at the surface of water, and also in other aqueous situations. Thus a *soap bubble* consists of a thin sphere of water in contact with the polar heads of soap molecules on both sides. Soap molecules entirely enveloped by water form a sphere, with the hydrophobic tails on the inside, away from the water. Such a structure is known as a *micelle*. Both ends of a soap molecule can satisfy their natural tendencies at the interface between *grease* and the surface of a ceramic. Ions in the latter attract the polar heads, while the hydrocarbon tails are compatible with grease. Soap molecules can thus ease their way into the interface, and ultimately cause the grease to become detached.

and it is monovalent. The acidic nature of the carboxyl group derives from dissociation of the hydroxyl part in water, to produce a negatively charged oxygen ion and a hydrogen ion. The latter immediately joins with a water molecule, to produce *hydronium*, as discussed in Chapter 6. The simplest series of organic acids can be regarded as consisting of an alkyl chain terminating with a carboxyl group. The general formula is RCOOH. In formic acid, R is simply a hydrogen atom. This compound is used in *dyeing* and the *tanning of leather*, and it takes its name from the Latin word *formica*, meaning ant. The earliest production of this acid involved the distillation of red ants, and the substance is also present in the stings of bees and wasps. A piece of blue litmus paper pushed into the nest of any of these creatures is rapidly covered with a host of red spots. *Acetic acid*, which is the next member in the series, can be formed from acetaldehyde in an analogous manner. It is also made directly by the continued oxidation of ethyl alcohol, and it is a major component of the *vinegar* that is produced when wine that is exposed to air becomes oxidized. The culprit in this transformation is the bacterium *Mycoderma aceti*, which possesses an enzyme capable of catalyzing the oxidation. Acetic acid is widely used as an *industrial solvent*, particularly in conjunction with *cellulose*, in the production of non-inflammable and pliable plastics.

The important group of substances in which an alkyl chain terminates with a carboxyl unit are known as the *fatty acids*. They display interesting chemical behaviour, which derives from a special feature of their structures. We saw earlier that an *alkyl chain* tends to avoid water, whereas the *carboxyl group*, because it can readily lose a hydrogen atom and thereby become charged, or polar, is compatible with that liquid. When fatty acids are present in water, they form globules known as *micelles*, in which the alkyl chains point inwards while the carboxyl groups face outwards towards the liquid. Similar structures encountered in the biological domain are extremely important. A still greater disparity between the two ends of a molecule is observed when the hydrogen at the carboxyl end of a fatty acid is replaced with a metallic atom such as sodium. The water-compatible end then shows an even greater affinity for water, and the resulting molecule can act as a mediator between water and fat. A good example is *sodium stearate*, which is the chemical name of common household *soap*. The alkyl chain has a convenient length for mixing with many greasy substances, while the polar unit has a marked attraction for water. This enables the soap to carry molecules of the grease off into the water. Soap also removes grease from ceramic objects. It does this because its molecules can work their way in along the grease–ceramic interface, the alkyl ends towards the grease and the polar ends facing the ions present in the surface of the ceramic. Water molecules are inevitably carried into the ionic interface, and permit the fatty film to float off.

An *ester* is an organic compound formed by replacing the hydrogen atom of an acid, organic or inorganic, by an organic radical. It is therefore the organic counterpart of an inorganic salt. Some examples are methyl formate, ethyl acetate, methyl chloride and ethyl sulphate. In connection with later examples of *esterification*, the totally organic varieties will be of particular interest. A standard reaction leading to the production of one of those is as follows:

$$\text{acid} + \text{alcohol} \rightleftharpoons \text{ester} + \text{water}.$$

This is similar to the inorganic reaction in which an acid and a base combine to produce a salt and water. A good example is the production of

Fatty acids

The *fatty acids* are monobasic organic acids, with the general formula RCOOH, where R is hydrogen or a group of carbon and hydrogen atoms. An example in which R is an alkyl chain is *arachidic acid*, which can be derived from peanuts. Its structure is

arachidic acid

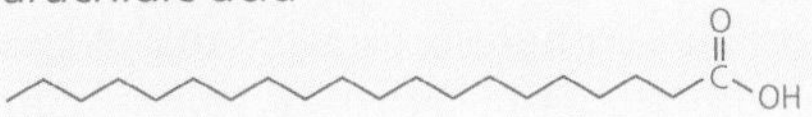

This white crystalline solid melts at 76.3 °C. Replacement of the carboxyl hydrogen by a metallic atom such as sodium increases the difference between the two ends of such a chain molecule; the aliphatic chain is shunned by water while the polar end is attracted to water. This is the basis of soaps, an example being

sodium stearate

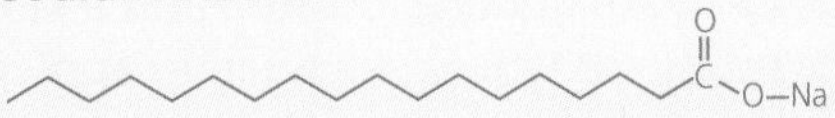

Such a molecule can act as a mediator between water and fat; whence its use as a cleansing agent. An *ester* is derived by replacing the hydroxyl hydrogen of an acid by an organic radical. The *glycerides*, a common constituent of the natural fats and oils known as *lipids*, are esters of glycerol with organic acids. The glycerol molecule itself has the following structure

glycerol

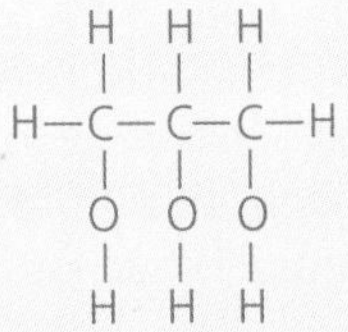

```
    H  H  H
    |  |  |
 H—C—C—C—H
    |  |  |
    O  O  O
    |  |  |
    H  H  H
```

and we see that it has three termini susceptible to esterification. Triolein, molecules of which have the structure

triolein

is common in many natural fats.

ethyl acetate from acetic acid and ethyl alcohol. The reaction is catalyzed by strong acids and is often pushed towards completion by using an excess of alcohol or organic acid. Esters are colourless liquids or solids that are insoluble in water, and they usually have a pleasant odour. This makes them popular vehicles for other strongly scented molecules in the *perfume industry*.

Kekulé's benzene ring

Probably the best known of all stories regarding chemistry is that associated with the name of Friedrich Kekulé von Stradonitz (1829–1896). Until his inspiring day-dream of snakes biting their own tails, all molecules were believed to have open chain structures. He realized that the molecule of *benzene*, C_6H_6, consists of a closed ring. Because the valence of carbon is nevertheless four, single and double carbon–carbon bonds must alternate around the hexagonal ring, and this gives two alternative possibilities:

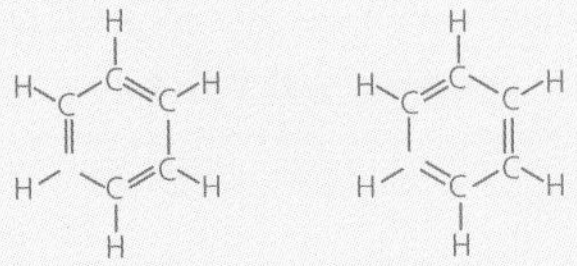

In fact there is a *resonance* between the two forms, and the three extra bonds are shared equally among the six pairs of atoms. This is often indicated by an inscribed circle:

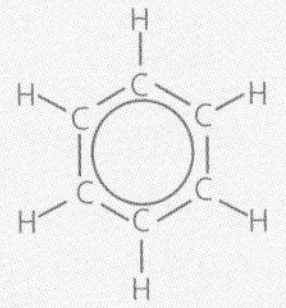

Monosubstitution products of benzene, such as C_6H_5Cl and C_6H_5Br, in which one of the hydrogens is replaced by a halogen atom, exist in only one form. This corroborates the *Kekulé model*; it shows that each hydrogen atom is equivalent to all the others. If a second hydrogen is replaced, however, no less than three *isomeric forms* are observed. This too is consistent with the Kekulé model. The carbons are numbered consecutively, starting at an arbitrarily chosen atom, and the first chlorine substitution is made at the hydrogen attached to the first carbon atom. There are three non-equivalent positions available for the second substitution. This can be made at positions 2, 3, or 4. The combinations 1+2, 1+3, and 1+4 give, respectively, *ortho*-dichlorobenzene, *meta*-dichlorobenzene, and *para*-dichlorobenzene. (These distinguishing prefixes are usually denoted by their initial letters.) These are clearly the only possibilities since 2+3 is the same as 1+2, while 2+4 is the same as 1+5, etc. The substituted carbon atoms in the three different isomers subtend angles of 60°, 120°, and 180°, respectively, at the centre of the benzene hexagon.

Particularly important examples of esterification involve the attachment of parts of fatty-acid molecules, known as *fatty-acid residues*, to *polyhydric alcohols*. *Glycerol*, also known as *glycerine*, is the most important of the latter. The name comes from the Greek word *glukus*, meaning sweet. The formal chemical name is 1,2,3-trihydroxypropane, the numbers indicating that there is a hydroxyl group attached to each of the carbon atoms of the basic propane molecule.

Natural fats and *oils* are the esters of glycerol with various fatty acids, and they are known as *lipids*. The main constituent of beef and mutton fat is *tristearin*, in which three fatty-acid chains, each comprising eighteen carbon atoms and thirty-five hydrogen atoms, are attached to the glycerol backbone. *Triolein*, which is found in *olive oil* and *lard*, is like tristearin but with one unsaturated bond located at the middle of each of the three chains. This means that the molecule contains six less hydrogen atoms. *Tripalmitin*, which is the chief constituent of *palm oil* and *human fat*, is another variation on the same theme, but with only fourteen carbons in each chain. These esters can be hydrolyzed in hot alkali solutions to liberate glycerol and fatty acids. If sodium hydroxide is used as the hydrolyzing agent, the products are

Fullerenes

While performing molecular beam experiments in 1985, Robert Curl, Harold Kroto and Richard Smalley – in collaboration with James Heath and Sean O'Brien – were surprised to find that they were producing molecules composed of sixty carbon atoms. The stability of these molecules clearly indicated that they were held together by covalent bonds, but no such structure was known in chemistry. Until that time, the only known allotropes of carbon were graphite and diamond. It soon transpired that the carbon atoms in C_{60}, as it was called, were arranged in the form of a truncated icosahedron, which itself resembles the type of ball used for soccer, with its panels of pentagons and hexagons (shown here against a blue background). The location of the structure's double bonds are indicated in the diagram with the white background, just one side of the molecule being shown in this case. The structure is also like a miniature version of the geodesic domes designed by the architect Richard Buckminster Fuller (1895–1993), a notable example being the American pavilion at Expo '67 in Montreal, shown in the colour photograph. This led to C_{60} being given the alternative name *buckminsterfullerene*. The original experiments had also revealed the presence of a molecule comprising seventy carbon atoms, and the emergence of even larger ball-shaped structures of the same type indicated that there is actually a whole family of *fullerenes*. By the early 1990s, Donald Huffman and Wolfgang Krätschmer had found a way of producing relatively large quantities of fullerene material, and new forms were discovered. There are, for example, cylindrical forms called *nanotubes*, which are in fact the strongest known material, and a two-dimensional version known as *graphene* was found in 2004. As is often the case, Nature was later shown to have anticipated the experimenters because fullerenes have been found to be present in candle soot. Very small quantities have even been found to occur naturally in such places as Shunga in Russia, Sudbury in Canada, and in New Zealand. The key to the molecular structure of a fullerene lies in all its carbon atoms displaying a coordination number of three, the bonds being of the strongly localized pi type. Such a structure is consistent with the rule embodied in a theorem by Leonhard Euler (1707–1783), namely that a closed arrangement must contain just 12 pentagons, no two of which may be located side-by-side, in addition to a number of hexagons. So it is the number of the latter which determines a fullerene molecule's size.

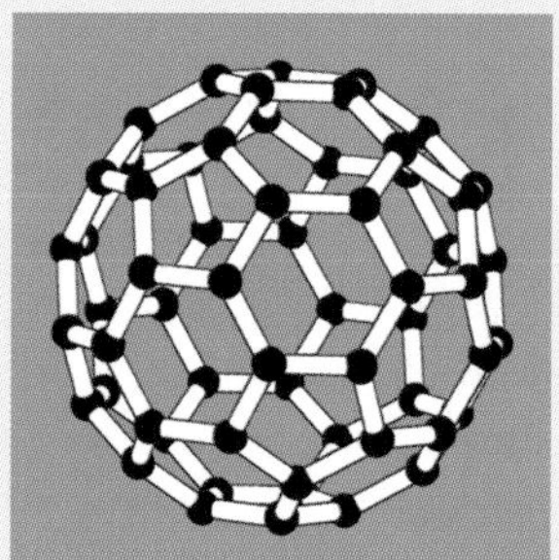

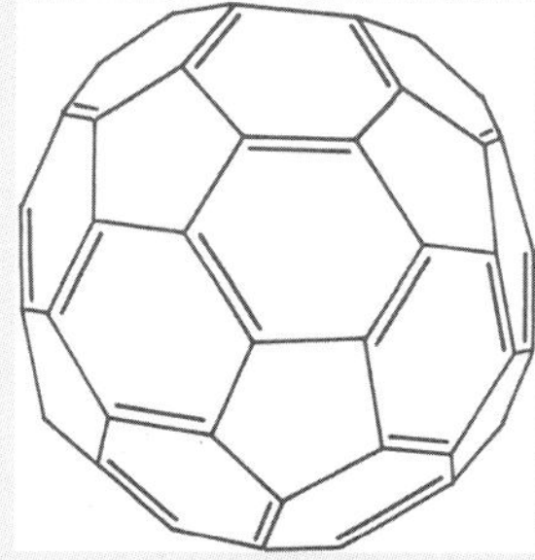

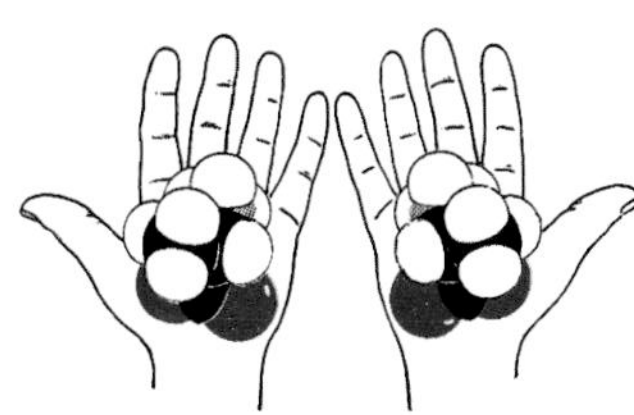

In 1848, Louis Pasteur (1822–1895) explained the different light-rotating properties of various types of *lactic acid* by assuming that the molecules of this compound can have two different geometrical forms, which are mirror images of each other. Such *optical isomers*, as they are called, are usually referred to by their handedness, left or right. This is illustrated here by space-filling models of left-handed and right-handed versions of the amino acid *alanine*.

glycerol and the sodium salt of the fatty acid. The production of common *soap* involves the boiling of animal fat and sodium hydroxide, in a process known as *saponification*, the product being the sodium stearate that was referred to earlier. It should be noted that the molecules of natural fats do not contain a polar unit, so they cannot mix readily with water. Fats flow easily when warmed, but form globules on cooling. Whence, for instance, the fatty feeling at the roof of the mouth after eating mutton.

The term *isomerism* has so far been applied exclusively to situations in which two chemically distinct compounds happen to have the same molecular composition, an example being the isomerism of ethyl alcohol and di-methyl ether. There is a second, and more subtle, type of isomerism that is just as important. It was theoretically established in connection with observations on certain optical properties of *lactic acid*, a compound present in sour milk. The electromagnetic waves of normal light vibrate in all possible planes containing the line of propagation, whereas in *polarized light* the vibrations are confined to a single plane. Certain substances, such as *tourmaline*, *Iceland spar* and *Polaroid*, are capable of converting ordinary light into polarized light. If two such polarizing substances have their planes of polarization lined up, light rays can pass through both. If, on the other hand, the planes are rotated so as to lie at right angles to each other, no light emerges from the second polarizer. Solutions of lactic acid are found to be *optically active*, in that they rotate the plane of polarization, the actual amount of rotation being measurable with two polarizers. No less than three different types of lactic acid were identified in this way: one rotated the light towards the right, one to the left, while a third was found to be optically inactive. Louis Pasteur, in 1848, suggested that optical activity in chemicals might be associated with an asymmetric arrangement of their atoms in space. In 1874, Jacobus van't Hoff and Joseph le Bel independently showed that the structure of lactic acid does indeed admit of two non-equivalent atomic configurations. In this compound there are four different groups all attached to a central carbon atom. Van't Hoff and le Bel suggested that these four groups are arranged roughly at the corners of a tetrahedron with the bonding carbon at its centre. There are two different ways of distributing four different groups among the corners of a tetrahedron: it is not possible to rotate one of these arrangements so as to bring it into exact coincidence with the other. Molecules based on these two configurations are found to be *mirror images* of each other, and this is the origin of the left-handed and right-handed optical activities. The optically inactive variant turned out to be simply an equal mixture of the two different forms. It has subsequently been found that some sources of this compound have a marked preference for one of the forms. Thus meat-extract lactic acid is *dextro* (right) active, whereas sour-milk lactic acid is an equal mixture of the *dextro* and *laevo* (left) forms. *Optical isomers* which are mirror images of each other are referred to as *enantiomers*, the term coming from the Greek word *enantios*, meaning opposite. An interesting variant of this *stereoisomerism*, as it is called, occurs in compounds containing bonds with restricted rotation, such as an unsaturated carbon–carbon bond. *Maleic acid* and *fumaric acid* are cases in point. The molecules of each of these isomers comprise four side groups, two of which are hydrogens and the other two carboxyl groups. In maleic acid, referred to as the *cis-form*, like groups are on the same side of the plane containing the double bond, while in fumaric acid, known as the *trans-form*, they are on opposite sides.

Colour changes in polydiacetylenes

The chemical compounds known as *disubstituted diacetylenes* are usually colourless. Their molecules have the general form R—C≡C—C≡C—R, where R, known as a substituent group, can be any of a wide range of organic radicals. Like acetylene itself, a compound of this class possesses an unsaturated region which is susceptible to modification by rearrangement of the interatomic bonds. The energy required to provoke such changes can be supplied thermally or by irradiation with ultraviolet light, X-rays, or gamma rays, and the individual molecules are thereby linked into chains. The reaction is indicated in the following structural diagram, in which it is revealed that half the triple bonds are still present in the composite molecule.

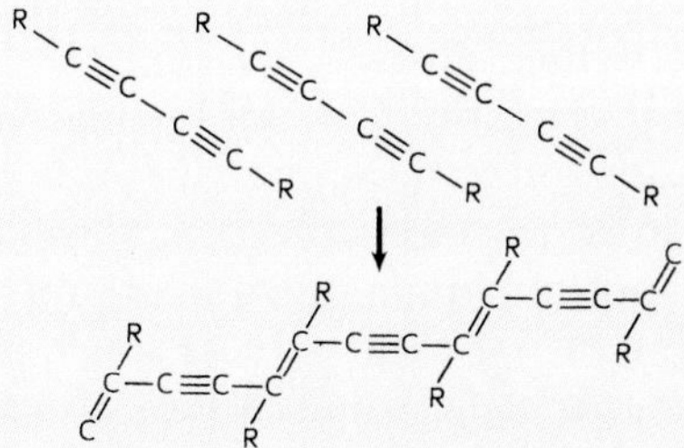

The joining of several smaller units into a chain is an example of *polymerization*. Interestingly, these larger molecules are coloured. The *polydiacetylenes* have highly conjugated backbones; although the single and multiple bonds are drawn in specific positions, the electrons from which they are formed are not so strictly localized. There is an interplay between the spatial configuration of the polymer molecule's backbone and the degree to which the backbone electrons are delocalized, and this in turn determines the colour of the compound. The substituent groups of the original molecules, referred to as *monomers*, become the side groups of the polymer, and they normally have little influence on the colour. But a family of polydiacetylenes first synthesized by Grodhan Patel in 1976 provide an interesting exception. Their side groups act upon the backbone and modify its shape. This alters the degree of *electron delocalization*, and thus changes the colour. A typical member of this interesting family is butoxy-carbonyl-methyleneurethane-diacetylene, now designated 4BCMU, in which R in the diagram above represents a rather long string of atoms having the molecular formula –$(CH_2)_4OCONHCH_2COO(CH_2)_3CH_3$. The structural formula of a section of this polydiacetylene, and also a photograph of a space-filling model, are shown below.

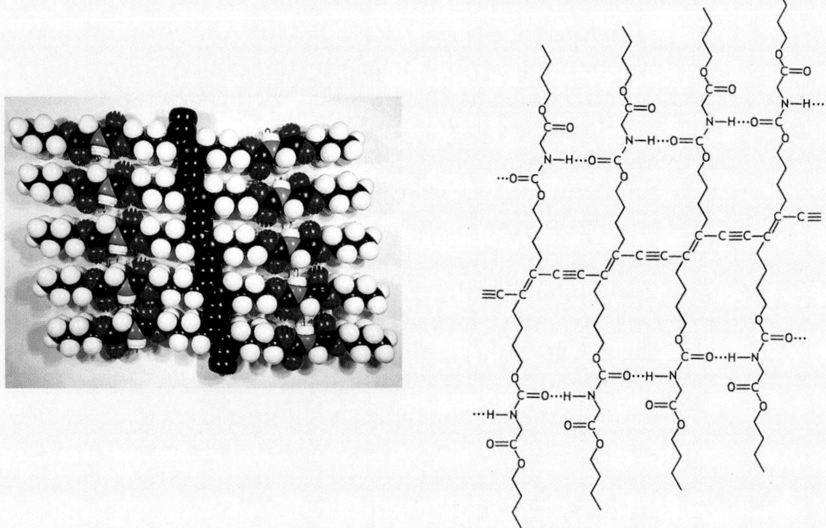

The special properties of 4BCMU derive from the additional presence of the *hydrogen bonds*, which are indicated by the usual dotted lines. The spatial arrangement of these bonds dictates the shape, and

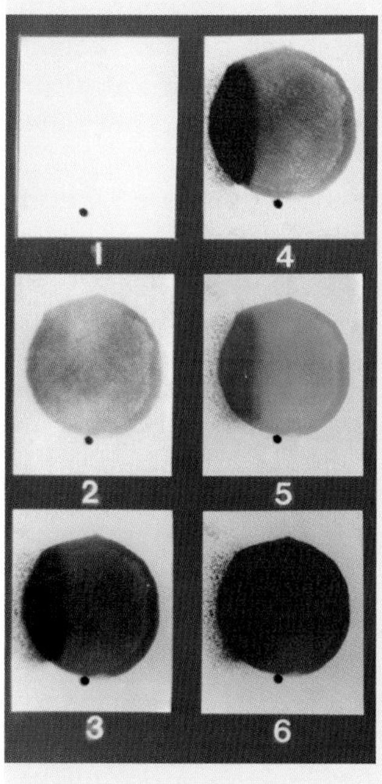

hence the colour, of the backbone. This is illustrated in the series of pictures shown above, the black dot serving as a reference marker. The first picture was obtained by spraying a five per cent solution of the monomer in ethyl alcohol onto a piece of paper, and permitting the solvent, to evaporate. A flat and compressed molecule of polydiacetylene appears blue because the backbone electrons are spread over a distance of some thirty repeat units. This is the situation that prevails when the polymer is in solid solution with its corresponding monomer, and it can be achieved by exposing the deposited monomer film to ultraviolet light for a fraction of a second (see picture 2). Further exposure increases the degree of polymerization, and a prolonged dose of radiation gives an almost metallic appearance, as shown at the left side of picture 3. If the unused monomer is removed, by heating to its melting temperature of 75 °C for example, the backbone remains flat but it becomes stretched. This increases the *electron localization* and the colour of the film changes to a light red (picture 4). Above about 145 °C, the hydrogen bonds break and the backbone rotates around the single bonds, producing a corrugated shape of the molecule and confining each backbone electron to just a few repeat units. Short wavelength light is then absorbed by the film and it appears yellow (picture 5). Upon cooling to room temperature, the hydrogen bonds are re-established, and the polymer returns to its flat stretched configuration. It thus recovers its red colour, as shown in the final picture.

In the compounds discussed thus far in this chapter, the carbon atoms have been arranged in chains, either singly or branched. A different configuration was discovered in connection with early researches on what have come to be known as *aromatic compounds*. In 1825, Michael Faraday reported the presence of an unidentified oil in a supply of compressed illuminating gas, and A. W. Hofmann and C. B. Mansfield isolated the new substance, *benzene*, twenty years later. It was chemically peculiar because it behaved in a manner most uncharacteristic of aliphatic compounds. Its molecular formula was shown to be C_6H_6, which would suggest a structure far less saturated than hexane, C_6H_{14}, in that it has eight fewer hydrogen atoms. It was, however, found to behave as if it were quite saturated, showing little propensity for forming addition products. The answer to this paradox was supplied by Friedrich Kekulé von Stradonitz. Reportedly during a day-dream, in which snakes were seen biting their own tails, he conceived the idea of carbon atoms arranged in *closed rings*. Actually, this does not immediately explain why benzene behaves as if it were saturated, because the hexagonal ring configuration would still have to involve three double bonds. Kekulé speculated, correctly as it transpired, that a double bond in a ring arrangement might behave differently than one in an open chain. The modern view is that a resonance is set up, and this shares the extra bonds equally among the six carbon atoms, keeping the ring flat, and causing the molecule to act as if it were saturated.

We have already noted that one of the main classes of food, namely fats, falls within the realm of organic chemistry. We come now to a second major group, the *carbohydrates*. These must not be confused with the similarly sounding hydrocarbons. It is also important to understand that these substances are not hydrates in the usual sense of the word; they are not compounds containing bound and identifiable water. Carbon forms no true hydrates. The name stems from the fact that these compounds contain hydrogen and oxygen in the same proportions as in water, namely two-to-one. The carbohydrate group includes such biologically significant substances as sugars, starches and celluloses, and sugars will be considered first since they are the simplest.

Sugars, the individual names of which all have the ending -ose, embody many of the principles discussed earlier in this chapter. The simplest members of this large group are referred to as *monosaccharides*, and they contain anywhere between three and more than seven carbon atoms. Except for the lighter examples, they find it energetically favourable to form ring structures, as first demonstrated by Charles Tanret in 1895. The most common of these compounds are the pentose sugars (general formula $C_5H_{10}O_5$), ribose, lyxose, xylose and arabinose, and the hexose sugars (general formula $C_6H_{12}O_6$), glucose, talose, mannose, idose, gulose, altrose, allose and galactose. Since they share a common molecular formula, members of these groups are clearly isomers. Several are also stereoisomers of one another. These sugars contain a CHO group and are therefore *aldoses*; they are sugars related to aldehydes. Other sugars are found to contain the group CO and are therefore *ketoses*; they are related to ketones. The most common example of this type of sugar is *fructose*. Because the differences between the various sugars are relatively subtle, it is customary to employ structural diagrams, particularly when distinguishing between *dextro* and *laevo* versions. The transition from the open-chain form of a sugar to the ring configuration involves the shifting of a hydrogen atom from the hydroxyl group lying closest to one end across to the carbonyl carbon at the other.

Monosaccharides can combine with one another to produce chains, the links of which are the individual rings. Anticipating something which we will encounter later, the joining process is an example of *polymerization*, and it belongs to the class known as *condensation* in that a water molecule is liberated during the joining process. Two-ring structures formed in this way are *disaccharides*, an example being the joining of glucose and fructose to form *sucrose*. Other important disaccharides are lactose, which comprises one glucose molecule and one galactose, and maltose, which is formed from two glucose molecules. The bond which holds the two rings in a disaccharide together is known as a *glycosidic linkage*. Glucose and fructose are both found in fruits and honey. Glucose is somewhat less sweet to the taste than fructose, and neither of them is as sweet as sucrose, which is also known as *cane sugar*. *Lactose*, or *milk-sugar*, is not found in fruits, but it occurs in the milk of all mammals, human milk containing about 7% and cow's milk about 5%. *Maltose* is formed during the breakdown of starch and is important in the brewing industry, while ribose, deoxyribose and mannose occur in different parts of the biological cell, and will be discussed in a later chapter, as will the way in which glucose becomes a source of biological energy.

The joining process does not stop at two ring units. Three monosaccharides joined together produce a *trisaccharide*, such as *raffinose*, while a string of between four and ten units is referred to as an *oligosaccharide*. Beyond this number the structure is called a *polysaccharide*, and several major classes exist. The main component of plant cell walls is *cellulose*, which consists of unbranched chains of glucose units, while the *pectins* present in most fruits are highly branched chains of galactose and arabinose. Two other important structural substances in plants are *arabans* and *xylans*, which are built up from the *pentoses*, arabinose and xylose. *Starch*, the food store of the

Sugars

Carbohydrates, one of the four types of organic compounds essential to life, are a large group composed only of carbon, hydrogen, and oxygen. Their name stems from the original belief, later proved fallacious, that they are hydrates of carbon. They comprise the *monosaccharides* and *disaccharides*, both of which are *sugars*, and the *polysaccharides*. The latter are subdivided into *starch* and *cellulose*, which are formed in plants, and the animal starch known as *glycogen*. The most familiar member of the family is *glucose*. It was first isolated in 1747, from raisins, by Andreas Marggraf (1709–1782). The name *glucose* was coined in 1838, by Jean Dumas (1800–1884), and the chemical structure was discovered by Emil Fischer (1852–1919) around the end of the nineteenth century. Sugar molecules can exist in both chain and ring forms, Charles Tanret (1847–1917) having shown, in 1895, that the latter structure is favoured by thermodynamics except for the smallest varieties. The open chain form of glucose is shown at the right of the following picture

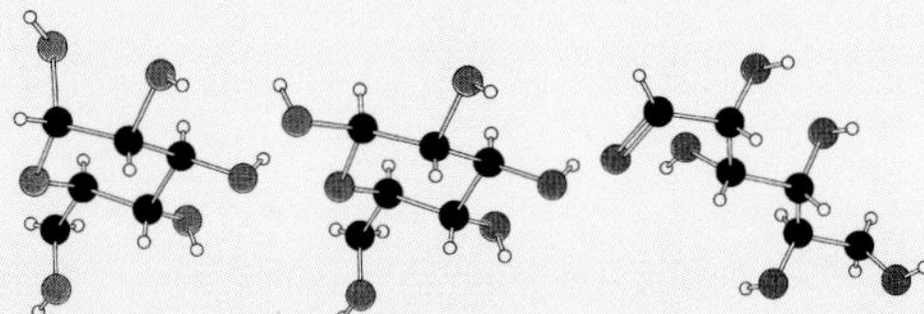

while the two *stereoisomers* of the ring configuration, *dextro*-glucose and *laevo*-glucose, are shown at the left and middle respectively. Although carbohydrates comprise only about one per cent of the typical animal body, they are the most abundant constituents of plant life, and several *plant sugars* are discussed in Chapter 18. One of the most common disaccharides of the animal domain is *lactose*, which occurs in the *milk* of all mammals. It comprises one molecule each of glucose and *galactose*, which are stereoisomers, as can be seen from their structural formulae. The laevo forms are as follows:

glucose *galactose*

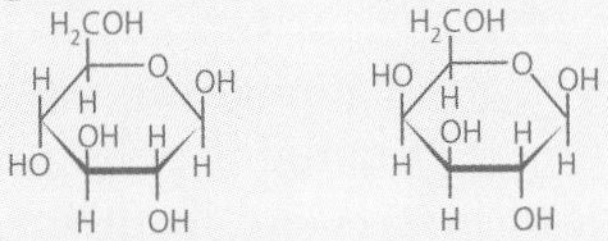

The two parts come together by shedding a total of one oxygen atom and two hydrogens, in the form of a water molecule. This is typical of the condensation type of polymerization (which will be discussed in Chapter 14), and the joining bridge is known as a *glycosidic link*. The *laevo* form of lactose is shown in the following picture:

laevo-lactose

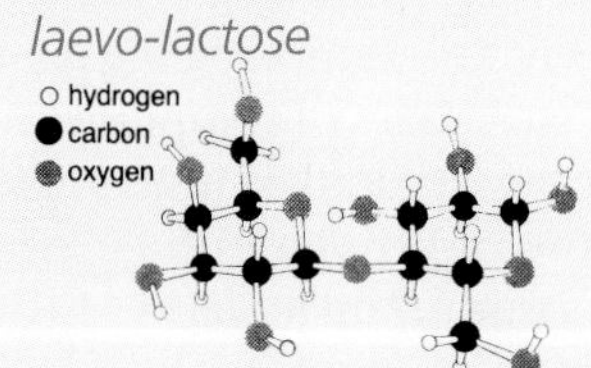

Glycogen, or *animal starch*, serves to store energy in living tissue. It is a polymer of glucose units. Unlike those other prominent biopolymers, proteins and nucleic acids, and indeed unlike most plant polysaccharides, glycogen displays considerable branching. In the following picture of a small part of a glycogen molecule, the arrows indicate joining paths to the rest of the structure.

glycogen (with branching)

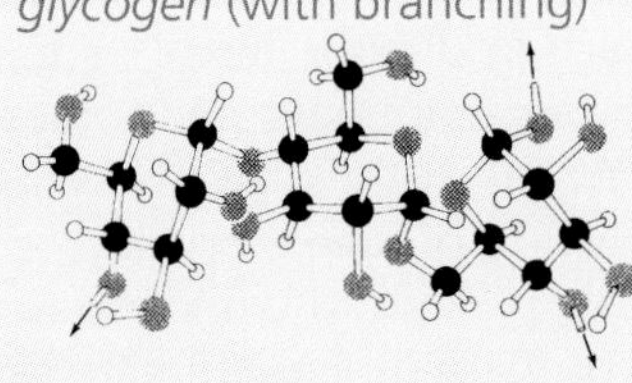

Purines and pyrimidines

Sugars are not the only *ring-shaped molecules* encountered in the biological realm. Other examples are found in the nucleic acids, the steroids, and in several molecules involved in energy conversion systems. The latter two groups are described in later chapters. *Nucleic acids* comprise sugar and phosphate units, together with two types of base: *pyrimidine* and *purine*. The pyrimidines have a single ring while the purines consist of a six-membered ring and a five-membered ring, which share a common carbon–carbon side. *Deoxyribonucleic acid*, generally the depository of the hereditary message, contains the pyrimidines *cytosine* and *thymine* and the purines *adenine* and *guanine*. In *ribonucleic acid*, which is most often a mediator of the genetic process, the pyrimidine thymine is replaced by another pyrimidine: *uracil*. Adenine is also a constituent of *adenosine triphosphate*, usually abbreviated to ATP, the ubiquitous energy transporter in biological systems. The structural formulae of the five bases are as follows:

adenine (purine) *guanine* (purine) 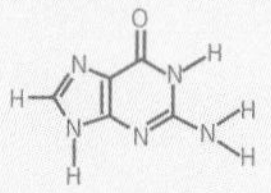 *cytosine* (pyrimidine)

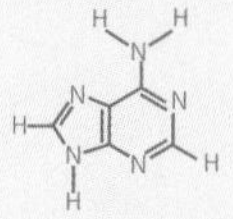

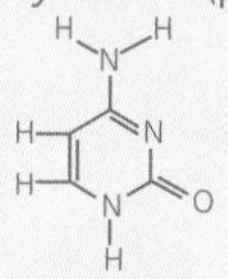

thymine (pyrimidine) 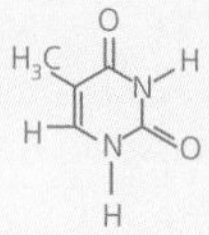 *uracil* (pyrimidine)

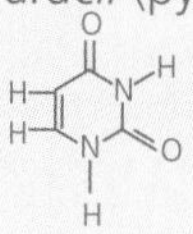

plant, consists of amylose, an unbranched chain of glucose units, plus amylopectin, in which the glucose chains are branched. *Glycogen*, which is the animal counterpart of starch, is similar to amylopectin, but it is even more branched, a dividing point occurring every 12 to 18 glucose units along any chain. Finally there is *chitin*, which is a nitrogen-containing polysaccharide that forms the basis of the horny exterior of insects and members of the Crustacea, such as the crab and the shrimp.

Organic chemistry gets excellent mileage out of its building blocks. All the compounds described thus far (except chitin) have involved atoms of just three elements, carbon, hydrogen and oxygen, and even so the choice has been highly selective. We now add a fourth, nitrogen. This element is usually found in association with hydrogen, an example being *ammonia*, NH_3, which is basic in the acid–base sense. Two classes of compound related to ammonia are the amides and amines. An *amide* is derived from an organic acid by replacing the hydroxyl portion of the carboxyl group by an *amino group*, NH_2. An *amine* is produced when this same group replaces a hydrogen atom in a hydrocarbon. The classic example of an amide is *carbamide*. It is in fact a special case because the starting material, *carbonic acid*, contains two hydroxyl units, and both of these are replaced by an amino group. The common name of carbamide is *urea*, and this was the substance enshrined by Wöhler as the first organic compound to be synthesized artificially. It is worth noting, in passing, that carbonic acid itself is a rather fundamental substance in that it is produced when carbon dioxide dissolves in water. Urea is found in the urine of all carnivorous mammals, and it represents the final stage in the decomposition of *proteins* by the digestive system. An adult human being excretes approximately 30 g of urea every day.

As examples of amines, we could hardly pick a more interesting group than the *amino acids*. There are about 20 of them in natural materials, and they lie at the very heart of the life processes. They take their name from the fact that they terminate at one end with a carboxyl group, and at the other with an amino group. These termini are acidic and basic, respectively. The coexistence of acid and base usually suggests the possibility of neutralization, but such a small linear molecule would have difficulty in biting its own tail, Kekulé fashion. This can in fact occur, the acidic part of the molecule reacting with the basic part to produce an *internal salt*, but it is relatively rare. However, there is the possibility that the acid end of one molecule could get together with the basic end of another, and it was Emil Fischer who first realized that the chains thus produced might be the basis of *protein structure*. Fischer's attempts to synthesize a protein failed, but he did succeed in producing *polypeptides* which were very similar to *peptones*, the primary decomposition products of proteins. Since each link in such a polypeptide can be any one of 20 different types, it is clear that these chains display great variety of composition and structure.

The words *amine* and *vitamin* have a historical link. In 1912, Frederick Hopkins found that a diet of purified casein, fat, carbohydrate and salts stunted the growth of young rats. When given a little milk, in addition, the creatures rapidly attained normal stature. Until this discovery, disease had invariably been attributed to the presence of an undesirable factor. Hopkins had demonstrated that trouble can also arise through the *absence* of something desirable: an accessory food factor, required in astonishingly small amounts. Around the same time, Casimir Funk postulated the existence of special substances, a deficiency of which could cause such afflictions as beri-beri, scurvy, pellagra and rickets, in humans. Believing them to be some sort of organic base, he called them *vital amines*, or vitamines. It was subsequently realized that they are not all amines, and the terminal letter was dropped, thus producing the present term vitamin. By 1915, especially through the efforts of E. V. McCollum and M. Davis, it was clear that these trace substances come in several varieties. A fat-soluble type, denoted by the letter *A*, was required to prevent *rickets* and other disorders, while the water-soluble *B* variety warded off *beri-beri*. These advances provided a rationalization of earlier observations on the occurrence and prevention of *scurvy*, the scourge responsible for greater loss of life amongst seamen than battle and shipwreck combined. In 1747, James Lind had discovered that a daily ration of lemon juice kept the disease at bay. Now, 170 years later, it became clear that citrus fruits too must contain a vitamin, designated by the letter *C*, and A. Harder and S. S. Zilva managed to produce highly concentrated samples of the active compound. In 1932, W. A. Waugh and C. G. King finally isolated the actual vitamin, and found it to be identical to the *hexuronic acid* extracted from cabbage by Albert Szent-Györgyi four years earlier. Together with W. N. Haworth, Szent-Györgyi succeeded in elucidating its chemical structure, and gave it the new name *ascorbic acid*. Numerous other vitamins have since been identified, and determination of their structures has often been followed by artificial synthesis.

As John Bernal once noted, it is the availability of materials which imposes the main limitation on the techniques of any age. In a developing society there is a constant demand for new substances, the ultimate source of which is frequently chemical synthesis. And the quest for innovation implies a severe test for a science. To win its spurs, it must not only explain the past; it must also correctly predict the future. Progress depends on advances in our understanding of the principles that govern the

Organic superconductors

The *superconductivity* displayed by some metallic crystals was explained in 1957 by John Bardeen (1908–1991), Leon Cooper and J. Robert Schrieffer. As described in Chapter 10, it depends upon a subtle attraction between pairs of electrons, which is mediated by the lattice of positive ions. This overcomes the inherent repulsion between two particles having the same electric charge, but the effect is weak and readily destroyed by thermal fluctuations if the temperature is more than a few degrees kelvin. Normal *electrical resistance* arises from the scattering of the electron waves by the *lattice vibrations*. In the superconducting state, the motions of all the *Cooper pairs* are mutually in phase (i.e. in step with one another) and a change in the momentum of one pair requires a compensating change in all the others. The energy needed to scatter a pair and create electrical resistance, in such a highly organized state, is much larger than the vibrational energy available in the lattice at low temperature. In 1964, William Little suggested that a quite different type of superconductivity might be possible in compounds comprising stacks of flat organic molecules. Noting that hydrocarbon molecules with loosely bound valence electrons are easily polarized, one part acquiring a positive charge while an adjacent part becomes negative, he postulated that this could lead to Cooper pair formation without the need for lattice distortion. Freed of the latter constraint, this new type of superconductivity should not be limited to very low temperatures, and Little speculated that the effect might even survive at or above room temperature for the most favourable compounds. That goal, which would have an enormous impact on electrical technology, has not yet been achieved. But the first organic superconductors have appeared. They consist primarily of stacks of tetramethyltetraselenafulvalene (TMTSF), the molecules of which are said to be *donors* in that they tend to give up their *valence electrons*. The molecular structure of TMTSF is:

H_3C Se Se CH_3
H_3C Se Se CH_3

Crystals are grown in which the TMTSF molecules are adjacent to *acceptor ions*, and they have paths of easy electrical conduction which are essentially molecular corridors. (See diagram below.) Along the favoured direction, an electron is scattered approximately once every 1000 molecules, which is ten to twenty times less frequently than in a typical metal. In fact, the superior conductivity of a metal is due only to its far more plentiful supply of conduction electrons. The pronounced directionality influences the way in which polarized light interacts with the surface of such a crystal. If the light is polarized parallel to the conducting direction, it is reflected and gives the crystal a characteristic metallic lustre: polarization in the transverse direction, in which the electrons are unable to move in response to the stimulating light

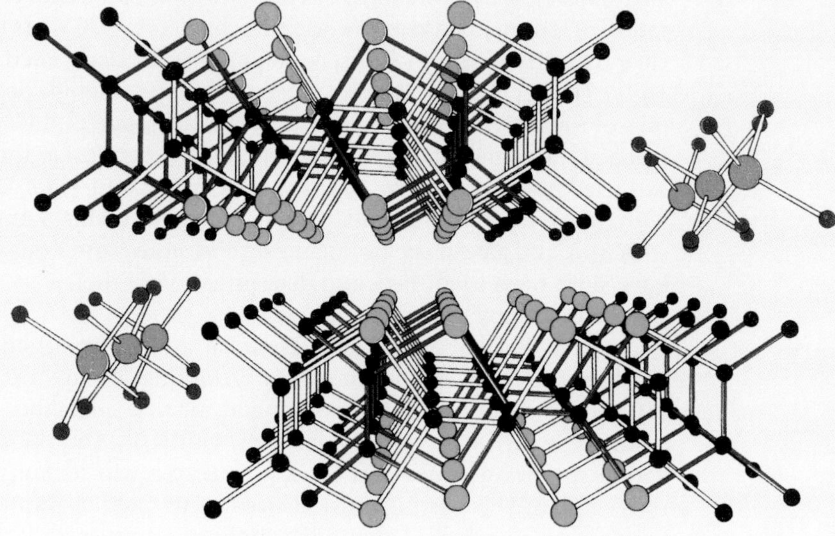

wave, produces a dull grey appearance. This is illustrated in two of the smaller pictures below (left and centre), which show a crystal comprising a tetracyanoplatinum complex with a structure similar to TMTSF. The third picture shows two crystals of tetrathiafulvalene-tetracyanoquinodimethane (TTF-TCNQ), the first organic compound shown to display metallic conductivity, mounted along mutually perpendicular directions. One of these directions gives parallelism between the plane of polarization of the incident light and the plane of easy electron motion, and thus produces opacity (i.e. zero transparency). In 1979, Klaus Bechgaard, Denis Jerome, Alain Mazaud and Michel Ribault reported that a compound based on TMTSF and PF_6 becomes superconducting at 0.9 degrees kelvin, at a pressure of 12 000 atmospheres. Just over a year later Bechgaard, working together with Kim Carneiro, Claus Jacobsen, Malte Olsen and Finn Berg Rasmussen, discovered that replacing PF_6 by ClO_4 produces a superconductor at atmospheric pressure, and again about 0.9 degrees kelvin. Future goals include verification that the Little mechanism is indeed being observed in these materials, now known as *Bechgaard salts*, and the development of compounds with higher superconducting transition temperatures.

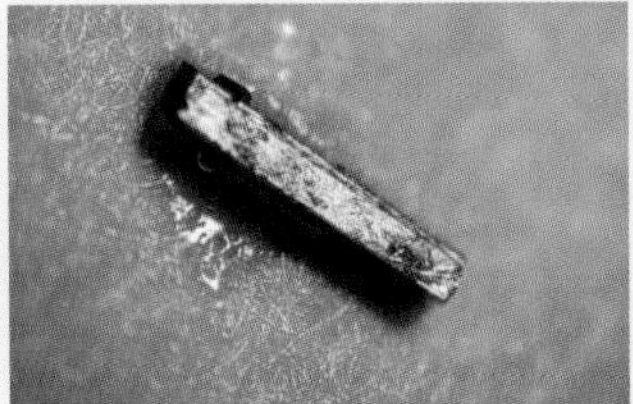

structure of molecules and their mutual interactions. The flat chemical formulae frequently found on the printed page are quite inadequate as a predictive aid; three-dimensional views are required to show how molecules shape up to do business with other molecules. The ball-and-stick models discussed earlier provide little help in this respect, and a quantum treatment of the electron wave states is mandatory. The electronic computer has played a vital role in this activity, and computer graphics technology has become a valuable tool in the design of new molecules and reaction routes (hence the term *computer-aided molecular design*). The modern organic chemist is rewarded for this theoretical effort by a harvest of novel compounds with properties and shapes that would earlier have been regarded as unattainable. In recent years, the growing inventory of artificial molecules has acquired such names as cubane, barrelene, basketane, prismane, tetrahedrane and propellane, none of which puts much burden on our powers of visual imagination. A recent addition to the list, by H. D. Martin and his colleagues, rejoices in the fanciful name pterodactyladiene; it comprises a strip of four four-sided rings, and its overall shape resembles a pair of outstretched wings. The advances in our understanding of reactions has permitted reduction, or even elimination, of the unwanted byproducts that have been a familiar frustration. It is thus appropriate that we close this chapter with a brief review of some of the events that have contributed to this revolution.

Oddly enough, one of the most significant advances involved compounds that do not formally qualify for inclusion in the organic family. These are the *boranes*, which contain only boron and hydrogen. Alfred Stock, had managed to prepare a number of such compounds, including $B_{10}H_{14}$, the molecular structure of which was found to resemble a basket. This was established by J. S. Kasper, C. M. Lucht and D. Harker in 1948. Now the boron atom has one less electron than that of carbon, and this appears to rule out hydrocarbon-like chain and ring structures. A clue was provided by Hugh Longuet-Higgins, who introduced the concept of *three-centre bonding*. This showed how a single hydrogen atom could provide a linking bridge between two borons even though only a single pair of electrons is involved. Following an exhaustive cataloguing of borane structures, William Lipscomb Jr. established the general result that the bonding in molecular frameworks is not, after all, limited by the *Lewis electron pair* concept described in Chapter 2. Together with Bryce Crawford Jr. and W. H. Eberhardt, he went on to show that a pair of electrons can also link three boron atoms, and this led to an explanation of even the most complex boron cage structures. This approach later showed why, in the *carboranes*, which are remarkably resistant to chemical and thermal stress, carbon atoms can enter into as many as five or six bonds.

In order to appreciate later theoretical advances we must consider certain experimental observations on molecular conformation, or structure. They were important because they showed that chemical reactions are critically dependent on molecular shape. A geometrical mismatch usually produces a dead end; it is like trying to put a left-hand glove on a right hand. The *stereochemical principle* is universal, as we should expect, and it provides the basis of our senses of *taste* and *smell*; molecules reaching the tongue or nose are discriminated according to their three-dimensional architecture. Some of the earliest studies were made by Odd Hassel in the 1930s, and he focussed attention on *cyclohexane*. Like benzene, this molecule comprises a ring of six carbon atoms, but it also has twelve rather than six hydrogens. It is thus saturated, and has no double or resonance bonds to keep the ring flat. It had long been realized that two non-planar forms of the molecule are possible: one shaped like a boat, and the other resembling a chair. Hassel probed the structure by the familiar approach of making it diffract the waves of a convenient radiation, in this case electrons. He found that one form can readily be flipped into the other, by thermal activation, and that the chair is the more stable. The significance of these results was appreciated by Derek Barton. In 1950 he showed how they implied a higher stability for certain *substituents*, which are new molecules formed by substituting one atom by that of a different element at certain positions. Barton went on to show how molecular conformation influences chemical reactivity. The door to more complicated molecules had been opened, and it became feasible to trace chemical pathways in their full three dimensional splendour. John Cornforth used radioactive labelling to follow the fate of specific atoms in enzyme-catalyzed reactions, and he was thereby able to give a surprisingly detailed picture of the stereochemistry of several compounds of biological significance. These included substances involved in the synthesis of *steroids*, a family to which *cholesterol* belongs. Access to the three-dimensional structure of a molecule implies that one can determine the *chirality*, which is what could be called its handedness. Vladimir Prelog explored this in a number of compounds comprising ring structures. Some of the rings contained nitrogen, while others were large and of importance to the *perfume industry*. Prelog was able to show how an atom or a group of atoms migrates across a ring, to a position four or five carbon atoms away.

The acquisition of stereochemical detail has led to sensitive control over chemical pathways. Georg Wittig synthesized alkenes by a method so tailored that a double bond could be introduced at a precisely specified point on the molecule. His techniques permit efficient production of a number of complex pharmaceutical compounds, including *vitamin A*, as well as artificial synthesis of insect *pheromones*. Natural versions of these latter substances are used by the creatures to attract members of the opposite sex, and the industrially manufactured versions are promising agents for pest control of specific species. The newly won expertise has also afforded improved production of such relatively prosaic compounds as alcohols, aldehydes, and ketones. Herbert Brown discovered new routes to these common substances, employing as intermediates the carboranes referred to earlier. The complexity of molecular architecture that can be synthesized in the test-tube has increased dramatically. Robert Woodward, in particular, showed how one can exercise stereochemical control over molecular conformation. In cases where a number of stereoisomers are possible, such control is vital if a mixture of wanted and unwanted molecules is to be avoided. Woodward developed methods for producing a single isomer at each step of the synthetic pathway, and his impressive list of artificial products included cholesterol, chlorophyll, the hormone cortisone, quinine and vitamin B_{12}.

Theories of molecular structure and reactions are taking the guesswork out of modern organic chemistry. They are based on *quantum mechanics*, of course, and much of the development work has made extensive use of the electronic computer. The fact that such calculations inevitably involve approximations is not a serious problem; the same is true of the digital approach to celestial mechanics, but this did not preclude Moon landings. To appreciate the motivation which led to these developments, we could not make a better choice than the following classic example: it is easy to combine the four-carbon chain of butadiene with the two-carbon chain of ethylene, to produce the six-carbon ring of cyclohexene, but two ethylenes do not readily interact to produce the four-carbon ring of cyclobutane. Both reactions are exothermic, involving the release of energy, so they should both be favoured thermodynamically, but the latter reaction has a prohibitively high energy barrier and other competing reactions intervene. In 1939, M. G. Evans pointed out the relationship between the transition state for the reacting four-carbon and two-carbon chains and the electronic structure of the six-carbon benzene ring. M. J. S. Dewar subsequently generalized this view into the idea of *orbital symmetry*, and it became clear that stability depends upon the *phases* of the interacting electron clouds, that is upon their ability to get into step with one another. As an example, we have the remarkable stability, first noticed by Erich Hückel, of cyclic molecules with backbones containing six atoms, ten atoms, fourteen atoms, and so on. There is an analogous rule for formal dinner parties at which the sexes must be alternated around the table and the host and hostess are to sit at opposite ends of the table; this can be achieved with six people, ten people, fourteen people, and so on, but four, eight or twelve people cannot be accommodated. Other stimulation came from observations on reactions which produced the reverse result, namely the opening of closed rings. In 1958, E. Vogel had noted that thermally induced ring opening is spatially specific, and that four-membered rings behave differently from those comprising six members.

In Chapter 10, we saw how electrons in a semiconductor can be transferred from one energy level to another. A similar transfer occurs between the electron clouds when two molecules interact, and this produces the *molecular orbital* that stabilizes the final atomic arrangement. But the question arises as to which of the several possible orbitals are actually involved. This problem was addressed by Kenichi Fukui in the mid 1950s, and he found that the most important reactions occur between the highest occupied orbital of the molecule which passes on some of its electrons, and the lowest unoccupied orbital of the molecule which receives those electrons. By analogy with their semiconductor counterparts, the two molecules are referred to as the *donor* and the *acceptor* respectively, and the two types of molecular orbital are known by the obvious acronyms *HOMO* and *LUMO*. Fukui called these the *frontier orbitals*, and he showed that the favoured reaction sites on molecules are the places where these orbitals have the largest amplitudes. Fukui and Roald Hoffmann independently discovered that the course of many reactions could be explained on the basis of the symmetry properties of the corresponding frontier orbitals, and that this principle can be applied without the need for detailed calculations of the electron states. We have thus come full circle, back to what was anticipated in the first two chapters: the essential features of chemical reactions can be traced to the nodal properties of orbitals, and their associated symmetries. Woodward and Hoffmann went on to discover the concept of the *concerted reaction*. If the electron clouds are rearranged in an uncoordinated fashion, old bonds being broken before new ones are formed, the reaction must pay the price of a higher free energy, and this means a lower probability. A far more favourable situation arises if all bonds are broken and reformed simultaneously, but this puts certain constraints on the symmetries of the participating orbitals. As Woodward and Hoffmann put it, the central concept of the principle of orbital symmetry conservation lies in the incontrovertible proposition that a chemical reaction will proceed the more readily, the greater the degree of bonding that can be maintained throughout the transformation.

Summary

Organic chemistry is the science of the compounds of carbon, the special properties of that element deriving from its high valence and preference for covalent bonding. Organic compounds tend to have low melting and boiling points, and therefore usually exist as liquids or gases at room temperature. Usually inflammable, but nevertheless slow to react, they frequently exhibit isomerism, a source of considerable structural variety. Similar carbon arrangements and terminal groups produce similar physical and chemical properties, and families of compounds, having the same general skeleton of atoms, are common. Examples are the alkanes, or paraffins, the alcohols, and the ethers. Interesting physical and chemical behaviour is associated with the molecular structures of the important fatty acids, one of which is related to soap. Aromatic compounds are distinguished by their cyclic structures, in which carbon atoms form a closed ring, as in benzene. Rings also occur in many of the sugars, the heavier members having several such arrangements. Three-dimensional networks of carbon-atom rings are the basis of the fullerenes, the simplest of which resemble footballs. The amino acids, with their combination of acidic and basic terminal groups, can link up to form long chains. Because a common amino acid can have one of 20 different side groups, these chains, the basis of proteins, show great variety. Progress in the determination and understanding of the three-dimensional structure and bonding of organic molecules has led to new organic compounds and new synthetic pathways.

13

There is no excellent beauty that hath not some strangeness in the proportion.

Francis Bacon (1561–1626) ***Of Beauty***

Strangeness in proportion: *liquid crystals*

Most substances can exist in three different states: solid, liquid and gas. The temperature, pressure and density collectively determine which form is adopted, and changes in the imposed conditions can produce the melting and boiling phase transitions. The solid state is usually crystalline, but some solids have the meta-stable glass structure. Differences between the three fundamental states are often depicted by simple diagrams in which atoms are represented by circles. For a gas, the circles are drawn randomly, with the distance between neighbouring circles somewhat larger than the circle diameter. The condensed states, crystal and liquid, are illustrated by arrangements in which the distances between the centres of neighbouring circles are comparable to the diameters, the difference between these forms lying mainly in the regular arrangement of the former and the relative randomness of the latter. We can go a long way with such models because real atoms are indeed roughly spherical. This is particularly true for noble gases such as argon and neon, because of their closed electron shells, but it is a reasonable approximation for all atoms. The question arises as to what happens if the atoms are replaced by molecules elongated in one direction. The answer is to be found in that peculiar intermediate state of matter: the *liquid crystal*.

A description of arrangements in terms of positions of centres of gravity is inadequate for elongated molecules. We require, in addition, knowledge of the orientations. Given this increase in the amount of information needed to describe the degree of order, we might speculate on the effects of varying position and orientation independently. We could keep all the long axes parallel, but permit the centres of gravity to disorder. Conversely, the latter could form a crystal lattice while the directions become disordered. Then again, both positions and orientations could simultaneously adopt varying degrees of disorder. It becomes apparent that many of the possible arrangements of elongated molecules cannot be fitted into the above classification of solid, liquid and gas. We must conclude, therefore, that there are more than just the three states of matter. The liquid crystal is a law unto itself; different from gas, liquid and crystal, but just as fundamental.

The existence of liquid crystalline phases was discovered in 1888, by Friedrich Reinitzer. His first observations were made during investigations of *cholesteryl benzoate*, the organic molecules of which are approximately eight times as long as they are wide. He observed that the solid state of this compound melts at 145 °C to give a liquid which, just above the melting point, is blue and quite cloudy. Raising the temperature further, he discovered that this cloudiness disappears quite sharply at 179 °C, to give a

Liquid crystal chemistry

The classic example of a mesogen is the compound 4,4′-dimethoxyazoxybenzene. The alternative chemical name is *p*-azoxyanisole, and it is usually referred to by the acronym PAA. Its structure is

p-azoxyanisole

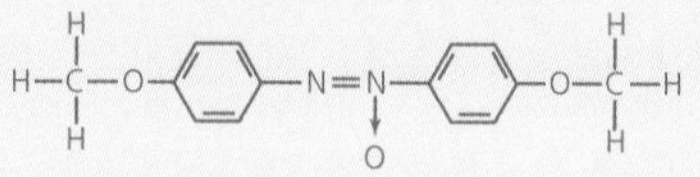

the arrow indicating affinity between a nitrogen atom and a nearby oxygen, because of the lone pair of electrons on the former. From a mechanical point of view, this molecule is essentially a rigid rod two nanometres long and half a nanometre wide. The two benzene rings are nearly coplanar. This compound has a crystal-nematic transition at 118.2 °C, and the nematic phase is transformed to the isotropic liquid at 135.3 °C. Mesogens displaying smectic behaviour are invariably longer than PAA. An example is ethyl *p*-azoxycinnamate

ethyl p-azoxycinnamate

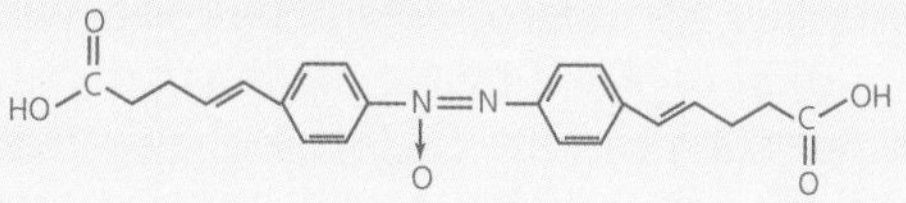

It is transformed from the crystal to the smectic A configuration at 140 °C, and the transition to the isotropic liquid occurs at 250 °C. This substance has historical significance. It was discovered in 1925 by Edmond Friedel (1895–1972) and used by him in the first demonstration of the layer nature of a smectic, which he accomplished by X-ray scattering. A similar experiment revealing the molecular arrangement in a nematic was first reported eight years later by John Bernal (1901–1971) and Dorothy Crowfoot (later Dorothy Hodgkin, 1910–1994). An appropriate example of a mesogen showing a cholesteric transition was the subject of the original study by Friedrich Reinitzer (1857–1927), *cholesteryl benzoate*. Its structure is

cholesteryl benzoate

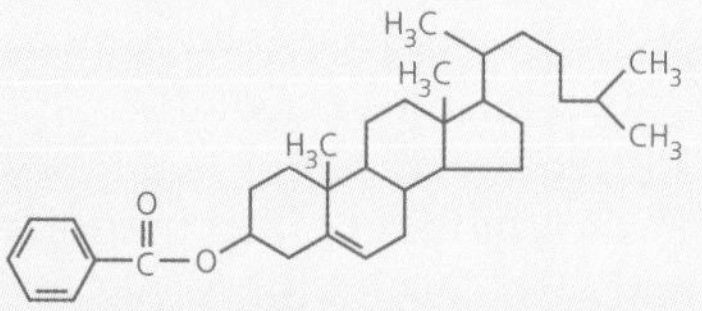

It should be borne in mind that such two-dimensional representations do inadequate justice to the actual three-dimensional structure.

transparent liquid. Reinitzer soon established that the appearance of the cloudy phase is a reversible phenomenon, but he was unable to shed any light on the physical origins of the cloudiness. Otto Lehmann succeeded in doing so, by studying specimens of the compound with the aid of *polarized light*. His experiments soon led him to conclude that the cloudiness stems from the elongated shape of the molecules, and that it is related to the degree of *orientational order*. The unusual combination of order and fluidity prompted him to name the new state crystalline liquid, although he later changed his mind and decided to call them liquid crystals. This change of heart is itself rather instructive, because it underlines the multiplicity of characteristics and the choice of emphasis.

One of the most important types of biological structure is the *lipid bilayer membrane*. It provides the enveloping layer for each cell, separating the latter's components from the exterior aqueous medium. This vital cellular 'skin' is a special example of a state of matter that lies between liquid and crystal: the *liquid crystal*. The molecules of liquid crystal *mesogens*, as they are called, invariably have elongated shapes. The bubble-like objects shown here are closed surfaces of lipid bilayer membrane, which were produced by hydration of lipids at low ionic strength. Such structures, which serve as experimental model cells devoid of their internal machinery, are known as *vesicles*.

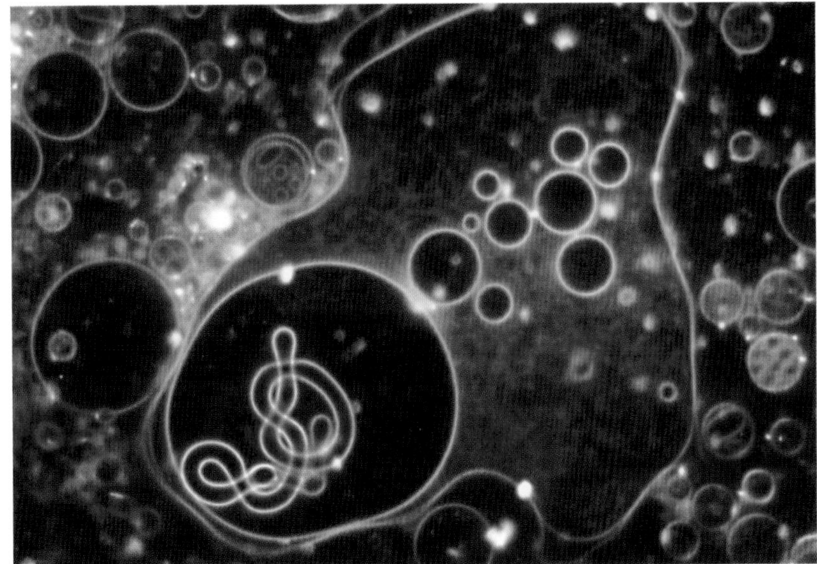

We should pause to consider why Lehmann concluded that a new state of matter had been discovered. A liquid composed of roughly spherical molecules permits the passage of light to an equal extent in all directions. This is so because a light beam travels through the same proportions of matter and free space, independent of its direction. But in an assembly of elongated molecules, with their axes all lying parallel, this proportion will depend on the direction of the beam. The amount of matter traversed determines what is known as the *refractive index*. As defined in Chapter 11, this is the ratio of the speed of light in a vacuum to its speed in the material in question. Substances for which the refractive index depends on direction are said to be birefringent, and numerous examples of *birefringence* in crystals have been discovered. Birefringent materials have an interesting influence on the polarization of light, that is to say on the plane in which the light wave can be thought of as oscillating. When a light beam passes through such a material, the plane of oscillation is turned through an angle that is proportional to the distance traversed. Observations on the polarization of light can readily be made with the aid of *Polaroid* films, of the type used in some sunglasses. This material permits the passage of light with only one particular plane of oscillation. If two such films are so arranged that their planes of accepted transmission lie at right angles, no light will be able to pass through both. If, however, birefringent material is placed between these two films, it will twist the plane of oscillation of the light emerging from the first film so that in general there will no longer be complete cancellation at the second, and some of the light will succeed in getting through. The polarized light technique has now become a standard tool in the observation of liquid crystals.

The task of classifying liquid crystals owes much to Georges Friedel, who in 1922 identified the three main types of arrangement. These have been given the names nematic, which comes from the Greek word *nematos* meaning thread; cholesteric, after the compound cholesterol, one of the first cholesterics to be identified as such; and smectic, invoking the Greek word *smectos*, which means soap. In a *nematic liquid crystal* the long axes of all the molecules are parallel, but the centres of gravity show no order, apart

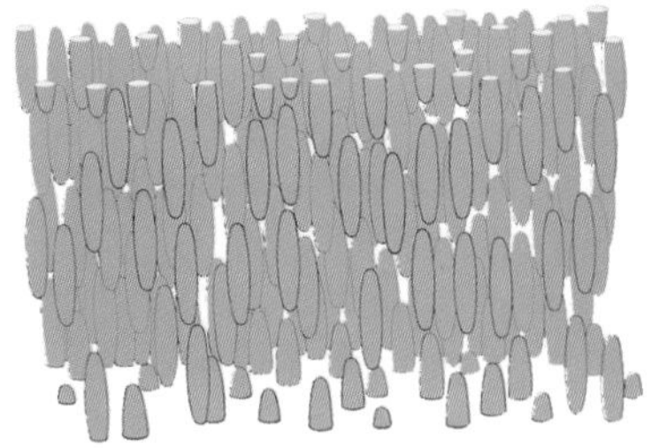
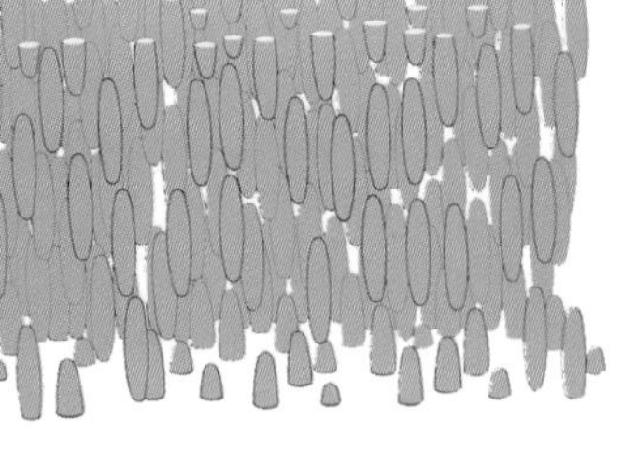
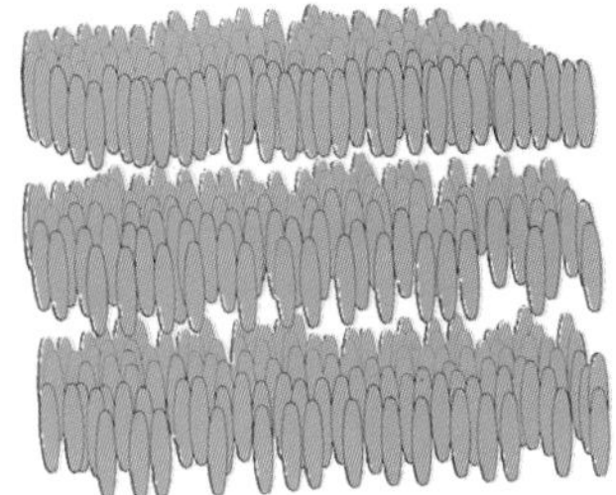
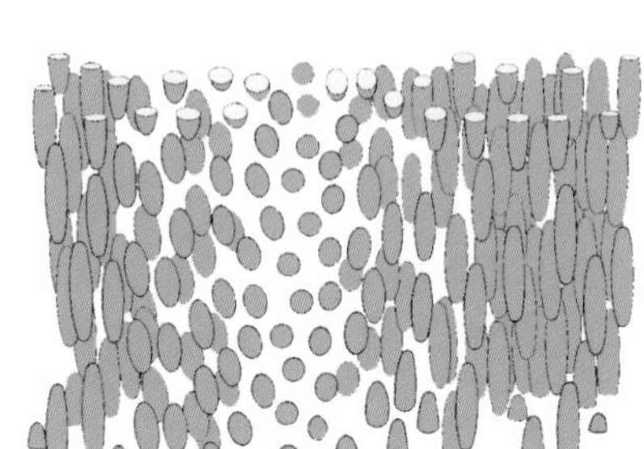

The three most familiar liquid crystal structures are the nematic, smectic, and cholesteric. In the *nematic* (top left), the elongated molecules all point in approximately the same direction, but their positions are otherwise disordered, rather like a travelling shoal of fish. The *smectic* form is characterized by its layer arrangement; the molecules resemble bottles on a series of shelves. The top right picture illustrates the smectic A type in which the molecules lie at right angles to the layers; several other smectic types have been identified. The *cholesteric* structure (bottom left) can be regarded as a nematic in which the director, or predominating direction of the molecular axes, gradually twists screw-fashion, as indicated in the bottom right picture. These pictures show only a small region of each structure and inevitably fail to illustrate the long-range features that arise from the deformability of liquid crystals. Thus the director in a nematic usually varies from place to place, giving the overall structure the appearance, when viewed through crossed polarizers, of twisting threads. Similarly, the layers of a smectic are readily bent and this gives the material a characteristically grainy appearance. The picture of the cholesteric structure grossly exaggerates the rate of twist; it shows just half the distance referred to as the pitch, and the actual pitch of many cholesterics is comparable to the wavelengths of visible light.

from that imposed by the condition that two molecules cannot physically interpenetrate. When we use the word parallel to describe the arrangement of axes, this is not meant in the rigorously mathematical sense because there must be a certain amount of thermal motion. A macroscopic analogy to the nematic configuration is a shoal of fish all swimming in the same direction. The *cholesteric* arrangement can be characterized by considering the molecular positions in a set of imaginary planes, equally spaced and oriented so as to contain the long axes of the molecules. The centres of gravity of the molecules in any one plane are disordered, but there is a preferred axial direction which varies in a systematic fashion from plane to plane. Between any two adjacent planes the preferred direction, known as the *director*, undergoes a rotation the magnitude of which is a characteristic of the material. After a certain number of planes have been traversed, the net rotation must reach 360°, and the director will again point in the original direction. The arrangement thus has *screw symmetry*, the distance between parallel directors being the *pitch*. It must be emphasized that the characterizing planes are *imaginary*; the molecules in a cholesteric liquid crystal are not actually arranged in planes, and the structure is essentially like a twisted nematic. In the *smectic* phase the molecules do lie in well-defined planes, rather like bottles standing on rows of equally spaced shelves. The directors in the various planes are usually parallel, though not always so, and each makes an angle to its plane that is characteristic of the material. The angle is frequently about 90°, but this is not necessarily the case. Indeed, for some materials this angle takes on different values depending on such physical conditions as temperature and pressure.

The three main types of liquid crystalline configuration are not independent. A nematic can be looked upon as a cholesteric in which the pitch has become infinite. Conversely, external stresses on a nematic can induce a cholesteric-like twist. Then again, if the density in a nematic is subjected to periodic variation, as occurs under the influence of longitudinal wave motion imposed parallel to the director, the structure is pushed towards the smectic configuration. The smectic arrangement is indeed equivalent to a

Multiphase mesogens

Many compounds adopt more than one liquid crystal arrangement, depending on the prevailing conditions. A typical example of a mesogen displaying both smectic and nematic modifications is the compound which rejoices in the name

N,N′-terephthalylidene-bis-(4-n-butylaniline), often mercifully shortened to TBBA. Its structure is as follows

N,N′-terephthalylidene-bis-(4-n-butylaniline)

This substance changes from crystal to *smectic B* at 112.5°C, and a transition to the *smectic C* form follows at 144.0 °C. When the temperature is raised above 172.5 °C, the *smectic A* structure is observed, while further heating to 198.5 °C produces the nematic structure. There is a final transition to the isotropic liquid at 235.5 °C. It is instructive to compare the behaviour of TBBA with that of the compound whose structure is as follows

bis-(p-6-methyloctyloxybenzylidene)-2-chloro-1,4-phenylenediamine

The two are superficially similar, but an obvious difference lies in the addition of a single chlorine atom, with an inevitable introduction of asymmetry. In the latter, initial melting of the crystalline phase produces a smectic C configuration, at 29 °C. But in this case further warming produces a cholesteric rather than a nematic, the transition occurring at 94.4 °C. The final transition to the isotropic liquid lies at 146.5 °C.

nematic modulated with a wavelength equal to the lengths of the individual molecules.

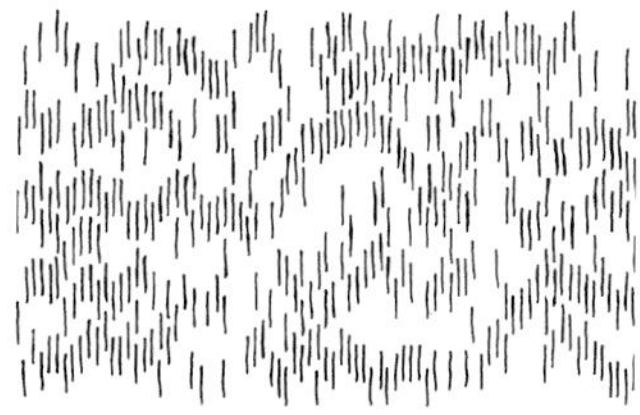

The various liquid crystal structures are not really independent. The cholesteric is like a *twisted nematic*, while the smectic A form is equivalent to a nematic with density modulation along the director. When a nematic nears the temperature at which it transforms to the smectic form, *density waves* appear as shown in this schematic picture, their amplitude increasing as the transition is approached.

Before considering further the different liquid crystal forms and the physical origins of their occurrence, it is necessary to introduce some terminology. Any substance which in a certain range of temperature exists as a liquid crystal is referred to as a *mesogen*. Liquid crystal phases are also referred to as *mesophases*, the two terms being synonymous. Until now we have considered only what are known as *thermotropic liquid crystals*, the name indicating that the liquid crystalline phase occurs when the temperature of the mesogen lies in the appropriate range. There is a second type, the *lyotropic liquid crystal*, which occurs only in mixed systems and in the appropriate range of composition. *Soap* has given its name to one type of liquid crystal configuration, and the soap–water system provides a good example of a lyotropic mesophase. A mixture of soap and water leads to the formation of the familiar slippery film when the concentration lies in the correct range. If there is either too much or too little soap, such films are not formed.

Liquid crystals are beautiful and mysterious, occupying a peculiar twilight zone between crystalline solid and isotropic liquid, the physical properties of the latter being, by definition, independent of direction. But they are not particularly rare. It turns out that almost one per cent of all organic compounds are mesogens, so liquid crystallinity is not one of Nature's oddities. However, many mesogens have transitions lying well removed from room temperature. Liquid crystals are particularly common in the biological

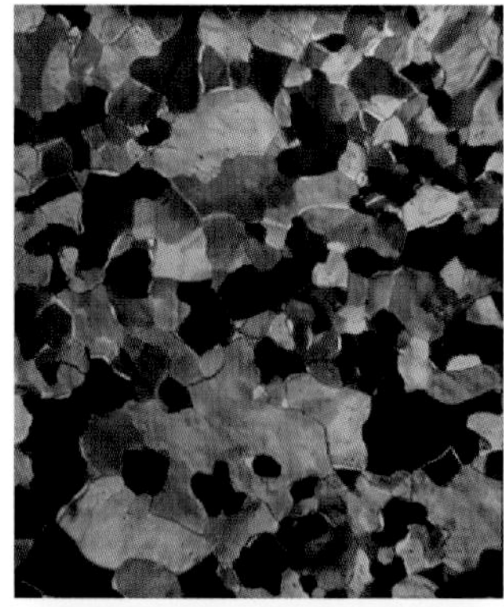

The layered arrangements in smectic liquid crystals give these a characteristic grainy appearance, usually referred to as a *mosaic texture*. The layers are quite flexible, and the *director* can display large variations over distances of the order of micrometres. Viewing through crossed polarizers gives preferential enhancement of colour for particular director orientations. The upper picture shows the grain structure typical of the *smectic* C structure, while the molecular layers of the *smectic* A, shown below, have been twisted into parabolic curves.

realm. Every *living cell* is bounded by a sort of molecular skin that is in this state: a *lipid double layer* that functions both as a containing bag and a selectively permeable gate. The *brain* in particular is rich in mesogens, these accounting for about 30% of the non-aqueous material. Several diseases are related to liquid crystal transitions, including such serious disorders as *arteriosclerosis* and *sickle cell anaemia*. In the latter, a mutation in the haemoglobin causes the *red blood cell membrane* to congeal, and lose some of its flexibility. This leads to some loss of water from the interior, causing the cell to shrink and adopt the characteristic sickle shape after which the disease is named. In the insect domain, liquid crystals often supply an attractive splash of iridescent colour, as in the wings of certain beetles. Not all insect colour is attributable to liquid crystals, however, much of it resulting from pigmentation or to simple interference effects.

The existence of liquid crystalline phases is related to mesogen chemistry. The prerequisite of elongated molecular shape is suggestive of the organic realm. Moreover, the units must be relatively stiff so we should expect mesogens to be reasonably rich in double bonds. Many of the most common examples are further stiffened by ring units. The forces between the elongated molecules are usually complex, and we must differentiate between lateral and terminal components. When these are not roughly equal in strength, thermal agitation can overcome the effects of one type while leaving the other intact. If the terminal forces are strong, for example, the lateral regularity of the crystal will be disrupted first, producing a nematic phase. With strong lateral forces, on the other hand, the smectic phase will be favoured, and the nematic arrangement will be seen only at a higher temperature, or not at all. In view of the great variety seen in organic chemistry, we should expect considerable variation in the detailed nature of the lateral forces. This is attested to by the multiplicity of smectic types. No fewer than eight distinct forms have been reported, and they do not exhaust the possibilities. They differ from one another in the angle that the director makes with the plane of the layer, and also in the arrangement of molecules within the layers.

The origin of the helical (screw-like) structure in *cholesterics* is intriguing. An early clue was provided by Georges Friedel, who found that addition of an optically active compound, such as lactic acid, can convert a nematic into a cholesteric. *Optical activity* is a consequence of what is known as *chirality*, a term used to indicate lopsidedness. A chiral molecule cannot be superimposed on its mirror image. It was subsequently found that all cholesteric liquid crystals are optically active. This implies that it is the asymmetric forces between chiral molecules which are responsible for the helical arrangement, and it shows why no single compound displays both nematic and cholesteric behaviour. Each of these configurations is observed separately in conjunction with the smectic arrangement, but they are never both associated with the same mesogen. A molecule either has or does not have chirality. It was earlier noted that the nematic can be looked upon as a cholesteric with infinite pitch; it is the logical extreme for the limiting case of zero chirality. Chirality is usually imposed upon a molecule by the presence of an asymmetrically appended atom or group of atoms. It is worth reiterating that a cholesteric is not a layered structure; the helical twisting occurs in a continuous manner.

Liquid crystal transitions are detected by the associated energies of transformation, some of which are rather small. The energy per molecule for a

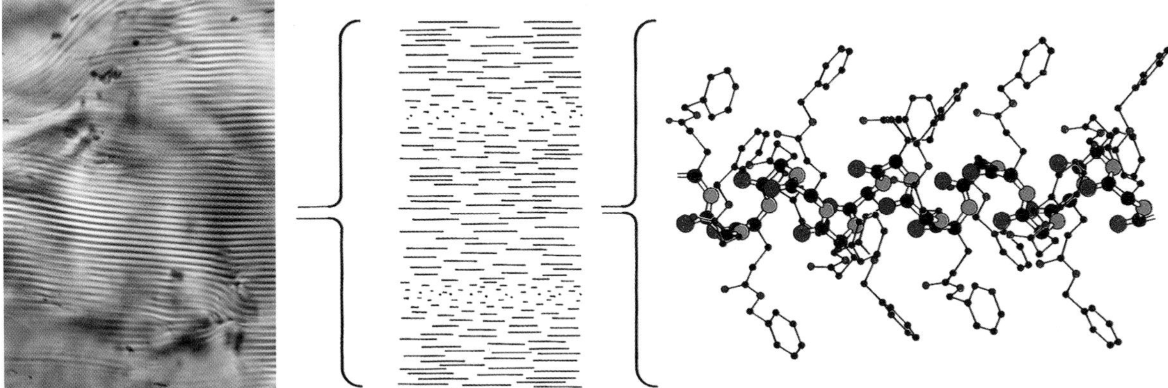

This composite picture illustrates several significant scales of distance in a *cholesteric liquid crystal*. Observation through crossed polarizers reveals a pattern of lines which arises from the molecular arrangement; the director gradually spirals up through the structure, corkscrew fashion, and each line corresponds to the situation in which the molecules are lying at right angles to the viewing direction. Three adjacent lines on the photograph define the pitch, as indicated by the middle diagram. The *lyotropic liquid crystal* consists of a 20% solution of the polymer polybenzyl-L-glutamate in dioxane. The polymer is based on the alpha-helix form of polypeptide, which is a common structural feature of *proteins*. This feature is emphasized by the exaggerated size of the backbone (right). (The standard colour convention is black for carbon, red for oxygen, and blue for nitrogen.) The width of the photograph is equivalent to about one millimetre. The cholesteric pitch is about a tenth of this, while the width of each polymer strand is about two nanometres.

typical transition from the nematic to the isotropic liquid, for example, is less than half that of the hydrogen bond. The first theory of liquid crystal transitions was put forward by Max Born in 1916. He treated the medium as an assembly of permanent electric dipoles and demonstrated the possibility of a transition from an isotropic phase to an anisotropic arrangement as the temperature is lowered.

A more realistic model was developed by Wilhelm Maier and Alfred Saupe in the late 1950s, following a suggestion made by F. Grandjean in 1917. They differentiated between the *repulsive forces* that prevent the molecules from overlapping one another and the so-called *dispersion forces* that arise because each molecule is able to induce electrical polarization of its neighbours. Maier and Saupe suggested that the latter are not at all negligible, a view endorsed by the experiments of J. Dewar and N. Goldberg, who probed the two factors independently. The Maier–Saupe model pictures each molecule as being located in a cylinder formed by its neighbours, and these impose both the repulsive and dispersive forces referred to above. By calculating the free energy as a function of the degree of order, Maier and Saupe were able to predict the transition temperature and obtained good agreement with experiment. Their theory nevertheless suffered from an obvious shortcoming; by treating the molecules individually it precluded investigation of correlation between the alignments of adjacent molecules. Even though long-range order is lacking in the isotropic liquid, a certain degree of nematic-like short-range order can be shown to persist just above the transition temperature. In 1937, Lev Landau had constructed a theory of transitions in terms of the degree of order, differentiating between cases in which this quantity disappears gradually or abruptly. Pierre de Gennes realized that the Landau approach is advantageous for liquid crystals in that it permits inclusion of spatial variations in order. He used it to predict the occurrence of fluctuations, the amplitude of which increases as the transition temperature is approached. His ideas were confirmed by the light scattering experiments of N. A. Clark, J. D. Litster and T. W. Stinson, in 1972.

There are three different ways in which a liquid crystal configuration can be deformed. They are characterized by reference to the director, the deformations being *bend*, *twist* and *splay*. The first two of these are reasonably straightforward, a twisted nematic taking on a configuration reminiscent of a cholesteric, for example. Splay is a fanning out and can be visualized by comparing the fingers of one's hand when mutually touching and

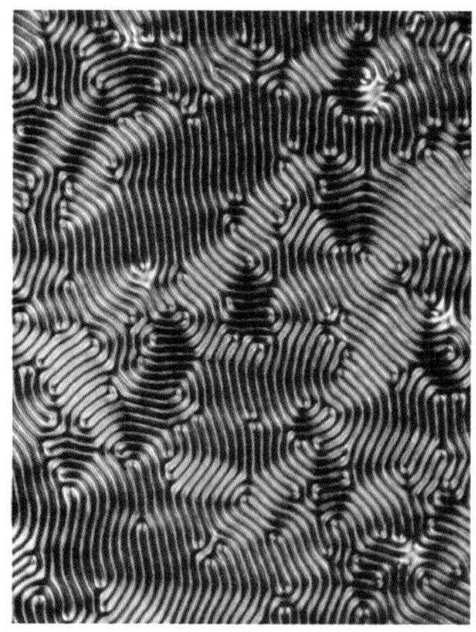

When a *cholesteric liquid crystal* is sandwiched between two parallel glass plates it will be forced to split up into zones of various director orientation unless the director is precisely parallel to the plane of the confining surfaces. The liquid crystal shown here is a mixture of methoxybenzylidene-n-butylaniline (MBBA) and the glue known as Canada balsam. The latter is not itself a liquid crystal but it possesses a pronounced *chirality*, or structural lopsidedness (handedness), and in sufficiently high concentration it converts the MBBA to a cholesteric. The pitch is about one micrometre, and is thus easily resolved by an optical microscope. The director does not lie parallel to the glass microscope slides, between which the material is held, and a complicated zone pattern has been produced.

when stretched apart. The force required to produce bending is considerably higher than for the other two deformations. Bend, twist and splay have it in common that the director changes direction in a continuous manner. Discontinuous changes are also possible, and were originally studied by F. Grandjean and Georges Friedel. Friedel referred to them as *noyaux*. These sudden disruptions to the molecular alignment were named *disclinations* by Charles Frank, in 1958, and they give rise to fascinating optical effects when viewed under crossed-polarization conditions.

Disclinations became the subject of intense study in the 1960s and 1970s, particularly by the Parisian school of Y. Bouligand, Jacques Friedel, M. Kléman, C. Williams and their colleagues. The defects have been given classifications that are consistent with those used for the crystalline imperfections, and we encounter familiar categories such as point and line disclinations. Although their centres always correspond to an abrupt change in direction, the long-range effect is such as to produce one of the above gradual deformations. Defects arise in pure liquid crystals, but a heterogeneous example provides an easier introduction. Let us consider an impurity particle in a nematic. If it has greater affinity for the ends of the molecules, some of these will attach themselves to it and stick out like the spikes of a sea urchin, or the quills of a rolled-up hedgehog. The director will be radially pointed toward the impurity, at all points, and the deformation will be purely of the splay type. Homogeneous defects occur as the reasonably long-lived remnants of mechanical disturbance. It is the tendency of a nematic to contain line disclinations that gives this type of liquid crystal the thread-like appearance after which it is named.

Particularly useful colour effects are seen in some cholesterics, variations being observed with changes of temperature and pressure. Because the structure is periodic along the direction of the helix axis, *Bragg interference* is possible for suitable wavelengths, just as in crystal structure determination. For the latter, the characteristic periodicities are comparable to atomic dimensions, so X-ray wavelengths are required. Typical cholesterics, on the other hand, have pitches of several hundred atomic diameters, so interference is obtained with wavelengths in the visible region of the electromagnetic spectrum. Strong Bragg reflection in the blue was the reason for Reinitzer's original observation of that colour in cholesteryl benzoate. The helical pitch of a cholesteric is determined by the relatively weak anisotropy of the intermolecular potential imposed by the chirality. It is critically dependent on purity and the prevailing physical conditions, and these same factors thus determine the observed colour. A mixture of equal parts of cholesteryl oleyl carbonate and cholesteryl non-anoate, for example, changes from red to blue when the temperature increases by just 3 °C. The reversal of the usual temperature trend should be noted; in this case we have red for cold, blue for hot. Because bending is relatively difficult, the main result of the increased availability of activation energy, as the temperature rises, is an increase in the amount of twist. This shortens the pitch, and thus decreases the wavelength most strongly reflected. The marked sensitivity to temperature change has led to several diagnostic applications. Exploitation began with J. L. Fergason's pioneering efforts in the early 1960s. In screening for breast cancers, for example, a cholesteric is smeared over the skin, and one detects the small increase in temperature that is characteristic of *tumour growth*. Other applications include detection of welding flaws, incipient fatigue fracture and malfunction in microelectronic circuits.

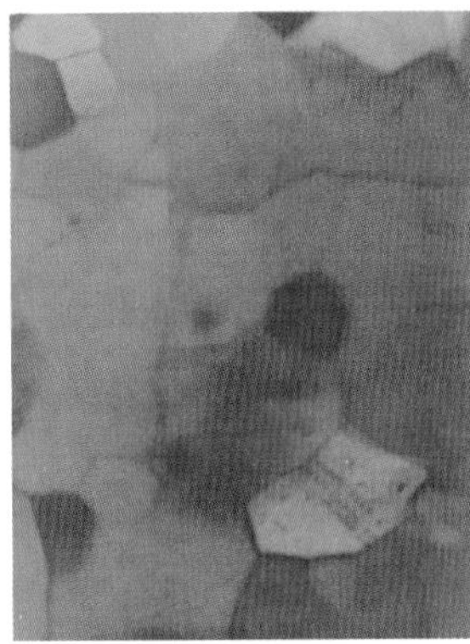
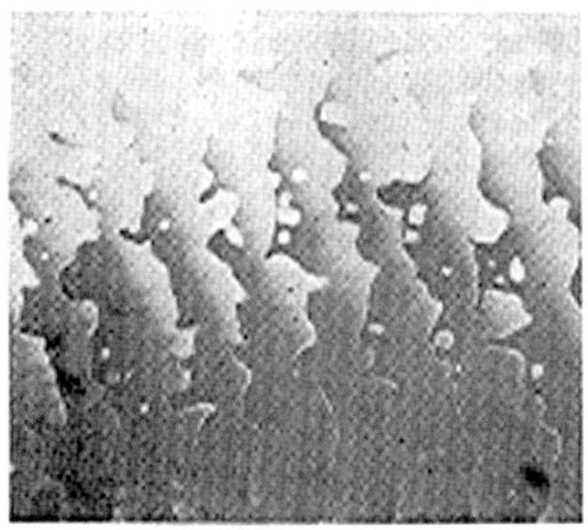

The liquid crystal structure which most closely resembles a normal crystal is the beautiful *cholesteric blue phase*. It shows many of the features often seen in true crystals, such as *grain boundaries* and *dislocations*. Arrays of the latter type of defect are faintly visible in the green-coloured grain near the middle of the picture. The origin of the blue phase structure is still uncertain. The black and white picture on the right shows a wedge-shaped specimen of the same blue phase, this geometry necessarily introducing a regular array of dislocations. A special observational technique was used in which the wavelength of the transmitted light varies along a particular direction; in this picture, it increases from bottom to top.

Television screens thin enough to be hung on a wall and *computer screens* well suited for portable machines are now routinely based on *liquid crystal displays*. The technique initially faced technical problems because of the prohibitively long times for molecular rearrangements, but that difficulty has long since been overcome. The experimental screen reproduced here, at its actual size, contained about twenty thousand twisted-nematic picture elements. It was one of the first such screens ever produced.

One of the major uses of liquid crystals has been in *display devices*. These frequently involve a nematic sandwiched between parallel glass plates made simultaneously transparent and electrically conducting by coating with tin oxide. For a typical sandwich spacing of 2 μm, an applied electrical potential of a few volts gives a field strong enough to cause both molecular alignment and impurity ion motion. In the so-called *dynamic scattering display*, the

Phospholipids

Particularly common among the membrane mesogens are the *phospholipids*. They have the general form

phospholipid

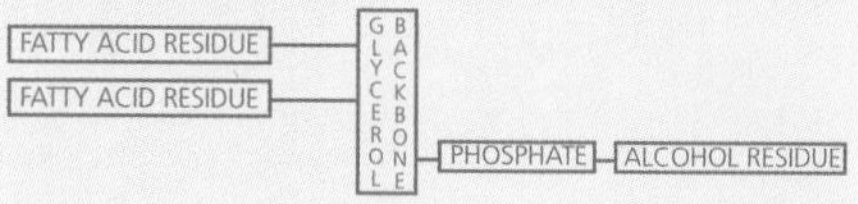

FATTY ACID RESIDUE
FATTY ACID RESIDUE
GLYCEROL BACKBONE
PHOSPHATE — ALCOHOL RESIDUE

The most common alcohols involved in these molecules are the following

choline

$HO{-}CH_2{-}CH_2{-}N^{+}(CH_3)_3$

serine

$HO{-}CH_2{-}CH(NH_3^{+}){-}COO^{-}$

ethanolamine

$HO{-}CH_2{-}CH_2{-}NH_3^{+}$

The phospholipids have a shape reminiscent of a tuning fork. An anthropomorphic analogy is also useful, because the molecules have easily identifiable head, shoulder, and leg regions. The most common variety is *phosphatidylcholine*, also known as *lecithin*. The dipalmityl version of this molecule has the structure shown below. It is present in the membranes of most animals and is particularly plentiful in *egg yolk*.

phosphatidylcholine

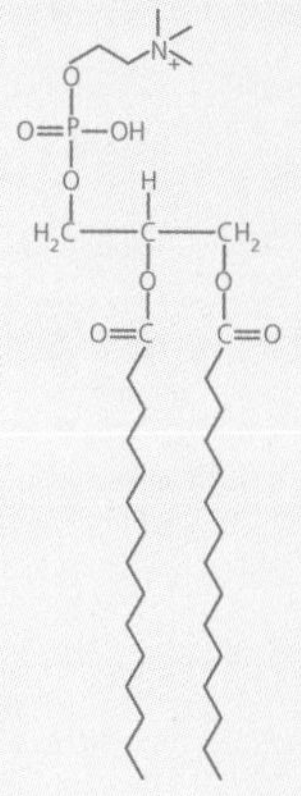

choline
phosphate
glycerol residue
fatty-acid residue

inner surfaces of the glasses are pre-treated so as to align the director parallel to the sides. Even a modest mechanical bias, such as can be achieved by stroking a glass plate with the finger, is found to cause an alignment that persists well into the bulk of the liquid crystal. Under zero field the sandwich appears transparent because the director is uniformly aligned. Application of the electric field causes motion of impurity ions, and the liquid crystal becomes turbid and opaque. In another type of device, crossed polars are added, one situated at each glass plate, and the inner surface alignment markings are mutually perpendicular. This causes the nematic to be twisted, the sandwich thickness corresponding to half the pitch. Light passing through the first polar has its plane or polarization turned through a right angle by the *twisted nematic*, so it is also able to pass through the

Minor membrane lipids

Not all lipids found in biological membranes are of the phospholipid type, and there is a second class of phospholipid that does not derive from the glycerine molecule. These are the *sphingolipids*, based on the sphingosine molecule. Their final form is nevertheless rather similar to the phospholipids. The two other types of lipid found in membranes are *glycolipids* and *cholesterol*. An important feature of the former is a sugar group at the hydrophilic end, and because sugar is compatible with water, this gives further directional stability to the molecule when it is in a membrane. A particularly important example of a glycolipid is *cerebroside*, which has the structure shown below:

cerebroside

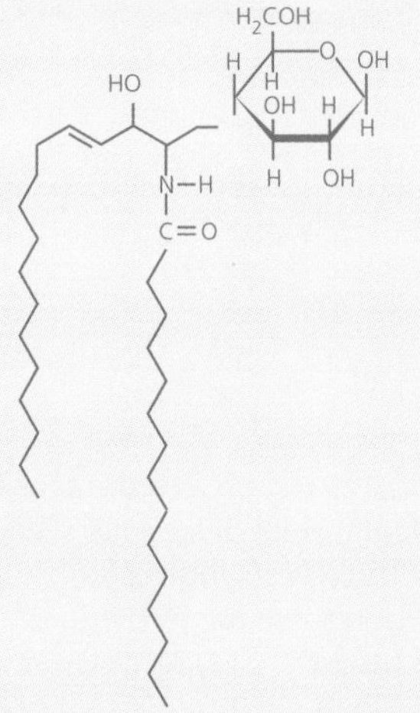

The important lipid cholesterol is a member of the chemical group known as the *steroids*, which also includes several *hormones*. Cholesterol itself has the form:

cholesterol

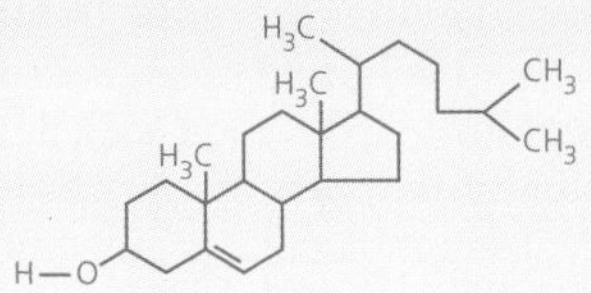

The entire molecule, apart from the hydroxyl group, is hydrophobic. One type of mesophase, the cholesteric, actually takes its name from cholesterol.

second polar and the device does not reflect. Application of the electric field causes the molecules to align themselves at right angles to the glass surface. In this orientation they have no influence on the polarization of incident light, which is thus reflected. Development of this technique owes much to the efforts of Wolfgang Helfrich.

In display devices of this type, symbols can be generated with appropriately shaped surface electrodes. The low operating voltages are ideal for portable units such as calculators and watches, which must incorporate their own power supplies, and power consumption is low because they do not have to generate light. Indeed, the stronger the light falling on them, the better the performance. This contrasts with the case of a cathode ray tube, for example, in which the appearance of the display is impaired by bright external illumination. A disadvantage arises from the relatively long

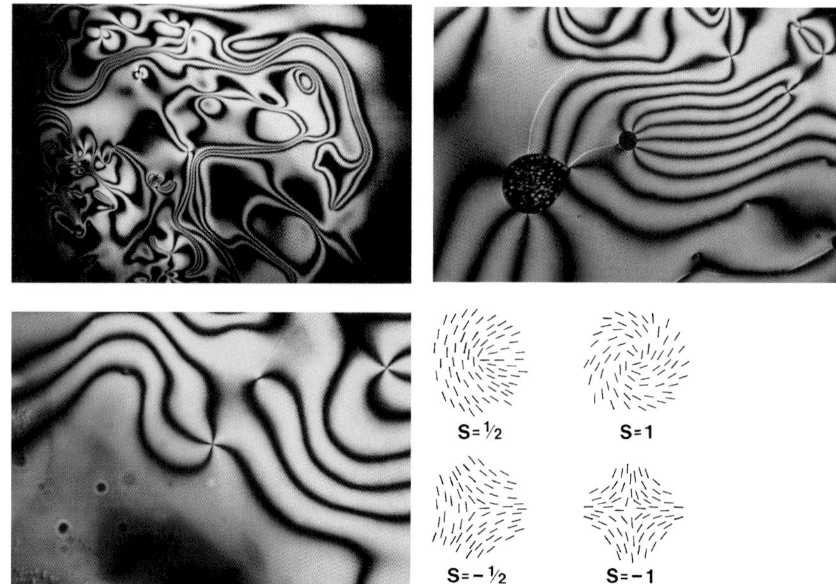

When a liquid crystal is viewed through *crossed polarizers*, the observed intensity depends on molecular orientation. Examples of orientational defects are seen in these three pictures of the nematic liquid crystal methoxybenzylidene-*n*-butylaniline. Although an optical microscope can not resolve the individual molecules, the nature of the defects can be inferred from the observed patterns. Lines known as brushes emerge from points referred to as singularities, and the strength (S) of a singularity is the number of emerging brushes divided by four. The strength of a closed surface is always two, so eight brushes should be visible. Examples are shown in the top right picture, in which an impurity particle is seen at the left and an air bubble at the centre. Eight brushes are associated with each of these features, two of them being mutually connected in the case of the bubble. The faint white lines are known as *inversion walls*; they are boundaries across which the *director* abruptly changes by 180°. A *disclination* of strength one is seen at the centre of the bottom picture, and another of strength a half appears above and to the right of this. The small diagram shows the idealized molecular arrangement around some common singularities.

relaxation time of a disturbed liquid crystal. Under the instantaneously imposed influence of an electric field, a typical nematic takes about one hundredth of a second (0.01 s) to respond, and recovery after removal of the stimulus occupies a somewhat greater period.

Nature must be given credit for the most fundamental use of liquid crystals. Certain mesogens form an essential part of the membrane that surrounds every living cell. In 1855, Carl Nägeli noted differences in the rates of penetration of pigments into damaged and undamaged plant cells and concluded that there must be an outer layer with its own special properties. He called it the *plasma membrane*. Its existence was confirmed in 1897 by Wilhelm Pfeffer, who also found it to be a universal barrier to the passage of water and solutes. Shortly thereafter, E. Charles Overton showed that this extreme view required modification. Although polar molecules certainly find it difficult to get through the membrane, non-polar groups such as alkyl chains and cholesterol have a relatively easy passage. Overton

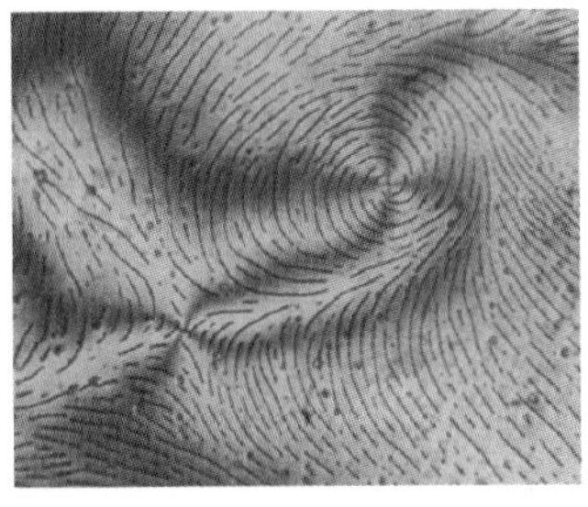

Dramatic confirmation of the implied presence of two *disclinations* is provided by this picture of a nematic liquid crystal observed between crossed polarizers. The two four-fold nodal points of the brushes each have a strength of one. In fact, because some of the brushes are seen to connect the two singularities, these disclinations must have opposite sign: +1 and −1. These conclusions are confirmed by the strings of liquid beads deliberately generated at the interface between the liquid crystal and another liquid, diethylene glycol. The beads reveal the *director field*, that is the map of directions adopted by the long axes of the molecules, and they show that the negative disclination lies to the left. Such a *disclination dipole* can spontaneously disappear, as the *turbulence* in the liquid crystal decreases, the positive and negative disclinations mutually annihilating.

concluded that the membrane exercised selective control through its *differential permeability* and that it was composed of *lipids*. In retrospect, it seems obvious that each cell must be surrounded by some sort of barrier. As Aleksandr Oparin stated in his classic volume *The Origin of Life*, in the mid 1930s, it is unlikely that there are living organisms, however primitive, that are not physically separated from the environment. Such an organism differs from its surroundings, by definition, so it must possess some structure that can maintain its special properties by providing a rampart against the outside world.

An important hint as to the actual molecular structure of the membrane was provided by Irving Langmuir's demonstration, in 1917, that lipid molecules tend to spread out into a layer one molecule thick, at air–water interfaces. Eight years later, E. Gorter and F. Grendel measured both the surface area of an *erythrocyte*, commonly known as a *red blood cell*, and the area of a reconstructed film formed by the lipids extracted from its membrane. The ratio was approximately 1:2, and they concluded that the membrane is two lipid molecules thick. In the mid 1930s, James Danielli, Hugh Davson, and E. Newton Harvey made accurate measurements of the *surface tension* of the plasma membrane and found this to be considerably lower than for most lipids. It was known that the addition of protein to oil, egg white to mackerel oil for example, lowers the surface tension, and this led Danielli, Davson and Harvey to propose a membrane model that is still regarded as essentially correct. The Gorter–Grendel lipid bilayer was envisaged as being located at the centre of the membrane, while the proteins formed a thin film at the lipid–water interfaces. This picture was subsequently refined by Davson and Danielli in 1954. The hydrophobic parts of the lipid were conjectured to lie at the bilayer interior, the hydrophilic regions of these

The wings of many insects are brightly coloured because they contain liquid crystals. The presence of such structures can be detected by observation through a Polaroid filter which selects for circularly polarized light. The electromagnetic oscillations in such light can be imagined as travelling in a corkscrew manner. If the handedness matches that of the molecular arrangement, the light is absorbed and the insect's wings appear black (left), while the opposite polarization produces reflection and reveals the wings in full colour (right). Albert Michelson (1852–1931) was the first to observe this effect, towards the end of the nineteenth century, and he correctly inferred the presence of a helical molecular arrangement, even though he was unaware that some *insect wings* contain liquid crystals.

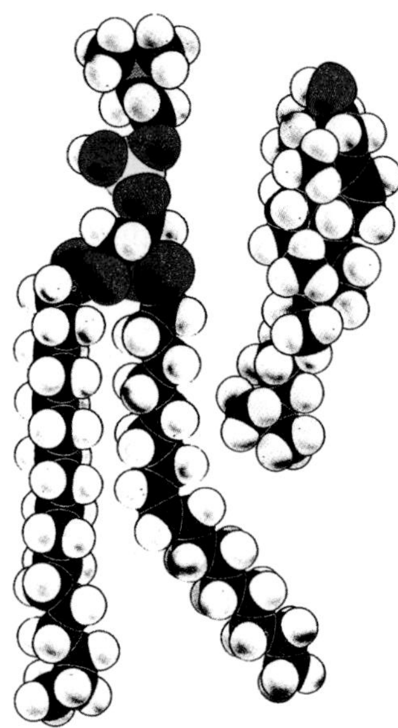

The liquid crystalline arrangement is vital to living systems. It provides the basis for the flexible and partially permeable skin that surrounds each cell. Such *biological membranes* contain members of several different chemical species, the two most important of which are *phospholipids*, here represented by dipalmityl phosphotidyl choline (left), and *cholesterol*. The oxygen-bearing ends of these two molecules, indicated by the red colouring on these space-filling models, are attracted to water, whereas the hydrogen-rich ends are repelled by that liquid.

molecules facing the water at either side, in the gaps between the proteins. They also proposed the existence of small pores through the membrane. Spectacular confirmation of the bilayer hypothesis was obtained in the mid 1950s, by J. D. Robertson, who found a way of staining membranes so as to permit direct observation in an *electron microscope*. He was able to resolve two parallel lines, corresponding to the two layers of molecules, although no distinction between protein and lipid was observable. Robertson was led to elevate the model to the status of a universal structure, always comprising a protein–lipid–protein sandwich: the so-called *unit membrane*. Proteins will be discussed in Chapter 15, where it will be shown that they too have hydrophobic and hydrophilic parts. The globular structure of a *protein* is generally attributed to the fact that a roughly spherical arrangement, with the hydrophobic parts hidden inside away from the water, corresponds to the lowest free energy. This is at variance with the Davson–Danielli–Robertson picture of unfolded proteins spread over the lipid exterior, and in 1966 S. Jonathan Singer and Donald Wallach independently postulated a quite different model of protein–lipid architecture. They visualized normal globular proteins dotted around the lipid surface, invariably penetrating into it and sometimes spanning the entire 6 nm width of the bilayer. Singer suggested that the latter variety of protein might have two distinct hydrophilic regions, one being attracted to each lipid–water interface.

The lipid portion of biological membranes is not composed of a single mesogen, and not all membrane molecules are mesogens. Of the three main constituents, lipids, proteins, and sugars, only the first form liquid crystal phases. The relative amounts of the different components show considerable variation, depending on the part of the organism in which the tissue is located. In the *myelin membranes* that sheath *nerve cells* the ratio of lipid to protein is about 10:1 by weight, whereas in the membranes that surround the *mitochondria*, which provide each cell with energy, it is approximately 1:3. The most common type of lipid in biological membranes is the *phospholipid*, the lesser members being *glycolipids* and *cholesterol*. It is appropriate to consider phospholipids because they hold the key to membrane structure. They are based on certain esters of the trihydric alcohol, glycerol. Attachment of a fatty-acid residue to each of the three hydroxyl positions on the glycerol molecule would produce a *neutral lipid*. Such lipids are present in the body as *fat*, stored in the *adipose tissue*. The *fatty-acid chains* are linear and can be defined by the number of carbon atoms they contain, together with the number of *unsaturated bonds*. The unsaturated (double) carbon–carbon bonds tend to be located roughly in the middle of the chains and give the latter a perceptible bend.

The neutral lipids, also known as *triglycerides*, do not form liquid crystal membranes; they lack a vital extra feature: a terminal *polar group*. The phospholipids differ from their neutral cousins in that one of the outer fatty-acid residues is replaced by one of a variety of groups, all of which incorporate a phosphate. It is clear that we are dealing with a multiplicity of families of molecules, in that there is considerable degree of choice in all three terminal groups. It is usually the case that one of the two fatty-acid residues is saturated while the other is unsaturated, the latter invariably being located at the middle position of the glycerol backbone. Why this should be so is not clear, and neither is it understood why the membranes in the *alveolar lining of the lung*, with their two saturated palmitic chains, should be a prominent exception.

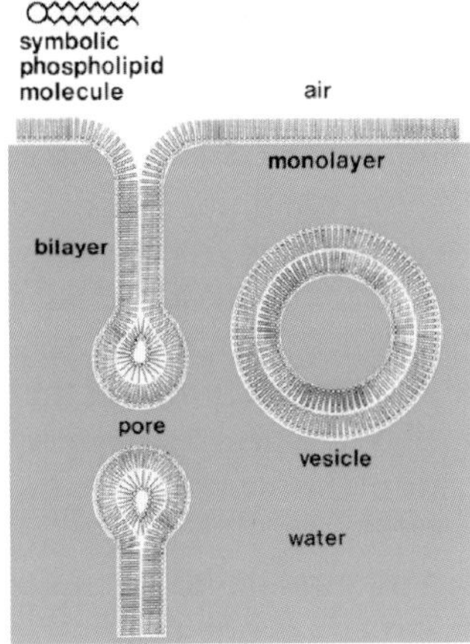

Because their hydrocarbon regions are strongly hydrophobic, *lipid bilayers* avoid developing edges. They thus have a tendency to form closed surfaces, rather like water-filled bubbles. These are the *lipid vesicles* which can be regarded as prototype cells. Indeed, the development of vesicle-forming molecules must have been a critical stage in the *evolution* of living organisms. In this schematic picture, which is not to scale, a vesicle is seen, as well as a hole in a bilayer. Such holes probably resemble the pores in a cell's *nuclear membrane*, through which the genetic information is transmitted.

One end of a lipid molecule is charged and therefore *hydrophilic*, while the other extremity has two fatty-acid chains which are *hydrophobic*. How are these dual tendencies satisfied in a cell? It is neatly accomplished by the molecules arranging themselves as a *bilayer*, with the head groups in the outer layer facing the extracellular water and the head groups of the inner layer facing inwards towards the cell's aqueous medium, known as the *cytoplasm*. Since each phospholipid molecule is approximately three nanometres long, the bilayer is approximately 6 nm thick. Depending on the temperature, and also the chemical composition of the surrounding aqueous medium, the molecules in each of the two layers are either ordered or disordered. In the disordered state the composite structure is reminiscent of the smectic liquid crystal phase. In this state the individual lipid molecules can readily migrate within their own layer. Lateral movement of molecules in the membrane was first demonstrated by David Frye and Michael Edidin in 1970. They studied *cell–cell fusion* induced by the Sendai virus, probing the subsequent distribution in the composite cell of proteins originally located on one or other of the participants. It was found that rapid mixing takes place at biological temperatures and, because this had occurred in the absence of an external energy source, Frye and Edidin concluded that it was due to lateral diffusion. Harden McConnell and Philippe Devaux, using what are known as *spin labels*, later established that lipids move around at a higher rate than proteins, as we should expect since they are smaller. It is found that the flipping of a molecule from one side of the bilayer to the other is a quite rare event. This is consistent with the recently established fact that many membranes are quite asymmetric in their lipid composition. The bilayer offers a remarkably high resistance to the passage of metallic ions, and most substances require the help of other membrane-bound bodies to pass through the membrane.

Interesting facts emerge if we examine the *fatty-acid composition profile* in different species and different organs. The profile for the *neutral triglycerides* varies from one species to another. This is hardly surprising since diets of various species can differ widely. For the *phospholipids* in a given species, the composition profile varies from one organ to another, but if we compare the profile in a given organ, there is remarkably little variation from one species to another. In the brain, for example, the most common fatty acid is of the oleic type irrespective of whether the animal is a mouse, rat, rabbit, pig, horse, ox or sheep, and in all cases palmitic acid and stearic acid are the next most common. This interesting observation is not particularly surprising if it is taken to indicate that the physical properties of a membrane are determined by its fatty-acid composition. The variation from one organ to another simply reflects the fact that there is a connection between physiological function and physical properties.

The *fluidity* of a phospholipid membrane is influenced by its cholesterol content; the role of that substance appears to be the lowering of the order–disorder transition temperature by disrupting the arrangement of lipid molecules. Physiological function derives from the membrane-bound proteins. Some of these appear to be confined to one side, but it is also known that others span the entire width of the bilayer. As we would expect, the mobility of such proteins in the plane of the membrane is low when the lipid molecules are in the ordered state, and considerably higher when the lipid is disordered. Since function might in some cases depend on the mobility of the proteins, it can be important for an organ's function that

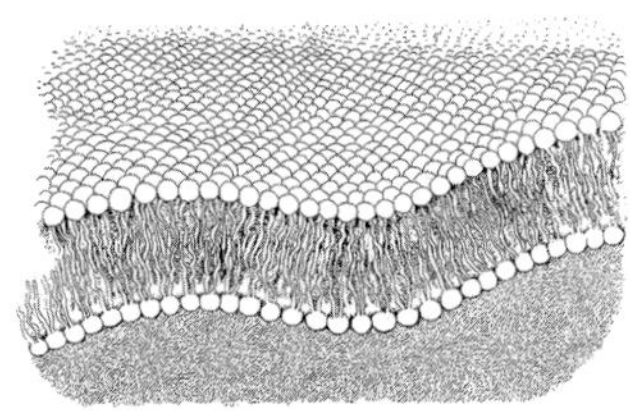

The phospholipid molecules from which biological membranes are primarily composed are *amphiphilic*: their two ends display contrasting affinities for water, one being attracted and the other shunned. These dual tendencies are both satisfied in the *lipid bilayer* arrangement, in which the polar regions are in contact with the water on either side while the hydrophobic tail regions are confined to the interior. The molecular arrangement is a special form of the *smectic* structure.

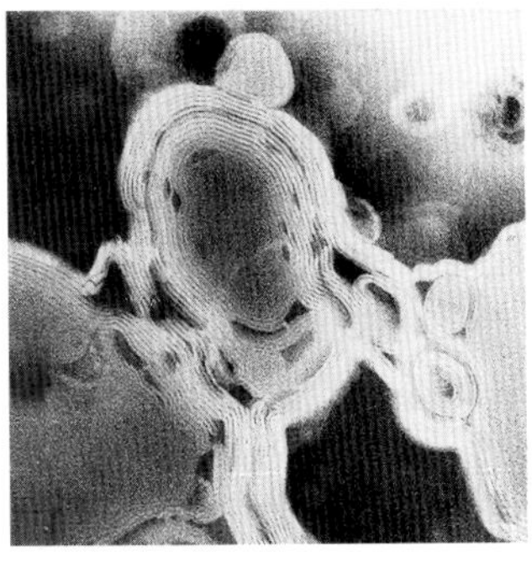

Phospholipid vesicles comprising a number of concentric bilayers have come to be known as *liposomes*. They can be made by subjecting a mixture of phospholipid and water to *ultrasonic radiation*. This transmission electron photomicrograph shows a number of multi-bilayer vesicles. Such structures have been used in tests on a number of cellular phenomena, and they are also employed as vehicles for introducing various chemicals to living cells. This type of study serves as a warning that excessive ultrasonic scanning of fetuses should probably be avoided.

the degree of lipid order falls within the desirable range. A dramatic example of this principle is seen in certain fish, which involuntarily adjust the composition of their membranes according to the water temperature. A similar thing is seen in some bacteria. S. Jonathan Singer and Garth Nicolson have synthesized the large amount of information on membranes, gained in recent years, into what is called the *fluid mosaic model*. Its chief features are that the phospholipid serves both as a solvent for protein and as a *permeability regulator*, the latter by adopting the bilayer configuration; that lipid–protein interaction is possibly essential for protein function; and that membrane proteins are free to migrate within the bilayer, unless restricted by specific interactions, but cannot rotate from one side of the membrane to the other.

In recent years, there has been much interest in membrane-bound units known as *vesicles*, or *liposomes*. Unlike cells, these structures often comprise several thicknesses of lipid bilayer, arranged like the layers of an onion. Their artificial formation was first achieved around 1960, by Alec Bangham, who studied how the lipid configuration varies with its concentration in water. Liposomes with a minimum diameter of about 30 nm can be produced by subjecting a lipid–water mixture to ultrasonic radiation. These structures are bound by a single lipid bilayer and are equivalent to diminutive cells with interiors devoid of everything but water, and their membranes purged of everything but lipid. They thus differ from the minimal configuration adopted by soap molecules: a single-layer sphere, with the tails inside and the polar heads facing the water, known as a *micelle*. Bangham and Robert Horne obtained cross-sectional pictures of liposomes in 1962, using electron microscopy, and resolved the individual layers. This showed that the bilayer itself can be observed, even in the absence of protein. Liposomes have become the standard test vehicle for experiments in which individual phenomena are studied, without the complexities of the normal cellular environment. In 1974 new use for them was found by Gerald Weissmann. Borrowing the Trojan Horse idea from Greek mythology, a team led by Weissmann used liposomes for packaging enzymes and delivering them to deficient cells.

Summary

Liquid crystals, or mesophases, are intermediate forms of condensed matter lying between the crystalline and liquid states, and they share characteristics with both of these extremes. Less ordered than a crystal, but more viscous than a liquid, their peculiarities stem from the elongated cigar-like shapes of their invariably organic molecules, and they give rise to spectacular optical effects. There are two classes: thermotropic mesogens, for which the transition from crystal to liquid crystal occurs upon changing the temperature, and lyotropic mesogens, with the transition caused by change of solvent concentration. Of the many distinct liquid crystal configurations, the most common are the thread-like nematic, in which the long axes of the molecules are aligned along a particular direction like a travelling shoal of fish, the stratified smectic, with its molecules arranged in layers like bottles on a rack of shelves, and the screw-like cholesteric, which is like a nematic with a uniformly twisting direction of orientation. Mesogens are widely employed in display devices and as temperature monitors. A smectic-like arrangement is prominent in the membranous skin of living cells, which is composed of lipid, protein, and cholesterol. Membrane lipid, the main variety of which is phospholipid, possesses the property of polarity at one end of its molecules. It adopts a double-layer configuration, the bilayer, in which the polar regions face outwards towards the extracellular liquid and inwards towards the cells' cytoplasm. Some diseases are associated with a liquid crystal phase transition.

14 Of snakes and ladders: *polymers*

What! will the line stretch out to the crack of doom?
William Shakespeare (1564–1616)
Macbeth

When Lewis Carroll's Walrus suggested that the time had come 'to talk of many things; of shoes and ships and sealing wax, of cabbages and kings', the intention was to reel off an impressively incongruous list of items. In point of fact, however, he was merely reciting an inventory of materials all of which are at least partly composed of *polymers*. And the list was by no means exhaustive, for he could have gone on to add paper and paint, rubber and rope, fur and flax, and wood and wool, and so on. The polymer domain is a veritable pot-pourri which displays greater variety than is seen in any other class of substances.

Of all materials used by Man, those based on polymer molecules have been around longest. *Wood*, for example, was pressed into service long before metals were being smelted, and wood is composed of cellulose, lignin and resins, all of which are natural organic polymers. Nevertheless, an adequate knowledge of polymer structure and properties has been acquired only during recent decades. The reason for the delay lies in the fact that these molecules are frequently very large and complex compared with anything encountered in metals and ceramics. Analysis of such huge aggregates of atoms, with molecular weights frequently running into the hundreds of thousands, has required the sophistication of modern instrumentation, and polymer science is still a relatively young discipline. The etymology of the word polymer is straightforward enough. The Greek word *poly* means many, and another Greek word *meros* means parts. The choice of this name stems from what has been found to be the defining characteristic of polymers; they are formed by the coming together of many smaller units. Some of these extend thread-like in one dimension, while others form sheets, and still others are arranged in three-dimensional networks. The individual units are called *monomers*, and most common polymers are composed of regular repetitions of just a few types of monomer, and often of only one. There is no formal restriction on the composition of a polymer, and examples like asbestos are not even organic. Because such inorganic polymers have appeared in earlier chapters, however, we will limit the discussion to molecules in which the element carbon plays a key role. Synthetic polymers of the organic type are usually associated with the word *plastic*, a term that is somewhat misleading in that it refers to a material which is not plastic in the physical sense. A plastic is best defined, therefore, as a material based primarily on polymeric substances, which is stable in normal use, but which at some stage during its manufacture is physically plastic. The early efforts in the field of organic polymers were of necessity directed towards naturally occurring polymers, but once the regularity of their structures had

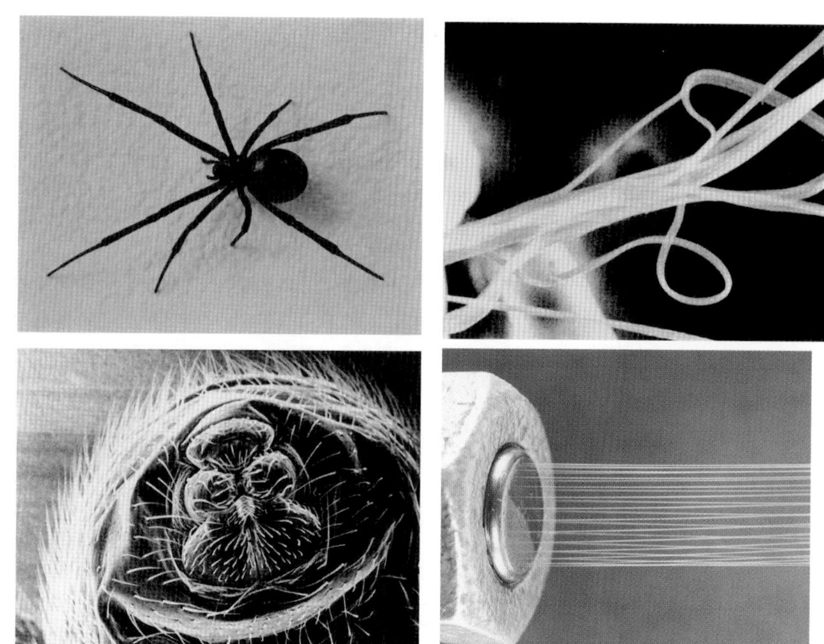

The first polymeric materials to attract scientific interest were silk and cobweb. In 1665, Robert Hooke (1635–1703) suggested that the products of the *silkworm* and *spider* could be imitated by drawing a suitable glue-like substance out into a thread. The industrial manufacture of *artificial fibre* does indeed resemble the natural process used by these humble creatures. A close-up view of the thread produced by a black widow spider (upper left) is shown at the lower left. The thread is extruded from an abdominal organ known as a *spinerette*, shown in the scanning electron photomicrograph at the upper right. The lower right picture shows cellulose acetate yarn emerging from an industrial spinerette.

been revealed, chemists began to tailor-make artificial polymers by stringing together combinations of monomers that do not occur naturally. This activity has now mushroomed into the vast industry that annually produces millions of metric tonnes of plastics, synthetic fibres, and related products.

In contrast to the technologies of metals and ceramics, the origins of which are antique and obscure, the mileposts of development in the polymer industry are so recent that the story is well documented. The first polymeric material to attract scientific interest appears to have been *silk*. Robert Hooke noted, in 1665, that this substance appears to be a type of glue, and he suggested that the product of the silkworm could be imitated by drawing a suitable glue-like material out into a thread. It transpired that he was more than 200 years before his time. Not surprisingly, the earliest technological use of materials of this type, in a rudimentary polymer chemistry, took advantage of naturally occurring substances. Among these were amber, rosin, gum arabic, asphalt and pitch, the latter two being valued for their waterproofing properties as early as 5000 years ago. Natural *resins* were familiar to the early Egyptians, who used them in varnishing sarcophagi. *Shellac*, which was in common use by the middle of the third century AD, is unique among the resins in that it is of animal origin. It is secreted by the insect *Tachardia lacca*, which is found on the lac tree *Butea frondosa*. Shellac is still used, alcoholic solutions having a consistency convenient for *varnishes, polishes*, electrical insulation and, at greater concentration of the polymer, *sealing wax*. Early explorers of the Central Americas were amazed by a substance used by the native Indians both for ball games and in the production of water-resistant boots. This elastic gum, now known as *rubber* and obtained primarily from the tree *Hevea brasiliensis*, was first made the object of scientific study in 1751, by Charles-Marie de la Condamine. Joseph Priestley must be given the credit for discovering its utility as an eraser of pencil marks, while the raincoat made of rubberized fabric is synonymous with the surname of its developer, Charles Mackintosh.

The requirements for *polymer formation* are not difficult to satisfy: there are numerous chemical groups which possess an appropriate number of *unsaturated bonds* that are free to provide links with neighbouring groups. And advances in synthetic techniques have made it possible to produce virtually any imaginable arrangement of the individual *monomers*. In these schematic representations of some of the more common forms, the coloured circles denote either single atoms or groups of atoms.

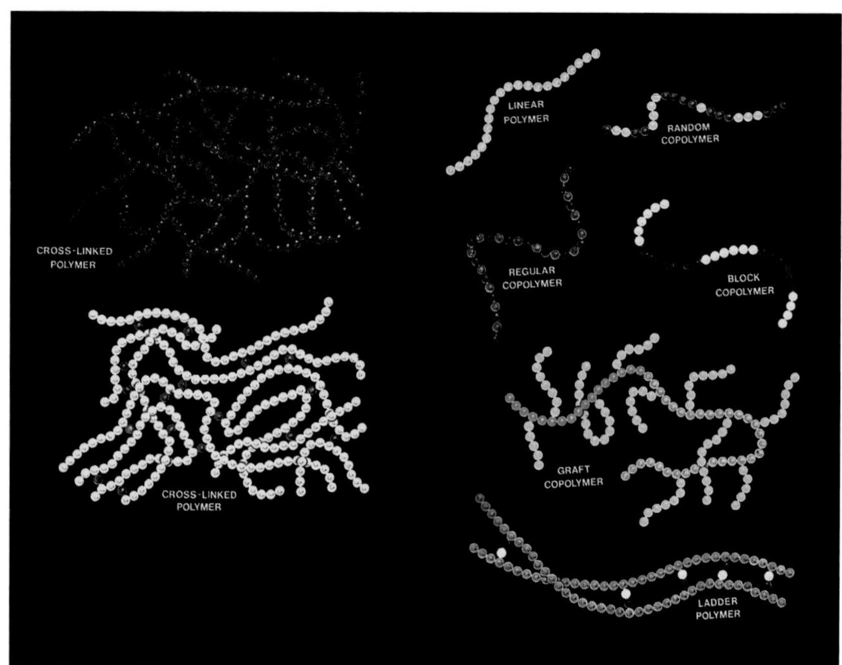

Nineteenth-century organic chemists often encountered substances that defied the normal classification as crystal or liquid, and they were unable to purify them to an extent sufficient for chemical identification. Some of these amorphous semisolids proved to have such desirable qualities that their preparation was systematically studied, and some were put into mass production. The subsequent explosive growth of the plastics industry was predicated on two developments: the elucidation of the principles of *polymer chemistry* and the extraction of the required monomers as byproducts of petroleum refining. Two events in the 1830s vie for recognition as the inception of the plastics era. One was the discovery, by Charles Goodyear in 1839, that rubber can be stiffened by the addition of sulphur and the application of heat, the process being known as *vulcanization*. The other, due to Christian Schönbein seven years earlier, was the demonstration that natural fibres can be dissolved in nitric acid to produce the mouldable product *cellulose nitrate*. Alexander Parkes was the first to attempt the commercial production of this plastic, in 1862, and he used castor oil and later camphor as a *plasticizer* to make it more pliable. In an unfortunate piece of timing, his Parkesine Company went into liquidation just one year before an event which was to prove decisive in establishing the plastics industry. With the unmerciful slaughter of the African elephant herds, the demand for ivory exceeded supply, and in 1868 the Phelan and Collender Company offered a $10 000 award to anyone who would assign them patent rights to a satisfactory substitute for *ivory* in the production of billiard balls. Unaware of the activities of Parkes, John Hyatt and his brother Isaiah solved the problem with a product which they named *celluloid*. It was in fact identical with the earlier material, cellulose nitrate plasticized with camphor, but the operation was improved by the use of heat and pressure during the moulding. There is no record of the $10 000 prize actually being paid out, and the tremendous success of their venture put the Hyatt brothers above such considerations. The number of applications of their product grew rapidly. Plastic dentures were soon widespread, and celluloid film was just what the embryonic photographic industry needed. And a veritable revolution was just around the corner. In 1880, Joseph Swan extruded cellulose nitrate through fine holes to produce threads that could be carbonized and used as electric light filaments. Then in 1884 came the real start of the artificial textile era, with the award of a patent, to Count Louis Hilaire de Chardonnet, for what was called *rayon*. This term is now applied to all man-made fibres based on cellulose, such fibres being known earlier as *artificial silks*. The Chardonnet process involved spinning a solution of cellulose nitrate in glacial acetic acid, and regeneration of the cellulose by reduction of the nitrate groups.

Cellulose nitrate suffered from a major disadvantage: it was inflammable. Each monomer of cellulose, the atomic structure of which will be discussed later, has three hydroxyl groups. Each of these can be replaced by a nitrate group, and the saturated version, *cellulose trinitrate*, is used as an explosive under the names *gun-cotton* and *cordite*. The non-saturated nitrates, while not as dangerous, must nevertheless be handled with caution, and this was a drawback of the early photographic films. In the nursery, celluloid dolls were a constant source of parental concern. A welcome alternative was provided by Otto Schweitzer, in 1857, who found that cellulose could be dissolved in an ammoniacal solution of copper oxide. Cellulose can then be regenerated from the solution by dilution with weak acids, or even water. This *cuprammonium rayon* process yields the finest filament of any synthetic fibre, but it is relatively expensive. A cheaper product, *viscose rayon*, was developed by Clayton Beadle, Edward Bevan and Charles Cross, in 1892. Charles Cross also introduced the principles of polymer chemistry to *paper manufacture*. The viscose process involves dissolving cellulose in sodium hydroxide and then adding carbon disulphide. The result is a solution of cellulose xanthate in dilute sodium hydroxide, called viscose, and regenerated cellulose is extracted from this thick brown treacle-like liquid by extruding it into a bath containing sulphuric acid and sodium sulphate. Viscose rayon went into commercial production in 1907, just four years after the cuprammonium variety. It appeared in a novel form, the thin transparent film known commercially as *cellophane*, in 1908. Finally, a fourth type of rayon deserves mention. In 1865, Paul Schutzenberger discovered the formation of *cellulose acetate*, a white fibrous material, by dissolving cellulose in acetic anhydride. The early difficulties of high production cost and difficulty in dyeing were overcome only much later, and large-scale manufacture was first achieved by the Dreyfus brothers, Camille and Henri, in 1919. Cellulose acetate is dissolved in chloroform to produce the dope used to paint cars and aircraft. The textile fibre is now marketed under the commercial name *Celanese*.

Although cellulose-based plastics dominated the early years of the industry, they did not have the field entirely to themselves. In 1851, twelve years after the discovery of vulcanization, Nelson Goodyear, the younger brother of Charles, found that incorporating a much larger amount of sulphur in rubber, before application of heat, produced a material as hard as the hardest wood. Because it resembled ebony, he called it *ebonite*, and it was invaluable to the infant telegraph and telephone industries. Another early plastic was *casein*. Discovered by Adolf Spiteller in 1890, it was first manufactured in Germany under the trade name *Galalith*, derived from the Greek words *gala*, meaning milk, and *lithos*, a stone. It was developed to satisfy the demand for white slates for school use. Casein is extracted from cow's milk, which contains about 3% of the solid, by precipitation through the addition of rennet. The production process is lengthy because of the need for hardening in formalin, which takes several weeks. Casein was earlier used for such items as buttons, umbrella handles, combs, pens and knitting needles, but it has now been priced out of the market.

All the plastics discussed thus far are based partly on naturally occurring substances. The first truly synthetic product was invented by Leo Baekeland in 1905. Formed by the irreversible reaction of phenol and formaldehyde, with the application of heat, and thus belonging to what we will later identify as the thermosetting class of polymers, it was given the trade name *Bakelite*. Baekeland filed a patent for his discovery in 1907, just one day before James Swinburne attempted to do the same thing. The synthetic polymer era had begun, and it rapidly gained momentum with numerous new plastics and, in 1934, the first completely synthetic fibre, *nylon*, invented by Wallace Carothers. This plastic made its commercial debut as the bristle material in Dr West's Miracle Tuft Tooth Brush, in 1938.

Fortunately, commercial exploitation of substances in our environment does not depend on an understanding of their atomic structure. Many of the above developments were achieved before the atomicity of matter was generally accepted. Even when X-ray methods were beginning to make inroads into the metallic and ceramic realms, polymers, with their long tangled strands, still defied precise structure determination. Some of the early scientific advances were secured through inspired induction, based on meagre experimental data. By 1920, one of the giants of polymer science, Hermann Staudinger, had elucidated the basic principles of polymerization and had paved the way to understanding the thermodynamics of these long-chain molecules. More recently, Paul Flory, through his analyses of the statistical aspects of *polymer conformation*, has done much to clarify the behaviour of materials belonging to this class. A useful analogy to a single polymer molecule is a piece of spaghetti. In their dry state, as usually purchased, pieces of spaghetti can be neatly stacked by laying them parallel to each other, to produce an arrangement that corresponds to a fully crystallized polymer. In fact, this situation has never been reached in practice, although the record degree of crystallinity in a polymer currently stands at about 95%. When it is being boiled, spaghetti is quite pliable, and it invariably forms a complex and tangled network which is analogous to the amorphous state of a polymer.

The special properties of polymers derive from the fact that they are composed of monomers, which have a tendency to link up and form long chains. The two important factors are *valence* and the *covalent bond*. The valence of an atom is determined by the number of electrons in its outermost electron shell. In the special case of covalent bonding there is a saturation effect, and valence can be looked upon as a measure of the number of arms that each atom stretches out towards its neighbours. In much the same way, it is the valence of the individual atoms which determines what is known as the *functionality* of a monomer. Just as it would be impossible to form a human chain from one-armed people, the formation of a polymer chain requires monomers with a functionality of at least two. If the functionality is greater than this, branching can occur, and this additional complication will be discussed later. The simplest possible

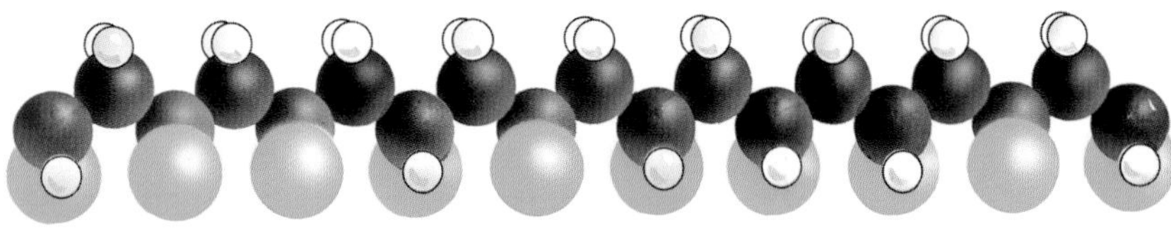

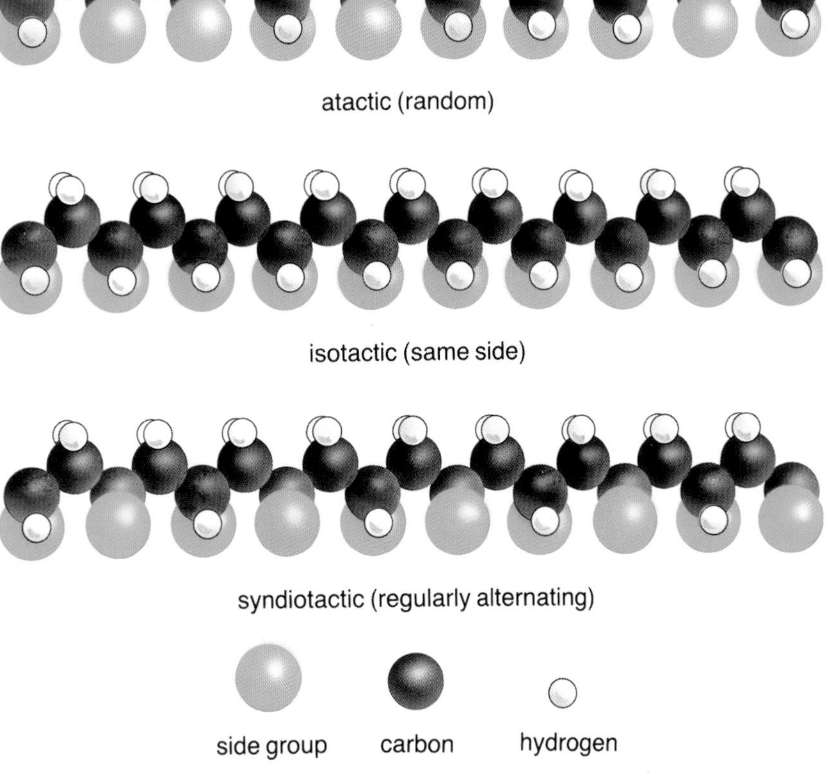

The distribution of the *side groups* along a polymer molecule is governed by thermodynamics, the most probable arrangement depending upon the strengths of the forces between the various types of atoms. When the side groups strongly interact with each other, and with their other neighbouring atoms, a regular pattern is established, as in the *isotactic* and *syndiotactic* arrangements. At the other extreme, when the side groups interact only weakly with their surroundings, the distribution displays the randomness of an *atactic polymer*.

polymer chain would be that for which the monomer is a single atom. Such chains are rare, the only common examples occurring in sulphur and selenium. More generally, the monomer is a molecule which can be as simple as a CH_2 group, or so complicated that it even contains rings of atoms.

A simple *linear polymer* is a chain molecule composed of bifunctional monomers, with monofunctional terminal units. If more than one bifunctional monomer is present, the chain is known as a *copolymer*. A copolymer in which groups of the same monomer are located adjacent to one another is referred to as a *block copolymer. Branching* of a polymer occurs when trifunctional (or higher functional) monomers are present. The multifunctional monomers are not necessarily different types of molecule. They can be variants of the bifunctional units. A special case of the branched polymer is the *graft copolymer*, in which one monomer appears in the main chain while groups of a second type are attached to this at trifunctional branch points. Finally, to complete this introduction to the general forms of polymer structure, there is the *cross-linked polymer*, in which the bridging portion is either a single monomer or a chain of monomers. In some cross-linked polymers the bridge is established by a single divalent atom. Deliberate cross-linking is usually achieved chemically by addition of atoms or radicals with the right functionality and bond strength, but it can also be produced physically by irradiation with ultraviolet light.

So far, we have discussed the different types of polymer in terms of the generalized monomers, without specifying their actual chemical composition. We must now consider which of the elements in the periodic table are actually found in natural and synthetic polymers. By far the most common of these are hydrogen and carbon, the latter being the polyvalent atom that frequently occurs in the polymer backbone. Let us consider, for example, linear *polyethylene*, also known as *polythene*, which is a polymeric paraffin having the following structure.

polyethylene

$$\begin{array}{ccccccccccccc} & \mathrm{H} & & \mathrm{H} & & \mathrm{H} & & \mathrm{H} & & \mathrm{H} & & \mathrm{H} & \\ & | & & | & & | & & | & & | & & | & \\ \mathrm{H}- & \mathrm{C} & - & \mathrm{C} & - & \mathrm{C} & -\cdots\cdots- & \mathrm{C} & - & \mathrm{C} & - & \mathrm{C} & -\mathrm{H} \\ & | & & | & & | & & | & & | & & | & \\ & \mathrm{H} & & \mathrm{H} & & \mathrm{H} & & \mathrm{H} & & \mathrm{H} & & \mathrm{H} & \end{array}$$

The repeating monomer is CH_2, and the monovalent terminal units are simply hydrogen atoms. Although the carbon atom has a valence of four, two of the bonds are used in binding the two hydrogen atoms, and this leaves two for the monomer as a whole. Polyethylene finds use in a variety of household applications, such as containers, tubing, and cable covers. It is so tough and wear-resistant that it can be employed in gear wheels and bearings. This highly useful substance, was discovered by accident in 1935, during high pressure experiments on ethylene–benzaldehyde mixtures carried out by Eric Fawcett and Reginald Gibson at the Imperial Chemical Industries laboratories. The production of polyethylene underwent a radical improvement through the discovery, by Karl Ziegler in the early 1950s, of very effective catalysts of the organo-metallic type. *Catalysts* play a vital role in most commercial polymerization reactions. A variation of the polyethylene theme is seen in *polytetrafluorethylene*, or PTFE for short, which is known by the trade names *Teflon* and *Fluon*. A fluorine atom replaces each of the hydrogen atoms in the polyethylene, which is quite straightforward since both are monovalent. The structure is as follows.

polytetrafluorethylene

$$\begin{array}{ccccccccccccc} & \mathrm{F} & & \mathrm{F} & & \mathrm{F} & & \mathrm{F} & & \mathrm{F} & & \mathrm{F} & \\ & | & & | & & | & & | & & | & & | & \\ \mathrm{F}- & \mathrm{C} & - & \mathrm{C} & - & \mathrm{C} & -\cdots\cdots- & \mathrm{C} & - & \mathrm{C} & - & \mathrm{C} & -\mathrm{F} \\ & | & & | & & | & & | & & | & & | & \\ & \mathrm{F} & & \mathrm{F} & & \mathrm{F} & & \mathrm{F} & & \mathrm{F} & & \mathrm{F} & \end{array}$$

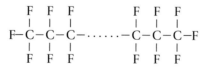

Invented by R. J. Plunkett of E. I. Du Pont de Nemours Company, in 1938, PTFE is particularly well known as the non-stick lining of modern frying pans and saucepans.

A slightly more complicated monomer occurs in the *vinyl polymer*, in which the repeating unit is

$$\begin{array}{ccccc} & \mathrm{H} & & \mathrm{H} & \\ & | & & | & \\ - & \mathrm{C} & - & \mathrm{C} & - \\ & | & & | & \\ & \mathrm{H} & & \mathrm{X} & \end{array}$$

X being a monovalent side group. In the case of *polyvinyl chloride*, which is also known as PVC and by the trade name *Vinylite*, X is chlorine. Discovered in 1838, by Victor Regnault, this plastic has a number of fortunate properties. Unlike polyethylene it can be high-frequency welded, and it is

Although most *polymer chains* show considerable flexibility they do not enjoy total freedom of movement. The presence of a double bond in the backbone, for example, can inhibit rotation around that interatomic link, and bulky side groups can give rise to the constraining effect known as *steric hindrance*. A well-known example of *geometrical isomerism*, arising from these factors, is seen in *natural rubbers*. In *gutta-percha*, all the carbon–carbon–carbon angles are equivalent and this gives an isotactic arrangement of the methyl side groups; the consequently straight backbone produces a polymer that is not very elastic. The other common form of rubber has a more kinked backbone, and a syndiotactic pattern of side groups. The marked elasticity of that latter form derives from the possibility of reversibly straightening out the convoluted backbone by the application of mechanical stress.

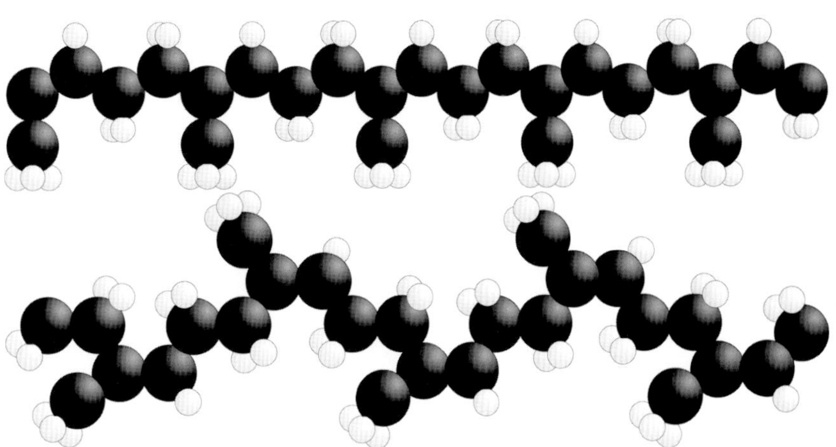

chemically resistant to petrol, various oils, and disinfectants. Naturally brittle, it can be softened by various *plasticizers*. It is used in containers, pipes and gutter fittings, water-resistant shoes, and even in flooring. Polyvinyl chloride came into general use in the 1930s. The introduction of a second type of monovalent side group raises the question of the positioning of these groups along the polymer chain. There are in fact three different possible arrangements which have been given the designations atactic, isotactic and syndiotactic. In the *atactic* case, the arrangement of the side groups is random. All the minor monovalent side groups are on the same side of the chain in the *isotactic* version, whereas their positions regularly alternate from one side to the other in the *syndiotactic* polymer. Side groups are not necessarily small and simple. In *polystyrene*, for example, one of them consists of a benzene ring. The structure of this transparent and hard polymer is as follows.

polystyrene

$$-CH-CH_2-CH-CH_2-CH-CH_2-$$

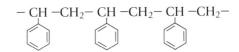

(each CH bearing a benzene ring)

It was elucidated by Staudinger and became a viable commercial product in 1937. Like polyethylene and polyvinyl chloride, polystyrene finds considerable use in household items and toys. An expanded version, which combines lightness with strength, is sold under the name *Styrofoam*. This plastic has become a standard material for packaging. Another polymer with bulky side groups is *polymethyl methacrylate*, which was discovered by E. Linnemann, in 1872. It was developed into a commercial plastic by R. Hill of Imperial Chemical Industries Ltd and came onto the market in 1933 under the name *Perspex*. Other trade names are *Lucite* and *Plexiglas*. Its first major application was in aircraft cockpit canopies, during the Second World War, and it continues to be the choice when transparency must be combined with strength. Its repeating unit has the following form.

polymethyl methacrylate

$$\begin{array}{cc} H & CH_3 \\ | & | \\ -C- & C- \\ | & | \\ H & COOCH_3 \end{array}$$

One of the most important mathematical concepts in polymer science is the *random walk*. First properly treated by Albert Einstein (1879–1955) in 1905, it relates the most probable distance between the two ends of a chain to the number of links, assuming that the joints between adjacent links are free to adopt any angle. The three chains shown here were generated at random by a computer, and they comprise 16, 64, and 256 links respectively. The starting and finishing points, in each case, are indicated by the green and red circles, which have radii equal to the length of one link. There is a clear tendency for the end-to-end distance to increase with increasing number of links, but this is only an average trend. As can be seen from the first picture (above left), the chain end was farther from the start after the eighth step than after the sixteenth.

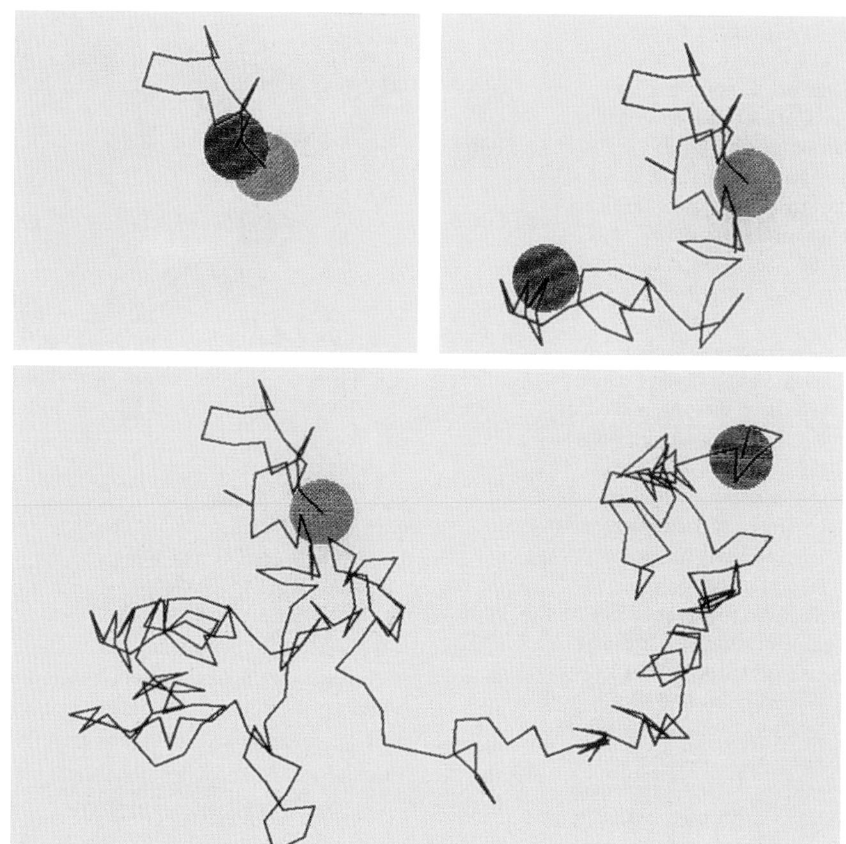

Natural rubber, the chemical name of which is *polyisoprene*, has a rather more complicated monomer, and it is interesting in that it introduces several new concepts. A single isoprene molecule has the structure $H_2C{=}C(CH_3)$–$C(H){=}CH_2$. In the common elastic variant, known as *cis*–1,4-polyisoprene, the repeating unit is as follows.

cis-1,4-polyisoprene

```
 H CH3 H  H
 |  |  |  |
-C-C=C-C-
 |        |
 H        H
```

Two of the backbone carbon atoms are joined together with a double bond. The single carbon–carbon bond is of the sigma type, around which the side-group atoms can easily swivel, while the double carbon–carbon bond displays the considerable resistance to rotation characteristic of the pi type. Another important feature of natural rubber is that the methyl group, CH_3, is so large that it interferes with the hydrogen atom on the next backbone carbon atom. Such interference is referred to as *steric hindrance*, and to avoid one group pushing against the other, the rubber molecule takes up a bent configuration. In *gutta-percha*, a much stiffer natural rubber that is a different form of polyisoprene, the methyl group and the hydrogen atom are on the opposite sides of the chain and therefore do not interfere. Two polymers that have the same chemical structure but different geometrical arrangements

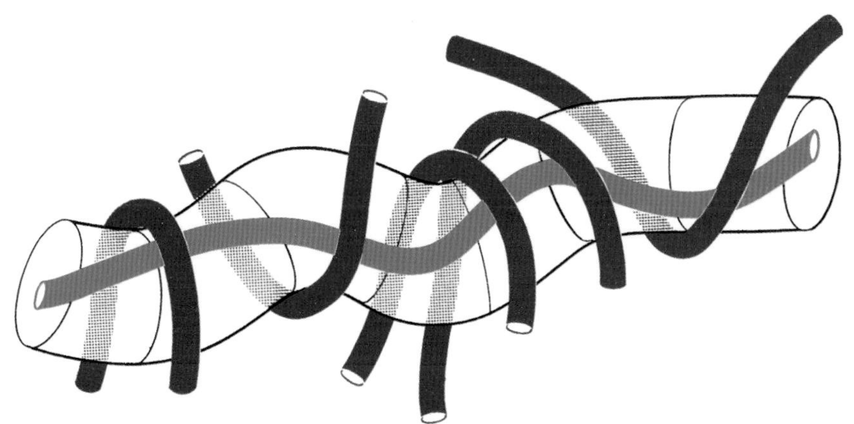

Diffusion of the individual molecules in a molten polymer is usually described by what is known as the *reptation model*, pioneered by Pierre-Gilles de Gennes and Sam Edwards. Each chain creeps, snake-like, along an imaginary tube defined by its neighbouring chains. The optimal average end-to-end length of a tube is governed by thermodynamics, and the ideal situation will prevail if the melt is given sufficient time. The overall motion of the polymer changes the conformation of the tube, and relatively rapid rates of flow cause departures from the equilibrium end-to-end distance. A chain can recover from such a disturbance, but this requires a *relaxation time*, the magnitude of which depends upon the temperature. These competing factors are responsible for the fact that a polymer can behave elastically under brief impact, and yet flow like a liquid given time.

are known as *geometric isomers*. The different arrangements frequently have a marked influence on mechanical behaviour. Because different configurations can occur, we must take advantage of the discriminating *cis–trans* nomenclature introduced in Chapter 12. In the examples just given, the elastic form of natural rubber is *cis*-1,4-polyisoprene, while gutta-percha is *trans*-1,4-polyisoprene. (The term *trans* indicates that certain side groups are on opposite sides of the backbone, while the *cis* form has both groups on the same side.) The *cis* version of rubber derives its elasticity from the kinked conformation that is forced on the chain by the steric hindrance between the impinging methyl and hydrogen side groups. Under tension, these kinks are straightened out, but the steric hindrance acts like a mechanical memory, and the rubber reverts to its kinked configuration when the load is removed. Gutta-percha, being devoid of steric effects, has no kinks to be straightened out. Tension is directly opposed by the backbone's covalent bonds, and the material is thus much stiffer. Interestingly, these two rubbers were both found in earlier versions of the golf ball: the elastic form at the centre and gutta-percha in the cover.

When the rubber is cross-linked, its elasticity is primarily determined by the constraints on the lengths of its interlink segments. A segment must terminate at its two defining cross-links, but there is a considerable degree of choice in the configuration that it can adopt between these two mandatory points. And these points are not fixed in space, because they are connected to other flexible chains. The problem is related to one solved by Albert Einstein in 1905. Known as the *random walk*, it concerns the average distance travelled in a journey of a given number of steps, if these have no mutual correlation. For such a drunkard's stagger, Einstein showed that the average distance is proportional to the square root of the number of steps. The concept might seem academic, but it turns out to be directly related to issues of practical consequence. In a single multifolded polymer chain, the average end-to-end distance determines the overall size of the molecular ball, which can be measured by light scattering. In a *cross-linked polymer*, the segments make a special type of random walk because of restrictions imposed by the end points. Sam Edwards and his colleagues have examined the consequences of these limitations for the material's elasticity. If the distance between the cross-links is comparable to the average random walk distance expected for the corresponding number of segments, there will be many different configurations of the latter which satisfy the terminal constraints.

When a molten polymer solidifies, numerous crystalline regions are nucleated almost simultaneously, and these grow outwards in the form of spheres. The individual regions are analogous to the grains in a metallic polycrystal, and they are referred to as *spherulites*. In this thin specimen of *polythene*, which was photographed in polarized light, the largest spherulites have diameters of about 0.1 millimetres.

The associated *entropy* will therefore be high. If there is considerable disparity between these two distances, on the other hand, the number of possible segment arrangements will be drastically reduced, and the entropy will be lower. In the extreme case in which the distance between two cross-links just equals the total segment length, there is only one possible configuration: a straight chain. As a piece of rubber is stretched, the distances between the various pairs of cross-links increase whereas the segment lengths remain constant. The above considerations apply, therefore, and stretching lowers the entropy. There is a consequent increase in free energy, and this opposes the applied stress. Further complexities arise if entanglements occur. These are links of primarily geometrical origin between chains, and they permit relative slip up to a limited amount. Such a material is weaker than the cross-linked variety at small extensions, but it becomes the stronger form when sufficiently stretched. *Entangled polymers* also display an interesting time-dependent effect. The material is elastic when a stretching force is first applied, but the structure gradually relaxes its configuration, and the elastic energy fades away. Pierre de Gennes, comparing the motion of an individual molecule to that of a snake, has called such relaxing migrations *reptation*.

Further insight into the nature of polymers can be gained by considering the different ways in which synthetic examples are fabricated. Broadly speaking such polymers can be divided into three classes: thermoplastic, thermosetting and non-mouldable. The important characteristic of the *thermoplastic* group is that they melt and resolidify with change in temperature, just as do simple materials composed of single atoms or small molecules. The interactions are very weak, so the molecules behave as independent entities and, if the temperature is high enough, slide past and over one another just like single atoms in simple substances such as copper. The higher viscosity of a polymer melt, compared with a metallic melt, derives from the necessity of cooperative motion in order to shift an entire polymer molecule. In *thermosetting polymers* chemical alteration is induced by the application of heat, and sometimes also pressure, so as to change an initially fluid-like material into a rigid solid. This is frequently achieved by inducing cross-linking between the linear molecules. One of the best known examples is the *vulcanization* of rubber by addition of sulphur and application of heat, as discussed below. *Non-mouldable polymers* represent an extreme in that they cannot be put into a fluid-like state without chemical decomposition occurring. Examples of natural polymers that are non-mouldable are *cellulose* and *asbestos*.

Polymer chains are produced by inducing chemical reactions between monomers so as to make them join to one another. The chemical reactions are invariably carried out in the presence of one or several *catalysts*, and the reactions themselves are of either the condensation or addition types. The polymerization process is possible because the monomers contain two or more functional units. Bifunctional monomers lead to the formation of linear polymers, while polyfunctional monomers produce branched polymers and three-dimensional network polymers. A typical example of a *condensation polymer* is *nylon*. The name, synonymous with a particular type of stocking, is actually generic for a family of polymers, also known as *polyamides*. Polymer formation by condensation is always a process of cutting up molecules and joining together parts of the fractions thereby

Condensation polymerization

A polymer produced by condensation has a chemical formula that is not merely a multiple of its constituent monomers. An example is *nylon*, which is nominally a *polyamide*. Synthetic polyamides are chemically related to the proteins and are structurally equivalent to an amino-acid condensation product. Industrial use of amino acids for this purpose would be inconvenient, however, and the practicable raw materials are diamines, $H_2N(CH_2)_nNH_2$, and dibasic organic acids, $HOOC(CH_2)_mCOOH$. The letters m and n indicate integers. The reaction involves condensation of the following type:

$$H_2N(CH_2)_nNH_2 + HOOC(CH_2)_mCOOH \rightarrow H_2N(CH_2)_nNHOC(CH_2)_mCOOH + H_2O$$

The liberation of a waste product, in this case water, is characteristic. A particularly common polyamide, nylon-6,6, has the following repeat unit:

nylon-6, 6

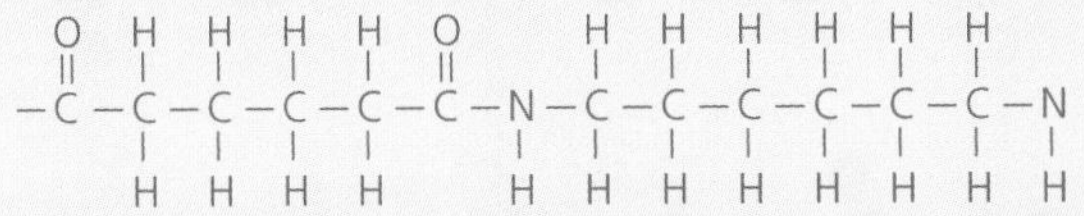

```
 O   H   H   H   H   O       H   H   H   H   H   H
 ‖   |   |   |   |   ‖       |   |   |   |   |   |
-C - C - C - C - C - C - N - C - C - C - C - C - C - N
     |   |   |   |       |   |   |   |   |   |   |   |
     H   H   H   H       H   H   H   H   H   H   H   H
```

The input, in this case, is a mixture of adipic acid and hexamethylene diamine. A good example of a *condensation polymer* having three links per repeat unit is *phenol formaldehyde*, usually referred to by its trade name, *Bakelite*. The repeating unit includes a benzene ring, and it has the following configuration:

phenol formaldehyde

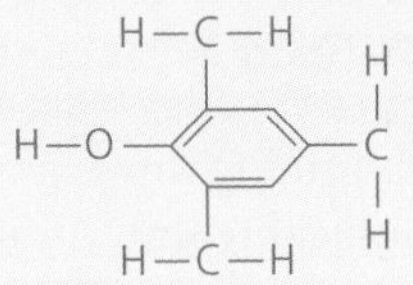

```
        H—C—H
                   H
          |        |
H—O—<  benzene  >—C
          |        |
                   H
        H—C—H
```

It forms a three-dimensionally connected network which produces a rather hard and brittle solid material. Bakelite was one of the most prominent structural plastics during the period up to the Second World War. It found use in radio cabinets, table tops, ashtrays, door handles and a host of other items.

produced. It invariably gives residual waste products, the water in the case of nylon fabrication being an example.

Addition polymerization is quite different from the condensation process in that it does not yield byproducts. It invariably involves the chemical interaction of molecules containing unsaturated regions. The term *unsaturation* refers to the fact that a molecule is associated with fewer than the maximum possible number of atoms, which are frequently hydrogens. Its characteristic is a multiplicity of bonding between the backbone atoms, and such bonds are said to be unsaturated if they are double or triple. It is usually the case that the reactivity of a molecule increases with the degree of unsaturation because the pi bond is particularly susceptible to attack. Moreover, the addition of an electronegative group usually makes a monomer even more reactive. For example vinylchloride, $CH_2{=}CHCl$, is more reactive than ethylene, $CH_2{=}CH_2$. On the other hand, if too many electronegative elements are present, the material will tend to be inert, an example of this being trichloroethylene, $Cl_2C{=}CHCl$.

Sulphur, the additive in the *vulcanization* process, is divalent. When it replaces hydrogen, it can set up links between the carbon atoms thus

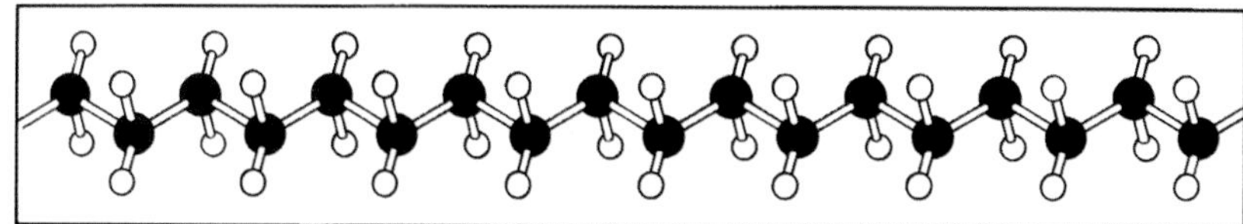

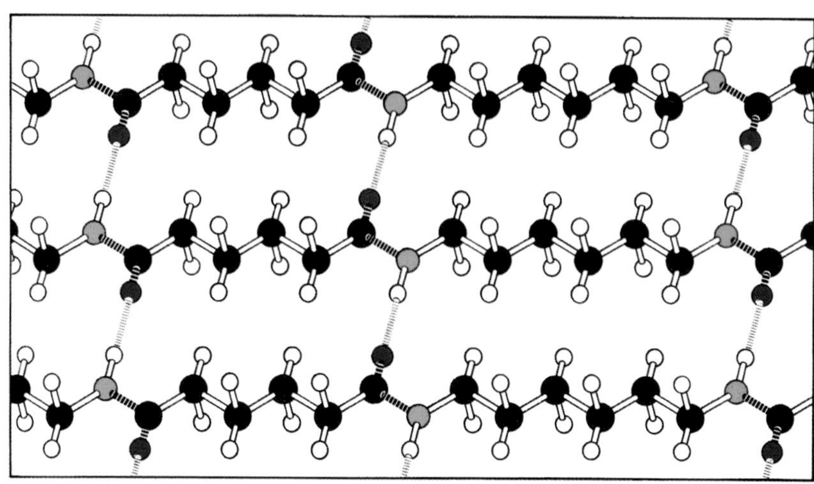

Cross-links between the individual molecular strands are an obvious source of rigidity in polymers. Such links can be of the strong covalent type, as with the *disulphide bonds* in *vulcanized rubber* and many *proteins*, but even the relatively weak *hydrogen bonds* prove adequate provided there are enough of them. This picture contrasts the situations in *polythene* (above), which does not show cross-linking, and *nylon*, with its high density of hydrogen bonds.

–C–S–C–, or, in the case of a disulphide bond thus –C–S–S–C. Depending on the actual amount of sulphur and the exact heat treatment, a range of end-products has been produced with mechanical properties varying from the *soft rubber* of a car tyre to the rigid solid of *ebonite*. Cross-linking through the agency of sulphur atoms is quite common in the biological realm. Among the more complex thermosetting plastics are the *epoxy resins*. The term is a combination of the Greek word for between and the abbreviation for oxygen. The latter appears in the chain backbone, as do benzene rings. The epoxies are usually self-hardening, by dint of their mildly exothermic reactions, which simultaneously produce linear chains and cross-linking.

One of the classical examples of *condensation polymerization* is seen in the production of *nylon*. Two solutions, one containing a diamine, the other a dibasic organic acid, react to produce nylon at their interface. In this demonstration, a solution of 1,6-hexane diamine is supported by a more dense solution of sebacoyl chloride. Each component supplies six-carbon monomers for the twelve-carbon repeats in the final nylon molecules. The polymer can be continuously pulled away from the interface in the form of a thread.

The mechanical properties of polymers are related to their structures. A widely accepted classification divides them into three types: rubbers, plastics and fibres. A *rubber*, or *elastomer*, is one that can be deformed to a high degree of strain in a reversible manner. The classical example of this type of behaviour is of course, that of *natural rubber*, which can be stretched up to several hundred per cent by simple tension but which will immediately regain its original shape when the tensile stress is removed. The other extreme is that displayed by a typical *fibre*. Such a material can be deformed reversibly by only a few per cent, but the intrinsic strength is considerably higher than that of a typical rubber. *Plastic materials* are intermediate between rubbers and fibres, both regarding deformability and strength. To appreciate the physical basis for the behaviour of these three general types we must delve deeper into their actual structures. The analogy between a polymer and a dish of spaghetti is again useful. When freshly cooked, the latter is in an almost fluid state, with the individual strands flowing past one another and constantly becoming tangled and untangled. If the spaghetti is left for a few hours, it becomes dry, and inspection reveals that it has a rather heterogeneous structure. In most of the bulk the strands are arranged in a more or less random tangle which is analogous to the amorphous state of a polymer. There are other regions, however, in which many spaghetti strands

Addition polymerization

The molecule of an *addition polymer* is simply a multiple of its monomer molecules. The *addition polymerization reaction* is frequently initiated by formation of a *free radical*, and it exploits the weakness of certain covalent bonds, such as the oxygen–oxygen bond in organic peroxides, ROOR, where R is some radical. The first stage is the splitting, or *homolytic cleavage*, of ROOR into two single RO · radicals, and the extra electrons on these free radicals can then interact with the pi electrons of the double bond of the monomer to be polymerized. The use of the dot symbol for the radical conforms with the *Lewis notation*. The polymerization then proceeds quite rapidly. An example is the formation of polyvinylchloride, for which the reaction sequence is as follows:

$$\mathrm{RO\cdot + CH_2{=}CHCl \rightarrow ROCH_2{-}CHCl\cdot}$$

$$\mathrm{ROCH_2{-}CHCl\cdot + CH_2 = CHCl \rightarrow ROCH_2{-}CHCl{-}CH_2{-}CHCl\cdot}$$

The growth process ends when two molecules which terminate with a free radical interact, as in

$$\mathrm{RO[CH_2{-}CHCl]_m\cdot + RO[CH_2{-}CHCl]_n\cdot \rightarrow RO[CH_2{-}CHCl]_{m+n}OR}$$

where *m* and *n* are whole numbers. Another example is synthetic rubber. Efforts to polymerize propylene, $CH_3{-}CR{=}CH_2$, were unsuccessful until Giulio Natta (1903–1979) discovered stereospecific catalysis in 1954. The propylene monomers are joined together by breaking half of the double bond, the CH_3 terminal becoming a side group. Natta realized that the previous failures had probably involved an atactic arrangement of the latter, and his suspicion that a catalyst of the type developed by Karl Ziegler (1898–1973) would produce an isotactic pattern

isotactic polypropylene

$$\begin{array}{cccccccccccccccccc}
 & CH_3 & & H & & CH_3 & & H & & CH_3 & & H & & CH_3 & & H & \\
 & | & & | & & | & & | & & | & & | & & | & & | & \\
- & C & - & C & - & C & - & C & - & C & - & C & - & C & - & C & - \\
 & | & & | & & | & & | & & | & & | & & | & & | & \\
 & H & & H & & H & & H & & H & & H & & H & & H &
\end{array}$$

proved to be correct. *Steric hindrance*, due to the methyl side groups, causes the polymer to coil into a helical conformation, giving it the desired compromise between rigidity and resilience.

lie parallel to one another in a crystalline array, and this too is observed in some polymers. Indeed, in many polymers both amorphous and crystalline regions coexist.

The factors that influence the arrangements observed in groups of polymer strands determine whether the material will be amorphous or crystalline. These factors are the stiffness of the individual strands and the interactions between them, and they are both determined by the chemical composition of the polymer. Stiffness is related to the degree of unsaturation in the backbone and to the possible effects of steric hindrance between side groups. Several factors determine the nature of the strand–strand interaction. There is the geometrical consideration of the fitting together of adjacent strands, which is favoured by small and regularly arranged side groups. Then there is the question of the interatomic forces between the strands. Whereas the covalent forces within the strand are of the strong primary type, the inter-strand forces are secondary and thus weak, except when cross-linking is present. The secondary forces involve either van der Waals or hydrogen bonds.

The ability of a polymer molecule to crystallize depends upon the flexibility of its backbone and the size and shape of its side groups. Regularity is present along the backbone, greatly facilitating the *crystallization process*, but the atoms along the chain must move cooperatively and this has an inhibiting influence. It is often the case, that the easiest path to crystallization involves the regular folding of the molecule back upon itself, as shown here.

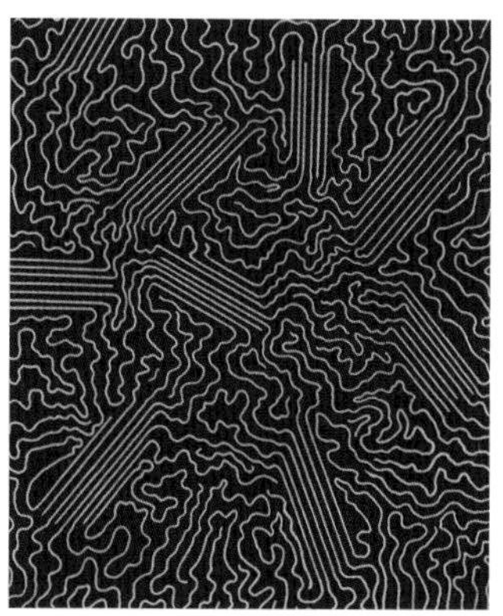

A fully crystalline polymer has never been produced, the maximum degree of crystallinity yet observed lying around 95%. Most common polymers consist of crystalline regions imbedded in an amorphous, or glassy, tangle of strands. This is usually a welcome compromise because the crystalline parts produce strength while the amorphous regions give the composite structure its flexibility.

A good example of an *amorphous polymer* is *atactic polystyrene*. It fails to crystallize because of the irregular arrangement of the bulky benzene rings that occur in its side chains. An example of the opposite extreme is *polytetrafluorethylene*, the PTFE or *Teflon* mentioned earlier. It is a simple linear polymer having a carbon difluoride monomer. The isotactic variant of polystyrene is also readily crystallized. If a polymer that is capable of crystallizing has fairly short individual chains, the degree of crystallization can be almost total. If the chains consist of several thousand individual units, on the other hand, only partial crystallization will occur. Such imperfect crystallization takes one of two characteristic forms. In one case the chains are found to fold back on themselves every 10 nm or so, producing an arrangement reminiscent of a Chinese cracker. Such structures are referred to as *lamellae*, and these tend to group together around nuclei to form what are known as *spherulites*. The nucleation of spherulites can be either homogeneous or heterogeneous, the latter occurring at some form of impurity. The final structure of a polymer showing this kind of behaviour is a collection of spherulites fitting fairly neatly to one another, with rather little amorphous material in between. As such the overall structure is not unlike the situation in a polycrystalline metal, because the narrow amorphous regions between the spherulites are analogous to *grain boundaries*. The spherulite structure is typical of polymers showing a high degree of crystallinity. In the second characteristic form of partial crystallization, usually referred to as the *fringed micelle structure*, the polymer molecules are neatly packed parallel to one another in some regions, while elsewhere the molecules are tangled in an amorphous mass. The dish of dried spaghetti, referred to earlier, is a particularly good macroscopic analogy to the fringed micelle structure. In this type of arrangement, crystallinity is not obtained only by individual molecules folding back on themselves, as in the case of the spherulite structure. Instead, an individual polymer molecule can pass through many different crystalline and amorphous regions. This form is typical of polymers showing a low degree of crystallinity.

Although the polymer molecules are lined up in a particular direction in one crystallite, different crystallites are oriented in different directions, so the overall structure is isotropic. If the solidified structure is subjected to deformation in one direction, the polymer chains tend to become aligned so as to produce a form of crystallization. The material is then said to be oriented because it will have developed a fibre axis along which the tensile strength is greatest. Alignment in this manner is sometimes achieved during the moulding of plastic items, by forcing the liquid polymer to pass through narrow constrictions. An even greater degree of alignment is achieved in fibres, such as nylon, by drawing them after the initial spinning process.

Thermosetting polymers that are amorphous tend to be rather brittle. They can be made less so by addition of substances composed of small molecules, which can occupy the interstices between the strands and function as a sort of molecular lubricant. Such substances are known as *plasticizers*, a typical example being *camphor*, which is added to cellulose nitrate to make *celluloid*. Another example is the addition of dioctylphthalate, which is used to plasticize polyvinyl chloride. This is the combination used in *plastic raincoats*. Unfortunately, such plasticizers, by dint of their small size, can gradually escape from the plastic so the product loses its flexibility with age. The cracking of old celluloid is a familiar example. A logical extreme of these ideas is found in many *paints, varnishes* and *adhesives*, in which thermosetting polymeric materials are suspended in organic solvents. The latter evaporate or dry, either at room temperature or with the application of heat, leaving behind a hard polymeric solid.

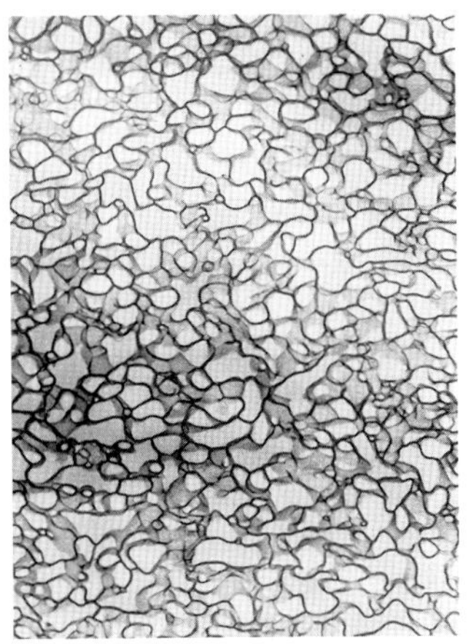

One trend in the search for new polymeric materials is the production of *composites*, in which one phase is incorporated in another. The examples shown here involve rubber–glassy composites. They were formed by synthesizing a *block polymer* within a monomer corresponding to one of the blocks: butadiene-styrene in *styrene*. The composite shown above is a 10% solution. The regularity of the rubber domains, and the thinner glassy regions, is a consequence of the *micellar structure* of the block polymer. When the concentration of the block polymer is higher, as in the 20% solution shown below, the rubber domains have numerous interconnections. Both composites have unusually high impact strengths.

Summary

Polymers are literally materials composed of many parts. Their structures consist of repeating groups of atoms, or monomers. The important organic members of this class, including artificial plastics, are usually long-chain molecules, sometimes branched or cross-linked. A typical linear polymer comprises a backbone along which side groups are arranged at regular intervals. Natural organic examples include proteins and cellulose. Polymers may be formed by condensation, which involves the splitting of smaller molecules and joining of some of the resulting fragments, or by addition, in which there are no waste byproducts. There are three broad classes: thermoplastic polymers such as polyethylene, which can be melted and resolidified and can thus be extruded into fibres; thermosetting polymers, in which irreversible changes occur on heating, as in the vulcanization of rubber; and non-mouldable polymers like Bakelite, which chemically decompose on heating. The physical state of a polymeric material, under given thermodynamic conditions, is governed primarily by chain flexibility, which is determined both by the backbone bonding and side-group size and shape. Polymers can crystallize, but never fully, the usual compromise being crystalline domains separated by amorphous regions.

15

Twist ye, twine ye! even so
Mingle shades of joy and woe,
Hope and fear, and peace and strife,
In the thread of human life.

Walter Scott
(1771–1832)
Guy Mannering

The vital threads: *biopolymers*

All molecules in a living organism serve a purpose, and we cannot say that a particular type is more important than any other. Nevertheless, it would be difficult to identify substances that play more varied and central roles than do the *biopolymers*. They contribute to the mechanical structure, and in the animal kingdom they provide the means of motion. They facilitate transport of smaller chemical units and mediate the chemical interaction between molecules, the biopolymers that serve this catalytic function being the proteins known as *enzymes*. There are biopolymers that form the immunological system and others that are involved in the mechanism of tissue differentiation. With so many functions to be fulfilled, we should expect to find many different types of these ubiquitous molecules, and this is certainly the case. It is believed, for example, that there are several hundred distinct enzymes within each cell of a higher organism such as a mammal. Even the task of perpetuating and translating the genetic message is entrusted to biopolymers.

The body of a typical *mammal* is mostly water – about 65% by weight in fact – but of the solid components no group is better represented than the proteins; they make up about 15% of the total weight. There are proteins in the skin, bones, tendons, muscles, blood and hair, and numerous proteins are constantly at work in the digestive system. In a *plant*, similarly, proteins govern the function of the living cells and are responsible for the manufacture of the cellulose walls which survive as fibrous tissue when these cells die. The name *protein*, coined by Gerhardus Mulder in 1838, is apposite because it is derived from the Greek word *proteios*, meaning of the first rank. Being polymers, proteins belong to a chemical family which also includes nylon, polyethylene, and rubber. A classification that groups one of the key building blocks of life with substances belonging to the inanimate world might seem puzzling, and the origin of the impressive versatility of proteins is deceptively simple. It lies in the nature of the protein monomer, an *amino-acid residue*. This has two parts, one of which is always the same and constitutes a building block in the polymer backbone, while the other is a *side group*. There are no fewer than 20 different common varieties of side group, and this is the source of the diversity. The polymer is known as a *polypeptide chain*. A different sequence of side groups produces a different polypeptide, and the number of possible combinations can be very large. Consider, for example, a chain consisting of 15 amino-acid residues. With a choice of 20 different side groups for each repeating unit, there are a total of $20^{15} = 3.3 \times 10^{19}$ possibilities. This recalls William Cowper's phrase: 'variety's the very spice of life'.

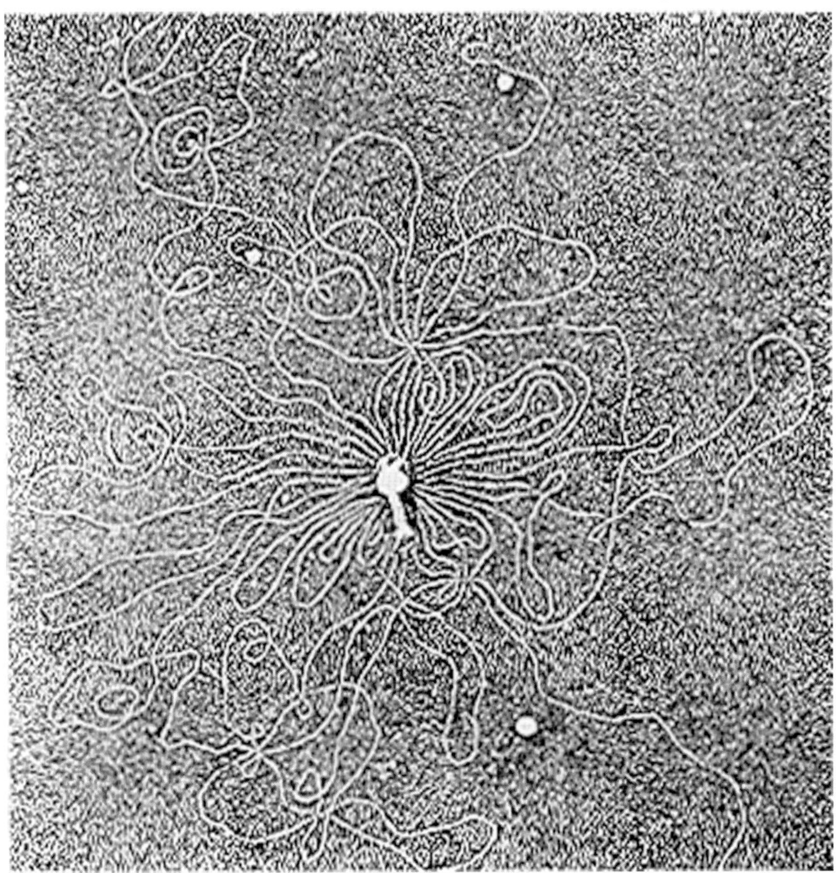

Viruses are microscopic agents which infect plants, animals, and bacteria. They are composed of the genetic material nucleic acid and a surrounding coat of protein or protein and lipid. The nucleic acid is deoxyribonucleic acid (DNA) or ribonucleic acid (RNA) in animal viruses, RNA in plant viruses, and usually DNA in bacteriophages, which are the bacteria-attacking variety. Lacking all but the most elementary cellular components, viruses are incapable of an independent living existence. A virus survives by injecting its nucleic acid into a cell and ruthlessly taking over the genetic apparatus, to produce numerous replicas of itself. Some viruses possess enzymes to help in the penetration of the host cell, while others achieve entry by purely mechanical means. This electron photomicrograph shows the coat of a *T2 coliphage* that has been ruptured by what is referred to as osmotic shock. This has split open the head of the virus, which has a diameter of about 100 nanometres, and released the single DNA molecule, the two ends of which are visible at the middle of the upper edge and the lower right-hand corner. When invading a cell, the virus attaches its tail to the cell membrane and squirts the entire 50-micrometre length of the DNA molecule into its prey by a sudden contraction. Its action thus resembles that of a hypodermic syringe.

The humorous point is often made that the monetary value of a human being, using current market prices for the constituent elements, is very little. The obvious fallacy in this approach is that it ignores the information content of the body. The real value is the sum of the prices of the various chemicals and the cost of arranging them in the correct configuration, and the former makes a negligible contribution to the total. This is hardly surprising. The same could be said of a television receiver. In view of the huge number of possible arrangements of amino-acid residues in a protein, the task of constructing a human body atom by atom, a sort of *ab initio* Frankenstein's monster, is a formidable proposition. Moreover, our 15-unit protein is a very modest affair compared with real structures, and it can easily be shown that the number of different possible proteins, having sizes encountered in terrestrial life forms, is greater than the number of particles in the Universe. It would seem that we have gone from one extreme to the other. If a polymer like nylon is relatively dull, because of its monotonous repetition, proteins might appear vulnerable by dint of their unmanageable variety; life based on these molecules might be precariously unique. Later, when considering proteins in action, we will see that they snatch security from what looks like dangerously wide scope for diversification.

Although each type of amino acid is chemically distinct, there is a strong familial resemblance between these molecules. They all have an acidic end, the *carboxyl group* COOH, and a basic end, the *amino group* NH_2, and since acids and bases tend to come together, neutralizing one another, structures

Essential and non-essential amino acids

The proteins that are ubiquitous in all living cells are composed of *amino acids*, of which there are twenty common types. *Autotrophic organisms*, such as most green plants, are able to manufacture all the amino acids they require, from simpler chemical compounds. In common with the other *heterotrophs*, the human being can synthesize only some of the necessary amino acids, and these are termed non-essential. Those that must be obtained through the diet are said to be essential. The amino acids not required from the environment are frequently produced by modification of one of the essential types. Under certain circumstances, this internal synthesis is inadequate and a normally non-essential amino acid must be extracted from food. An example is the special need for arginine by growing children. The division between the two classes is not strongly specific to a particular species. The following list, for adult humans, is equally applicable to birds, rats, and several insects.

Non-essential	Essential
alanine	histidine
arginine	isoleucine
asparagine	leucine
aspartate	lysine
cysteine	methionine
glutamate	phenylalanine
glutamine	threonine
glycine	tryptophan
proline	valine
serine	
tyrosine	

can be formed by joining amino acids head-to-tail fashion in a chain. An amino-acid residue is the group of atoms that remains after a molecule of water has been removed: H from the amino end and OH from the carboxyl end. When a link is added to the chain, it is the terminus of the latter which loses an OH group, while the approaching amino acid sheds a hydrogen atom. It was Emil Fischer who first showed, in 1904, that this is the basis of protein structure. The formation of a polypeptide is an example of *condensation polymerization*. The two hydrogen atoms and the one oxygen atom, shed by each amino-acid linkage, form one water molecule, the polymerization byproduct. It is an important feature of polypeptides that they never exhibit branching. They are always simple linear structures, although the chains can become cross-linked by the formation of secondary bonds. The chain has both carbon and nitrogen atoms in its backbone, but this does not make it unique. The same is true of nylon, for example.

Some of the most salient features of polypeptides derive from peculiarities of the *peptide bond*, and are an interesting example of principles discussed earlier in connection with covalent bonding. Each repeating unit contains two nitrogen–carbon single bonds, one carbon–carbon single bond, one double bond between the *carbonyl carbon* and the oxygen atom, one carbon–hydrogen single bond, one nitrogen–hydrogen single bond, and finally one bond between what is referred to as the *alpha-carbon*, denoted by Cα, and the side group. We would expect the carbon–oxygen double bond to be of the pi type and the carbon–nitrogen single bond to be of the sigma type. It turns out, however, that it is energetically more favourable for part of the pi orbital of the former to be transferred to the carbon–nitrogen bond, producing a single pi-like bond stretching from the oxygen atom across the carbonyl carbon atom and over to the nitrogen atom. Because a pi bond

Amino-acid polymerization

All proteins consist of one or more linear polymer chains known as *polypeptides.* Each monomer is the residue of an amino acid, of which there are twenty common varieties. Relatively rare types of amino acid are found in special locations, usually where the protein interacts with a nucleic acid genetic messenger. The polypeptide chain grows by sequential addition of amino-acid residues, this being an example of condensation polymerization. The reaction is as follows

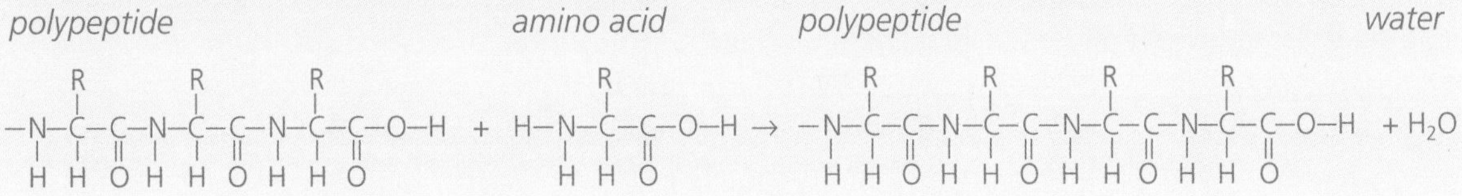

```
polypeptide                                        amino acid          polypeptide                                  water

      R         R         R                            R                      R         R         R         R
      |         |         |                            |                      |         |         |         |
 —N—C—C—N—C—C—N—C—C—O—H  +  H—N—C—C—O—H  →  —N—C—C—N—C—C—N—C—C—N—C—C—O—H  + H2O
   |  |  ||  |  |  ||  |  |  ||              |  |  ||            |  |  ||  |  |  ||  |  |  ||  |  |  ||
   H  H  O   H  H  O   H  H  O               H  H  O             H  H  O   H  H  O   H  H  O   H  H  O
```

the letter R representing one of the twenty common, or few rare, side groups. As with all polymerizations of the condensation type, there is a reaction by product, in this case water.

prohibits rotation, all three atoms must lie in the same plane. Moreover, the alpha-carbon atoms are found to lie in that same plane, because this happens to be the lowest-energy direction for the hybridized single-bond orbitals. The polypeptide polymer thus becomes a chain of units each of which is constrained to lie in a single plane. There is nevertheless considerable flexibility because the C–Cα and N–Cα bonds are both of the single sigma type and therefore permit essentially free rotation. Indeed, the main limitation to rotation about these bonds is imposed by the side groups.

The 20 different common amino acids are divided into two main groups, the distinction being whether the side group is polar or non-polar. The non-polar examples, which are hydrophobic, can be readily distinguished because their extremities usually consist of hydrogen atoms bound to carbons. There are nine *non-polar amino acids,* and they range from the simple *glycine,* in which the side group is a single hydrogen atom, to the complicated *tryptophan* with a side group containing nine carbon atoms, one nitrogen atom, and eight hydrogen atoms. By convention, each amino acid has been given a three-letter symbol, such as gly for glycine, ala for *alanine,* and trp for tryptophan. The other main group, those containing *polar side groups,* can be further subdivided into three classes. The first contains *aspartic acid* (asp) and *glutamic acid* (glu). At physiological pH, both develop negative charges at their extremities and are therefore acidic. As a consequence of the charge, they are also hydrophilic. A second class consists of *lysine* (lys) and *arginine* (arg). These develop a positive charge at physiological pH and are therefore basic and hydrophilic. The final class contains amino acids such as *glutamine* (gln) and *cysteine* (cys), which are all neutral at physiological pH. Whether a particular residue is polar or non-polar is of crucial importance to the *conformation* of the polypeptide, when it is in aqueous solution. The non-polar residues, particularly the more bulky of them, avoid water by tending to lie at the interior of the protein. Conversely, the polar residues prefer to be located on the outside. A change of solvent causes some of the residues to move either towards the outside, or inside, leading to subtle changes of conformation that are frequently of great physiological importance. The catalytic action of some enzymes appears to be influenced by such changes of chemical environment.

Links between polypeptide chains

The three-dimensional structure of a *protein* is determined by the sequence of amino acids along its polypeptide chain or chains. These influence conformation in various ways. The chemical nature of the various amino-acid side groups dictates the way in which the polymer folds, minimizing the free energy to arrive at its thermodynamically most favoured state. Further stabilization can be achieved by bonds which link points on a single chain or adjacent parts of different chains. The strongest of such bonds is the *disulphide* or *cystine bridge*. It is formed when two cysteine side groups approach, lose their hydrogen atoms, and produce a sulphur-sulphur bond.

cysteine side groups *cystine bond*

```
   O   H   H   O   H           O   H   H   O   H
   ||  |   |   ||  |           ||  |   |   ||  |
 - C - C - N - C - C -       - C - C - N - C - C -
       |           |               |           |
   H - C - H       R           H - C - H       R
       |                           |
       S - H                       S
                                   |
   H - S                           S
       |                           |
   H - C - H       R           H - C - H       R
       |           |               |           |
 - C - C - N - C - C -       - C - C - N - C - C -
   ||  |   |   ||  |           ||  |   |   ||  |
   O   H   H   O   H           O   H   H   O   H
```

The bond strength is about 0.35 attojoules, about ten times larger than that for a hydrogen bond. The most common form of the latter links a carbonyl oxygen with the hydrogen atom attached to the backbone nitrogen atom, for example:

hydrogen bonds

```
      R   H       R   H
      |   |       |   |
  - C - C - N - C - C - N -
    ||  |       ||  |
    O   H       O   H
    .           .
    .           .
    .           .
    H   H       H   H
    |   |       |   |
  - N - C - C - N - C - C -
        |   ||      |   ||
        R   O       R   O
```

A third common link is the salt bridge, an arrangement which is possible when (basic) amino and (acidic) carboxyl side groups are in close proximity. An example is the interaction between aspartic acid and lysine.

aspartic acid *lysine*

```
      |                                            |
  H - N                        H                   N - H
      |              //O       |                   |
  H - C - CH2 - C              H - N+ - (CH2)4 -   C - H
      |              \O-       |                   |
  O = C                        H                   C = O
      |                                            |
```

The stabilizing force is the electric attraction between the two opposing charges. It is not a bond in the usual sense in that there is no sharing of electrons, and the attraction survives in spite of considerable variation in the charge–charge distance because electric forces are long-ranged.

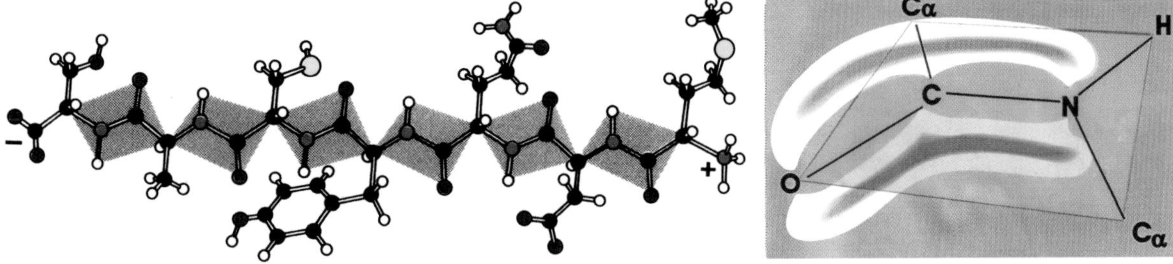

The *primary structure* of a protein is a linear chain polymer, or several such chains, the individual monomers being *amino-acid residues*. The link between each successive pair of monomers involves a bond between a carbonyl carbon atom and a nitrogen atom, but although the chain itself is flexible, this type of bond is quite rigid. The stiffness arises from the presence of an extended pi bond which stretches from the nitrogen atom across the carbonyl carbon and on to the carbonyl oxygen atom, as shown in the smaller picture. This *amide link*, as it is called, is sufficient to hold six atoms in a single plane, and the chain flexibility derives entirely from the mutual rotations that are possible at the joints between adjacent amide links. As is clear from the chain shown, the rotations occur at the other carbon atoms, known as the alpha-carbons, and these are also the sites of side-group attachment.

Before moving on to the three-dimensional structure of proteins, further remarks should be made about the polypeptide chain. The chemical formula for the amino-acid molecule makes no reference to the surrounding solvent. The molecule is generally ionized when in aqueous solution. Just what happens depends on the availability of hydrogen ions, that is on the pH. In an acidic solution there are many free hydrogen ions, and the amino group becomes positively ionized. In a basic solution, on the other hand, the carboxyl group is negatively ionized, because a hydrogen ion is removed from the molecule in an effort to make up for the lack of such ions in solution. At neutral pH, that is for values around 7, both the amino and carboxyl groups are ionized, and the molecule is said to be dipolar or *zwitterionic*. Because both ends are so readily ionized, the molecule is highly soluble in water. This mode of ionization does not favour the spontaneous establishment of the peptide bond, because the latter involves removal of a hydrogen from the NH_2 end, whereas in the zwitterionic form this terminal becomes NH_3^+. The stringing together of a number of amino acids to form a polypeptide chain requires the mediating agency of a catalyst, and in the living cell this job is done by the *ribosomes*. These are partly composed of protein, and this introduces an important dimension to the story, for we see that proteins are not just the products of a cell's machinery. In some cases, they are also the machine tools. Another interesting consequence of the aqueous ionization of the amino and carboxyl groups is that a polypeptide has a well-developed sense of direction. Irrespective of how many units the chain contains, one end is always an amino terminal and is therefore positively charged, while the other end is always a negatively charged carboxyl group. We will later see that the sequence starts at the amino end. Thus the polypeptide asp–gly–arg, for example, is not the same as the polypeptide arg–gly–asp.

The sequence of amino acids in a protein defines what is known as its *primary structure*. The question arises as to how it transforms from a linear polymer molecule to a three-dimensional structure, given that branching does not occur. There are several factors. One comes from *steric effects* due to the side groups, and another is the establishment of various bonds, the strongest type being the *disulphide bridge*. This is formed when two cysteine residues approach each other and lose their hydrogen atoms. The two sulphur atoms link together, and the bond strength is about 0.35 aJ. This is an oxidation reaction. The disulphide bond is also referred to as a *cystine bond*. The weaker kind of secondary link that binds polypeptide chains together is the *hydrogen bond*, in which the hydrogen atom attached to the nitrogen atom on one chain forms a bond with the carbonyl oxygen of another chain. In practice, since neighbouring polypeptide chains often lie parallel to each other, hydrogen bonds frequently occur in groups. The strength of

Ionized forms of amino acid

Amino-acid molecules are usually ionized when they are in aqueous solution. The form of the ionization depends upon the pH. In an acidic solution, there is an abundance of free hydrogen ions, and the amino end of the molecule tends to lose an electron in an effort to neutralize the surrounding positive charges. It thereby becomes positively charged, the structure becoming

```
     H    R
     |    |
H — N+ — C — C — O — H
     |    |    ||
     H    H    O
```

In a basic solution, with its low concentration of positively charged hydrogen ions, the carboxyl end readily loses a proton, acquiring a negative charge in the process. The molecule then takes the following form:

```
          R
          |
H — N — C — C −− O
     |    |    ||
     N    H    O
```

At neutral pH, for which the numerical value is around 7, both termini are charged

```
     H    R
     |    |
H — N+ — C — C −− O
     |    |    ||
     H    H    O
```

and the molecule is said to be dipolar or zwitterionic. With two charged ends, this form is highly soluble in water. It does not favour the formation of a peptide bond, however, because the latter requires the loss of a hydrogen atom from the NH_2 end rather than capture of such an atom to produce NH_3^+.

the hydrogen bond is only about 0.03 aJ. A third type of link which can stabilize a protein is the *salt bridge*. This can be established when amino and carboxyl side groups are in close proximity. An example is the interaction between aspartic acid and lysine. The stabilizing force is the electric attraction between the two opposing charges. It is not a bond in the usual sense in that there is no sharing of electrons, and the attraction survives in spite of considerable variation in the charge–charge distance.

An interatomic bond can always be broken if sufficient energy is available, as, for example, through heating. Hydrogen bonds are weaker than cystine bonds, so they are ruptured first as the temperature of a protein is raised. This causes partial unwinding, and the protein is said to be denatured. A *denatured protein* has not necessarily been destroyed; if the temperature is lowered, it can regain its original three-dimensional conformation, with the same hydrogen bonding. This is expected because the cell builds its proteins one amino acid at a time, and the finished chain must be able to fold up into its correct shape. The latter is uniquely determined by the sequence of amino-acid side groups. Thus although understanding of protein function requires knowledge of spatial configuration, investigation of primary structure, that is the amino-acid sequence, is an indispensable first step.

Protein structure can also be permanently disrupted. Examples from everyday experience include the chemical denaturation of milk protein by curdling it with vinegar, the thermal denaturation of egg or meat protein

In the context of *protein structure*, an amino-acid molecule can be imagined as comprising four parts. There is a detachable hydrogen atom at one end and a detachable hydroxyl unit at the other. The part remaining when these termini are gone is referred to as a *residue*, and it is this portion that contributes to the protein structure. The residue itself consists of the backbone unit and a *side group*, and protein variety derives from the fact that there are twenty different common types of the latter. The side groups are broadly classified by their polarity, and the polar varieties are further characterized by their pK values, that is to say by the pH at which they have a 50% probability of being dissociated.

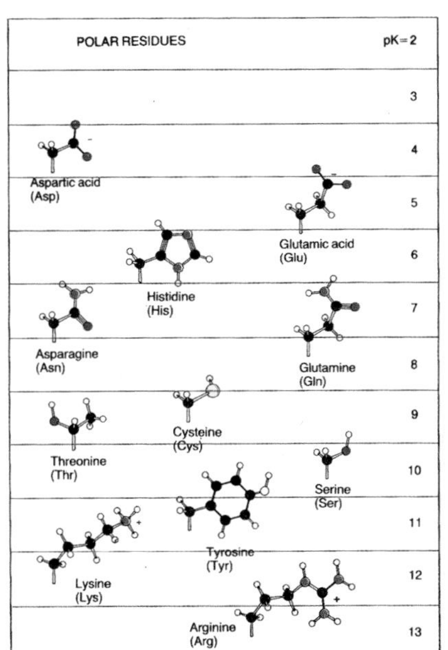

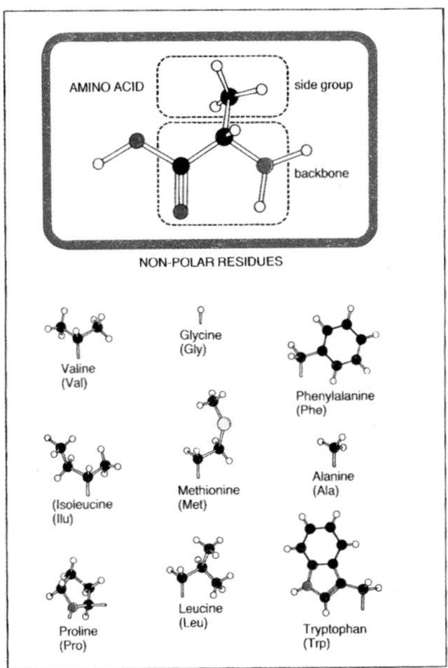

when it is cooked, and the mechanical denaturation of egg white when it is beaten. In all these cases the process is irreversible; like Humpty Dumpty, the protein cannot be reassembled. In the mid 1950s Christian Anfinsen studied milder reactions in which proteins were reversibly denatured and discovered important facts about the development of three-dimensional protein structure. The object of his investigations was *bovine pancreatic ribonuclease*. Werner Hirs, Stanford Moore and William Stein had shown this to be a 124-residue single-chained polypeptide, having eight cysteines which pair off to produce four cystine bonds. Now there are 105 ways in which eight cysteines can pair, and only one corresponds to the natural and enzymatically active version of the protein. The remaining 104 combinations were referred to as being scrambled. Cystine disulphide bonds are irreversibly oxidized by performic acid, each sulphur collecting three oxygen atoms. They can be reversibly cleaved and reduced by *beta*-mercaptoethanol. Anfinsen found that denaturation in *beta*-mercaptoethanol and a strong urea solution was reversible; removal of these agents by dialysis permitted the eight cysteines to re-establish the correct four cystine bonds, and enzymatic activity was restored. Removal of the *beta*-mercaptoethanol had allowed the sulphurs to re-oxidize by forming disulphide bridges. He went on to show that if this re-oxidation occurs in the presence of the urea, only about 1% of the enzymatic activity is regained. Most of the protein molecules were being reassembled in scrambled versions. The urea apparently disturbed the folding of the polypeptide chain. Finally, Anfinsen discovered that reintroduction of traces of *beta*-mercaptoethanol enabled the aberrant molecules to gradually unscramble themselves. He interpreted these results as indicating the dominant role of *thermodynamics* in protein folding. The three-dimensional structure of a native protein in its normal physiological environment, consisting of solvent at the correct acidity, ionic strength and temperature, is the one for which the free energy is lowest. As Anfinsen put it: in terms of natural selection through the design of macromolecules, a protein molecule only makes stable, structural sense when it exists under conditions similar to those for which it was selected – the physiological state.

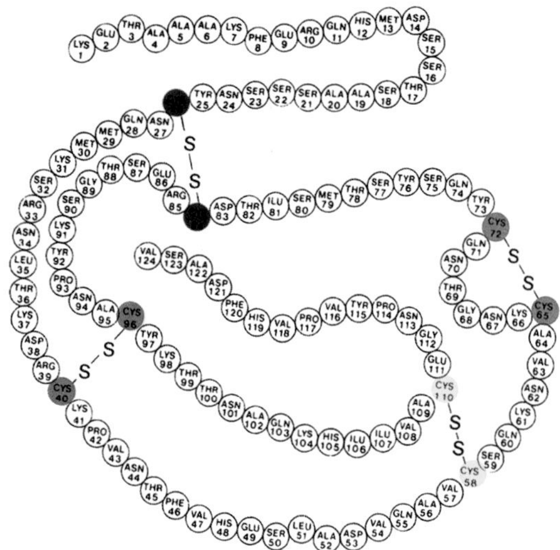

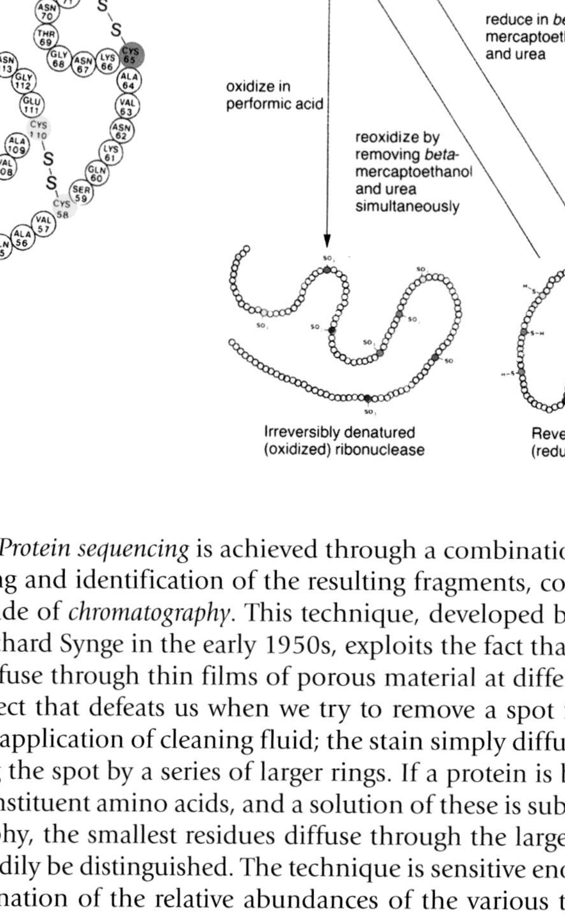

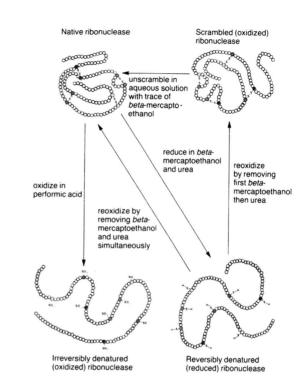

Although it is not yet possible to predict the three-dimensional structure of a protein from a knowledge of its amino-acid sequence, it is clear that the latter unambiguously determines the most stable configuration of the molecule. And the work of Christian Anfinsen (1916–1995) has established that the observed form of a protein is that determined by thermodynamics. *Ribonuclease* consists of a single polypeptide chain, of 124 amino-acid residues, its structure being stabilized by four *disulphide bridges*. Anfinsen showed that this protein can be reversibly denatured in a solution of *beta*-mercaptoethanol and urea, the former compound disrupting the bridges. The urea apparently influences the conformation of the backbone, because its delayed removal from the denaturing solution allows the protein to adopt a scrambled arrangement in which bridges form between the wrong pairs of cysteine residues. The thermodynamically correct structure can be recovered from such a metastable state by gentle warming in a dilute solution of *beta*-mercaptoethanol. This permits the disulphide bonds to rearrange themselves.

Protein sequencing is achieved through a combination of chemical chopping and identification of the resulting fragments, considerable use being made of *chromatography*. This technique, developed by Archer Martin and Richard Synge in the early 1950s, exploits the fact that different molecules diffuse through thin films of porous material at different rates. This is the effect that defeats us when we try to remove a spot from soiled clothing by application of cleaning fluid; the stain simply diffuses outwards, replacing the spot by a series of larger rings. If a protein is broken down into its constituent amino acids, and a solution of these is subjected to chromatography, the smallest residues diffuse through the largest distances and can readily be distinguished. The technique is sensitive enough to permit determination of the relative abundances of the various types of residue. The process is repeated with weaker *lysing* (i.e. disrupting) *agents*, to produce larger fragments, and these too are analyzed by chromatography. By using different agents to produce different distributions of fragments, it is possible to derive the entire *amino-acid sequence* unambiguously. Such investigations are facilitated by certain enzymes, which preferentially combine either with the amino or acid end of the polypeptide. Frederick Sanger completed the first sequencing of a protein in 1954. Using such enzymes as *pepsin* and *trypsin*, he sequenced the entire length of 51 amino acids in the protein *insulin*, finding it to be composed of two chains, one with 30 residues and the other with 21, linked together by two cystine disulphide bonds. Since this first success, many other proteins, with increasingly higher molecular weight and complexity, have been sequenced.

A means of unambiguously predicting the three-dimensional conformation of a protein, starting only with its primary structure and a knowledge of the interatomic forces, could be likened to finding the Rosetta Stone of protein structure. In spite of some progress, through the individual

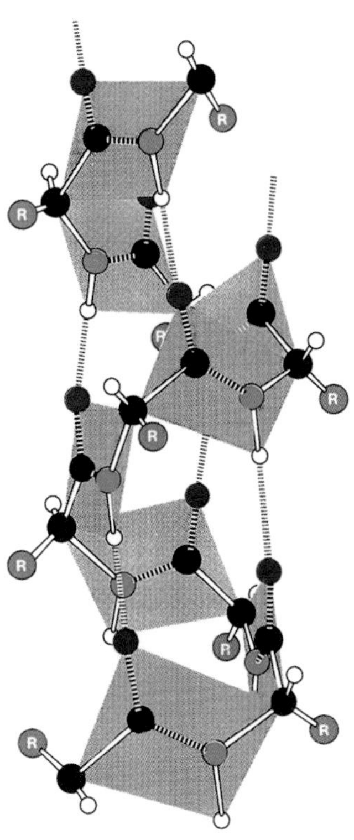

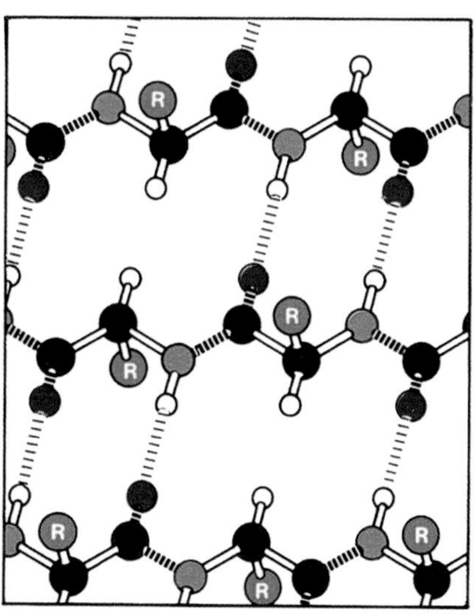

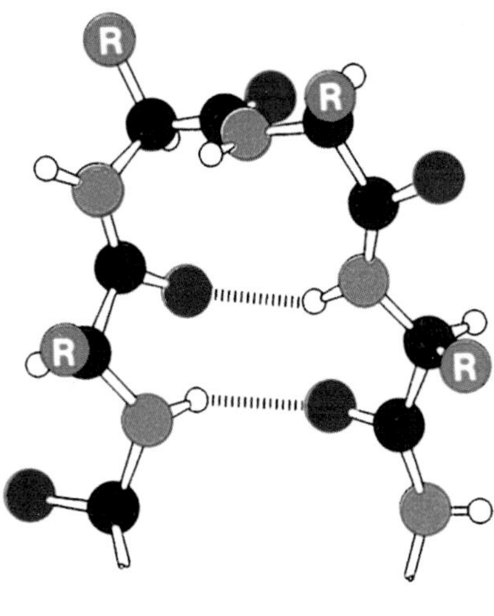

A key feature of the *polypeptide chain* is the oxygen and hydrogen atoms, none of which are involved in the bonding responsible for the primary structure. These can pair off and produce the hydrogen bonds which play a major role in determining a protein's *secondary structure*. There are several particularly common arrangements. The chain can adopt the form of a single helix, the oxygen of one amide unit linking with the hydrogen of a unit farther along the chain. A particularly stable helix, the *alpha helix*, results when the linking involves every third amide unit. This structure (left) is found in many proteins, and it is the main constituent of the *keratin* in hair fibrils. (Collagen, which is described in Chapter 18, involves a triple helix.) Another common structural feature of proteins is the *beta sheet* (middle). When separate polypeptide strands lie adjacent and parallel, but run in opposite directions, their oxygen and hydrogen atoms can form hydrogen bonds, producing a pleated sheet configuration. A typical example of this arrangement provides the structural basis of *silk*. When the beta sheet is present in proteins it usually occurs in patches that are much smaller than that shown in the diagram. A special limited form of the beta sheet often occurs where a chain bends back on itself; the two sides of such a loop effectively run in opposite directions, permitting hydrogen bonds to form (right) as indicated.

efforts of Henrik Bohr, Cyrus Chothia, Christopher Dobson, Alan Ferscht, John Finney, Hans Frauenfelder, Martin Karplus, Cyrus Levinthal, Michael Levitt, Per-Anker Lindgaard, Gopalasamudram Ramachandran, Frederick Richards, Barry Robson, Harold Scheraga, Peter Wolynes and others, this is not yet possible, and one must still resort to structure analysis by X-ray scattering. Such activity is almost as old as X-ray crystallography itself, the pioneers being William Astbury, John Bernal, Dorothy Hodgkin, Kathleen Lonsdale and J. M. Robertson. Two major difficulties are encountered: getting the large protein molecules to form good crystals, and overcoming the notorious *phase problem*. The latter arises because X-rays cannot be focussed in the same manner as light rays. One must photograph a *diffraction pattern* and then reconstruct the object from it. Unfortunately, only the amplitudes can be recorded, not the phases, so reconstruction of an unambiguous picture of the object presents difficulties. Nowadays, one measures the differences between the diffraction patterns of the protein with and without the attachment of a heavy atom. The latter dominates the scattering and alters all the phases. This heavy atom must not change the structure of the molecule. Hence the name of the technique: *heavy atom isomorphous replacement.*

Another approach that has scored several spectacular successes, though still relatively young, involves building scale models. It requires knowledge of atomic sizes, interatomic bonding, a grasp of relevant biophysical and biochemical concepts and, not the least, intuition and inspired guess work. Possessing these prerequisites in good measure, Linus Pauling initiated studies in the late 1930s aimed at revealing the basic structural units of proteins. The belief that different proteins might contain identical building blocks stemmed from the fact that the diffraction patterns of such things as hair, finger nail, porcupine quill, muscle, skin, connective tissue, and even the flagellae of some bacteria, are remarkably similar. It was known that a polypeptide resembles a charm bracelet, when stretched out; a closely packed one with an amino-acid side group hanging from every link. The problem lay in finding a way for this chain to fold into a structure which, through its simplicity, could be a common feature of proteins. Because the side group sequence varies from one protein to another, it seemed that the backbone must fold without these appendages getting in the way. Together with Robert Corey, Pauling discovered a polypeptide structure that is consistent with all the experimental evidence and attractively simple. The backbone is simply coiled up into a helix. The distance between adjacent turns of the structure, the pitch, is such that it could just be spanned by a hydrogen bond. The overall structure resembles a tube around which the amino-acid side groups spiral. It was subsequently established that this *alpha helix*, as it is called, is indeed the *secondary structure* of many sections of polypeptide in proteins.

If all proteins consisted of a single alpha helix, the only variation being in length and variety of side groups, elucidation of protein structure would be trivial, and these molecules would hardly be capable of the great variety of functions they perform. Secondary structure is by no means the whole story. The physiological properties of a protein emerge only at the third and fourth levels: the tertiary and quaternary structures. A good analogy to tertiary structure is what happened to the helical cord of one of the old-style telephones when bunched up. Some sections remain fairly straight, but there is considerable kinking where the helix folds back on itself. It is at this level of structure that the nature of the side groups, and their actual sequence, becomes significant. Folding brings certain side groups into close proximity, and side group size and chemistry determine the configuration of the resulting structure. A mutation which changes the primary sequence of a protein, substituting one side group for another, for example, can have a profound influence on *tertiary structure. Quaternary structure* arises when a protein consists of more than one polypeptide chain. A classic example is *haemoglobin*, which has four chains. The tertiary structures of some proteins are actually quite simple and do not involve folds or even sharp changes in direction. They appear primarily in structural tissue. Such structures often contain many helices packed together. Nature solves this packing problem by arranging the helices in a hierarchy of superstructures. In *alpha keratin*, the basic subunit of hair, nail, horn and similar substances, the alpha helices come in groups of seven, with six helices wound around a central helix. These seven helices form a single strand, and the strands themselves are arranged in similar groups of seven. Corroboration for these superstructures, which were proposed independently by Francis Crick and by Linus Pauling

Another labour-saving device

In Chapter 12, we noted that the positions of the carbon atoms in molecular diagrams is often conveyed implicitly, because there are usually so many of them. Jane Richardson developed a similar labour-saving convention for secondary-structure motifs in the depiction of proteins. In her scheme, the alpha-helix is merely indicated by a spiral-formed ribbon, with no attempt to show the individual atoms, while the beta-sheet is replaced by an analogously atom-free pair – or set – of flat ribbons. By equipping the latter with arrow heads, Richardson was able to indicate whether the beta-sheet belonged to the parallel or anti-parallel variety. Three examples of protein illustrations using her convention are shown here. The first picture (with the alpha helices coloured blue) shows a molecule of Lipolase (a lipase), which catalyzes the reaction in which a triglyceride is cleaved to glycerol and a free fatty acid. The second picture (with the helices in red) shows a molecule of Novamyl (an amylase), which breaks down starch into smaller fragments. Finally, the third picture (with helices in yellow) shows a phytase, which catalyzes the reaction in which inorganic phosphate groups are liberated from phytic acid. The Richardson convention for depicting alpha helices is sometimes further simplified by merely indicating them as cylinders. An example appears in the final chapter of this book.

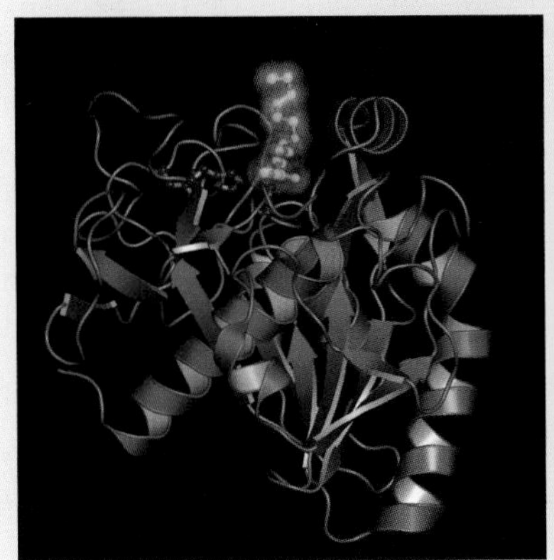

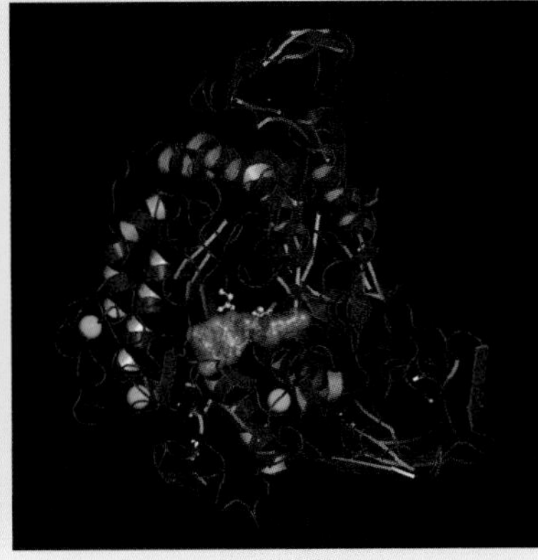

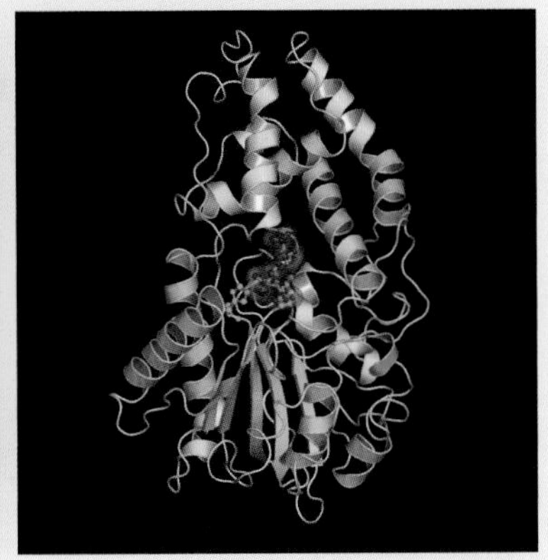

and his colleagues, was supplied by X-ray diffraction. The same technique, in the hands of Max Perutz, provided support for the alpha-helix model.

Although the original investigations of amino-acid sequence were both ingenious and inspired, one aspect of the work required little fantasy. The tools for cutting proteins into smaller sections were already at hand; the body uses such agents in its digestive system. They are in fact other proteins, known as *enzymes*, and they collectively chop the ingested protein into individual amino acids. The latter are augmented by other amino acids that the organism itself produces. Other enzymes then reassemble these units into the proteins required by the organism. Although Walter De La Mare was not aware of these processes, his verse nicely sums them up: *It's a very odd thing – as odd as can be – that whatever Miss T eats turns into Miss T.*

The question arises as to how the units are reassembled in the desired order. The actual functioning of proteins will be discussed later in this chapter, but it is already clear that these impressive molecules must be put together according to a specific plan. What acts as the blueprint? The digestive action of enzymes such as pepsin was known well before they were used to help unravel protein structure. Using pepsin to break down cells by destroying their proteins, in 1869, Friedrich Miescher was surprised to

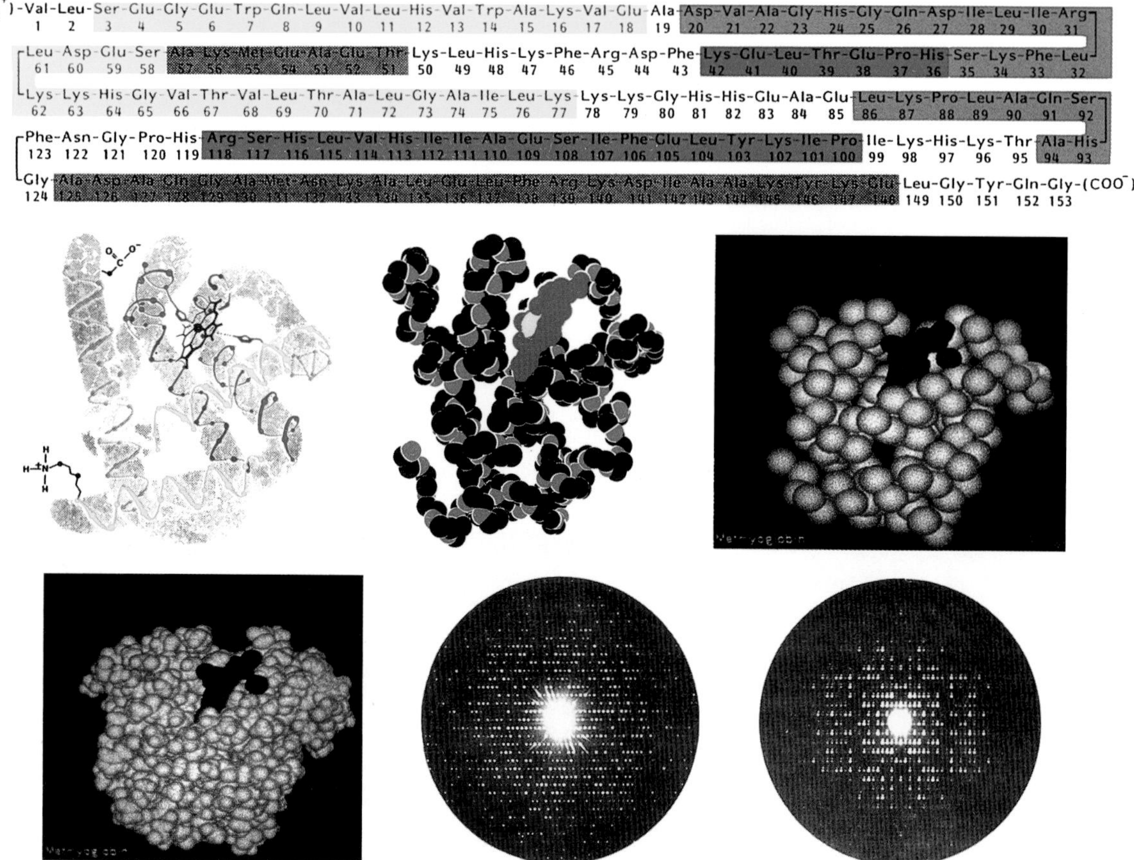

The *tertiary structure* of a protein is determined by its content of such secondary units as helices and sheets, and by the way in which these are packed together. *Myoglobin*, the first protein for which the three-dimensional structure was determined, is a bun-shaped molecule 4.5 nanometres wide, 2.5 nanometres thick, and with a weight of 17800 daltons. Of its 153 amino-acid residues, 121 are involved in *alpha helices* and there are eight such sections. These are indicated by colours in the table, which lists the amino-acid sequence. The molecule is compact, with very little free space in its interior, and it provides a sort of capsule for the *haem group*, which is the molecule's biochemically active site. Four views of the molecule are shown, all from the same angle. The first indicates only the alpha-carbon atoms and it reveals the arrangement of the helices, these being coloured according to the convention used in the table. The haem group is shown in red. The second picture shows all the carbon (black) and nitrogen (blue) atoms in the polypeptide backbone, and the haem group appears in green, with its central iron atom in yellow. The first of the less detailed computer-generated pictures shows each amino-acid residue as a single sphere, while the second comprises all the non-hydrogen atoms. The haem group appears in red. The structure of myoglobin was established through the X-ray diffraction studies of John Kendrew (1917–1997) and his colleagues. However, an X-ray film can record the amplitudes but not the phases of the diffracted waves. This was overcome by taking diffraction pictures with and without a heavy atom attached to each of the myoglobin molecules. Such an atom so dominates the X-ray scattering that it shifts the phases of the various waves, producing the differences apparent in the two diffraction patterns. The one on the left was produced with a normal crystal, while the composite picture on the right shows the patterns of normal and heavy-atom-labelled crystals superimposed but slightly out of register.

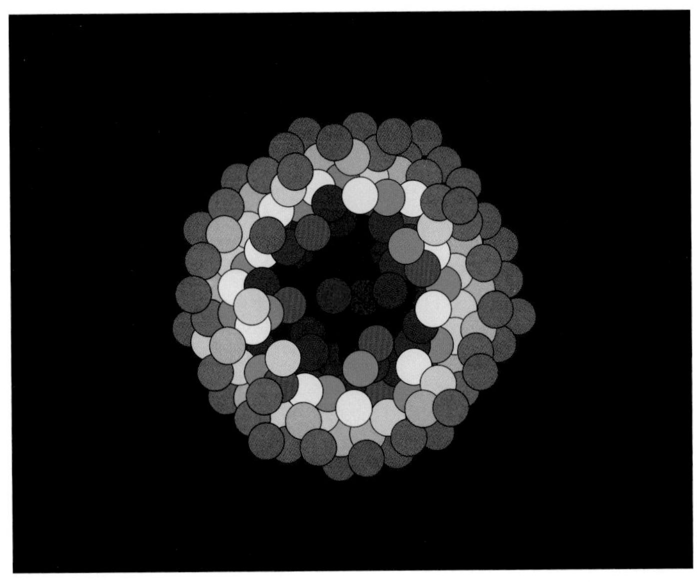

Proteins are not static structures. *Thermal agitation* causes their constituent atoms to vibrate about their equilibrium positions, thereby permitting the overall structure to explore many different configurations. Some of these are conducive to interaction with nearby molecules, and it is in this manner that proteins belonging to the class known as *enzymes* perform their useful functions. The atoms of the symbolic protein shown here have been coloured so as to indicate the relative amplitude of their vibrations, the violet, blue and green colours indicating smaller amplitude while the yellow, orange and red colours correspond to successively higher amplitude. Not surprisingly, it is the atoms lying at the protein molecule's surface which have the largest vibrational amplitudes because they are subject to the weakest positional constraints. And it is naturally these surface atoms which are candidates for interacting with other molecules. Bacteria have been found in the Antarctic whose proteins have been selected by evolution for the unusually large vibrational amplitudes of their surface atoms, this apparently compensating for the lower temperatures in which they exist.

find a residue that had withstood attack. The substance was distinctly non-protein; it contained far more phosphorus than is usually found in protein. Finding it to be associated with cell nuclei, he called the new substance nuclein, but later changed this to *nucleic acid* when he found it to have pronounced acidic properties. Miescher discovered that the thymus glands of young calves and salmon sperm were particularly rich sources of his new substance. Not long after, his senior colleague Felix Hoppe-Seyler found a similar yet distinct substance in bacteria, yeast and plants. Its molecular weight was comparable to the animal variety, but its atomic composition was slightly different. They concluded, erroneously as it turned out, that one variety of nucleic acid is always associated with animal cells while the other type is found only in plant cells. Albrecht Kossel became interested in the structure of nucleic acids and broke down specimens by hydrolysis. He found that the two varieties were indeed different. The animal type contained sugar, phosphoric acid, the *purine* bases *adenine and guanine*, and the *pyrimidine* bases *cytosine* and *thymine*. The plant variety was different in that thymine was replaced by another pyrimidine: *uracil*. Shortly after the completion of this analysis, Phoebus Levene showed that the sugars too are different in the two varieties of nucleic acid. To begin with he found that both sugars are somewhat unusual, containing five carbon atoms rather than the more common six. He also found that the sugar of the animal type lacked an oxygen atom and was the compound *deoxyribose*. In the plant variety, the sugar turned out to be *ribose*. The two nucleic acids have since come to be known by their acronyms *DNA*, for *deoxyribonucleic acid*, and *RNA*, for *ribonucleic acid*. Levene also discovered how the different units are connected together. A purine or pyrimidine base is attached to the sugar, and this in turn is joined to the phosphate unit. Such a composite came to be known as a *nucleotide*, but no information could be gained as to how the individual nucleotide monomers were joined together to produce the entire nucleic acid.

The belief that DNA is associated only with animal cells, and that RNA is just as inseparable from plant cells, persisted until 1914 and the discoveries

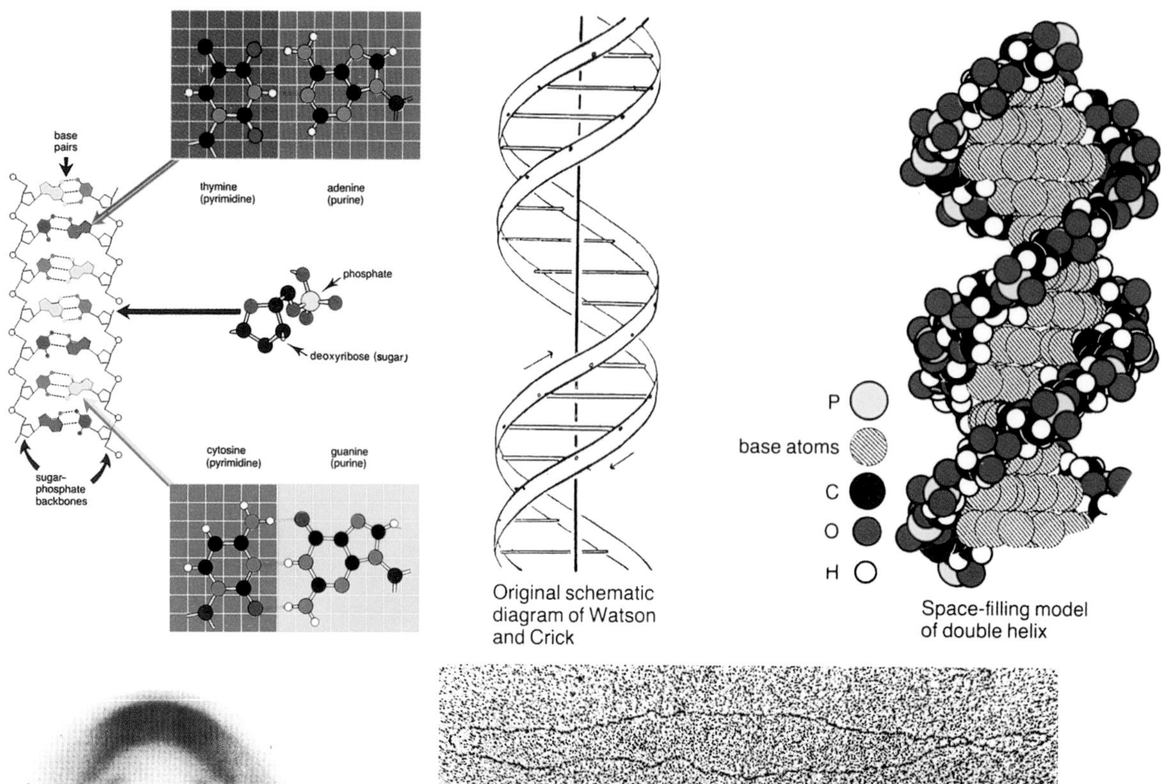

Original schematic diagram of Watson and Crick

Space-filling model of double helix

Electron photomicrograph of DNA of simian virus SV40

Rosalind Franklin's X-ray diffraction pattern

As first described in 1953 by James Watson and Francis Crick (1916–2004), *Deoxyribonucleic acid* (DNA) is a *ladder polymer* consisting of two sugar–phosphate backbones joined by pyrimidine–purine rungs. The diagram at the left side is schematic in that the molecule is shown unfolded and with the bases rotated into a plan view. Thymine occurs only in conjunction with adenine, and cytosine only with guanine, and this base pairing produces rungs of the same length. This can be seen by comparing the positions of the sugar carbon atoms, shown in grey, relative to the underlying grid. In the actual structure of DNA, the backbones are twisted into a *double helix*, as shown in the original Watson–Crick diagram (top centre), while the basal planes that form the rungs of the ladder lie at right angles to the helix axis (top right). The electron photomicrograph of *simian virus DNA* does not resolve such details, and the molecule appears as a thin continuous line. As is well known, a vital step in the search for DNA's molecular structure was the production of the definitive X-ray diffraction pattern – reproduced here – by Rosalind Franklin (1920–1958). It is probably no exaggeration to call this the most important such picture ever taken.

of Robert Feulgen. He found a dye that would colour DNA while not affecting RNA, and another pigment that would preferentially stain RNA. He was thus able to monitor the whereabouts of each of the nucleic acids in their respective cells. Curiosity led him to apply both stains to a single cell, and he uncovered the fact that DNA and RNA are found together in both plant and animal cells. The distinction between animal and plant cells, when based on nucleic acid content as least, had been shown to be fallacious. Together with his student, Martin Behrens, Feulgen established the real distinction: DNA resides in the *nucleus* whereas RNA primarily floats around in the *cytoplasm*.

Techniques for revealing a cell's *chromosomes* by staining were already well advanced at this time, and the most exciting of Feulgen's discoveries was that DNA is actually contained in those nucleus-bound structures. To put this early DNA research in perspective, it is necessary to anticipate Chapter 16 and give a succinct description of the most salient features of a *cell*. These units are membrane-bounded globules of the liquid cytoplasm, in which float various *organelles*, themselves enclosed in membranes. One of the most obvious organelles is the nucleus, whose bounding membrane contains holes so as to make its aqueous content continuous with the cytoplasm. Inside the nucleus are the chromosomes and also a smaller body called the *nucleolus*. During cell division the chromosomes are seen to double in number and divide themselves equally between the two daughter cells. Jean Brachet and Torbjörn Caspersson independently discovered more sensitive dyes that permitted detection of even the smallest amounts of nucleic acid. They established that DNA is found in both animal and plant cells, but only in the nucleus, and indeed only right in the chromosomes. (It subsequently transpired that DNA also occurs in certain specialized structures found elsewhere in the cell, as will be discussed in later chapters.) RNA was also shown to be in both animal and plant cells, and most of it is in the cytoplasm. There is some RNA in the nucleus, however, particularly around the nucleolus, and traces of RNA are even seen associated with the chromosomes.

It might seem surprising that biologists did not leap at the idea of nucleic acids carrying the *hereditary message*, following these beautiful demonstrations. Chromosome multiplication during cell division had already been discovered, and the importance of the nucleus was also well appreciated. But DNA is not the only molecule present in the nucleus, and it had already been established that this organelle also contains protein. Moreover, protein's active role in life's processes was already common knowledge, whereas nothing particularly noteworthy was associated with either form of nucleic acid. Even if the latter were involved in perpetuation of cellular line, by dint of their presence in the nucleus, protein was expected to play the dominant part. The protein found in the nucleus is rather special. A large proportion of its amino acids are arginine which, because of their NH_2 side groups, are markedly alkaline. DNA and RNA are both acidic, and alkalis and acids tend to combine and neutralize each other. This is why protein and nucleic acid are found in conjunction in the nucleus. The combined structure is referred to as *nucleoprotein*, while the special form of protein is known as *histone*. An account of the events that led to identification of DNA as the hereditary material must be deferred until the next chapter. It will transpire that the role of nucleic acid is to direct the manufacture of proteins; DNA is in fact the blueprint referred to earlier. But what is there about the structure of this molecule that enables it to direct such an important task? And how are the instructions remembered in spite of the molecule-splitting events that occur during cell division? It is to these questions, which clearly lie at the very heart of life itself, that we now turn.

By the late 1940s, fairly detailed information regarding the physical characteristics of DNA was beginning to accumulate. Seen in the *electron microscope*, the molecule appeared as a long, thin thread. Its diametre was only 2 nm, whereas its length, which varied from one species to another, was many thousand times greater. *X-ray diffraction patterns*, obtained from specimens in which a large number of molecules were arranged parallel to one another, were taken by Maurice Wilkins, Rosalind Franklin, Raymond Gosling, and their collaborators. Certain features of these patterns were rather disconcerting. They did not display the anticipated variation for DNA extracted from different species. Even more puzzling was the appearance of strong X-ray reflections corresponding to a distance around 3 nm, considerable greater than would be expected from the dimensions of a single nucleotide. The width of a nucleotide is a few tenths of a nanometre, and close packing of nucleotides would be expected to produce characteristic distances of about the same magnitude. The results of biochemical studies by Erwin Chargaff, J. N. Davidson, and their colleagues, were particularly intriguing. They determined the relative proportions of sugar, phosphate, purine and pyrimidine in DNA from various species. The amounts of deoxyribose and phosphate were approximately equal. This was not surprising since there are equal amounts of these substances in each nucleotide, but it showed that nothing was being lost during consolidation of many nucleotides into a single DNA molecule. The amount of a particular base varied from one species to another, and this presumably had something to do with the nature of the hereditary message being stored in the molecule. The really fascinating findings were certain regularities of distribution. The proportions of pyrimidine and purine appeared to be equal. Moreover, although the amounts of adenine and guanine varied from one species to another, the amount of adenine always seemed to be about the same as that of thymine, while a similar equality between the amounts of guanine and cytosine was also noted.

The successful incorporation of this diverse information into an acceptable model of the DNA molecule was achieved by James Watson and Francis Crick, in 1953. They suggested that DNA must be an example of something that was discussed in the preceding chapter: a *ladder polymer*. The vital result which led them to this model was that a coplanar hydrogen-bonded arrangement of adenine and thymine has exactly the same overall length as a similar combination of guanine and cytosine. Such *base pairs* can thus act as the rungs of the ladder, the sides of which are composed of sugar and phosphate units in regular alternation. No other base combinations produce rungs of an acceptable length, the observed pairings securing their uniqueness by excluding all other possibilities. The equality in the amounts of adenine and thymine, and also of guanine and cytosine was now clear, as indeed was the equality in the amounts of purine and pyrimidine, since each base pair involves one of each of these types of molecule. Building a scale model of their arrangement, Watson and Crick found that *steric hindrance* causes the ladder to twist into a helical conformation: the now-famous *double helix*. The spacing between adjacent base pairs is a few tenths of a nanometre, as expected, but two successive bases are offset from each other by an angle of just over 30°. The pitch of the double helix is such that there is parallelism between bases spaced eleven base pairs apart, and this corresponds to 3.4 nm, the origin of the mysterious X-ray reflections referred to earlier. The beauty of the double helix lies in its eminent suitability as a hereditary perpetuator, as Watson and Crick immediately appreciated. If the two strands are separated, each can act as a template for the construction of a new strand, which, because of the rigid control imposed by the base pairing, cannot help being perfectly complementary to the original strand. Because each of the original strands can function in this way, the result of such division is the production of two new double

The genetic code

The amino-acid sequences of an organism's proteins are specified by the genes, the information being stored in nucleic-acid polymer molecules. In *eukaryotes*, the *genetic message* is stored in deoxyribonucleic acid (DNA), which is located in the nucleus, and it is transferred outside that cellular organelle by messenger ribonucleic acid (messenger-RNA). The polymer backbones of both these types of nucleic acid consist of alternating ribose (sugar) and phosphate units, and one base is attached to each sugar. The bases come in five varieties. In DNA, there are the two purines adenine (A) and guanine (G) and the two pyrimidines cytosine (C) and thymine (T). The latter is replaced by uracil (U) in RNA. The genetic code is based on triplets, three consecutive bases specifying one amino-acid residue. The following table lists the amino acids corresponding to each triplet, as well as the stop signals. It is noteworthy that the code displays considerable redundancy, several triplets coding for each amino acid.

First base (5′ end)		**Second Base**					Third base (3′ end)
		U	C	A	G		
	U	phe	ser	tyr	cys	U	
		phe	ser	tyr	cys	C	
		leu	ser	stop	stop	A	
		leu	ser	stop	trp	G	
	C	leu	pro	his	arg	U	
		leu	pro	his	arg	C	
		leu	pro	gln	arg	A	
		leu	pro	gln	arg	G	
	A	ile	thr	asn	ser	U	
		ile	thr	asn	ser	C	
		ile	thr	lys	arg	A	
		met	thr	lys	arg	G	
	G	val	ala	asp	gly	U	
		val	ala	asp	gly	C	
		val	ala	glu	gly	A	
		val	ala	glu	gly	G	

helices, each of which is a perfect copy of the original. (A proposal along the lines of two *complementary molecules*, as a basis for the hereditary material, had already been put forward by Linus Pauling and Max Delbrück, in 1940, but an actual model did not emerge at that time.) The *genetic message* was clearly encoded in the arrangement of base pairs, and this must vary from one species to another. Since the structural complexity of an organism should be roughly related to the number of instructions contained in its nucleic acid, we could say that Watson and Crick had discovered the ladder of life upon which the various species have made their evolutionary ascents.

The information stored in the DNA specifies the types of amino acid occurring in proteins, and the sequences in their primary structures. *Collinearity*, that is that the order of DNA instructions is identical to the order of amino-acid side groups on the corresponding protein, was assumed and later verified experimentally. Important issues now arose, the most obvious being the nature of the *genetic code*. The four bases, denoted by A, T, C and G, are the genetic alphabet, but what are the words? Progress with this point had to await the solution of another problem. Proteins are too large to pass through the holes in the nuclear membrane, and because they are found outside this organelle, they must be produced in the cytoplasm. But the instructions for that fabrication are given by the DNA, which is inside the nucleus and never leaves it. How was the message being transmitted from nucleus to cytoplasm? An obvious candidate was RNA, which had been found in both cytoplasm and nucleus. Investigations of RNA had revealed a water-soluble form and another variant associated with particles having diameters of approximately 2 nm. These particles, detected by electron microscopy, are the *ribosomes*, and they contain RNA and protein in roughly equal proportions. Neither form of RNA seemed to fit the bill, however, and in 1959 Jacques Monod and François Jacob suggested that the mysterious messenger must be a compound that is broken down, and the resultant fragments ultimately dispersed in the cytoplasm, fairly rapidly after its information has been delivered. Confirmation of the Monod–Jacob hypothesis soon followed. Matthew Meselson, Sydney Brenner and Jacob himself exploited the fact that, when a virus invades a cell, it does so by injecting some of its own genetic material. Using the colon bacterium, *Escherichia coli*, as a target, and the virus known as T4 bacteriophage, and employing radioactive labelling to identify the different RNA products, they were able to establish that a messenger variety of RNA was indeed being formed. Independent verification of the existence of *messenger-RNA* was provided by Sol Spiegelman and Benjamin Hall, and it involved a neat piece of polymer chemistry. It had earlier been shown that the hydrogen bonds holding the base pairs together in DNA can be disrupted by gentle heating. In effect, this unzips the double helix, and the process is reversible because lowering the temperature causes the structure to zip up again. Spiegelman and Hall reasoned that since the putative messenger RNA was to be a transcription of a corresponding length of DNA, its base sequence must be complementary to one of the two DNA strands. As such, it must be capable of recombining with that strand to produce a *DNA–RNA hybrid*. Once again, the trick was to keep a check on the various strands by using different radioactive labels.

There is no room here for a detailed description of the numerous experiments which contributed to the detailed picture we now have concerning the protein-making activities of RNA. A brief summary of the essential steps must suffice, and this is an ideal point at which to introduce several definitions. The formation of two new DNA double helices from an original one is referred to as *replication*. The genetic instructions contained in the sequence of bases in the DNA are transmitted to the extra-nuclear region by messenger-RNA. The vital transcription is accomplished by the enzyme *RNA-polymerase*, which reads the message by using one strand of the DNA as a template and builds the RNA from the complementary bases. There is a minor complication here. The bases cytosine, adenine, and guanine are all used by both DNA and RNA. In RNA, however, the DNA base thymine is replaced by a different pyrimidine, uracil. This single-ring compound resembles thymine but lacks the latter's methyl group. Like thymine, it forms a complementary base pair only with the purine, adenine. The message-reading process is known as *transcription*. The messenger-RNA, having emerged from the nucleus, uses its information to direct the stringing together of amino-acid residues into a polymer, a process known as *translation*. This is accomplished with the aid of two other forms of RNA,

In the *replication* of DNA, the two complementary strands of the molecule unwind and the hydrogen bonds that hold the *nucleotide bases* together in pairs are ruptured. This enables the strands to separate, and they each serve as a template for the growth of a new complementary strand. Thus one of the strands of each daughter DNA molecule is new, while the other is derived from the parent DNA molecule. The process is controlled by a number of proteins which variously twist, cut, and rejoin the strands. The *genetic message* encoded in the base sequence of a DNA molecule is passed on to a second molecule, *messenger-RNA*, by a process known as *transcription*. This starts at specific sites, called *promoters*, and the new strand is assembled by a protein known as *RNA-polymerase*. The message is made available by the sequential unwinding of short lengths of the DNA, only one of the strands of that molecule actually being transcribed. Just as in replication, the fidelity of the genetic information is preserved through base pairing, but with the slight modification that an adenine base on DNA gives rise to a uracil base on the RNA rather than a thymine. In this schematic diagram of the two processes the deoxyribose–phosphate backbones of DNA are coloured black and magenta, while the ribose–phosphate backbone of RNA is shown in the purple.

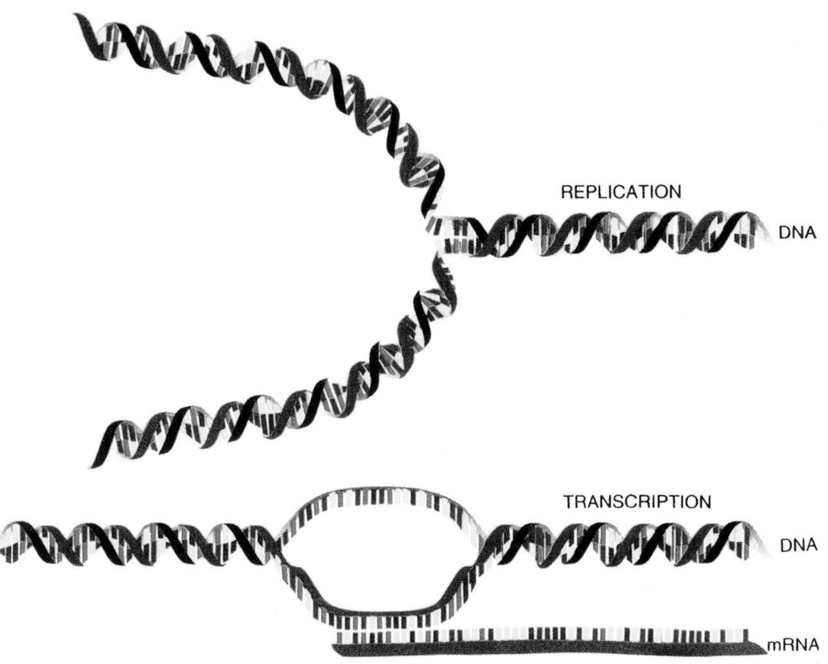

transfer-RNA and *ribosomal-RNA*, the latter working in conjunction with proteins in the ribosomes. Each ribosome acts as a sort of construction machine, capable of reading the instruction on a particular piece of messenger-RNA and acquiring the appropriate amino acid to add to the growing protein. Its source of these building blocks are the transfer-RNA molecules, each of which must latch on to the right amino acid and be ready with it when required by the ribosome.

To understand how these molecules can perform in such a sophisticated fashion, we need look no further than the base-pairing mechanism. Although, unlike DNA, RNA is sometimes single stranded, this does not prevent base pairing from taking place. To see how this works, let us consider the following situation. Imagine that the five consecutive bases, numbers 10 to 14, on a single strand of RNA, are GUCCA. Suppose also that the five consecutive bases, numbers 25 to 29, are UGGAC. Pairing is possible between bases 10 and 29, 11 and 28, and so on. This means that the RNA can form a loop by bending back upon itself so as to permit hydrogen bonding across five consecutive base pairs. Had the second group of bases been located at positions 15 to 19, this would not have been possible, because it would have required the RNA to fold back upon itself over a single inter-base distance. Such a hairpin bend would involve a prohibitively high energy. A single strand of transfer-RNA can develop loops at various points so as to produce a *clover-leaf pattern*, with leaf diameters determined by the cross-linking complementary bases. Such superstructure formation enables transfer-RNA to recognize the correct amino acid. It must also be able to recognize the instruction, known as a *codon*, stored on the relevant piece of a messenger-RNA, but that task is easier because base pairing can be exploited. Recognition is achieved by the transfer-RNA having the requisite complementary bases, this unit being referred to as an *anticodon*. There are 20 different brands of transfer-RNA, each capable of transporting the appropriate amino acid and presenting it at the ribosome when called for.

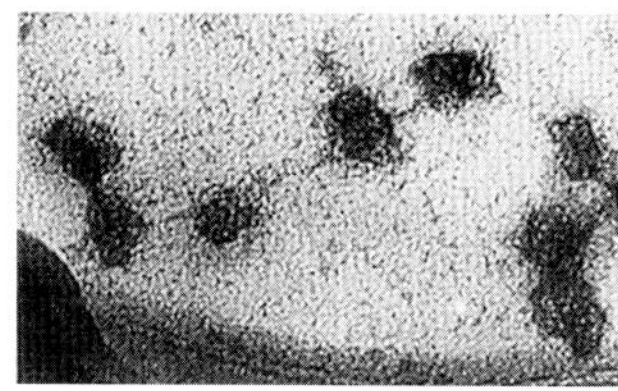

After the genetic message has been transcribed from the DNA to the messenger-RNA, the latter leaves the nucleus and its coded instructions are translated into the corresponding proteins. The message includes start and stop signals, and this divides it into self-contained passages: the genes. One gene codes for one protein, and the latter is synthesized with the aid of a ribosome, which functions rather like a machine tool. Short stretches of another nucleic acid, *transfer-RNA*, are held in specific geometrical shapes through pyrimidine-purine base pairing, and this enables them to latch on to particular amino acids as indicated in this schematic diagram. The amino acids are transported to the ribosome and the latter is able to read the sequence of bases on the messenger-RNA, three at a time, and attach the appropriate amino acid to the growing protein. The diagram indicates that several ribosomes can be simultaneously at work along a single stretch of messenger-RNA, and evidence for such a *polysome*, as it is called, was obtained in 1963 by Cecil Hall, Paul Knopf, Alexander Rich, Henry Slater, and Jonathan Warner. An example is shown in the electron photomicrograph.

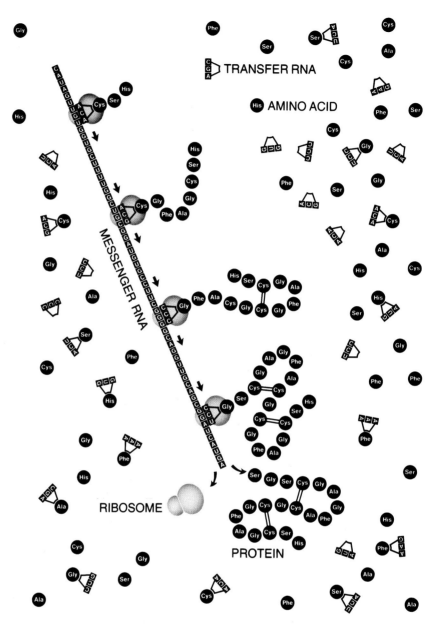

Energy is required during the polymerization process that consolidates the individual amino-acid residues into the protein. As first demonstrated by Paul Zamecnik, Mahlon Hoagland, and their colleagues, the energy required to polymerize the amino acids is supplied by ATP, part of that molecule actually becoming attached to the major fragment of the amino acid. (As will be discussed in the following chapter, *adenosine triphosphate, ATP*, is the universal currency of free energy in biological systems, and it transfers this energy in two separate units, each approximately twice that of the hydrogen bond, by successively being hydrolyzed to adenosine diphosphate, ADP, and adenosine monophosphate, AMP. The other byproduct of the hydrolysis is pyrophosphate.) The resulting compound molecule is known as an *activated amino acid*, the job of joining the two parts together falling on amino-acid activating enzymes. There is one of these enzymes for each type

This planar view of an unwound transfer ribonucleic acid (tRNA) molecule indicates the cloverleaf structure formed by base pairing, the unpaired bases (some of them unusual) producing four loops. The largest loop contains between eight and twelve bases, while two of the others have seven each. The three-base anticodon is indicated by the shading. The native three-dimensional structure, determined by Alexander Rich and his colleagues, resembles a twisted L. In the early 1980s, T. Inoue and Leslie Orgel reported that a new strand of RNA can be formed alongside an existing strand, its base sequence being complementary to that of the original, without the help of a protein enzyme. In this pioneering experiment, the original strand contained exclusively cytosine bases, and the daughter strand possessed only guanine bases, despite the fact that nucleotides containing the other three bases were present in the surrounding aqueous medium. Shortly thereafter, Thomas Cech and Sidney Altman made the astonishing discovery that RNA molecules can themselves function as catalysts, for a limited variety of reactions. This solved the chicken-and-egg conundrum of molecular biology: if DNA is required to dictate the formation of proteins, some of which are needed in the replication of DNA, which came first? It now seems that neither DNA nor protein can claim primacy in this respect, and that the earliest – more primitive – stages of molecular evolution took place in what has been referred to as an *RNA world*.

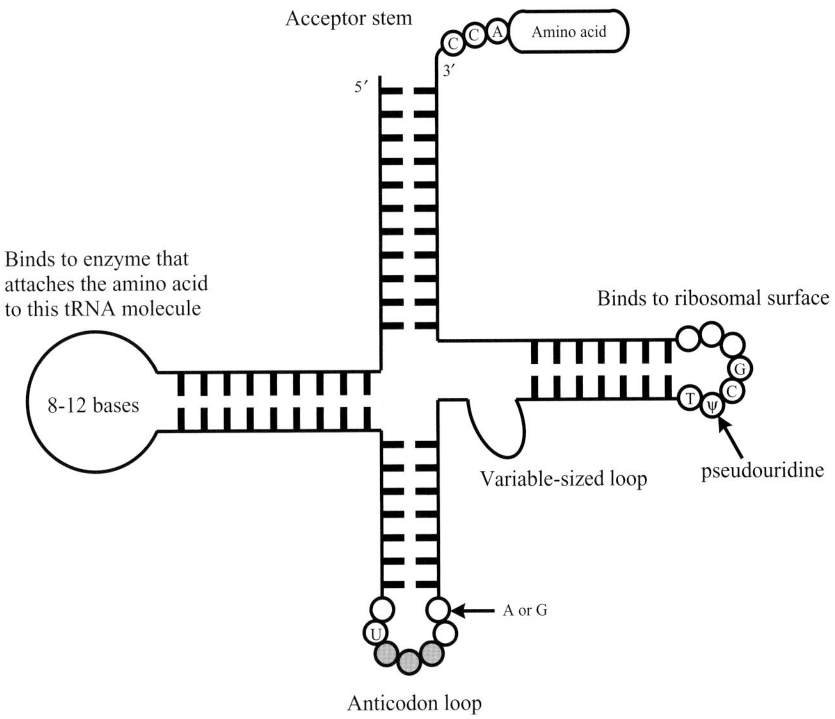

of amino acid. The complex-forming reaction is

$$\mathrm{H_2N \cdot \overset{\displaystyle R}{\overset{|}{C}H} \cdot COOH + ATP + enzyme \rightarrow}$$
$$\mathrm{H_2N \cdot \overset{\displaystyle R}{\overset{|}{C}H} \cdot CO \sim AMP \cdot enzyme + pyrophosphate}$$

the AMP being joined to the amino acid by an energy-rich bond, as indicated by the wavy line. The polymerization of RNA, from individual nucleotides, involves similar activated nucleotides as precursors.

Transfer-RNA is obviously a molecule with remarkable abilities. With the help of the activating enzyme, it selects the right amino acid and brings it to the messenger-RNA, where it recognizes the correct codon by matching this with its own anticodon. There remains the question of how the amino-acid residues are actually pushed together to form the polypeptide. In the language of mechanical engineering, some sort of machine tool is required. This is the role played by the ribosomes. These RNA–protein complexes, which are plainly visible in an electron microscope, are essentially two globular masses having opposing curved faces which dovetail neatly into each other, a sort of spring arrangement holding them together. There is a groove, which enables the ribosome to slide along the messenger-RNA, and there is another depression into which the transfer-RNA fits. The amino acid is thereby brought into the correct position, and a swing of the hinge causes it to be joined to the free end of the growing polypeptide. One could loosely compare the two parts of the ribosome with a hammer and an anvil, at which are forged the links of life. The ribosome works its way along the messenger-RNA, with the growing polypeptide dangling from it, and being lengthened by one amino-acid residue for every codon that is read. Only a small region of the messenger-RNA is occupied at any instant by a particular ribosome, and there is nothing to prevent another ribosome being simultaneously active at another part of the messenger-RNA. Electron microscope evidence that this occurs was obtained by Alexander Rich and his colleagues in 1963, and the necklace-like arrangement of several ribosomes all attached to the same strand of messenger-RNA is called a *polyribosome*, or simply a *polysome*.

Two new amino acids were discovered in 1981, both apparently unique to proteins that associate with RNA. G. Wilhelm and K. D. Kupka found amino citric acid in the ribonucleoproteins of calf thymus, bovine and human spleen, and the bacteria *Escherichia coli* and *Salmonella typhi*. (The former is named after its discoverer, Theodor Escherich, who isolated it in 1885.) There is a variety of such proteins, but all are found in combination with RNA. Tad Koch and John van Buskirk subsequently discovered *beta*-carboxyaspartic acid in *E. coli* ribosomes. Another unusual amino acid, *gamma*-carboxyglutamic acid, had earlier been observed in ribosomes. A striking characteristic of all three of these uncommon polypeptide components is their unusually high acidity. They thus bear large negative charges, a feature also displayed by RNA itself. This could mean that these three amino acids electrically repel RNA, possibly keeping it away from certain regions of the protein molecules.

But what precisely is a *codon*? That it has something to do with the arrangement of bases in the DNA has already been made clear, but what arrangement corresponds to which amino acid? And how many bases are there in each codon? We can follow the reasoning of George Gamow, who speculated on this problem of information transfer in 1954. There are only four bases, A, T, C and G, so a single-base codon could code for only one of four amino acids. Even two-base codons would not be able to cover the twenty different common amino acids, because there are only 16 different ways of arranging four types of base in groups of two: AA, AT, AC, AG, TA, TT, and so on. If a codon contained three bases, however, there would be more than enough different arrangements, 64 of them in fact, to cover the 20 types of common amino acid. Gamow's suggestion was therefore that the genetic message is written in three-letter words. In such a system, assuming that there is no overlapping of codon triplets, the protein ribonuclease, with its 124 amino-acid residues, would require a messenger-RNA with 372 bases.

Confirmation of the three-base hypothesis was provided by the beautiful experiments of Francis Crick, Sydney Brenner, Leslie Barnett and Richard Watts-Tobin in 1962. Their approach was made possible by the existence of a special type of mutant. *Mutations* alter the hereditary message by changing the sequence of bases in DNA. Most mutations take the form of a simple substitution of one base for another. The codon containing the altered base still has the correct number of bases, but its new sequence either corresponds to a different amino acid or it is complete nonsense. In the latter case, the part of the protein corresponding to this codon is simply omitted. There is a group of mutagenic agents, known as the *acridines*, which appear to have a more profound influence on their targets. Genes which carry acridine-induced mutations completely lose their ability to produce protein. Crick and his colleagues concluded that the entire genetic message is disrupted after the point of mutation. They referred to this as a shift of reading frame and reasoned that it must arise either from the deletion or addition of a base. Such a special type of mutation would have a devastating effect on the way in which the base sequence is read. Whatever the number of bases in a codon, deletion or addition of a single base would cause all codons thereafter to be read out of step. They studied the rII gene of the T4 bacteriophage, which had been completely mapped out in earlier studies by Seymour Benzer.

As a preliminary, Crick and his co-workers had to establish how one differentiates between addition and deletion mutations, and they then tried these in various combinations. If there is one deletion and one addition, for example, the code will get out of step at the first mutation but into step again at the second. This would produce a protein whose structure was correct except for those amino acids corresponding to the codons lying between the two mutations. Likewise, two additions and two deletions will produce a protein that is correct for all codons lying before the first mutation and after the last. They also investigated the effect of multiple mutations of a single type, that is either addition or deletion but not both. They found that DNA with two deletions malfunctions in much the same way as one with a single deletion. For three deletions, however, the result was similar to that for equal numbers of deletions and additions: the code appeared to get back into step after the last mutation. The conclusion was inescapable: Gamow had been correct; codons consist of three consecutive bases and the genetic message is read in non-overlapping base triplets. As the jubilant Crick remarked: 'two blacks do not make a white, but *three* do'. The results of these elegant experiments implied two other important facts. To begin with, the generally held belief that the code is read in only one direction, starting from a fixed point, had been validated. The second point concerned the nature of the erroneous codons lying in the region between the first and last mutations. Because genes containing such garbled sections nevertheless produce single whole proteins, with a number of incorrect amino-acid side groups, almost any triplet of bases must correspond to one or another amino acid. Since there are 64 different possible arrangements of three bases, the code must contain a certain degree of *degeneracy*, that is built-in redundancy, with some amino acids being coded for by several different base triplets.

The direction in which a protein grows was established by Howard Dintzis in 1961. He exposed haemoglobin-producing cells, known as *reticulocytes*, to radioactively labelled leucine for periods shorter than it takes to synthesize the entire protein. By splitting the final chain into fragments, with the enzyme trypsin, he was able to measure the amount of radioactivity at various positions. The carboxyl end had the most, and there was a steadily decreasing activity as the amino end was approached, that terminal having the least. The radioactive leucine at the carboxyl end had obviously been decaying for the shortest period, indicating that the carboxyl end was synthesized last. Dintzis also found that an entire 574-residue four-chain haemoglobin molecule takes a mere 90 s to assemble. DNA and RNA strands also have directionality. If the five carbon atoms in each backbone ribose are numbered in a consistent manner, the phosphates on either side are linked to the fifth and third members. Each nucleic acid strand therefore has a 5′ end and a 3′ end, according to the accepted nomenclature. It was established that messenger-RNA is translated in the direction running from the fifth carbon towards the third carbon in each unit. Other exciting developments were reported in 1961. Jerard Hurwitz warmed a mixture of DNA, nucleotides, and the appropriate enzymes and obtained messenger-RNA.

The test-tube experiments of G. David Novelli went one stage farther. He artificially manufactured the enzyme *beta*-galactosidase from a mixture of DNA, nucleotides, ribosomes and amino acids.

Great strides had been made, but the genetic code was not yet cracked. The alphabet and the length of a word were now known, but there remained the questions of the dictionary and punctuation. It transpired that a vital prerequisite had been discovered by Severo Ochoa and Marianne Grunberg-Mango in 1955: an enzyme capable of synthesizing RNA in a test tube. (An enzyme capable of doing the analogous thing for DNA was discovered by Arthur Kornberg and his colleagues, around the same time.) Marshall Nirenberg and J. Heinrich Matthaei used this molecule, known as *polynucleotide phosphorylase,* to catalyze the formation of an RNA polymer containing only one type of base: uracil. They then used this *polyuridylic acid,* called poly-U for short, as messenger-RNA, making it react with a mixture of all 20 common amino acids. The resulting polypeptide contained only phenylalanine. The genetic dictionary had been opened, and the first word, UUU, meant phenylalanine. Moreover, the method of obtaining other words was reasonably obvious. Addition of a small amount of adenylic acid produced an RNA polymer whose bases were still predominantly UUU, but with a small chance of being AUU, UAU or UUA. With an equal mixture of uridylic and adenylic acids, AAU, AUA, UAA and AAA were added to the list, and so on. This approach received an important boost from H. Gobind Khorana's successful synthesis of DNA strands with specific base sequences. One of the punctuation marks was discovered in this way, the three triplets UAA, UAG and UGA all meaning *stop*. The *start* signal is apparently more complex. It involves either of the triplets AUG or GUG, but additional structures are required on the adjacent stretches of nucleic acid, without which these base sequences give methionine and valine, respectively. Khorana ultimately succeeded in artificially producing an entire gene.

Two points should be made regarding the degeneracy alluded to earlier. One is that it minimizes the deleterious effects of mutation, in that a change of one base still allows a triplet to code for the same amino acid in some cases, but not in all. Secondly, as Ulf Lagerkvist emphasized in 1978, many codons are read with sufficient accuracy if two out of three bases are correct. Because misreading by this method could threaten the accuracy of protein synthesis, he suggested that it is strictly confined to codon families in which two-out-of-three reading can be used with impunity. Final mention should be made of the genetic material found in the subcellular structures known as *mitochondria,* the cell's power plants. In animal cells, they are the only other organelles that contain DNA, apart from the nucleus, and there is evidence that they use a different genetic code. Alex Tzagoloff and his colleagues found that the triplet CUA codes for threonine in the mitochondrion, whereas it usually gives leucine. Moreover, the normal termination triplet UGA corresponds to tryptophan in this organelle. Finally, mitochondria have ribosomes that resemble those of bacteria. These observations lend credence to the idea put forward by Lynn Margulis that mitochondria originally led an independent existence as primitive *aerobic bacteria,* which developed symbiotic ties with their present hosts, ultimately becoming absorbed and integrated by them. The normal bacterial genetic code is identical to that of cells with separate nuclei (i.e. *eukaryotes*), however, so the above differences must be attributed to degeneration during the postulated independent period of development.

The sole function of the genetic machinery is to specify the amino-acid sequences which constitute primary protein structure, and the latter ultimately dictates the spatial configuration adopted by a protein. We return now to the question of the biological functions of these crucial molecules. Some, like those found in the connective tissue of animals, play an essentially structural role, while others carry out such specialized tasks as the regulation of molecular transport across cell membranes. A particularly important class of proteins are the *enzymes*. There are many hundred different types, and they have it in common that each catalyzes a specific chemical reaction. By convention, the substance on which the enzyme acts is referred to as a *substrate*, and the name of the enzyme is formed by adding the suffix -ase to the name of the substrate and the function performed. Thus isocitrate dehydrogenase catalyzes the removal of hydrogen from isocitrate, while phosphohexoisomerase catalyzes the isomeric change in configuration of a specific sugar phosphate. The biochemical tasks performed by enzymes include the splitting of molecules into smaller fragments, and the reverse process, including the stringing together of many similar units to form a biological polymer. There are also enzymes for oxidation, for reduction, for transfer, and for modification of three-dimensional structures. The word enzyme was coined by Willy Kuhne, in 1878, and comes from the Greek words *en*, meaning in, and *zyme*, which means yeast.

The idea that chemical reactions in living systems depend on *catalysts* was first put forward in 1835 by Jöns Berzelius. It met opposition, Louis Pasteur's demonstration of the role of living cells in fermentation and putrefaction enjoying more support. The key discovery was made in 1897 by Hans Büchner and Eduard Büchner. Desiring to produce a cell-free extract for medicinal applications, they ground yeast and sand in a mortar and obtained an apparently inert juice. Anticipating that its properties might be destroyed by antiseptics, they stored the juice in sucrose and were amazed to find that this resulted in fermentation. Enzymatic action had been shown to depend not on living cells as such but on certain cellular products. About a decade earlier, Svante Arrhenius had analyzed the kinetics of catalyzed reactions and concluded that a catalyst–substrate complex must be formed as an intermediate stage. Evidence of such a complex was obtained by Cornelius O'Sullivan and F. W. Thompson. They heated a solution of the yeast enzyme *invertase* and its substrate sugar. The excess protein coagulated and precipitated, while the invertase was found to be recoverable. The substrate, by its presence, had apparently protected the enzyme from the heat, indicating the formation of a complex with its own distinct physical properties. Formation of a catalyst–substrate complex was put on a quantitative footing by Leonor Michaëlis and Maud Menten in 1913. Catalytic rate is approximately proportional to the amount of substrate present when this is quite small, whereas the rate levels off to a roughly constant value, independent of substrate concentration, when the latter is high. This is readily understandable on the basis of saturation of available catalytic sites when essentially all enzymes are involved in a complex at any instant. Definitive confirmation of the *Michaëlis–Menten model* was obtained by Britton Chance, in the 1940s. He used very rapid mixing and flow techniques to study the complexing of peroxidase with various substrates, monitoring the concentrations of the components by the intensities of their different characteristic colours. Peroxidase catalyzes the reaction between hydrogen peroxide and many reducing agents, to form water and an oxidized substance.

The fact that enzymes are proteins was established in 1926 by James Sumner, who managed to isolate and crystallize a pure specimen. The mode of operation of these biological workhorses had been the subject of an inspired guess in 1894, when Emil Fischer proposed that enzyme and substrate fit together like lock and key. The complementary shapes of the two molecules were envisaged as producing a jigsaw-like match, excluding all but one specific substrate. Although this idea was almost dogma for many decades there were difficulties, such as the apparent competition between different substrates in certain enzymatic reactions. Moreover, some processes were found to involve the sequential binding to the enzyme of several molecules of different types, and the order of attachment was apparently not arbitrary. This led Daniel Koshland to reject the rigid template idea, in 1963, and to replace it by an *induced-fit model*. He emphasized the deformability of proteins and anticipated that an enzyme can change its shape, to a certain extent, so as to fit the molecular contours of the substrate. And in this respect one must bear in mind that the deformability will increase with increasing temperature; proteins are dynamic structures. An important elaboration was proposed in 1965 by Jean-Pierre Changeux, Jacques Monod and Jeffries Wyman. Referred to as the *allosteric model*, it accounts for the cooperative functioning of the different subunits of an enzyme consisting of several parts, each of which can operate on a substrate molecule. The classic example of an *allosteric enzyme* is *haemoglobin*, which consists of four units grouped in two pairs.

The induced-fit and allosteric theories can account for such finer details of enzymology as the existence of *coenzymes*. These non-protein molecules come in two varieties, some promoting catalysis and others inhibiting it. Several coenzymes have been found to be related to certain *vitamins*. A coenzyme need not lie adjacent to the substrate during the reaction, and most of these secondary agents act at specific regulatory sites on the enzyme that are reasonably remote from the active region. The typical enzyme is pictured as a globular protein molecule, on one side of which is a cleft or groove having the approximate shape of that part of the substrate on which it is to act. A slight change of shape enables it to make a precise fit, the change possibly being provoked by a coenzyme. If it is a cutting enzyme, the partial enveloping of the substrate is followed by its being split at a specific site and the two fragments are then released. The key to this functioning, and indeed to most biochemical processes, is *specificity*. It derives from the precise arrangement of atoms in proteins, which in turn follows from the chemical structure and *interatomic forces*. Arieh Warshel has suggested that local envelopment of the substrate isolates it from the aqueous environment, and that this modifies some bonds from covalent towards ionic, making them more susceptible to cleavage.

Some of the factors controlling shape, such as steric hindrance and the formation of disulphide and hydrogen bonds, were discussed earlier. Two other important factors are the aqueous medium in which the enzyme resides and formation of *salt bridges*. The hydrophobic or hydrophilic tendencies of some of the amino-acid side groups cause these to prefer the inside or outside positions, respectively, of the folded molecule, but a change in pH of the surroundings can affect the balance and cause subtle geometrical changes. When the net charge on a protein is zero, the molecule will not migrate under the influence of an electric field. The pH value for which neutrality prevails is known as the *isoelectric point*, and it is a characteristic of

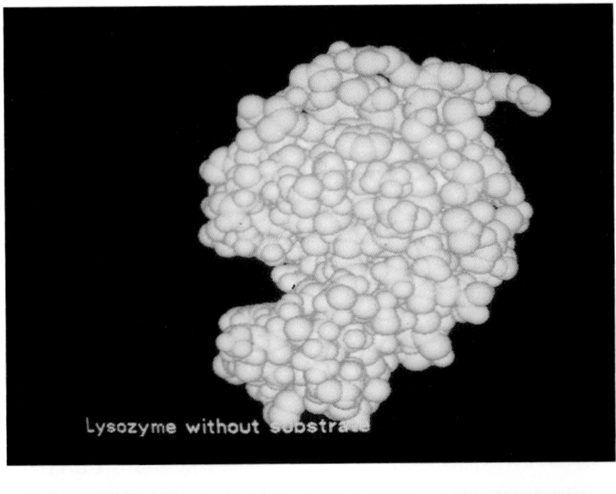

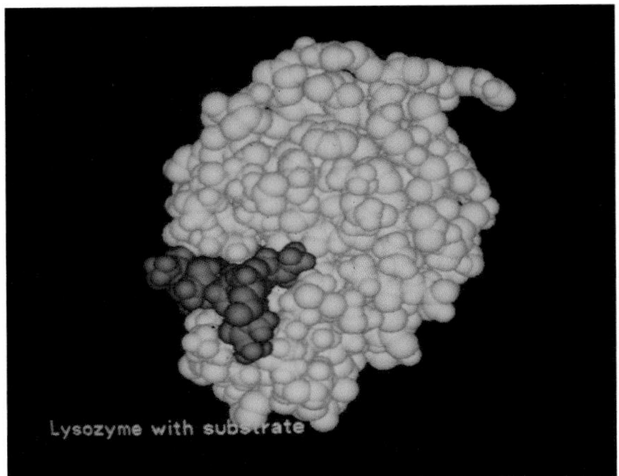

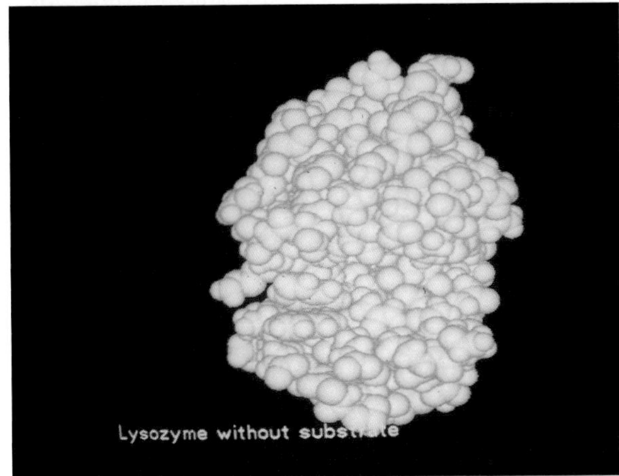

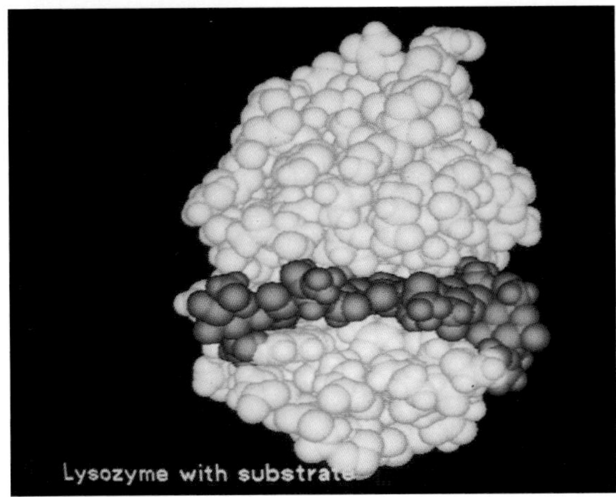

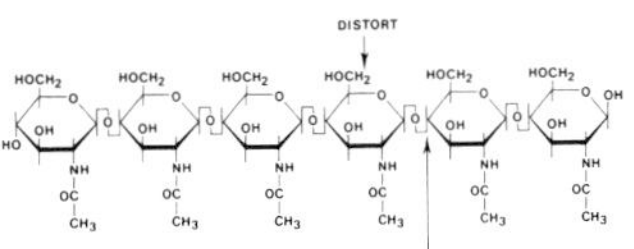

In 1922, Alexander Fleming (1881–1955) discovered a substance in his own nasal mucus capable of breaking down, or lysing, certain bacteria. He showed the substance to be an enzyme, called it *lysozyme*, and found that it is widely distributed in Nature. This protein molecule, 14 500 daltons in weight, consists of 129 amino-acid residues in a single polypeptide chain, the structure of which is stabilized by four cystine (disulphide) bridges. Its three-dimensional configuration was elucidated in 1965, through the X-ray studies of David Phillips (1924–1999) and his colleagues. They found the structure to comprise much less alpha helix and rather more beta sheet than myoglobin and haemoglobin, and discovered that the molecule has a well-defined cleft running across its middle. The enzyme attacks bacteria by cleaving certain bonds in the polymers in their cell walls. These polymers are similar to cellulose in that they involve sugar monomers, the latter being *N*-acetylglucosamine (NAG) and *N*-acetylmuramic acid (NAM). NAG and NAM monomers alternate in bacterial cell-wall polysaccharides whereas the chitin in crustacean shells, also a lysozyme substrate, contains only NAG residues. The cleavage is achieved through distortion of the polymer when it is positioned in the lysozyme cleft; one of the sugar six-carbon rings is twisted from its stable 'chair' form into a 'half-chair' configuration, and the resulting redistribution of electrons makes the adjoining *glycosidic linkage* susceptible to rupture. These computer-generated views of lysozyme show the molecule with and without a hexa-NAG substrate, the molecular structure of which is indicated by the inset. The upper pictures show the molecule from the side while the lower ones show it from the front.

a given protein. Proteins with a high proportion of basic side groups, such as haemoglobin, have high isoelectric points, while those with a preponderance of acidic side groups have low isoelectric points, an example being pepsin.

The first enzyme for which the three-dimensional structure was elucidated was *lysozyme*. This substance was discovered in 1922 by Alexander Fleming. While suffering from a cold, he let a few drops of nasal mucus fall on a bacterial culture, and he later found that it had locally destroyed the bacterial cells. It was subsequently discovered that lysozyme breaks down the polysaccharide cell walls of bacteria by splitting a specific bond between two adjacent amino sugars known as N-acetylglucosamine and N-acetylmuramic acid. The structure was determined through X-ray diffraction by David Phillips and his colleagues in 1965, and it was found to comprise a cleft into which the polysaccharide chain fits. Inhibition of an enzyme can result from the binding of an extraneous molecule to a site somewhat removed from the active region, this attachment causing a distortion of the enzyme which spoils the latter's neat fit with the substrate. More direct inhibition occurs when a bogus molecule competes with the real substrate. A key that does not quite fit a lock can cause the latter to jam. In an analogous manner, an alien molecule can masquerade as a substrate and block the active site of an enzyme. The enzyme is then said to be *poisoned*. An example is provided by the synthetic *antibiotic* sulphanilamide. Certain bacteria require para-aminobenzoic acid for synthesis of folic acid, an indispensable *growth factor*. Sulphanilamide competes for the catalyzing cleft on the relevant enzyme and thus interferes with the vital reaction. Sulphanilamide does not harm us because we cannot synthesize folic acid; it is obtained directly from the diet. This substance, commonly known as a *sulpha drug*, saved many lives during the Second World War, combat troops carrying it in powder form and sprinkling it directly into open wounds to prevent infection.

The substrates acted upon by enzymes show great variety, and when they are themselves proteins, the corresponding chemical reactions involve one protein acting upon another. Perhaps the most impressive enzymes are those that manipulate the hereditary material itself, and we will come to these *restriction enzymes*, as they are called, later. Before this, it will be appropriate to consider the many other important interactions between proteins and nucleic acids. We have already seen that these two types of biochemical act together in the ribosomes; they are also found in conjunction in the *chromosomes*, in *viruses*, and in what are known as *operons*. The transfer of information that enables a section of DNA, a gene in fact, to direct the construction of a protein has already been described. Nothing has been said, however, of how the rates of transcription and translation are controlled. In the early 1960s, François Jacob and Jacques Monod demonstrated that many organisms lacking the enzymes to handle certain substrates could be stimulated by those same substrates into producing the missing catalysts. Jacob and Monod studied the colon bacterium *E. coli*. This bacterium uses lactose, a disaccharide comprising one galactose unit and one glucose, as a source of carbon. The hydrolysis of lactose is catalyzed by *beta*-galactosidase, and an *E. coli* cell growing on lactose normally contains thousands of molecules of this enzyme. A cell grown on an alternative source of carbon, such as glucose or glycerol, is found to contain very little *beta*-galactosidase. Two other proteins are synthesized concurrently with the enzyme: galactoside permease, which mediates transport of lactose across the bacterial cell membrane,

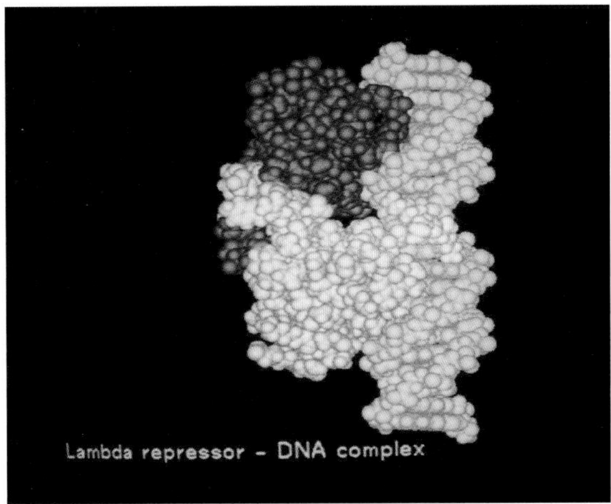

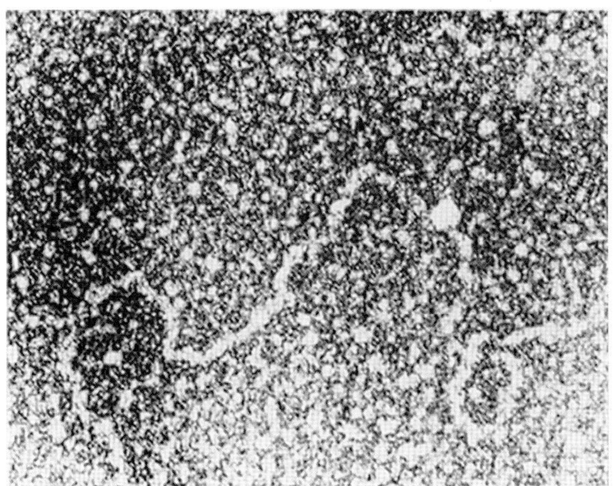

The expression of a gene involves the biochemical synthesis of the protein molecule coded for in the gene's nucleotide base sequence. This process is generally under the control of other proteins, known as *repressors* and *promoters*, and these have their own corresponding genes. The colour picture, which was generated on a computer, shows two of the key parts of the *lambda repressor*, in yellow and blue, attached to part of the DNA double helix. The sugar–phosphate backbones of the latter, and also the nucleotide bases, appear in light blue. Part of the repressor, which is produced in the bacterial virus known as *lambda phage*, is seen to protrude into the wider groove of the double helix. This enables it to probe the bases, and to recognize a specific site on the DNA by making particularly intimate contact with a region having a shape complementary to its own surface. The electron photomicrograph shows another example, the *lac* repressor, bound to a twisted stretch of DNA.

and thiogalactoside transacetylase, the function of which remains unclear. Especially revealing was the fact that production of all three proteins could be stimulated by molecules closely resembling lactose in their chemical and spatial structure. Because the three proteins are rapidly synthesized in response to lactose, it was obvious that the information required for their manufacture was already present in the bacterium's genetic machinery. It was also clear that, in the absence of the normal substrate, or something which can mimic it, the enzyme-producing mechanism was switched off, as if to spare the cell from futile effort.

These conclusions led Jacob and Monod to the idea of the *repressor* and the *operon*. Their model postulated the existence of three sections of genetic material located on the chromosome: a *regulator gene*, an *operator site* and a set of *structural genes*. The operator site and structural genes are collectively referred to as an operon. They sometimes lie side by side in the chromosome, but the regulator gene is often physically separated from the structural genes. We must not be misled by the terminology; an operon carries a set of genetic instructions but, like all DNA, it is passive. The regulator gene produces a repressor which can interact with the operator and effectively switch it off. Because the operator site is adjacent to the structural sections, which are under its control, the binding of the repressor molecule ultimately suppresses transcription of those structural genes. Walter Gilbert and Benno Müller-Hill showed that the repressor of the lactose operon, now known as the *lac repressor*, is a protein. It was found to bind to DNA carrying the lactose operon, but not to DNA lacking that gene sequence. Repressor proteins function in a manner not unlike the enzymes discussed earlier. They do not catalyze breaking or joining reactions, but they do have active sites, with which they cover and block regions of DNA. Their functioning, which is critically dependent on three-dimensional structure, can be compared to that of enzymes which require coenzymes. The relationship between DNA and the repressor is analogous to that of a substrate and an enzyme, while the substance that triggers the repressor, producing critical allosteric changes of shape, can be compared with a coenzyme. Repressors supply the means for exercising delicate control over genetic expression.

The DNA of an *E. coli* cell has about 4×10^6 base pairs, and the molecule is a closed loop about 1 mm in length. A human cell contains about 2 m of

Restriction enzymes

Restriction enzymes, cleave nucleic acid strands at sites specified by particular base sequences. Some cut the sugar–phosphate backbones at a position removed from the recognition site, while others make an incision at the locality recognized. *Eco* P1 is found in *Escherichia coli* and makes its cut between 25 and 27 bases away from the recognition position.

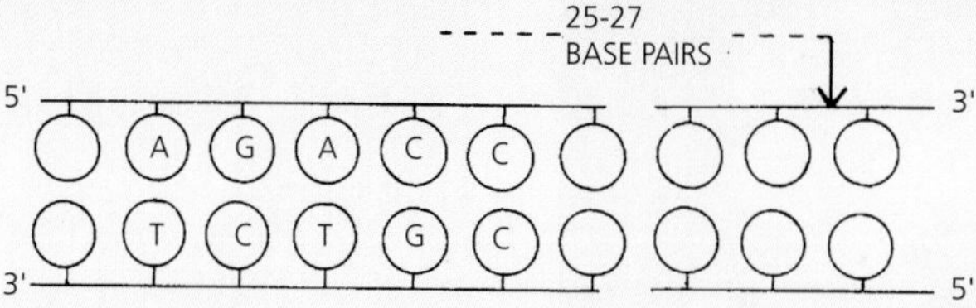

Eco P1

The arrow indicates the position of the cut, and the identities of the bases (circles) that are not specified are arbitrary. The horizontal lines represent the backbones. *Hae* III belongs to the second, more specific, group. It is produced by *Haemophilus aegyptius,* and it makes its cut at the symmetrical position as follows:

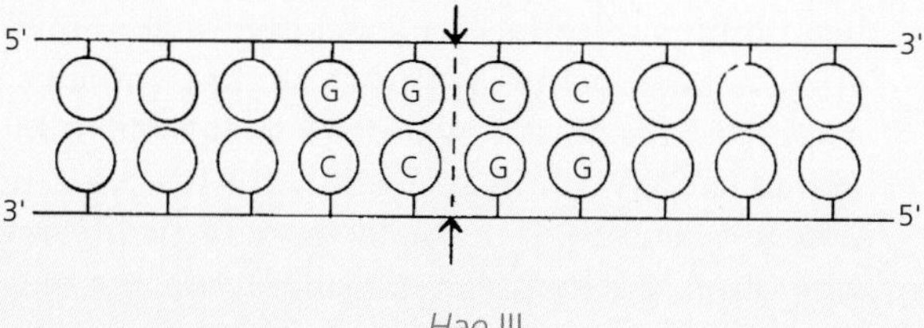

Hae III

This yields fragments that have blunt ends. Restriction enzymes of the second group which separate base pairs produce termini with a few unpaired bases, and these are called sticky ends because of their ability to combine with a complementary terminus. An example is *Bam* H1, found in *Bacillus amyloliquefaciens.* It gives fragments with four unpaired bases:

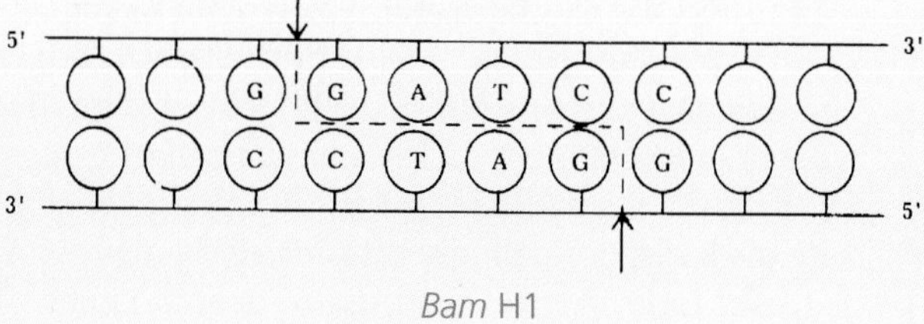

Bam H1

Bacteria possess immunity against their own restriction enzymes, this being conferred by methyl groups attached to certain bases at the recognition sites.

DNA comprising roughly 6×10^9 base pairs, and this is distributed among the 46 units that constitute the 23 chromosome pairs. These chromosomes, visible by optical microscopy during cell division, have a total length of only 0.2 mm, so there must be a high degree of packing of the basic double helical material. In fact, these numbers indicate a packing factor lying in the region of 10 000, and this has usually been taken to imply a hierarchy of folding. We turn now to the question of how this compaction is achieved and first note some definitions. The protein found in the chromosomes is known as *histone,* and the complex of histone and nucleic acid is a special kind of *nucleoprotein* called *chromatin*. During cell division, each chromosome divides into two *chromatids*. Research on the packing problem aims at describing the arrangement of histones and DNA in a chromatid. When the cell is not dividing, this structure apparently becomes unwound, so that the genetic message can be transcribed and translated.

The first evidence of regularity in chromatids was obtained by Maurice Wilkins and Vittorio Luzzati around 1960, by X-ray diffraction. They observed a reasonably well-defined repeat distance of about 10 nm. Chemical support for this conclusion was provided by Dean Hewish and Leigh Burgoyne in 1973. They discovered a DNA-cleaving enzyme, known as a *nuclease*, in the nuclei of rat liver cells and found that it splits chromatin into fragments having a certain minimum size. They concluded that the histones somehow confer a regular pattern of protection against the nuclease. At about the same time, Ada Olins and Donald Olins obtained striking electron microscope pictures of a ruptured nucleus, which revealed the chromatin as a beads-on-string arrangement, and Pierre Oudet later dubbed the beads *nucleosomes*. It was soon appreciated that nucleosomes were the origin of the regularity seen by Wilkins and Luzzati and that they were composed of histone in conjunction with the DNA thread. This association was apparently the origin of the protection observed by Hewish and Burgoyne. Histones come in five varieties, designated H1, H2A, H2B, H3 and H4. It was found that only the latter four types are present in the nucleosomes. In fact, they form an octamer comprising two molecules of each variety, as demonstrated by Roger Kornberg and Jean Thomas.

Study of the structure of the nucleosome unit was greatly facilitated by production of crystallized specimens by John Finch, Aaron Klug, and their colleagues. They obtained X-ray pictures from which it appeared that the nucleosome core has a flat oyster-like shape, 6 nm thick and 11 nm in diameter. They showed that the DNA must be wound twice around the core with a pitch of about 3 nm. This was consistent with the discovery, by Kornberg and Marcus Noll, that approximately 200 base pairs are associated with each nucleosome. Confirmation was provided by the neutron-scattering studies of John Pardon, Brian Richards and E. Morton Bradbury, which clearly showed the DNA as lying around the histone molecules. It became clear that what appeared to be a beads-on-string arrangement actually involved a string on beads. Finch, Klug, and their associates proposed that the individual nucleosomes in a chromatid are themselves arranged in a helix with a pitch of about 11 nm and a diameter of 300 nm. They called this structure a *solenoid*, and it is reminiscent of a prediction by Wilkins that the DNA double helix in a chromatid is wound into a larger helix. Indeed, the solenoid arrangement actually involves a three-fold hierarchy of helices; and even this is not the end of the story. The coiling of the DNA around the nucleosome cores gives a packing factor of seven, while the formation of the solenoid gives a further factor of six. This still leaves unaccounted for a factor of about 250.

In 1978, Leth Bak, Francis Crick and Jesper Zeuthen argued that there is another level of coiling intermediate to the solenoid and the visible chromatid. They called it a *unit fibre*, obtained evidence of its existence with an electron microscope, and showed that it gives another packing factor of about 40. This leaves a final factor of just 6 for the packing of the unit fibres into the chromatid. If all these results stand the test of time, it seems that Hamlet, in his famous soliloquy, might more accurately have spoken of

The *chromosomes* of higher organisms are composed of *chromatin*, a composite of protein and nucleic acid. When chromatin is observed in its unwound form, as in the electron photomicrograph shown here, the protein appears as globules arranged along the nucleic acid like beads on a string. The protein component consists of individual molecules known as *histones*. There are five types of these, and they are found in a wide range of eucaryotic organisms. Groups of eight histones, two of each of four of the known types, form what are known as nucleosomes and these are the globules seen in the photomicrograph. In the chromosomes, the composite is wound up into a compact structure in which the nucleic acid spirals around the *nucleosomes*, which themselves spiral into a helix. The fifth type of histone might arrange itself along the inside of this latter helix, known as a solenoid, keeping the nucleic acid windings in place. The nucleotide bases have been omitted except for the few indicated where the molecule ends.

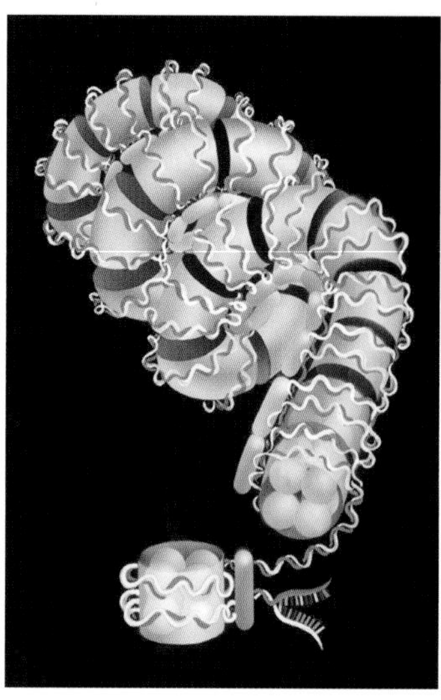

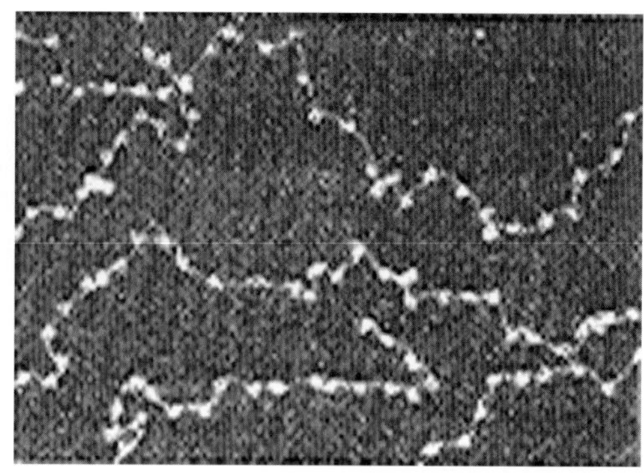

'... these mortal coiled coiled coiled coiled coils'. There remains the question of the H1 histone, which Kornberg and Noll had shown to be only loosely associated with each nucleosome. Experiments performed by Fritz Thoma and Theo Koller, and independently by Ingrid Eicken, Walter Keller, Ulrike Müller and Hanswalter Zentgraf, have revealed that H1 has a state of aggregation that depends on the ionic strength of the aqueous medium. This has led Klug and Kornberg to propose a modified solenoid model in which the H1 histones lock the pairs of DNA coils in place and possibly link up to form a secondary helical structure lying inside and parallel to the solenoid of nucleosome units.

Perhaps the most startling of the nucleic-acid–protein complexes are viruses. These sub-cellular objects, which lie at the frontier between living and non-living structures, show the genetic machinery in its most ruthless form. Lacking the usual life-supporting systems of the cell, and being thus incapable of an independent existence, they survive by commandeering a cell's hereditary mechanism. They are the ultimate parasites. The origins of viral science are usually traced to Edward Jenner's discovery, in 1798, that immunity to smallpox can be acquired by deliberate infection with the milder disease cowpox. The term *vaccination* comes from the Latin name for cowpox: *Variolae vaccinae*. In 1892, Dmitri Ivanovsky found that an agent capable of attacking tobacco leaves was small enough to pass through a filter that removed all bacteria. He erroneously concluded that the infection was the result of a simple chemical reaction. Seven years later, after confirming these observations, Martinus Beyerinck concluded that the mysterious agent was not just a very small cell but a more fundamental unit that can reproduce only as a cell parasite. Borrowing from the word virulent, he called the object a *virus*. The leaf-mottling agent studied by Ivanovsky is now known as the *tobacco-mosaic virus*, or *TMV*. Numerous other examples were soon found to be associated with a host of diseases afflicting plants and animals. William Elford succeeded in measuring their sizes, using a remarkably uniform and fine collodion filter, and the smallpox variety was found to be about 100 nm in diameter. Electron microscopy later permitted resolution of viruses smaller by a factor of 10, such as the one that causes *foot-and-mouth disease*.

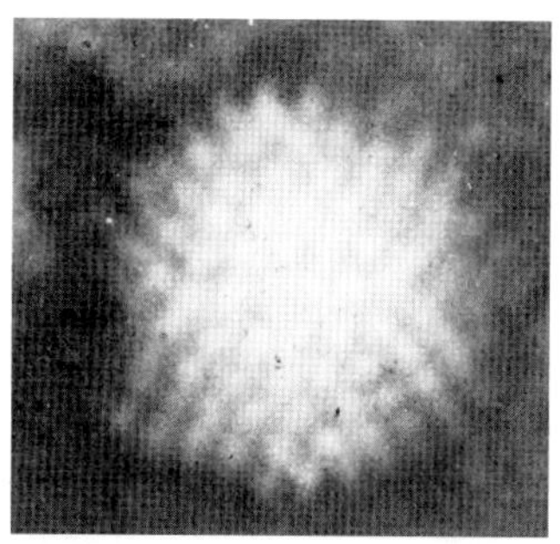

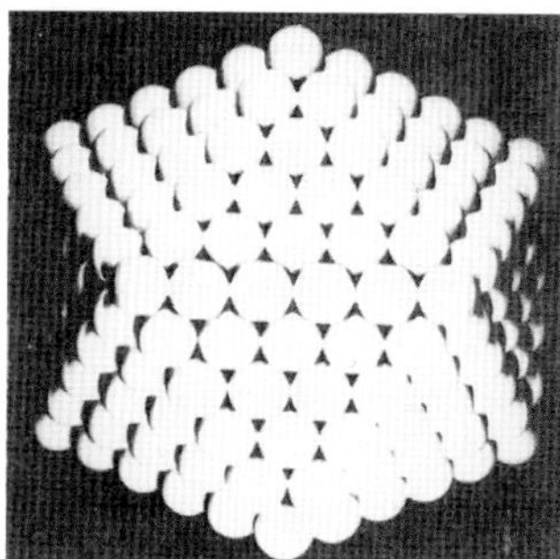

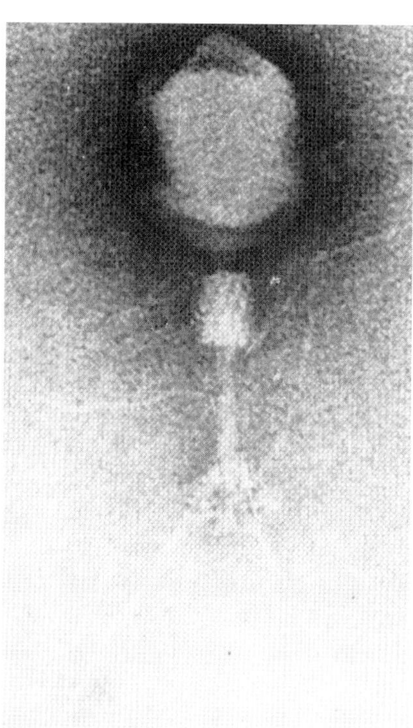

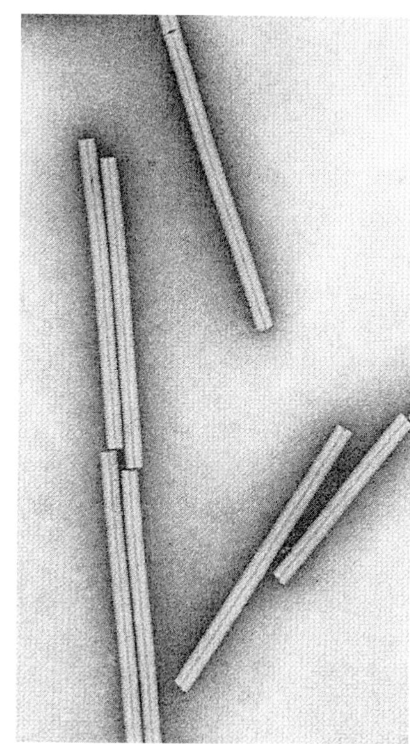

Observed by electron microscopy, *viruses* have remarkably geometrical shapes, as befits their intermediate status between the animate and inanimate domains. This derives from a biochemical simplicity that often involves just one type of protein and one nucleic-acid molecule, the former providing a container for the latter. In the *adenovirus* (upper left), a perfect icosahedron about 80 nanometres wide, the globular proteins form the flat outer faces by adopting a triangular lattice arrangement. This is simulated by the ball model (lower left). The *T2 bacteriophage* (centre) has a more complicated structure, the icosahedral head, cylindrical tail sheath, tail spike and tail fibres all being visible here. The tail fibres attach onto the cellular membrane of a victim and the nucleic acid contained in the head is injected into the prey via the tail sheath. The *tobacco-mosaic virus* (right) consists of a single 6400-nucleotide-long strand of RNA, representing about four genes, packed between the helical turns of an 18-nanometre diameter protein coat comprising 2130 identical subunits. The RNA–protein complex, stabilized by hydrophobic interaction, is susceptible to changes in the aqueous medium and tends to disintegrate at low acidity levels.

One of the most remarkable events in the history of virus research, because it came so early and won recognition so late, was Peyton Rous's observation of the transmission of carcinogenic activity by a cell-free filtrate prepared from avian connective-tissue tumour. Injected into healthy chickens, it induced cancer and became known as the *Rous sarcoma virus*. Four years later, in 1915, Frederick Twort noticed that part of a colony of *Staphylococcus* bacteria, being cultured on agar jelly, took on a glassy appearance. Trying in vain to generate further colonies from the affected area, he soon realized that the organisms had been killed. He found that filtering did not remove the mysterious agent and concluded that is was a virus that preyed on bacteria. Shortly thereafter, these observations were confirmed by Felix d'Herelle, who also established that no simple chemical compound was involved. He called the virus *bacteriophage*, the ending coming from the Greek word *phagos*, which means to consume. Ironically, the much-vaunted work of Wendel Stanley, in the mid 1930s, led virus research off on the wrong track. His efforts were commendable enough, involving as they did the boiling down of a ton of tobacco plants infected with TMV. From this, he extracted sufficient pure virus to establish that this could be crystallized, a fact which he took as an indication of similarity to simple chemical molecules. Stanley believed in fact that viruses are composed solely of protein. We now see, in retrospect, that the earlier and unheralded work of Max Schlesinger was of greater significance. By exposing *E. coli* cultures to bacteriophage, and centrifuging the cultures after the bacteria had been destroyed, he obtained a virus concentrate which chemical analysis showed to be half protein and half DNA.

Much of the virus story is better told in the context of cellular science and will thus be deferred to the next chapter. It is appropriate, however, to consider a few of the more recent developments here. In 1945, Rollin Hotchkiss discovered that the DNA of an invading virus actually joins up with the cell's own DNA when it takes over its victim's genetic machinery. In the early 1950s, electron microscopy revealed regular arrangements of substructures deployed around the viral surface, and these were correctly surmised to be the protein component. By observing viruses from different angles, Robert Horne was able to demonstrate that they invariably have a simple polyhedral shape, and this revealed the basis for the easy crystallization seen by Stanley. It was reasonably clear that the DNA must lie inside, protected by the protein sheath, and waiting to commit its act of piracy. The rationale of virus structure was clarified by Francis Crick and James Watson in 1957. Noting that the small size puts a limit on the amount of DNA that a virus can contain, they argued that it would be able to code for only a few proteins. They therefore suggested that the outer coat, known as a *capsid*, must be composed of a number of identical units symmetrically packed. These protein units are referred to as *capsomers*. In 1955, Heinz Fraenkel-Conrat and Robley Williams succeeded in reassembling the protein and DNA from dissociated viruses into new viruses possessing full potency. This exposed the essential simplicity that underlies the cold efficiency of these structures. The assembly of viruses was subsequently studied by Donald Caspar, John Finch, Rosalind Franklin and Aaron Klug, who used their observations as a source of information regarding the DNA–protein interaction. It appears that the DNA fits into a groove on the protein surface, the snug fit stretching over a length that spans about a dozen base pairs.

Study of numerous viruses has shown that two structures are particularly common: a helical arrangement, as seen in TMV, and an icosahedral form, examples of which include *adenovirus*, *polyoma virus* and the *herpes virus* that causes cold sores. The *poliomyelitis virus* has a less common dodecahedral structure. The simplest varieties have as few as three genes, while the most complex can have more than 200. Not all viruses contain DNA. Some, such as the Rous sarcoma variety, contain RNA instead. In such cases, as predicted by Howard Temin in 1964 and verified by Temin and David Baltimore six years later, the flow of genetic information is contrary to the usual direction, passing from RNA to DNA with the aid of an enzyme referred to as *reverse transcriptase*.

The newest addition to the genetic arsenal, by date of discovery at least, is the *viroid*. The first evidence of this agent of disease had appeared in the

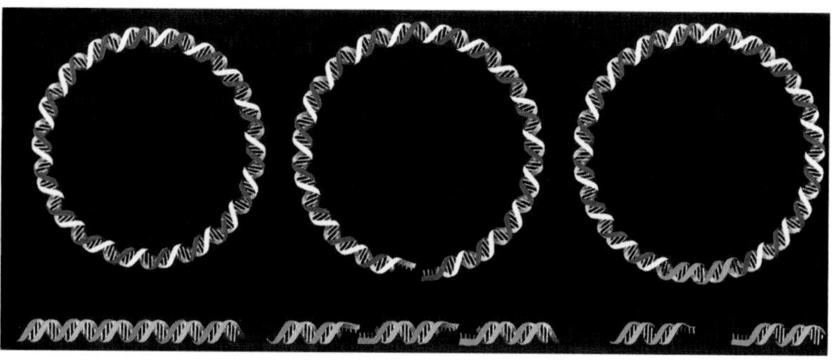

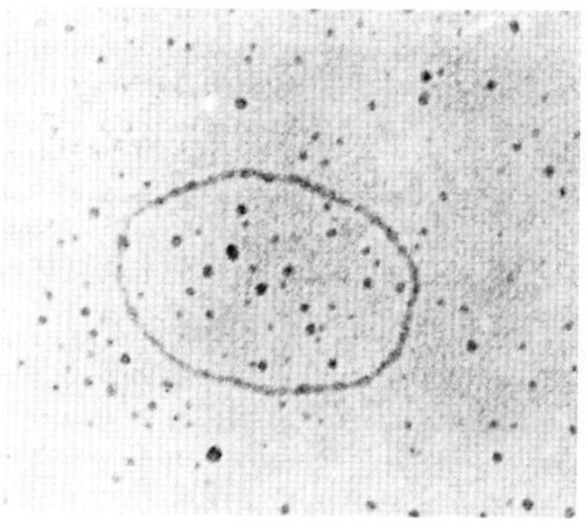

The advent of genetic engineering was triggered by two key discoveries: *restriction endonucleases* and *plasmids*. A restriction endonuclease is an enzyme that acts as a precise atomic-level scalpel; it is capable of cutting double-stranded DNA at a specific site, at which the nucleotide base sequence displays two-fold rotational symmetry. The most useful varieties are those which leave a number of unpaired bases on one of the strands. Such cleavage is said to produce a '*sticky end*', and it can recombine with any other DNA fragment produced by the same enzyme. In the example shown here, the restriction endonuclease is *Eco* RI, and the symmetry of its cut is apparent. (By analogy with passages like the famous 'able was I ere I saw Elba,' which read the same both forwards and backwards, such cleavages are said to be palindromic.) Plasmids are small mobile genetic elements in the form of closed loops of DNA. They have been found in many bacteria, and are readily extracted from and inserted into these organisms. Genetic manipulation is achieved by using a specific restriction endonuclease to cut a desired stretch of DNA from a convenient source, and inserting it into a plasmid that has been cleaved by the same enzyme. The resulting chimera, as it is called, is then introduced into a fresh bacterium. This is done by soaking the organism in a warm calcium chloride solution to make its membrane permeable. The electron photomicrograph shows a roughly 6000-nucleotide-pair plasmid extracted from the yeast bacterium *Saccharomyces*. There are between about 50 and 100 such plasmids in each bacterium, and the length of the loop, here made visible by a combination of protein attachment and heavy-metal shadowing, is approximately 1.7 micrometres.

1920s, in the form of spindle-tuber blight in potatoes. Plant pathologists soon established that no bacteria or other micro-organisms were involved in the infection, and it was assumed that the trouble was caused by a virus. In 1962, William Raymer and Muriel O'Brien discovered that the spindle-tuber agent could be transmitted from potato to tomato, young plants of the latter developing twisted leaves at a surprisingly rapid rate. The highly infectious entity seemed to be able to permeate an organism very quickly. It also seemed to be rather light, because subjecting a solution to a centrifugal force 100 000 times as large as gravity did not cause the agent to settle to the bottom of a test tube. Such a force would have produced a sediment of even the smallest virus. Together with T. O. Diener, Raymer went on to show that the mysterious factor became inactivated in the presence of the RNA-digesting enzyme *ribonuclease*, whereas it was unaffected by other enzymes known to break down DNA. Their conclusion was that the source of infection was a relatively short length of RNA which, unlike a virus, was not encapsulated in a protein coat.

Electron microscopy of spindle-tuber viroids, by José Sogo and Theo Koller, showed them to be rod-like with a length of about 50 nm. Their width was similar to that of double-stranded DNA, but it was subsequently found that this was caused by complementary base pairing across opposite sides of a single closed RNA loop. Viroids thus bear a structural resemblance to transfer-RNA. Hans Gross and his colleagues worked out the entire nucleotide sequence of this viroid, finding the 359 bases to be distributed in a rather suggestive manner: 73 adenine, 77 uracil, 101 guanine and 108 cytosine. Recalling that adenine pairs with uracil, and guanine with cytosine, this distribution would permit a high degree of cross-linking provided that the individual bases are in favourable positions. This proves to be the case, about 90% of the bases being involved in pairing. A rod shape is thereby produced, with the length quoted earlier. Viroids are the smallest known agents of infectious disease, and they appear to be located predominantly in the nuclei of the cells of the afflicted organism. Different varieties have now been identified in connection with about a dozen plant diseases, and there are indications of a link with certain animal diseases, possibly including one that affects the human nervous system. Just how they function is still not clear, but it is suspected that they interfere with the control of genes that code for certain *growth hormones*.

Viroids are not the only example of extra-chromosomal genetic material found in living cells. There is also the *plasmid*, and it takes the form of an isolated and closed loop of double-stranded DNA. Like the viroids, plasmids have no protein sheath, and they cannot survive outside their host cell. They multiply independently of the cell's chromosomal DNA, the multiplication products being divided in the usual way during cell division. Unlike the chromosomal DNA, the genetic material contained in a plasmid does not appear to be essential to a cell's life cycle, but it can confer certain favourable characteristics which might help a cell to survive. Plasmids are found in bacterial cells. As will be described in the next chapter, Joshua Lederberg discovered that such cells, or *procaryotes* as they are called, can pass on genetic information between themselves by a sort of mating process known as *conjugation*. Noting that specific genetic traits could be communicated independently, he concluded that they must be coded for by genetic material that is isolated from the main procaryotic chromosome. In 1952 he coined the term plasmid, to describe such an independent fragment. Plasmids account for only a small fraction of the genetic material in a cell, usually not more than a few per cent, but it is they alone which happen to carry the information required for conjugation. Moreover, they enable their host cells to resist a variety of toxic agents, including *antibiotics*. During conjugation they take their desirable qualities with them, thereby permitting the recipient to survive while the donor perishes. Viruses and viroids are unwanted parasites, but plasmids are most welcome guests.

Plasmids took on industrial significance in the 1970s with the development of what became known as *genetic engineering*. The general aim of this enterprise is the production of useful substances by bacteria containing modified genetic machinery: farming at the cellular level. It was made possible by the discovery of certain enzymes which are able to cleave DNA at specific sites, and these were used to insert desirable genes. The enzymes are known as *restriction endonucleases*, or simply as *restriction enzymes*, and their existence was first postulated by Werner Arber and Daisy Dussoix in 1962. From observations on genetic variations in certain bacteriophages, they concluded that there must be enzymes capable of cutting unprotected DNA. It transpired that the particular enzymes they were studying were not particularly specific. The site at which each enzyme recognized a particular sequence of bases was found to lie somewhat removed from the cutting

The cause of the dreaded *acquired immune deficiency syndrome* (*AIDS*) is the *human immunodeficiency virus* (*HIV*), which is about a hundred nanometres in diameter. Its outer surface consists of a phospholipid bilayer membrane which provides symmetrically distributed anchoring sites for eighty molecules of the glycoprotein called gp120. These are arranged at the corners of twelve pentagons and twenty hexagons, with an extra protein at the centre of each hexagon (left picture). An inner shell of protein units, in the shape of an icosahedron, envelopes the central core. The latter, shaped rather like the bottom of a champagne bottle, contains the genetic material of HIV. The virus attacks a particular type of *white blood cell* (or *lymphocyte*) known as a T4 cell, either merging with the latter's membrane or becoming engulfed by it (right picture, which is a transmission electron micrograph). Once inside its victim, the virus disintegrates, releasing its RNA and the enzyme reverse transcriptase. This enzyme copies the invading genetic material into DNA, which becomes incorporated into the target cell's nucleus. This act of piracy enables the virus to become a permanent component of the T4 cells, and these unwittingly spread the virus when they respond to some other infection.

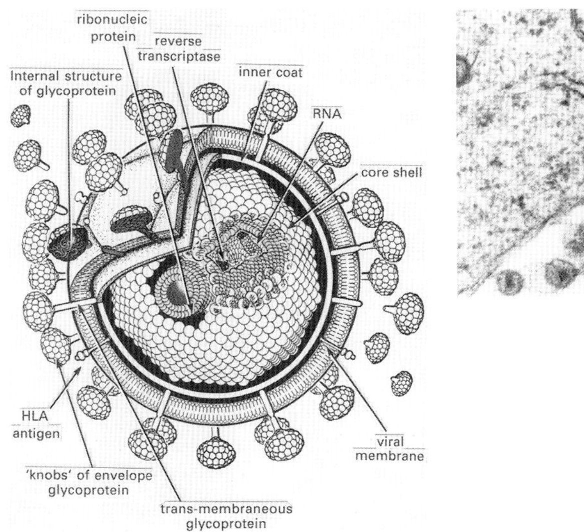

point. Other examples later studied by Hamilton Smith and Kent Wilcox had the desired accuracy, recognition and incision occurring in the same location. Restriction enzymes of this latter class can further be grouped into two types: those that cut only the two sugar–phosphate backbones, and those that also separate a number of nucleotide base pairs. An example of the first type is *Hae* III which recognizes a particular palindromic sequence of guanines and cytosines and makes its cuts at the symmetric position. The other type produces termini having a few unpaired bases, each of which can readily combine with a complementary terminus. Such *sticky ends,* as they became known, are particularly useful to the gene splicer. An example is *Bam* H1, and it produces sticky ends with four unpaired bases.

We should not be surprised at the number of base pairs cleaved by such an enzyme. The pitch of the DNA double helix comprises about ten base pairs, but the number of pairs accessible from one side is no greater than four or five. Genetic engineering is carried out in the following manner. Bacteria are burst open with detergents, which disrupt their membranes, and their plasmids are recovered after separation by centrifuging. The plasmids are cleaved at specific sites by a particular restriction enzyme and are mixed with gene segments from other organisms that have been snipped out of chromosomes by the same enzyme. Having complementary sticky ends, the various lengths of DNA can combine to produce a composite containing genetic material from two or more different organisms. Such composite loops are called *plasmid chimeras,* after the mythical beast with the head of a lion, the body of a goat, and the tail of a serpent. The joining reaction is catalyzed by the enzyme DNA-ligase. The first such *recombinant DNA* was produced by Paul Berg, David Jackson and Robert Symons in 1971, and it was a chimera of the five-gene DNA of the simian virus SV40, a length of plasmid DNA from *E. coli,* and several genes from the virus known as bacteriophage lambda. By 1981, genetic engineering had already advanced to the stage where such medically important substances as *insulin* and *interferon* were being produced routinely in considerable quantities. Goals for the future include manipulation of the amino-acid composition in mass-produced protein, to produce a predominance of the essential amino acids.

An obvious question arises in connection with the particular restriction enzymes that a bacterium produces: how does its own DNA survive being chopped into pieces? It turns out that the bacterial DNA does indeed have the desired immunity, this being conferred by methyl groups that are attached to certain bases at the recognition sites. Consider, for example, the action of the *Hind* III restriction endonuclease of *Haemophilus influenzae*. It produces a four-base sticky-end cleavage. The corresponding sequence on this bacterium's own DNA has two methylated adenines, and these provide immunity by sticking out into the larger groove of the double helix and physically precluding recognition.

Cleavage and splicing could be called the precision machining and riveting activities of the genetic engineer. Another important aspect of the enterprise can be regarded as the reading of the blueprint. It is the determination of the base sequence of a length of DNA, and here too restriction enzymes play a key role. Two similar approaches were developed independently by Walter Gilbert and Frederick Sanger, both exploiting the fact that molecules diffuse under the influence of an electric field, at a rate proportional to their size. Gilbert, together with Allan Maxam, cleaved DNA at specific sites, with a restriction enzyme, and labelled the ends radioactively. They then divided the sample into four batches, treating each with a chemical that attacks the sugar–phosphate backbone at only one of the four types of base. In each case, the chemical was so dilute that only a small, and random, fraction of the corresponding base sites were cleaved. The solution containing the various fragments was then placed at the lower end of a film of gel across which a voltage was applied, and in a given time the fragments rose to heights that were inversely proportional to their sizes. One can distinguish fragments differing by only a single base; a resolution of 0.5 nm. An autoradiographic print is obtained for each batch, and when the four ladder-like pictures are placed side by side, the positions of the different bases can be read off directly. Sanger's method is slightly different in that the splitting into fragments is achieved by what are known as *base analogues*. These resemble the actual bases found in DNA but differ from them sufficiently to preclude pairing. The Sanger technique involves making complementary copies, and the copying process is brought to a halt by the counterfeit bases. In 1977, Sanger and his colleagues completed the first determination of the entire base sequence of a complete DNA-containing virus: the 5347-base genetic blueprint of the bacterial virus ΦX174.

This chapter has been concerned primarily with the interplay between two types of polymer which collectively hold the key to life's processes: the biopolymers known as nucleic acids and proteins. Since there must have been a time in the remote past when the terrestrial environment was too hostile to permit any form of life, we might ask how things started. Aristotle believed that life could spring from inanimate matter, if the latter was suitably influenced by a 'soul', and the sixteenth-century physician Paracelsus claimed to have produced an embryonic man by correct manipulation of appropriate elements. (Such a *homunculus*, or little man, also appears in Goethe's classic; Faust's assistant, Wagner, conjuring up a 'pretty manikin, moving, living, seeing' in a glass vessel known as an alembic.) René Descartes and Isaac Newton were among later advocates of the idea that life can be spontaneously generated. The opposing view, that only life can beget life, received strong support in the middle of the seventeenth century, when Francesco Redi showed that maggots appear in decaying meat only if this is

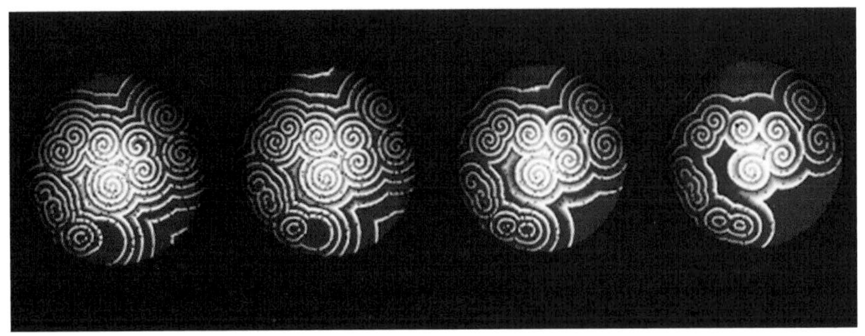

In 1920 Alfred Lotka (1880–1949) predicted that some chemical reactions might display stable oscillations, and the effect was indeed observed by William Bray a year later. The suggestion elicited ridicule, and Bray's discovery was overlooked. Forty years later, however, Boris Belousov reported such oscillations in mixtures of organic fuels and water. The fuels were citrate and malate, which are both involved in the energy-producing reactions in living cells. Then in 1970, Albert Zaikin and Anatol Zhabotinsky discovered the recipe for a chemical reaction that not only pulses but also sends sharp waves of colour out across a surface, like grass fires across a field. The mixture contained bromide and bromate, which are two forms of bromine compounds, the organic fuel malonate, and phenanthroline, an organic complex of iron that simultaneously serves as a catalyst and a colour indicator. The dye is red when its iron atom is reduced to the ferrous form, with two electrons in its outer orbital, and blue when it is oxidized to the ferric form with three outer electrons. The concentration of bromide ions in solution determines the oxidation state of iron, so this concentration is indicated by the waves of colour, seen at four different stages of development in this picture. Similar wave patterns are observed to travel along the membranes of nerve cells, and they have been conjectured to provide the basis of the *circadian rhythm*, otherwise referred to as the *biological clock*. Manfred Eigen has suggested that an initially uniform solution of a few amino acids and certain short lengths of nucleic acid might start to throb in an analogous manner, and that this could have provided the *origin of life*.

not hermetically isolated from the surroundings. Two hundred years later, Louis Pasteur appeared to have settled the issue when he exposed a boiled solution of organic materials to air from which all micro-organisms had been filtered. No life appeared in the container, and Pasteur's painstaking experiment earned him the prize offered by the French Academy of Sciences for such a rigorous proof.

We have already seen that experiments such as those performed by Jerard Hurwitz and G. David Novelli indicate a subcellular basis of life, and laboratory studies by Stanley Miller, in 1952, lent very strong support to the fundamental chemical view. Miller filled a large flask with gases assumed to be present in the early terrestrial atmosphere: hydrogen, methane, ammonia and water vapour. He boiled the mixture, to approximate the conditions of the remote volcanic past. Finally, he simulated the effects of lightning by subjecting the primordial brew to 60 000 V electric discharges. The result was dramatic; subsequent chemical analysis showed his mixture to contain four different kinds of amino acid, as well as some simple organic polymers. Later experiments by Leslie Orgel added an important dimension to the story. He slowly froze an aqueous solution of hydrogen cyanide, ammonia, and other simple organic compounds, in order to concentrate them into close conjunction, and discovered that he had produced not only amino acids but also adenine, one of the four nucleic acid bases. The concept of spontaneous generation had made a spectacular comeback.

John Bernal suggested that the surfaces of clay minerals may have provided catalytic sites for the early polymerization of organic building blocks, and both Ephraim Katchalsky and Graham Cairns-Smith have shown that the common clay *montmorillite* has a crystal structure that is ideally suited to play the role of template for amino-acid association. More recently, Stephen Bondy and Marilyn Harrington have demonstrated that another clay mineral, *bentonite*, preferentially catalyzes the polymerization of left-handed amino acids and right-handed sugars. These are indeed the forms found in Nature.

Let us now consider these life-giving processes at the fundamental physical level. The consolidation of monomers into polymers with specific sequences of either nucleotide bases or amino-acid residues represents a decrease in *entropy*, but this is no problem. *Thermodynamics* merely requires that the overall entropy of a system and its environment shall not decrease. The increase of information that accompanies the development of a biologically specific molecule is more than offset by the acquisition of free energy from the surroundings. But thermodynamics is unable to predict the emergence of self-sustaining structures, because, as Alan Cottrell has

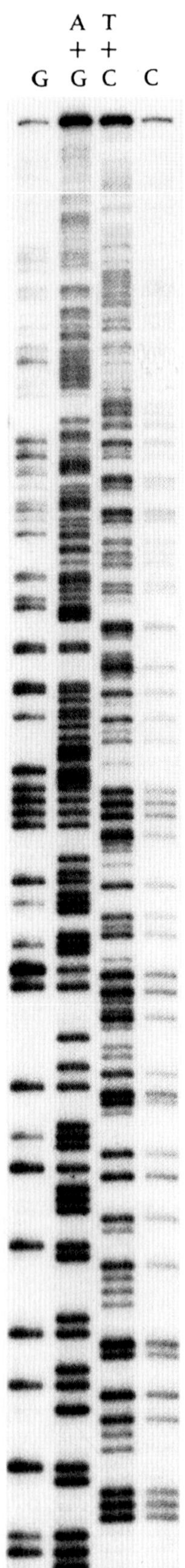

emphasized, this branch of science does not encompass such concepts as quality and purpose. At the thermodynamic level, nucleic acids and proteins are mysterious engines. Studying organized structures, of which biopolymers are good examples, Ilya Prigogine found that highly non-equilibrium situations can be generated, and can survive, as long as the rate of energy influx exceeds a certain threshold value. Intriguing manifestations of the spontaneous generation of pattern have been discovered, starting with a reaction studied by B. P. Belousov in 1958. Ions of the metal cerium catalyzed in water the oxidation of an organic fuel by bromate. Surprisingly, this reaction was found to have no stable steady state; it oscillates with an almost clock-like precision, waves of alternating colour systematically passing through the container. A prediction of such chemical oscillation, by Alfred Lotka in 1920, came before its time, and met only with ridicule. A. M. Zhabotinsky and A. N. Zaikin later developed a number of systems which showed the effect more dramatically. These phenomena are examples of what is now referred to as *self-organization*.

Manfred Eigen has proposed that life must have sprung from similar *pattern generation* within an originally homogeneous system. In his picture, we are not confronted with the old puzzle about the chicken and the egg, when considering whether priority must be accorded to nucleic acid or protein. He was able to demonstrate that they evolved together. His model favours short stretches of RNA as the original nucleic acid, and these were postulated as being able to recognize certain amino acids without the help of enzymes. Making estimates of the degree of interaction between the different components, Eigen was able to show that the system would display *collective oscillations*. Assuming that mutations to the components occur randomly, he found that the system would evolve by a sort of *natural selection* at the molecular level. The genetic code was seen as being quite primitive initially, capable of specifying only four amino acids. As Eigen and Ruthild Winkler-Oswatitsch have emphasized, evolution appears to be an excellent example of a game of chance, but one played according to strict chemical rules. At the molecular level at least, life is predicated on inequality; in the social life of the biopolymer the guiding philosophy is Herbert Spencer's – *the survival of the fittest* – rather than that of Karl Marx. Eigen's ideas received dramatic endorsement in the early 1980s when

Sequencing the nucleotide bases on a strand of DNA has become a routine exercise. The 5′ end is first labelled with the radioactive isotope ^{32}P and methyl groups are then attached to a limited number of the guanine bases by reacting the specimen with dimethyl sulphate for a few minutes. Subsequent treatment with piperidine causes cleavage at the methylated sites, and the different lengths of segment are now separated by making them diffuse through a polyacrylamide gel, under the influence of an electric field. The field strength is about 50 volts per centimetre, and the smallest fragments diffuse the farthest. Three similar treatments with formic acid, a hydrazine and sodium chloride mixture, and hydrazine alone, produce cleavages at adenine and guanine, thymine and cytosine, and only cytosine, respectively. By running four diffusion columns simultaneously, therefore, the entire base sequence can be read off. In this *autoradiographic gel*, the uncleaved fragments are at the top and this is why bars appear in all four columns. Reading from the bottom, the base sequence starts: AGGCCCAG . . . The specimen is a 432-base fragment of *Escherichia coli*, located about a thousand bases from the origin of replication.

The polymerase chain reaction and genetic fingerprinting

Each multicellular organism starts its existence as a single cell and grows thereafter by the process of cell division. This produces first two cells, then four, then eight and so on. The genetic material in the nucleus of each cell is duplicated during each division, so the increase in the number of copies of this material keeps pace with the increase in the number of cells. One could refer to that increase as a chain reaction, borrowing from the similar term applied to the nuclear process described in Chapter 1, though the chain reaction is admittedly a rather slow one. The replication of the genetic material during each cell division involves the enzyme *DNA polymerase*, as described in the present chapter. This is able to synthesize new stretches of the genetic material, provided that there is a plentiful supply of DNA building blocks in the form of deoxynucleoside triphosphates, the synthesis naturally occurring at body temperature. As the science of molecular genetics progressed, it became clear that it would be enormously advantageous if multiple copies could be produced of any strand of DNA in which there happened to be a particular interest. Kary Mullis solved this problem in an astonishingly straightforward manner in 1983. He had been specializing in the production of *oligonucleotides*, that is to say stretches of single-stranded DNA with specific sequences of bases. When the base sequence of such an oligonucleotide is precisely complementary to the sequence at one end of a DNA stretch one wishes to replicate, the oligonucleotide is referred to as a *primer*. Mullis realized that the replication process could be mimicked in a test tube – and greatly speeded up – by prising the complementary strands of DNA apart through the simple expedient of raising the temperature. This is variously referred to as *melting* or *denaturation*. When the sample had cooled again, the replication process would repeat itself, provided that there were copious supplies of primer molecules and deoxynucleoside triphosphates. The diagram illustrates the process in schematic form, the complementary strands being indicated in red and green. The attachment of a primer to the complementary sequence of bases on the strand to be replicated is referred to as *annealing*. The fact that a chain reaction takes place is clear, even though only three successive stages are shown, and one should note that DNA stretches of the desired length soon predominate. There was one technical difficulty, and this was soon overcome: the DNA polymerase would itself be vulnerable to denaturation because of the elevated temperature. But there are organisms whose polymerase molecules are clearly able to withstand the near-boiling temperatures in hot springs because they are able to grow and multiply under such extreme conditions. An example is the so-called

Taq polymerase present in *Thermus aquaticus*, which thrives in hot pools such as the one known as Morning Glory in Yellowstone National Park, shown in the photograph. Starting with a single stretch of DNA, a hundred billion copies can be fabricated within a few hours by the polymerase chain reaction (PCR). The target stretch may be pure, but it may also be a small part of a complex mixture of genetic material. PCR makes an important contribution to the technique known as *genetic fingerprinting*, which was pioneered by Alec Jeffreys and his colleagues in 1985. In order to understand what is entailed, one must first note that it is advantageous for some stretches of DNA to be present in multiple copies – known as *tandem repeats* – in the genome because this speeds up the rate of production of the corresponding pieces of cellular machinery. This is the case for the stretches coding for ribosomal RNA, transfer RNA and histones, for example. There is a clear evolutionary advantage associated with tandem repeats because they permit faster cell division. The origin of other repeats, such as those referred to as *satellite DNA*, is less obvious; they may be the outcome of slippage along its template strand of the new strand being formed during replication. Some satellites occur in very short versions, known as *microsatellites*, and they often cause genetic diseases. There are also regions known as *minisatellites* which have up to fifty repeats of stretches composed of up to about a hundred base pairs. The interesting fact is that whereas the actual base-pair sequence in a particular minisatellite is conserved from individual to individual, different people have different numbers of repeats. And examination of the abundance of repeats for various minisatellites permits one to build up a profile for an individual that is actually more specific than his or her ordinary fingerprints. Identification in criminal cases is an obvious – and now familiar – application, while genetic fingerprinting's most spectacular victory to date was probably the verification that bones found buried on the outskirts of Sverdlovsk were indeed those of the final Russian royal family, murdered in 1918. This was established through comparison with genetic material provided by Philip, Duke of Edinburgh, another member of the same family tree. The final outcome was that Nicholas II (1868–1918), his wife and children finally received a proper burial.

Thomas Cech and Sidney Altman, together with their respective colleagues, discovered that RNA molecules can indeed function as powerful and selective catalysts.

A number of technical advances achieved in 1980, and subsequently, made sequencing of the entire genetic information in an organism – that is to say, the organism's *genome* – increasingly feasible. The basic methods invented by Frederick Sanger and by Walter Gilbert and Allan Maxam were described earlier in this chapter. These were soon being exploited by David Botstein, Ronald Davis, Mark Skolnick and Ray White, in a method known as *restriction-fragment length polymorphism* (*RFLP*). This technique is made possible by the way genes are inherited. Because we receive only one copy of each chromosome from each parent, and because there is a swapping of parts of the material in each pair of chromosomes (as will be described in the next chapter), the chromosome that is passed on carries a new combination of genes. But the number of such swaps is far fewer than the number of genes, so there is a high probability that a given gene will remain adjacent to its original neighbour on the new hybrid chromosome. If an easily detected stretch of DNA can be found that is a neighbour of a disease-causing gene, therefore, it will serve as *marker* for tracing the inheritance of that gene. There is the additional requirement that the said stretch should come in several variants, because the region of DNA immediately adjacent to the disease gene is then likely to differ in a readily detectable manner from the corresponding region associated with the normal version of the disease. James Gusella succeeded in isolating such a marker for *Huntington's disease* in 1983, that severe brain disorder resulting in involuntary writhing movements of the limbs.

A year later, the complete 170 kb (that is to say 170 thousand base-pair) DNA sequence of the *Epstein–Barr virus* was published. There then followed two further technical breakthroughs which were to speed things up dramatically. Kary Mullis realized that the requirement of an approximately constant body temperature, which sets a limit on the rate at which reactions take place, does not apply in the test tube. Moreover, the *DNA polymerase* in certain hot-spring organisms can survive and function at much higher temperatures, enabling Mullis to replicate large amounts of genetic material in what has come to be known as the *polymerase chain reaction*. Shortly thereafter, Leroy Hood and Lloyd Smith produced the first automatic DNA sequencing machine.

J. Craig Venter, Claire Fraser and Hamilton Smith were able to publish the entire 1.8 Mb sequence of *Haemophilus influenzae* in 1995, the first independent organism to be thus analyzed, and only a year was to pass before the 12 Mb sequence of the yeast *Saccharomyces cerevisiae* was similarly made public. This organism was found to possess slightly under 6000 genes. Fred Blattner and Guy Plunkett established no record with their report of the 5 Mb sequence of *Escherichia coli*, a year later, but they had at least revealed the hereditary make-up of a famous genetic work-horse. In 1998, the 97 Mb sequence of the nematode worm *Caenorhabditis elegans* was reported by John Sulston, Robert Waterston and their colleagues. The estimated number of genes in this organism was 19 099, which is about 6000 more than are found in another favourite of geneticists, *Drosophila melanogaster*, alias the common fruit fly (180 Mb). The first plant sequence to be published was the 125 Mb of *Arabidopsis thaliana*, which is estimated to possess about 25 000 genes.

The first human chromosome to be sequenced was, not surprisingly, the shortest, namely chromosome 22, its 34 Mb stretch being published just before the turn of the millennium. The quest for sequencing the entire *human genome* was naturally an undertaking of considerable proportions, with large numbers of collaborators pooling their resources. The historic announcement of the job's completion came in June 2000, with simultaneous publication of the compendious documentation in the prestigious journals *Nature* and *Science* the following February. The leaders in this monumental achievement were Venter, introduced above, and Francis Collins.

Perhaps the biggest surprise to emerge from this work has been the indication that the human genome comprises only about 30 000 different genes. This is an increase of only 50% over the above-mentioned nematode worm, and it suggests that complexity of body structure is not well correlated with gene number. But as Jean-Michel Claverie has stressed, the relationship between *gene number*, N, and *structural complexity* might not be a simple one. As was discussed earlier in this chapter, individual genes are activated by special promotor proteins, and each of them can thus be regarded as being in either its activate or inactive state. This leads Claverie to suggest that complexity might be related to the quantity 2^N, rather than to N, say. If this is the case, there would be a 2^{10000} fold increase in complexity in proceeding from the nematode to the human, a factor with which we may feel more comfortable.

The dependence on gene number might be even more recondite, because many aspects of structure are determined by more than one gene. There has been a certain amount of loose talk about individual genes for specific diseases, but it is certainly the case that many diseases are related to more than one gene. The devastating mental condition of *autism*, for example, has been estimated to be provoked by faults in up to half a dozen genes. It is true that some single-gene diseases have been identified, an example being *Huntington's disease*, but such simple inheritance is probably the exception rather than the rule.

The science of genetics still faces worthy challenges, therefore, and there is still the difficult issue of predicting a protein's three-dimensional structure from its primary amino-acid sequence – what some have referred to as the second half of the genetic code. The reward for solving that latter problem would be great, given that it would open the way to designing proteins ab initio for a specific function. That quest provides the engine for much computer-based activity in pharmaceutical laboratories the world over.

Summary

Many important biological molecules are polymers. In proteins, each monomer is one of 20 different amino-acid residues; whence protein diversity. The hydrophilic or hydrophobic tendencies of the various side groups, their steric interactions, and cross-linking determine the three-dimensional structure of the protein. Spatial conformation governs enzymatic function, in which protein and substrate interact like lock and key. In eucaryotes, the repository of genetic information is a ladder polymer, deoxyribonucleic acid, or DNA, which steric hindrance twists into a double helix. The rung-like cross-links are purine–pyrimidine base pairs. The DNA directs the construction of proteins, three consecutive bases coding for one amino-acid residue. The DNA is contained in the chromosomes, in nucleic-acid–protein composites called nucleosomes. Gene regulation is governed by promotor and repressor molecules, which are also proteins. Nucleic-acid–protein composites are also the basis of viruses, in which crystal-like arrays of proteins form the outer shell. Restriction enzymes cut nucleic acid strands at specific sites, identified by a particular sequence of bases, permitting manipulation of the genetic message. Such sub-cellular surgery, and exploitation of the transporting capability of small loops of nucleic acid called plasmids, is the basis of genetic engineering. The entire human genome had been elucidated by mid 2000, and it produced the surprising result that a human being possesses only about 30 000 different genes, a mere 50% increase over the number found in a nematode worm.

16

A million million spermatozoa,
All of them alive:
Out of their cataclysm but one poor Noah
Dare hope to survive.
And among that billion minus one
Might there have chanced to be
Shakespeare, another Newton, a new Donne,
But the One was Me.

Aldous Huxley (1894–1963)
The Fifth Philosopher's Song

The essential bag: *the cell*

Although biopolymers are more sophisticated than their simpler cousins, such as polythene and rubber, this alone does not give them an obvious reason for existing. Nucleic-acid codes for proteins, some of which contribute to the formation of further nucleic acid, but if these products were free to float away through the surrounding water, no advantage would be gained by the original molecule. The essential extra factor was discovered over 300 years ago. In 1663, using one of the earliest microscopes, Robert Hooke observed that *cork* has a structure reminiscent of a honeycomb, with pores separated by diaphragms so as to produce a collection of little boxes, which he called *cells*. Their universality was not appreciated until 1824, when Henri Dutrochet concluded that entire animals and plants are aggregates of cells, arranged according to some definite plan. He also suggested that growth is caused by an increase in either cell size or cell number, or both. The idea of the cell was further elevated in 1838, when Matthias Schleiden proposed that it is the basic structural unit in all organisms. By the following year, Schleiden and Theodor Schwann had perceived that cells are capable of independent existence, and two decades later Rudolf Virchow speculated that cells can only be produced by other cells. The concept of an indivisible unit had thus been introduced into the realm of biology, and it foreshadowed the arrival of the atomic and quantum ideas in physics and chemistry. It is this division into compartments, each with its own set of hereditary molecules, that provides the competitive basis of life.

Recognition of the cell as the common theme of living systems was no mean achievement. Cells come in a great variety of shapes and sizes. The yolk of an ostrich egg is a single cell with a diameter of several centimetres, while a typical *animal cell* measures about 0.001 cm across. Some *bacterial cells* are smaller than this by a factor of 30. Cell shapes include the polyhedron of a typical plant cell, the roughly spherical shape of an egg cell, the octopus-like nerve cell, the tadpole-like sperm cell, the lozenge-like blood cell, and so on. There is also great diversity in the degree of aggregation. The bacterium is a single cell, while a fully grown human comprises approximately one million million cells. But in spite of this variety of form and dimension, the purpose of every cell is essentially the same, and this is intimately connected with what the existence of cells does to the distribution of biopolymers. It separates them into elementary groups, which are forced to carry on an independent struggle for existence. The cell must use the materials and energy available in its environment to survive and reproduce. In pursuit of this goal, it can adopt one of two strategies. It can function as a separate creature or throw its lot in with others, so as to form a multicellular organism.

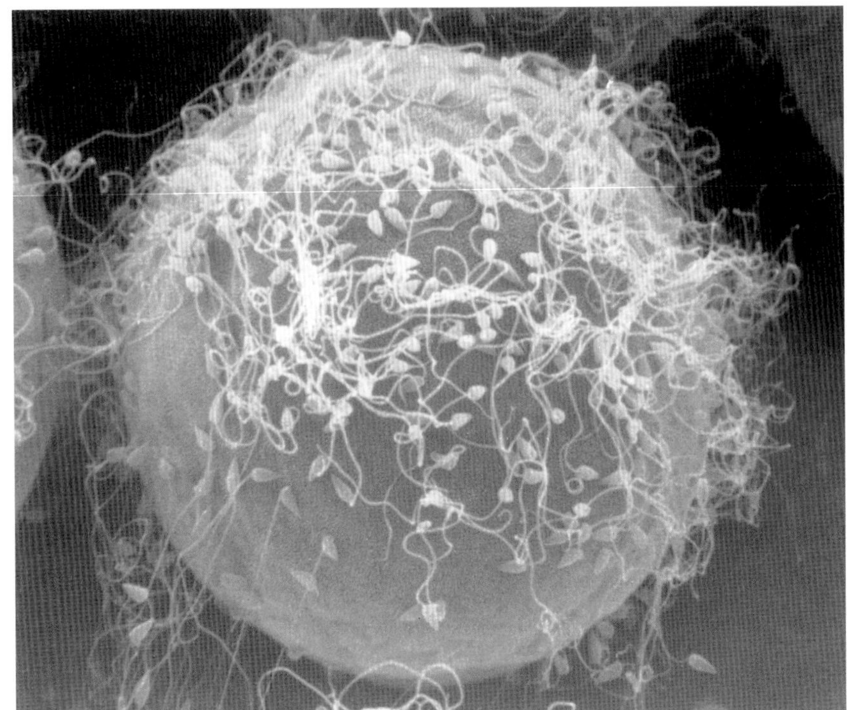

Polyspermy, the fertilization of an egg cell by more than one *spermatozoon*, is prevented by the egg's double membrane system. The outer coat, known as the *vitelline layer*, contains receptor sites and reacts to the sperm cell's penetration by transient changes in the concentration of sodium and calcium ions. A new membrane, which develops within about one minute, prevents the entry of other sperm. This scanning electron photomicrograph shows a sea-urchin egg surrounded by the numerous spermatozoa which failed to achieve the first and only penetration.

The latter alternative expresses the safety in numbers idea at the cellular level, but this applies only to the total organism, not to the individual cells, millions of which die every day in the typical animal.

A major division of living things is made on the basis of how they obtain their energy. *Autotrophic organisms* are independent of external organic compounds for their energy requirements. The *chlorophyll molecules* in *green plants* enable them to convert the energy of light photons directly into the energy of chemical bonding required to synthesize large organic structures from small inorganic molecules. *Heterotrophic organisms*, such as animals, are incapable of this and must eat to survive. They use the Sun's energy in an indirect fashion, either by consuming plants or by devouring other animals that have fed on plants. The price paid for this parasitic existence is a more complex metabolism, and this meant a later emergence in the evolutionary picture. Bacteria span this division; there are *autotrophic bacteria* and *heterotrophic bacteria*. The metabolic processes in plants and animals are quite similar, but the animal must possess additional machinery with which it can hunt for nourishment. Hence the obvious external differences. Plants are static because they simply let the energy come to them, and their structural requirements are modest: plenty of surface area for the trapping of sunlight and for the elimination of oxygen following the breaking down of water, and a vascular system for transporting the latter, together with dissolved nutrients, up from the roots. In keeping with their stationary existence, plants are fairly rigid structures, their stiffness deriving partly from the cellulose in their cell walls and partly from the pressure within the cells. Animal structures are quite different. They involve moving parts. They had to develop an *alimentary system* for consuming food and eliminating waste, a means of *locomotion* to enable them to forage, and a *nervous system* to coordinate movement. In the larger animal organisms, moreover, the movement of oxygen and carbon

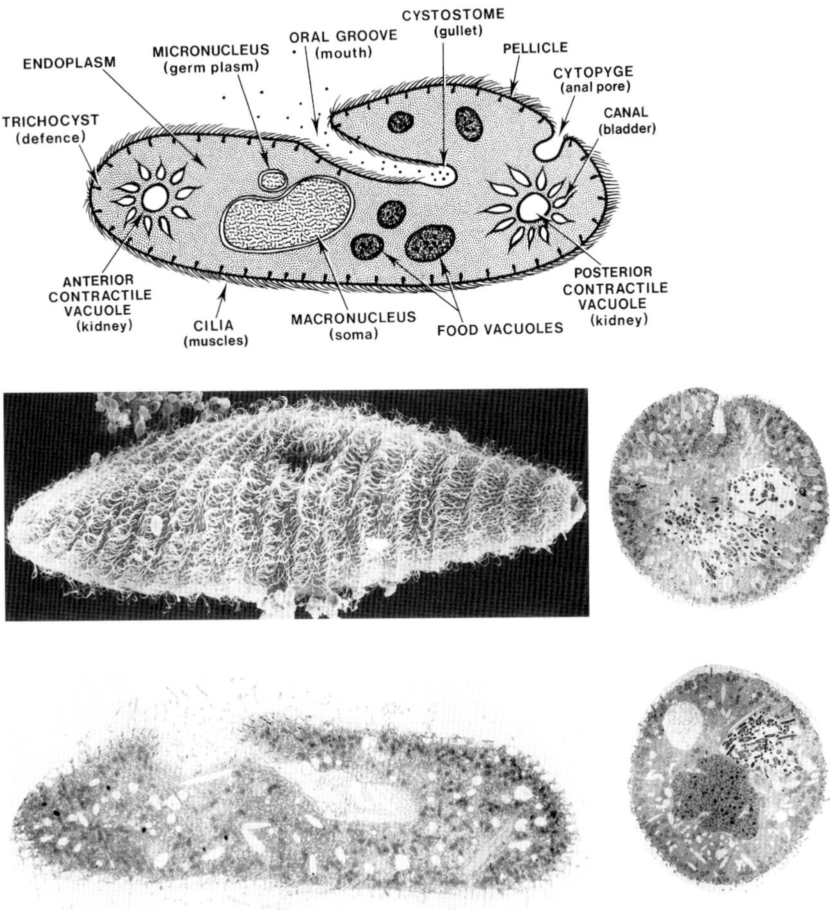

Paramecium tetraurelia is a member of the single-celled group (or phylum) known as the *Protozoa*. The latter differ from bacteria in that they have at least one well-defined nucleus. The *Paramecium* genus belongs to the Ciliophora class, which is characterized by the possession of cilia and a double nucleus (macronucleus and micronucleus). The *cilia* are visible as the hair-like covering seen on the scanning electron photomicrograph (middle left) and they propel the creature by a coordinated waving motion. The single longitudinal sectional view and the transverse sections, all taken by transmission electron microscopy, show several of the features indicated by the schematic diagram (above). Identifiable characteristics of the roughly 200-micrometre long protozoon are the oral cavity (lower left and upper right); food vacuoles and trichocysts (upper right and lower right); and the macronucleus and a contractile vacuole (lower right).

dioxide by simple diffusion is too slow, so a vascular system is required for the rapid transport of these molecules through the body. Hence the lungs, heart, arteries and veins of such creatures.

These classifications do not always lead to clear distinctions. There are organisms that seem to be both plant and animal. The single-celled *Euglena gracilis*, which appears as the green scum covering stagnant ponds, gets its colour from the pigment chlorophyll. It maintains this plant-like hue when basking in the sunshine, absorbing energy, but in the absence of light the green disappears and the creature swims about capturing food in a distinctly animal-like fashion. Neither is degree of aggregation always a reliable basis for classification. The slime mould *Dictyostelium discoideum*, which thrives in unclean bathrooms, can exist both as a single amoeboid cell or as a multicellular organism. It is capable of slug-like motion and the formation of fruiting structures which shed spores. In multicellular organisms, cell specialization is pronounced, individual organs consisting of large numbers of similar cells. The alimentary canal of a higher animal, for example, is a tube in which cellular differentiation produces an oesophagus, a stomach, a duodenum and an intestine. Even in unicellular organisms the various metabolic events do not happen in a random fashion. The transport of material within the cell frequently occurs by a process known as a *cyclosis*, in which the cellular liquids, together with various suspended bodies, are

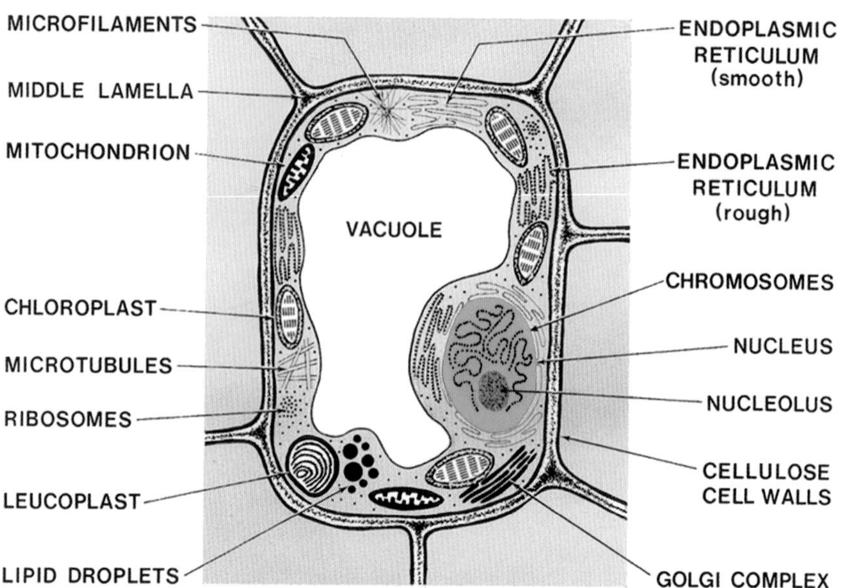

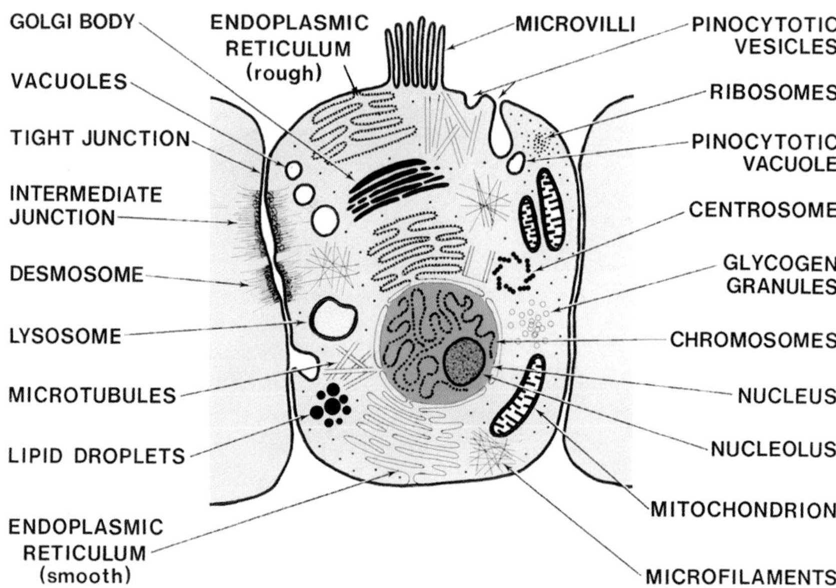

The similarities and differences between the cells of plants and animals are indicated by these rather schematic diagrams. The plant cell distinguishes itself by the presence of the *vacuole*, *chloroplasts*, and a relatively rigid wall structure lying outside its cytoplasmic membrane. In the animal cell, the *cytoplasmic membrane* marks the outermost boundary, and this gives the cell its characteristic flexibility. The presence of movable surface protrusions such as *microvilli* (shown here) and *cilia*, in some cells, is another animal speciality.

made to flow in a directed fashion by rhythmic contractions. In the freshwater protozoon *Paramecium*, food enters the cell at a fixed point, which is in fact a primitive mouth: the *cytostome*. The food is packaged in small envelopes of membrane, and these follow a fixed route around the cell, their contents gradually being acted upon and modified. They ultimately discharge waste products at another specific point: the *cytopyge*. The prefix 'cyto' in these and other terms denotes cellular.

The cell, this primary unit of life, is a most remarkable structure by any yardstick. It is capable of consuming chemicals and changing them into new compounds, which it uses either to develop its own structure or to carry out its various functions. It possesses the means of extracting the energy required in its vital chemical reactions, either directly from sunlight or indirectly from

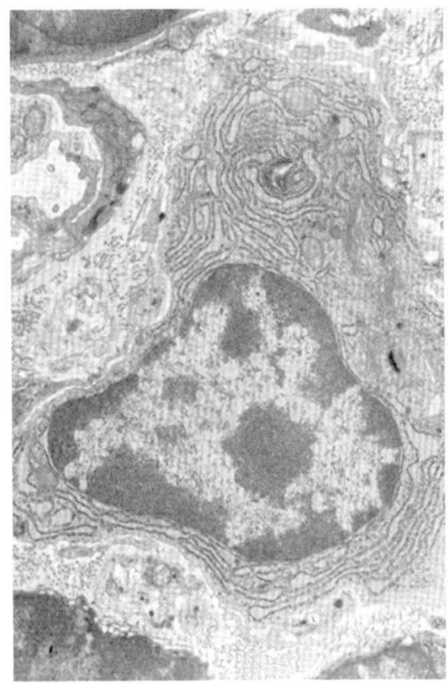

Transmission electron microscopy has proved to be a powerful tool for the study of the cell interior, and many cellular features were discovered with this technique. This view of part of a cell from the intestine of a bat shows the roughly five-micrometre-diameter *nucleus*, surrounded by the *rough endoplasmic reticulum*. The small dots, just discernible at this magnification, are the *ribosomes* which participate in the assembly of proteins.

food that contains stored energy. Moreover, since all cells ultimately die, they must be able to reproduce themselves by the process of *cell division*. Indeed, such division is vital to the overall growth of an organism because this is accomplished by an increase in the total number of cells rather than by cellular enlargement. Finally, and most impressively, the cell is capable of developing into different forms depending on its position in an organism. This last feature is the above-mentioned *cellular differentiation*, and we will be taking a closer look at it in the next chapter. Anything capable of such sophisticated behaviour would be expected to have a rather complex structure, and this is certainly the case. Optical microscopy, and more recently electron microscopy, have revealed that the cell is essentially a closed bag of fluid, known as *cytoplasm*. The outer skin consists of a membrane capable of discriminating between various chemical compounds, transmitting some and excluding others. The evolution of molecules capable of forming such a membrane was clearly a critical milestone in the development of life; with the appearance of membrane material, Nature had acquired the means of separating space into an inside and an outside.

All cells of a given type have approximately the same size, and roughly the same characteristic shape. Cells of one kind contain about the same proportions of a variety of internal structures known as *organelles*. In *eucaryotic cells*, the most familiar of these organelles is the nucleus, which itself is bounded by a membrane. The *nucleus* was discovered by Robert Brown in 1883. The nuclear membrane contains pores wide enough to allow the passage of fairly large molecules, so there is continuity between the regions of cytoplasm that are inside and outside the nucleus. The nucleus contains the genetic material, deoxyribonucleic acid, usually referred to by its acronym DNA, the ladder polymer that carries the hereditary information required to ensure survival of the cell lineage. The nucleus is the major control centre of the cell, and its contents direct the manufacture of proteins. The other type of cell, the *prokaryote*, has no nucleus; its genetic material lies in the cytoplasm, and is located approximately at the cell's centre. The two major groups of procaryotes are *bacteria*, first observed by Antonie van Leeuwenhoek in 1683, and *blue-green algae*. In the modern classification, the former are referred to as *Schizomycophyta*, while the blue-green algae, which are bacteria possessing the faculty of photosynthesis, are termed *Cyanophyta*. In all other organisms, the unit of structure is the eukaryotic cell.

Four other important types of organelle in eukaryotic cells are the mitochondria, the endoplasmic reticulum, the Golgi body, and the lysosomes. Plant cells are unique in that they also have *chloroplasts*, which contain the chlorophyll that is essential for converting the energy of solar irradiation. These are all bounded by membranes, but their shapes and functions vary considerably. The *mitochondrion*, discovered in muscle tissue in 1857, by Rudolf von Kölliker, and first isolated by him 30 years later, is actually enclosed by two membranes: a smooth outer membrane and an inner membrane that is folded into a ripple pattern of what are known as *cristae*. The purpose of the latter, discovered through electron microscopy by George Palade and Fritjof Sjöstrand in 1954, appears to provide support for groups of molecules involved in the extraction of energy from partially metabolized foodstuffs. The mitochondria are indeed the power stations of the cell; they are the sites of the respiratory process, as first suggested by B. Kingsbury in 1912, and the energy they produce is stored in the chemical compound *adenosine triphosphate*, usually referred to by its acronym *ATP*.

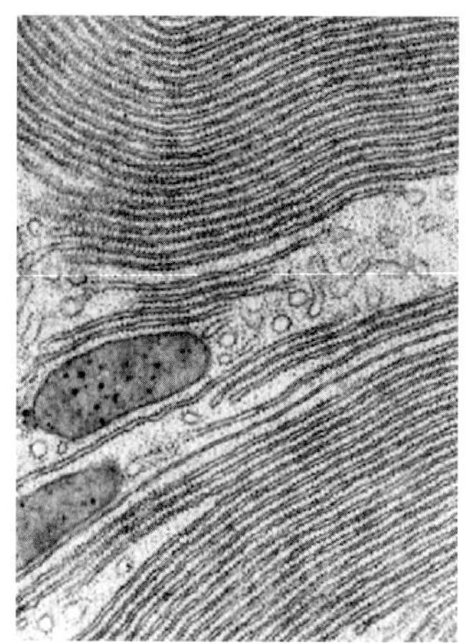

The high resolution provided by transmission electron microscopy reveals the origin of the uneven surface in the *rough endoplasmic reticulum*. These lipid membranous sheets, seen here in cross-section, are found to be dotted with the small globular bodies called ribosomes. Ribosomes are involved in protein synthesis, and they are themselves composed of proteins. In the absence of these agents, the endoplasmic reticulum is said to be smooth. The parallel arrangement of the membranes is believed to provide channels for directing the transport of various molecules. This specimen was taken from an acinar cell of the pancreas.

Chloroplasts and mitochondria are the only other types of organelle, apart from the nucleus, known to contain DNA. But the genetic code used by the mitochondria appears to be somewhat different from that obeyed by nuclear DNA. This has strengthened the suspicion, first expressed in 1890 by Richard Altmann, that mitochondria were once independent bacteria that entered a symbiotic relationship with eucaryotic cells and were ultimately absorbed and incorporated by them. Mitochondria do show a striking resemblance to the bacterium *Paracoccus denitrificans*.

ATP was discovered in 1929, by Karl Lohmann, and after its chemical structure had been determined it was synthesized by Alexander Todd, in 1948. By that time, Fritz Lipmann had already demonstrated that ATP is the major transporter of chemical energy within cells. Even before that, in 1935, Vladimir Engelhardt had shown that muscle contraction requires the presence of this compound, and Herman Kalckar demonstrated two years later that its synthesis is implicated in cell respiration. We will return to ATP later in this chapter.

The *endoplasmic reticulum* and the Golgi body are involved in what might be called the manufacture and transport functions in the cell. The former, discovered by electron microscopy and given its name by Keith Porter in 1953, consists of a fine system of interconnected surfaces which branch out from the wall of the nucleus into the cytoplasm. Porter and Palade later showed that these sheet-like extensions are usually dotted with a dense coverage of the smaller bodies known as *ribosomes*. The latter, which are also found floating in the cell's cytoplasm, are the locations of protein manufacture. The *Golgi body*, named after Camillo Golgi, who discovered it in 1898, consists of a series of flat disc-like bags surrounded by membrane, the outer edge of each disc being rather thicker than the centre. The central regions of the discs, known as *cisternae*, are continuous, whereas the edges are perforated. As demonstrated by Palade, membrane-bounded capsules containing newly fabricated protein periodically break away from these ragged edges. The Golgi body thus has the function of packaging proteins prior to their transport to other parts of the cell, or indeed to regions outside the cell. *Lysosomes* are small membrane-bounded bags containing enzymes capable of breaking down a variety of substances. They were discovered by Christian de Duve in 1952. The cell uses these to break down organelles that have become redundant. Indeed, they appear to provide a mechanism of selective tissue destruction, like that which occurs in developmental processes such as *metamorphosis*. Lysosomes also help the cell digest substances that have been taken in from the outside. In Chapter 18 we will consider yet another membrane-bounded organelle: the *vacuole*, which gives a plant cell rigidity by exerting an internal pressure.

Our list of structures found in the eucaryotic cell is by no means complete. There are, for example, the hollow cylindrical bodies known as *centrioles*, which are usually located in the vicinity of the nucleus, and which play a key role in the process of *cell division*. They typically have a diameter of about 0.15 μm, and a length three times larger, and they are commonly observed in pairs lying mutually at right angles just outside the nuclear membrane. Then there are the *microfilaments*. These are fairly straight fibres consisting of proteins, two of which resemble the *actin* found in muscle fibres. These filaments appear to occur in two varieties: one with a diameter of about 5 nm, and the other twice as thick. They are frequently observed to lie at the inner surface of the cell membrane and are seen to form clusters around

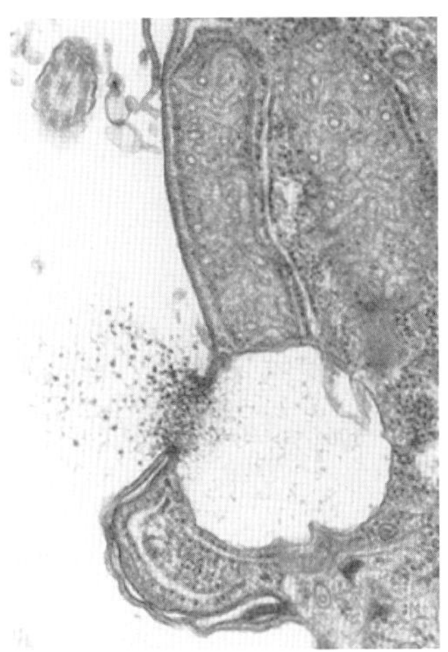

A living cell frequently secretes substances by enveloping them in a membrane *vesicle,* which releases its contents by fusing with the cytoplasmic membrane. Although the process (known as *exocytosis*) takes only about a millisecond, it can be arrested by rapid freezing and studied by transmission electron microscopy. This example captures the instant of discharge from *Tetrahymena,* a ciliated protozoon related to *Paramecium.* The driving force for the almost explosive ejection derives from the elongated shape of the vesicle prior to fusion with the cell's outer membrane. The final shape of the vesicle is seen to be a sphere, with a diameter of about a micrometre.

the newly formed partition, or *septum,* during cell division. There are also thicker filaments known as *microtubules,* which are rather straight fibrous tubes. Their outer diameters are roughly 24 nm, which makes them about a hundred times as wide as a water molecule. These objects were first observed in the nineteenth century, but their structure was not properly investigated until the 1960s. They were given their name by Myron Ledbetter, Keith Porter and David Slautterback in 1963, and Martin Chalfie and J. Nicol Thomson later showed the individual microtubule lengths to lie in the range 10–25 μm. They have a simple form with no evidence of branching, and they are quite resilient, returning to their original shape after deformation. It has been found that they are formed by aggregation of strands known as *protofilaments.* There are usually thirteen of these to each microtubule, give or take a couple. Each protofilament is composed of many identical protein subunits. The latter are actually dimers, consisting of two rather similar proteins: *alpha*-tubulin and *beta*-tubulin, each comprising about 500 amino-acid residues. It seems that microtubules serve the cell in a variety of ways. They operate in conjunction with the centrioles during cell division, forming a sort of scaffolding along which the chromosomes separate. In addition, they appear to function in a manner similar to the endoplasmic reticulum in facilitating transport of material from one part of the cell to another. Finally, microtubules located at the cell surface, bounded by the cell membrane and causing this to protrude outwards, are the basis of the arm-like cilia that some cells use as a means of locomotion. The effect is not unlike that produced by the novice camper, tent poles pushing against the canvas from the inside.

In 1929, Rudolph Peters speculated that a tenuous network of fibrous molecules might be present in a cell, and that this could coordinate enzymatic activity. Studies of the flow – or rheological properties – of cells also hinted at an underlying structure, because the interior appeared capable of both viscous flow like a liquid and elastic deformation like a solid. Some attributed this behaviour to a cytoplasm that is in the colloidal gel state, but the majority, invoking the earlier observations of microfilaments and microtubules, began thinking in terms of what became known as the *cytoskeleton.* Around 1970, exploiting the large depth of view provided by an electron microscope operating at 1 000 000 V, Ian Buckley, Keith Porter and John Wolosewick discovered an irregular three-dimensional network of interlinked fibres. This was connected to the outer membrane and appeared to permeate the entire cell. Because it resembled the bar-like, or trabecular, structure of spongy bone, they called it the *microtrabecular lattice.* It was found to incorporate the various organelles, microfilaments, and microtubules, arranging them in a reasonably organized superstructure as Peters had anticipated. Subsequent studies showed that this structure is not static; it changes in response to mechanical deformation of the cell's shape, and also to alterations in the cellular environment. The scaffolding appears to become dismantled under certain conditions, the microfilaments and microtubules totally, and the microtrabecular strands partially. But when the deforming stress is removed or the environment restored to its previous condition, the structure is reassembled in its former pattern. Particularly dramatic effects of this type are provoked by the drug *cytochalasin B,* the changes again being reversible.

The cell surface itself consists of a membrane. This is reasonably smooth at the molecular level, but on a grosser scale it is often covered with a dense

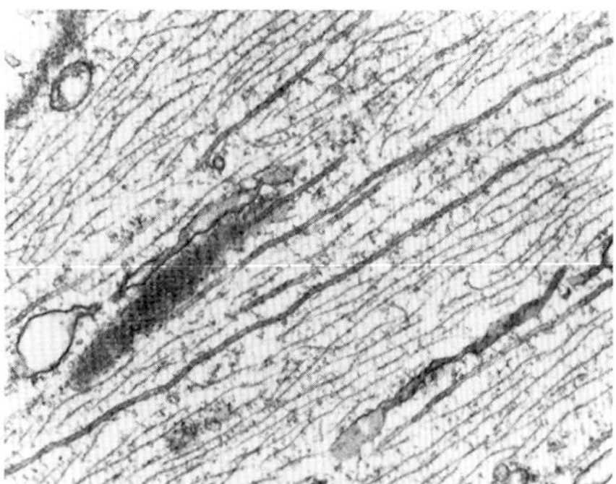

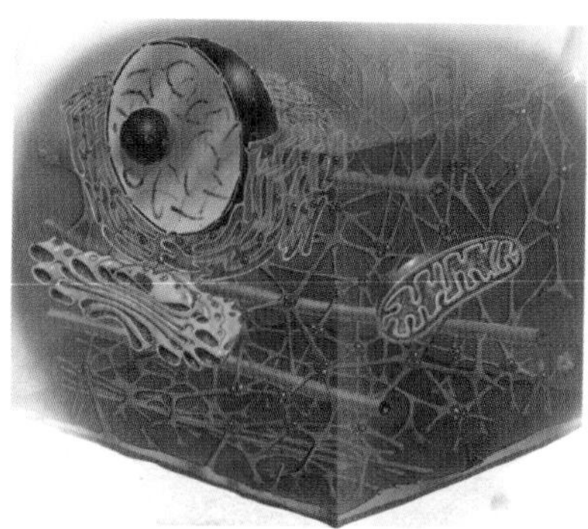

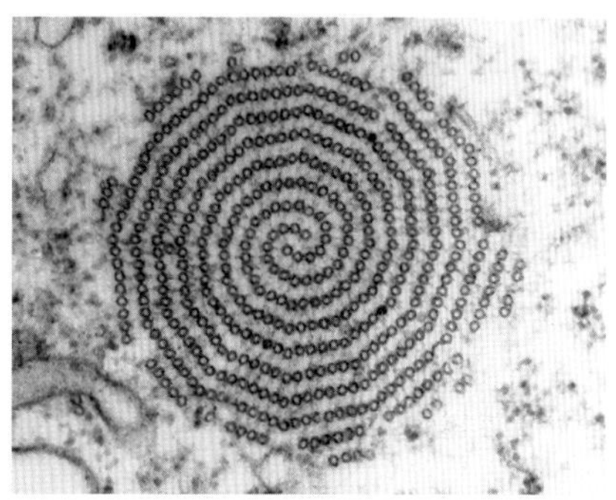

The electron microscope has revealed that what were previously taken to be relatively open spaces in the cell, containing only the cytoplasm, ribosomes, and small molecules such as amino acids, are in fact threaded with a three-dimensional network of *microtubules* and *microfilaments*. The former are hollow structures, composed of *tubulin* subunits, and with a diameter of about 24 nanometres, while the latter are somewhat thinner and simpler. The diagram indicates the way in which the network holds the various organelles in place, the cut-away view showing a *mitochondrion* (right), *Golgi body* (lower left) and the *nucleus* surrounded by the *rough endoplasmic reticulum*. The red objects are *ribosomes*. The electron photomicrographs show the thread-like structures in two different specimens: a nerve-cell axon seen from the side (upper picture); and in the special arrangement present in the bundle of microtubules known as an *axoneme*, seen in cross-section (lower picture). Together with a surrounding membrane, the axoneme forms the *flagellum* which propels such cells as spermatozoa and protozoa, this picture showing the axoneme structure in *Echinosphaerium nucleofilum*.

pattern of projections, reminiscent of fingers and known as *microvilli*. This outer membrane is also capable of forming much larger indentations, which are referred to as *invaginations*. These are used by the cell to surround and capture food. The roughly spherical envelopes thereby formed are referred to as *pinocytotic vesicles* when they are used to ingest fluid, and *phagocytotic vesicles* when the food is solid. Some cells, known as *phagocytes*, specialize in this mechanism and are used by certain multicellular organisms for defence against invading bacteria. Contact between the walls of adjacent cells shows variation in the degree of intimacy between neighbouring membranes. In the loosest arrangement, between *epithelial cells*, the adjacent membranes are separated by a distance of about 30 nm, the intervening space being filled by fluid. At several sites, epithelial cells are cemented together by a dense packing of fibrous material. Such a junction is referred to as a *desmosome*. In the tightest arrangement, known as a *zonula occludens*, the adjacent membranes are in actual molecular contact.

Not everyone, asked to name the largest organ in the human body, would think of the skin. There is an understandable temptation to look upon it as merely a sort of organic bag, which acts as little more than a conveniently elastic envelope. In much the same way, it is easy to underestimate the importance and complexity of the cell membrane. Until recently it was regarded as a reasonably inert but flexible container, with just the right blend of strength and permeability for containing its life-giving components. We now know that the cell membrane, through its intricate structure, actively participates in the life of the cell and dictates many of its essential functions. The compositions of cellular membranes and of those associated with organelles show considerable variation. They variously comprise six distinct types of chemical: *phospholipids, proteins, water, cholesterol, sugars*

Aerobic and anaerobic respiration

The joining of phosphate and ADP to produce molecules of ATP, each of which thereby stores 0.06 attojoules of chemical energy, is the net product of a sequence of processes collectively termed respiration. There are two basic forms. Aerobic respiration requires oxygen, and it is by far the more common. Oxygen is provided by simple diffusion or, in the higher animals, by the breathing and circulation functions. The overall chemical reaction is deceptively simple

glucose		*oxygen*		*carbon dioxide*		*water*
$C_6H_{12}O_6$	+	$6O_2$	$\rightarrow$	$6CO_2$	+	$6H_2O$

It produces 2.28 attojoules in the oxidation of one molecule of glucose. There are many individual chemical steps, and the energy is released in conveniently small packets rather than in one big bang. The anaerobic form does not require oxygen absorption. It occurs naturally in yeasts and a few other primitive organisms, and also in the tissues of higher species when deprived of oxygen. In the former, it involves incomplete breakdown of glucose by the reaction

glucose		*ethyl alcohol*		*carbon dioxide*
$C_6H_{12}O_6$	$\rightarrow$	$2C_2H_5OH$	+	$2CO_2$

This produces much less energy than the aerobic process: about 0.25 attojoules per glucose molecule. Moreover, it produces ethyl alcohol, which is toxic to most living tissue. Yeast can survive in a moderate concentration of alcohol, and its anaerobic reaction is exploited in the production of that compound in brewing and carbon dioxide in baking. If the oxygen supply in an aerobic system becomes seriously depleted, the process is arrested at an intermediate stage, and anaerobic glycolysis takes over, giving the lactate that appears in overworked muscles, for example. This is still a useful source of energy, which can be released by further oxidation, when this becomes possible. The anaerobic process also occurs in tissue that is too remote from the air supply to allow satisfaction of its oxygen requirements by simple diffusion. An example occurs at the centre of a large apple.

and *metal ions*. These are found in different proportions, the phospholipids, proteins and water always being present, while the cholesterol, sugars and metal ions are sometimes missing. The sugars are invariably linked to lipids or proteins, existing as *glycolipids* and *glycoproteins*, respectively. The sugar regions of these molecules stick out from the membrane surface into the surrounding aqueous medium, with which they are compatible. The ratio of protein to lipid shows considerable variation, depending on the organism in which the cell is located. In human *myelin*, there is between five and ten times more lipid than protein. At the inner mitochondrial membrane, on the other hand, there is about three times as much protein as lipid.

Membrane proteins typically have diameters in the 5–10 nm range. Some are imbedded in one side of the membrane lipid bilayer, while others penetrate right through its 5 nm thickness. The latter are referred to as *integral membrane proteins*, and they usually have hydrophobic amino acids in the central regions of their folded structures, these amino acids making contact with the interior of the bilayer. These proteins play vital roles in mediating passage of electrons, ions, and small molecules through the membrane, a different protein being responsible for each specific conduction or regulating function. In some membranes, the proteins are packed as densely as $10^{17}/m^2$, while in others they are sparsely dispersed. They are also sometimes distributed in patches. A closely packed network of microtubules often

The key, and only essential, component of the biological membrane is a *phospholipid bilayer*, with its tail units in the membrane interior and its head groups facing inwards towards the cytoplasm and outwards towards the cell exterior. The other features, which are present to a degree which varies with cell type, are *cholesterol* (the small darker molecules imbedded in the bilayer), *proteins* (the large globular molecules, one of which is shown spanning the width of the bilayer), and *polysaccharides*, the chains of which may be attached to proteins or lipids. The transmission electron photomicrograph (upper right) of the membrane of a human erythrocyte, or red blood cell, actually distinguishes the two layers of the membrane, even though their total width is only about five nanometres. But it does not resolve individual proteins. The preparation was fixed in glutaraldehyde and imbedded in polyglutaraldehyde. The bonds between the lipid heads and the surrounding water are stronger than those between the lipid tail groups. If a cell is rapidly frozen and chipped with a sharp knife, therefore, it is possible to split the two halves of the bilayer apart. This technique permits examination of the arrangement of proteins, which appear as surface roughening in the freeze fracture photomicrographs (below), and the different patterns observed in these two views of the same cell type support the belief that the lipid bilayer is in a fluid state, permitting lateral protein motion.

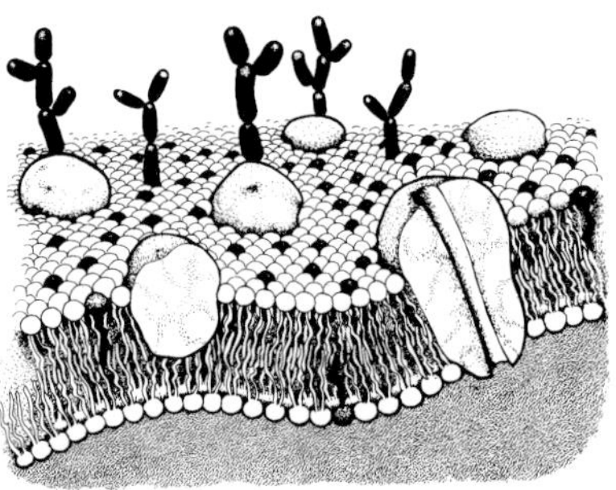

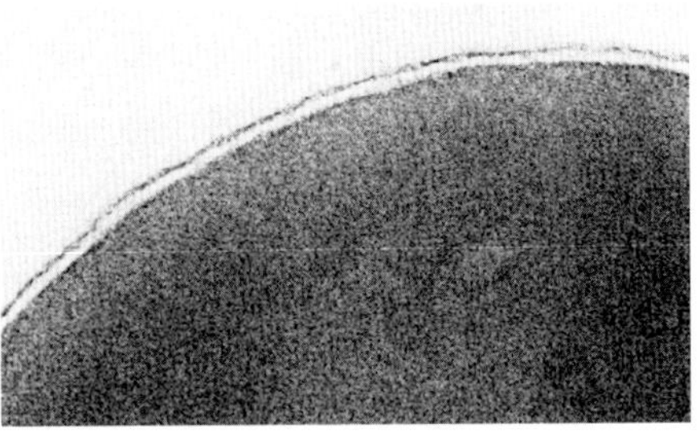

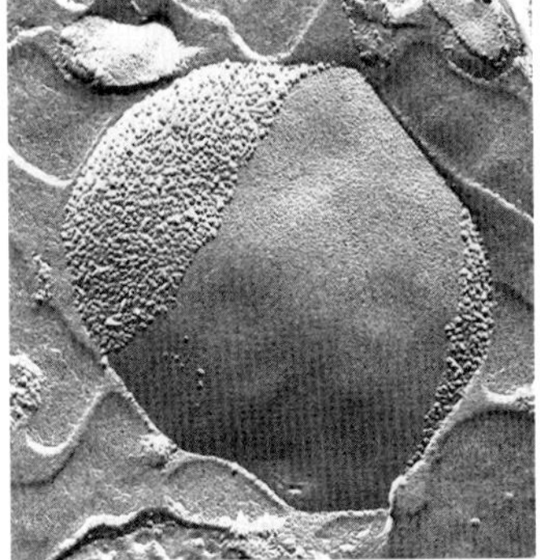

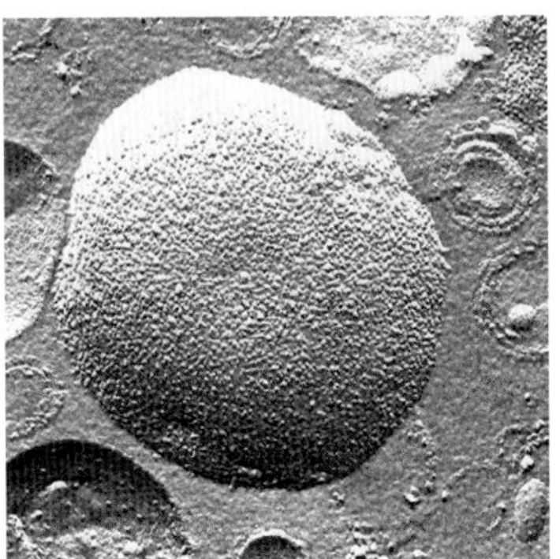

lies at the inner surface of the membrane. Its role is not clear; it could act as a strengthener, but it might also help give the cell surface a particular shape, as in the dimpled form of the *red blood cell*.

For animal cells, the *cell membrane* marks the physical boundary, but this is not the case in plants. The limiting layer of the *plant cell* is the *cell wall*. It consists of cellulose, hemicellulose, and lignin, all of which are organic polymers, and these lie outside the cell membrane. *Bacteria* also have layers external to their cell membranes, and they are broadly classified according to their response to a stain discovered by Christian Gram in 1884. Those stained by his method, said to be Gram-positive, have walls that can be as much as 80 nm thick, and these can represent up to a quarter of the cell's dry weight. These bacteria have a single membrane, and the surrounding walls contain teichoic acid, which is apparently not encountered elsewhere in Nature. *Gram-positive bacteria*, such as *Staphylococcus* and *Streptococcus* are susceptible to the bactericidal action of penicillin and are also attacked by lysozyme. *Gram-negative bacteria*, such as *Escherichia coli* and those which cause typhoid, have thinner walls, usually measuring about 10 nm. But these comprise two lipid bilayer membranes, both of which contain transport-mediating proteins. It is this double arrangement which presumably gives such bacteria their higher resistance to chemical invasion. The cellular membrane itself is known by several aliases: *cytoplasmic membrane, plasma membrane* and *plasmalemma*.

The processes of life include performance of mechanical, electrical and chemical work, and these functions depend on maintenance of organized structure. To accomplish these tasks, every cell needs both a source of energy and a means of husbanding this valuable commodity. Of the different forms, it is the spatially localized chemical variety which is a particularly appropriate energy currency in the cell, and the coinage is *ATP*. Energy is required to form molecules of this compound, and it is released again when the molecules are broken down by *hydrolysis*. The ATP molecules are thus ideal for mediating energy transfer. The ultimate energy source for all living things is of course the Sun. Only plants possess molecular machinery capable of converting the radiant energy carried by light photons into electron flow. Electrical current is a transient phenomenon, however, and energy storage requires use of this electricity to produce chemical rearrangement. The cells in plant leaves contain *chloroplasts*, in which ordered molecular structures enable charge motion to split water into hydrogen and oxygen. The hydrogen remains chemically bound while the oxygen is liberated. The splitting process also produces ATP, and the energy contained in this compound is used in uniting the hydrogen with ingested carbon dioxide to produce *glucose*. These processes are collectively termed *photosynthesis*. Its product is used by both plants and animals for the reverse process, *respiration*, in which chemically stored energy and oxygen interact to produce ATP, and also water and carbon dioxide. In its most direct form this process actually uses glucose, or its polymers such as glycogen, but additional routes employed particularly by animals involve energy storage in fats and proteins.

The driving force for these processes arises from the universal tendency towards a state of lower free energy. But the reactions require a mediating agency. Glucose does not spontaneously burst into flame when exposed to oxygen, but it can be made to burn and release its chemically stored energy if it is sufficiently heated. Because an increase in reaction rate by appreciable heating is not feasible for a biological system, the *energy barriers* must be overcome by circumvention: they are bypassed via the lower hurdles associated with the catalytic activity of enzymes. This is one of the most outstanding properties of life. The price paid for this speeding up is confinement to specific chemical routes: the *metabolic pathways*. The overall result is nevertheless the same: energy is extracted and made available through the formation, diffusion and breakdown of ATP. In principle, there is not a lot of difference between keeping warm by burning wood, the cellulose of which is related to glucose, and eating sugar to acquire energy internally. Around 1500, Leonardo da Vinci compared human nutrition with the burning of a candle, and he was close to the truth; the candles of that period were made from animal fat. Respiration is carried out in the mitochondria, which, like chloroplasts, have a characteristically ordered microstructure. Each of these organelles is shaped like an elongated egg, two or three times as long as it is wide, and with a length of about 4 μm. Correct functioning of both types of organelle depends on electron transport, and it seems that these structures require properly designed electrical circuits, just like their technological counterparts. The basic chemicals consumed in photosynthesis are liberated again in respiration, and conservation of one element in particular is emphasized in what is referred to as the *carbon cycle*. Energy also flows in both directions, but the overall trend is degradation from the directed nature of solar radiation to the randomized emission of heat from the living organism. The modification is accompanied, however, by the local increase in free energy associated with production of the ordered structures that are so vital to life.

The important *ATP molecule* is a linear arrangement of three types of unit that we have encountered previously. The first is the purine *adenine*. This is joined by a nitrogen–carbon bond to the sugar, ribose, the compound thus formed being known as *adenosine*. If a phosphate group is joined via

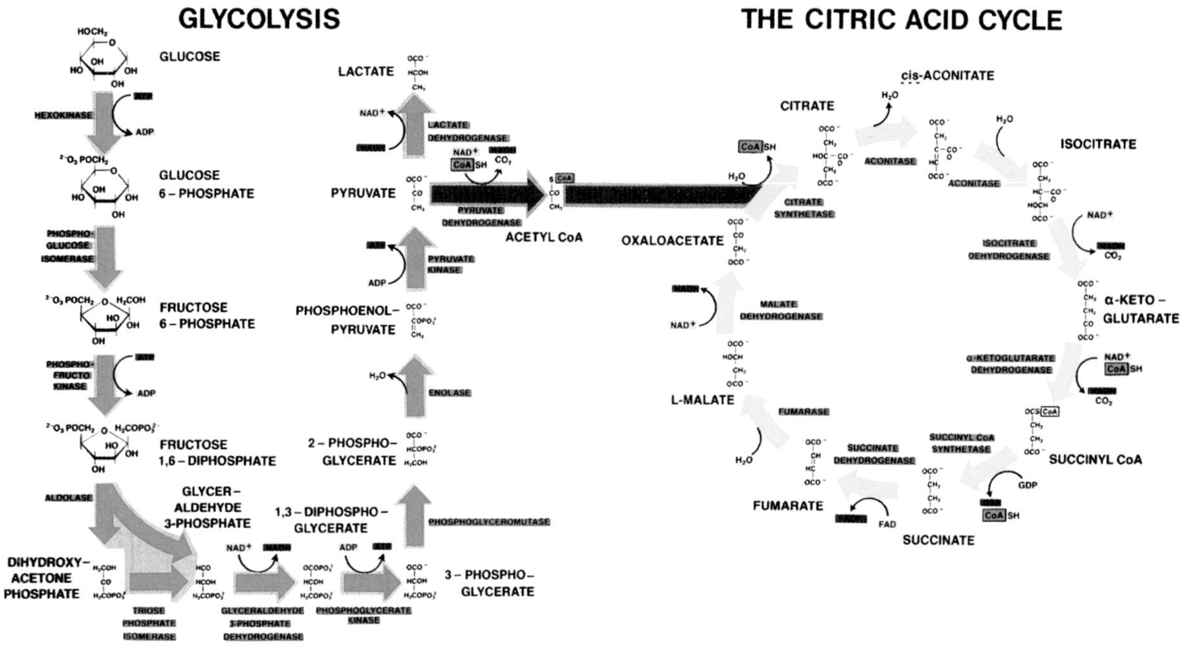

GLYCOLYSIS

THE CITRIC ACID CYCLE

The breaking down of organic molecules (catabolism) to smaller chemical units, and the associated liberation of energy, which is ultimately transported by ATP, is referred to as *respiration*. The cell uses a number of such reaction pathways, the classic example being *glycolysis*, the splitting of glucose molecules. This chain of reactions feeds, via the compound acetyl-CoA, into the *citric-acid cycle*. The latter involves a closed series of reactions and thus does not exhaust the supply of participating compounds. Other catabolic routes, such as that which breaks down fats, also feed into the citric-acid cycle. The colour code indicates the participating enzymes in blue, and shows an energy gain for the cell by a transition from yellow to red. The compound *cis*-aconitate is a short-lived intermediate between citrate and isocitrate. GDP and GTP are the acronyms of guanosine diphosphate and guanosine triphosphate, respectively, and the energy stored in the latter is readily passed on to ADP, which is converted to ATP with the help of the enzyme nucleoside diphosphokinase. Of the many participating enzymes, only coenzyme A (CoA) actually becomes chemically bound during the performance of its task.

an oxygen atom to the fifth carbon atom of the ribose, a molecule of the nucleotide *adenosine monophosphate, AMP*, is produced. (It must be emphasized that the above description refers only to the structure of the AMP, not to the anabolic pathway by which it is formed; the latter is more complicated, and involves the intermediate production of *inosine monophosphate*, which contains a purine double ring arrangement.) We have already seen that the polymerization of nucleotides, with alternating sugar and phosphate links, produces a *nucleic-acid backbone* of DNA or RNA. In the present case, however, the polymerization is of a different type. Further phosphate units can be tacked on to the first unit, through *pyrophosphate bonds*, to produce *adenosine diphosphate, ADP*, and ultimately *adenosine triphosphate, ATP*. In each case, the attachment of a phosphate group is referred to as *phosphorylation*. The vital features of the process are that it is endothermic, free energy being expended to form the bond, and that the phosphorylated molecule is metastable. These pyrophosphate bonds have a particularly high

ATP: the cell's energy coinage

Life's processes require a steady input of energy, and those organisms which are unable to obtain this directly from sunlight must satisfy their requirements by consuming food in which energy is stored in a chemical form. They then transform it to such other forms as the electrical energy of nerve impulse transmission and the mechanical energy of muscle contraction. The coupling between these different types of energy is not direct. It is mediated by *adenosine triphosphate*, known by its acronym ATP. Molecules of this compound comprise five parts: a purine (adenine); a sugar (ribose); and three phosphate units.

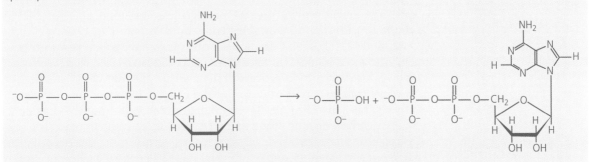

The molecule is metastable in that the breaking of a phosphorus–oxygen bond to produce *adenosine diphosphate*, ADP, and inorganic phosphate is accomplished by the liberation of about 0.06 attojoules of free energy. There are several factors which favour this reaction. The splitting of ATP permits a higher degree of hydration, and it also relieves the electric repulsion between the negatively-charged oxygen atoms. In some cases, the splitting is taken one stage further, ADP being hydrolyzed to *adenosine monophosphate*, AMP, again with liberation of about 0.06 attojoules of energy. The turnover of ATP is very rapid, a given molecule of this compound usually surviving no more than a minute or so after its manufacture. This high rate of consumption is illustrated by the following figures for an adult human. The daily energy intake is about 2400 Calories, or roughly 10^7 joules, and this is equivalent to the hydrolysis of approximately 1.7×10^{26} molecules of ATP. One molecule of this compound weighs about 8×10^{-25} kilograms, so the daily consumption could be stored in roughly 140 kilograms of ATP, which is well in excess of the normal body weight. At any instant the human body possesses a concentration of ATP that could be provided by the oxidation of just a few grams of glucose.

space-filling model of ATP

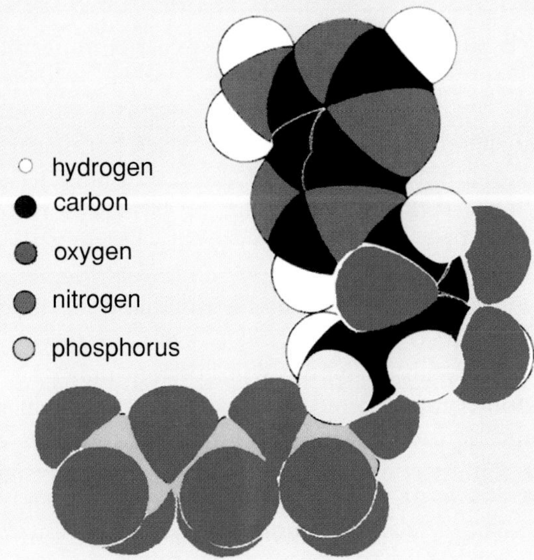

energy, about 0.06 aJ per molecule in fact. This is roughly double the energy of a hydrogen bond. ATP functions as a *coenzyme*, one of the phosphate groups being readily transferred to another substance, with the help of an enzyme. This is accompanied by transfer of a large fraction of the pyrophosphate bond energy. The most common reaction hydrolyzes ATP to ADP and inorganic phosphate, but further hydrolysis of ADP to AMP and phosphate is also exploited under certain circumstances. The energy is available for such tasks as chemical synthesis and maintenance of chemical gradients. It is also employed in initiation of the metabolic processes which ultimately lead to production of further ATP molecules. The importance of ATP in cellular function was first established by Fritz Lipmann, in 1941.

Around 1770, Joseph Priestley demonstrated that animals consume oxygen, and about a decade later Antoine Lavoisier suggested that *metabolism* involves oxidation by a process equivalent to burning. The net effect of *aerobic respiration* is redistribution of the carbon, hydrogen and oxygen in carbohydrates, proteins and fats, and oxygen taken in from the outside to produce carbon dioxide, water and ATP. Because movement of oxygen and hydrogen atoms is involved, we must reconsider the processes of oxidation and reduction. A compound or radical is said to be oxidized when an oxygen atom becomes bound to it, while reduction is the loss of oxygen. Another form of *oxidation* is the removal of hydrogen from a compound or radical, and addition of a hydrogen atom thus causes *reduction*. Finally, because life's reactions are carried out in an aqueous environment, in which hydrogen ions are present, removal of an electron is equivalent to loss of a hydrogen atom. This is guaranteed by the tendency of atoms to prefer the electrically neutral state. The third form of oxidation is therefore loss of an electron, and electron acquisition produces reduction. There is an inevitable complementarity in these processes; during the transfer of an oxygen, a hydrogen or an electron, an oxidizing agent is reduced while a reducing agent is oxidized. The two-part nature of the process is emphasized by use of the term *redox reaction*. Reference to donation and acceptance of electrons is reminiscent of our earlier discussion of *semiconduction*, and the resemblance is not superficial; some of the molecules involved in the respiratory process are indeed to be regarded as *organic semiconductors*. The processes involved in respiration are essentially the sequential passing of a hydrogen atom or an electron along a series of these carrier molecules: a chain of redox reactions. The route thus taken is like a bucket brigade, in which energy is tapped off at convenient points. It is hydrogen, or an electron, which must move towards oxygen, and not vice versa, because the oxygen is so much heavier; Mohammed must go to the mountain. It is worth noting the similarity between the energy producing processes in the living organism and those exploited in a battery. In each case, neutral atoms are dissociated into electrons and positive ions, which travel independent paths at different speeds, ultimately to be reunited with a member of an oppositely charged species, in an attempt to restore electrical neutrality.

A metabolic pathway has directionality because the various interatomic bonds encountered in it have different energies. Proceeding along the path, the binding energies progressively increase, and the water molecule is a natural termination point because its internal links are unusually strong. The bond between hydrogen and oxygen has an energy of 0.77 aJ. (attojoules). This is 12% higher than the energy of a hydrogen–carbon bond and 18% higher than that of a hydrogen–nitrogen bond. Over some stretches of a pathway, the moving entity actually is a hydrogen atom, but in other regions an electron is passed along the chain, detachment of the hydrogen ion initiating the electron motion. The latter variant is equivalent to transfer of a hydrogen atom because the electron is reunited with a hydrogen ion, recovered from the aqueous solution, when its journey is completed. If the hydrogen is to be moved over a considerable distance by diffusion, rather than shuffled across a relatively short path, it is transported by one of the dinucleotide *electron carriers*. One of these is *nicotinamide adenine dinucleotide*, usually abbreviated to NAD^+, the superscript denoting that it is an electron acceptor in the oxidation of energy-supplying molecules. This important chemical vehicle was known earlier by the name diphosphopyridine nucleotide, or DPN^+. A near relative of NAD^+, also active as an electron carrier, is *nicotinamide adenine dinucleotide phosphate*, $NADP^+$, formerly known by the acronym TPN^+. The other major electron carrier is *flavin adenine dinucleotide*, or *FAD*. When NAD^+ oxidizes an energy-rich molecule, it acquires a hydrogen ion and two electrons, becoming reduced to *NADH* in the process. In contrast, FAD accepts both hydrogen ions, and their associated electrons, when it oxidizes a molecule; it thus becomes $FADH_2$. These reduced electron carriers pass on their passengers, acting as reducing agents by becoming electron donors.

In a plant, the main source of energy is the glucose produced in the leaves. Animals obtain their energy by eating either plants or other animals that have fed on plants. In some unicellular organisms, the food is acquired by the direct engulfment known as *phagocytosis*. The cell walls of bacteria are too rigid to permit this mechanism, and these creatures acquire their nourishment by diffusion. In the higher animals, energy-giving compounds reach the individual cells only after a series of steps in which the food eaten is broken down by *enzymes*. In the stomach, *pepsin* breaks down protein into fragments known as *peptones*. The pancreatic enzymes, *trypsin*, *amylase* and *lipase* act in the duodenum upon peptones, starch and fat, respectively, to produce peptides, which are polymers of a few amino acids, maltose and a mixture of fatty acids and glycerol. Finally, in the intestine, *peptidase* chops up peptides to produce individual amino acids; *sucrase* works on sucrose to give glucose and fructose; *maltase* converts maltose to glucose; and lipase breaks down more fat to fatty acids and glycerol. The final soup of amino acids, sugars, and fatty acids is absorbed through the wall of the intestine, whence it passes into the blood stream to be distributed to the various regions of the organism. Energy can be extracted from all these end products, but under normal circumstances most of the cell's requirements are satisfied by the splitting of sugar. It is thus appropriate that we consider the details of this vital process.

Release of the chemical energy stored in glucose occurs in two stages. The first is known as *glycolysis*, and its individual steps were elucidated in the late 1930s, particularly through the efforts of Gustav Embden and Otto Meyerhof. Major contributions were also made by Carl Cori and Gerty Cori, Carl Neuberg, Jacob Parnas and Otto Warburg. In these initial stages of *respiration*, glucose is split, and the resultant fragments modified, by a system of enzymes and coenzymes. The process is anaerobic, and the metabolic pathway consists of an initial common route followed by branching into various possible ways of recycling NADH. One alternative, used by animals, some fungi and bacteria, produces lactate. Another, the *fermentation process*, employed by yeasts and a few fungi and bacteria, gives ethyl alcohol. In each case, *pyruvate* is produced at the penultimate stage of the reaction. Glycolysis produces a relatively small amount of energy, in the form of ATP, but it is an important power source for short period demands that exceed the available oxygen supply. The sprinter's muscles are rich in lactate at the end of a race. The *glycolytic*, or *Embden–Meyerhof, pathway* also feeds directly into a second series of reactions that use pyruvate as a starting point. In *aerobic respiration*, these reactions ultimately produce carbon dioxide and water, and liberate for use the majority of the original energy which is still held in the molecular fragments. The nature of this latter metabolic pathway was a biochemical mystery, because it was not clear how the reactions could avoid exhausting the supply of whatever compound is required to initiate the chain. This would abruptly terminate the respiratory release of energy. The problem was ultimately solved in 1937, following contributions by Albert Szent-Györgyi, Franz Knoop and Carl Martius, and by Hans Krebs who showed that the series of reactions is cyclic. The supply of the compound oxaloacetate required to initiate this *citric-acid*, or *Krebs, cycle* is actually replenished by the reactions. The enzymes required for glycolysis float freely in the cytoplasm, so this part of the breakdown process releases its modest amount of energy throughout the cell. The enzymes responsible for the citric-acid cycle, on the other hand, are attached to the inner membranes of the *mitochondria*, so these final stages of the process are much more localized. The citric-acid cycle is not an isolated chain of reactions; it is linked to the *urea cycle* that is involved in amino-acid degradation, and compounds are fed off into this latter pathway even under anaerobic conditions. In most cells, where oxygen is available, chemical transfer around the full citric-acid cycle is possible, with consequent release of energy.

The initial stages in the breakdown of glucose are rather like the pulling apart of a doughnut by a pair of hands. Two phosphate units become attached to the glucose molecule, having been split away from ATP, which thus becomes ADP. This therefore involves the investment of energy, but it is later repaid with interest. The first phosphate addition is catalyzed by the enzyme *hexokinase*, and another enzyme, phosphoglucose isomerase, converts the compound molecule to fructose-6-phosphate. The second phosphate is then added to produce fructose-1-6-diphosphate, the enzyme in this case being phosphofructokinase. As usual, the numbers appearing in the names of these compounds indicate the locations on the molecule at which the radicals become attached. The two hands are now in place, ready to break the doughnut. The enzyme that provides the muscle is aldolase, and the two fragments are dihydroxyacetone-phosphate and glyceraldehyde-3-phosphate. A six-carbon ring has been split into two three-carbon molecules, which are in fact isomers. Only the latter isomer can be utilized directly by the cell, but the enzyme triose phosphate isomerase is able to convert the dihydroxyacetone-phosphate into the usable form.

The stage is now set for the important transport and energy-producing steps to be initiated. They require the presence of water, inorganic phosphate, ADP, and the coenzyme NAD^+, which is to transport hydrogen. The next step is the breaking down of water to produce oxygen, which binds a phosphate unit to the glyceraldehyde-3-phosphate, and hydrogen, which becomes attached to the NAD^+. The reaction is catalyzed by the enzyme glyceraldehyde 3-phosphate dehydrogenase, and it involves chemical subtlety. The compound produced by this reaction is 1-3-diphosphoglycerate, this name deriving from the fact that phosphate groups are attached to the

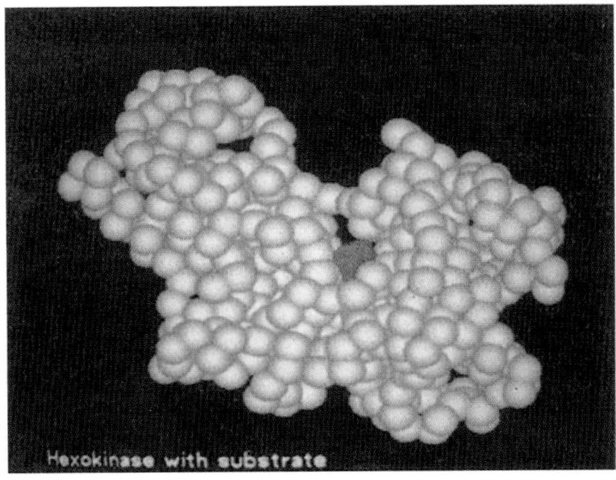

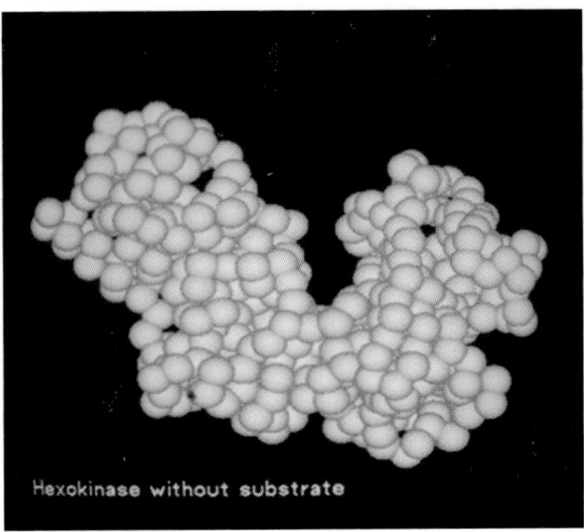

The enzyme *hexokinase*, which catalyzes the attachment of a phosphate unit to a glucose to produce glucose 6-phosphate, undergoes a marked change of shape when a molecule of glucose becomes bound to its centrally-located cleft. The substrate, coloured red in this computer-generated picture, fits neatly into the base of the cleft. The two lobes of the enzyme, clearly seen in the right picture, close around and envelop the glucose, as if to get a better grip on it. Functional colouring is used in the picture, and each sphere in the enzyme represents an amino-acid residue.

first and third carbon atoms. But the two bonds are not equivalent because the environments of the carbon atoms differ. It turns out that the phosphate at the first position is bound by an energy-rich bond, and this can react with ADP to give ATP. The final products are 3-phosphoglycerate and ATP. We are still some way from the end of this tortuous pathway. The next step involves the isomeration of the 3-phosphoglycerate to 2-phosphoglycerate, the change being a shift of the phosphate group from the third carbon position to the second. This job is efficiently done by the enzyme phosphoglyceromutase, and the resulting molecule is then ready to be split by the enzyme enolase, to release water. The phosphoenolpyruvate thus formed possesses an energy-rich bond.

Up to this point, with nine individual reactions carried out, the energy balance sheet looks rather bleak. Two ATP molecules had to be used in order to get the reaction going, and until now the pay-off has only been two ATP molecules. The account is already in the black, however, because the oxidation of glyceraldehyde-3-phosphate gave two molecules of NADH, and these later yield further molecules of ATP. Returning to the phosphoenolpyruvate, another molecule of ATP is now formed, this tenth reaction being catalyzed by the enzyme pyruvate kinase. The enolpyruvate produced in this reaction spontaneously transforms, with the movement of a hydrogen atom and rearrangement of bonds, into its isomer, pyruvate. We have come a long way from the original glucose molecule, eleven individual steps in fact, and a considerable amount of energy has already been made available to the system. This completes the *glycolytic pathway*. As noted earlier, this part of the metabolic process does not require the input of oxygen, and in animal cells under such anaerobic conditions the pyruvate is then reduced by NADH to form *lactate*. If oxygen is available, however, further reactions occur which ultimately lead to the complete breakdown of pyruvate into carbon dioxide, water and, most importantly, energy. We now turn to those processes.

The reaction between pyruvate and oxygen to produce *acetic acid* and carbon dioxide requires the cooperation of several enzymes, and of secondary helpers known as *cofactors*. One of these, discovered by Fritz Lipmann in 1950, is *coenzyme A*, usually abbreviated to CoA. It contains an adenine nucleotide and a terminal sulphur–hydrogen group. The reaction also requires the cofactor *thiamine, pyrophosphate* and *lipoic acid*. The first of

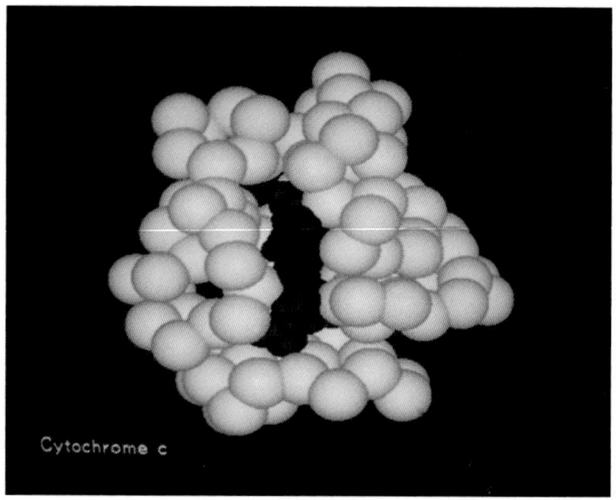

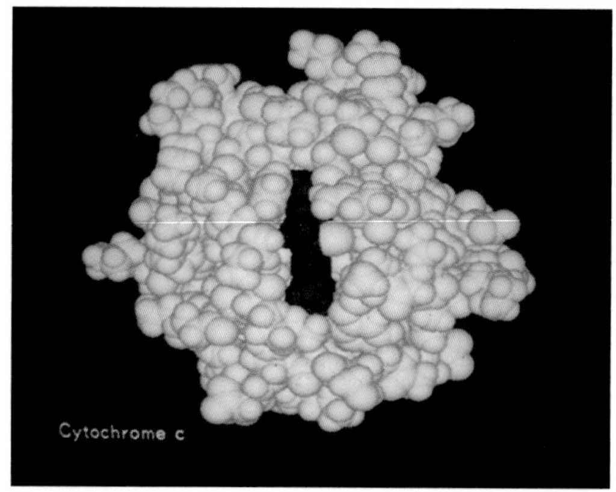

A centrally-located *haem group* provides the active site in the *cytochrome c* molecule, just as in myoglobin and the four subunits of haemoglobin. But in this case the task of the haem's iron atom is to mediate the transport of an electron rather than an oxygen atom. Functional colouring is used in this computer-generated diagram; the 104 amino-acid residues, arranged in a single polypeptide chain, have all been given the same blue-white colour while the haem group is shown in red. The left picture shows each amino-acid residue as a single sphere, while the right one shows all the atoms in the molecule.

these is commonly known as *Vitamin* B_1, and its importance was recognized by Rudolph Peters in the 1920s, when he found that animals given a diet deficient in this compound accumulate pyruvate in their blood streams. The reaction produces NADH which can later be oxidized with the further production of energy, while the three-carbon compound, pyruvate, is converted to the two-carbon acetyl-CoA. Oxidation of the two remaining carbon atoms occurs indirectly. In fact, the next step appears to conflict with our ultimate aim, because it involves the production of a six-carbon compound. The acetyl-CoA joins with the four-carbon compound, oxaloacetate, to produce the six-carbon *citrate.* This is an anabolic reaction, and as such needs energy, which is conveniently supplied by the energy-rich bond of the acetyl-CoA. It requires citrate synthetase, and the reduced product, CoASH, is liberated. The citrate is then converted to its isomer, isocitrate, with the aid of the enzyme aconitase, *cis*-aconitate appearing as a short-lived intermediate compound. The isocitrate is oxidized to produce α-ketogluterate, a molecule of NAD^+ being reduced to NADH in the process. The enzyme in this case is isocitrate dehydrogenase.

α-Ketogluterate is a five-carbon compound, so we have again begun the process of breaking molecules down to smaller units. It is structurally similar to pyruvate, and it too can make use of the combination of lipoic acid, Vitamin B_1, NAD^+, and CoASH. This results in the production of a compound having only four carbon atoms, the composite succinyl-CoA, and it also produces an energy-rich bond. A further entry is made on the credit side of the energy balance sheet because a molecule of NADH is also produced. Indeed, we have almost an embarrassment of riches, because the energy-rich bond can now be employed to convert a molecule of ADP to ATP. This occurs while succinyl-CoA is being transformed to succinate, by the enzyme succinyl-CoA synthetase, and the ATP is produced after the intermediate formation of a related compound, guanosine triphosphate. Since molecules of NADH will later be oxidized to produce further ATP, our faith in the latter as a viable energy coinage is being vindicated; the cash register is now ringing merrily. And there is more to come. The next reaction, catalyzed by succinate dehydrogenase, converts succinate to fumarate with a further transfer of hydrogen. In this case, however, it is passed on not to NAD^+ but to another carrier, *flavoprotein,* to which we will return shortly. There then follows the conversion of fumarate to L-malate, with the help of the enzyme fumarase, and this is succeeded by conversion of L-malate to oxaloacetate, through the enzyme malate dehydrogenase. We can now appreciate just

Operons: a possible mechanism of differentiation

François Jacob and Jacques Monod (1910–1976) suggested that two operons can combine to produce a control mechanism leading to differentiation. The system has two possible states, and it is reminiscent of the electronic circuit known as the flip-flop. Let us suppose that we have two anabolic pathways each depending on its own chain of enzymes. In the first, enzyme a_1 produces molecule A_1, and this is then modified by enzyme b_1 to produce molecule B_1. The chain continues with enzyme c_1 acting upon B_1 to produce molecule C_1, and so on. Imagine that there is a second chain involving enzymes a_2, b_2, c_2 producing molecule A_2, and taking it through a series of modifications, B_2, C_2, and so on. Jacob and Monod pointed out that for some such chains the product C_2 might interfere with the action of enzyme a_1, and therefore have control on the first pathway. Similarly product C_1 might inhibit enzyme a_2. The two pathways thus exercise mutual control, and a steady state is possible with either one or the other chain in operation. Moreover, the initial conditions determine which of the chains operates at any time. If the supply of C_1 is plentiful at the outset, the second pathway will be suppressed while the first functions. If, on the other hand, the system starts with a plentiful supply of C_2, the enzyme a_1 will be inhibited and it will be the second pathway which is operational.

how frugal this system is, because oxaloacetate is none other than the compound originally required to combine with acetyl-CoA and produce citrate. Thus instead of a continuous breaking down, as occurs in glycolysis, the processing of citrate occurs in a cycle, which sloughs off energy and carbon dioxide during every transit.

Oxidation of one molecule of NADH yields three molecules of ATP, so the entire process, from glucose, through the Embden–Meyerhof pathway, and around one rotation of the Krebs cycle, leads to the production of as much as 38 molecules of ATP. Since each of these can produce 0.06 aJ, a total of 2.28 aJ is made available to the cell. The complete oxidation of a glucose molecule would ideally liberate 5.15 aJ, so the overall efficiency of the processes is approximately 45%, a level rarely attained in our technologies. The best steam locomotives were about 10% efficient. It must be emphasized, however, that the above figures refer to the ideal situation, and ignore the loss which robs the cell of perhaps as much as half of the expected energy. The energy transfer processes are not tightly coupled, and this appears to be the source of the leakage. As impressive as the above numbers for glucose are, they are nevertheless surpassed by those applying to *fatty-acid oxidation*. Full catabolism of a single molecule of palmitate, for example, produces 129 molecules of ATP. One gram of glucose produces 0.21 moles of ATP when fully oxidized, whereas the same weight of palmitate yields 0.53 moles of ATP. Because the energy of ATP is available for biosynthesis, we see that diets which advise avoidance of excess fat make good sense.

There remains the question of how NADH passes its hydrogen atom on so as ultimately to reduce an oxygen atom and realize its potential as an energy source: how its currency equivalent is cashed in. This does not happen directly, the hydrogen, or its equivalent, being passed along a series of molecules before it finally reaches an oxygen atom. These molecules are collectively referred to as the *respiratory chain*. The first two members actually pass on hydrogen, while the remaining five provide a conduit for an electron. The various links in the chain have it in common that they are fairly large molecules, only part of which is actually involved in the transport process. In this respect they are reminiscent of other large functional proteins. They are tightly bound to the mitochondrial membrane and their

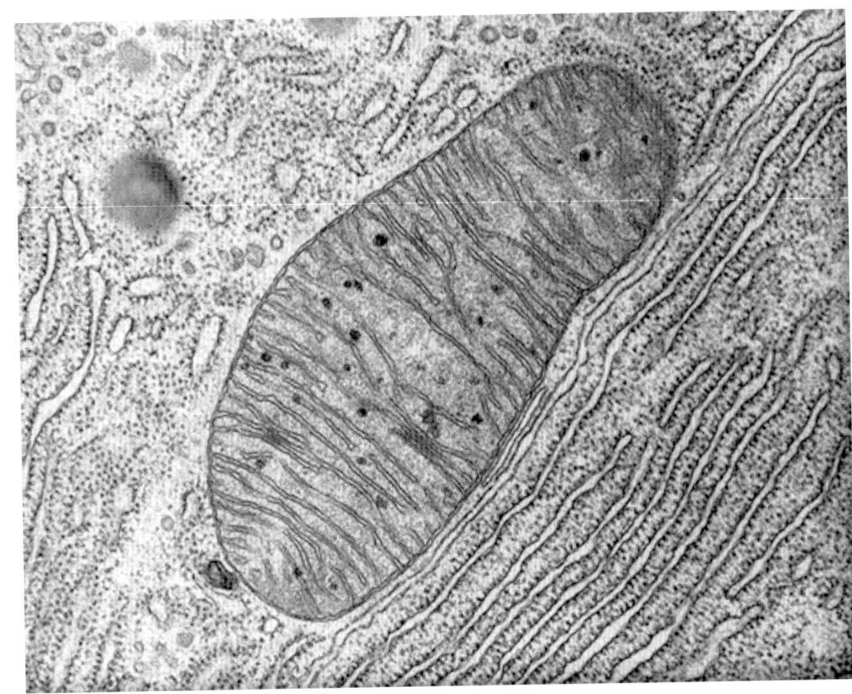

Mitochondria are the organelles responsible for generating the cell's energy requirements, by breaking down organic molecules such as glucose. Their numbers vary, and more than two thousand have been counted in a single rat liver cell. This transmission electron photomicrograph shows, in cross-section, a single mitochondrion in a cell from the pancreas of a bat. Its length is about two micrometres, and the picture reveals the double membrane structure, with the inner lipid bilayer multiply-folded into the sheets, or lamellae, known as *cristae*. At higher magnification, the latter are found to be dotted with approximately 9-nanometre-diameter projections. Efraim Racker (1913–1991) demonstrated that removal of these bodies arrested the production of the energy-transporting compound ATP, and thereby established that they are the sites of the respiratory enzymes. Part of the endoplasmic reticulum, and its associated ribosomes, are also visible in the picture.

transport of hydrogen or electrons ultimately leads to conversion of molecules of ADP to ATP by the enzyme *ATP synthase*. The NAD^+ molecule comprises two nucleotides joined together at their phosphate groups, and it is the nicotinamide end which carries the hydrogen. The hydrogen is passed on to the next molecule down the line, *flavin mononucleotide*, or *FMN*, which is attached to a large protein, the combination being known as *flavoprotein*. The active site on this molecule is located near the middle of three adjacent rings, and FMN is reduced when it accepts hydrogen, and subsequently oxidized when it passes that atom on down the line.

The third carrier in the sequence is *quinone*, also known as *coenzyme Q*. It is common in numerous biological systems, a ubiquity reflected in its other alias, *ubiquinone*. This compound possesses a fatty-acid section which probably buries itself into the hydrophobic part of the mitochondrial membrane, anchoring itself in the appropriate place. The going obviously gets tougher at this point and, like a traveller lightening the load by casting off all except the absolute essentials, the ubiquinone splits the hydrogen atom into an electron and a proton, shedding the latter into the surrounding water. Thus oxidized, it passes the electron on to the first of the four remaining carriers, all of which are *cytochromes*. These proteins take their name from the colouration that derives from the iron atom at their active sites. In this respect they are comparable to *haemoglobin*. The first cytochrome, cytochrome b, contains a ferric ion, Fe^{3+}, and this is reduced to a ferrous ion, Fe^{2+}, by the electron removed from the hydrogen atom. The remaining members of the chain, cytochrome c_1, cytochrome c and cytochrome oxidase, sequentially transmitting the electron like a team of basketball players passing the ball, finally deliver it to the oxygen, which in an animal has been brought from the lungs by haemoglobin. The final reaction of the respiratory chain then takes place: combination of transmitted electrons, hydrogen ions acquired from the local aqueous medium, and ingested oxygen, to produce water. This is journey's end, the hydrogen gained by NAD^+ during glycolysis and

Electron carriers

Living organisms obtain their energy by breaking down chemical compounds in a series of processes known as *respiration*. There are two types: *aerobic respiration*, in which ingested atmospheric oxygen is converted to carbon dioxide, and *anaerobic respiration*, which utilizes chemically combined oxygen. In each case the first stages of the respiratory chain involve the transport of hydrogen or electrons to oxygen, the latter thereby becoming reduced. There are a variety of carriers, which function consecutively, passing the mobile charge across the membrane of the mitochondrion. The following pictures show the oxidized and reduced forms of the major carrier molecules. The active part of the structure is highlighted by the red background colour, in each case. *Nicotinamide adenine dinucleotide* (NAD^+) is a composite molecule, as its name suggests. It comprises a molecule of adenosine monophosphate (AMP) joined, via an oxygen atom, to nicotinamide mononucleotide (NMN). The latter, which contains the active site, has the green background.

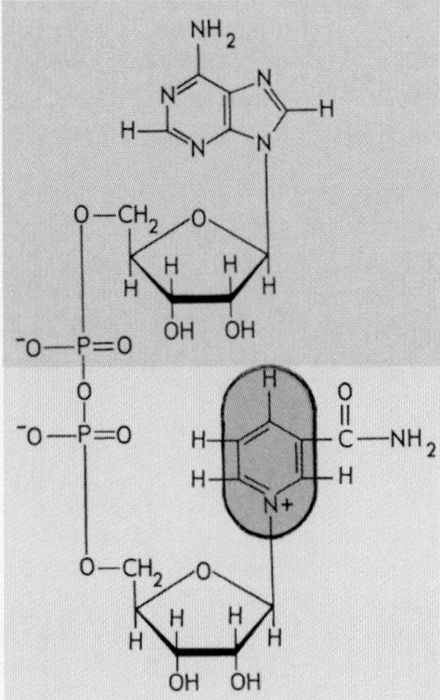

oxidized nicotinamide adenine dinucleotide (NAD^+)

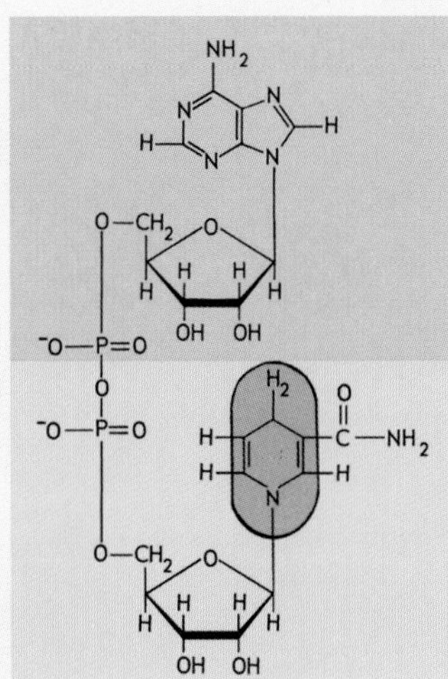

reduced nicotinamide adenine dinucleotide (NADH)

The flavins, *riboflavin mononucleotide* (FMN) and *flavin adenine dinucleotide* (FAD), differ by the AMP unit that is part of the latter, larger, molecule. These carriers are indicated by the yellow and orange backgrounds respectively. FAD is an alternative to NAD as the initial carrier in the respiratory chain, while FMN provides the second link in both cases.

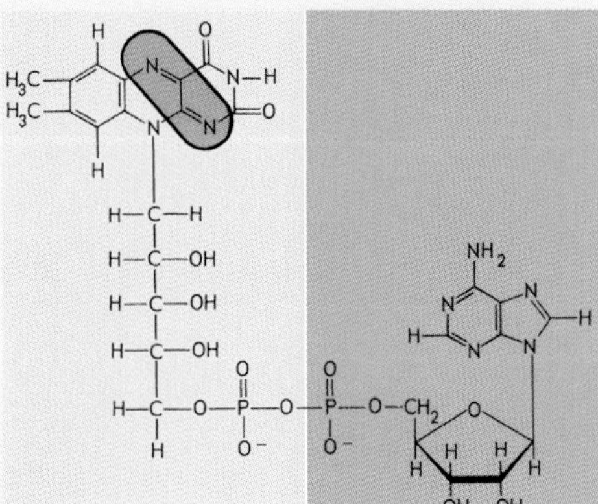

oxidized flavin adenine dinucleotide (FAD)

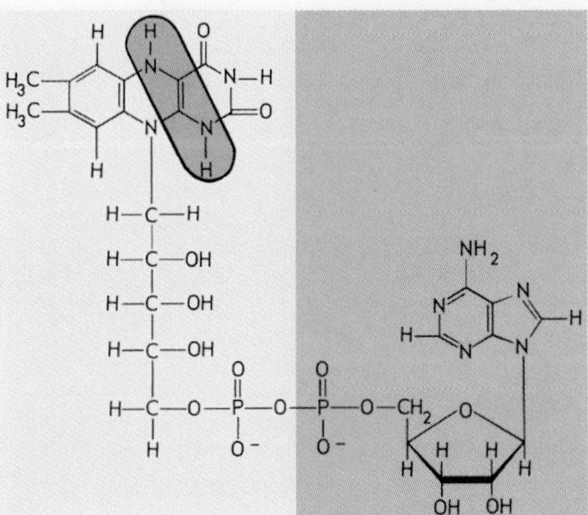

reduced flavin adenine dinucleotide ($FADH_2$)

The third carrier, *coenzyme Q*, also known as *ubiquinone*, is the only mobile molecule in the chain. It is believed to diffuse across the mitochondrial membrane, ferrying electrons from FMN to the cytochrome molecules that provide the end station of the journey. The oxidized and reduced forms of coenzyme Q are as follows:

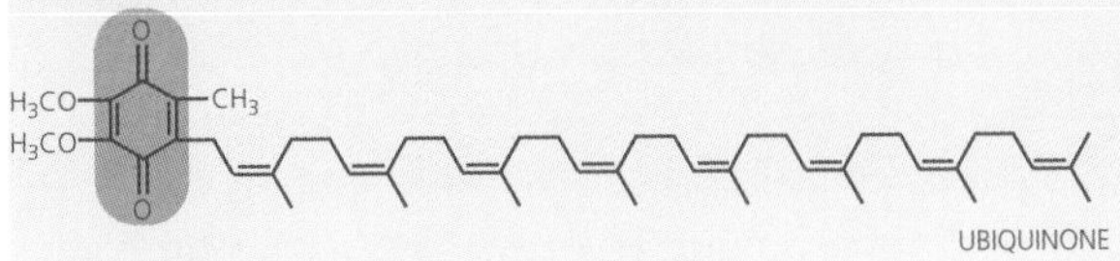

oxidized coenzyme Q (ubiquinone)

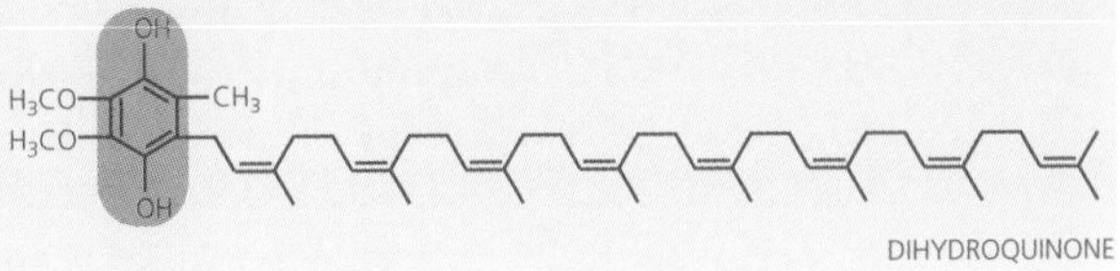

reduced coenzyme Q (dihydroquinone)

The *cytochromes* are a closely-related family of carriers, each of which consists of a protein with an attached (prosthetic) group. The latter is the active part of the structure, in each case, and it is a haem group similar to that found in haemoglobin. The oxidation of cytochrome, by attachment of molecular oxygen to the central iron atom, and its subsequent reduction, is the final step by which oxygen participates in the respiratory process. The diagram below shows a typical cytochrome haem; it is haem a, which is found in cytochrome a.

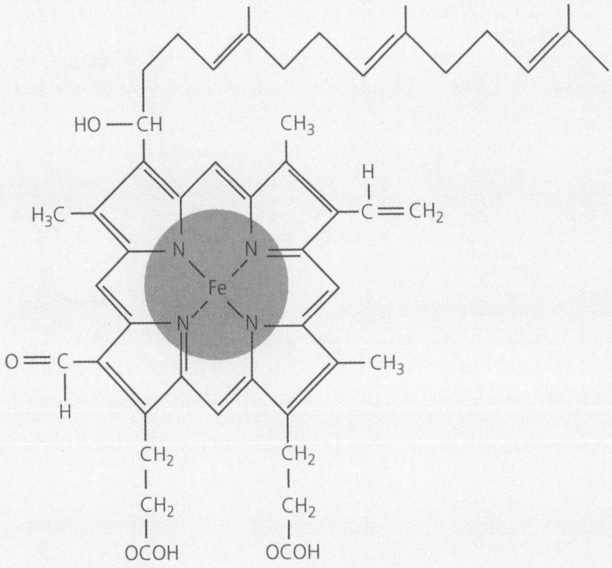

heme a

A single respiratory chain typically handles between a few and several hundred electrons per second, depending on how active the cell is. Mitochondria are egg shaped, with a length of about 2 micrometres, and the average cell contains several hundred of these organelles. We have already seen that the optimal adult intake of 2400 Calories, or roughly 10^7 joules, is sufficient to produce 1.7×10^{26} molecules of ATP. And this rate of energy consumption was shown to be equivalent to about 120 watts. The potential difference across the mitochondrial membrane has been measured to be 0.14 volts, so the total electrical current flowing in the body at any instant is an impressive 850 amperes or so. One ampere is equivalent to 6×10^{18} electrons per second, so our daily energy supply causes the transfer of about 4×10^{26} electrons across respiratory chains. This indicates that 2 or 3 electrons are transferred for each ATP molecule produced.

the citric-acid cycle finally having reached its target oxygen atom. We must bear in mind, however, that this has been accomplished by proxy; the hydrogen ion captured by the oxygen is not the one actually shed by the NADH. Although these final steps take many lines to describe, they occur within one thousandth of a second (0.001 s). Each chain carries several hundred electrons per second. Since passage of a pair of hydrogen atoms down the line ultimately produces three molecules of ATP, not all these steps are energy liberating. Indeed, the degree of physical linkage between the electron carriers and the ATP-fabricating enzyme, ATP-syntase, remains to be elucidated. The complexity of this sequence of events suggests vulnerability, but everything happens automatically so long as there is an adequate supply of the raw materials. The delicate mechanisms can easily be disturbed, however, as in the event of *cyanide poisoning*. Cyanide prevents oxidation of reduced cytochromes, and the flow of energy is rapidly terminated, resulting in seizure and death. Left alone, the mitochondria work in a steady and continuous manner, producing up to 75% of the cell's energy requirements.

The vital role of cytochromes in respiration was demonstrated by David Keilin in 1923, and he later succeeded in reconstructing part of the transport chain. The importance of iron was discovered already in 1912, by Otto Warburg, but it took him another 20 years to establish the structure of the haem molecule, in which the iron atom is surrounded by nitrogens and carbons like a spider sitting at the centre of its web. The haem units are carried by the cytochromes, which appear to be universal for respiratory chains. Some bacteria are able to reduce a nitrate, even though preferring oxygen, and others such as *Desulphovibrio* work anaerobically on a sulphate, giving off the pungent *hydrogen sulphide* as a waste product. All these forms nevertheless employ cytochromes, and Richard Dickerson has traced the evolution of cytochrome-bearing organisms back to what appear to have been primordial *photosynthetic bacteria*. Identification of the mitochondrion as the site of oxidation was achieved as recently as 1949, by Eugene Kennedy and Albert Lehninger, the latter demonstrating two years later that at least part of the respiratory chain transports electrons during *oxidative phosphorylation*.

A new aspect of the story was revealed by electron microscopy when Humberto Fernández-Morán observed small lumps dotted over the mitochondrial membranes. Efraim Racker went on to show that ATP production is arrested when these F_1 projections, as they are now called, are removed, even though the remaining mitochondrial membrane is still able to transfer electrons to oxygen. In the early 1950s, it was believed that the phosphorylation is directly coupled to the redox steps, both processes being managed by the same enzymes. This view, propounded by E. C. Slater and others, has since been modified by David Green, who suggested the existence of membrane-bound macromolecules having at least two alternative configurations, differing considerably in their conformational energies. He postulated that electron transport flips such a unit into its high energy state, the subsequent relaxation providing sufficient free energy for phosphorylation of ADP, presumably by an enzyme that is an integral part of the macromolecule.

The theory was put on a firmer footing by Paul Boyer, first in 1964 through his suggestion that ATP synthesis involves a structural change in the *ATP synthase* itself, and about a decade later when he discovered that the energy-requiring step in the synthesis simultaneously accomplishes two things: binding of ADP and inorganic phosphate and the release of ATP. It should be stressed, however, that the ATP molecule being released during a given synthetic step is not necessarily composed of the ADP and inorganic phosphate that become bound. Boyer's ideas were subsequently to receive impressive endorsement. In 1981, John Walker succeeded in determining the DNA sequence of the genes which direct fabrication of ATP synthase, and by 1994 he and his collaborators had determined the actual molecular structure of the F_1 part of this enzyme – the part which Racker had shown was so important. Speculation soon arose that the functioning of the synthase complex involves rotation, possibly analogous to rotation of the flagellum possessed by some bacteria. This was confirmed by the chemical studies of Richard Cross, by the spectroscopic studies of Wolfgang Junge and by the microscopic studies of Masasuke Yoshida. Let us now return to the issue of electron transport.

The proteins of the *electron transport system* can be regarded as semiconductors, and the *redox chain* involves a series of steps in each of which an electron is passed from a donor to an acceptor. The distance between adjacent electron trapping centres can be as large as 3 nm. This rules out electron motion by systematic shuffling of temporary chemical bonds, and it suggests that *tunnelling* is important. This is a purely quantum phenomenon, without classical counterpart. We would normally expect a particle to be able to pass through a barrier only if it has sufficient energy to thus penetrate it. But the rules of motion are different at the subatomic level. Because of the Heisenberg principle discussed in Chapter 1, there is an inherent ambiguity in the location of an electron; we could never be quite sure which side of the barrier it is on, whether it is still waiting to start its journey or has already arrived at its destination. And this gives the electron a finite chance of tunnelling through a barrier even when lacking the energy to penetrate it in the classical sense. The situation is complicated, however, because the *tunnelling probability* depends on the molecular configuration that defines the barrier. And we have already seen that this configuration is subject to deformation. Temperature fluctuations, for example, could enable a protein instantaneously to adopt a conformation favouring *electron tunnelling*. John Hopfield and Joshua Jortner have independently shown that tunnelling can occur even in the absence of such thermal activation. The situation can be compared with a person walking on a see-saw. In the classical case, the see-saw flips over when the person passes the fulcrum, but the quantum regime demands a ghost-like person whose position is intrinsically uncertain. The situation in the electron-transporting protein is still subject to calculable probabilities, however, and one can work out the chance of the conformational flip implicated in ATP production.

An important contribution to the story was made in 1961, by Peter Mitchell. Like most strokes of enlightenment, is seems so obvious in retrospect. Mitchell noted that energy dissipation implies directionality, and the topological prerequisite of the latter is automatically provided by the mitochondrial membrane; it divides space into an inside and an outside. The above mechanism of electron transport was not challenged, but the manufacture of ATP is attributed to a general process rather than one specifically linked to the respiratory chain. The crux of the Mitchell theory lies in the fact that electrons readily pass down the respiratory chain whereas hydrogen ions do not. This differential transport sets up a hydrogen ion, or proton, gradient, and this is equivalent to what Mitchell called a *chemiosmotic*, or *electrochemical*

ATP synthesis

The very important reaction in which adenine diphosphate (ADP) and inorganic phosphate (P_i) are consolidated into adenine triphosphate (ATP) is catalyzed by ATP-synthase. In 1964, Paul Boyer postulated that the process involves a structural change in that enzyme. The ATP molecule is metastable; there is an energy barrier which prevents its spontaneous fission into ADP and P_i, with the release of the stored energy. About a decade later, Boyer demonstrated that the energy-requiring step in the synthesis involves the release of a complete ATP molecule from the enzyme, together with the simultaneous binding of an ADP molecule and a P_i group. As mentioned earlier in this chapter, electron microscopy had revealed that part of the enzyme (now designated F_0) is embedded in the mitochondrial membrane, whereas a somewhat larger portion (F_1) sticks out into the surrounding cytoplasm. Ephraim Racker (1913–1991) succeeded in isolating the F_1 fragment in 1961, and John Walker and his colleagues later determined the DNA sequences of the genes that code for the entire enzyme. Their subsequent X-ray diffraction investigation – completed in 1994 – established the three-dimensional structures of the various parts of the F_1 portion, and this is shown in the diagram at a more macroscopic level, in order to bring out the relationship with the membrane and the F_0 part. It turned out that the F_1 part consists of seven units, the proteins of three of which (designated α) are identical both with respect to primary and tertiary structure. The primary structures of the proteins of three of the remaining four units (designated β) are also identical, whereas their instantaneous tertiary structures are distinctly different. Those three conformations are referred to by the words *tight*, *loose* and *open*, these terms indicating the various binding, compaction and release stages first discovered by Boyer. It should be noted from the diagram that it is rotation of the centrally located seventh protein (designated γ) which provokes the β units into adopting their different conformations. The rotation of this seventh unit is caused by the rotation of the membrane-bound F_0 unit, this, in turn, being due to the flow of hydrogen ions, a mechanism that could be very loosely compared with the action of the wind on a freely rotatable propeller. The photograph, taken by Feng Yan and colleagues, shows the proteoid roots of a white lupin plant growing in a solidified agar medium supplemented with the pH indicator bromocresol purple. The yellow colour indicates acidification (that is to say, decrease in pH) of the medium resulting from the activity of a plasma membrane proton pump, which, as we have just seen, is associated with the movement of hydrogen ions from one side of the membrane to the other.

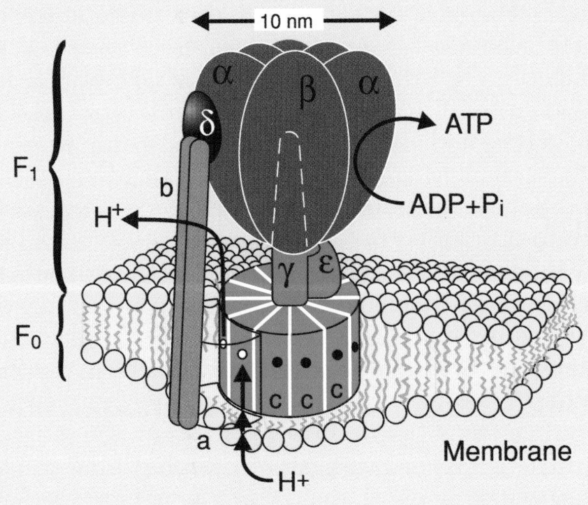

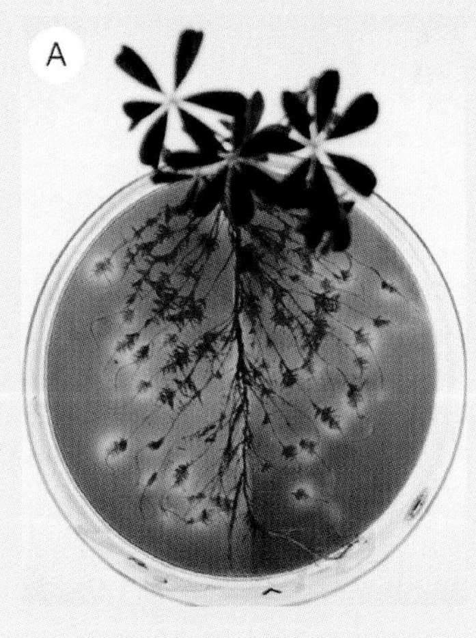

force. Comparing the situation with that which provokes electron flow, he coined the term *proticity*. The electron flow across the membrane caused by oxidative respiration, or as we shall see by photosynthesis, sets up a proton gradient, and consequently a voltage across the lipid bilayer. This gradient provides the energy required by membrane-bound ATP synthase to form ATP from ADP and inorganic phosphate. In one variant of Mitchell's theory, the ATP synthase did indeed span the width of the bilayer, as was later comfirmed. The proton–oxygen interaction was conjectured to occur on the side having the higher hydrogen ion content (i.e. low pH and positive voltage), and the ADP–phosphate reaction proceeded on the other. Another version envisaged a protein possessing an internal channel down which protons flow, this leading to the linkage of ADP and inorganic phosphate to produce ATP. This is indeed what is now believed to take place in that part of the ATP synthase complex that is buried within the membrane. The *chemiosmotic theory* has been endorsed by numerous experiments, many of them performed by Mitchell himself, in collaboration with Jennifer Moyle.

Just what is done with the energy made available through respiration depends on the type of cell. If it is a *red blood cell*, some of the energy is used by the haemoglobin molecules in their oxygen transporting function. In *nerve cells*, the chemical gradients of sodium and potassium which provide the basis for nerve impulses must be maintained, while the typical *plant cell* uses energy to link sugar units and produce cellulose fibres. And every cell requires a constant source of energy to carry out the synthesis of proteins. In connection with *cell division*, there is also the energy needed in duplication of the genetic material, although this is a relatively modest consumer of that commodity. We have already discussed the molecular aspects of these functions; there remain the more macroscopic aspects, and it is to these that we now turn. Before getting down to details, we should consider the exciting series of breakthroughs that provided so much insight into life's most fundamental phenomena.

Our knowledge of life's evolutionary processes owes a great deal to two momentous advances during a six-year period in the middle of the nineteenth century. One was the *theory of natural selection*, which provoked much public debate. Commonly referred to as the principle of *survival of the fittest*, it should rather be interpreted in terms of reproduction. In a genetically uneven population, certain individuals can contribute to the succeeding generation a disproportionately large number of offspring that survive to fertility. The composition of the population is thereby changed, and the species evolves. This idea occurred to Charles Darwin in 1842, and independently to Alfred Wallace a few years later, and in 1859 Darwin published *The Origin of Species*. The second major breakthrough, achieved in relative obscurity by Gregor Mendel in 1865, gave us the science of *genetics*. His success was based on pure breeding lines and sufficient observations on conflicting characteristics to provide good statistics. Using pea plants, and recording such features as stem height, flower colour and seed shape, he studied the effect of crossing plants with various combinations of distinguishing traits. He noted, for example, that when a pure-bred tall plant is crossed with a pure-bred dwarf plant, their progeny are invariably just as tall as the tall parent; there is no tendency to produce a compromise plant of intermediate height.

A further surprise was seen in the third generation. When second-generation tall plants are crossed, about one quarter of the third generation are dwarfs that are just as short as their dwarf grandparents. Mendel had established the importance of distinguishing between an organism's *phenotype*, which is the sum of its visible characteristics, and what would now be called its *genotype*. Hereditary features can remain latent or concealed, being carried by an individual even though they are not expressed. Mendel's brilliant analysis of these results yielded the basic laws of genetics. Inheritance comes in indivisible units, now known as *genes*. Once again we have the principle of the ultimately indivisible entity; we encountered this in the quantization of energy, in atomic theory, and with the living cell. The idea that reproduction involves a sort of mixing of the blood had been exposed as a fallacy. Mendel also realized that domination of some characteristics over others, said to be *recessive*, and the possibility of re-emergence of the latter, implies that genes must come in pairs. He saw heredity as involving an element of chance, even though the outcome of a particular pairing is not totally capricious.

A frequently used illustration of these principles is the domination of brown eyes (B) over blue eyes (b) in humans. Of the three possible combinations of eye colour genes, BB, Bb, and bb, only the last produces blue eyes because the brown-eye gene dominates. Mendel had concluded that each parent passes on only one gene, and we must consider what happens in the case of mating of the pure lines, BB and bb. The offspring can only carry the gene combination Bb, and they are thus bound to have brown eyes. If brown-eyed parents produce a blue-eyed child, they need not be surprised, for they have proved that they are both of the Bb type. It is sometimes stated that blue-eyed parents, on the other hand, cannot produce brown-eyed offspring, but this is not correct. The distribution of eye pigment is regulated by more than one gene, and certain gene combinations do indeed permit blue-eyed parents to have a brown-eyed child, although such an event is rare. There are even people with one eye of each colour. Sex, in the sense of gender, is determined by a somewhat different mechanism. We have already seen that *chromosomes* come in pairs, and the two members of each pair are said to be mutually homologous. In the human, the male chromosomes of one *homologous pair* have different size and shape, the larger being designated X and the smaller Y. The chromosomes of the corresponding pair in the female are identical, and are of the X type. *Sexual differentiation* hinges on the fact that males have one active X chromosome and one active Y chromosome, whereas females have two of the X type, one of which is always inactive. The mother will always contribute an X chromosome to her child, whereas the father gives either an X or a Y. The latter case results in the XY combination, and the child is another male. Mendel succeeded in establishing both the existence of genes and the rule of their combination, known as the *law of segregation*, because of his wise choice of characteristics. A phenotypic feature such as the human face, varying as it does from, say, Shirley Temple's to that of the late Genghis Khan, is not determined by a single gene; a study of inherited human physiognomy would have told Mendel very little. He also proposed a second law, that of *independent assortment*, in which he argued that genes for different characteristics have no influence on one another. In this case, Mendel's choice of features was unfortunate because it prevented him from observing another important phenomenon: *gene linkage*. Independent assortment would be the invariable rule only if each chromosome carried a single gene. This is never the case: a chromosome usually comprises numerous genes, which are thereby linked.

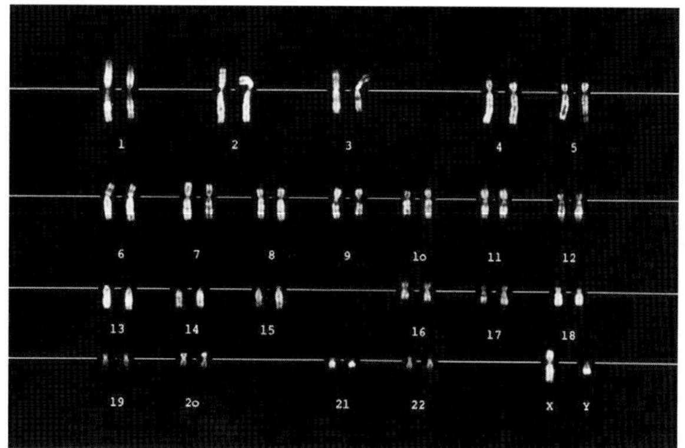

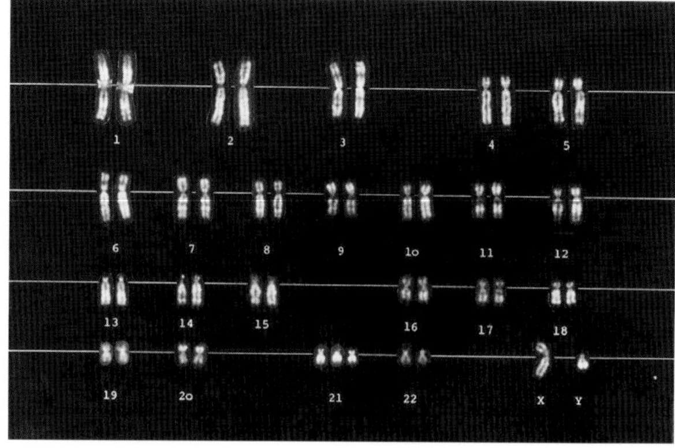

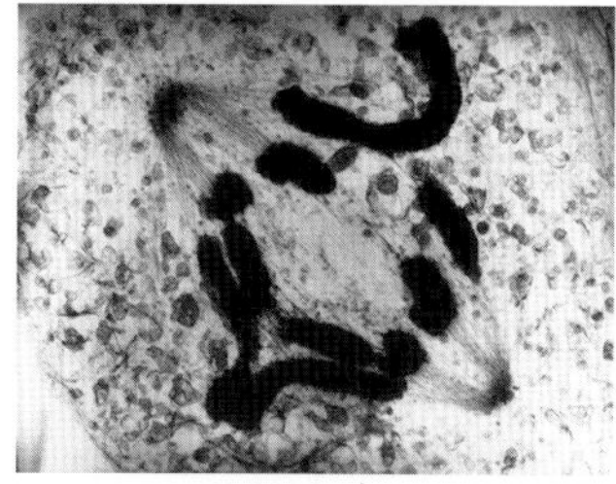

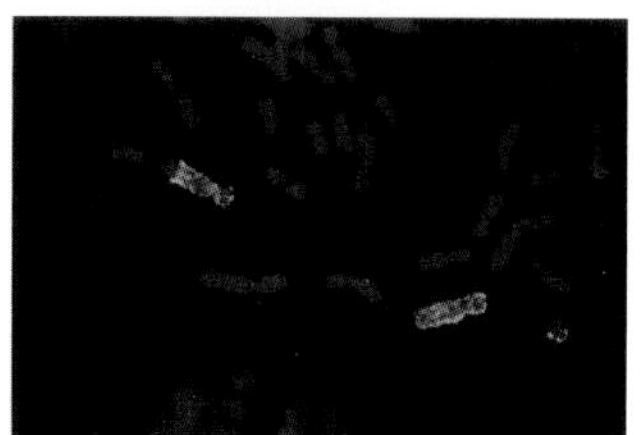

Mitosis is the process by which the nucleus divides into two parts, one part going to each of the daughter cells during cell division. Each chromosome duplicates, and one copy appears in each of the new nuclei. This photomicrograph (left) shows the chromosomes separating towards the two extremities of a dividing cultured mammalian cell, by moving along the guiding microtubules which are also visible. This stage of mitosis is known as anaphase. There are 23 pairs of chromosomes in human somatic (body) cells, one pair being the sex chromosomes: XX for the homogametic (female) sex and XY for the heterogametic (male) sex. The top two pictures show the full chromosome complement of two human males, prepared by the staining technique known as Q-banding. The left-hand picture shows a normal specimen, whereas the right-hand picture reveals the additional twenty-first chromosome characteristic of a patient with *Down's syndrome*, otherwise known as *mongolism*. The colour photograph was produced during a genetic project on autism, carried out by Margareta Mikkelsen, Niels Tommerup and the present author. It shows that a *translocation* has occurred between chromosomes 5 and 18, such that parts of each of one copy of these chromosomes have been transferred to the other. This was detected by labelling chromosomes 5 with a fluorescing marker, and one can see (just to the left of centre in the picture) that one copy of that chromosome has remained intact. But the other copy has become split into a major portion (seen on the right) and a minor portion (lower right), these both now being attached to parts of the original copy of chromosome 18. The victim of this chance event has been autistic since birth.

Mendel's laws were discovered before the existence of atoms and genes had been established. There was therefore a long way to go before a connection with anything at the atomic level could become apparent. Let us survey the historical developments that ultimately led to the link between Mendel's peas and biopolymers. In the latter half of the nineteenth century, steady improvements to the optical microscope permitted cells to be investigated at progressively finer levels of detail. *Cell division* was discovered, as was the fusion of two cells during *fertilization*. Staining their specimens to increase the visibility of microstructure, biologists observed thread-like features, a few micrometres in length, in the cell nucleus. They were named *chromosomes*, after the Greek for colour bodies, and were found to occur in pairs. These structures were observed to undergo changes during cell division. In the resting cell, the threads are more or less randomly distributed, and in this state they are now referred to as *chromatin*. The chromosomes appear during the early stages of the cell cycle, and immediately prior to division, each becomes a double thread. The two threads, known as *chromatids*, separate during division, each becoming a new chromosome, so the two daughter

MITOSIS

1 The somatic, or body, cells of a eucaryotic organism multiply by a process known as *binary fission*, dividing into two daughter cells each with the full complement of chromosomes, and hence of the genes. The *chromosomes* are present in homologous pairs, and their number is thus said to be *diploid*. The two copies of a gene, located at the same relative position, or locus, on homologous chromosomes are said to be *alleles*. The idealized cell shown here has two homologous pairs, the green chromosomes being inherited from the male parent, say, and the yellow from the female parent.

2 By the stage known as *prophase*, each chromosome appears as a double thread consisting of two *chromatids*. There are thus two copies of each allele. This is illustrated here for just one gene location. The alleles have been given different colours to indicate the possibility that they are different, as when dominant and recessive versions are both present. The *nucleolus* and *nuclear membrane*, both indicated in black, are becoming less visible. The *centrosome*, located just outside the nucleus, has split in two.

3 In *metaphase*, the nucleolus and nuclear membrane are gone and the chromatid pairs are attached to individual microtubules at locations referred to as *kinetochores*. The chromatids are coiled up, but this is not indicated on this diagram, which merely keeps track of the fates of the various chromatids.

4 The cell division has entered *anaphase*, and the two members of each chromatid pair are moving with the help of the microtubule scaffolding, towards the respective ends of the now elongated cell. The *microtubules* appear to mediate the motion by a combination of growth and selective degeneration. It should be noted that the existence of homologous pairs is without significance for the mitotic process; each member of a pair gives rise to two chromatids and these are separated during anaphase, one going to each of the daughter cells.

5 In *telophase*, the two diploid sets of chromosomes uncoil, elongate and gradually disappear within the two new nuclear membranes. An incipient membrane known as the *septum* appears at the equatorial plane of the elongated cell. Its growth is related to the microtubule spindles, which are now breaking up.

6 A new nucleolus appears within each new nucleus. The dividing septum has now developed into two lipid bilayer membranes, which are continuous with the original cytoplasmic membrane. The fission process is thus complete and one cell has been replaced by two, with identical sets of chromosomes. The time occupied by the mitotic process varies widely, but it usually lies between about a half and three hours.

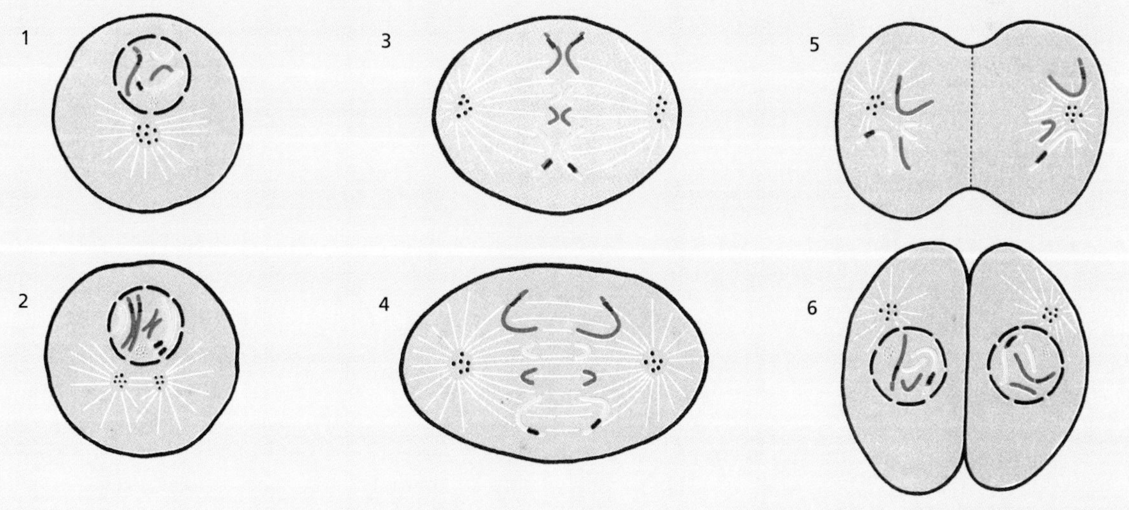

MEIOSIS

1. *Meiosis* involves two successive cell divisions but only one chromosome duplication process. It thereby produces four cells each having only the *haploid* (single) set of chromosomes: one of each homologous pair. When the chromosomes first become visible in prophase, they appear as a tangle of single threads. This is known as the *leptotene stage*, and the chromosomes have in fact already doubled. By a process known as *synapsis*, homologous chromosomes then pair up, with corresponding loci adhering. This is the *zygotene stage*, and each associated chromosome pair is called a *bivalent*.
2. The bivalents shorten and get thicker (not shown here), and the double-thread structure of each chromosome becomes apparent. Each bivalent thus comprises four chromatids. This is the *pachytene stage*.
3. The two chromatids of one member of a homologous pair now separate from the two belonging to the other member, except at certain points where interchanges of homologous chromosome units have occurred. These points are the *chiasmata*, and with this crossing over the division has reached the diplotene stage. The *kinetochore* (or spindle attachment) of each original chromosome has not yet duplicated, so the chromatid pairs are also still joined at these points.
4. In the first anaphase and telophase, pairs of chromatids move to opposite poles of the dividing cell. Because of the segment exchange during *crossing over*, the separating chromatids are in general hybrids of the original homologous pair. Random choice governs the direction in which each separating pair moves towards one of the poles.
5. There are now two cells, each with a set of recombinant and segregated chromosomes. This first division may be followed by a resting period, but the second stage sometimes occurs immediately.
6. No chromosome duplication precedes the second anaphase. The two members of each chromatid pair move towards opposite poles. Again, the direction of separation is a matter of chance. This random distribution of chromatids provides the basis of *Mendel's law of independent assortment*.
7. Because the chromosomes have doubled only once in two cell divisions, each of the four granddaughter cells possess only half the normal complement of these genetic molecules. This is the haploid number, and the cells are *germ cells*, or *gametes*. The combined effects of crossing over and *independent segregation* (of the chromatids during the two anaphase stages) make it extremely unlikely that a gamete contains genetic elements from only one of the parents of the organism in which the meiosis occurs.

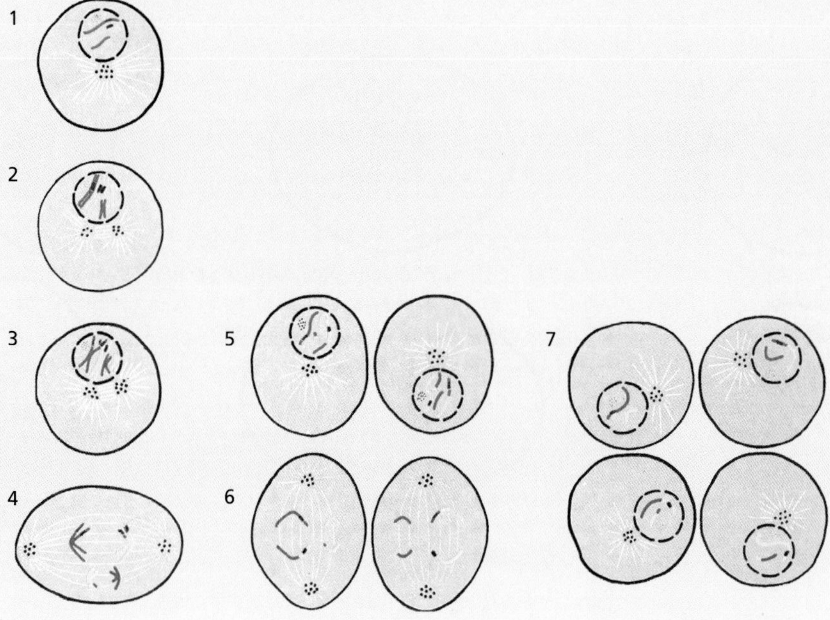

cells contain single strands. It became apparent that this type of behaviour made chromosomes candidates as Mendel's carriers of heredity, and in 1887 August Weismann argued that the number of chromosomes must remain constant. He proposed the existence of two different types of cell: *germ cells*, which are involved in the reproduction process; and *somatic cells* (also known as body cells), which differentiate to produce the various tissues. He showed that a fertilized cell, now called a *zygote*, must produce both germ cells and somatic cells. The former contain only one set of chromosomes, the *haploid* number, while the somatic cells contain the full complement of two sets, the *diploid* number. This prevents the obvious nonsense of steadily increasing numbers of chromosomes in successive generations, and Weismann suggested that the production of germ cells involves a reduction division, *meiosis*, in contrast to the normal mode of division, *mitosis*, which preserves the chromosome number. The term mitosis had been coined by Walther Flemming in the early 1880s.

The difference between mitosis and meiosis is clearly fundamental so we must consider these processes in greater detail. Before mitosis, the nucleus of the resting cell is visible, and it is seen to contain both the nucleolus and the chromatin, which is in its uncoiled state. The resting, or *interphase*, period is terminated by gradual disappearance of the *nucleolus* and the *nuclear membrane*, and the chromatin is suddenly seen to consist of parallel pairs of chromatid strands. The cell then enters *prophase*, in which the daughter chromatids coil up into a short thick configuration. At the same time, the *centrioles*, which until now have been located just outside the nuclear membrane, split apart so as to produce two poles between which stretch a number of *microtubules*. The chromatid pairs are seen to have a connecting point at their centres, and these points now become located on the centriole tubules. In *metaphase*, the chromatid pairs separate and slide along the microtubules towards the two poles of the now elliptically shaped cell. The separation is completed during *anaphase*, and the cell is then ready to divide. The two poles move apart, stretching the cell, which is now in *telophase*, and a separating membrane, the *septum*, gradually forms. The two daughter cells move apart, while the chromosomes unwind. Nuclear membranes and nucleoli become visible in each new cell. This is the process by which each somatic cell fissions, and it is important to note that the chromosome pairing has no special significance for the mitotic mechanism.

Cell division by meiosis is more specialized and occurs only in the reproductive organs. It involves cells known as *oocytes*, and the interphase stage resembles that of mitosis. After this, however, the two homologous members of each pair of chromosomes become closely associated and actually adhere at one point. This pairing process is known as *synapsis*, and the associated pair is called a *bivalent*. Following this so-called *zygotene stage*, the bivalents coil up, and each chromosome is seen to be double, the bivalent now consisting of four chromatids in close association. It is here that the sexual mixing occurs because, in a process known as *crossing over*, sections of the individual chromatids interchange places. When the cell subsequently divides, each daughter cell contains the full diploid complement of chromosomes, but these are now of the mixed variety. There then occurs a second division, into a total of four cells, but in this case the homologous chromosomes of the daughter cells separate without first doubling into two chromatids. The four granddaughter cells thus contain only the haploid chromosome number, namely one member of each homologous pair. These are the *germ cells*, or *gametes*. In the female, meiotic division produces cells of unequal size, and only one of the four gametes survives as a viable *ovum*. In the male, on the other hand, all four gametes survive, and these are the *spermatozoa*. *Sexual reproduction* is advantageous to a species. The mixing of genes during crossing over ensures that each offspring is genetically unique and different from either parent. Every fusion of gametes to produce a zygote is a new throw of the evolutionary dice, and although the probability might be small, there is always a chance that the offspring will be significantly better adapted to the environment than its mother or father. This chance is further increased by *mutations*, which change the activity of a single gene in an unpredictable manner.

The molecular mechanism of mitosis in *eukaryotes* is still not completely elucidated. That microtubules are involved is well established. In 1889, B. Pernice noticed that the toxic alkaloid, *colchicine*, chemically isolated six years earlier, seems to promote cell proliferation. This substance is obtained from the meadow saffron *Colchicum autumnale*, a plant famous in the treatment of joint pain. It was subsequently realized that colchicine, far from provoking cells to divide, actually arrests them in metaphase. Independent studies by Pierre Dustin and O. J. Eigsti, and Gary Borisy and Edwin Taylor, in the 1960s, established that colchicine interferes with microtubule assembly. *Tubulin* is not a symmetric molecule, so microtubules assembled from these units have polarity. Barry Kieffer, Robert Margolis and Leslie Wilson suggested that one consequence would be a unidirectional sliding when two oppositely polarized microtubules are in contact. They also noted that tubule assembly is favoured at one end whereas the other end actually tends to break up under certain conditions. These two effects were combined into a model for the separation process during anaphase. Darrell Carney and Kathryn Crossin found that colchicine stimulates DNA duplication, and that this tendency is reversed by the anti-tumour drug *taxol*, which is known to promote microtubule assembly. It thus appears that local disruption of the *cytoskeleton* is a prerequisite for DNA duplication, while microtubule reassembly is required if the chromatids are to separate. Theodore Puck added another dimension, in 1977, when he suggested that information is transferred from the cell surface to the nucleus by the cytoskeleton, and that the unregulated growth characteristic of cancer is a consequence of disorganization of the microtubule–microfilament assembly. Finally, there remains the mysterious role of the *nucleolus*. The genes in the chromosomes carry the total genetic information, but this nebulous little intra-nuclear organelle might determine which parts of the message are to be read. Interestingly, William MacCarty suggested as early as 1936 that the nucleolar material might prove to be the most important marker of malignancy.

The early observations of chromosomes left biologists with an apparent paradox; there were not enough of them to account for all the phenotypic features. Rats have 21 pairs, Mendel's pea plants seven, fruit flies only four, and there is a species of round worm that has but a single pair. Moreover, the number of chromosomes did not seem to reflect the position of a species on the evolutionary ladder. *Homo sapiens* has 23 pairs, but the potato has one more than this. There is a species of crayfish that has as many as 100 pairs, and the one-celled rhizopod has many hundreds of pairs. The riddle was resolved by Thomas Morgan, in 1926, who suggested that genes are not separate entities, but are instead grouped together in the chromosomes, each of the latter containing a large number of these hereditary units. The

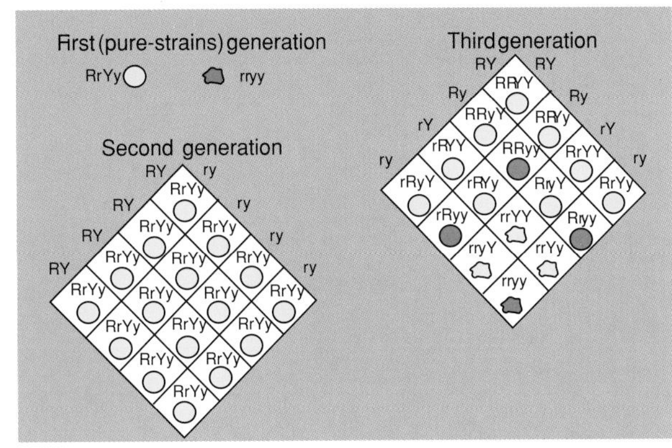

The genetic traits used by Gregor Mendel (1822–1884) for his studies of inheritance were felicitously chosen, because they are dictated by genes present on two different chromosomes. The features he focussed on were shape and colour, and he was careful to start by breeding like with like, in order to obtain pure strains. The dominant shape gene (R) produces round peas, whereas the recessive version (r) produces peas with a wrinkled appearance. Similarly, the dominant colour gene (Y) gives yellow peas while its recessive counterpart (y) produces a darker green colour. There are two copies of each homologous gene in a given pea, of course, because there are two copies of each chromosome, so as indicated in the diagram, the second generation offspring cannot avoid having the gene combination RrYy. This can only produce round, yellow peas. But as also indicated, the third generation will display a greater variety of gene combinations, and this will lead to production of round yellow peas, wrinkled yellow peas, round green peas and wrinkled green peas in the proportions 9:3:3:1. This prediction is in very good agreement with the statistics actually obtained in Mendel's experiments, and it originally led him to believe – erroneously, as it turned out – that such *independent assortment* is always the case.

genes corresponding to the characteristics on which Mendel had made his observations were all located on different chromosomes, and it was for this reason that they appeared to sort independently. In 1900, Hugo de Vries noticed that changes in the appearance of primroses are not gradual, but occur suddenly as a spontaneous mutation. Shortly after Morgan's suggestion, Hermann Müller and L. G. Stadlerr discovered that the *mutation rate* is greatly increased by exposing gamete cells to *X-rays*. It had meanwhile been observed that certain chromosomes, such as those in the cells of the salivary glands of certain flies, are large enough to permit observations of internal structure. Alternating light and dark bands were seen to traverse the chromosomes, and it was soon discovered that these too can be changed by irradiation with X-rays. Morgan's hypothesis had been verified; these bands were Mendel's hereditary units: the *genes*.

The location of the genes and the ultimate target of their influence had been revealed, but other questions remained unanswered. It was still not clear what a gene looked like at the atomic level because at that time genes had been observed only through optical microscopes, and optical wavelengths span many atomic diameters. Moreover, it was not obvious what the immediate function of a gene could be. As early as 1917, before enzymes were isolated, Sewall Wright had suggested that skin colour in humans is determined by a series of enzymes which are controlled by genes. This was a bold speculation at the time, and it was not verified until 23 years later. The common bread mould *Neurospora crassa* can survive on an austere diet of glucose, inorganic salts and one of the B vitamins, biotin, because it can internally manufacture the enzymes needed to produce its other biochemical requirements. George Beadle and Edward Tatum irradiated specimens of this mould with X-rays and ultimately succeeded in producing a mutated strain incapable of manufacturing Vitamin B_6. When fed a diet augmented with this vitamin, the spore survived. They later produced another mutant deficient in the ability to produce Vitamin B_1. Beadle and Tatum concluded that their mutant strains were suffering from defective enzymes; Wright's gene–enzyme link had been proved correct.

This left the question of the exact nature of a gene. We must return to the year 1928, and experiments by Frederick Griffith on pneumonia bacteria, *Diplococcus pneumoniae*. Of the various forms of this microbe, Griffith was particularly interested in two: a virulent variety having an exterior consisting of a polysaccharide coating, referred to by the letter S because it forms smooth-surfaced colonies, and a harmless type that has lost its capsule and

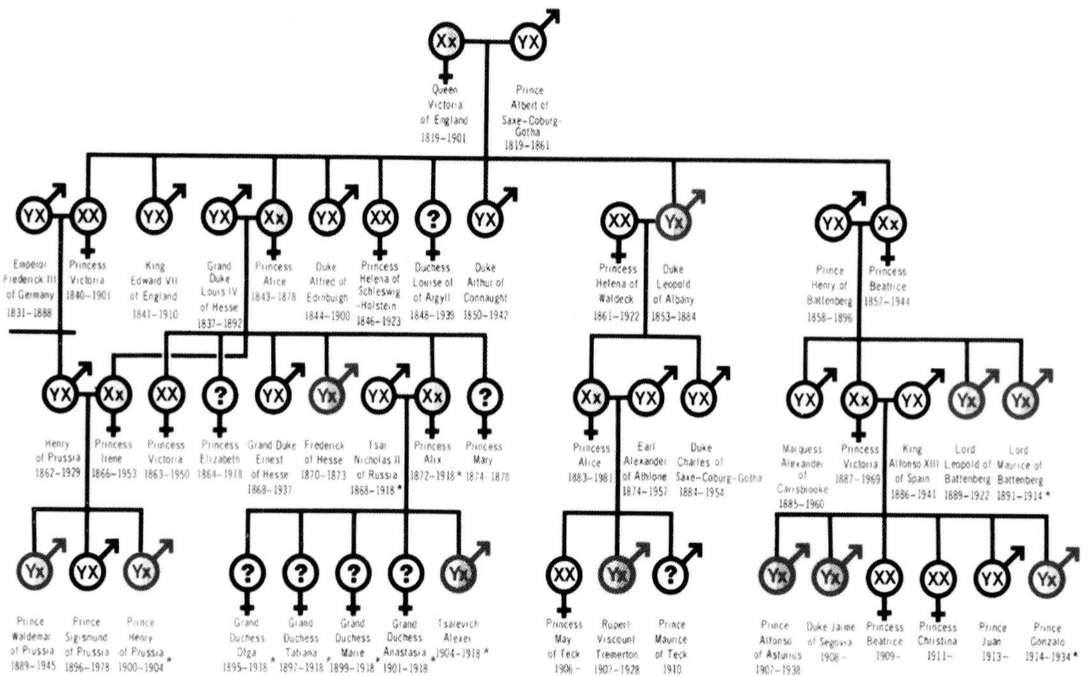

Haemophilia is a recessive trait carried on an X chromosome. Queen Victoria (1819–1901) was a carrier, and the disease came to expression in one of her four sons, namely the Duke Leopold of Albany (1853–1884), while two of her five daughters were known carriers. (An asterisk indicates that the person did not die a natural death.) It is sometimes said that the disease is carried by some female members of an affected family, but comes to expression only in the male line. A detailed study of the Queen's family tree reveals that this is not necessarily so; a female could indeed be haemophiliac, if there were sufficient inbreeding.

therefore produces rough-surfaced colonies, the so-called R form. The R form is a mutant lacking an enzyme required in production of the polysaccharide. Griffith discovered that if he injected mice with both heat-killed S type and untreated live R type bacteria, the mice developed the disease and perished. The live but harmless bacteria had somehow managed to get together with the virulent form to produce a lethal hybrid. The nature of this transformation was discovered about 15 years later by Oswald Avery, Colin MacLeod and Maclin McCarty. They repeated the Griffith experiment with various parts of the S type removed. Prior removal of the polysaccharide coat had no effect, and neither did removal of the protein capsule. But if the DNA was either removed or denatured, the lethal activity disappeared. It was furthermore demonstrated that this activity could be disrupted by the enzyme deoxyribonuclease, which breaks down DNA. Thus it became apparent that the transformation in Griffith's experiments involved penetration of some of the lethal DNA into the harmless variety, which then caused a genetic transformation of the latter, turning it into a dangerous copy of the S type. The chemical identity of Mendel's hereditary units had been established; genes are composed of the biopolymer DNA.

Bacteria are fairly complex systems, and skepticism persisted regarding the result of Avery and his colleagues. Then in 1952 came evidence from a system which is so simple that all doubts were dispelled. As we saw earlier,

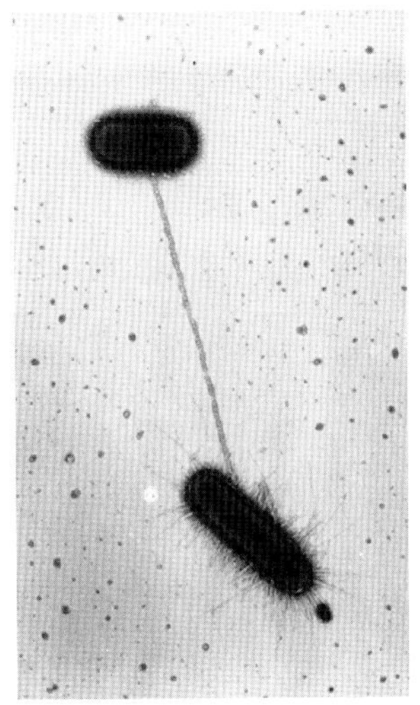

Bacteria, such as the *Escherichia coli* specimens shown in this transmission electron photomicrograph, do not reproduce sexually; they multiply by simple cell division. Genetic variation is nevertheless made possible by the process of *conjugation*, in which DNA is passed from one individual to another. Such an event is revealed in this picture. The cell seen left, covered with numerous appendages called Type 2 (or somatic) pili, is the donor. Contact with the appendage-free recipient cell is mediated by the structure known as an F pilus. The formation of this bridging tube is controlled by genes contained in a separate infectious genetic element: the F plasmid. The numerous small bodies seen surrounding the F pilus are special virus particles which use this connecting passage to infect the donor. The Type 2 pili play no role in *conjugation*, but they are required by the bacterium in its colonization of the intestine, its normal habitat. They also confer on *Escherichia coli* the ability to cause diarrhoea, and vaccines against this malady have been prepared from these pili.

viruses are nothing more than DNA, or sometimes RNA, surrounded by a coat of protein. Alfred Hershey and Martha Chase used viruses that infect bacteria. These are known as *bacteriophages*, or simply *phages*, the phages in question being T2 and T4 which attack a particular variety of bacteria, the colon bacillus, *Escherichia coli*. These viruses survive and multiply by taking over the genetic apparatus of their victims, forcing them to mass-produce new viruses. The goal was to establish which part of the virus did the actual damage, protein or DNA. The protein in these viruses contains sulphur but no phosphorus, whereas the DNA contains phosphorus and no sulphur. Hershey and Chase therefore introduced radioactive phosphorus and radioactive sulphur into their viruses by growing the latter on bacterial cultures containing these labelled elements. When the viruses were turned loose on a fresh supply of bacteria, it was easy to keep track of the two components and it became clear that it is the DNA that causes the hereditary takeover. It is not difficult to see how the chemical nature of DNA qualifies it for the hereditary role. We have seen how the two strands of this biopolymer can split apart, and how a new complementary strand can be grown on each of the originals, which act as templates. This is the process underlying the sudden appearance of chromatid pairs during late interphase.

Maintenance of the diploid number of chromosomes is accomplished by production of gamete cells having only the haploid number. When two of these fuse together, in sexual reproduction, the resulting zygote cell thus has the full diploid complement. This process presents no problem for multicellular organisms. The desired specialization of some cells to form the oocytes that are capable of meiosis is achieved by *tissue differentiation*. But this leaves the question of genetic variation in *prokaryotes*, which increase their numbers simply by cell division and growth. Unlike their eucaryotic superiors, they do not have a nucleus, and their genetic material is distributed in the cytoplasm as filaments of DNA. An auspicious start on the question of reproduction in prokaryotes was made in 1953 by Salvador Luria and Max Delbrück. Through observations on virus-resistant strains of *E. coli*, they demonstrated that the chief agency of hereditary change in these lowly creatures is mutation, just as in higher organisms. Then a few years later, Joshua Lederberg and Edward Tatum discovered just how genetic variation in bacteria is accomplished. It had long been suspected that even in these unicellular organisms some sort of mating, or *conjugation* as it is called in this case, must occur. It was also appreciated, however, that conjugation must be a relatively rare occurrence. If carried out directly, by using an optical microscope, a search for conjugating bacteria would be like looking for a needle in a haystack. They therefore used a genetic test, employing various strains of E. coli, billions of which are present in the colon of every human. They ingeniously arranged their experiments in such a way that only *recombinants*, that is cells having gene combinations clearly differing from those of either of the initial strains, could survive in their cultures. Conjugation in prokaryotes was established, and it became clear that at least some of the hereditary variation in these organisms is accomplished by fusion of the DNA strands of two individuals, and interchange of genes by a process resembling crossing over during the meiotic cycle in eucaryotes.

It might seem that all life's secrets now lay exposed. The nature of genes, and what they control, was clear. This included the mechanism whereby genetic instructions stored in DNA are passed on, through the various types of RNA, to produce proteins. And, through determinations of their three-dimensional structures, it had even been revealed how functional proteins carry out their enzymatic activities. It could appear that there remained only pedestrian detail. Nothing would be farther from the truth, however, for we have come to the mightiest problem of them all: *developmental biology*. In multicellular organisms, what is it that causes cellular differentiation? It is paradoxical that our knowledge had been so full at the microscopic level, beyond the range of unaided human vision, and yet so incomplete concerning what is manifestly apparent to the naked eye: that cells develop in different ways so as to produce hand and heart, and stem and seed. In spite of this diversification, every somatic cell appears to carry in its nucleus the same chromosomes, consisting of the same genes, which are composed of DNA having precisely the same sequence of base pairs. Something must dictate that only part of the information is translated, allowing different cells to produce the various tissues, organs, and systems of the animal or plant. And the differentiating mechanism was doubly mysterious because it did not appear consistent in its demands for precision; the fingers of a hand and petals of a flower are subject to rigid numerical constraints, while the alveoli of a lung and branches of a tree are merely required to follow the correct general arrangement. In the following chapter, therefore, we must turn to the important issues of *cellular differentiation* and *tissue differentiation*, and their relevance for the overall structure of an organism.

Summary

The cell is the fundamental unit of all living organisms. It consists of a flexible membranous bag, of limited permeability, containing an aqueous cytoplasmic solution, in which are suspended a variety of organelles. The latter, also membrane-bounded, carry out the processes essential to the cell's survival. In the eucaryotic cells of higher organisms, the nucleus contains the genetic message, encoded in deoxyribonucleic acid, or DNA, which directs protein formation. This is catalyzed by the ribosomes and occurs at the endoplasmic reticulum. The Golgi apparatus is responsible for the encapsulation of fabricated protein and its transport to other regions of the cell. Respiration, or the extraction of energy by the breaking down of food, particularly glucose, is carried out in the mitochondria, the product being energy-rich molecules of adenosine triphosphate, or ATP.

Lysosomes, present in large numbers in the cytoplasm, remove dead material through their liberation of enzymes. Rod-like aggregates of proteins with well-defined morphology, the microtubules and microfilaments, mediate both cell locomotion and cell division. Although differing in composition, the cytoplasmic and organelle membranes are all basically a fluid-like phospholipid bilayer containing proteins and steroids such as cholesterol. The surface of the cytoplasmic membrane also has carbohydrate-containing substances, glycolipids and glycoproteins, which participate in intercellular recognition and the binding of regulatory molecules such as hormones. There are two modes of cell division in eucaryotic organisms. Mitosis produces new somatic cells, thereby causing the body to grow, whereas meiosis produces the germ cells used in reproduction. The process of crossing over during the latter mode leads to the exchange of genes between homologous chromosomes, this, coupled with the occasional mutation of a gene, underlies genetic variability. Variability is also seen in procaryotic organisms, being achieved by a simpler, more direct route.

17

The evolution of the horse was certainly a most tortuous process. None of your seven day nonsense! Seven days' labour wouldn't evolve one primitive earthworm.

John Mortimer (b. 1923)
A Voyage Round my Father

Mortal coils: *the organism*

There are interesting parallels between the structuring of multicellular organisms and societies. We see the same elements of aggregation, communication, commitment and differentiation, and although some regret the decline of the protean ideal, specialization was a minor penalty to pay for the development of modern society. After all, more people have flown in aeroplanes than ever held a pilot's licence, and more have enjoyed the benefits of surgery than ever wielded a scalpel. And how many of us would be meat eaters if we had to do our own butchering? There are even similarities in the way societies and multicellular organisms subordinate the individual to the point of dispensability. A nation survives the death of any citizen, however prominent, and a mature organism hardly seems to notice replacement of its individual cells. Moreover, it now appears that the collective structure has a vested interest in the mortality of its members; the multicellular organism follows a policy of *programmed death*, and a steady turnover increases the chance of a favourable mutation. This too might have its social counterpart; what happens to individual members after they have procreated is of minor importance, and old worn-out citizens are replaced by fresh young individuals.

When considering the question of *developmental biology*, it is important to bear in mind the ideas embodied in the *Darwin–Wallace theory*. *Mutations* are usually detrimental, but they occasionally help a species in its fundamental aims of survival and reproduction. The important point here is the element of chance; organisms and their component structures are not designed to fulfill a predetermined purpose – they are the product of a series of random modifications. We do not have eyes because of an anticipated need to read; we can read because we have eyes. The same is true at the cellular level. It would be difficult to conceive of a more pragmatic entity than the cell: it does precisely what it is capable of doing under the prevailing conditions, no more and no less.

A key issue in developmental biology – earlier referred to as *embryology* – was identified by Aristotle in the fourth century BC. Speculating about how the different features of an embryo are formed, he realized that there are two contrasting possibilities: either every feature is present at the very outset, and merely grows as the organism grows, or the generation of different features is triggered at various stages during growth. Aristotle had thus perceived a philosophical schism – *preformation* contra *epigenesis* – and he opted firmly for the latter. Preformation was still enjoying much support as recently as the end of the nineteenth century, one view maintaining that a miniscule but complete human being – known as a *homunculus* – resides within the

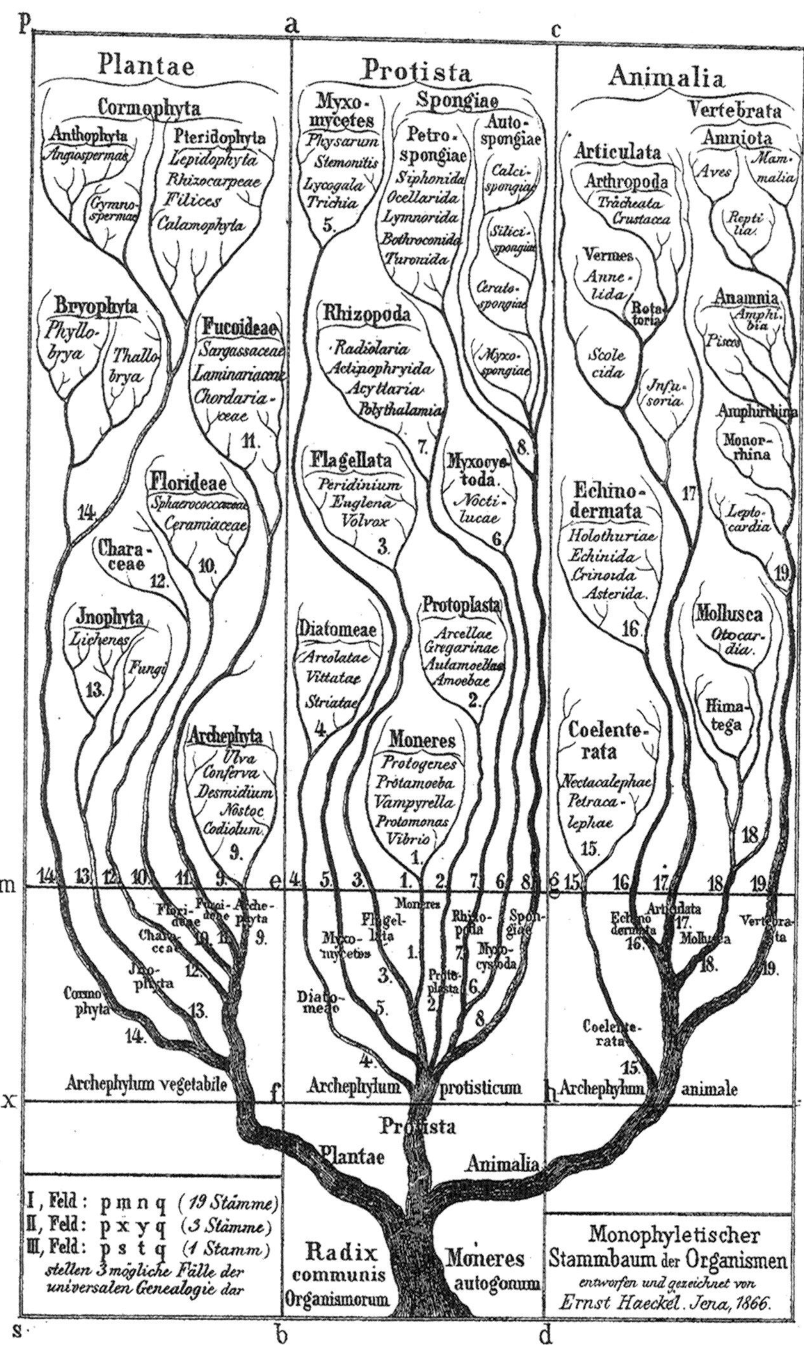

Although Ernst Haeckel (1834–1919) made some bad errors, such as his maintaining that 'ontogeny [the course of development during an individual's lifetime] recapitulates phylogeny [evolutionary history]', and his statement that 'politics is applied biology', many of his more sober efforts deserve recognition. He was one of the first to support Darwin's evolutionary ideas, for example, and it was he who coined such terms as *ecology*, *phylum* and *phylogeny*. But he is best known as the person who made the first attempts to place the various species in a scheme representing their evolutionary relationships – what would now be called an evolutionary tree. One of these commendable efforts, from 1866, is reproduced here, and it is immediately clear that Haeckel was strongly influenced by the current ubiquity of plants and animals; they claim two of his tree's three main branches. But he was close to the truth in regarding bacteria (Moneres) as being the original form of life, and in his perceiving that unicellular eucaryotic organisms (Protista) represent a separate series of developments, and that they did not serve merely as progenitors of multicellular life.

head of every human sperm. Optical microscopy had exposed the existence of cells, as noted in the previous chapter, but there were doubts as to whether the resolving power of these instruments was sufficient to reveal the tiniest details. Such reticence was prudent; even modern electron microscopes are hard pressed when aiming to resolve atomic arrangements in the molecules that dictate inheritance. In a sense, adherents of the preformation doctrine focussed on the wrong level of structure; the genetic material is indeed present and complete the moment egg–sperm fusion produces the *zygote*, but the various tissues develop only subsequently. And the manner in which genes direct the structure and time of generation of those tissues is rather complicated, because of the interplay between a number of factors. The process is nevertheless deterministic.

We now know that *tissue differentiation* is caused by four distinct processes at the cellular level: *cell multiplication, cell movement, change of cell character* and *cell–cell signalling*. And we also know which contributing cellular mechanisms are controlled by the genes, namely their spatial organization, their change in form, their differentiation, and their growth. It is this multiplicity of processes and factors that inevitably leads to a complex scenario. It is not possible to study the arrangement of tissues in the mature organism and draw reliable conclusions about the rules governing their formation. Such induction by working backwards – the *top-down* approach – fails. One must observe the successive stages of development experimentally, and construct a *bottom-up* picture of events. Only then can one appreciate the remarkable fact that complexity can be the result of basic processes that are relatively simple. Lewis Wolpert has drawn comparisons with the Japanese paper-folding art *origami*, in which an infinity of different patterns can be generated from a few simple folding rules, any specific pattern being determined by which rule is applied at which stage in the sequence. The human and the chimpanzee are composed of the same cell types, but the latter's numbers and spatial arrangement differ in the two species. Let us now consider the underlying processes in greater detail.

A fascinating example of *differentiation* is seen in the slime mould *Dictyostelium discoideum*. This feeds on bacteria, and when supplies are abundant, individual cells can survive independently, moving about and hunting their prey. When the bacteria are in short supply, or disappear completely, some of the *Dictyostelium* cells secrete a cyclic form of AMP, and this provokes the remaining cells to migrate toward those emitting members of the colony. When they reach the *cyclic-AMP* source, the cells cluster to form a single slug-like creature known as a *pseudoplasmodium*, which can crawl around in search of nourishment. Differentiation sets in, and individual cells specialize, some remaining as base material while others contribute to the formation of a *stalk*. Cells at the end of the stalk are further differentiated into a *fruiting structure* which, when ripe, sheds *spores*. These hatch to become new slime mould amoebae. Although the cells of higher organisms never enjoy such a period of independent existence, they display the same general behaviour. In a tree, for example, the majority of cells in the trunk are dead, but they still play a vital role in supporting the entire organism, enabling the branches to reach out and permit seeds to fall sufficiently far from the roots that they have a chance of survival. Similarly, and notwithstanding their apparent sophistication, we could say that the diverse activities of differentiated tissues in higher animals are directed towards a single goal: fusion of one of the organism's *gamete cells* with the gamete of another individual of the species. The underlying principle is sacrifice by the many for the ultimate benefit of the potent few.

Differentiation obviously has a spatial dimension: cells in different parts of an organism develop in different ways. In a typical mammal, for example, this specialization produces about 200 different types of tissue. There is also a less obvious temporal dimension. Proper development of an organism requires that certain cells must die at the correct time. This commonly occurs in the modelling of the limbs. Mammalian fingers and toes are joined during early development, but subsequently separate when the connecting

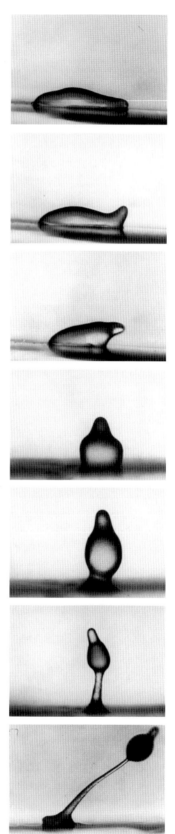

A remarkable example of *cellular differentiation* is observed in the humble slime mould *Dictyostelium discoideum.* The amoebae of this species are formed from spores, and they replicate and hunt bacterial food as free individuals. When threatened with starvation, however, they seem to sacrifice their independent existence for the common goal of survival. Moving along the concentration gradient of the chemical formerly known as *acrasin*, which was identified as cyclic-AMP (adenosine 3,5-monophosphate) in 1967, the amoebae aggregate around a few of their number which initiate emission of this attractant. They thereby form a multicellular slug-like creature which crawls about, forms a stalk, and ultimately a *fruiting body* that sheds new *spores* to complete the life cycle. Cellular differentiation in this process appears to be based on time of arrival at the aggregate, known as a *pseudoplasmodium*, and the cells can revert to the amoeboid state if they are removed from the mass at any stage prior to formation of the mature fruiting body. The two sequences of pictures show successive stages of the life cycle as viewed from the side (right) and from above (left).

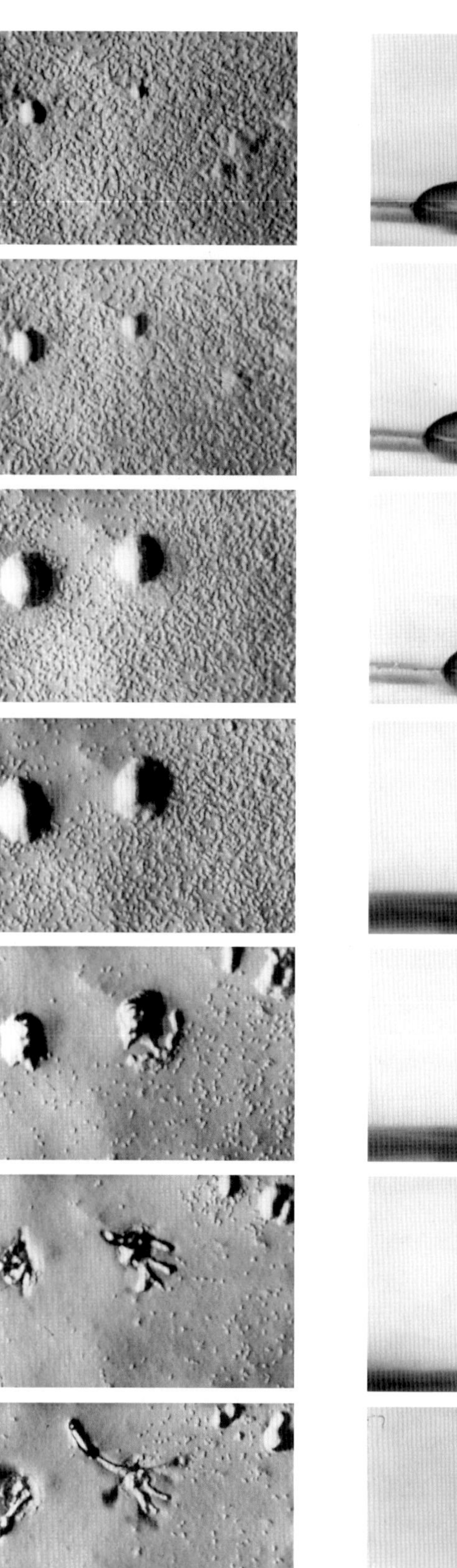

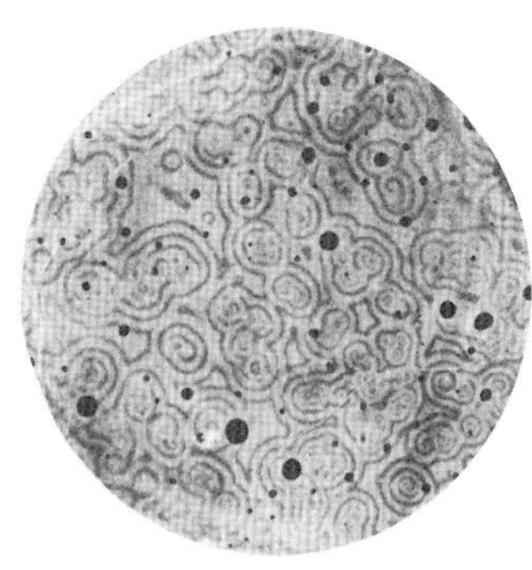

The *cyclic-AMP* that provokes aggregation of *Dictyostelium discoideum* has been found to be emitted in waves. At the onset of the consolidation process the cells become segregated into territories, and cells within each territory are attracted towards its centre, moving in the direction of increasing cyclic-AMP concentration. A cell also appears to respond to this compound by releasing additional cyclic-AMP, thereby relaying the signal. This picture, termed a fluorographic image, detects cyclic-AMP that has been radioactively labelled with the tritium isotope of hydrogen (i.e. ^{3}H). The observed wave pattern bears a striking resemblance to that produced by the Zhabotinsky–Zaikin reaction illustrated in Chapter 15.

tissue disappears. We can gain insight into the nature of the differentiation problem by considering the various factors which control cell shape. To begin with there is the physical state of the bounding membrane, which consists predominantly of lipids, steroids such as cholesterol, proteins and – on the external surface – sugars. The proteins display great variety, a single cell perhaps containing about 1000 different types. Other proteins influence shape from the cell interior, by aggregating to form *microtubules* and *microfilaments* which then prod the membranes at their inner surfaces.

The genetic instructions in the various cells could be expressed differentially, not all the protein blueprints being used at a given time. There must be a mechanism whereby genes are switched on and off. Some proteins, such as those involved in energy metabolism, are ubiquitous, and the corresponding genes must always be active. Other genes, producing proteins in highly differentiated cells, are switched on only in special regions of the organism. It seems reasonable to expect that proteins govern cellular differentiation. Chromosomes contain both DNA and proteins belonging to the special class known as *histones*. By the mid 1960s, Alfred Mirsky and his colleagues had established that gene activity can be altered by making chemical changes in the histone molecules. We have already seen that substitution of a single amino-acid residue can change the folded-up, or tertiary, structure of a protein, and have a profound influence on the way it functions. More recent research has probed the three-dimensional configuration of the histone–DNA complex. The histone units are arranged in the nucleosome groups described in Chapter 15, and the *nucleosomes* form a chain which is coiled up in a helical conformation. The DNA double helix is wrapped around the circumferences of the histones in a continuous fashion, passing from one of these multiple proteins to the next in the chain. The DNA superstructure is thus a helical configuration of helically arranged helices.

Just how this determines which genes are transcribed to the *messenger-RNA* is now becoming clear, because some regions of the DNA must remain hidden by such coiling. Other regions are prevented from passing on genetic information by proteins bound to the DNA. Insight has also been gained in studies of giant chromosomes, such as those in the salivary glands of the larvae of *Drosophila melanogaster*, the common *fruit-fly*. These special thickened structures are formed by the parallel arrangement of numerous identical chromatids, this being the result of repeated replication without cell division, so as to produce what are known as *polytene chromosomes*. Chromosomes are crossed by alternating light and dark bands. The latter contain nearly all the DNA and histone, and are in fact the genes. Some chromosomes contain sections that are blown out into what are known as *puffs*. Wolfgang Beermann found that the location of these puffs varies from one tissue to another. Together with Ulrich Clever, he also demonstrated that the puffs are not only rich in DNA and histone, but also contain a high concentration of RNA. Moreover, very little RNA is found in regions of the chromosomes other than the puffs. Finally, together with Jan-Erik Edström, Beermann and Clever established that puffs are genes actually in the process of producing messenger-RNA.

The mechanics of gene transcription are thus becoming clear, but this still leaves the question of control. What is the origin of the signals that command the histones to arrange themselves in a certain configuration, exposing a particular set of genes in a differentiating cell? It seems reasonable to suppose that such signals come from outside the cell. This implies diffusion of controlling substances through the intercellular regions. The concept of external chemical control is not difficult to accept. A dramatic, and tragic, example was seen with the drug *thalidomide*. Originally used as a tranquilizer, it was frequently prescribed to pregnant women, until it was found to cause deformities. These often took the form of seal limbs, in which the upper and lower arms were so stunted that the hands emerged directly from the shoulders. The existence of signal substances received strong endorsement from the discovery of organic compounds known as *hormones*. In animals, these are produced in minute quantities in the *endocrine glands* and transported in the blood stream to other locations, where they have a profound influence on cell function. If development does involve transmission and reception of extracellular substances, it appears that there must be both start and stop signals. Indeed, there must be something which tells the entire organism when to stop growing. Replacement of cells in tissues that can be regenerated, such as in the healing of a cut finger, must involve a mechanism which ensures that cell renewal terminates precisely when the job is completed; just the right number of new cells appear, restoring the tissue to its correct numerical strength.

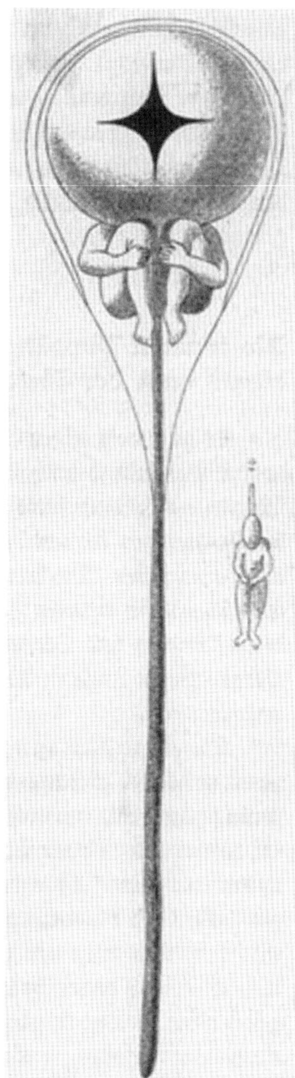

Some of those who believed that the entire organism was present at birth in microscopic form, and that subsequent development merely involved growth in size, assumed that a miniature person – a so-called homunculus – was present in the head of each sperm. The female body was perceived as providing suitably nourishing surroundings for the early stages of this development, but not as contributing anything by way of actual tissue. Nicolas Hartsoeker (1656–1725) captured this extraordinary conjecture in a drawing published in 1694.

Finding ways in which development of individual cells can be externally controlled presents no problem. One factor is the actual transport of the signal molecules, referred to as *morphogens*. In travelling from source to target, these molecules either diffuse or are circulated in the blood stream, and can use a combination of both mechanisms. The larger the organism, the greater are the number of target cells, and the longer is the transportation distance. A signal substance might have no influence on cells more remote than some critical distance, because it has become too thinned out, and thus fallen below a certain threshold concentration. These ideas support the suggestion, originally made by Hans Driesch just before 1900, that differentiating cells respond to some form of *positional information*. This concept received support from independent studies of the colouring patterns of butterfly wings, by B. N. Schwanwitsch and Fritz Süffert in the 1920s. More specifically, the positional information was envisaged as a *chemical gradient*, by C. M. Child around the same time. In the 1970s, Lewis Wolpert suggested that differentiation through variations of morphogen concentration could be a universal mechanism, and he emphasized the role of a superimposed time factor; development is also determined by the length of time a cell spends in a particular zone. Wolpert compared the basic situation to the French flag, in which a slight change in position can cause a radical change in colour if a boundary is traversed. The intercellular spaces, through which the signal molecules must ultimately diffuse, have shapes and transmission characteristics which are determined by the protein components of the local cell membranes. If these proteins are produced by a mechanism which is under the control of the signal substances, this could lead to *self-regulation*. Edward Tangl claimed, in the nineteenth century, to have observed regions of contact between the cytoplasm of adjacent cells. Firm evidence of such local continuity was obtained in 1959 by Edwin Furshpan and David Potter, who detected electrical transmission between nerve cells of a crayfish.

Yoshinobou Kanno and Werner Loewenstein later showed that intercellular continuity is a general phenomenon, not limited to the nervous system. The structure of these *gap junctions*, as they are called, is becoming clear. Donald Caspar and David Goodenough found them to consist of a protein

During embryonic development, the precursor of the brain and spinal cord appears as a layer of cells called the neural plate. It is visible as the dark line in this picture taken by John Gurdon of the one-day-old embryo of an African clawed frog, *Xenopus*. The roughly one millimetre diameter embryo contains about 60 000 cells at this stage.

cylinder with inner and outer diameters of about 2 nm and 8 nm respectively. These cylinders just span the *lipid bilayer*, and intercellular continuity is established when cylinders in adjacent cell membranes become aligned. Nigel Unwin and Guido Zampighi showed each cylinder to be a hexagonal arrangement of six proteins, each comprising about 250 amino-acid residues. The hollow centre of the hexagon contains only water, not lipid, and will thus permit transmission of molecules whose diameter does not exceed 2 nm. Malcolm Finbow, John Pitts and James Simms demonstrated that gap junctions can pass nucleotides, sugar phosphates, amino acids, various enzymatic cofactors and other small molecules, whereas RNA and proteins are selectively stopped. This is just what we should expect; if the hereditary messenger molecules were free to travel from one cell to another, the multicellular state would lose its *raison d'être*.

It is important to bear in mind that the multicellular state is not a prerequisite for structural differentiation. The tropical seaweed *Acetabularia* is a single-celled alga which reaches a height of several centimetres. It has a well-defined stalk, a funnel-like cap, and an anchoring organ, known as the *rhizoid*. The single nucleus is found in the latter. We have also seen that the single-celled *Paramecium* possesses several specialized structures. The concept of positional information seems just as valid for these primitive forms of life. Even for organisms comprising many cells, the nature of the chemical signals that cause differentiation, and the location of their origins, are becoming clear. When they have developed, the mammalian endocrine glands exercise control over the growth processes. This becomes apparent when a gland malfunctions. In *Cushing's disease*, which is traceable to a faulty adrenal gland, the limbs are thinner than normal, while the abdomen is distended. Insufficient activity of the *thyroid gland* produces the condition known as *myxoedema*, in which the face becomes puffy and the limbs swollen. This cannot be the whole story, however, because differentiation has already set in well before these glands have finished developing. We

Homeobox genes

As has been discussed earlier in this book, genes code for the composition and structure of proteins, and when the proteins in question have evolved to interact with the genetic material itself, the corresponding genes are able to direct the overall development of the host organism. This is the case with *homeobox genes*, which all contain a roughly 180-base stretch known as a *homeobox*. The latter is therefore able to code for a stretch of protein about 60 amino-acid residues in length, this being referred to as a *homeodomain*. The homeodomain has a helix-turn-helix motif that is typical of protein segments that bind to DNA, and thereby regulate *transcription* (see the molecular diagram overleaf, in which the helices are drawn in the labour-saving fashion described in Chapter 15). Homeobox genes thus encode *transcription factors*, and the surprising fact is that a single one of these is able to switch on a whole cascade of other genes, and thereby direct the formation of a large portion of the organism, such as an entire limb. A mutation in a homeobox gene therefore has devastating potential for producing a malformed body part. This is referred to as *homeotic transformation*, or simply *homeosis*. Let us take the fruit fly *Drosophila* as an example. As with all insects, this fly's body is composed of three major parts: head, thorax and abdomen. The thorax consists of three segments, T1, T2 and T3, each of which includes a pair of legs, so insects have six legs in all. In the majority of insect orders, both T2 and T3 also include a pair of wings, a good example being the four-winged honeybee. But in flies, only the T2 segment is equipped with wings, the T3 segment having instead a pair of balancing organs known as *halteres*. The absence of wings on T3 is due to the suppressing influence of a gene called *Ultrabithorax* and in the 1970s, Edward Lewis introduced a double mutation into this gene and, in effect, suppressed the suppression. The upshot was a fruit fly with two pairs of wings, instead of the normal one pair

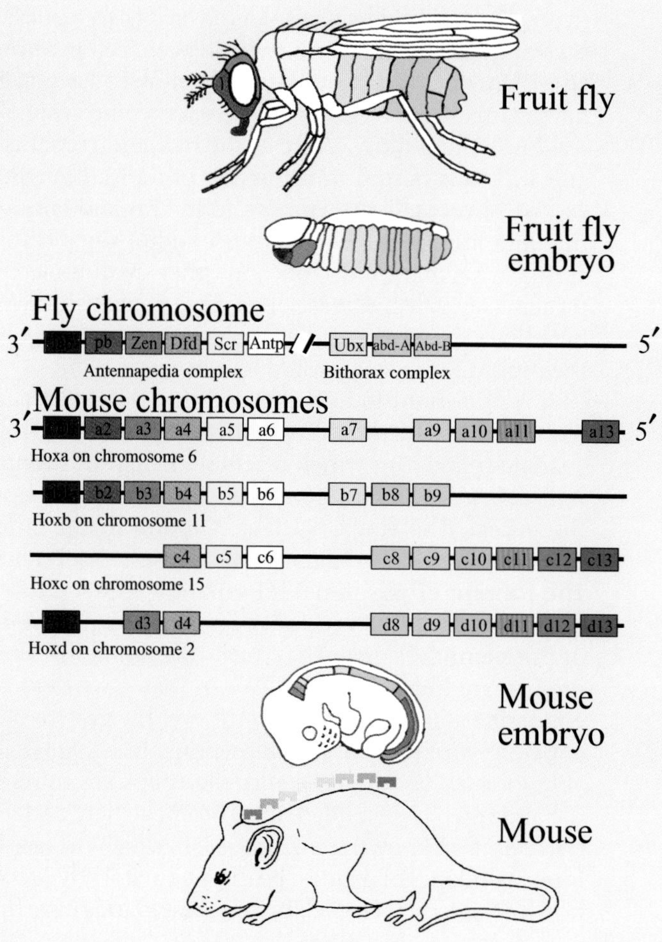

(see the photograph). A similar effect of mutation to another gene, known as *Antennapedia*, produces an even weirder fly with a pair of legs growing out of its head, where its antennae are usually located. It is now known that homeotic genes come in clusters (see the remaining diagram), the original cluster probably having been formed by gene duplication and divergence. Genes produced in this manner are known as *paralogs*, and the corresponding genes in the different clusters are known as a *paralogous subgroup*. It is noteworthy that the genes in a cluster are arranged in the same order on the chromosome as are the body segments that they influence (see the diagram). There is strong evidence that the homeobox genes have been conserved during evolution; this is hardly surprising, given the central role that they play. Thus the mouse has such genes on four of its chromosomes (see the diagram) they being referred to as *Hox genes* (Hoxa, Hoxb, Hoxc and Hoxd) in this case. And in the human, analogously, there are the HOXA, HOXB, HOXC and HOXD clusters.

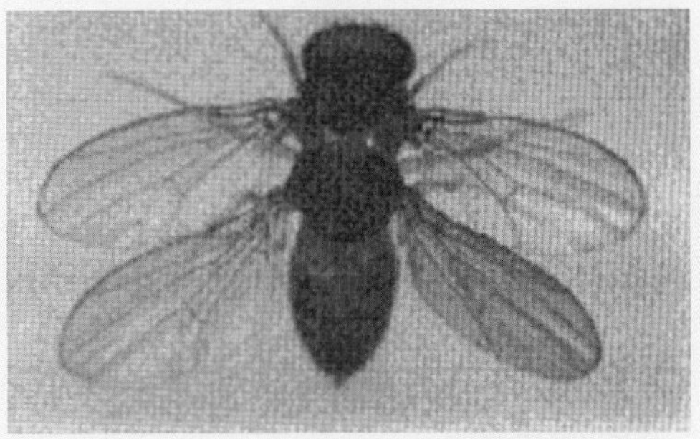

must therefore consider the multicellular organism during the early stages of its development: *embryogenesis*.

An early success, already implicit in what we have been discussing, was verification of August Weismann's thesis that inherited characteristics stem exclusively from the parents' *germ cells* – an egg from the mother and a sperm from the father – and not from the *somatic cells* that comprise the remainder of the body. Weismann further postulated that subsequent cell division would be asymmetric, the two cells produced by fission of a single cell displaying mutual differences. Wilhelm Roux destroyed one of the two cells produced by the first cleavage of a frog zygote, and the surviving cell developed into a half-embryo. It must be borne in mind, however, that Roux had left the destroyed cell *in place*. When the zygote has subdivided only a couple of times, all cells, known as *blastomeres* at this early stage, are probably of a single type. Studies of the developing sea-urchin, *Echenius esculentus*, for example, confirm this. In 1892, Hans Driesch *separated* the embryo at the four-blastomere stage and found that each was capable of developing into a complete sea-urchin. A chance event of this type gives rise to human *identical twins*. The difference between this result and the

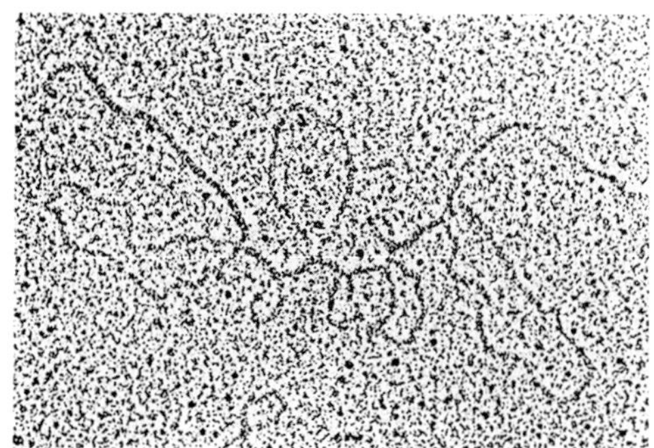

One of the most significant recent developments in biological research has been the discovery that the genetic message is often subject to editing prior to translation. The hereditary mechanism is thus more complicated and dynamic than had previously been suspected. The organization of an individual gene in a higher organism is fundamentally different from that of a gene in a bacterium. The gene of a higher eucaryote is discontinuous; segments which will ultimately be translated are interspersed between apparently dispensible segments that are eliminated during a post-transcription editing process. These protein-coding and non-coding sequences are now called *exons* and *introns*, respectively, and examples of split-gene organization are seen in the electron photomicrograph. A single strand of DNA incorporating the gene for the egg-white protein ovalbumin has been allowed to hybridize with the single strand of ovalbumin messenger-RNA, the molecule from which the protein is translated. The complementary base pairs of the corresponding exons in the two molecules can combine and produce a double strand, but this leaves the introns in the DNA unpaired so these single-strand segments appear as the loops which are visible in the picture. The introns vary in length from 251 base pairs to about 1600.

one obtained by Roux had revealed the fact of *regulation*: the capacity of an embryo for normal development despite early removal of some of its cells.

Driesch found, however, that the third division, or *cleavage* as it is called, produces differentiation. Each blastomere produces two unlike cells, so the eight-blastomere embryo is divided into two regions: a light-coloured group of four cells called the *vegetal hemisphere*, and a darker group known as the *animal hemisphere*. Cells in the latter region are destined to become the *ectoderm*, or outer part of the embryo, while the vegetal hemisphere will produce the *endoderm*. The ectoderm forms the *skin, nervous system* and *sense organs*, while the endoderm gives rise to the *alimentary system*. At the next cleavage, a small region near the pole of the vegetal hemisphere is composed of the very small cells known as *micromeres*, which initiate development of the *mesoderm*. This develops into *muscle, blood, sex glands*, and other internal structures. After a few more divisions, a small hollow ball of cells, the *blastula*, is formed, and the cells at its inside are already distinct from those on the outer surface.

Further growth leads to relative movement of differentiated cells and a cleft forms, producing a shape rather like that of a cowrie shell. This is the *gastrula*, in which the different types of cell are essentially in the relative positions that they will occupy in the final organism. The movement is known as *gastrulation*, and it essentially turns the blastula inside out, through an opening known as the *blastopore*. In many mammals, the latter subsequently develops into the *anus*. Gastrulation brings the various types of cell into transitory contact, and experiments performed by Hans Spemann and Hilde Mangold in the 1920s established that they exert the mutual influence on each other's subsequent development known as *induction*. The outcome seems to be that the various cells become committed to a particular mode of development, even though they appear indistinguishable at this stage. Comparing the embryos of different mammals at similar fractions of their *gestation periods*, we find that they look quite similar. A chicken embryo three days after conception, for example, resembles a human embryo at the three-week stage. As the gestation period proceeds, however, the differences between embryos of different species become more pronounced, and a human *foetus* at fifteen weeks is clearly recognizable as an incipient person.

As we have already seen, the differentiation process often involves cell movement. Some primitive organisms are capable of reassembling themselves after they have been broken up, by squeezing them through a sieve. Just after 1900, H. V. Wilson demonstrated that *hydra* and *sponges* can accomplish this even if cells originating from different parts of the organism have

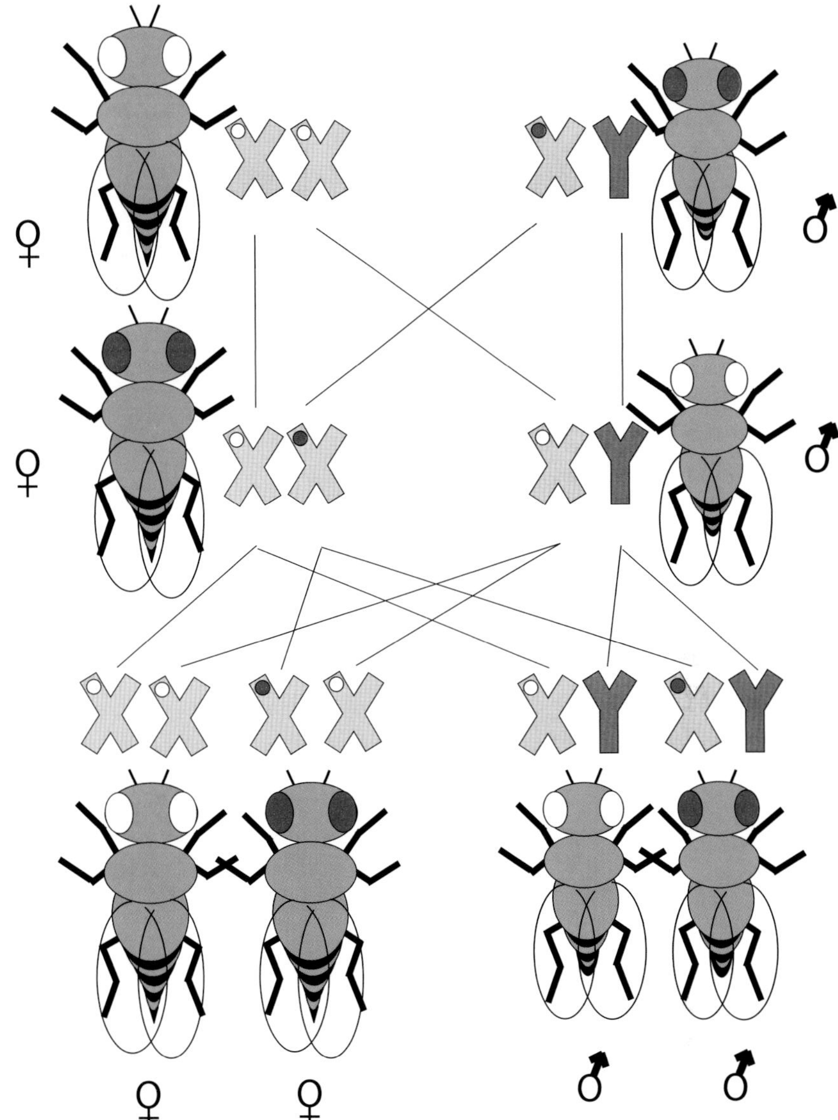

The diagram in the previous chapter illustrating Mendel's classic experiment with peas, through which he discovered the underlying principles of genetics, embodied his assumed *law of independent assortment*. As was explained in that chapter, he was fortunate that the characteristics he chose to study did in fact obey that law; the shapes and colours of his peas were dictated by genes lying in different chromosomes. When characteristics are governed by genes that happen to lie close to each other on the same chromosome, there is a high probability that these genes will remain close to one another despite the crossing over that occurs during meiosis. And if a gene is linked to one of the sex chromosomes, in which crossing over does not occur, the linking of the characteristics always survives meiosis. This is true of the gene that determines eye colour in the fruit fly *Drosophila*, as shown here. There are two variants of that gene, only one of which – the dominant variant – produces the red coloration. The gene lies on the X chromosome, so the red eye coloration can appear in flies of either sex, as indicated in this highly schematic diagram.

been thoroughly mixed. The sorting process involves relative movement between cells of different types, and aggregation of similar cells when they find each other. Observed under a microscope, the cells are seen to have numerous finger-like projections, the rhythmic motion of which propels the cell along. The clustering process must involve cell–cell recognition, which will be discussed later. Cellular movement induced by a gradient of the chemical concentration of a stimulating substance is known as *chemotaxis*. In one of its modes, this process is the result of changes in the *cytoplasmic membrane*. The provoking chemical generates a weak spot on the latter, possibly through disruption of the underlying microfilaments, and the membrane bulges outward into what is called a *pseudopod*, or false foot. The cytoplasm flows into this, producing directed movement. In a gradient of the stimulating compound, pseudopods – also known as *filopodia* – will preferentially develop either toward or away from the region of high concentration, biasing the cell into a well-defined direction of

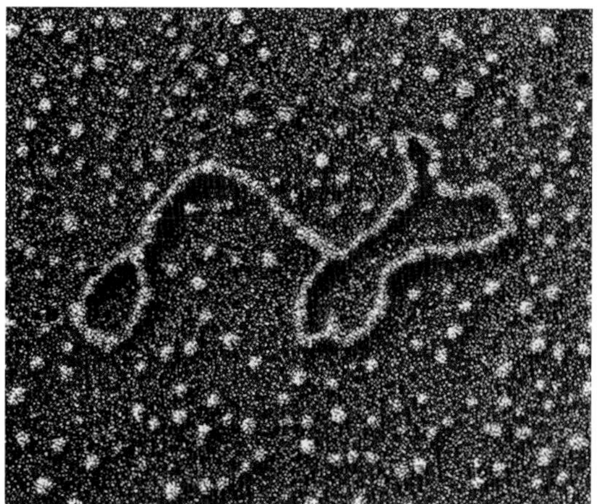

Until recently, it was believed that *genetic variation* in higher organisms arises solely from the combined effects of *mutation* and the reshuffling of chromosomes during *sexual reproduction*. During the late 1940s, following numerous studies of the variegated seed pigmentation of specimens of the common corn plant *Zea mays* (right) that had undergone chromosome breakage, Barbara McClintock (1902–1994) concluded that certain genetic elements can be moved from one chromosomal region to another, and thus provide another source of diversity. Such *transposable genetic elements* have since been discovered, and are referred to as transposons. An example is shown in the electron photomicrograph (left): a transposon has been inserted into the double-strand DNA of the plasmid pSC105. The composite was then denatured and the complementary bases were subsequently permitted to recombine. Because the *transposon* has inverted-repeat termini, this produces single-stranded terminal loops joined by a double-stranded segment.

chemotactic travel. Other forms of chemotaxis appear to involve relative movement of various components of the cytoskeleton. Interestingly enough, two of the implicated biopolymers are *actin* and *myosin*, both of which are known to be involved in muscle function.

The cell surface plays a vital role both in the interactions between cells and in their response to substances in the cellular environment. We have already noted the importance of gap junctions in cellular communication, and we now turn to the more general aspects of the question, highlighting three types of membrane component: *glycolipids*, *glycoproteins* and the proteins known as *immunoglobins*. The first two, believed to participate in *intercellular recognition*, can possess combinations of as many as five sugars, including glucose, galactose and mannose. The sugars are of course water soluble, so they naturally adopt positions at the membrane–water interface. Because up to about 20 individual sugar units can be appended to the main part of the molecule, and because branching is also possible, the variety is enormous. This provides considerable scope for cell surface specificity. In 1970, S. Roseman suggested that specific adhesion between cells, vital to tissue formation, occurs because adjoining cell surfaces carry complementary structures. Of the numerous possibilities, a neat fit between a polysaccharide unit on one membrane and an enzyme such as *glycosyl transferase* on the other is particularly attractive.

There will be more to say about recognition mechanisms later, but meanwhile let us return to the developing embryo. We had left it at the stage of gastrulation, during which the basic layout of the body had been established. The inner region of the mesoderm mediates formation of the *notochord*, which has the form of a flexible rod stretching from head to tail. (We humans have only a vestigial tail, of course: the *coccyx*.) The nervous system's *brain* and *spinal cord* will later lie along the same route. Two other portions of the mesoderm form the *somites*, deployed on either side of the notochord, and it is they which will subsequently develop into the vertebrae, muscles and skin. During the next major stage, known as *neurulation*, the ectoderm folds itself into a tube – the *neural tube* – which differentiates into the components of the nervous system. (In the malformation known as *spina bifida*, the folded tube fails to close properly upon itself.) There then follows *organogenesis*, during which the somites differentiate into the above-mentioned structures, as well as into the other organs, and in some species this is followed by *metamorphosis*, the classic example being the process by which a tadpole is transformed into a frog. It involves an example of something mentioned earlier, namely the selective removal of certain cells.

The arrangement of structural features along an animal's long axis – that is to say its antero-posterior axis – has been found to be determined by a set of genes specifying positional identity. Originally discovered in the fruit fly *Drosophila*, in which they are called the *homeobox genes*, they were subsequently found to have counterparts known as *Hox genes* in mammals. The homeobox gene takes its name from a 180-base-pair stretch of the DNA sequence, which codes for a 60-amino-acid portion of DNA-binding protein known as a *transcription factor*. That portion comprises two alpha helices separated by a beta turn – a familiar DNA-binding motif. A mutation in some homeobox or Hox genes provokes a *homeotic transformation*, in which one body part is replaced by another. In *Drosophila*, for example, one such transformation produces a fly with twice the normal number of wings while another creates a weird creature with legs growing out of its head instead of antenae.

Four chromosomes in the mouse are found to include a number of Hox genes, namely chromosomes 6, 11, 15 and 2, the corresponding clusters of *homeotic genes* being collectively designated *Hoxa*, *Hoxb*, *Hoxc* and *Hoxd*. There is some similarity in the homeotic genes on different chromosomes, in which case they are referred to as *paralogs*, and are said to form a *paralogous subgroup*. There appears to be redundancy because some paralogs can act as a back-up if another member of their subgroup is non-functional. In the mouse, two subgroups involve all four chromosomes, namely the subgroup comprising Hoxa4, Hoxb4, Hoxc4 and Hoxd4, and also the one made up of Hoxa9, Hoxb9, Hoxc9 and Hoxd9. It is because of the existence of subgroups that gaps have been left in the gene designations in a cluster on a given chromosome. There are no Hoxa8 or Hoxa12, for example. (The designations are naturally numbered from the 3′-end of the DNA sequence.) The multiplicty of clusters, which nevertheless display considerable similarities, suggests that they arose during evolution through the dual processes of gene duplication and divergence.

The role of a specific homeotic gene can be studied by the *gene knockout method*, which exploits the possibilities provided by the *recombination techniques* we considered in Chapter 15. It is found that the low-numbered genes (that is to say those located at the 3′-end) are activated most anteriorly. The anterior border of the region dictated by a given homeotic gene is found to be sharply defined, whereas the gene's influence fades away more gradually toward the posterior region. It is also clear that genes located more posteriorly have the ability to turn off those lying toward the creature's head region.

It has been found that there is considerable similarity between homeotic clusters in different species of vertebrates – humans and mice, for example. Jeffrey Innis and Douglas Mortlock have discovered a mutation in the HOXA13 gene in a family with hereditary abnormalities of the thumbs and big toes, some of the female members additionally having uterine abnormalities that adversely affected their fertility. It transpired that the mutation was coding a change from the amino acid tryptophan into nonsense or a stop codon, and this produced a HOXA13 protein lacking its terminal 20 amino acids. Analogously, mutations in the mouse hoxa13 gene lead to smaller digits and, in males, malformed penises. This coupling of limb and genital abnormalities may explain why so many vertebrates have limbs

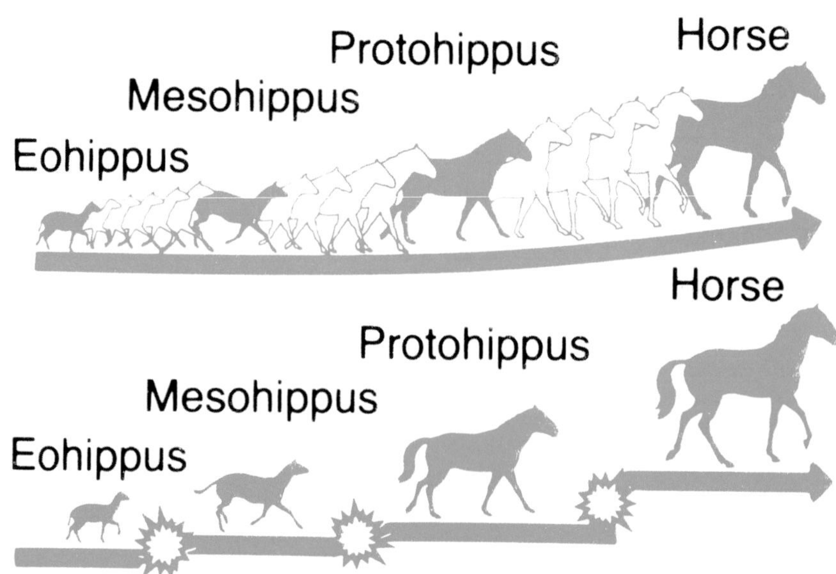

The biologist's concept of *heredity* has undergone frequent revision since the great achievements of Gregor Mendel (1822–1884) and Charles Darwin (1809–1882) around the middle of the nineteenth century. The discovery of the central role played by the *chromosomes*, the identification of DNA as the hereditary perpetuator, and the elucidation of the atomic arrangement in that molecule were all major mileposts. The demonstrations that genetic elements are movable, that certain sections of the cell's DNA are present in several copies, and that many genes are split and subject to editing, collectively amount to another significant step. They are suggestive of potentially dramatic genetic changes remaining silent in the gene pool, waiting for certain factors to provoke sudden evolutionary changes like those conjectured in the lower part of this schematic illustration.

with five digits; individuals with deviant numbers of fingers or toes may have died off because of their reduced reproductive capability.

The overall structure of an organism is dictated by certain genes, therefore, and in an animal this includes the spatial arrangement of such components as the limbs and internal organs. As will be discussed in the next chapter, the same general principles are also at work in the plant. Genes thus dictate the structure of the self. But what about the all-too-frequent invasions of self from the outside – from viruses, bacteria, fungi and parasites. Genes also dictate the repulsion of these non-self aliens, through a machinery whose complexity reflects the variety of the challenges to the organism's integrity. Collectively, the ability of plants and animals to fight infection is known as *immunity*, of course, and about 1% of human cells are involved in one or other aspect of the *immune response*.

In plants, immunity is conferred by such features as the waxy surface, which prevents entry and development of pathogens, and the thick *cuticle*, which blocks penetration by fungal germ tubes. It is also provided by protoplasmic substances that inhibit development of pathogens, and by the acquired immunity that follows recovery from an acute disease. These plant mechanisms have their counterparts in animals, which have impervious skin and antiseptic body fluids such as stomach acid. But the immune response in the animal body can additionally call upon a far more extensive arsenal of cellular and molecular weapons, including the sophisticated and very important *adaptive mechanism*. At the core of this impressive ability lies a *recognition mechanism* used by the cell in defence against alien substances. Paul Ehrlich, in 1885, suggested that resistance to external disease-generating agencies derives from certain molecules floating in the blood stream. Such an invading body is now called an *antigen*.

Ehrlich imagined the antigen as having a surface feature of specific shape, equivalent to a sort of key. The protecting molecules were postulated to have shapes complementary to these keys, so that they could function as locks. These *antibody* locks were conjectured to float in the body fluids, meet an invading antigen, and latch on to its keys so as to cancel their effect. (Plants

In the conclusion to his *The Origin of Species* (1859), Charles Darwin (1809–1882) noted that: . . . the chief cause of our natural unwillingness to admit that one species has given birth to other and distinct species is that we are always slow in admitting great changes of which we do not see the steps. And in his Chapter 6, which addresses possible difficulties for his theory, a section entitled *On the Absence or Rarity of Transitional Varieties*, he suggests that: . . . when two or more varieties have been formed in different portions of a strictly continuous area, intermediate varieties will, it is probable, at first have been formed in the intermediate zones, but they will generally have had a short duration. These opinions attest to the difficulties facing those who search for the elusive intermediates. Fossilized remains of such transient forms have nevertheless been found, an example being the eight-toed *Acanthostega* (a reconstruction of which is shown here), discovered by Jennifer Clack, and *Livoniana*, identified by Per Ahlberg. Ahlberg noted that several features of that jaw bone are intermediate to those found in certain fish and the early land-based tetrapods; it is a true missing link.

do not produce antibodies.) *Cell–cell recognition* could occur in a similar manner if each cell has a series of both locks and keys, on its bounding membrane. The Ehrlich theory has been endorsed by subsequent developments. Small foreign molecules, known as *haptens*, usually do not elicit an immune reaction, but can do if they become attached to a macromolecular carrier. The activated part of the latter is then said to become a *haptenic determinant*. Proteins, nucleic acids, and polysaccharides are prominent among the other antigens.

Before discussing the molecular-level aspects of immunity, however, we ought to consider its macroscopic features. The grossest distinction is between the immune system's *cellular* and *humoral* components, the latter being the collective designation of circulating substances in the blood and tissues. The cells in question are *white blood cells* (as distinct from the oxygen-transporting red variety we will encounter in a later chapter) and the tissue cells known as *mast cells* and *macrophages*. The white blood cell varieties are the *granulocytes*, which mainly stay in the blood, and the *lymphocytes*, which can pass into the surrounding tissue. The most important humoral substances are the antibiotic agents *lysozyme*, which is produced by macrophages and which kills bacteria by rupturing their protective exteriors, and *interferon*, which blocks reproduction of viruses. Another prominent humoral agent is the series of substances collectively known as *complement*, members of which participate in a cascade of chemical reactions when stimulated by infection. Macrophages and granulocytes are *phagocytes*, which acquire this label because they can eliminate alien matter by engulfing and digesting it. We must take a closer look at the structure of antibodies, but let us first round off our discussion of the cellular level.

Adaptive immunity comes in two forms, active and passive, the latter type being acquired from such sources as maternal milk and vaccination. Active antibody formation is the job of the lymphocytes, which come in two varieties, namely the *B-cells* produced in bone marrow and the *T-cells* produced in the thymus. The latter come in several versions, two of which are especially important: the *helper T-cells*, which recognize alien antigens and trigger the response of the B-cells, and the *killer T-cells* that destroy foreign matter in tissue, which the B-cells cannot access. The issue of self vis-à-vis non-self is particularly vexing in the case of *organ transplantation*, the tissue of the transplanted organ being recognizably different from that of the intended host because of the considerable variation in some tissue proteins, from person to person. Some of the proteins from a donor may be perceived as antigens by the host's immune system.

The main source of tissue protein variability appears to be concentrated along a stretch of a single chromosome, comprising upward of 50 genes, this being known as the *major histocompatability complex* (*MHC*). (The prefix *histo* denotes something related to the body's tissues; formally and historically, *histology* has been the study of tissue, but the subject now concerns itself primarily with the structure of cells and their interactions.) MHC is structurally altered by infection. The immune system then no longer recognizes it as belonging to host tissue, and it becomes susceptible to attack by T-cells. Peter Doherty and Rolf Zinkernagel made an interesting discovery regarding the specificity of cell-mediated immune defence. They asked themselves how T-cells can specifically identify infected cells, and thereby kill only those targets. It transpired that this requires a dual recognition process, in which there is simultaneous atomic-level docking of proteins on the

Acanthostega, shown reconstructed in the previous illustration, lived about 365 million years ago, during the Late Devonian period. Like the equally old *Ichthyostega*, it was a very primitive tetrapod, despite its having retained a fish tail and fin rays. In 2006, Edward Daeschler, Neil Shubin and Farish Jenkins Jr. announced their discovery of Late Devonian fossils found in river sediments on Ellesmere Island in Nunavut, Arctic Canada that clearly predate *Acanthostega* and *Ichthyostega*. They called the roughly 380 million year old animal *Tiktaalik rosea*, and drew particular attention to the transitional nature of the skeleton of the pectoral fin, seen on the right (the animal's left) of this picture. Its distal part – that is to say, its extremity – is adapted for flexing upwards, which suggests that it was being used to prop up the animal, and there is a marked resemblance to tetrapod digits. The bones are nevertheless components of a fin, so *Tiktaalik* reveals what was happening when fins were becoming limbs. As Per Ahlberg and Jennifer Clack remarked, when commenting on this remarkable discovery: 'this really is what our ancestors looked like when they began to leave the water.'

T-cell surface with both the antigen and the MHC molecule on the surface of the alien invader.

Antibody proteins are now known as *immunoglobulins* (and formerly as *gamma-globulins*). Five classes have been identified (each comprising several sub-varieties): IgA, IgD, IgE, IgG and IgM. *Immunoglobulin G*, for example, has a molecular weight of about 150 000 Da (daltons). Rodney Porter showed that this can be cleaved into three roughly equal fragments by the enzyme *papain*, and he found that only one of these has a shape regular enough to promote crystallization. At about the same time, in 1959, Gerald Edelman showed that the molecule consists of two kinds of polypeptide chain, one heavy and one light, and Porter subsequently reconstituted immunoglobulin G from four chains, two of each variety. When electron microscope observations by Michael Green and Robert Valentine revealed a Y-shaped molecule, it was not difficult to arrive at a structure for this antibody. The light chains are located only on the upper branches whereas each of the heavy chains spans an upper branch and the stem. A disulphide bond holds the two parts of the stem together, and other hydrogen bonds shape the two strands of each upper branch in much the same way as similar bonds provide the architecture and specificity of transfer-RNA. Both heavy and light chains have portions with the structural variability which enables different antigens to be nullified. It is this dual structure which endows the molecule with such diversity. A given cell contains genetic instructions for just one of perhaps a thousand different arrangements of each chain; this

Taxonomy: the classification of organisms

Organisms are classified in a hierarchical series of groups, the lowest level of distinction being that of species. *Taxonomy*, the science of such classification, was founded by Carl von Linné (1707–1778), also known as Carolus Linnaeus. It involves identification, description, and naming, and was traditionally based on anatomy. The later introduction of biochemical data has improved the accuracy of differentiation. The system is not rigid; it is subject to modification as more is learned about organisms and their fossil history. Experimental taxonomy aims at discovering how patterns of variation evolved. The taxonomic hierarchy, in which a group at any particular level can comprise several groups at the level lying immediately below, is as follows:

KINGDOM
PHYLUM
CLASS
ORDER
FAMILY
GENUS
SPECIES

In the plant kingdom, phyla are frequently referred to as divisions. For most animals, and many plants, a species is a group of individuals able to breed only among themselves, but reproductive isolation is not an inflexible criterion, and a comprehensive definition of species has not yet been found. Horses and donkeys belong to separate species; they can interbreed to produce mules, but these offspring are infertile. A similar result can be produced by even more widely separated species such as lions and tigers. The progeny of some interspecies pairings are fertile, however, and this underlies the fact that taxonomic distinctions are not ideally precise. The standard method of naming species involves the binomial nomenclature, which is usually in Latin. The second part of the name is peculiar to the species while the

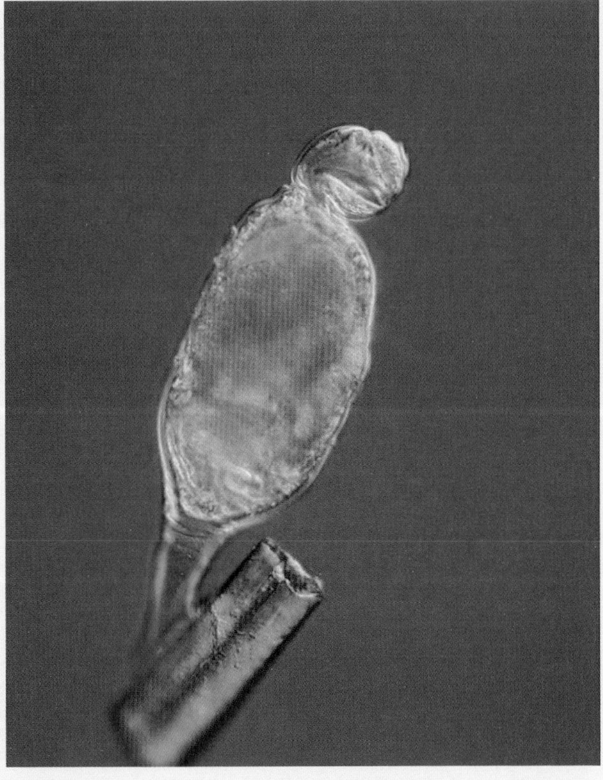

first (generic) part designates the genus to which it belongs. Thus the wolf, *Canis lupus,* and the dog, *Canis familiaris,* are different species belonging to the same genus. When a subspecies, or variety, is unambiguously identifiable, a trinomial naming system is used. An example is *Homo sapiens neanderthalensis*. New species are discovered regularly, of course, and the same is true of the higher taxonomic levels, though less frequently. There are nevertheless occasional major surprises, as when a new class – or even a new phylum – is discovered. This what happened in 1991, when Peter Funch and Reinhardt Kristensen came across a new aquatic species in the Northern Kattegat, near the port of Frederikshavn, Denmark, and named it *Symbion pandora* (see the colour picture, which was taken with a light microscope). The species name was chosen to reflect the remarkable fact that a larva is visible inside the main creature, waiting to be released and continue the life cycle, in a manner reminiscent of Pandora's box. The creature was found living in the mouth appendages of the lobster *Nephrops norvegicus*, with which it has a symbiotic relationship. The researchers' subsequent studies revealed that *S. pandora* belongs to a hitherto unknown genus (*Symbion*), and to an equally novel family (Symbiidae). But this was not the end of the surprises, because it then transpired that *S. pandora* belongs to a new order (Symbiida) and a new class (Eucycliophora). Finally, in 1995, Funch and Kristensen were able to announce that it is indeed the first discovered member of a new phylum, which has been named Cycliophora. *S. pandora* is tiny and primitive. It consists of a bell-shaped funnel terminating in a circular mouth, a trunk and a stalk, the entire body being about 350 micrometres in length and just over 100 micrometres in width.

gives about a million possible combinations. This, then, is what lies at the molecular heart of *adaptive immunity*.

The *clonal selective theory* of antibody formation, put forward by F. MacFarlane Burnet, Niels Jerne, Joshua Lederberg and David Talmage, in the 1950s, holds that each cell is genetically committed to produce a single type of antibody, which becomes bound to the surface membrane. If a maturing cell encounters the antigen complementary to its antibody, it dies; the ability to produce the required range of biochemicals has not yet been acquired. When a mature cell encounters its antigen, it synthesizes large amounts of the corresponding antibody, and it is also stimulated to divide, providing numerous *clones* of itself to carry on the fight against the intruding substance. Those clones that survive after the invader has been vanquished confer immunity against future attack. It should be noted that this behaviour guarantees that the tissue will not try to reject itself, a situation which suggested itself to Ehrlich and which he called *horror autotoxicus*. Unfortunately, as first discovered by Ehrlich's contemporary Karl Landsteiner, such a state of affairs can exist. Examples include the *autoimmune diseases* known as *diabetes mellitus* and *myasthenia gravis*, a crippling affliction of the nervous system. During the latter half of the twentieth century, another apparently autoimmune disease, *polymyalgia rheumatica*, became increasingly prevalent. It attacks all the body's muscles, and is thus extremely painful. Its cause is still a mystery, some experts inclining toward an explanation that sees increasing environmental pollution as the culprit.

The structural variation displayed by the polypeptide chains of antibodies must reflect great diversity in the genetic composition of those chromosomal regions responsible for the immune response. This brings us back to the cell's nucleic acids. In 1892, August Weismann suggested that cellular differentiation is caused by progressive loss of genes. We now know that this is not the case; inactivation of genes results not from loss but from *repression* by certain proteins. Recent developments have exposed surprising

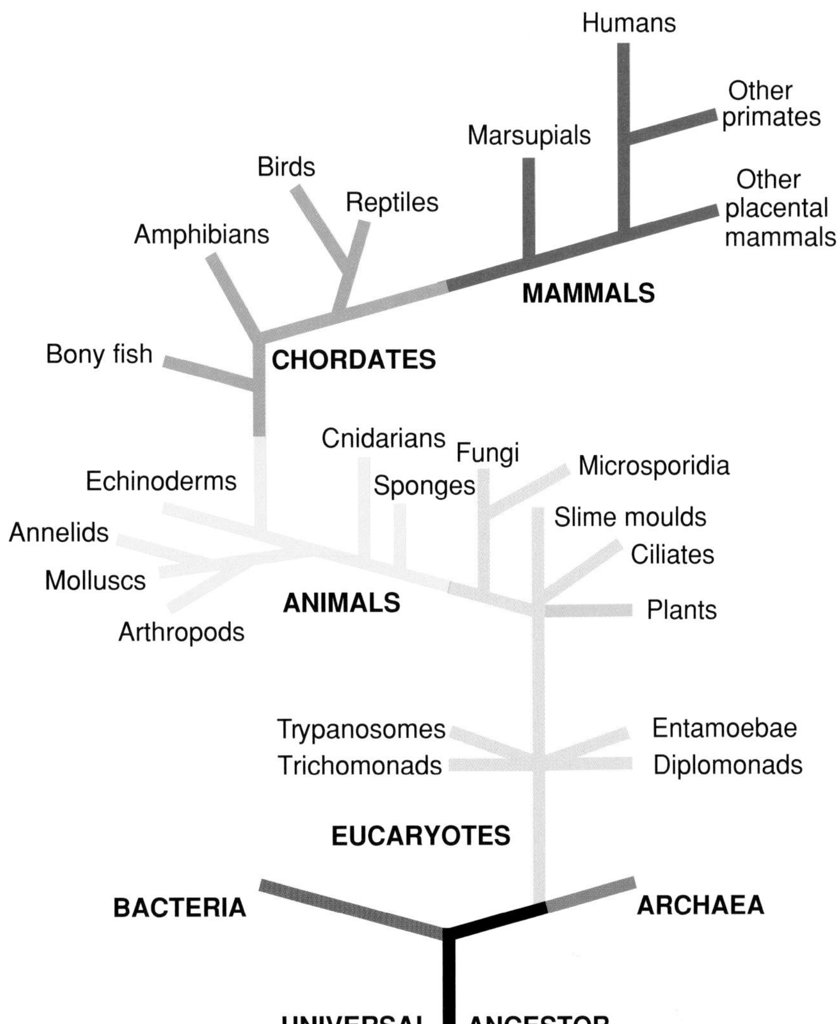

This modern version of the evolutionary tree is based on the sequencing of 16S-tRNA, and on one developed by Carl Woese in the 1980s. It is instructive to compare this with Ernst Haeckel's tree reproduced at the start of the present chapter.

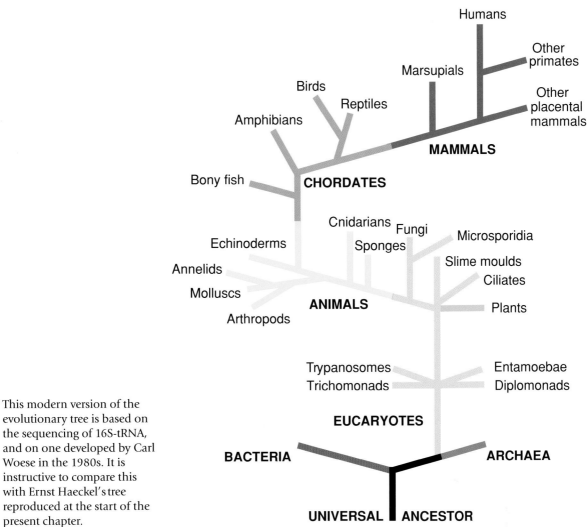

properties of the hereditary material. Regarded collectively, they reveal that the genetic machinery is more complex and dynamic than previously appreciated. But these discoveries might have been anticipated, because it had long been known that eucaryotic cells have a surfeit of nucleic acid. The amount of that material contained in the single chromosome of a typical procaryote comprises about 4 million bases. With three bases per *codon*, this means that roughly 1.3 million amino acids can be specified. Assuming the average protein to be composed of 400 amino-acid residues, the genetic material of the procaryote is sufficient for about 3000 proteins, which seems a reasonable number. In the higher eucaryotes such as mammals, birds and amphibians, the chromosomes collectively possess as many as 3000 million base pairs. If all this information were used, it would be sufficient to dictate fabrication of three million different proteins. This greatly exceeds the observed variety of proteins in higher eucaryotes. As we noted in Chapter 16, the number of different proteins in the human being lies around 30 000. The mechanisms whereby the genetic message is made to produce proteins in procaryotes and eucaryotes are quite different. In the former, *transcription* and *translation* occur at the same time and place, whereas these processes are both temporally and spatially separated in eucaryotes. In the latter, the DNA is first transcribed to *messenger RNA*, which then leaves the nucleus before its recorded instructions are used as a blueprint for protein manufacture. It is now clear that the RNA originally transcribed from the DNA must be subject to a considerable amount of editing before it emerges from the nucleus.

By using *restriction enzymes*, and the techniques which have now made genetic engineering feasible, a number of laboratories simultaneously obtained evidence that parts of the transcribed message are edited out. By using the terminology advocated by Walter Gilbert and his colleagues, the DNA was found to be composed of alternating segments referred to as *introns* and *exons*. Groups led by N. H. Carey, Pierre Chambon, Richard Flavell, Philip Leder and Susuma Tonegawa showed that only the base sequences complementary to the exon segments occur on the emerging messenger-RNA. It was found that introns can be present in stretches of DNA that code for a single protein; whence the term *split gene*. The editing mechanism was subsequently revealed by the electron microscope studies of Madeleine Cochet, Frank Gannon, Jean-Paul Le Pennec and Fabienne Perrin. The superfluous segments of the transcribed nucleic acid, referred to as *precursor messenger-RNA*, form closed loops on the single-stranded molecule. These are then cut out, and the transcribed exons are spliced together, the excised material floating away and presumably being degraded. This editing mechanism appears to be universal for the higher eukaryotes, most genes being of the split type. Among the notable exceptions are the genes which code for the histones, and, as demonstrated by Frederick Sanger and his colleagues in 1981, those contained in human mitochondria. The latter discovery is intriguing because it reveals a difference from the genes of yeast mitochondria, which contain numerous introns. Tonegawa, Leder and others have shown that the immunoglobulin genes are split, and splicing is probably the source of the great diversity displayed by those antibody molecules.

There remain the questions of the origin and purpose of the introns. Indeed, there appears to be a considerable amount of DNA that has no specific function. In some species, including such unlike organisms as lilies and salamanders, there is 20 times as much nucleic acid as there is in humans. It cannot contribute to the fitness of its host organism, and thus confers no *selective advantage*. This surplus is now referred to as *selfish DNA*, and Francis Crick and Leslie Orgel have called it the ultimate parasite. It possibly arises as a natural imperfection of the replication mechanism, subject as it is to random mutation. An organism can gain by duplication of advantageous genes, while the rate of elimination of unwanted sequences, possibly made redundant during evolutionary advance, might be rather slow. As Crick and Orgel have emphasized, selfish DNA owes its existence to the ease and necessity of DNA replication. This extra DNA might in fact have a subtle purpose. Roy Britten and Eric Davidson have suggested that it could be used for control purposes, segments coming into play if they happen to lie in the right place. T. Cavalier-Smith has proposed that the excess material could be used by the cell to slow down its development or achieve a larger physical size, and M. D. Bennett has indeed found a striking connection between the minimum generation time of herbaceous plants and the DNA content of their cells.

The close genealogical relationship between humans and several ape species is revealed in the similarities of the banding patterns seen in their respective chromosomes, when these are appropriately stained (upper part of the figure). One chromosome from each of five of the twenty-three chromosome pairs is shown, for the human (H), chimpanzee (C), gorilla (G) and orangutan (O). The largest difference is seen in the second chromosome, which is composed of two parts in the chimpanzee, gorilla and orangutan. The lower figure is an ancestral tree for the same species, this time differentiating between two species of chimpanzee. Each of the transverse lines represents a chromosomal change. The human, gorilla and chimpanzee probably shared an ancestor in whom the microscopic genetic organisation of the chromosomes was similar to that found in the present human. Indeed, eighteen of the twenty-three modern human chromosomes are virtually identical to those of our common hominoid forebear.

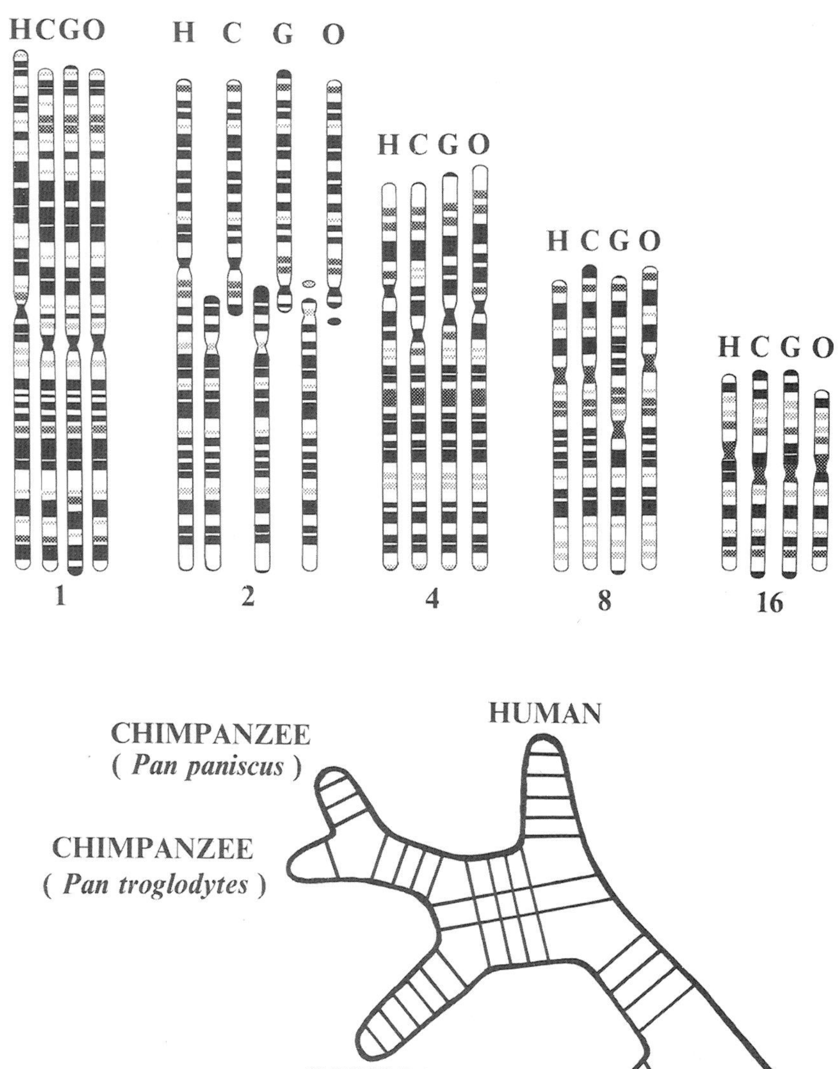

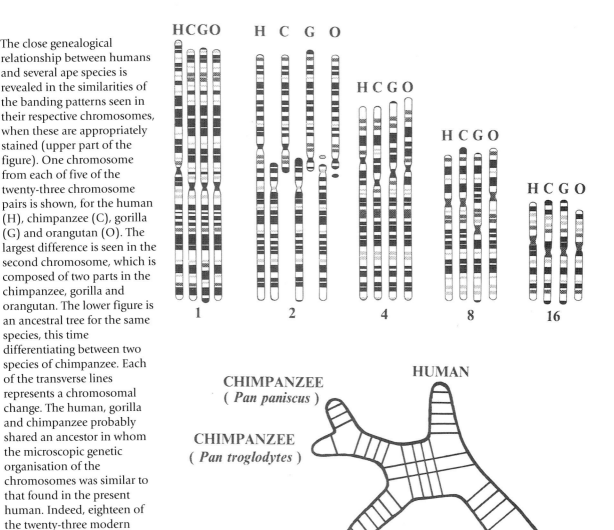

Complexity in the genetic material is not limited to the existence of introns and selfish DNA. It is now apparent that genetic elements can be transposed from one chromosomal region to another. Until now we have touched on only two sources of biological diversity: mutation and homologous recombination during meiosis. There is another factor, however, and it has an even greater potential for genetic variation. In the late 1940s, Barbara McClintock stumbled on a mysterious phenomenon while studying the common corn plant *Zea mays*. The kernels of this species are frequently mottled, patches of colour showing up against a uniform background. These tints change kaleidoscopically from one generation to the next. Monitoring the distribution of pigmentation in plant lines that had undergone repeated cycles of chromosome cross-over, McClintock discovered that the pigment genes were being switched on and off by what she called *controlling elements*. Even more remarkably, these switches appeared at different regions of the chromosomes, as if they were being physically moved from one place to another. The cells seemed to be achieving their own genetic engineering. McClintock's suggestion that certain chromosomal elements are transposable met with much scepticism. Working with a plant not unlike that studied by Gregor Mendel, her studies too appeared destined for obscurity.

About 20 years after her first reports, however, Elke Jordan, Michael Malamy, Heinz Saedler, James Shapiro and Peter Starlinger discovered *transposable elements* in *Escherichia coli*. They observed that certain mutated sequences of DNA contained extra segments up to 2000 nucleotides in length. These had somehow become inserted into the original nucleic acid by non-homologous recombination. It has since become clear, particularly through electron microscope observations by Stanley Cohen and Dennis Kopecko, that the *transponsons*, as they are now called, are duplicated stretches of genetic information which are moved about as closed loops. These can maintain their integrity because of links formed by sets of complementary bases. Whether or not such transposition is relevant to cellular differentiation remains a moot point, but there is another possibility that is just as exciting. McClintock herself noted that transposition could provide a means for the rapid evolution of control mechanisms, particularly when several genes must be switched simultaneously. Moreover, because physical shifting can bring together genes with no ancestral relationships, it could cause a sudden evolutionary change; *jumping genes* might provoke evolutionary leaps. The essentially discrete nature of genetic information at the molecular level ensures, of course, that evolution must always occur in steps, but McClintock's work indicates that the steps might be relatively large. Such *hopolution*, as one could call it, would require reappraisal of the Darwin–Wallace ideas, advocating as they do a gradual evolutionary ascent. And it suggests that the commendable hunt, by Richard Leakey and his colleagues, for the undiscovered intermediates between ape and human – the *missing links* – must expect to identify only a limited number of these. *Homo sapiens*, like every other species, will enjoy only an intrinsically limited run in creation's theatre, and our replacements might be surprisingly alien.

This major issue is, of course, the bone of contention between those who support and those who oppose the idea of evolution, so let us delve deeper into the matter. A genetic reshuffle in a single member of a species does not produce a new type of organism; this requires a more widespread modification of the *gene pool*. A particularly favourable situation arises in small highly interbred groups, as with some species of deer and horse. The offspring of a dominant male and a female harem will mate indiscriminately, giving a reasonably high probability of *homozygous transmission* of a mutation, that is through a cell line in which both chromosomes of a homologous pair have identical genes for the feature in question. In such an event, the two sets of genes in the fertilized ovum, the *zygote*, would both carry the altered genetic information. It is an interesting fact that the eight species of horse physically resemble one another, but have surprisingly different numbers of chromosomes. One type of zebra has 32, while the rare, and almost unpronounceable, Przewalski's wild horse has 66.

The apparent total lack of missing links is often cited by those antagonistic to evolution, favouring as they do the biblical explanation of spontaneous and simultaneous creation of all creatures. It is not surprising, therefore, that palaeontologists have devoted much effort to finding fossilized traces of the elusive intermediates. The hunt is certainly not limited to discovering links between apes and humans. An even more important search has been directed toward the assumed transition between fish and animals with four legs terminating in digits, that is to say *tetrapods*. There must have been a time when certain fish began to crawl up onto land, this perhaps having occurred in shallow pools that periodically dried out. This mode of locomotion would be reminiscent of what mudskippers achieve in our time. Those involved in this search point to the fact that fish, amphibians, birds and mammals all have four limbs, even though those limbs display considerable variation in structural detail. So a central idea is that certain fish started to crawl with their four lateral–ventral fins, and that these gradually evolved into the legs that are the eponymous characteristic of the tetrapods. In the case of birds, two of those legs further evolved into wings, the assumed intermediate being the *dinosaur* known as *Archeopteryx*.

The challenge lay in discovering fossils of fish that had crawled along the bottom of a river or a lake, using their fins, either as an exclusive mode of locomotion or by way of augmenting their swimming. And the search seemed to have been fruitful in 1938 when some South African fishermen discovered the multicoloured body – just over a metre and half long – of a mysterious fish and took it to Marjorie Courtenay-Latimer, a museum curator. Her colleague, J. L. B. Smith, confirmed that it was a hitherto unknown species. This was the *coelacanth*, which appeared to be an ancient species that had somehow survived to the present era. The event was a false alarm, however, because live specimens were also discovered, and they were clearly *not* capable of crawling. The coelacanth, despite being intrinsically interesting, had proved to be a red herring.

One particularly credible explanation of the lack of putative intermediates in the fossil record is that relatively few individuals may have been involved, and that their presence on Earth could have been rather short-lived; they may have been replaced by yet more evolved forms relatively quickly. Moreover, evolution could have suffered numerous failed attempts, the underlying mutations producing life-forms that were simply not viable. When the attempt did not fail, on the other hand, the intermediate may have been quickly replaced – on the geological time scale, that is – and what appears to have been a sudden, discontinuous change may really have involved a rapidly evolving series of intermediates. Fish in the Devonian period were believed to be particularly good candidates as the progenitors of crawling creatures. *Eusthenopteron*, for example, had the required pectoral fin (that is to say, the forelimb) but lacked the feet and toes. In the 1930s, Erik Jarvik and his colleagues discovered fossils of a creature that bore a striking resemblance to *Eusthenopteron*, and called it *Ichthyostega*. The latter, a tetrapod, clearly had toes. Jennifer Clack discovered the fossil of another promising tetrapod, *Acanthostega*, and in 1990 her colleague, Michael Coates, was astonished to find that it had *eight* such digits. Five digits has always been taken to be the rule, since no tetrapod now living has more than that number, so here indeed was the hint of a short-lived intermediate. Clack and her colleagues then re-examined *Ichthyostega* and found that Jarvik had in fact overlooked two supernumerary toes. He had also missed something else at least as important, namely that *Ichthyostega*'s legs were not for walking; Clack was able to put forward strong arguments supporting the view that they were in fact paddles.

So although the discoveries of *Ichthyostega* and *Acanthostega* had clearly given a massive boost to the idea that evolution had occurred, it was

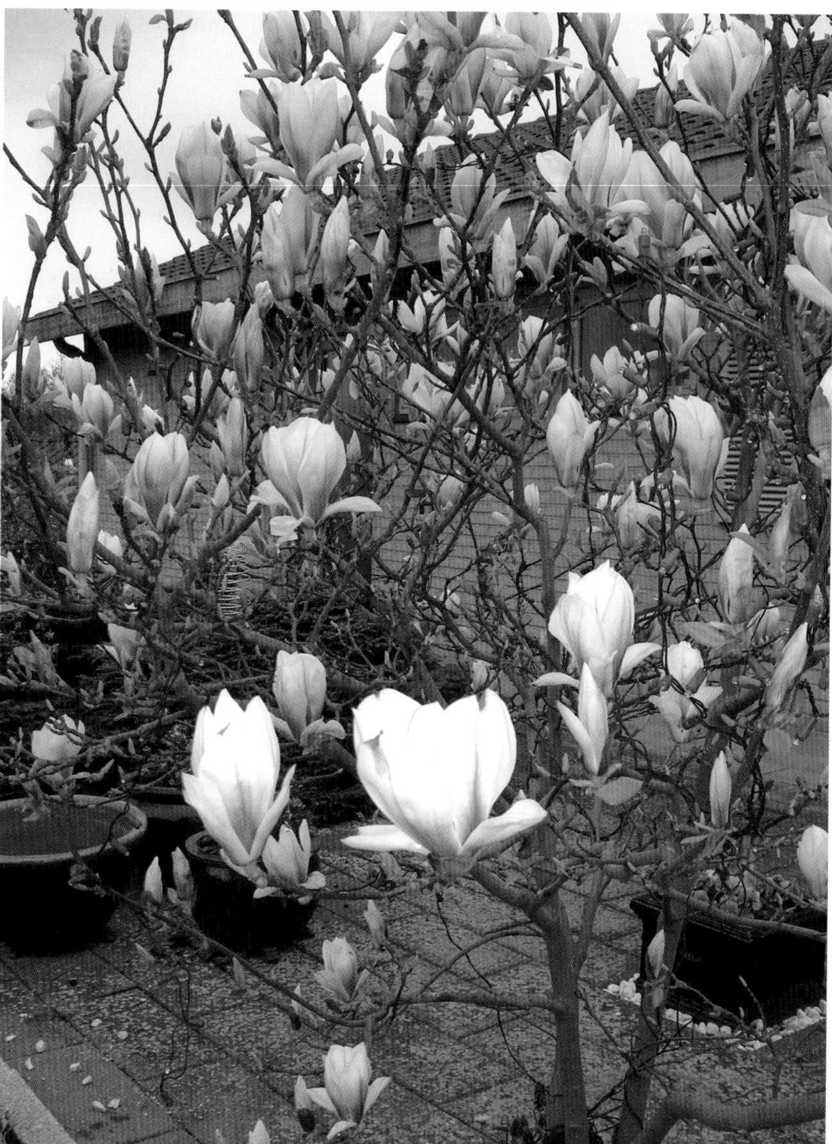

The presence of a magnolia tree, with its rather sophisticated appearance, in many a domestic garden could lead one to believe that it is a relatively recent product of the horticulturist's art. The surprising fact is, however, that the magnolia has not changed its form and size during the last one hundred million years. There were magnolia trees like the one shown here when dinosaurs roamed the Earth!

now apparent that there was still another chapter in this story. As mentioned above, the assumption had long been that the crucial transition had involved a fish crawling onto land on its fins, but now the researchers had to explain why toed limbs were evolving when creatures were still confined to their aquatic environment. Observations made in 1994, in the Red Hill area of Pennsylvania by Edward Daeschler, endorsed that different possibility. Examining Devonian sandstone deposits in that region, Daeschler discovered strata with distinct green coloration, as well as fossils of the earliest tetrapods, and concluded that underwater plants were far more numerous during the Devonian period than had previously been assumed. He consequently put forward the idea that digits evolved at the extremities of some fins, enabling their possessors – the tetrapod *Hynerpeton*, for example – to escape predators such as *Hyneria* by grasping the stems of

This beautiful picture taken by Bruce Holst of the Marie Selby Botanical Garden in Sarasota, Florida shows one of the table-top mountains of the Guayana region of Bolivar State, in southern Venezuela, which are known locally by their indigenous name *tepui*. This particular example is known as the Aparamán tepui, and is one of a chain known as Los Testigos, which means the witnesses. The summit of the Aparamán tepui is 2 100 metres above sea level, and its sides are vertical around the entire perimeter. The top is thus isolated from the surrounding terrain, as it has been over geological time. Flora have been discovered on the plateau which have not been observed elsewhere on Earth, this adding to the evidence found by Charles Darwin (1809–1882) on the Galapagos islands and documented in his *The Origin of Species* (1859). The occurrence of such isolated places shows that the novel *The Lost World*, by Arthur Conan Doyle (1859–1930), was not particularly far-fetched.

aquatic plants, thereby propelling themselves deeper into the submarine undergrowth.

There then remained the task of demonstrating that one or other of these toed creatures had made the transition from water to dry land. It had been noted that tetrapodomorph fish and early tetrapods with limbs had very similarly constructed mouths and noses, so it is likely that they breathed in similar ways, gulping air as well as using their gills in water. *Ichthyostega* and *Acanthostega*, for example, both had grooved gill skeletons like the fish that preceded them. So in searching for intermediates that represented a sort of half-way stage, one had to look for differences that were more subtle. Per Ahlberg chanced upon precisely such a fossil and subjected it to what is known as a *cladistic analysis*, this comparing the positions and appearance of numerous anatomical features with those of well established species. He named the 360 million year old fossil *Livoniana* and showed – by noting the positions of its teeth, and the paths through the jaw bone taken by certain blood vessels – that it was a true intermediate between a fish and a tetrapod. In 2002, Ahlberg collaborated with Min Zhu, Wenjin Zhao and Lianto Jia in a study of the 355 million year old fossils of a creature named *Sinostega pani*, found along the Chinese coastline, and noted its overall resemblance to *Acanthostega*. They thus concluded that tetrapods had effectively colonized much of our planet by the end of the Devonian period, 354 million years ago. When one adds this evidence to related finds in Russia, Australia, Latvia, the USA and most recently in Belgium, one can appreciate why the case for intermediates is now generally regarded as well established.

In this chapter, we have considered a few of the fascinating things that occur when all goes according to plan. Let us close with a brief look at the situation when things appear to go wrong; when restraint gives way to anarchy, as in the various forms of *cancer*. One of the most striking characteristics of the *malignant cell* is its lack of *contact inhibition*. When grown in culture, normal cells in a limited area stop dividing when they have filled the available space, but cancer cells do not confine themselves to such a monolayer. They pile up on top of one another in an uncontrolled fashion. This implies breakdown in intercellular communication, possibly because

Chaos and Extinctions

One of the most remarkable scientific developments in the latter part of the twentieth century was the discovery of *deterministic chaos*. Systems that are completely described by well-formulated – and often surprisingly simple – interactions between their components are nevertheless capable of displaying temporal developments that are chaotic. It is impossible, even in principle, to predict the system's evolution further ahead than a rather small amount of time. The reason for this lies in what is known as *non-linearity*, which can be taken to mean lack of simple proportion. In Nature, very few relationships do have simple proportions, but chaotic effects came to light only recently because science did not previously have the means of coping with the complexities of the phenomena. The advent of fast electronic computers was the key breakthrough, and many of the fundamental truths about the chaotic state have emerged through *computer simulations*. The essence of this activity is calculation of the new value of a variable, given a record of its immediately previous value, and a suitable prescription for going from one to the other. The pioneers of this originally esoteric enterprise, in the early 1960s, included Stephen Smale, Michael Hénon and his student Carl Heiles, and Boris Chirikov. They were joined within a decade by Mitchell Feigenbaum and by Robert May. Feigenbaum was interested in what is known as *logistic mapping*, this being described by the prescription

$$\lambda x_{OLD}(1-x_{OLD}) \rightarrow x_{NEW}$$

where x is simply a variable (between 0 and 1) whose new value, x_{NEW}, is to be calculated from its immediate previous value, x_{OLD}. Feigenbaum explored the variation in time of the product of this prescription, over a large number of repetitions, for values of the parameter λ ranging from 0 to 4. (If λ exceeds 4, x_{NEW} rapidly goes to minus infinity.) Feigenbaum discovered that as λ increases, x_{NEW} shows an increasing tendency to *bifurcate*, that is to say randomly adopt values in two different ranges of size. And he was astonished to observe a radical change when λ reached 3.57. The calculated value of x_{NEW} started to vary chaotically (within certain limits), even though the prescription for computing it was completely specified. Robert May was particularly interested in the implications of deterministic chaos for biological systems, including its influence on population genetics. As a typical example, we can imagine x_{NEW} to be the size of the population of a particular species in a specific year, given that x_{OLD} was its size the previous year. The above prescription certainly looks reasonable because the new population size would be expected to depend on how many individuals there had been present the previous year (see the prescription's first term, after the parameter λ, which can be regarded as indicating what fraction of an offspring one sexually mature adult can participate in producing per year, on average). And the second term can be rationalized as indicating that the competition for limited resources will become increasingly severe as the population size increases, the term becoming zero when x_{OLD} reaches unity. This is reminiscent of the oft-stated fact that no food will be producible if the entire Earth's surface is covered with people, standing shoulder to shoulder. Working together with George Oster and James Yorke, May found that x_{NEW} tends toward zero for λ less than unity, whereas it levels off to a constant value when λ lies between 1 and 3. For even greater values of λ, x_{NEW} starts to oscillate wildly, in a manner suggestive of boom–bust cycles. Ultimately, when λ increases beyond that value of 3.57, x_{NEW} varies chaotically, and can suddenly become zero. This is a very profound result; it indicates that a species can become extinct despite the apparent adequacy of resources and number of breeding adults, for no other reason than that the underlying dynamics have exceeded a certain threshold of complexity.

of changes in the cytoplasmic membrane. Just after 1900, V. Ellermann and O. Bang transferred *leukaemia* from one bird to another with a sub-cellular extract, and Peyton Rous accomplished a similar result with tissue tumours in chickens. In each case the transmitting agent was a virus. In Chapter 15 we saw that the normal direction of genetic flow, from DNA to RNA, can be reversed by the enzymes known as RNA-dependent DNA polymerases, or *reverse transcriptases*. This is a way in which viral RNA can be inserted into a host's genetic material. Robert Huebner and George Todaro have suggested that cancer-generating, or *oncogenic*, virus fragments are present in all normal cells, but that they are repressed. A malignancy was postulated to arise upon failure of this *repression*, presumably by another agency such as radiation or chemical stimulation. It is possible that the resulting mechanism is relatively simple because some of the known carcinogenic viruses, such as Simian Virus 40 and polyoma, contain only enough nucleic acid to code for five to ten proteins.

In several respects the malignant cell appears to have lost its ability to differentiate; cancer cells seem more primitive than their benign cousins. This raises the interesting issue of *senescence*. While sophisticated organisms displaying a high degree of specialization have well-defined life spans, primitive forms are apparently able to put off death indefinitely. An extreme example is seen in the mammalian *red blood cell*. It is highly differentiated, even to the extent that it has no nucleus, and it survives only for about 120 days, after which it wears out owing primarily to decreased enzymatic activity. The principle holds not only for individual cells but also for highly differentiated organisms. From studies of longevity statistics, Alex Comfort has established that whereas the average human life span has steadily increased, the upper limit remains virtually unchanged around 90 years; even if we have a favourable pedigree, and look after ourselves, we have a hard time getting more than 20 years beyond the biblical three score years and ten. For a number of higher cell types, Leonard Hayflick and Paul Moorehead have discovered that there is a limit to cell division which lies around 50 doublings. After this the cells degenerate and die out. Together with Audrey Muggleton-Harris and Woodring Wright, and performing cellular surgery in which old nuclei were transferred to young cells, and vice versa, Hayflick has also established that the time-keeper of this inevitable decline resides in the nucleus.

Zhores Medvedev and Leslie Orgel have proposed that the limits are imposed by the accumulated mistakes in the transmission of the genetic message. Up to a point these were conjectured to be tolerated, but the cell ultimately reaches a point of no return, a sort of error catastrophe. A more detailed picture has been given by Ronald Hart and Richard Setlow. They find that cells are able to repair damage to their DNA caused by ultraviolet radiation, to an extent proportional to the lifetime of the species. Thus humans live about twice as long as chimpanzees, and the *DNA repair rate* in their cells is about twice as high. But in both species the repair ability declines as the cells approach their reproductive limit. Only two mammalian cell types that avoid the inevitability of death are known: cancer cells and germ cells. A particularly famous example of the former are the numerous surviving cultures of *HeLa cells*. (It is standard medical practice to label cells with only the initial letters of a patient's name, so as to avoid divulging the person's identity.) The HeLa cells are abnormal in that they contain anywhere from 50 to 350 chromosomes. The immortality of cancer and germ cells might be achieved by an essentially similar mechanism. As Hayflick has emphasized, the input of genetic information when an *oncogenic virus* invades an animal cell might resemble the genetic reshuffle when a sperm cell fuses with an egg; the joining of two gametes might effectively reset the biological clock. A human being might be simply a sophisticated machine which a germ cell uses to produce further immortal germ cells.

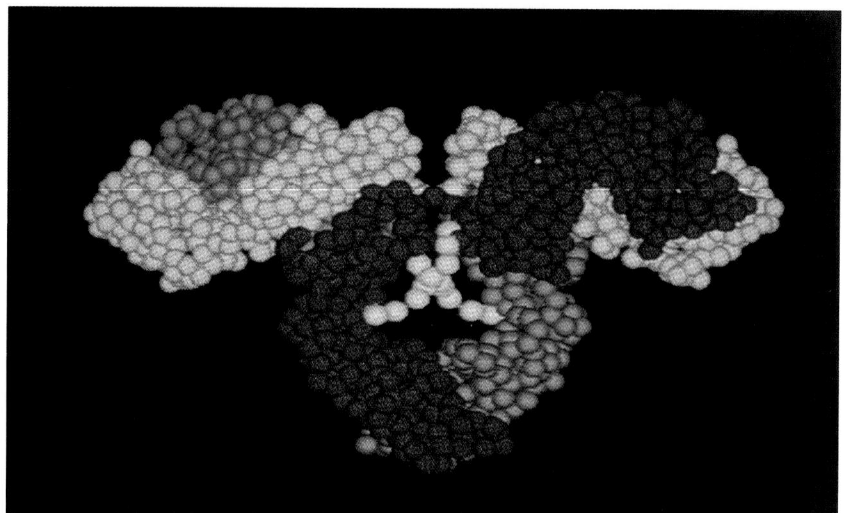

Lymphocytes, the blood's white cells, are produced in the bone marrow and multiply in the *lymph nodes*, spleen, and thymus. They protect the body's tissues against foreign molecules and organisms by producing *antibodies*, which come in an impressive variety of forms. This diversity permits the lymphocytes to discriminate between an apparently unlimited number of alien substances, referred to as *antigens*, and to disarm the invaders by binding the antibodies to them. Each antibody molecule is Y-shaped, and it is a composite structure of two heavy and two light protein chains, the former running the length of the Y while the latter are found only in the upper branches. The entire structure is held together by disulphide bonds. Both pairs of chains help the antibody to recognize and latch on to the antigen, whereas the subsequent immunologic task is left to the heavy chains alone. The staggering diversity of the antibody molecules stems from the fact that the upper extremities of all four chains, logically designated the variable regions, display great variety of amino-acid sequence. The lower parts of the two heavy chains are constant, and generally belong to one of just five classes. This explains why the vast number of different antigens that can be recognized are nevertheless disposed of by one of only five types of mechanism. The computer-generated picture of a typical antibody molecule shows the amino-acid residues as spheres. The heavy chains are indicated in red and blue, and the light chains in yellow. The variable regions lie at the outer extremities of the upper branches on either side. If the variable and constant parts of a particular antibody were coded for by adjacent sequences of DNA, it would be hard to explain why only one sequence is subject to genetic variability. William Dreyer and J. Claude Bennett suggested that the sequences occur at separate locations, there being many stretches corresponding to the different types of variable region and just a few constant chain sequences. They also proposed that the DNA sequences were brought together during the shuffling and splicing now known to occur after transcription. These ideas have been largely verified by Philip Leder, David Swan and Tasuku Honjo, and the variable regions found to be translated from separated stretches of DNA, allowing for a multiplicity of combinations.

The nature of Hayflick's time-keeper is now well established, and one of these is located at the extremity each chromosome. It is known as a *telomere*, and it consists of multiple repeats of the DNA base sequence TTAGGG. Each telomere can initially be as long as 15 000 bases in the foetus, but upwards of 200 bases are lost during each cell division. This is in approximate agreement with the above-mentioned limit of about 50 cell divisions, and it was found that a cell dies when the telomere length falls below a critical value, that destruction process being known as *apoptosis*. The importance of the telomere is therefore clear: if it were not present, critical chromosome damage would occur even during cell division in the foetus. It transpires that there is a mechanism whereby the part of a telomere lost during a cell division can be restored, thereby resetting the time-keeper as it were. The restoration process is under the control of an enzyme known as *telomere terminal transferase*, or simply *telomerase*, which consists of protein and RNA. Not surprisingly, it does its job by adding TTAGGG blocks to the end of each chromosome. Telomerase is found in three types of cell: foetal cells, adult germ cells and tumour cells. The ageing process in normal cells can thus be attributed either to a lack of telomerase or to the fact that it is present but inactive. This raises the prospect of determining a cell's fate by manipulating telomerase activity, and that has obvious implications for the control of both senescence and cancer.

Another possible method of control has been suggested by research on the mechanism that triggers the various stages of mitosis that we considered in the previous chapter. Leland Hartwell, studying baker's yeast, *Saccharomyces cerevisiae*, in the early 1970s discovered a series of genes involved in cell cycle control. These are known as the *CDC genes*, and Hartwell found that one of them, CDC28, controls a cell's progression through the earliest stage of the cell cycle, that is to say from the previous mitotic division to the point of DNA replication. (It is worth noting that the cells of the nervous and muscular systems in animals do not divide after they have differentiated, so CDC28 control does not apply to them.) Hartwell also discovered a system of molecular checking, which verifies that the various steps of the process have been properly carried out, this thereby ensuring that those steps occur in the correct sequence. A few years later, Paul Nurse discovered the similar gene cdc2 in another type of yeast, namely the so-called fission yeast *Schizosaccharomyces pombe*. He showed that it controls the two later stages of the cell cycle, that is to say the stage from DNA synthesis to chromosome duplication and the stage from chromosome duplication to cell division. About a decade later, Nurse isolated and identified the corresponding human gene, now designated CDK1. The similarity between the yeast and human genes indicated that CDK function has been conserved throughout evolution.

There remained one puzzling fact: the amount of CDC and CDK in the respective organisms appeared to be constant during the entire cell cycle. How could they be carrying out their regulatory functions while displaying such lack of variation? Timothy Hunt found the final piece of the jigsaw in the early 1980s, when he discovered proteins known as *cyclins*. Each of these binds to a specific CDK molecule, thereby controlling the latter's level of activity. And in this case there *was* the expected variation in concentration; the level of cyclin gradually increases throughout the cell cycle, starting at zero immediately after cell division and reaching a maximum just before the next mitosis. The letters CDK stand for *cyclin-dependent kinase*, these kinases being the proteins coded for by the gene, those product molecules exercising control by the familiar process of phosphorylation, when they are themselves activated by the corresponding cyclin. Not surprisingly, the combined efforts of Hartwell, Nurse and Hunt have raised hopes of external control of cell division, and through that of combating cancer by arresting cell proliferation.

Summary

Cellular differentiation, by which tissues, organs and systems are generated, has now been largely elucidated. In the embryo, everything is not preformed and present in miniature at the outset, as previously assumed; cells undergo modification as they divide and multiply. The process is highly nonlinear; a small change in a cell's position in the organism can lead to a radical change in the tissue to which it is contributing. Tissue differentiation is caused by four distinct processes: cell multiplication, cell movement, change of cell character and cell–cell signalling. Spatial organization and differentiation are controlled by the genes. The arrangement of structural features along an animal's long axis is determined by the Hox genes, which code for portions of a DNA-binding protein known as a transcription factor. Mutations in this gene provoke a homeotic transformation, one body part being replaced by another. The upshot can be bizarre, as in the cases of supernumerary limbs and grossly displaced body parts. The immune system comprises the white blood cells and the tissue cells known as mast cells and macrophages. Immunity ultimately stems from molecules with great adaptive power because their structures are composites of units which themselves display considerable variability. Variability at a grosser scale is produced when certain stretches of the genetic material are shifted from one chromosomal site to another, and this may be the mechanism underlying evolution. The changes thus occur in a manner that is essentially discontinuous, but the transition from one species to its successor may nevertheless be reasonably smooth. Those opposed to Darwinism cite the lack of life forms intermediate to known species, but the fossils of such creatures have now been found. Most cell types appear to be programmed for ultimate death, known as apoptosis. The stretch of DNA called a telomere at the end of each chromosome loses part of its length during each cell division, and apoptosis occurs when the telomere length falls below a critical value, this limit being reached after about fifty divisions. The two exceptions are cancer cells and germ cells, their immortality stemming from the enzymatic activity of telomerase, which maintains telomere length at its original value.

18

How vainly men themselves amaze
To win the palm, the oak, or bays;
And their uncessant labours see
Crown'd from some single herb or tree,
Whose short and narrow verged shade
Does prudently their toils upbraid;
While all flowers and all trees do close
To weave the garlands of repose.

Andrew Marvell (1621–1678)
The Garden

A place in the Sun: *the plant*

The common aims of plants and animals are the conversion of energy to their own purposes and the perpetuation of their species. All energy in the solar system ultimately comes from the Sun as light emission, and organisms that can use this source directly belong to one of the *autotrophic groups*, which are independent of outside sources of organic substances. These direct solar energy converters, or *phototrophs*, are almost exclusively *plants*, the rare exceptions being found among the *bacteria*. An example of this latter esoteric group is *Halobacterium halobium*. But not all plants are autotrophic. A major division, or phylum, of the plant kingdom are the *mycophyta*, or *fungi*, which include mushrooms, moulds and yeasts. These *heterotrophs* are all either *saprophytes*, obtaining nutriment in solution from the decaying tissues of plants or animals, or *parasites*. The latter group are important agents of disease, mainly in other plants but occasionally in animals.

We will not be concerned here with such plants, and neither will we consider all phototrophic plants. The *algae*, which are simple photosynthetic plants, were formerly classified as belonging to a single taxonomic division. They are now differentiated according to such characteristics as pigment, cell-wall structure and assimilatory products, and have been recognized as occupying separate divisions. The smallest variety are the *Bacillariophyta*, also known as *diatoms*. These unicellular plants thrive in both salt water and fresh water, and the organic remains of their ancestors probably produced the Earth's *petroleum* deposits. They are present in *plankton*, and ultimately provide the basis of all aquatic life. Other types come in a variety of colours, and include the green *Chlorophyta*, blue-green *Cyanophyta*, golden-brown *Chrysophyta*, olive-brown *Phacophyta*, yellow *Xanthophyta*, and the red *Rhodophyta*. The algae are ubiquitous in water and in damp places, and the mobile varieties move about with the aid of *flagella*. These project from the organism as fine threads, up to about 60μm long, and perform a lashing or undulating movement.

The division of phototrophic plants that will be our main interest here are the *Spermatophyta* or *seed plants*. They are the dominant flora of our era, including trees, shrubs, herbaceous plants and grasses. As we shall see later, they are divided into two classes: *Gymnospermae*, in which the seeds lie fairly exposed and relatively unprotected, and *Angiospermae*, which have their seeds enclosed in an ovary. Although the Earth's seed plants comprise a great variety of species, with a wide range of size and shape, they have several basic features in common: roots, stem, leaves and reproductive systems. These units act collectively so as to realize the twin goals of viability

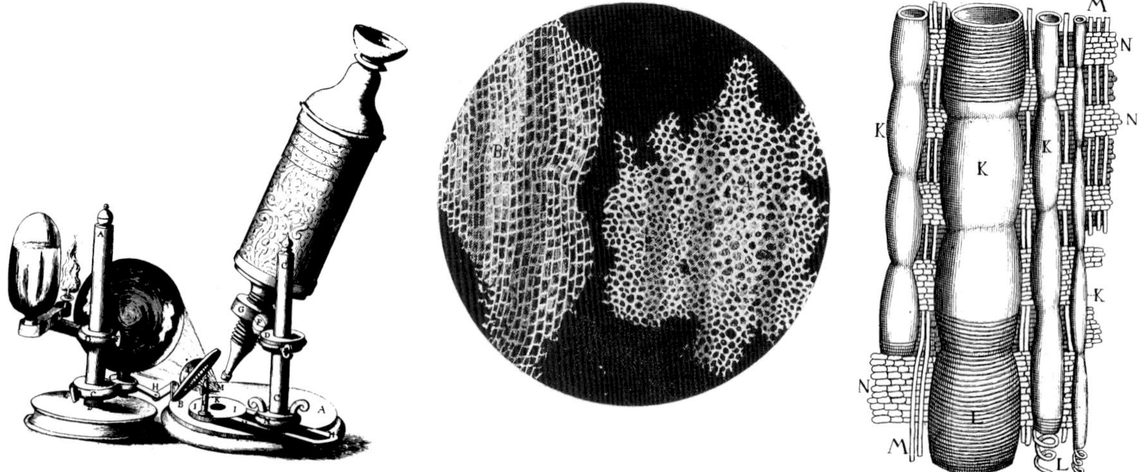

The pioneering application of optical microscopy to botany by Robert Hooke (1635–1703), in 1663, reaped an immediate reward: the discovery of the cell. The etchings show the microscope with which he achieved this breakthrough, and his impressions of what he observed when he first focussed his instrument on a cork specimen. The practitioners of the new technique rapidly grew in number and the right-hand picture indicates how far the field had progressed within a single decade. It shows a section of a wood specimen studied by Marcello Malpighi (1628–1694), as it appeared in his *Anatomia Plantarum*, published in 1673. The various types of cell, including those responsible for liquid transport, are clearly depicted.

and continuity, and they are driven by the most important of all terrestrial reactions: *photosynthesis*.

Knowledge of the way in which a plant functions is surprisingly recent. Earlier than 300 years ago, our forebears were content to accept plants as part of Nature's bounty, and before 1630 there seems to have been little curiosity about their origin. Since plants grow in soil, it is not surprising that this was assumed to be their sole source of nutriment. In that year, Johann van Helmont planted a 2 kg willow sapling in 100 kg of soil. Five years later the plant weighed 85 kg, even though the soil had lost only 50 g. Van Helmont concluded that the extra weight must somehow have been converted from the only other factor in his experiment: the regular watering of the plant. Almost 100 years passed before the next significant quantitative observations, and even these seem to have initially obscured the main facts. Stephen Hales detected the liberation of water from plants in 1727. Covering a plant with a bell-shaped glass vessel, he found that water condensed on the inner surface of this container. His discovery that the weight of this water was roughly equal to that fed to the roots seemed to indicate that the plant was functioning as a sort of siphon. He later noticed, however, that a plant which was being starved of water nevertheless continued to emit this substance, but it simultaneously lost weight. The water being given off during *transpiration*, as the process is called, does not simply pass through the plant as would water through household plumbing. At some point, it was now clear, the water must actually be an integral part of the plant structure. Not having the advantage of water's visibility, gases were discovered to be associated with plant function only several decades later. Around 1770, Joseph Priestley discovered that plants give off a gas that supports animal respiration. He also found that this sustained combustion, and the gas later became known as *oxygen*. Ten years after this, Jan Ingenhousz demonstrated that oxygen is liberated only by a plant's green parts, and only when these are in sunlight. He went on to show that the liberation of oxygen is accompanied by absorption of carbon dioxide. Indeed, he even realized that the carbon in that compound is used by the plant to manufacture some of the components on which its structure is based.

By the beginning of the nineteenth century, plant study had become an impressively exact science. Sensitive measurement by Theodore de Saussure,

Nitrogen fixation

Nitrogen, one of the essential elements in all amino acids, is abundant in the Earth's atmosphere, where it exists in the molecular form, N_2. Its conversion to biological purposes is very difficult because this molecule has a triple bond with a strength of 1.57 attojoules. Only a few strains of micro-organisms, such as certain soil-inhabiting bacteria and some blue-green algae, possess the molecular machinery necessary to change the bond configuration and permit attachment of other elements such as carbon and hydrogen. These difficult processes are collectively termed nitrogen fixation. Some nitrogen-fixing bacteria live symbiotically with leguminous plants such as peas, beans, and clover, forming root nodules in which the fixation occurs. Other species live independently in the soil. The Rhizobium bacteria, for example, contain the powerful reducing agent ferredoxin, an electron carrier that is also important to photosynthesis. It mediates rearrangement of the bonds in nitrogen and water molecules to produce NH_4^+ and hydrogen ions, the changes requiring absorption of a considerable amount of energy. All life depends upon these few varieties of nitrogen fixers.

in 1804, revealed that the weight of *carbon dioxide* absorbed by a plant, when bathed in sunlight, is not equal to the weight of oxygen liberated. His suggestion that the deficit must come from water was tantamount to the discovery of *photosynthesis*, by which carbon dioxide reacts with water so as to produce the plant's structural carbon, with liberation of oxygen into the atmosphere. It was soon appreciated that plants function essentially like machines, converting the simplest of compounds into more complicated organic molecules. We recall that carbon dioxide is not classified as an organic compound, so a useful characterization of the plant kingdom could be based on the fact that its members do not require the input of organic carbon. Autotrophic organisms such as algae that live on the surface of water have their essential requirements immediately at hand: the water itself and its dissolved minerals, sunlight and carbon dioxide. Thus blessed by providence, they have been under no pressure to evolve, and their members have changed but little from their primordial ancestors. In order to secure a footing on land, plants had to become more sophisticated. They required root systems to burrow down to the water level, a transport system to bring this liquid up to the sites of photosynthesis, and a stem to hold the leaves up in the sunshine.

Let us therefore consider the various parts of a typical plant in some detail. A good starting point is the structure devoted to liquid transport. The *vascular system* of a plant has the dual tasks of conducting water up to the leaves and distributing the *glucose* produced by photosynthesis to the rest of the structure. The cells responsible for water conduction are known collectively as the *xylem*, while the sugar conductors are the *phloem*. The two structures are located quite close to each other in the plant, the xylem always being nearer the centre than the phloem. In most parts of the plant, the vascular system is hidden below the outer surface, but it does become visible in the leaves, particularly the underside, where it appears as a system of *veins*. The common use of this latter term emphasizes similarity to the vascular system of higher animals, but the analogy should not be taken too far. Thus, although the blood system also has two directions to flow, the arterial flow away from the heart and the venous flow towards it, the counter-flowing systems in a plant are at no point in actual contact. Moreover, the vascular system of a plant has no counterpart of the animal heart. Its liquid flow is achieved in a simpler manner, through *transpiration*. The gases involved in

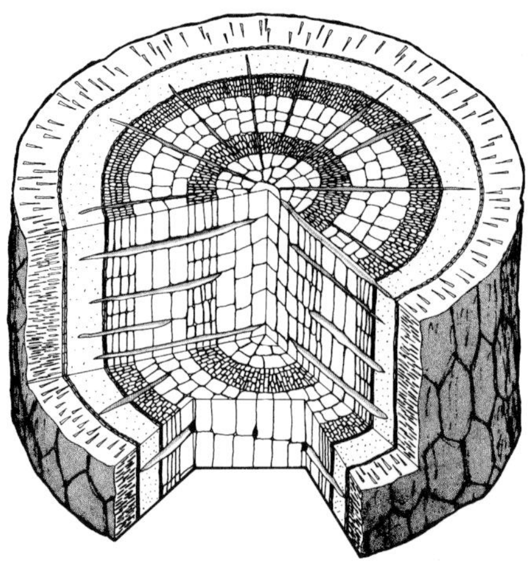

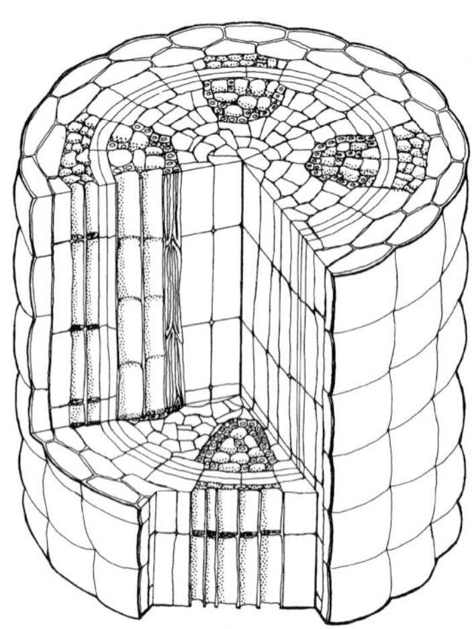

Typical features in woody (left) and herbaceous plant stems are revealed by the various common sections. The transverse section, which is that produced by the lumberjack when felling a tree, shows the concentric arrangement of such layers as the bark, phloem, cambium, and xylem. This section also shows the clustering of the liquid transporting cells into vascular bundles, in the herbaceous plant. The radial longitudinal section, which is that produced when a log is split into smaller sections for firewood, exposes the radial ray cells. This section also reveals the fact that the majority of the cells are elongated parallel to the direction of the stem's axis. The other type of cut illustrated here is the tangential longitudinal section.

the input and output of photosynthesis reach the sites of that reaction by penetrating through the spaces between cells, in the rather loosely packed arrangement at the underside of the leaves. Plants can be grossly divided into two groups: *herbaceous plants* and *woody plants*. In the former, which have soft and pliable stems, the vascular system is arranged in discrete groups of phloem and xylem cells, known as *vascular bundles*. The woody stem, on the other hand, has a continuous system of *vascular cells*, these being arranged all around the circumference of an approximate circle.

Until now, we have considered only the gross features of a plant, but the differences between its parts really become clear at the cellular level. Cutting through the stem of a herbaceous plant, or the trunk of a tree, we observe that the structure consists of concentric rings. Inspection of the section with a simple optical microscope would further reveal the cells, which are not uniform in size and shape, but show considerable variation. In a typical herbaceous plant, the outermost single layer of cells constitute what is known as the *epidermis*, the function of which is to prevent loss of water and give protection to the underlying cells. Immediately below this layer lies the *cortex*, the cells of which are larger and less tightly packed than those of the epidermis. Moving inwards from the cortex, we then encounter the vascular bundles. Finally, inside these lies the *pith*, which consists almost solely of dead cells. The structure of a tree is somewhat more complicated. Instead of the epidermis there is the *bark*, consisting of a corky tissue of dead cells, these having been produced by a growth layer known as the *cork cambium*. Inside this lie, in turn, the phloem, cambium, xylem and pith. The last-mentioned is also referred to as the *medulla*. The other obvious difference between the stem of a herbaceous plant and the trunk of a tree is that in the latter the xylem surrounding the pith is far more prominent. It is this which is referred to as *wood*.

A plant increases its thickness by continual cell division, which adds new cells both inside and outside the *cambium*. In a tree, the new cells formed on the inside simply become new xylem, while those formed on the outside of the cambium are the phloem. In a herbaceous plant, new xylem and phloem are produced only at certain points around the ring of cambium: the vascular bundles referred to above. The remaining growth simply gives rise to new pith and cortex. Both the xylem and the phloem are heterogeneous, consisting of cells of various types and states. The cells that are still

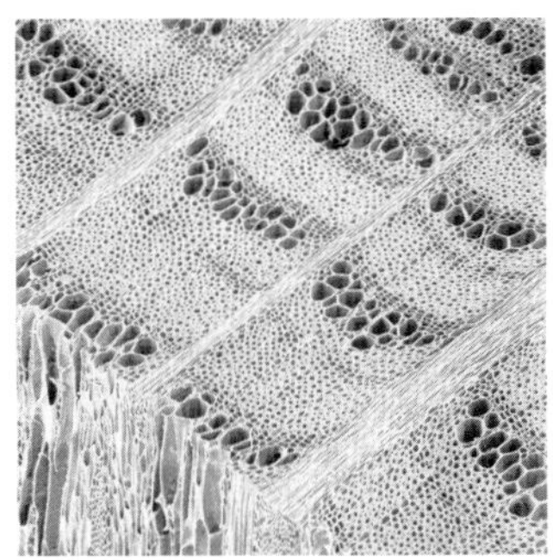

Because vessels have their long dimensions parallel to the axis of the stem, they appear as elongated tubes in radial and tangential sections and as roughly circular holes in transverse sections. The opposite is true of ray parenchyma cells, which radiate out horizontally from the axis. In this scanning electron photomicrograph of *Knightia excelsa,* the transverse section faces towards the top. The width of the picture corresponds to about two millimetres.

alive are known as *parenchyma cells,* and they have thin walls. They occur in both xylem and phloem. Parenchyma cells are usually loosely packed, with the intercellular space containing air. One of their main functions is the storage of *starch,* the plant's principal reserve food material. In the xylem there is another prominent type of cell, the *tracheid,* which is a fully developed cell, but one which has died and from which the cytoplasm has disappeared. Tracheids are invariably tightly packed, and the space between them is filled with a complex material that we will consider later. They are usually elongated, and their walls are relatively thick. In the phloem, the counterpart of the tracheid is the *sieve cell,* but this is found only in the Gymnospermae. Another type of cell is the *fibre,* which is present in the xylem of Angiospermae and the phloem of both classes. Their main contribution is to the mechanical strength of the plant. Finally, in the Angiospermae, there are *vessel members* and *sieve-tube members*; the former are present in the xylem and the latter in the phloem. Plant cells communicate with one another over limited regions in the areas of their mutual contact. This takes place through what are known as *pit pairs.* A pit is a place in a cell boundary at which much of the normal wall material is missing, and it appears as a physical depression. The pits occur in complementary pairs, one in each of the adjacent cells, and all that remains at such a pit site is a relatively thin membrane through which liquid can diffuse from one cell to the other. In the vascular system, these structures are of a special type known as *bordered pit pairs.* The border is a thickened region, the added strength of which compensates for the missing material at the middle of the pit.

The actual arrangement of cells within the two important regions of the vascular system is conveniently discussed by taking the tree as an example. Both xylem and phloem consist of a network of cells arranged with their long dimensions both parallel with and transverse to the axis of the trunk. This produces a structure containing vertical and horizontal channels through which liquid can flow. In the phloem of the Angiospermae, sieve tubes are formed from rows of sieve-tube members arranged end to end and connected by walls that are dotted with holes, these walls being known as *sieve plates.* In the Gymnospermae, there is a somewhat similar arrangement of sieve cells, the contacts between which are established through the above-mentioned pit pairs. The xylem has a more complicated structure which consists mostly of tracheid cells, arranged with their long axes parallel to the axis of the trunk, but also includes some parenchyma cells. Some of these latter have their long axes parallel to those of the tracheid cells, but others, known as *parenchyma rays,* lie at right angles to this, radiating out from the centre of the pith towards the bark. Many of the parenchyma ray cells share pit pairs with tracheids and sieve cells, the structures in this case being referred to as *cross-field pits.* When vessel members are present, these form channels in the xylem referred to as *vessels.* Their function is to transport *sap* from the roots up through the trunk to the leaves, a task otherwise performed by the tracheids. Vessels are comparable in structure to the sieve tubes in the phloem, and they are formed by the end-to-end arrangement of vessel members. The walls at the joining areas between the cells lack cellulose, and this renders them susceptible to attack by certain enzymes. The end regions ultimately disappear, to leave what are called *perforation plates.* In view of the rather unusual terminology it should be emphasized that both sieve plates and perforation plates actually contain or consist of holes. An idea of the relative proportions of different types of

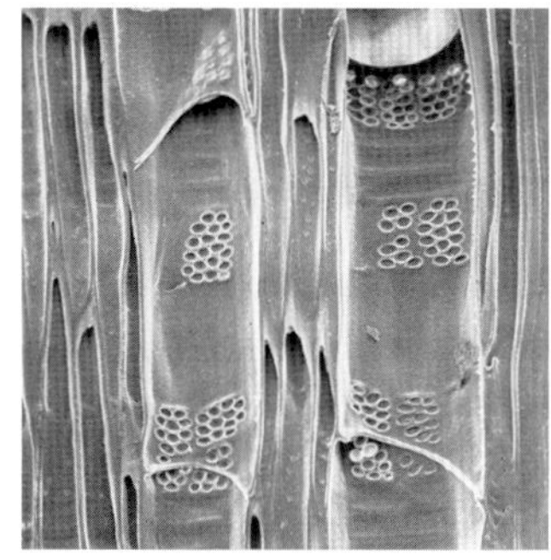

Vessel members, which are arranged vertically, interact with the horizontally-lying ray cells through cross-field pits. These structures, examples of which are seen here in a specimen of black poplar (*Populus nigra*), thus provide the connections in a system of thoroughfares that can loosely be compared with the elevator shafts and corridors of a multistorey building. The width of this scanning electron photomicrograph is equivalent to about one third of a millimetre.

cell in wood is given by the fact that tracheids account for over 95% of the total weight in a typical coniferous specimen.

The final degree of detail concerns the actual structure of an individual plant cell. Although the following description refers specifically to wood, it is worth noting that the main constituent of the tracheid belongs to a larger category of materials known as *natural fibres*. Thus much of what follows would also apply to leaf fibres such as *hemp* and *sisal*, to bark fibres like *jute* and *flax*, and to seed fibres such as *cotton* and *kapok*. A typical tracheid in a *hardwood* has a length somewhere between 0.25 mm and 1 mm, while its diameter is smaller by a factor of about 10. In *softwoods* the average tracheid length is about 3.5 mm, and in one species, *Araucaria*, it reaches as much as 11 mm. In an animal cell, everything of interest lies at or within the cytoplasmic membrane. This is not true of the plant cell; the wall is outside the membrane, and when the cell dies, the wall is all that is left. The walls can be compared to the skeleton of an animal, and they are laid down in two stages: primary and secondary. The former is present when the cell is still growing, while the latter develops only when growth is about to terminate and differentiation sets in. The total wall thickness is typically about one tenth of the diameter of the cell, for cells generated in the early part of the growing season, while late wood has a somewhat thicker wall. The walls are found to consist of several layers. Working from the outside we first have the intercellular material, also known as the *middle lamella*. Inside this lie, in turn, four different layers: the primary wall, designated by the letter P, the outer layer of the secondary wall, S1, the middle layer of the secondary wall, S2, and the innermost layer of the secondary wall, S3. Finally, we come to the *helical thickening*, HT, and the *warty layer*, W. The various walls are all composed primarily of cellulose, arranged in the form of thin threads known as *microfibrils*.

Cellulose is a natural polymer, the structure of which will be described later. The difference in the arrangement of the various walls lies in the configuration of the microfibrils. In the primary wall these are irregularly and rather loosely disposed, the interfibril space being filled with *lignin* and *hemicellulose*, both of which are organic polymers. The microfibrils in the S1 layer lie neatly parallel in a direction almost perpendicular to the long axis of the cell. In S2, which is by far the thickest of the layers, they are tightly packed and lie almost parallel to the cellular axis. It is this part of the structure which gives the cell most of its mechanical strength. The structure of S3 is rather similar to that in S1, both with respect to thickness and direction of microfibril alignment. Lignin is present in the narrow spaces between the microfibrils. The helical thickening is a feature observed on that surface of the S3 which faces towards the centre of the cell. It is not always present, but it is particularly prominent in the tracheids of some of the *conifers*, especially those of the *Douglas fir*. The warty layer takes its name from the small protrusions which are apparently formed just before the death of the cell's *protoplast*. The latter is the membrane-bounded region lying within the various cell walls. It is the actively metabolizing part of the plant cell.

When the primary wall is being formed, the cell is stretching out in the direction of its long axis. At this time it still contains its cytoplasmic interior, including the *nucleus*, the *Golgi body*, and the extensive membrane system of the *endoplasmic reticulum*. The latter is probably continuous with the surface membrane, and from this composite are extruded the growing cellulose

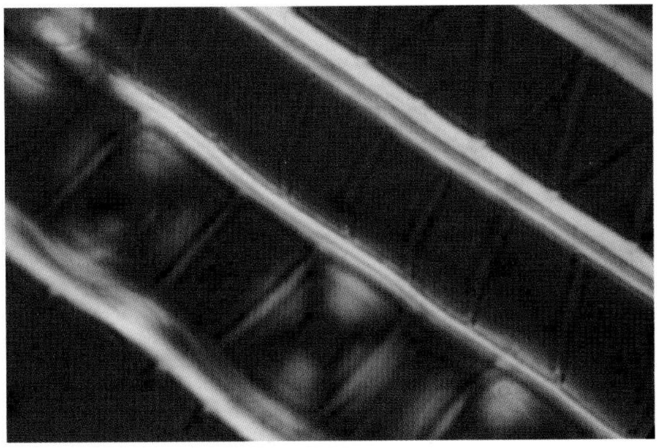

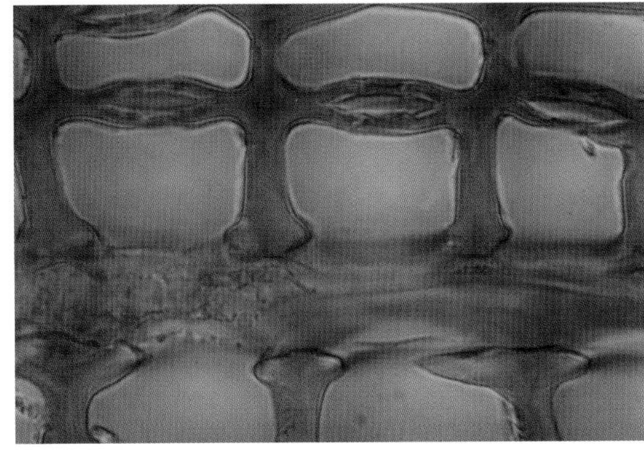

Adjacent tracheid cells interact through the thin membranes which span the areas of what are known as bordered pits. These two photomicrographs show examples with different viewing directions. In the longitudinal section of yew (*Taxus baccata*), on the left, they are seen face on; observation in polarized red light reveals the pits as circles at the centres of orange and blue-green sectors. The transverse section of Scots pine (*Pinus sylvestris*), on the right, was coloured by phloroglucinol-hydrochloric acid, which specifically stains lignin light red. Bordered pits are seen in cross-section in this specimen, appearing on the radial cell walls. Each picture has a width corresponding to about ninety micrometres. Helical thickenings are visible in the yew specimen, while non-lignified ray cell walls in the pine wood are seen spanning the width of the right-hand picture.

molecules. At the same time, there is deposition outside the cell of hemicelluloses, *pectin*, and a small amount of lignin, some of this material being passed on from the Golgi apparatus in the membrane-bounded packets known as *vesicles*. The cell also contains one large *vacuole*, which plays a decisive role in the rigidity of the plant. By selectively permitting the passage of inorganic salts, it develops a pressure that has been estimated to vary from 8 atm to 18 atm, and it thereby presses the cytoplasm out against the cell wall. Because the same thing is happening simultaneously in the surrounding cells, this holds the walls rigidly in place. The pressure developed by the vacuole is known as the *turgor pressure*. The secondary walls form on the inside of the primary wall after elongation of the cell is complete.

Because microfibrils are parallel packings of cellulose molecules, and because the microfibrils represent such a large proportion of the fibrous material, it is clear that cellulose is the most prominent structural molecule of every plant. It is thus appropriate that we examine *cellulose* in atomic detail. As was established by the X-ray investigations of Kurt Meyer and Herman Mark, in the 1920s, it is a linear polymer molecule, composed of glucose monomers, and it does not display branching. The individual cellulose molecules are usually very long indeed, a typical number of monomers lying in the tens of thousands. It is not surprising to find that *glucose* is the building block for cellulose, because it is the product of photosynthesis, which will be described later. There are two interesting points regarding the arrangement of the links in the polymer chain. One is that the individual monomers have what is referred to as the *chair conformation*, so the molecule is saw-toothed rather than ribbon-like in appearance. The other peculiar feature is that successive monomers in the chain are rotated through 180°, so that the bulkier side groups point in alternatingly opposite directions. This atomic arrangement is the key to the great tensile strength of cellulose, which exceeds that of steel if specimens of equal thickness are compared. The molecule is straight, and the covalent bonding, both in the six-membered rings and between the monomers, makes for a stiff chain. Moreover, the alternating arrangement of the side groups allows adjacent cellulose molecules to fit neatly together. This permits lateral bonding through a mixture of *hydrogen bonds* and *van der Waals bonds*. The orderly packing of the cellulose molecules leads to a degree of crystallinity in the microfibrils that can be as high as 80%, and the overall structure is quite reminiscent of the *fringed micelle texture* often seen in synthetic polymers. It is interesting to note that whereas isolated cellulose molecules are highly attractive to water,

Plant sugars and cellulose

In addition to glucose and galactose, which were discussed in Chapter 12, there are many other sugars commonly found in plants. The best known is fructose, which is unusual in that it comprises a five-membered ring. It is nevertheless a hexose; it has six carbon atoms. Its structural formula is shown below, together with those of several other plant sugars. Arabinose also has a five-membered ring, and, like xylose, it contains only five carbon atoms and is thus a pentose. Fucose and rhamnose both lack a hydroxyl group, while galacturonic acid has the unusual feature of a carbonyl oxygen.

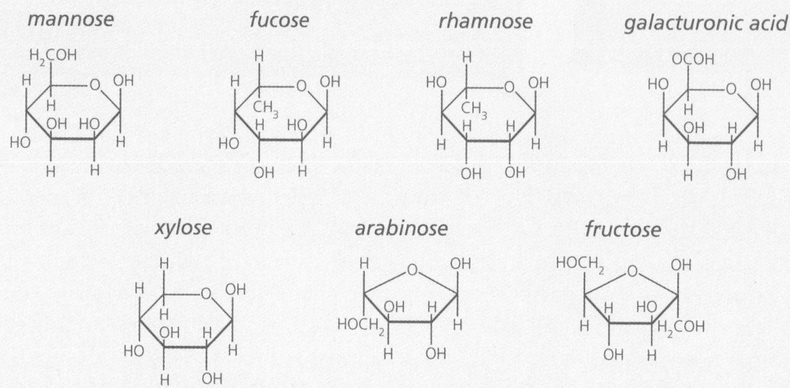

In 1811, Constantine Kirchhoff discovered that starch can be broken down by acids, to produce a sweet syrup from which sugar can be crystallized. It subsequently transpired that this product is identical to the sugar found in grapes, honey, and indeed the urine of diabetics. But another common plant sugar, that obtained from sugar cane, sugar beet and maple tree sap, is different. This is sucrose, the familiar domestic form, and it is a disaccharide. It comprises one molecule of glucose and one of fructose, joined by a glycosidic link. A molecule of sucrose is shown in the following diagram.

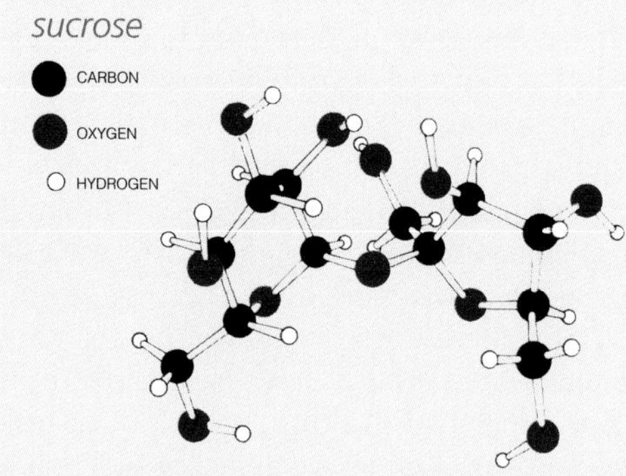

The two most abundant plant polysaccharides are cellulose, the main structural polymer, and amylose, also known as starch, which the plant uses to store energy. Sections of these important molecules are shown in the following diagrams.

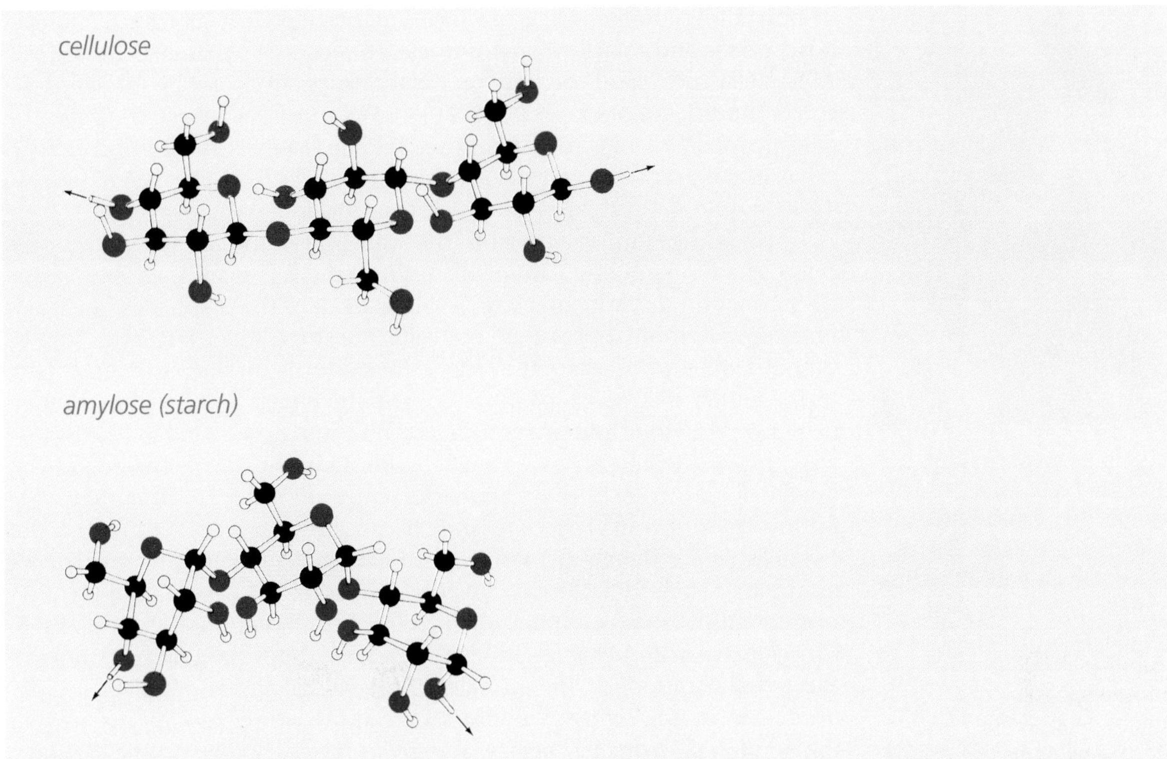

The monomer is glucose in both cases, the structural differences arising from the way in which the glucose units are arranged. The arrows indicate the junctions with the rest of the molecule, in each case.

because of the presence of so many hydroxyl groups, a microfibril is insoluble because those same groups are no longer available for reaction. They are already interacting with each other to form the links between the cellulose molecules.

The middle lamella contains molecules related to cellulose, known as *hemicelluloses* and *pectins*, and also *lignins*. Like cellulose itself, these are carbohydrates. They distinguish themselves from cellulose because, in the context of plant tissue at least, they show only relatively minor degrees of crystallization and polymerization. Actually the hemicelluloses are a family of molecules that comprises a number of different sugars, including glucose, xylose, galactose, fucose, mannose and arabinose. The chemistry of pectins is still an active area of research, and their structures have not yet been fully established. It is known that they all contain a high proportion of polygalacturonic acid, which has a polymeric structure. Pectins also comprise such sugars as rhamnose, galactose and arabinose. The situation regarding the lignins is even less clear, because of the sheer complexity of these compounds. It has been firmly established that the chief building block of the structure is phenylpropane, but just how the several phenylpropane molecules and their various oxidation products are linked together is still a matter of much discussion. As an example of the relative abundances of the different components, *Douglas fir* contains about 57% cellulose, 28% lignin and the remainder is hemicellulose and pectin. Regarding the sugar

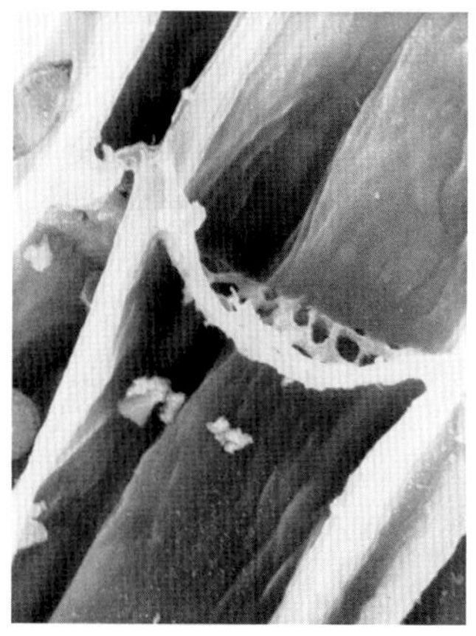

Sieve-tube members are elongated cells arranged end to end. They can function as fluid conduits in the phloem because their interconnecting regions, known as sieve plates, are dotted with holes. This scanning electron photomicrograph of a longitudinal section of *Trochodendron aralioides* shows the pores in a sieve plate separating two cells. The width of the picture is about twenty micrometres.

composition of the cell walls of various plants, it is clear that glucose, arabinose, xylose and galactose predominate. In *sugar cane*, these represent 13%, 25%, 26% and 9% of the total by weight, respectively, while for *wheat* the corresponding numbers are 18%, 21%, 25% and 12%. In woody tissue, the conspicuous extra sugar is mannose. Taking Douglas fir again, as a typical example, the following sugar composition has been measured: glucose 44%, mannose 11%, galactose 5%, xylose 3% and arabinose 3%.

It was earlier emphasized that the wall of a plant cell lies outside the cytoplasmic membrane. Up to this point, little has been said about what goes on within that organic bag. To a great extent the organelles, and their function, resemble those of animal cells. But there are nevertheless significant differences, and again they suggest superiority on the part of the plant. In the earlier discussion of biopolymers, the *amino acids* used by human beings were divided into two groups: essential and non-essential. The former were thus named because they must be found in the diet. Plants require no uptake of amino acids: they can manufacture all of the 20 varieties needed for protein synthesis. This is possible because they possess all the enzymes required in the metabolic pathways of amino-acid synthesis. A domestic example of such self-sufficiency is the cultivation of *cress* on damp cotton wool; it absorbs only certain minerals from the water. Plants are also self-sufficient regarding *vitamins*. Indeed, they are an important dietary source of these vital compounds, for animals. *Carnivorous plants*, also known as *insectivores*, are an interesting exception to the above principles. Species such as the familiar *pitcher plant*, *sundew* and *Venus flytrap*, possess enzymes which enable them to digest trapped insects, thereby augmenting their supply of amino acids and minerals. This ability enables them to thrive in relatively barren soil. Charles Darwin was the first to demonstrate that some insectivores can digest protein, using egg white as a test material. At least 500 species of carnivorous plants have now been identified.

Considering that plants are also capable of photosynthesis, which enables them to obtain carbon from the atmosphere, their independence is impressive. But they nevertheless have a major inadequacy: they cannot directly convert atmospheric nitrogen for their own purposes. Such a facility, called *nitrogen fixation*, is possessed by certain bacteria found in the soil. Known as *Rhizobium*, they reside in *root nodules* and enjoy a symbiotic existence with their host plants, supplying nitrogen and obtaining carbon, by way of sugar, in return. Not surprisingly, in view of its essential presence in proteins, nitrogen heads the list of every plant's mineral requirements. For the average green plant, the ideal quantity is about 15 mmol/l (millimoles per litre) of ingested water, and the element is obtained from salts containing both nitrate and ammonium ions. Somewhat smaller amounts of several other minerals are also mandatory: potassium, for maintenance of the small voltages which occur across cell membranes; calcium, which influences membrane permeability; phosphorus, required in nucleic acids, ATP, and membrane lipids; sulphur, which provides the cross-links in proteins and is also important in coenzymes for carbohydrate and lipid metabolism; magnesium, the central atom in the chlorophyll molecule; and iron and copper, which occupy the analogous active sites in several of the molecules in the *electron transport chain*. Even this long list is incomplete because there are about an equal number of other elements required in trace amounts that can be as low as 10 nmol/l (nanomoles per litre) of water.

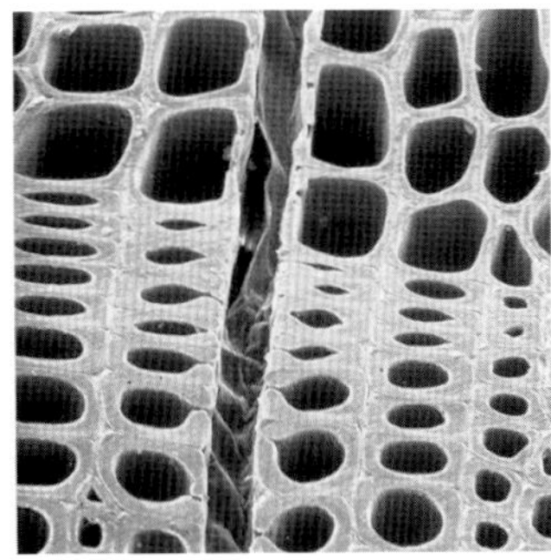

The abrupt change in cell wall thickness which produces the familiar annual ring pattern is clearly visible in this transverse section of Monterey pine (*Pinus radiata*). The late-wood cells, seen in the lower part of this scanning electron photomicrograph, are also somewhat smaller than their early-wood counterparts. The cells are tracheids and a ray vertically traverses the picture, the width of which corresponds to about a third of a millimetre.

Unlike animals, plants grow throughout their lives, the main sites of dimensional increase being located at the extremities: the tip of the roots and the shoots. In some plants, there is a well-defined secondary growth, but such a stage does not occur in many herbaceous species. It is also absent in *monocotyledonous species*, that is species having only one leaf-forming section in their embryos. When the establishment of the basic structure of the plant is followed by a period of secondary growth, the latter lasts as long as the plant lives. Secondary tissue is laid down either side of the cambium, producing secondary xylem and secondary phloem, which respectively push the primary xylem and primary phloem farther away from the cambium. During the early part of the annual growth season, the new secondary xylem consists of large cells with relatively thin walls. As the season proceeds towards autumn, the cells get smaller and their walls thicker, and this produces the familiar ring pattern, which can be seen clearly in transverse section. These are of course the *annual rings*, from which the age of the tree can be determined. In general, the regions of a plant in which the cells are multiplying most rapidly are known as the *meristems*, and the terminal growth points are referred to as *apical meristems*. The cambium is simply an extension of the apical meristem parallel to the axis of the plant.

As with animals, *cell differentiation* and *specialization* in plants appear to be under the control of chemical signal substances, some of which have been isolated and identified, including *indoleacetic acid* and *gibberellic acid*. The former promotes differentiation of xylem, while the latter appears to play an analogous role in phloem. The ratio of the amounts of these two substances present in the cambium seems to be important in the control of that growth region. Plants have a built-in mechanism which prevents excess accumulation of growth hormones. Such natural safeguards are ineffective against artificial compounds which sufficiently resemble the hormones that they can mimic their action, while eluding biological control. This is the basis of some *herbicides*, a particularly potent example being 2,4-dichlorophenoxyacetic acid. But it is toxic only to *dicotyledonous plants*. Monocotyledonous varieties like tulips, palms, and cereals are unaffected. This differential effect is exploited in the control of *weeds* in lawns and agricultural crops.

Indoleacetic acid is one of a family of plant hormones known as *auxins*. Their discovery followed an early observation by Charles Darwin and his son Francis that seedlings can be dissuaded from bending towards the light if their extremities are covered by an opaque cap. Frits Went detected that such shoot tips secrete a growth-promoting substance, in 1927, and this was subsequently found to be auxin. Applied to plant cuttings, auxin increases both the number of roots and the proportion of successful grafts, and has had considerable commercial significance. Other families of growth hormones have now been discovered. The *cytokinins*, the most prominent of which is zeatin, are related to the purine base adenine which we have encountered in connection with nucleic acids. These compounds promote leaf growth and appear to delay leaf deterioration and detachment. They thus influence senescence, the natural ageing process. Cytokinins also promote cell enlargement and seed germination, which they appear to achieve by increasing nucleic-acid metabolism and protein synthesis. The opposite trend is caused by another hormone known as *abscisic acid*,

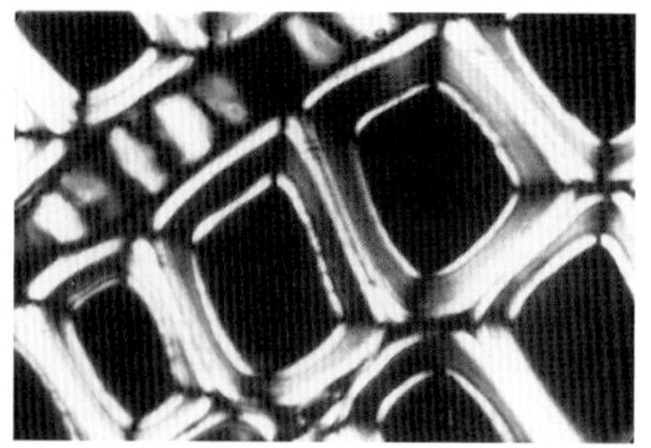

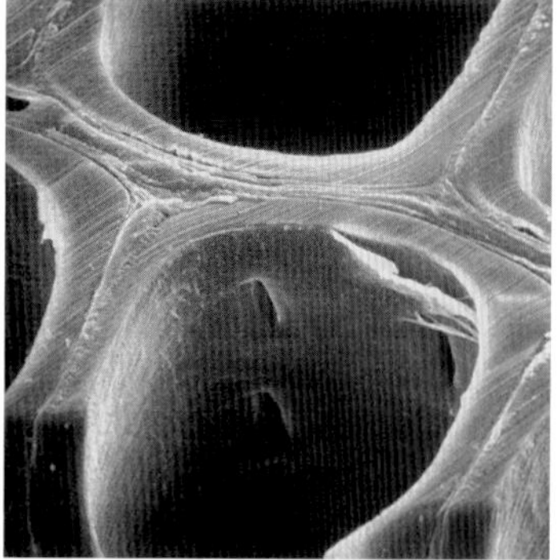

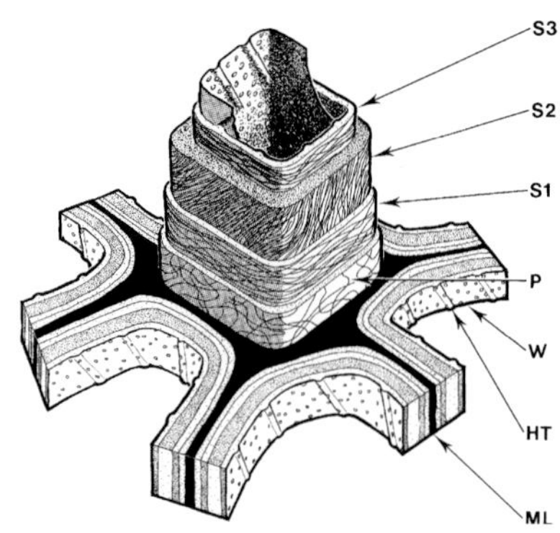

The wall of a wood cell is a quite complex structure. No less than seven concentrically-arranged components are distinguishable in many specimens. The letters identify these on the rather idealized diagram. The middle lamella (ML) is composed of pectin, hemicellulose, and lignin. The remaining layers, in which cellulose is also present, are known as the primary wall (P), the outer, middle, and inner secondary walls (S1, S2, and S3), the helical thickening (HT), and the warty layer (W). These features can be distinguished in the middle picture, which is a scanning electron photomicrograph of Monterey pine (*Pinus radiata*). The transverse sectional cut has caused partial separation of the S2 and S3 layers. The warty layer is barely discernible on the inner surface. When viewed between crossed polarizers, in an optical microscope, the various wall layers are more readily identified because the cellulose molecules in their microfibrils lie at different angles to the incident light. The layers thus rotate the plane of polarization by different amounts, and this gives them varying degrees of optical density. Four layers can be seen in the transverse cross-section of *Pinus radiata* on the left. Working inwards from the intercellular region, these are ML, a combined P and S1, S2, and S3.

which can suppress growth to such an extent that it can actually induce dormancy.

Unlike higher animals, plants must fend for themselves from the earliest stage of their development. And the germinating seed must grow in the correct directions in order to reach its vital requirements. Gravity causes water to seek the lowest possible level, and the *geotropism* of growing main roots enables them to reach downward toward moisture. Plant stems, on the other hand, are said to be negatively geotropic because they elongate in the upward direction. *Phototropism* is the tendency for growth toward the light that is essential for photosynthesis. A full explanation of these mechanisms is still not available, but it does seem clear that various growth hormones are involved. Auxin appears to be redistributed toward the lower side of a root, and because it actually inhibits growth of root cells, this causes a horizontal growing root to curve downward. Auxin's influence on shoots is apparently just the opposite, so they are stimulated to grow upward. Phototropism, which causes indoor plants to bend toward a window, has been attributed to a greater concentration of auxin on the side facing the light, but other hormones are believed to be implicated simultaneously. There is another, more subtle, growth trait in plants. When the growing organism is deprived of light, as in the case of a seedling which has not yet reached the soil surface, the stem elongates at an unusually high rate and the leaves remain small and primitive. Moreover, chloroplast and chlorophyll generation are arrested. All four effects are reversed when the plant reaches the light. This set of tendencies, known as *etiolation*, is of obvious benefit to the juvenile plant, and H. A. Borthwick and S. B. Hendricks have traced the behaviour to a control molecule they called *phytochrome*. They postulated that this blue-coloured chromoprotein is capable of adopting two conformations, and thereby explained why red light promotes leaf growth whereas wavelengths near the infrared (heat) region have an inhibiting influence. The Sun's infrared radiation heats the soil but no light penetrates to the underlying regions, so the shoots develop leaves only when they reach the surface.

The question of cellular differentiation was discussed in the previous two chapters, and little need be added here except in respect of the apparently major role played by *microtubules*. The presence of these structures in plants was discovered in the early 1960s by Myron Ledbetter and Keith Porter. Microtubules form remarkably regular arrays in the vicinity of the cytoplasmic membrane, lying parallel to the direction of the cellulose

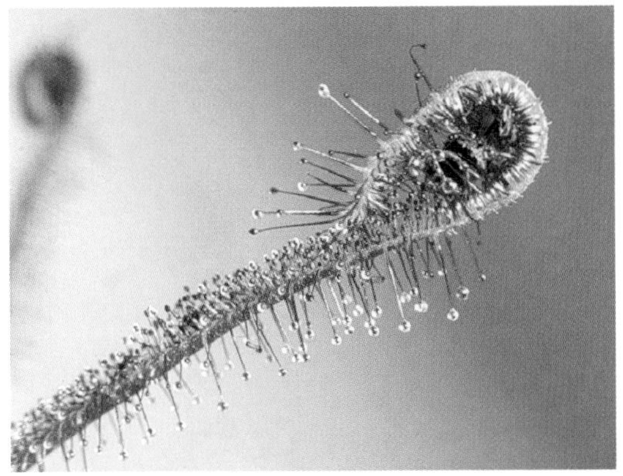

Plants are capable of synthesizing all the amino acids required by their metabolism, so they are usually not meat eaters. Some species can extract nutriment from animal remains, however, because they possess the enzymes necessary for breaking down chitin and protein. The insectivorous plants have developed a range of strategies for trapping their prey. The leaves of the sundew (*Drosera*) shown on the left, are covered with sticky tentacles. These are particularly interesting in that they play a dual role: they have conduits which supply the glands that secrete the digestive enzymes, and the digestion products are transported to the leaf by the same route. The leaf of the Venus flytrap (*Dionaea*), pictured on the right, consists of two lobes connected by a midrib. The upper surface of each lobe is equipped with three sensory hairs. If any of these is touched, it transmits an electrical signal to the midrib region and the lobes close. The rows of interlocking teeth, arranged along the margins of the lobes, block the escape of the prey.

microfibrils. It is possible that they actually marshal the latter into the correct direction. As noted earlier, microtubules are disrupted by *colchicine*, and in the presence of this compound the microfilament arrangement in a growing plant cell becomes uncharacteristically haphazard.

The products of the differentiation include the elaboration of the root and branching systems, and the development of the leaves and reproductive centres. The continual growth of the *roots* ensures that the extremities will be very fine structures, measuring a few tens of cells across, and this guarantees that the vascular system is close to its source of water. In the roots, the xylem and phloem lie side by side and not coaxial with each other as is the case farther up the stem. This enables moisture to enter the xylem, without traversing the phloem. In the leaf, the pressure of evolution has led to the production of special cells which help to support this delicate but vital structure. The vascular bundles themselves are highly branched, and they are surrounded by a strengthening and supporting tissue that consists of *sclerenchyma fibres* and *collenchyma cells*. Most importantly, the *leaf* contains *mesophyll cells*, which are further differentiated into *palisade* and *spongy cells*. The former are very rich in chloroplasts, and they are the main sites of photosynthesis. The spongy mesophyll contains fewer chloroplasts, and its cells are rather loosely packed, with quite large air spaces between them. These spaces communicate directly with the atmosphere, through openings on the lower side of the leaf known as *stomata* (singular: stoma). It should be noted that although the non-green parts of a plant do not participate in photosynthesis, they nevertheless require a gas exchange system with the atmosphere in order to carry out *respiration*. The rather open packing of the cells at the underside of leaves, so necessary for the exchange of gases, is also the main cause of water loss by *transpiration*. One could say that this is the penalty the plant pays for its *aerobic metabolism*, but transpiration is in fact useful. The enhanced water flow that it provokes increases transport of necessary minerals up to the leaves, and it also helps to keep the plant cool by evaporation.

In spite of the diversity of outward form seen in the plant world, the basic aim remains the same: water, carbon dioxide and sunlight must rendezvous in the chloroplasts and produce glucose by photosynthesis. Having described the various structures of a plant that enable it to realize this goal, we now turn to the molecular aspects of the contributory processes. It will already be clear that water transport plays a vital role in the functioning of

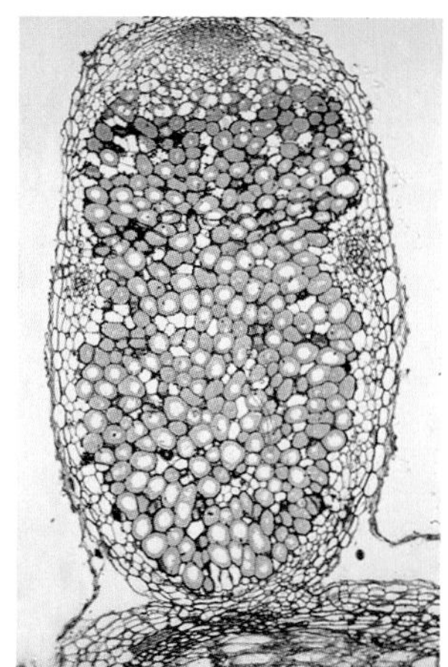

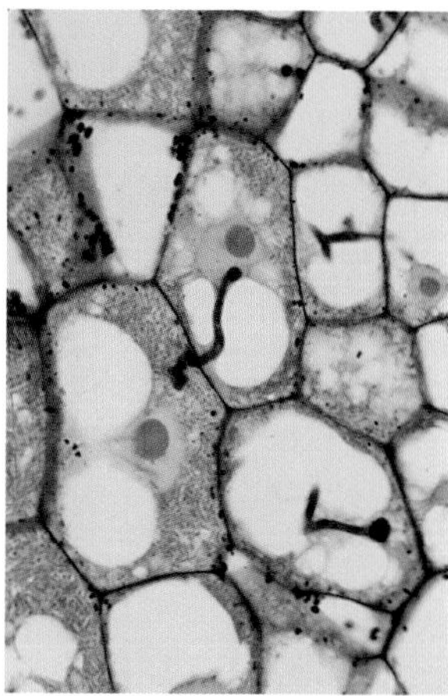

Nitrogen fixation is accomplished by certain bacteria, which enjoy a symbiotic association with leguminous plants such as clover. The bacteria are present in the cells of root nodules, which are visible as roughly spherical swellings in the top picture. The host plant shown here is lesser yellow trefoil (*Trifolium dubium*). The lower left picture shows a longitudinal section of a nodule. The reddish colouring seen on a freshly-cut nodule derives from the presence of leghaemoglobin, an oxygen carrier capable of maintaining the optimum oxygen concentration for efficient nitrogen fixation. The bacteria enter the root through a root hair, causing it to curl up and develop an infection thread which is usually tube-shaped. The bacteria reside in this thread and its gradual elongation provides access to the inner part of the cortex. It is here that the bacteria provoke the cell division and proliferation that ultimately produces the nodule, and the young cells in the vicinity of the apical meristem (i.e. the growing extremity) are penetrated by infection threads. In the final picture, which has a width corresponding to about 140 micrometres, the infection threads appear as dark worm-like objects. The bacteria, in this case *Rhizobium,* have a rather similar appearance, but are much smaller. They are just about visible in the regions adjacent to the cell walls. The bacteria are not in direct contact with the host cell's cytoplasm, being surrounded by a membrane provided by the host itself.

any plant, and it is interesting in several respects. There are several physical and chemical agencies which can provoke flow. The most obvious is pressure difference, and this is what causes a liquid to flow through a drinking straw when suction is applied to one end. Suction is, of course, simply a lowering of pressure, and an analogous situation exists in the leaves due to the evaporation of water molecules into the atmosphere. This causes a decrease relative to the atmospheric pressure that acts on the water near the roots, so the liquid tends to flow from root to leaf. We should note, however, that the maximum possible pressure difference that could be set up in this way is one atmosphere; it would correspond to the highly unlikely situation of such rapid transpiration that the pressure at the leaves actually fell to zero. One atmosphere (1 atm) of pressure is capable of supporting a column of water with a height of approximately 10 m. We are thus confronted with the problem of explaining how it is that trees having heights well in excess of this, such as the giant *Californian redwood, Sequoiadendron giganteum* contrive to get sap to their uppermost leaves. These trees can attain heights of more than 100 m. Other forces must be at play.

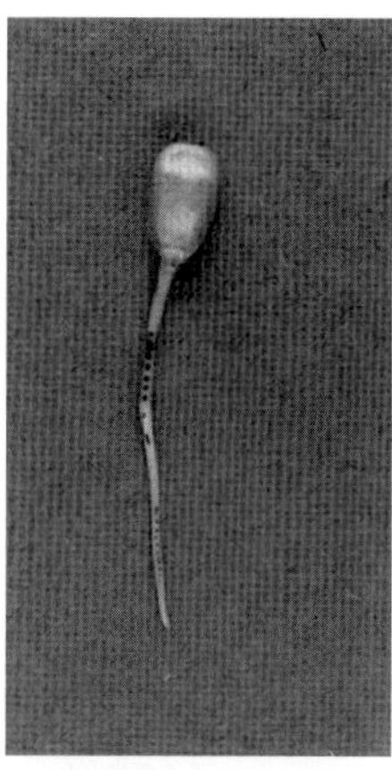

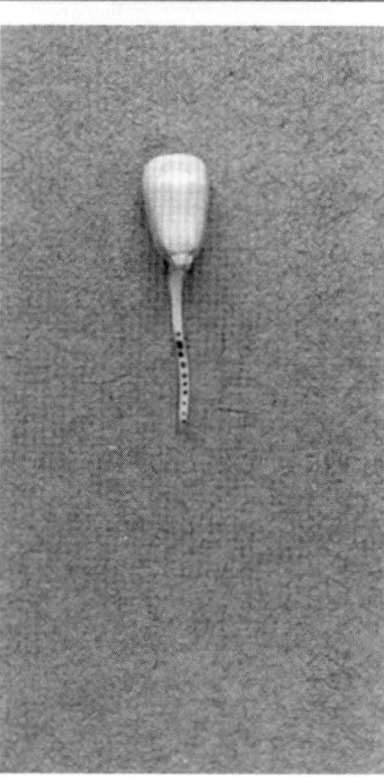

Plant growth is not uniform throughout the organism; the dimensional changes preferentially occur at the extremities. This can be checked by suitably marking a growing plant's surface. Ten points at one-millimetre intervals, were drawn on the root of this corn seedling, and the second picture shows the situation fourteen hours later. It is evident that the tip region experienced the largest growth during this period, and the lowest original spot has been stretched out into twelve smaller spots. Observations must be made over a much smaller period if one wishes to determine the instantaneous growth rate.

A particularly powerful cause of water transport, *osmosis*, arises because of differences in solute concentration. A membrane is said to be *semipermeable*, under a given set of conditions, if some molecules can readily pass through its microscopic pores while others are prevented from doing so by their size. If such a membrane is used as the dividing wall between a region containing a pure solvent, such as water, and another region containing a solution of a solute in that solvent, and if the solute particles are too large to get through the membrane, the pure solvent flows preferentially in the direction of the solution, as if in an effort to dilute it. The rate and amount of this flow are determined by the strength of the solution, and the flow continues until it is counterbalanced by some other agency. This latter is frequently hydrostatic pressure, and it can arise in the following type of situation. Imagine that a concentrated solution of common salt is completely enclosed in a bag made of such a membrane, and this bag is then placed in pure water. The water flows into the bag, and the solution in the latter becomes continually more diluted. The entry of water into the bag increases the pressure, however, and there must come a point at which this equals the effective chemical pressure that is driving the water inwards. In that situation, equilibrium will have been attained and the flow of water will cease. The osmotic process is at work across the boundary of the cell. The cytoplasmic membrane, and also several of the organelle membranes, contain proteins which are able to selectively transfer inorganic ions such as potassium or sodium across the membrane in a particular direction. Such a *transport mechanism* is said to be of the *active* type, as distinct from normal diffusion, which is known as *passive transport*. The direction of selectivity is usually such that the region inside the membranous envelope is rich in potassium and deficient in sodium, while the external region has a high concentration of sodium but is depleted of potassium. There appears to be a protein pump for these ions, and a second protein works on chlorine to produce a higher concentration of its ions on the inside of the membrane.

Here, then, is one of the main uses of the energy made available by respiration. The ATP molecules liberated by that process provide the fuel for the active pumps, and these in turn maintain the gradients of the various inorganic ions. Finally, the osmotic pressure produced by these ionic gradients causes water to flow. In the case of the vacuole, the inward flow of water produces a hydrostatic pressure in that organelle which can reach, as we saw earlier, up to 18 atm. It is this that produces the *turgor pressure* which holds the plant rigid. In the event of excessive transpiration, as during a period of unusually high temperature, the turgor pressure falls below its optimum value, and the plant wilts.

The osmotic passage of water into a cell would leave the region outside depleted of that liquid if it were not replaced by the supply coming up from the roots. Thus there is a second driving force which must be added to that caused by transpiration. But there is yet another agency that pulls the water up through the vascular system, and it turns out to be the most important of all. The *cohesion theory of sap ascent* was put forward in 1894, by Henry Dixon, but it won widespread support only much later. In the earlier discussion of *capillarity*, a description was given of the way in which a very narrow column of liquid can be supported by the cohesive forces between the molecules of the liquid and the atoms in the wall of the tube. The stronger these forces are, the longer is the column that can be supported. In the case of the plant,

Plant hormones

Hormones play a prominent role in plant growth and development. Their transport and distribution are said to be of the active type, which means that their movement is controlled by cellular components and is not merely dictated by diffusion. Cell growth and division is directed primarily by the interplay between auxins and cytokinins. The most common of the auxins, which promote growth by increasing the rate of cell elongation, is indoleacetic acid, the molecular structure of which is as follows, the obvious carbon and hydrogen atoms being implicit:

indoleacetic acid

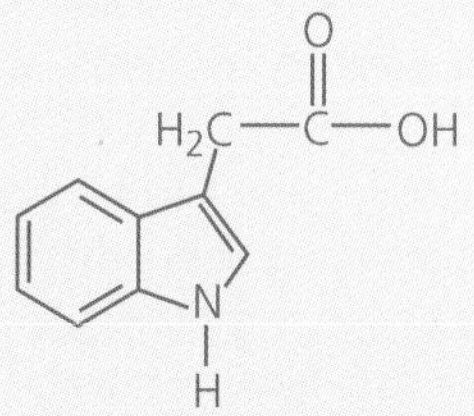

O
H_2C—C—OH
N
H

Auxins are widespread in living organisms. In addition to their ubiquitous presence in plants, they have been isolated from fungi, bacteria, and human saliva and urine. The cytokinins stimulate cell division, this action requiring the presence of auxins. They are chemically identified as purines, and the molecular structure of the most common type, zeatin, has the familiar double-ring structure:

zeatin

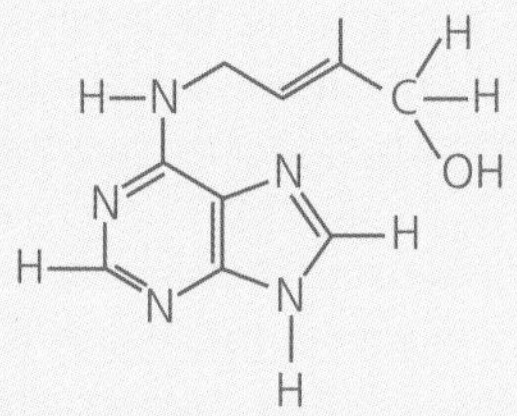

H
H—N C—H
N OH
N —H
H— N N
H

The gibberellins are usually present in minute quantities, and accelerate shoot and stem elongation. They were first discovered in the fungus *Gibberella fujikuroi* which causes abnormal growth when it infects rice plants. Dwarf peas treated with these hormones rapidly attain normal stature, and gibberellins have been used to increase the rate of growth of celery, producing a more tender stalk. This family of hormones comprises over thirty distinct chemical compounds, with molecular structures resembling those of the steroids of several animal hormones. Abscisic acid is a growth-inhibiting compound which counteracts the promoting hormones. It is present in most plant organs and its action, apparently reversible, is believed to lie in control of nucleic acid and protein synthesis. Its molecular structure is as follows:

abscisic acid

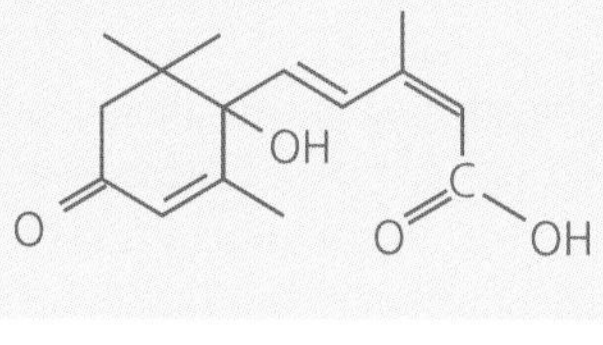

OH
C
O O OH

the important forces are those between the water and the atoms in the walls of the vessels. As we have seen, these walls consist primarily of cellulose, and the important chemical units in this context are the hydroxyl groups. Combining with nearby water molecules, these groups can form the mixed type of bond, a combination of hydrogen and ionic bonding, that exists in the water itself. The ultimate limitation on the height of a capillary

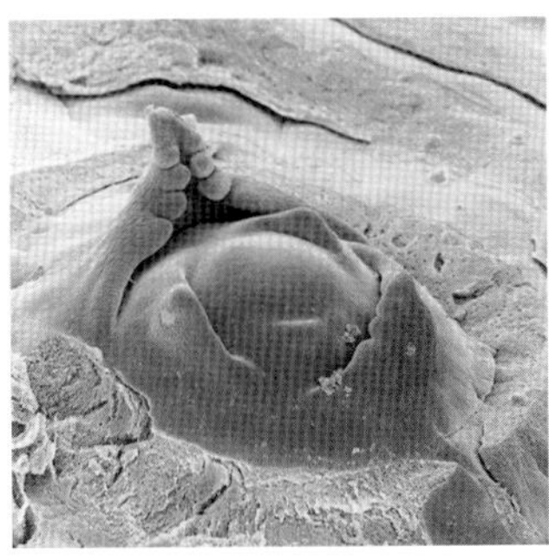

The point at which the new leaves of a growing plant appear is called the apex. The example seen in this scanning electron photomicrograph of a celery plant is about a millimetre in diameter. The emerging leaves are first seen as ridges, and they appear at equal time intervals if the temperature is constant. One interval is termed a plastachron.

column of water is determined by the *tensile strength* of the water itself, and the magnitude of the water–cellulose interaction is such as to permit this limit to be approached. Although it seems paradoxical, wet objects can be dried more effectively with a damp towel than a dry one. This arises from the helpful attraction between the absorbed water molecules and those already present in the towel fibres. Similarly, the moisture already inevitably present in the youngest cells provides the foothold for the rising water capillaries. It has indeed been demonstrated that water travels predominantly via the vessels and tracheids produced during the latest growth season: the outermost annual ring. Here, then, is the real reason why there are such large trees as the giant redwoods.

At the end of its journey, that water which is not lost by transpiration reaches the *palisade cells*, and its arrival brings us to the most important of all terrestrial chemical reactions: *photosynthesis*. More than 1000 years passed, following the original suggestion of de Saussure, before any real inroads were made on the problem of discovering exactly what the mechanism entails. Before going into the historical development of the story, we must consider certain facts which reveal a surprising complexity. In one of the input molecules, carbon dioxide, there are twice as many oxygen atoms as carbon atoms. In carbohydrates such as glucose, however, there are equal numbers of carbon and oxygen atoms. Adding the fact that photosynthesis liberates oxygen, it seems natural to conclude that the process involves removal of an oxygen atom from carbon dioxide and its release into the atmosphere. Such an interpretation would be quite fallacious. As demonstrated by Ruben Kamen, who used radioactive oxygen to label that element first in water and then in carbon dioxide, the oxygen liberated in photosynthesis is derived exclusively from water. Cornelis Van Niel had already hinted at something of this sort, in the 1930s, when he suggested that the main process occurring in photosynthesis is the light-driven production of a reducing agent. Apart from the participation of light, this is reminiscent of respiration, about which we should recall a few basic facts. *Oxidation* of a compound can occur in one of three ways: by addition of oxygen, or through the loss of hydrogen or an electron. *Reduction*, the reverse process, involves loss of oxygen, or gain of hydrogen or an electron. A *reducing agent* is a compound that causes another to be reduced, and the reducing agent itself becomes oxidized in the process. The other relevant aspect of the photosynthesis story is the participation of a number of molecules which pass hydrogen atoms or electrons from one to another. Indeed, some of these molecules are *cytochromes*, just as in the *respiratory chain*, while another molecule, NADP, is a cousin of an earlier acquaintance, NAD^+. Finally, and not surprisingly, considering what has been stated regarding its universality, there is ATP. We will soon see how these two energy-bearing molecules drive the reaction which produces glucose, the so-called *dark reaction*. First, however, we must consider how the photosynthetic process produces these two molecules in what is known as the *light reaction*.

Confirmation of the idea originally put forward by Van Niel came in 1939, with the work of Robert Hill, who experimented on chloroplasts that had been extracted from leaves by mashing these up, making a solution from the debris, and finally separating the chloroplasts with the aid of a centrifuge. He found that chloroplasts are capable of producing oxygen, when bathed in light, but only in the presence of a substance that could readily be reduced. It was subsequently found that the reducible substance actually

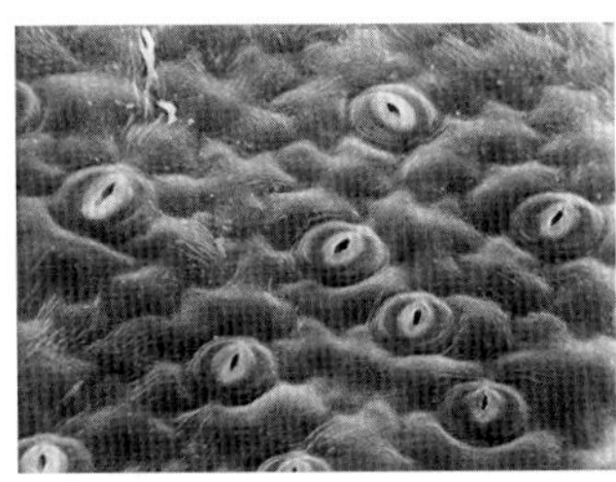

Stomata are openings in the epidermis on the lower side of the leaf, and they are bounded by two specialized guard cells. These cells are capable of changing shape, thereby regulating the gas exchange between the mesophyll and the atmosphere. This scanning electron photomicrograph shows stomata and the kidney-shaped guard cells in the leaf of a Christmas rose plant (*Helleborus*), the width of the picture spanning about a quarter of a millimetre.

present in palisade cells is nicotinamide adenine dinucleotide phosphate, *NADP*. The process which takes place in the chloroplasts, in the presence of light, is reduction of this coenzyme to produce $NADPH_2$. This is known as the *Hill reaction* and, like so many other biological processes, it requires the help of another molecule: *ferredoxin*. The latter is a protein with a single iron prosthetic group, which, in its reduced form, is the most powerful of all biological reducing agents. It was later discovered, by Daniel Arnon and Albert Frenkel, that a second energy-storing process is at work in the chloroplasts. It requires the presence of electron carriers such as the cytochromes, and the hydrogen carrier, *quinone*, and it produces ATP from ADP and inorganic phosphate. From what was stated earlier regarding the formation of ATP, and because the reaction requires light, it is not surprising that the process is known as *photophosphorylation*. It was also found that the Hill reaction and photophosphorylation occur not only simultaneously but cooperatively, with the participating molecules actually coupled together.

What is accomplished by the light energy in photosynthesis is thus reasonably clear, but there remains the question of just how these changes are brought about. Because photosynthesis is limited to the green parts of a plant, and since that colour derives from the pigment chlorophyll, it seems obvious that this must be the vital molecule. Strong support for this conclusion comes from optical experiments. Measurement of the rate of photosynthesis at various wavelengths yields what is known as an *action spectrum*, and some wavelengths are found to be more effective at producing the reaction than others. If a beam of white light is passed through a solution of chlorophyll, the red and blue wavelengths are absorbed particularly strongly, and the light that emerges is rich in the complementary wavelengths, the greens and yellows. The strongest absorption of light by chlorophyll occurs for just those wavelengths which produce the greatest activity levels in photosynthesis, that is the peaks of the action spectrum. The satisfaction engendered by this demonstration was subsequently tempered by the discovery that chlorophyll comes in several forms, each with its own characteristic peaks in the absorption spectrum. The most prominent version of the molecule, known as *chlorophyll a*, consists of two main parts: a bulky head group and a thinner hydrocarbon tail. Apart from the fact that it is branched, the tail resembles those of the *phospholipids*, and it burrows down into one of the many membrane bilayers situated within the chloroplast to give the chlorophyll an anchoring point. The head group is the business end of the molecule, and it is based on a cyclic arrangement of carbon and nitrogen that is known as the *porphyrin ring* system. At its centre, straddling the space between four nitrogen atoms, lies a doubly charged magnesium ion. *Chlorophyll b* is like chlorophyll a except for certain differences in the porphyrin ring, while *chlorophyll c* lacks the hydrocarbon tail group.

It has become clear that chlorophyll molecules do not function individually during photosynthesis. When this pigment is extracted from chloroplasts, it loses its photosynthetic abilities. Additional evidence came from the work of Robert Emerson and William Arnold, who investigated the efficiency of the process by exposing chloroplasts to brief flashes of light. They found that the maximum photosynthetic yield corresponds to one quantum of light energy being used by about four hundred chlorophyll molecules. The cooperative functioning of groups of molecules of this size can be compared with the increased efficiency of a radio telescope constructed from

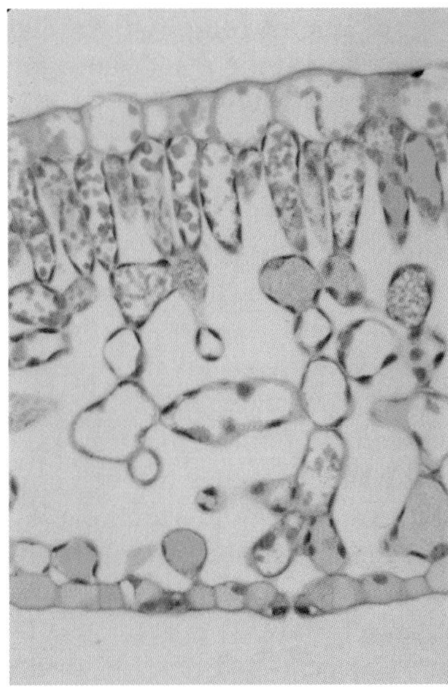
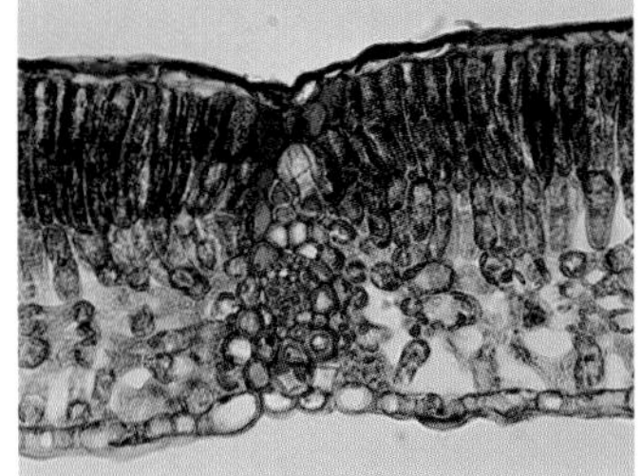
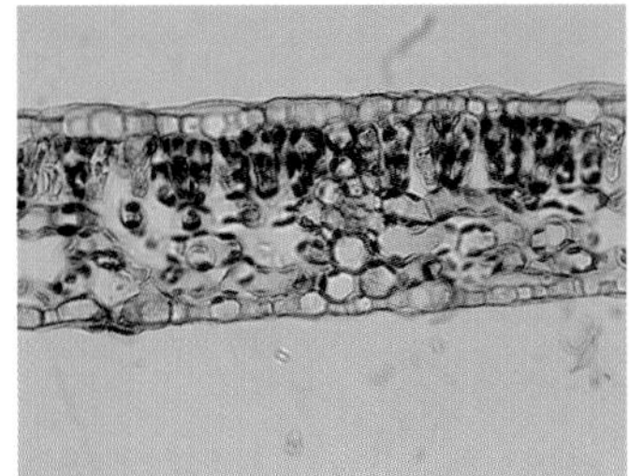

The open structure of the leaf interior is apparent in these cross-sectional views of specimens of beech (*Fagus silvestris*). In the first picture the mesophyll, or photosynthetic part of the leaf, is visible in the upper layer; the chloroplasts are the coloured spots arranged along the cell walls. A stoma, or pore, through which contact is made with the atmosphere, can be seen near the middle of the lower surface. The upper right and bottom right pictures show, respectively, sections of leaves from the exposed crown of the tree and from the shady inner region. The widths of both pictures, which were taken through a blue filter, correspond to about a third of a millimetre. The differences between the two specimens reveal the plant's response to varying environmental conditions, in this case light intensity. The leaf from the light part has a mesophyll region consisting of two layers of elongated palisade cells whereas the shaded leaf has only a single layer of bag-shaped palisade cells. Such structural variations arise during cellular differentiation, each leaf becoming committed to a subsequent pattern of development at a certain stage of its growth.

a large number of individual antennae. Finally, there is the evidence from electron microscopy, which has exposed the presence of structures with diameters of between 10 nm and 20 nm in chloroplast membranes.

The electron microscope has revealed much more about the internal composition of *chloroplasts*, which are about 10 μm in length. Each chloroplast contains numerous membranes stretched out parallel to the long dimension, and arranged in pairs known as *thylakoids*. In disc-shaped regions of the latter, the two membranes separate and envelope groups of approximately 400 chlorophyll molecules, rather like slices of green cheese lying between two pancakes. These are the functional units discovered by Emerson and Arnold, and they are known as *quantasomes*. In the higher plants, adjacent quantasomes line up with each other to form a stack known as a *granum* (plural: *grana*), which resembles a pile of coins. The regions between the grana are known as *stroma*, and they are crossed by the empty parts of the thylakoids. *Starch grains*, when present, are seen in the stroma.

Valuable information on the nature of the coupling between the Hill reaction and photophosphorylation, and an explanation of the multiplicity of chlorophyll types, was gained by Emerson in 1958. More than 20 years earlier, he had discovered that the liberation of a hydrogen atom from water requires two light quanta rather than one. Probing the influence of wavelength on the energy balance sheet, he perceived a strong hint of cooperativity. It was already well known that the photosynthetic response is rather poor at the long-wavelength (low-energy) red end of the spectrum, while the efficiency, measured in molecules of carbon dioxide converted per standard number of light quanta, is much higher at the short-wavelength (high-energy) blue end. Emerson discovered that when chloroplasts are simultaneously illuminated with both red and blue light, the conversion efficiency of the red wavelengths is greatly increased. Indeed, in the presence of blue light, the red wavelengths are just as effective as the blue. Robert Hill and D. S. Bendall suggested that the *Emerson effect*, as it has come to be called, must be caused by two different groups of chlorophyll units acting in tandem.

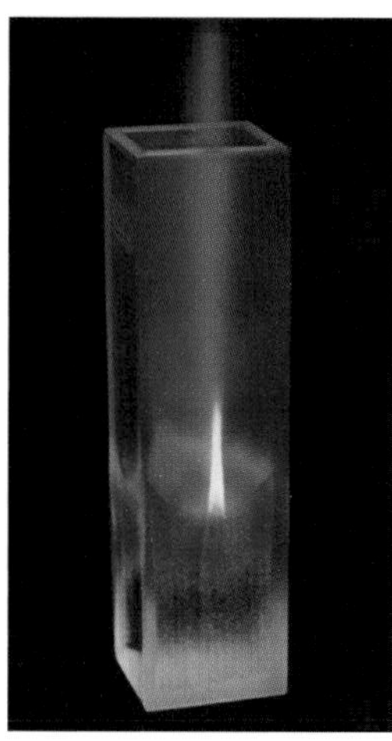

Photosynthesis involves the absorption of light by the pigment chlorophyll, and transmission of the captured energy to other molecules capable of converting water and carbon dioxide into glucose and oxygen. The natural green colour of the pigment is visible at the bottom of the container, which holds a solution of chlorophyll in ether. A beam of blue light, made visible by the scattering from airborne dust, enters the solution from above. Some of the energy absorbed from the light is dissipated as heat, while the remainder is re-emitted as fluorescence in the red region of the spectrum. Such fluorescence actually occurs in plant leaves too, but it is much weaker; most of the absorbed energy is passed on to the synthesizing enzymes.

A complete picture of the photosynthetic process is now at hand. An electron associated with the magnesium ion on one chlorophyll molecule is raised to an excited energy state by the absorption of a light quantum, and the excitation energy is then passed on in the form of hydrogen transport across a *quinone* molecule, with subsequent electron transport down a *cytochrome chain*. This chain, in turn, is coupled to another group of chlorophyll molecules which, by receiving the energy, become more readily excited by another light quantum. It is apparently this second excitation process, occurring within a few hundred picoseconds, which is linked to the Hill reaction, while both processes contribute to the production of ATP. A somewhat analogous situation occurs with the test-your-strength device which is a common feature of amusement parks. If sufficient energy is imparted to the lever, the slider reaches the bell and part of the mechanical effect is converted to sound energy, which could in principle be used to do further work. The gradual conversion of the kinetic energy of electrons to the potential energy of chemical bonds can be compared to the way in which a hot potato loses heat when passed around a ring of people. It was earlier thought that *ATP production* occurs at specific sites on the energy transport chains but Peter Mitchell, in 1961, put forward a radically different picture. As discussed in Chapter 16, he suggested that the main outcome of the different transport processes is the setting up of a hydrogen ion gradient. This is equivalent to what he called a *chemiosmotic force*, which then causes ATP production.

It is clear that Nature was confronted with several problems in ensuring that the various reactions involved in photosynthesis proceed as described. To begin with, several chlorophyll molecules have to lie sufficiently close to one another that they can function cooperatively. And it is equally vital that electrons being passed from one region to another, during the various stages, are prevented from recapture near the site of their excitation; the overall goal of the process is charge separation, and it would be arrested rapidly if that occurred. So the underlying rationale is placement of the participants in close proximity while keeping them away from substances that would cause premature oxidation or reduction. As we have seen so often before, such constraints call for a conglomerate of protein sub-units, in which are embedded the smaller molecules involved in the actual reactions. The challenge lay in determining the structure of this *photosynthetic reaction centre*. And the problem was exacerbated by the fact that this centre normally resides within a membrane. Proteins thus located usually owe their stability to the hydrophobic nature of the membrane's interior, and removal from the membrane would be expected to cause denaturation, that is to say disruption of the structure. This problem was solved by Johann Deisenhofer, Robert Huber and Hartmut Michel, who devised a method of extracting reaction centres from the photosynthetic membrane while retaining just enough of the lipid to ensure that denaturation did not occur. And when they subsequently crystallized a large number of the extracted centres, and subjected the resulting crystal to X-ray diffraction, they were pleased to find that the remnant lipid did not interfere with the determination of the centre's structure. Their efforts were rewarded with a breathtaking picture of numerous alpha helices, which were seen to provide scaffolding for the smaller molecules that participated in the reactions described above. So detailed were their results that Deisenhofer, Huber and Michel were able to follow the route taken by the moving electrons,

Chlorophyll

Chlorophyll is the green pigment present in virtually all higher plants. Molecules of this compound are located in the chloroplasts, where they convert the fleeting energy of incident light to the more persistent commodity, chemical energy. There are several modifications of the compound, the version known as chlorophyll *a* being the only one common to all plant organisms. Its structure is shown in the diagram, the hydrocarbon tail serving to anchor the molecule in place on the photosynthetic membrane. In chlorophyll *b*, the CH_3 methyl group at the upper right is replaced by an aldehyde group, CHO. Chlorophyll *c* lacks the hydrocarbon tail. The active end of each chlorophyll molecule consists of a porphyrin ring at the centre of which is attached a magnesium ion, Mg^{2+}, linked to four nitrogen atoms. This part of the molecule bears a strong resemblance to the haem parts of the cytochrome involved in the respiratory chain, and the oxygen carriers, haemoglobin and myoglobin, which are found in the blood and muscles, respectively. Cytochrome is discussed in Chapter 16, and the oxygen carriers in Chapter 19, the magnesium being replaced by iron in all three molecules. Chlorophyll molecules show a pronounced absorption of light at the red end of the visible spectrum; they thus tend to predominantly reflect the complementary colour and appear green. The maximum absorption occurs at slightly different, and reasonably well-defined, wavelengths for the different molecular variants, whereas the absorption characteristic of a typical plant covers a much broader range of wavelengths. This led Bessel Kok to suggest that the various types of chlorophyll function cooperatively as a pigmentary unit.

chlorophyll

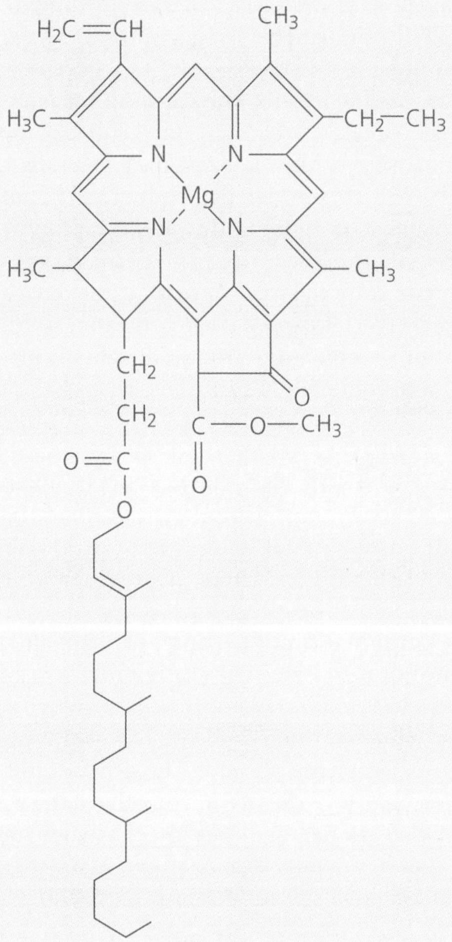

The chloroplast is the site of photosynthesis. Chlorophyll molecules are present in the parallel layers of membrane collectively known as a granum. Several such grana are visible in the left sectional view of a chloroplast recorded by transmission electron microscopy and are indicated in the schematic picture to the right. The length of the entire organelle is about two micrometres.

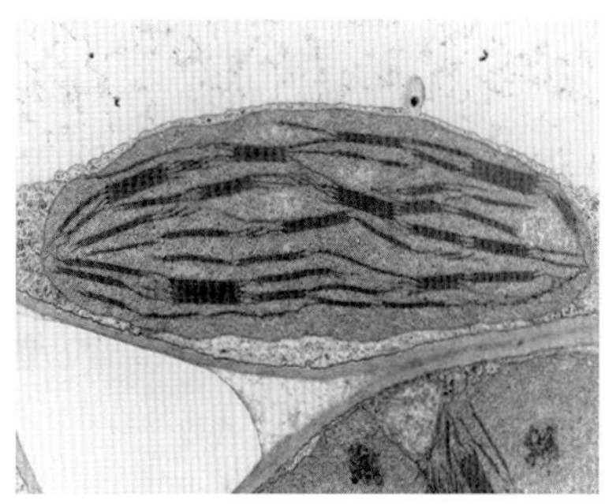

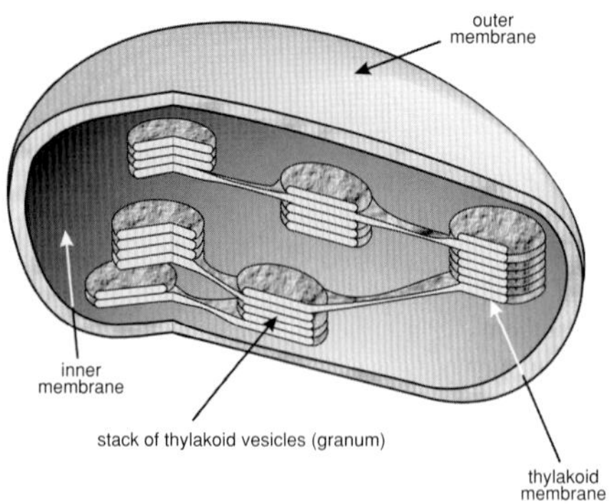

after the original excitation event. What they had achieved was surely the ultimate victory for the technique pioneered by Max von Laue, Walter Friedrich, Paul Knipping, William Bragg and Lawrence Bragg over seven decades earlier.

There remains the question of how the energy converted by chlorophyll from sunlight into $NADPH_2$ and ATP is used to convert carbon dioxide into glucose. Our understanding of this final piece of the jigsaw puzzle owes much to the efforts of Melvin Calvin and his associates, around 1950. Exposing photosynthesizing algae to short bursts of carbon dioxide that contained radioactively labelled carbon, and by an ingenious analysis, they managed to work out the various steps of the process. As with the earlier work of Hans Krebs on the citric-acid reactions, the solution turned out to be a cyclic process. In the Krebs process, it is oxaloacetate which completes the cycle, while in the case of glucose synthesis it is *ribulose bisphosphate*. The combination of carbon dioxide with this molecule produces intermediate sugar acid. The product immediately combines with water, to split into two separate molecules of *phosphoglyceric acid*, these reactions being catalyzed by the enzyme ribulose bisphosphate carboxylase. These reactions are said to fix carbon dioxide, in that they convert it from a gas to part of a solid. This part of the process does not require energy input. In fact, it actually gives off energy, so carbon dioxide will automatically be fixed as long as there is a plentiful supply of ribulose bisphosphate. It is the conversion of phosphoglyceric acid to glucose which requires the energy that has been harvested from sunlight by the chlorophyll. $NADPH_2$ and ATP now act in unison on the phosphoglyceric acid to produce *triose phosphate* and *orthophosphoric acid*. The massive formation of triose phosphate enables the chloroplast to reverse the early steps of *glycolysis* and consolidate these three-carbon sugars into the six-carbon sugar, glucose. The ribulose bisphosphate is regenerated in a series of steps that include the combination of triose phosphate and fructose phosphate into ribulose monophosphate. The entire process is known as the *reductive pentose phosphate cycle*.

The synthesized glucose is of course available to the leaves themselves, but most of it is passed down the plant via the phloem. There are several ways in which it is used by the non-green parts of the plant. The cells of the storage regions contain organelles known as *plastids*, or *amyloplasts*, in which *starch* is synthesized from the glucose. Starch is readily broken down in the digestive systems of animals, and it is of course a staple food. Much of the glucose is used in the production of cellulose, which is not digested in the human alimentary canal. It is, on the other hand, valuable in our diet because it provides roughage. It can easily be demonstrated that glucose is a source of

Photosynthesis

The determination of the three-dimensional structure of the photosynthetic reaction centre, in 1984, when taken together with the results of earlier kinetic and spectroscopic investigations, provided a detailed picture of the primary photosynthetic reactions. The atomic-level picture of the reaction centre was established through the X-ray studies carried out by Johann Deisenhofer, Robert Huber and Hartmut Michel, and it is shown here with the alpha helices drawn according to the convention described in Chapter 15. The reaction centre consists of four sub-units, two of which have five membrane-spanning helices apiece. These two units are in intimate mutual contact, and it is they which contain the photochemically active groups. A third sub-unit is located at the inner surface of the membrane, and it is anchored to the above two units by a single helix. It contains no active sites. Finally, the fourth unit is the cytochrome molecule, which is also bound to the dimer, but at the latter's other extremity, on the outer surface of the membrane. It contains four haem sites. The photochemically active chain consists of two dimerized chlorophyll molecules, two monomeric chlorophylls, two pheophytin molecules (these resembling the chlorophylls, except that they lack the latter's central magnesium atom), two quinone molecules (one located at each of the two first-mentioned sub-units), and finally a single iron ion, this being located on the two-fold symmetry axis of the overall structure (as is the above-mentioned chlorophyll dimer). The initial photon-induced excitation of the chlorophyll dimer makes the latter strongly reducing, and this excitation is followed within three picoseconds by charge separation. The latter involves an electron being transferred to one of the pheophytins, and, 200 picoseconds later, the electron being passed on to one of the quinones. By the time 300 nanoseconds has elapsed since the initial excitation, the chlorophyll dimer has been re-reduced by the bound cytochrome, the first chain reaction being completed around the 100 microsecond mark with the electron being transferred from one quinone to the other. The subsequent absorption of a second quantum of light energy (that is to say a second photon) produces a similar chain of reactions, again via the active sites on the first-mentioned sub-unit, but this time the end product is reduction of the other quinone to produce hydroquinone. The latter leaves the reaction centre and is replaced by a new quinone molecule from the surrounding membrane. The electrons are ultimately returned to the cytochrome region of the reaction centre by other electron carriers. Finally, the gradient of hydrogen ions (that is to say, protons) is used to turn the rotor of the ATP-synthase enzyme, as described in Chapter 16, thereby converting the harvested light energy to energy-rich ATP molecules. The various steps are illustrated ina highly schematic fashion in the second diagram.

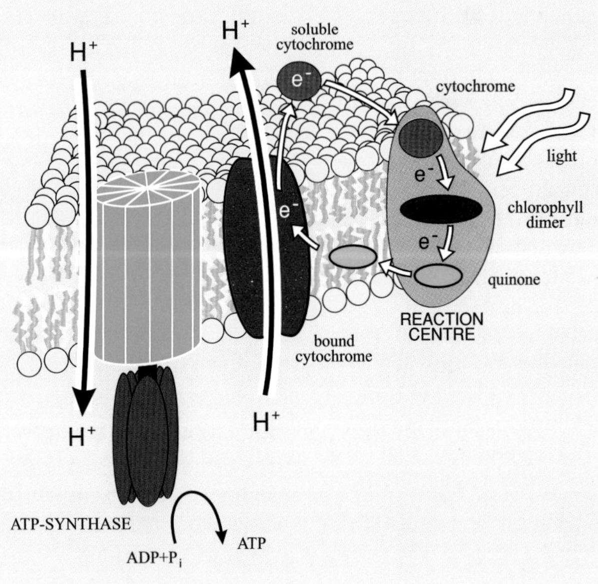

Biological mutation is a sudden change in the molecular structure of the nucleic acid that constitutes the genetic material, and the most important mutations occur during or immediately prior to meiosis, because such changes can be inherited by the progeny. Chloroplasts are vital to survival of a plant, and mutations which influence the development or function of these organelles are usually lethal. In these three specimens of a strain of barley known as Bonus, each batch of shoots derives from a single seed pod; it is apparent that not all the offspring carry a particular mutation. Mutations can occur in both the nuclear DNA and the chloroplast DNA. The examples shown here are all recessive nuclear gene mutants which were deliberately induced. The first mutant (top left) is known as Albina 36. It blocks the early development of chloroplasts, and was caused by irradiation with X-rays. Xantha 44 (top right) is a mutant which blocks a step in chlorophyll synthesis. It was chemically induced by the compound *p-N*-di-(beta-chloroethyl) – phenylalanine a derivative of mustard gas. A striped appearance characterizes barley carrying the Striata 12 mutant (bottom left), caused by purine-9-riboside. The final picture shows a naturally occurring mutation in dandelion (*Taraxacum*). The amount of chlorophyll is noticeably reduced in some parts, but this plant is still producing enough photosynthesizing tissue to survive.

energy. When burned in oxygen, it gives off heat, liberating carbon dioxide and leaving behind water. Although the thought might offend the epicurean sensibilities of the chef and the *maître d'hotel*, the fact is that when we eat vegetables we are basically just consuming glucose, cellulose and starch.

With the steadily increasing insight afforded by research on photosynthesis, it is not surprising that many have set their sights on imitating the process artificially. Success in such a venture would have an impact on the energy problem which could hardly be exaggerated. Several possible strategies have been suggested none of which would involve ATP as an intermediate, as in the plant. We have seen that the natural mechanism has two stages: water is oxidized to oxygen, (positive) hydrogen ions, and (negative) electrons, the energy being derived from light photons; and the electron is raised to a higher energy level, by further photon absorption, from which it is able to reduce carbon dioxide. In one of the artificial schemes, the excited electron reduces water so as to liberate hydrogen gas. In another, advocated by Melvin Calvin, the oxygen atom liberated during the first of the above stages is trapped by complexed manganese atoms, the resultant compound becoming a strong oxidizing agent. Both approaches require a molecular complex which, like chlorophyll, can transiently retain the energy absorbed from the sunlight. This sensitizer reacts with a carrier which, through becoming excited, can reversibly donate and accept electrons. In a system devised by George Porter and his colleagues, the positively charged sensitizer reacts with a sacrificial electron donor, releasing hydrogen ions which are then reduced to hydrogen gas by the carrier. The sacrificial donor is designed to decompose, thereby preventing the reverse reaction. In another approach, Michael Grätzel, John Kiwi and Kuppuswamy Kalyanasundaram split water with the aid of tris-(2,2′-bipyridine)-ruthenium. Molecules of this compound involve a central ruthenium atom that is bound to one nitrogen on each of six surrounding ring structures. They thus bear a superficial resemblance to chlorophyll. The remaining difficulty with this process is

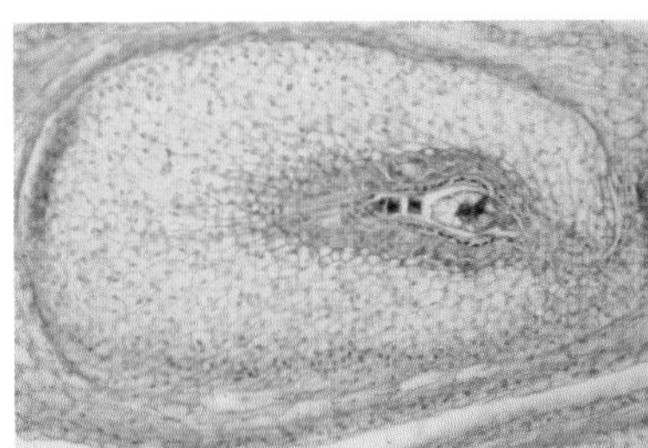
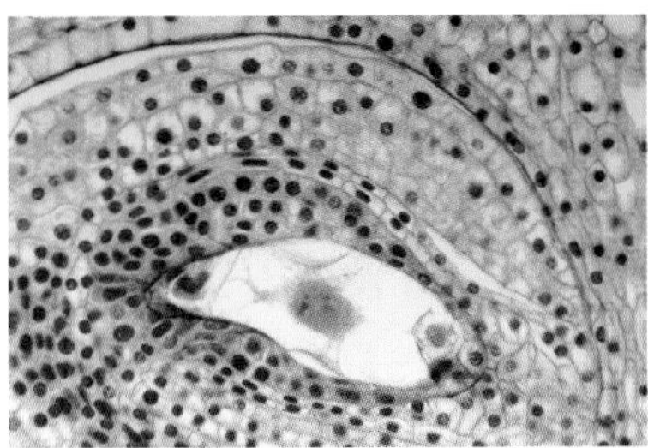
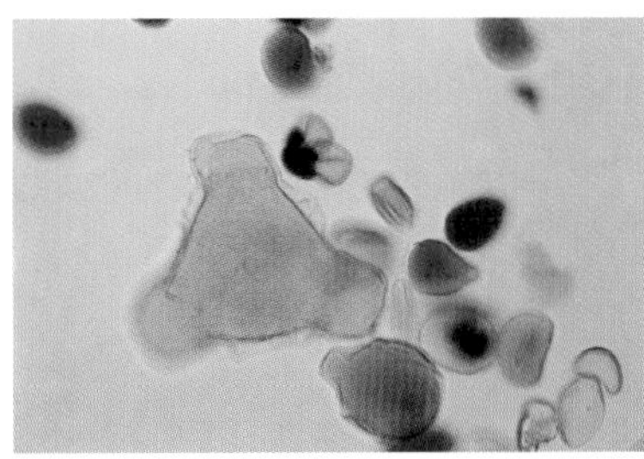

Mature embryo sacs are visible at the centres of two ovule specimens, from a daisy (*Bellis perennis*), pictured at the left, and a crown imperial (*Fritillaria imperialis*) pictured in the middle. A central body, the nucellus, encloses the sac, and is itself surrounded by the cellular layers known as integuments. The latter differentiate and provide the protective coat of the mature seed. Following pollination, a pollen tube grows and gradually protrudes through an opening between the integuments called the micropyle. This runs in from the right in both pictures. An egg cell and two other cells are located at the junction of the embryo sac and the micropyle, and a sperm nucleus passes through the pollen tube, fuses with the egg nucleus, and produces the diploid zygote that later divides to form a multicellular embryo. Another sperm nucleus fuses with the two nuclei at the centre of the embryo sac, forming the triploid endosperm. The latter divides to produce the nutritive tissue, which is consumed by the embryo either immediately or later depending on the type of plant. The entire process is called double fertilization. The final picture reveals the considerable variation of shape and size displayed by mature pollen grains. Each has one or more thin areas, or pores, through which the pollen tubes will appear during germination. *Oenothera drummondii,* commonly called evening primrose, produces large triangular grains with a pore at each corner, while the grains of the dwarf mountain pine tree (*Pinus mugo*) are easily recognized because they are equipped with two large air sacs. Pollen grains are produced by the stamens. When mature, they are released from the pollen sacs and transmitted to the stigma, primarily by the wind or insects.

physical separation of the oxygen and hydrogen. Although a working system has not yet been achieved, it has been estimated that up to 15% of the incident solar energy should ultimately be convertible, compared with the 1% attained by plants. It is to be hoped that such artificial energy trappers will soon find their place in the Sun.

There has long been fascination with behaviour exhibited by chloroplasts which is almost suggestive of intelligence. In dim light they position themselves face-on, so as to maximize exposure, but when the radiation is sufficiently strong they swivel around and meet the potentially harmful rays edge-on. Michael Blatt, Karin Klein and Gottfried Wagner discovered, in 1980, the origin of this movement. And it was an exciting revelation. Using electron microscopy, they observed filaments of *actin*, attached to chloroplasts and running out into the surrounding cytoplasm. Treated with another filamentous protein, *heavy meromyosin*, they were found to induce a configuration of the latter that is strikingly similar to that seen in animal muscle. As will be discussed in the following chapter, muscle action is indeed based on mutual movement of actin and myosin filaments. There is even evidence that another feature of muscle contraction, the presence of calcium ions, is also imitated in chloroplast movement. Finally, the actual light sensitivity was found to be provided by *phytochrome*. In Chapter 16, it was noted that actin, at least, is also implicated in the movements that accompany cell division. It is possible that the mechanism is a universal feature of cellular motion.

The other important activity of a plant, apart from its energy-converting functions, is *reproduction*. Much has already been described regarding the cellular aspects of this process, so further comment will be limited to those features that are peculiar to common plants. The dominant flora of the present

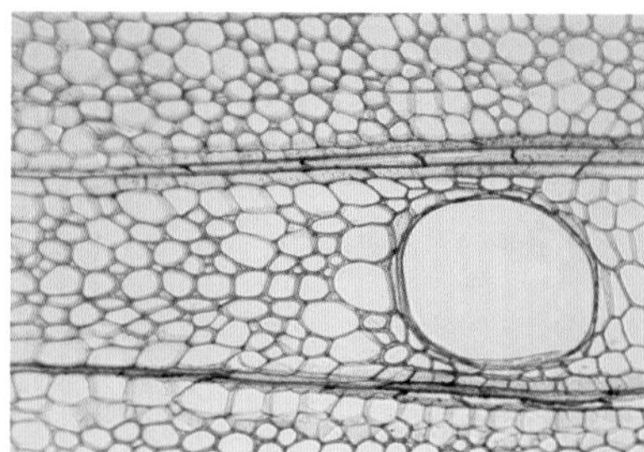

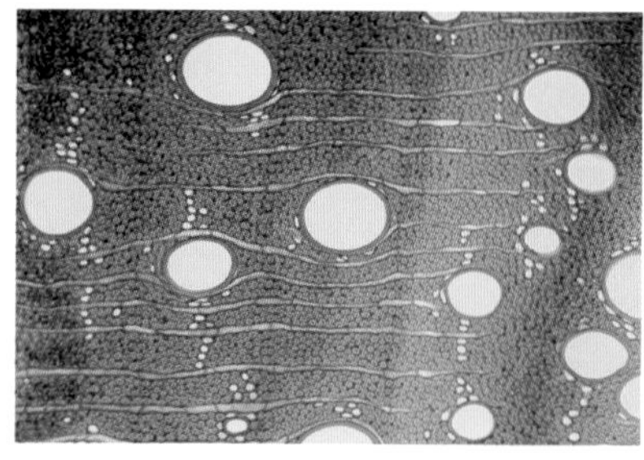

Carpenter's timber has a surprisingly wide range of density, and this arises from the different degrees of porosity displayed by wood microstructures. The extremes are illustrated in these transverse sections of the lightest and heaviest known examples. Balsa (*Ochroma lagopus*), shown at the left, has a density of only 0.04 grams per cubic centimetre (based on its oven-dry weight and volume). Lignum vitae (*Guajacum officinale*), seen right has a density of 1.4 grams per cubic centimetre. This thirty-five-fold difference is reflected in the striking contrast between the two pictures, each of which has a width equivalent to about a millimetre. The larger holes are vessels, seen in cross-section. Balsa wood is used in marine engineering, to promote buoyancy, and it is popular among the builders of model aircraft. Lignum vitae is very hard and resistant to biological attack. Capable of being highly polished, it is prized by the furniture craftsman.

time, that is *trees, shrubs, grasses* and various *herbaceous plants*, belong to a division of the plant kingdom known as *Spermatophyta*. The cells of these organisms are grossly divided into two differentiated types. The diploid somatic cells make up the body of the plant, which in this case is called the *sporophyte*. The haploid male *gametophyte*, which is called a *pollen grain*, produces two gametes. Not being mobile, they are dependent on the wind, insects, or other creatures, which transfer them to the female gametophytes, each of which is contained within an *ovule*. The pollen is actually caught on the *stigma*, which is separated from the ovules by a tube known as the *style*. The pollen sends down a tube through the style to meet the ovule, and through this it discharges its gametophyte nuclei. There then follows the special double fertilization process, with one male germ nucleus fusing with a female germ nucleus to produce the diploid zygote, which will later divide and multiply into many sporophyte cells. The second male germ nucleus fuses with two of the smaller products of the meiotic division, namely the (haploid) *polar nuclei*. This produces the *primary endosperm nucleus*, which is remarkable for the fact that it is triploid, containing three sets of chromosomes. The primary endosperm nucleus undergoes many mitotic cell divisions and produces the *endosperm tissue* which surrounds and protects the growing embryo. At the same time, the fertilized ovule is dividing and differentiating into the *seed* and its coat. Finally, the surrounding ovary undergoes development and differentiation, ultimately becoming the *fruit*. The *Spermatophyta* are divided into two large groups: the *Gymnospermae* and the *Angiospermae*. The former are relatively primitive and include the *conifers*, such as pine, spruce, yew, larch, cedar and fir. Their xylem never contains vessels. Reminiscent of the young lady who was said to wear her heart on her sleeve, the *Gymnospermae* carry their ovules relatively unprotected on open structures known as *megasporophylls*, such as the *fir cone*. The *Angiospermae*, which include flowering plants and *deciduous* (leaf-shedding) *trees*, distinguish themselves by having their ovules enclosed within an ovary.

The importance of plant life to mankind could hardly be exaggerated. Plants provide us with virtually our entire diet; the major part directly, as grains, vegetables and fruits, and the balance indirectly through animal life which subsists on grain and grass. They also give us a variety of natural fibres, paper and structural timber. Just as significantly, plants play the vital role of generating oxygen, and any widespread threat to the vast tropical forests would thus be a matter of grave concern. The carpenter's trade is made interesting by the variety of wood types and by the heterogeneity of wood structure. Species variations in cellular composition produce a range of density, from the ultra-light *balsa* to the very heavy *lignum vitae*. Hardness too shows considerable variation, the Gymnospermae tending to be soft and the other end of the scale being occupied by such woods as oak,

The lignin, pectin, and hemicellulose mixture in the middle lamella region between wood cells is a plastic. It is therefore capable of being moulded. This is often exploited in furniture manufacture, the wood being subjected to steam heat prior to shaping. A similar effect can be obtained through use of certain chemicals. The extreme deformation in the example shown here was achieved after prolonged exposure to ammonia, which causes softening of lignin.

teak and ebony. The xylem and phloem of newly cut timber obviously contains a great deal of moisture, and until this has evaporated, the wood is subject to dimensional changes: whence the need for *seasoning*. The art of cutting wood lies in maximizing exposure of such features as rings, knots and rays, while not jeopardizing strength through conflict with the directionality imposed by the grain. Particularly striking examples of the beautiful effects obtainable in transverse radial sectioning are seen in *Australian silky oak* and *cabinetmaker's walnut*. The middle lamella of wood being a natural thermo-mouldable plastic, timber can be bent to surprisingly high degrees of strain after prolonged exposure to steam heat. This extends its range of application into areas which would otherwise be the monopoly of metals. The substructure of wood actually aids in its protection against decay. The vascular system of interconnecting channels, when devoid of water after proper seasoning, provide conduits for impregnation with rot-resisting fluids.

Usefulness aside, plants have one other commendable feature: their aesthetic quality. Plant life constitutes a major part of the environment and it is fortunate that Nature's competitive principles have engendered such diversity of size, shape, scent, texture, taste and hue. But external appearance notwithstanding, we could reasonably claim that there is an equal beauty in a plant's internal structure. An appreciation of the intricacies of photosynthesis, germination, transpiration, capillarity and the like, gives added depth to the sentiments captured in Joyce Kilmer's couplet:

I think that I shall never see
A poem lovely as a tree.

Summary

All higher plants, whether herbaceous or woody, have three features in common: roots, a stem and leaves, the cells of the latter containing chloroplasts, the site of photosynthesis. This most important of all terrestrial reactions uses the energy in sunlight to combine carbon dioxide and oxygen into glucose. It involves photon-induced electron transport and is mediated by chlorophyll. A vascular system of interconnected tubular cells transports water from the roots to the leaves, through the xylem, and glucose from the leaves to the rest of the plant, via the phloem. Growth occurs in the cambium, a limited cylindrical layer coaxial with the centre of the stem, trunk, or limb. It adds new phloem cells on its outer surface, and new xylem cells toward the inside. The cells have prominent walls consisting of several layers of microfibrils, these being composed of bundles of cellulose molecules. The intercellular space is filled with lignin, hemicellulose and pectin. Stem rigidity is provided by the turgor pressure produced by the vacuoles in the living cells, and in woody plants this is augmented by the stiffening effect of successive layers of dead xylem. Cellular differentiation, and the directional trends of geotropism and phototropism, are controlled by hormones, the auxins being particularly prominent. Certain artificial compounds which mimic hormonal action, while evading the normal biological controls, are effective herbicides. The Gymnospermae, including the softwood conifers, have exposed and relatively unprotected reproduction systems. In the Angiospermae, including the hardwood deciduous trees, these systems are contained in flowers, and the fertilized ovule becomes a fruit which encloses one or more seeds.

19

'Unting is all that's worth living for – all time is lost wot is not spent in 'unting – it is like the hair we breathe – if we have it not we die – it's the sport of kings.
Robert Smith Surtees (1803–1864) ***Handley Cross***

Mandatory hunter: *the animal*

The most vital possessions of the plant are the chloroplasts in the cells of its leaves. Through these, the organism can convert, for its own purposes, a fraction of the solar energy that falls upon it. Sunlight is remarkably uniform in its intensity, and a plant has little to gain by moving about. It can thus afford a relatively stationary existence, the only motion being that required to reach regions beyond the shade of its immediate environment and its competitors. Because they have no chloroplast-bearing cells, animals have paid the penalty of having to develop several specialized functions in order to satisfy their energy requirements. They must have a means of locomotion and feeding, and some form of coordination of these faculties, however primitive, is needed. And as insurance against the unsuccessful forage, they should be able to store digested energy. Moreover, unless an animal is so small that the normal process of diffusion is adequate, it must have a circulatory system to distribute dissolved gases and chemical compounds to its various parts. Finally, because correct operation of these components depends on a certain degree of structural integrity, the animal body requires connective tissue and, if it is sufficiently large, a rigid framework to support it. Hunting is said to be the sport of kings. But in the broader sense of the word – the unceasing quest for consumption of energy-bearing material – hunting is the common lot of all creatures, great and small. This chapter is a description of that remarkable composite of muscle, nerve, collagen, fat, blood, bone and brain: the *animal*.

A reasonable appreciation of the role of *muscle* in limb movement can be gained from the simplest of observations. The muscle that flexes as the joint hinges feels stiff, and tension is clearly an important factor. Medieval anatomical dissections, performed in surreptitious defiance of papal prohibition, were unable to shed light on the origin of the force. They did lead to a detailed charting of the various sinews and tendons, however, and one volume of *De Humani Corporis Fabrica (On the Fabric of the Human Body)*, published by Andreas Vesalius in 1543, was an exquisitely detailed muscular atlas. Further progress came only with the advent of microscopy, which showed each muscle to consist of hundreds of thousands of tightly packed fibres. Each *fibre* is enveloped by a membrane, and the electrical insulation thereby provided allows the inside to maintain a negative potential of about one tenth of a volt (0.1 V) relative to the exterior. The better resolving power of improved versions of the microscope revealed that each of these fibres is composed of one or two thousand smaller units, known as fibrils, these too lying parallel to the length of the muscle. A typical *fibril* has a diameter of about 1 μm. Finally, electron microscope studies demonstrated

Before the Renaissance, human anatomical studies were subject to the taboos that naturally accompanied vitalism; the interior of the body was the vessel of the soul, and was thus out of bounds. The grosser features of animal anatomy were familiar through the eating of meat, of course, but detailed investigations did not become common until around the end of the fifteenth century. Leonardo da Vinci (1452–1519) set new standards of observation and documentation, and he was followed by Andreas Vesalius (1514–1564), who is generally regarded as the founder of modern anatomy. These pictures of the human nervous and muscular systems appeared in the latter's *De Humani Corporis Fabrica*, published in 1543.

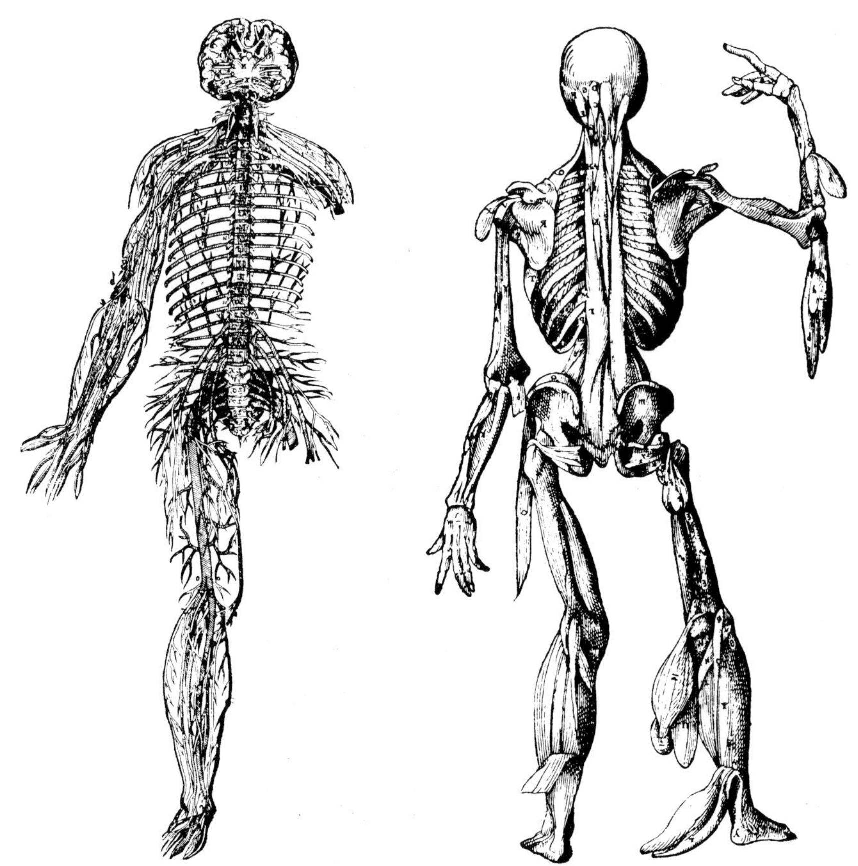

that each fibril comprises hundreds of *filaments*. Work at the beginning of the twentieth century, by H. H. Weber and his colleagues, had demonstrated the presence in muscle of a protein that was given the name *myosin*. This was subsequently shown to be heterogeneous, and a second protein, *actin*, was also identified. The two substances were apparently associated, and thus collectively referred to as *actomyosin*. In 1939, Vladimir Engelhardt and Militsa Lyubimova showed that actomyosin can enzymatically split the energy-bearing compound ATP. A particularly exciting series of discoveries were made by Albert Szent-Györgyi and his colleagues in the early 1940s. First they showed that mixing pure actin and pure myosin, in a concentrated salt solution, produced a great increase in viscosity. This indicated a strong interaction between the two substances. Upon addition of ATP, the viscosity was reduced to its original level, indicating that the actomyosin was being dissociated into actin and myosin. A subsequent experiment produced an even more dramatic result. Actomyosin preparations, rich in actin, were dissolved in a dilute solution of potassium, magnesium, and calcium ions, the thread-like molecules being oriented by flow. Upon addition of ATP, the threads contracted! The underlying basis of muscle function had been revealed to be an ATP-mediated interaction between myosin and actin. Another interesting fact emerged during subsequent chemical analysis: the common form of actin is a globular molecule, but in the presence of ATP and the above ions, it polymerizes into long threads.

The structure of the two main proteins involved in muscle function is now well known. A myosin filament is approximately 16 nm thick, while an actin

The Earth and the human body: composition

The human body, like all other living organisms, consists predominantly of the lighter elements. Oxygen is the most abundant of these, but only because the living cell contains so much water in its cytoplasm. In terms of the dry weight, there is actually three times as much carbon as oxygen. The following tables contrast the compositions of the human body and the Earth, the entries being given as percentages of the whole by weight, in each case.

The Earth

O	**Si**	**Al**	**Fe**	**Ca**	**Na**	**K**
46.5	28.0	8.1	5.1	3.5	2.8	2.5
Mg	**Ti**	**H**	**C**	**Cl**	**P**	**S**
2.0	0.58	0.20	0.20	0.19	0.11	0.10

The human body

O	**C**	**H**	**N**	**Ca**	**P**	**K**
65.0	18.0	10.0	5.0	1.5	1.0	0.35
S	**Na**	**Cl**	**Mg**	**Fr**	**Zn**	
0.25	0.15	0.15	0.05	0.004	0.004	

filament measures only 6 nm. It was primarily the work of Hugh Huxley and Jean Hanson, in 1953, that exposed the connection between these filaments and muscle contraction. Observed at high magnification, a longitudinal section of striated muscle fibril displays a regularly alternating series of light and dark stripes, known as the I-bands and A-bands respectively. A thin and very dark line runs along the centre of each I-band. This is the Z-line. The repeat length of the pattern, which is the distance between two successive Z-lines, is referred to as a *sarcomere*. In a relaxed muscle, this measures about 2.5 μm. The vital observation concerned the dimensional changes of the sarcomere and the I-band upon contraction.

Both decrease in length, whereas the A-band stays the same. Substructure can be resolved within the A-band, in the form of a central light strip known as the H-zone. This is found to decrease in width as the muscle contracts. Huxley and Hanson interpreted these observations in terms of a *sliding filament model*, the mechanism being analogous to a piston moving in a confining cylinder, as in a cycle pump. The pistons are actin filaments, while the cylinders are bunches of myosin filaments stacked parallel to one another. The latter maintain a constant length during the contraction process, and they define the width of the A-band. Very high resolution photomicrographs reveal that the actin filaments entering an A-band, from opposite ends, actually overlap each other at the centre of that band when contraction is at its maximum. This accounts for an observation made by Archibald Hill, early in the twentieth century, that a stimulated muscle can exert less tension when contracted than when extended, the prerequisite assumption being that the vital forces arise through interactions between the actin and myosin filaments. At the atomic level, these cannot be quite smooth, and it is natural to think in terms of *cross bridges* that span the gap between the two types of molecule. Sequential formation and breaking of these bonds would lead to mutual sliding. Confirmation of this idea, and endorsement of the sliding filament model, was provided by Andrew Huxley and Rolf Niedergerke who demonstrated a direct relationship between tension and the number of myosin–actin cross bridges in a single muscle cell.

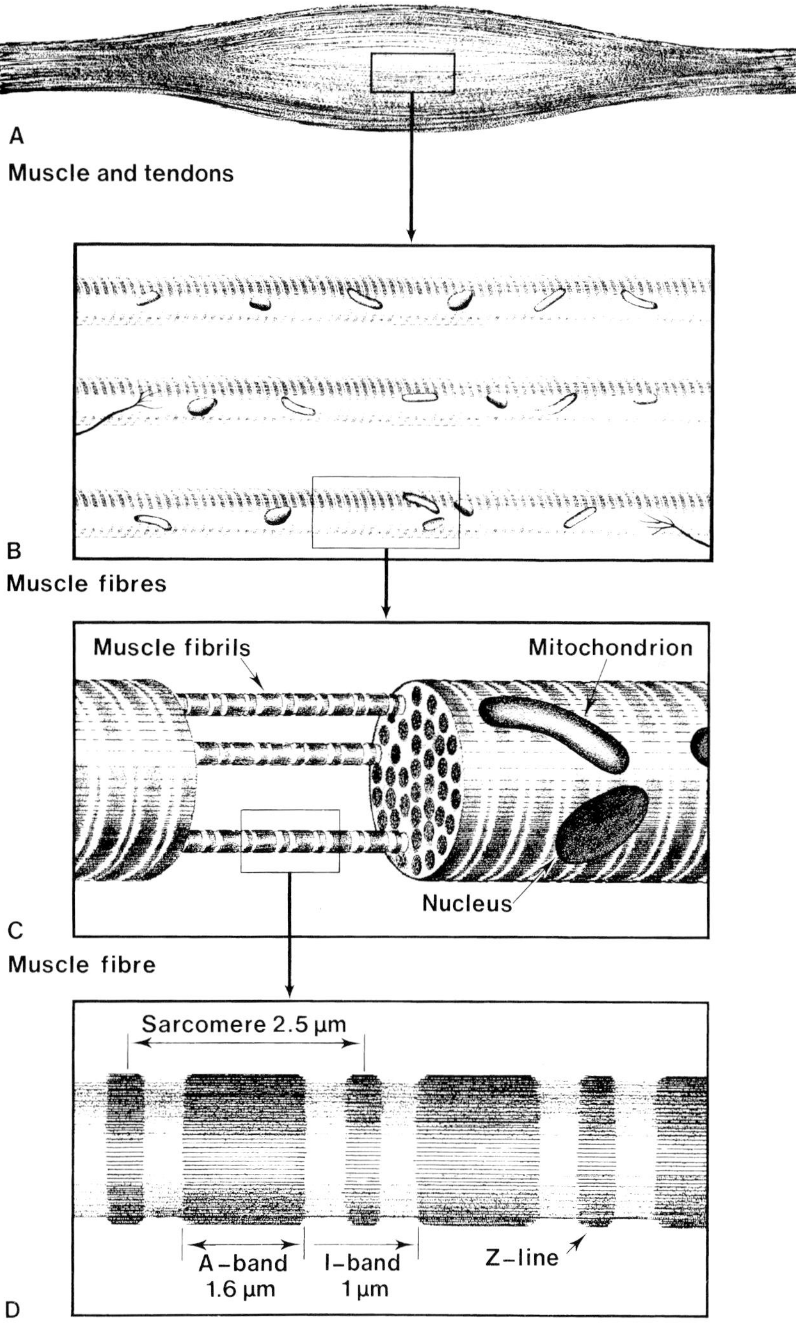

The structure of muscle was gradually elucidated as the power of microscopes increased. The optical instrument revealed first the presence of fibres and later the fact that each fibre comprises numerous fibrils. Finally, the electron microscope permitted resolution of the many filaments of which each fibril is composed.

The nature of the cross bridges was gradually elucidated through investigation of the finer structural detail of myosin and actin. In 1953, Andrew Szent-Györgyi found that the former is actually composed of two fragments, light and heavy *meromyosin*, and electron photomicrographs revealed that the light version is straight while the heavy type is shaped like the head of a golf club. Hugh Huxley went on to show that the two components are positioned so as to give myosin the appearance of a bunch of golf clubs, with the heads all pointing in the same direction and staggered along the

Electron microscopical studies were crucial to the elucidation of muscle action. They revealed both the existence and the arrangement of the filaments, and indeed the structure of the latter. A muscle fibril contracts and expands through the sliding of one type of filament through the spaces produced by the triangular arrangement of another type of filament. These filaments are myosin and actin, and they are referred to as the thick and thin components, respectively. The myosin molecules adopt an arrangement reminiscent of a bunch of golf clubs, the head units successively making and breaking contact with the actin double helix during the mutual sliding of the molecules. The actin molecule (seen to be composed of globular units in the diagram) has been found to have a quite complex structure, with relatively narrow tropomyosin helices (shown in black) running alongside but not quite in its grooves. These in turn are acted upon by a troponin complex (indicated by the blue globules) which respond to calcium ions by slightly shifting the tropomyosin strands farther into the actin grooves, so as to permit the myosin heads full access to the actin's globular units. The energy used in achieving the sliding motion is supplied by ATP.

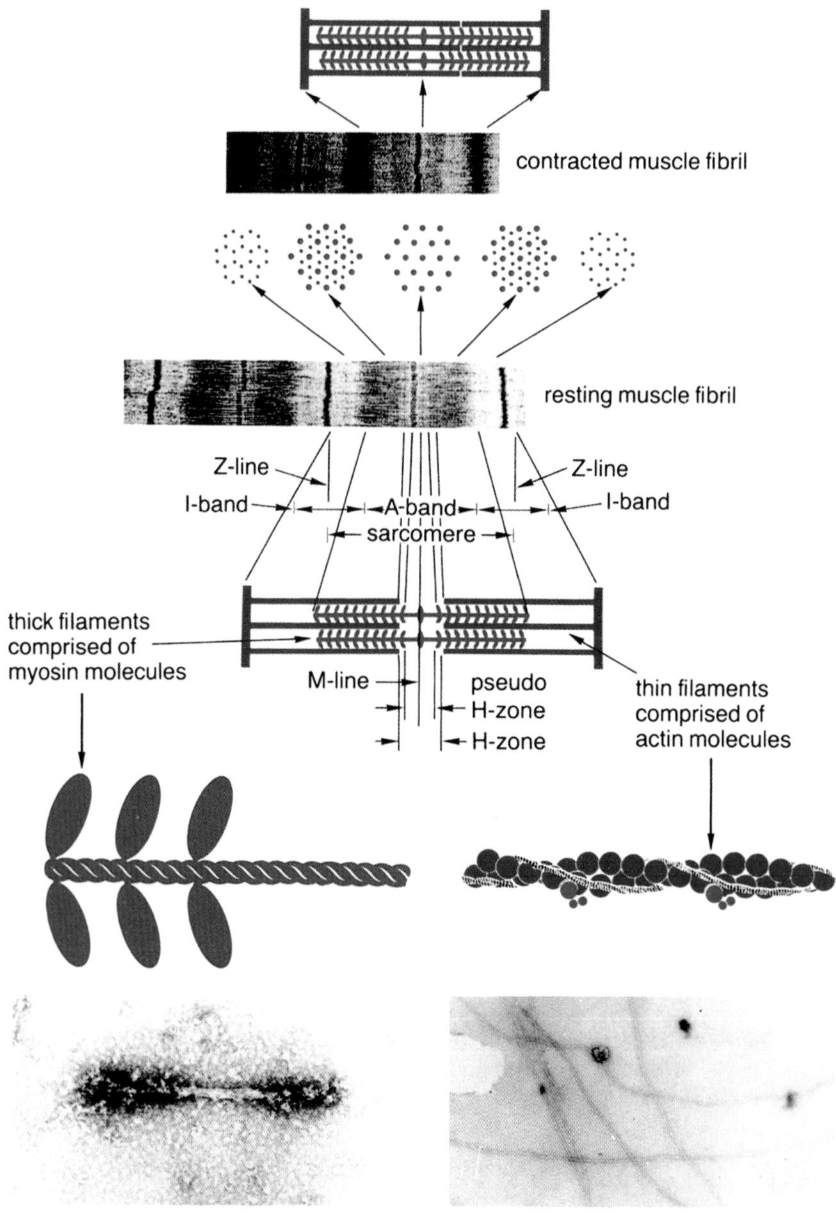

axis. It was already known that actomyosin can enzymatically split ATP, and when it was subsequently found that it is the heavy meromyosin that actually accomplishes this, the role of the myosin club heads as part of the cross bridges was firmly established. Meanwhile, Jean Hanson and Jack Lowy, using high resolution electron microscopy, discovered that the actin filaments are composed of two strands, twisted rope-like into a helix, each consisting of globular units strung out like beads in a necklace. There were clear indications of the presence of other molecules. In the mid 1940s, Kenneth Bailey had isolated *tropomyosin* from muscle filaments, and Carolyn Cohen, Kenneth Holmes, Alfred Holtzer and Susan Lowey showed that it was basically alpha-helical in structure. Hanson and Lowy suggested that the long tropomyosin molecules might lie along the two grooves in the actin helix.

An unlucky Olympian?
In the 1972 Olympic swimming championships, the 400-metre individual medley event was won by Gunnar Larsson. He covered the distance in 4 minutes 31.981 seconds, just 0.002 seconds faster than Tim McKee. One could question whether the use of such small timing margins is realistic; they are comparable to variations in the duration of processes in the nervous system. The sound of the starter's pistol sends nerve impulses from the ears to the brain, which then transmits further impulses to the limbs. These signals travel at about 18 metres per second, and would thus cover a distance of some 36 millimetres in 0.002 seconds. This is considerably less than the range of human height. Assuming that the nerve impulse velocity is precisely the same in all individuals, the taller competitor would seem to be at a disadvantage. There is, of course, the compensating advantage of superior reach, but it might have been more appropriate to award gold medals to both Larsson and McKee.

Another minor component, *troponin*, was discovered by Setsuro Ebashi and Ayako Kodama, and it was found that this too is located on the actin, troponin molecules occurring at regular intervals every seventh globule along the actin helix.

The smallest entities involved in muscle contraction are individual ions, dissolved in the cytoplasm of each muscle cell. As early as 1883, Sidney Ringer had demonstrated that calcium ions are essential to the functioning of frog heart muscle, and in the late 1940s L. V. Heilbrunn and F. J. Wiercinski showed that these are the only common ions capable of causing contraction when injected into isolated muscles. The way in which calcium ions actually trigger muscle contraction is rather subtle. The accepted model, dating from 1972, was put forward by Hugh Huxley and John Haselgrove, and independently by David Parry and John Squire. The tropomyosin molecule is envisaged as normally interfering with the interaction between the actin and the club-like bulges of the myosin. At a critical calcium ion level, there is a slight change in the configuration of the troponin molecule, and this allows the tropomyosin to move deeper into the groove of the actin helix, cancelling the repressive effect and allowing the actin and myosin to interact. The control mechanism is thus one of physical blocking and unblocking of active sites. Subtle *allosteric changes* in a protein molecule, caused by alteration of the ionic content in the surrounding aqueous medium, is indeed a common means of physiological control.

The pleasant sound for which the *humming-bird* is named is not emitted vocally. It is caused by the rapid beating of that creature's wings, and the rate of reversal of the relevant muscles is about 200 per second. This is almost as high a frequency as that of the musical tone middle C, which lies at 261 cycles per second. Archibald Hill pointed out that the time lapse between stimulation by the associated nerve and mechanical response of the muscle is too short to permit diffusion of chemical substances from the exterior membrane to the innermost filaments. Electron microscope investigations have revealed structures lying roughly at right angles to the filaments, spanning the width of each muscle cell. These are the *transverse tubules* and the *sarcoplasmic reticulum*, which are not mutually connected. Although some details remain to be clarified, it appears that these two cellular components are responsible both for rapidly supplying calcium ions to cause contraction and for quickly scavenging these same ions from the filaments to permit relaxation. This cyclic process can continue as long as ATP is available. When the supply of the latter is cut off, as in anoxia or death, the muscles assume a state of rigor, or extreme stiffness.

The functioning of muscle fibres is just one case of coordinated motion at the cellular level. Other examples, such as the rhythmic waving of the *cilia* and the systematic sliding of the *mitotic spindles* during cell division, have been mentioned earlier, and it is also a fact that myosin and actin are much in evidence in the cell. It is tempting to speculate as to the generality of the principles operative in muscle. Normal cellular pH and ionic strengths are certainly conducive to polymerization of both myosin and actin, and both ATP and membrane-controlled calcium ions are ubiquitous.

Until now, nothing has been said regarding the stimulus which causes liberation of calcium ions in muscle. It comes from a *motor nerve*, which makes contact with the membrane surrounding the muscle cell, and this is an appropriate point at which to address the question of *nerve function*. In 1791, Luigi Galvani discovered the essentially electrical nature of neuromuscular

The wings of the hummingbird beat so rapidly that they emit the audible frequency for which this creature is named. The tone lies around the musical middle C, which corresponds to 261 hertz (i.e. cycles per second). Archibald Hill (1886–1977) was the first to realize that such high cycling rates cannot be achieved unless there is a direct coupling of the motor nerves to the innermost reaches of the muscle cells.

function. He observed the twitching of frog muscle when stimulated by artificial or atmospheric electricity, and he also detected electrical discharge from living nerve and muscle tissue. Some 60 years later, Hermann von Helmholtz showed that the signal velocity in a frog's sciatic nerve, far from being immeasurably large, as previously assumed by Isaac Newton and others, was a modest 30 m/s. In the latter part of the nineteenth century, the individual efforts of experimenters such as C. Matteucci, Emil Du Bois Reymond, L. Hermann, and Julius Bernstein established that nerve signals are transmitted as well-defined pulses of short duration. The pulses appeared to be uniform, in contrast to the different durations used in the Morse code. Regarding the physical structure of the nerve itself, two events single themselves out. One was the development, by Camillo Golgi around 1875, of a chemical means of rendering entire cells visible in an optical microscope. This was accomplished by selective staining so that only a small fraction of the cells in a piece of tissue became coloured at one time. The other was achieved by Santiago Ramón y Cajal, who used the Golgi technique to demonstrate that nerve cells, or *neurons* as they are now called, are independently functioning units. He made exquisitely detailed drawings of networks formed by the octopus-like neurons, showing how a 'tentacle' of one cell would seem to reach out and touch the body of another. Close contact between nerve cells had been observed as early as 1897, in fact, by Charles Sherrington. He referred to such a junction as a *synapse*, and in 1921 Otto Loewi showed that information is transported across the small gap between cells by chemical means rather than by electricity. The chemical substance involved was called a *neurotransmitter*, and Loewi subsequently identified it as the compound *acetylcholine*.

The long protuberances emerging from a nerve cell, the tentacles referred to above, are now known as *processes*. There are two types: the relatively short *dendrites*, of which there are many per cell, and the *axon*. Each neuron has a single axon, and this is usually quite elongated. Indeed, axons may have lengths up to 5000 cell diameters. The axons of some neurons are sheathed by *myelin*, which appears to act as a sort of insulating cover. The flow of electrical information in the neuron is from the dendrites, in towards the cell body, and thence out along the axon, ultimately to pass to another neuron, via a synaptic junction, or to a muscle. It is axons of the latter type which are myelinated, the insulation being periodically interrupted at sites known as *nodes of Ranvier*. It is at these nodes that brief pulses of ionic diffusion occur, as we shall see shortly, and they give the axon its vital property of electric excitation.

Although it is long and thin, like the wires in an electric circuit, an axon, seen as a prospective conduit, would seem to be at a disadvantage. Its membranous skin, primarily composed of *lipid*, is electrically insulating, and even the aqueous axoplasm is only weakly conducting. Indeed, if the axon were to transmit signals in the usual way, by currents flowing parallel to its major axis, these would become severely attenuated within a few millimetres. The secrets of neuromuscular function were already beginning to emerge in the early years of the twentieth century. In 1902 E. Overton showed that frog muscle loses its excitability if the sodium ions are removed from the surrounding aqueous medium. It was already known at that time that the inner region of the nerve cell is rich in potassium ions, and Overton made the remarkably prophetic suggestion that excitation might arise through sodium and potassium exchange. Just 50 years were required to vindicate

The relative slipping of filamentary molecules appears to be a widely used device in animal motion. It provides the basis for the undulating movement of the flagellum, by which free cells such as spermatozoa and protozoa propel themselves, and it also underlies the coordinated flapping of the cilia which removes foreign particles from the lining of the respiratory passages. The filaments are of various thickness and composition, just as in muscle. The electron photomicrograph shows, in cross-section, the arrangement of microtubules in a contractile region of the ciliate *Pseudamicrothorax dubius*. Each microtubule is composed of a helical arrangement of tubulin molecules, the overall diameter being about 25 nanometres. The dark-field optical photomicrograph shows four bacteria, each with its single helical flagellum at its rear end. Such flagella are much simpler in construction than the ones present in the ciliates. It should be noted that the four bacteria are all swimming in roughly the same direction. Their motion is governed by a chemical memory mechanism that persists for a few seconds, and it biases their travel in the direction of increasing concentration of nutrients.

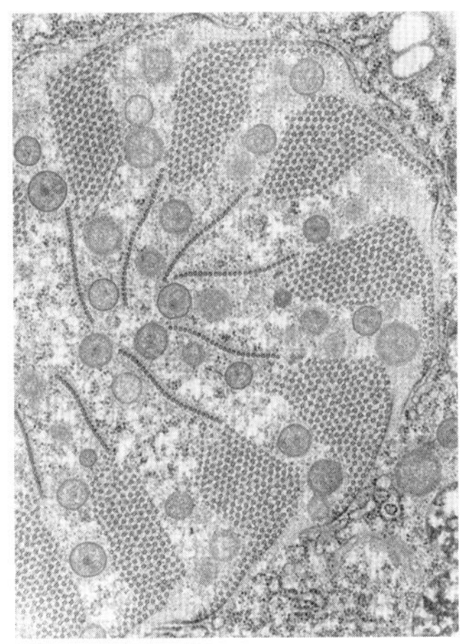

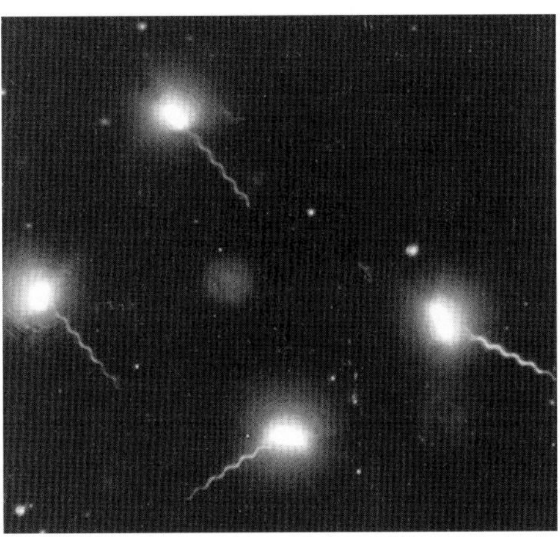

this inspired guess. The year 1902 also brought the important *membrane hypothesis* of Julius Bernstein, who attributed the impulse to a transient change in the ionic permeability of the nerve membrane. Another important step was the discovery, by Edgar Adrian and Keith Lucas in 1916, that a nerve responds to stimulation in an all-or-nothing manner: signals below a certain strength cannot make the nerve 'fire', but signals stronger than this *threshold* value are transmitted with virtually no attenuation. Adrian also demonstrated that the impulses all have a standard amplitude and that the only variability is in the frequency of their bursts, which can be as high as 1000 per second. Because of its characteristics, the nerve impulse has come to be known as the *action potential* or spike potential.

An important step was taken in 1933, when J. Z. Young found that the swim reflex of the squid, *Loligo*, is controlled by neurons having an axonal diameter of 0.5 mm. Such *giant axons*, being relatively easy to experiment upon, became the standard object of nerve studies. They were used in the mid 1930s by K. S. Cole and H. J. Curtis, who found that the resistance of the axon membrane falls precipitously during the action potential whereas the capacitance (i.e. the ability to hold electric charge) remains virtually unaltered. This ruled out a simple electrical short circuit, and Cole and Curtis suggested that the electrical changes were caused by a leakage of ions through sites which represented as little as 1% of the total membrane area. The most intense study of the ionic transport associated with the action potential was made in the next three decades by the school led by Alan Hodgkin, Andrew Huxley and Bernard Katz. The axon membrane is maintained at a *resting potential* of about 0.1 V, this being negative on the inside. It arises because of imbalance across the membrane in the concentrations of sodium and potassium ions. There are about ten times as many sodium ions outside the membrane as there are inside, while the gradient of potassium ions, in the opposite direction, is even larger. The membrane is slightly leaky to both types of ion, but particularly so to potassium, and metabolic energy is required to maintain the chemical gradients, as will be discussed shortly. Because of these concentration imbalances, the resting nerve is like a loaded fuse.

In 1952, Hodgkin and Huxley were able to present the following picture of the *action potential*. For a stimulus greater than Adrian's threshold, there is first a flux of sodium ions through the membrane toward the inside. This decreases sharply after about 1 ms and, at the same time, there is a rapid

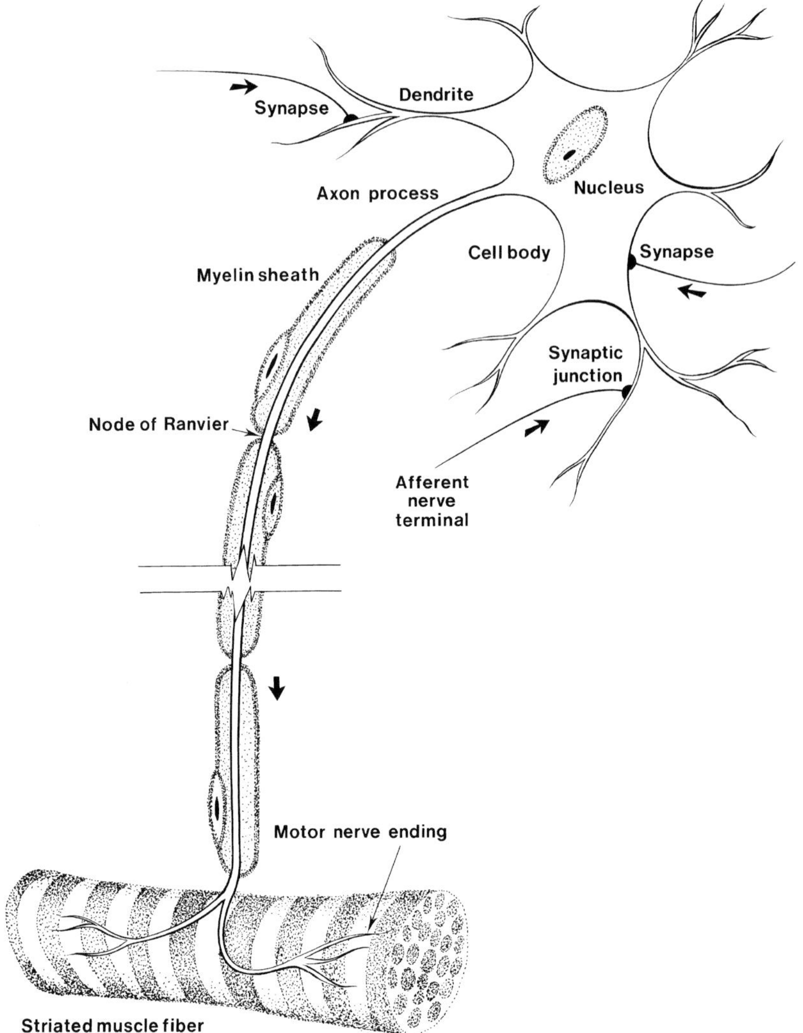

Muscle action is triggered by the electrical signals received from a motor nerve. The nerve cell has one highly elongated extension, or process, which can have a length of many centimetres, and this terminates in intimate contact with the membranes of the muscle cells. It is known as an axon and it is surrounded, over most of its length, by myelin, which acts as a sort of electrically insulating layer. The myelin is occasionally interrupted, at sites known as nodes of Ranvier, at which the electrical signal receives a boost. A nerve signal travels much faster along such a myelinated axon than it would if the sheath were absent. Nerve cells can also receive signals, from other nerve cells, by way of synaptic contacts to their dendrites, or minor extensions.

increase of potassium flux towards the outside, this latter pulse being less pronounced than its sodium counterpart, but lasting somewhat longer. The counteracting fluxes, being temporally offset from each other, cause the potential across the membrane to swing first from its negative resting value to a positive potential of about 0.04 V and then back again to the initial −0.1 V. There is a slight overshoot on the negative swing, with a duration of about 3 ms, during which period the nerve cannot be made to fire. That period is now referred to as the *refractory period*. As Hodgkin and Huxley pointed out, the spike owes its existence to an instability: the voltage dependence of the sodium conductance has a region of *positive feedback*, in which depolarization (i.e. a decrease in the magnitude of the negative potential difference across the membrane) causes an increase in the permeability to sodium ions which, in turn, leads to a further depolarization. Just how the sodium and potassium ions get through the membrane became steadily clearer, with growing evidence that they pass along *ion channels* that are pipe-like holes through proteins that lie in, and span the width of, the nerve membrane. Hodgkin and Richard Keynes have shown that the ions move along single-file fashion, indicating that the channels are of near-atomic dimensions.

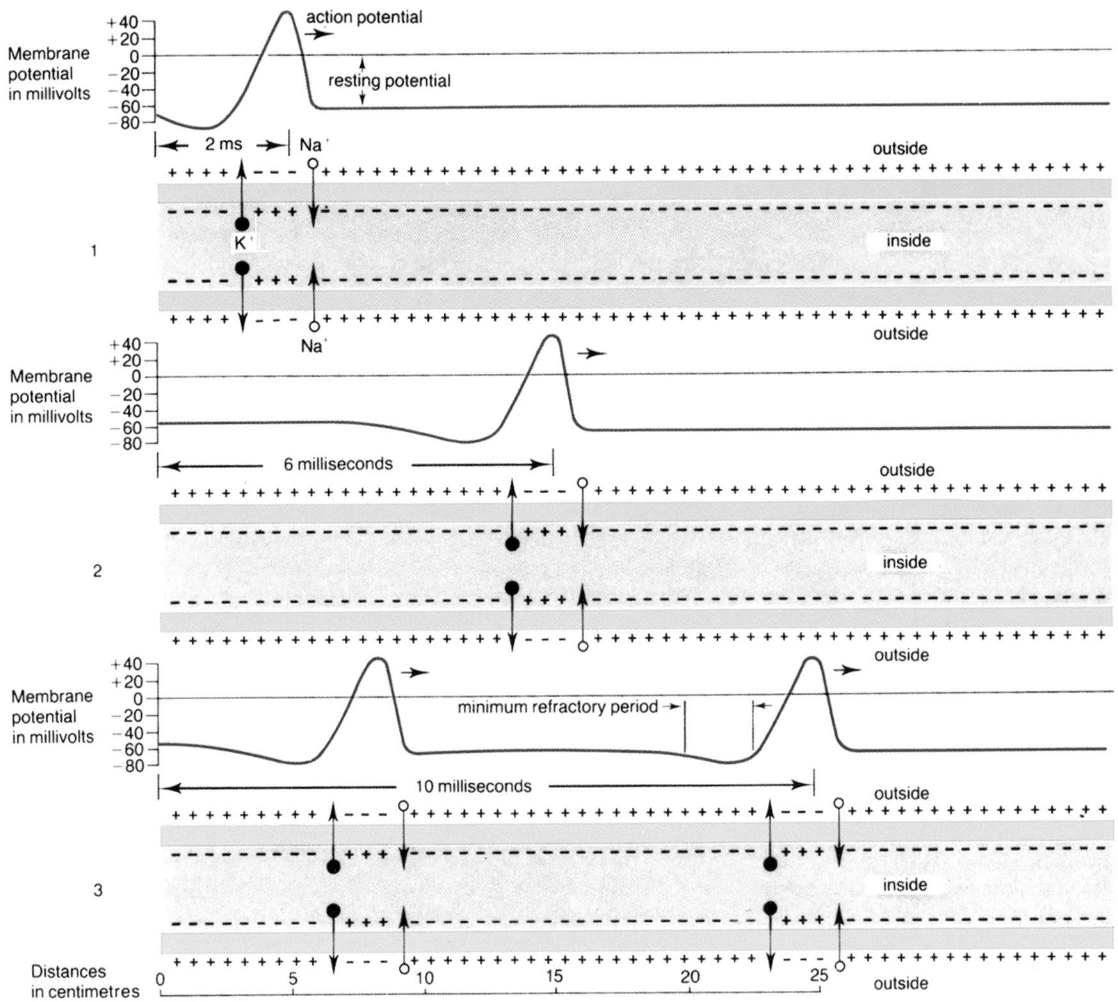

A major milepost was reached in 1952, when Alan Hodgkin (1914–1998) and Andrew Huxley showed that the nerve impulse, or action potential as it is called, is driven by a sudden influx of sodium ions across the nerve membrane, here indicated in yellow, followed within about a millisecond by an outflux of potassium ions. This causes the voltage (or potential) across the membrane to swing from about −0.1 volts to about +0.04 volts, and back again, all within approximately two milliseconds. This pulse behaviour is indicated by the red line which shows the voltage distribution along the nerve axon at three different instants, mutually separated by four milliseconds. The pulse travels at about twenty metres a second, and is generated only if stimulated by an initial voltage drop exceeding a certain threshold value.

Independent endorsement was supplied by determination of the structure of the powerful *nerve poisons*, *tetrodotoxin* and *saxitoxin*. The molecules of both these compounds have charged groups which mimic the sodium ion by entering the sodium channel, the poisonous effect arising from physical blockage of the latter.

The indications were thus that the ions move through the channels unaccompanied by the water molecules that are strongly bound to them in solution. Such water molecules are referred to collectively as a *hydration shell*, their actual number depending upon the surface charge density of the ion. This is larger for sodium than it is for potassium, because of the sodium ion's smaller radius, and that fact is reflected in its hydration shell comprising about five water molecules, compared with the potassium ion's three.

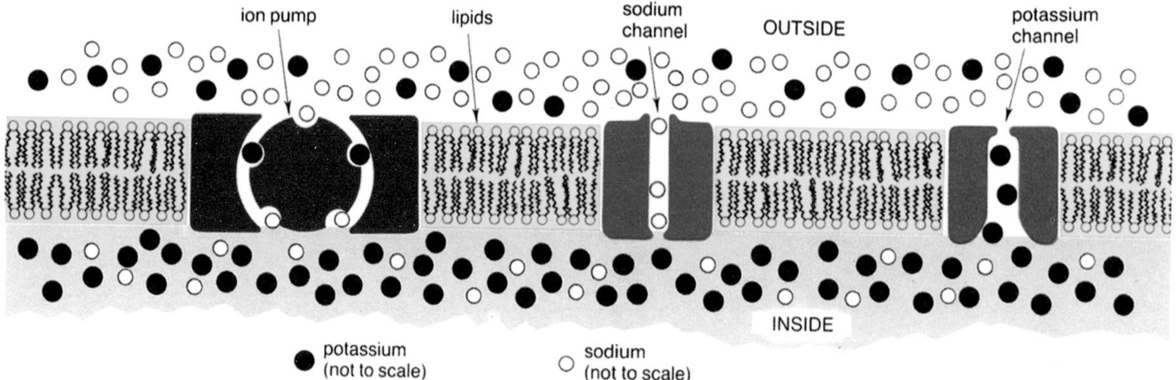

As in all biological membranes, the nerve membrane basically consists of a lipid bilayer dotted with proteins, the latter belonging to the special variety which span the entire bilayer width. There are three key types: one (discovered by Jens Skou in 1957) powered by ATP, which selectively transfers sodium to the outside and potassium to the inside, while the other two (whose existence was predicted in 1952 by Alan Hodgkin, 1914–1998, and Andrew Huxley) mediate the diffusion of sodium and potassium. These three types of protein are indicated in red, blue, and green respectively in this highly schematic diagram. The action of the ion pump achieves what is referred to as active transport, and thereby keeps the nerve in a primed state, ready to fire. The sodium and potassium channels, shown with idealized 'gates', permit the ionic movements that produce the nerve impulse. The effectiveness of some poisons derives from the ability of their molecules to fit into the sodium or potassium channels and block the flow of ions.

The water molecules in a hydration shell are not permanently attached to the ion. They periodically move away, and are immediately replaced by other water molecules because of the strong electrostatic attraction of the ion. The question arose, therefore, as to how that attraction could be overcome, permitting the ion to slip alone through the channel. The publication of the three-dimensional atomic structure of a potassium channel, by Roderick MacKinnon and his colleagues in 2003, resolved the mystery. Such channels are proteins, of course, and the pipelike hole through the potassium channel was found to be lined with the oxygen atoms that are a standard feature of these molecules. The remarkable discovery was that the oxygen atoms lining the hole were arranged in such a way that they would mimic the oxygen atoms in a hydration shell; the pipe was, in effect, a pseudo-hydration shell. It was thus clear that the energy required to tear the moving ion away from its hydration shell in the water surrounding the channel would be much smaller than had been anticipated.

During the action potential, the flow of ions is in the direction dictated by the principles of *electrodiffusion*: from a region of high concentration towards a place where the concentration is lower. This is true for both sodium and potassium, and such diffusion is termed passive. The processes which maintain the nerve in its loaded state, with its concentration imbalances of both sodium and potassium are, on the other hand, 'uphill' mechanisms. (We encountered analogous processes for plant membranes in the previous chapter.) Such *active transport*, as it is called, obviously requires the expenditure of energy. Indeed, perhaps as much as a quarter of all energy consumed by a human being is used in maintaining the concentration gradients in the neuromuscular system. There must be a mechanism whereby the active transport occurs, an *ion pump* as it has been called, and in this case too, knowledge of the molecular details is steadily increasing. In the early 1950s, Alan Hodgkin and Richard Keynes had shown that sodium ions flow into a nerve cell when it conducts a signal, and that ATP is consumed when such ions are transported in the opposite direction. They also found that the latter process is inhibited when ATP synthesis is disrupted. Demonstration of the essential validity of the pump hypothesis came in 1957, when Jens Skou discovered an enzyme that hydrolyzes ATP (i.e. splits it, to liberate its stored energy) only in the presence of sodium and potassium, and traces of magnesium. The enzyme, given the cumbersome but logical name Na^+,K^+-ATPase, simultaneously transfers sodium ions to the exterior and potassium ions to the interior. It subsequently transpired that three sodium

Although the idea of ion channels through the nerve cell membrane (and indeed through the membranes of other types of cell) quickly became popular, it took several decades to obtain atomic-scale evidence of their existence. The experimental challenge lay in extracting numerous channels from the membrane, without disrupting their structure, and then inducing them to crystallize. And demonstrating that certain parts of the channel were responsible for its voltage sensitivity required inspired coupling of electrical measurements with the techniques of molecular biology. This mammoth task was finally accomplished for one type of potassium channel in 2003, by Roderick MacKinnon and his colleagues.

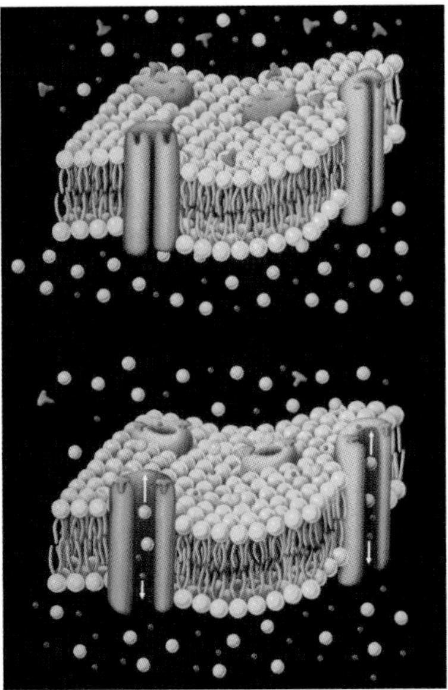

ions are actively transported out of the cell for each pair of potassium ions redeployed in the opposite direction. The actual atomic arrangement in the ion-transporting enzyme still remains to be determined.

The two main processes involved in the transmission of a nerve impulse along a neural axon are thus essentially understood. The active transport system maintains the axon in the ready-to-fire state, keeping the fuse loaded so to speak, and it rapidly restores this condition after firing. Moreover, as Hodgkin and Huxley also demonstrated, the counter-directed bursts of sodium and potassium ion flux, and the resulting double reversal of voltage across the nerve membrane, has just those characteristics that would be required to propagate a solitary wave of disturbance along the axon at the experimentally observed velocity. The question now arises as to what occurs when the nerve impulse has travelled the length of the axon and reached the synapse. We have already seen that information is transmitted across the synaptic gap by acetylcholine. Bernard Katz discovered that this compound is released in discrete packets of an apparently standard size. Once again we encounter the concept of a quantum of action. The unit in this case is approximately 10 000 molecules of the compound, and subsequent investigations revealed that the packaging is in bags of phospholipid bilayer membrane. These *vesicles*, as they are called, reside in the vicinity of the *presynaptic membrane*, and they release their charge of acetylcholine into the *synaptic gap* by fusing with that surface.

Individual neurotransmitters are doing this, in a random manner, all the time, but such individual haphazard events produce a depolarization of voltage at the *postsynaptic membrane* of only about 0.0005 V. When a nerve impulse arrives at the synapse, several acetylcholine-bearing vesicles fuse with the presynaptic membrane in unison, and this triggers an increase in postsynaptic membrane permeability, large enough to elicit a strong response. Several control mechanisms are in evidence. The number of vesicles close enough to the presynaptic membrane to be involved in the triggered fusion appears to be limited by structural proteins residing in the adjacent cytoplasm. These tubular structures seem to marshal the vesicles

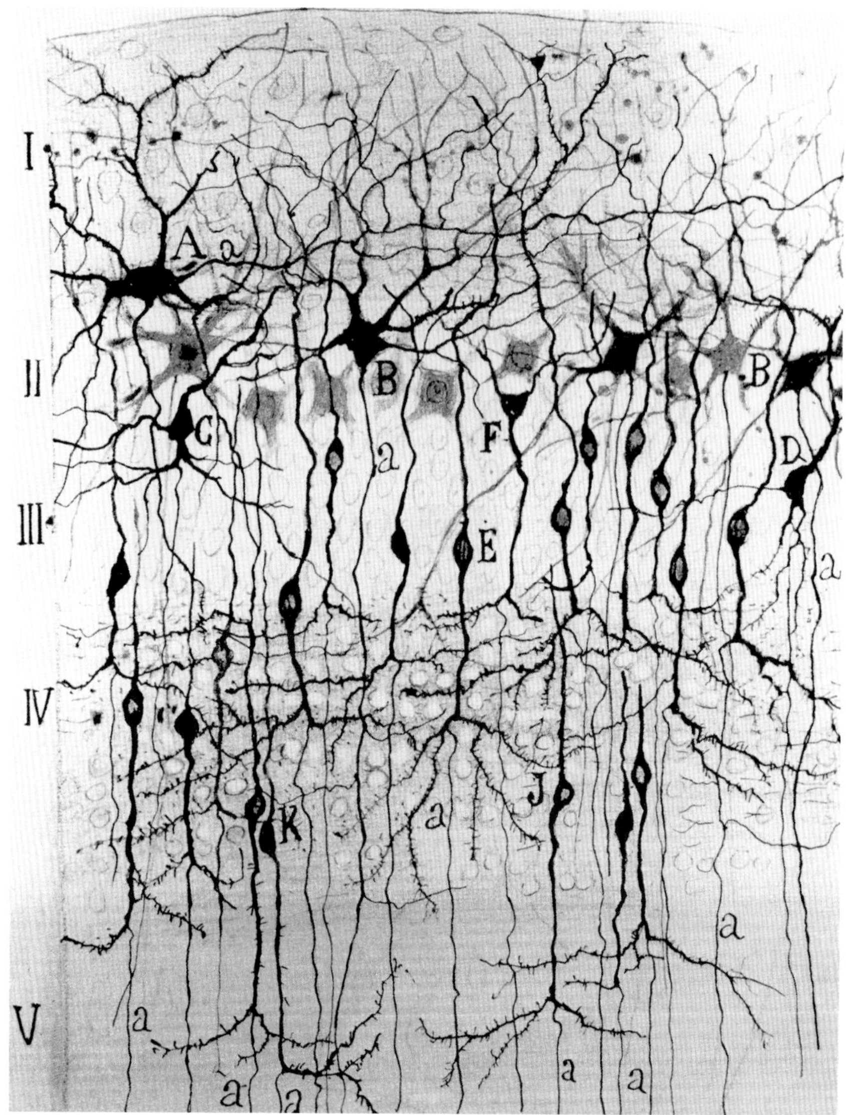

Study of the arrangement of nerve cells in brain tissue was greatly facilitated by a staining technique developed by Camillio Golgi (1844–1926). It stains only a few cells, but those that do become coloured are revealed in their entirety, and this produces a microscopic view that would otherwise be hopelessly cluttered. Santiago Ramón y Cajal (1852–1934) employed Golgi staining in his extensive investigations, drawing by hand what was visible in the microscope. An example of his meticulous work is shown here, and features the stratification into distinct layers that he discovered (although he did miss a sixth layer discovered more recently). The lower case letters **a** indicate the axons of the various neurons, amongst which are one with a horizontal axon (**A**), large stellate neurons (**B**), small pyramidal neurons (**C** and **D**), a superficial-layer fusiform neuron (**E**) and deep-layer fusiform neurons (**J** and **K**). In contrast to most other types of cell, neurons do not divide after birth, so the nervous system's complement of these cells must serve a lifetime. New synapses can be formed between existing neurons, however, and such generation is particularly vigorous in young children.

into an orderly queue. The spike-invoked fusion is mediated by calcium, the actual mechanism of release of these ions presumably having much in common with the analogous process in the muscle cell described earlier.

Bernard Katz and Ricardo Miledi found that within about 0.33 ms after the fusion of a single vesicle at the presynaptic membrane, approximately 2000 ion channels are opened in the postsynaptic membrane, the channels somehow permitting passive diffusion of both sodium and potassium. Once again, molecular details are becoming available, particularly through the efforts of Nigel Unwin and his colleagues. This information is tending to endorse the view put forward in the mid 1960s by Jean-Pierre Changeux and Arthur Karlin. They proposed the existence of a receptor molecule with functionally (and probably structurally) distinct parts: an *acetylcholine receptor* which induces a conformational change that causes channel opening. If acetylcholine survived indefinitely, the channels would remain open and postsynaptic membrane excitability would be nullified. The compound is,

Nerve cells and synapses

Charles Sherrington (1857–1952) coined the term *synapse* for the junction between two nerve cells and noted that the existence of such junctions effectively joins the individual neurons into a vast interconnected mesh, which he famously likened to an *enchanted loom*. Information is carried across one of these junctions by chemicals known as neurotransmitters, one particularly common variety being acetylcholine. Bernard Katz and Ricardo Miledi showed that this compound is released in discrete packets of a roughly standard size. The packaging unit is a phospholipid vesicle, capable of holding about ten thousand molecules of the neurotransmitter, and the waiting vesicles are marshalled up to the presynaptic membrane by microtubules. All these features are shown in the two diagrams. The arrival of a nerve impulse causes some of the vesicles to fuse with the presynaptic membrane. The neurotransmitter molecules thereby released diffuse across the synaptic gap and activate ion-channel proteins on the postsynaptic membrane. The electron photomicrograph shows vesicles caught in the act of fusion with the presynaptic membrane. In the 1940s, Donald Hebb (1904–1985) suggested that the selective restructuring of synapses might underlie the long-term learning process.

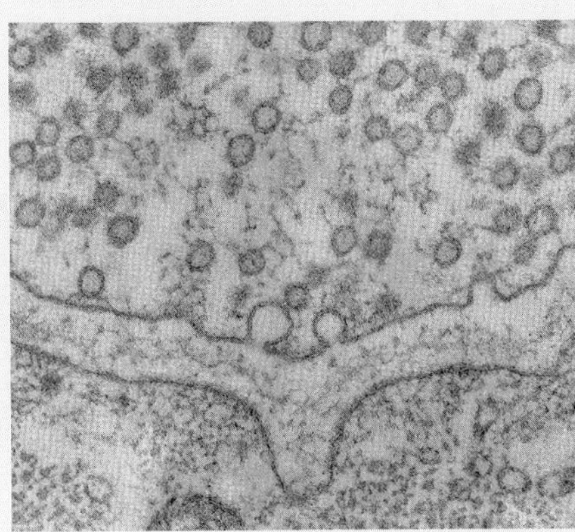

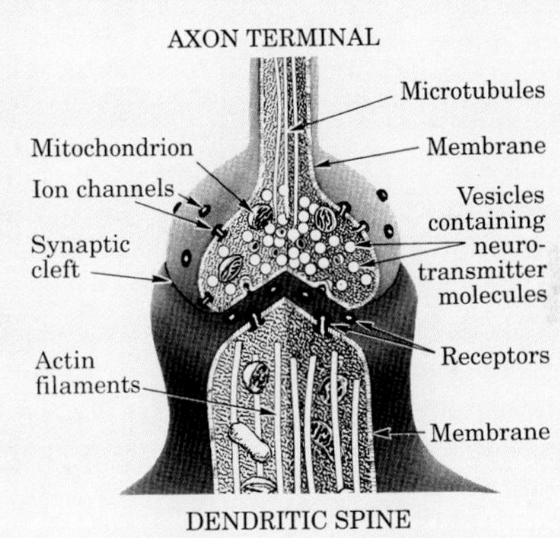

in fact, broken down by the enzyme *acetylcholinesterase*, discovered in 1938 by David Nachmansohn, and Katz and Miledi have estimated that about a third of all acetylcholine is hydrolyzed before it ever reaches its target receptors. Even those molecules that do succeed in contributing to the post-synaptic voltage change are broken down within a millisecond or so. This is important because the responses of the nervous system would otherwise be unacceptably sluggish.

Acetylcholine is not the only neurotransmitter; several dozen different examples have now been identified. Particularly prominent amonst these are the *excitatory neurotransmitter glutamate* and the *inhibitory neurotransmitter gamma-aminobutyric acid* (*GABA*). Some neurotransmitters, such as *adrenaline* and *noradrenaline*, also serve as *hormones*. Knowledge of the principles of nerve conduction has led to pharmacological advances. Enhancement of function can be secured by use of drugs which inhibit acetylcholinesterase. This is the basis of the treatment of the eye disease *glaucoma* by *neostigmine*, for instance. Among the more nefarious exploitations of this branch of science has been the development of powerful nerve gases for military use.

The neurons are part of the vast nervous system of the animal, and this is normally divided into two parts: the *central nervous system*, which in the vertebrates comprises the *brain* and the *spinal cord*, and the *peripheral apparatus*. Not all the neurons in the latter are directly dependent on the brain, some,

Vision is probably the best understood of the senses. The chemical compounds active in the light-sensitive cells have been identified, and the neural connections between the retina and the visual region of the brain have been traced and mapped in full three-dimensional detail. The primary event of the visual process is the change of shape of a molecule of 11-*cis*-retinal, induced by absorption of a light photon, and this causes a conformational change in a rhodopsin molecule. The latter type of molecule is present in the membranous discs of each retinal rod cell (see diagram left, and cross-sectional electron photomicrograph, middle). Modification of the shape of a rhodopsin molecule releases numerous molecules of a small chemical transmitter substance, believed to be either calcium ions or cyclic guanosine monophosphate, and this in turn causes the change in sodium permeability of the cellular membrane which results in the generation of a nerve impulse. The right-hand pictures show retinal rod and cone cells, at different magnifications, as observed by scanning electron microscopy. Both types of cell have a diameter of about one micrometre. The picture reproduced at the larger magnification was taken from a specimen of the mudpuppy (*Necturus*).

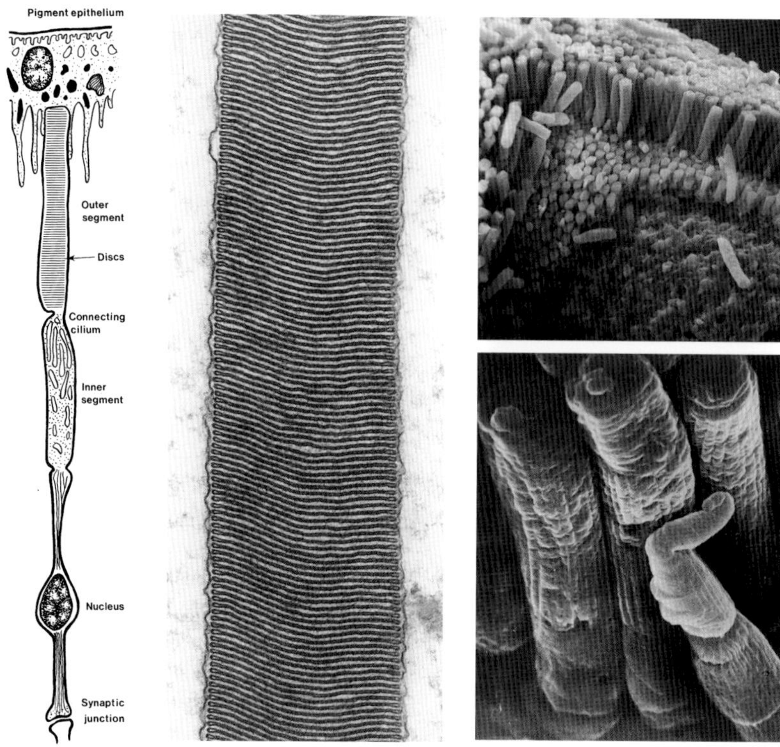

known as the *autonomic ganglia*, serving as simple but reliable controllers of the internal organs, including the *heart*. One differentiates between nerves which carry information towards the central nervous system, the *afferent nerves*, and the *efferent nerves* which transmit messages from the central nervous system to executive organs such as *muscles* and *glands*. Afferent information originates at the various *sense organs*, which are divided into five types: touch, hearing, taste, smell and sight. To this traditional classification we could add balance. In all cases, there are special protein *receptor molecules* present in the membranes of the *receptor cells* of these organs. They act as transducers, converting energy into nerve impulses, and as such they can be differentiated into four groups: *thermal receptors, mechanoreceptors, chemoreceptors* and *photoreceptors*. The *sense of pain* also depends upon the presence of mechanoreceptors, which detect actual or merely threatened damage to tissue.

The sensations of heat and cold are the easiest to understand because they depend on the simple principles of chemical reaction rates. As we have already seen, reactions proceed at a rate that is determined by temperature, hotter implying faster. For a typical biological process, a rise in temperature of 10 °C produces a doubling of reaction rate. *Thermal sense* is provided by nerve endings that spontaneously generate action potentials, the rate of these being determined by temperature. The nerve cells involved in the *tactile sense* are of various types. In the more primitive of these, pressure causes deformation of the nerve membrane which, in turn, alters the sodium conductivity. When the pressure is sufficiently large, an action potential is evoked. If the pressure is sustained, the active transport process adjusts itself so as to restore the membrane potential to its initial resting value, and the nerve, said then to have *adapted*, ceases to emit signals. Release of the pressure produces a new change of deformation, and another series

Biochemistry of vision

The key molecule involved in vision consists of two parts: the protein opsin, and the attached prosthetic group 11-*cis*-retinal. The latter is not produced by the body, and must be acquired through the diet. It is not consumed in its operative form, but is ingested as the precursor all-*trans*-retinol. The structure of this important molecule, commonly known as Vitamin A, is

all-trans-retinol (Vitamin A)

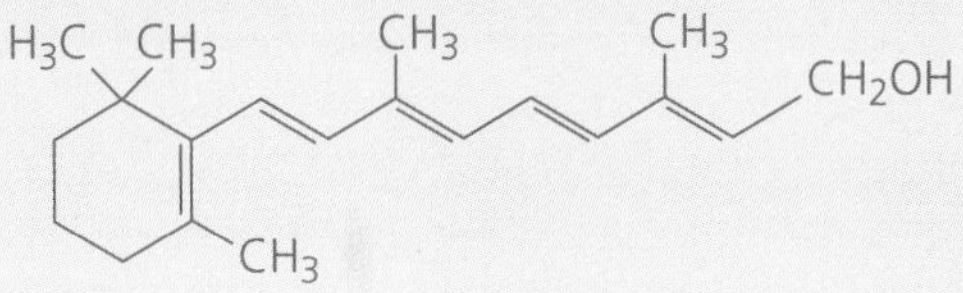

H3C CH3 CH3 CH3 CH2OH CH3

all obvious carbon and hydrogen atoms not being shown, by the usual convention. The structure of 11-*cis*-retinal is

11-*cis*-retinal

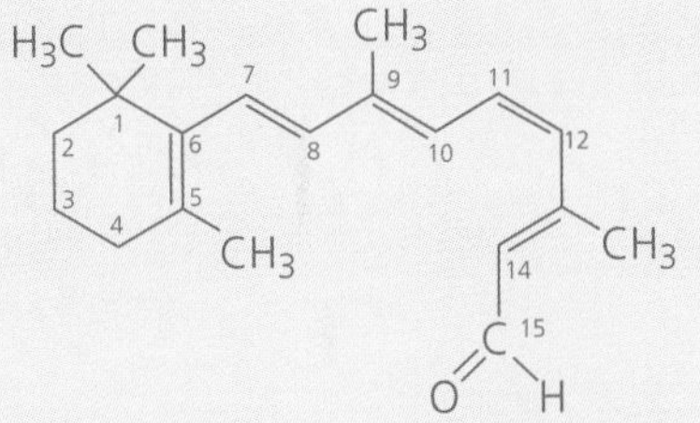

H3C CH3 CH3 CH3 CH3 O H

the number 11 denoting that the *cis* configuration occurs at the eleventh carbon atom numbering from the end with the ring structure. This group is attached to the lysine side-chain of the opsin via what is referred to as a Schiff base, thus

11-cis-retinal lysine side chain *protonated Schiff base*

$$\mathrm{R{-}C(=O)H} + \mathrm{H_2N{-}(CH_2)_4{-}opsin} \rightarrow \mathrm{R{-}CH{=}N^{+}H{-}(CH_2)_4{-}opsin} + \mathrm{H_2O}$$

When a light photon falls on the 11-*cis*-retinal, its energy is consumed in altering the geometry of the molecule, to produce all-trans-retinal. This isomerization, as it is called, involves the swivelling of the eleventh carbon–carbon bond through 180°. The structure of the resultant isomer is as follows:

all-trans- retinal

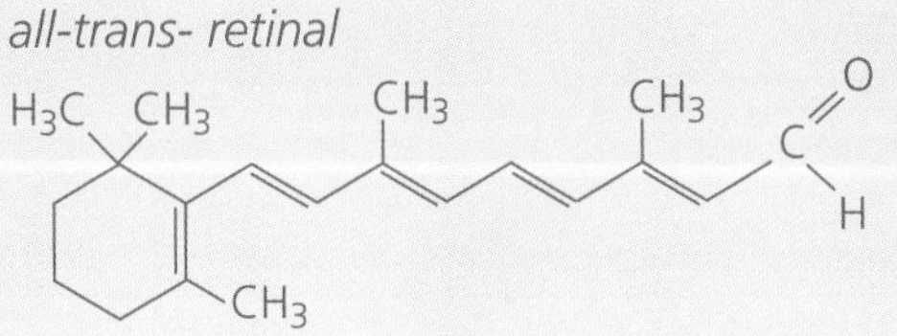

H3C CH3 CH3 CH3 O H CH3

of action potentials. In other mechanoreceptors, the primary interaction is with the hair-like cilia, and in this case the microtubules lying immediately below the cell surface appear to be involved in the generation of the action potentials. A process of this type is involved in the *audioreceptors* of the ear. The *sense of balance* in the human being derives from similar receptors in the three *semicircular canals* of the inner ear, which are mutually perpendicular. They are filled with fluid that contains crystals of calcium carbonate. These

Blood

Following the discovery of blood circulation in 1628, by William Harvey (1578–1657), the focus of interest shifted to its actual composition and structure. It consists of *plasma*, which is an aqueous solution of protein, salt and glucose that also transports other substances around the body; *red cells* (also known as erythrocytes), which account for almost half of the mass and transport oxygen through the agency of their haemoglobin molecules; *white cells* (also known as leukocytes), which play a key role in immunity; and *platelets*, which produce clotting when required. The haemoglobin molecule consists of four polypeptide chains: two identical alpha chains and two identical beta chains, comprising 141 and 146 amino-acid residues respectively. A haem group is attached to each of the four units, and this gives the molecule its oxygen-transporting capability. Each haem sits just inside a cleft between adjacent alpha-helical sections, and relatively small movements of the latter permit an oxygen molecule to approach and become loosely bound to the haem's iron atom. The picture, based on Max Perutz (1914–2002) and his colleagues' determination of the molecule's structure, differentiates between the two types of chain and indicates the haem groups in red. Haemoglobin can be roughly compared to a cluster of four myoglobin molecules, but it is capable of displaying cooperative phenomena not seen in its lighter cousin. The scanning electron photomicrograph of red blood cells reveals their characteristically dimpled shape. Each erythrocyte is about five micrometres in diameter and contains many millions of haemoglobin molecules. There are about five litres of blood in the circulatory system of an adult human. The fact that this vital bodily fluid also participates in the immune response is underlined by the existence of *blood groups*, discovered in 1900, and of the *rhesus factor*, discovered in 1940, both by Karl Landsteiner (1868–1943). Blood transfusion was a hazardous undertaking before Landsteiner's discoveries; administering an incompatible blood type to a patient could trigger a life-threatening immune response. The following table should only be used as a rough guide, therefore, because many anomalous special cases exist.

Recipient blood type	Donor plasma	Donor white blood	Donor red blood
O+	any O, A, B or AB	O+, O−	O+, O−
O−	any O, A, B or AB	O−	O−
A+	any A or AB	any A+ or A−	any A+ A− O+ or O−
A−	any A or AB	A−	any A− or O−
B+	any B or AB	any B+ or B−	any B+ B− O+ or O−
B−	any B or AB	B−	any B− or O−
AB+	any AB	any AB+ or AB−	any AB+ AB− A+ A− B+ B− O+ or O−
AB−	AB	AB−	any AB− A− B− or O−

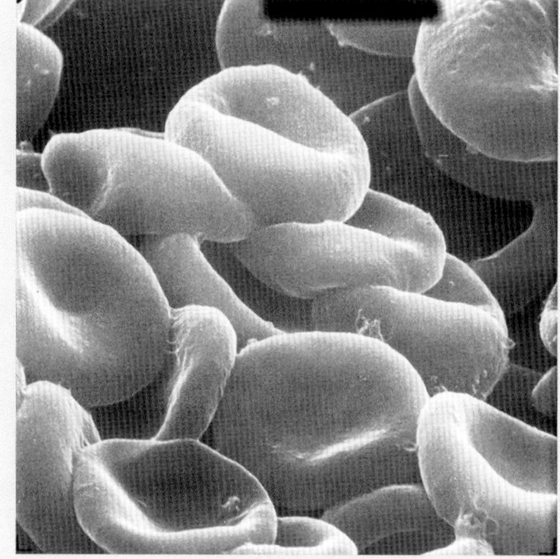

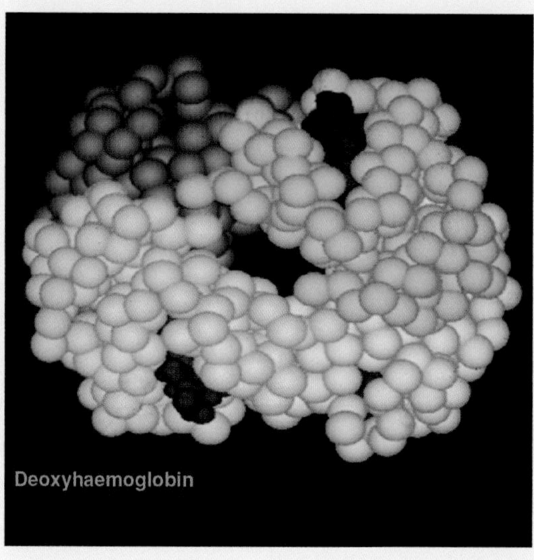

Haem

Although the oxygen-transporting molecules haemoglobin and myoglobin consist primarily of protein, the actual binding of oxygen depends upon the presence of a non-protein unit known as a haem group. Each subunit of haemoglobin, and each molecule of myoglobin, contains such a group. It is a ring structure with an iron Fe^{2+} ion sitting at the centre like a spider on its web. The ring is the organic compound protoporphyrin, and the iron is linked to four nitrogen atoms. When the haem is in place in the carrier molecule it makes a fifth bond to one nitrogen of a nearby histidine unit, while a sixth link can be provided by a water molecule which is readily displaced by a molecule of oxygen.

ferroprotoporphyrin (IX)

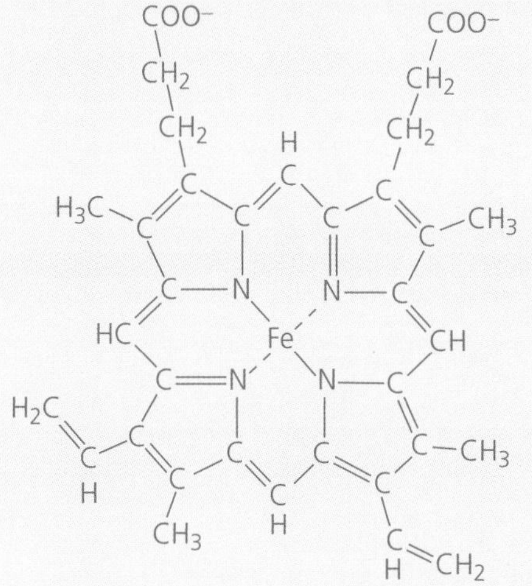

This prosthetic group, with slight modifications in its structure, also plays a central role in the electron transport involved in respiration, and in photosynthesis. In the former, it is present in cytochrome, which, unlike the blood molecules, contains iron in both the Fe^{2+} and Fe^{3+} charge states. In the chlorophyll that is vital to photosynthesis the iron is replaced by Mg^{2+}, and some of the side groups are modified.

otoliths, as they are called, provide directional sense by pressing more or less firmly against the *cilia* in the various canals, the brain being capable of synthesizing the three sets of messages into an orientation with respect to the direction of gravity.

The chemoreceptors which provide the sensations of taste (formally known as *gustation*) and smell (*olfaction*) function by changing the resting potential of their associated membranes. There are four different types of receptor in the taste buds of the tongue, corresponding to the four *basic tastes*: sweet, sour, bitter and salty. The chemical structures of the receptor molecules and the manner in which they function remain to be determined. But one could guess that they resemble those of the olfactory system, the understanding of which was greatly increased through the efforts of Richard Axel and Linda Buck in the 1980s. They discovered that there is a family of about 1000 genes which code for different olfactory receptor molecules, this thus being about 3% of the total human gene complement. Each receptor molecule is present in the membrane of a receptor cell, of course, and Axel and Buck showed that a given type of cell contains only one type of receptor molecule. Furthermore, a given type of receptor molecule can detect

Aldol cross-links

Collagen is able to make its vital contribution to the structure of the animal body because it combines strength with flexibility. These qualities derive from the covalently-bonded chains of the collagen molecules, and cross-links provide additional strength in much the same way as they do in vulcanized rubber. In the case of collagen, the intramolecular bridges within each triple helix are provided by aldol condensation between pairs of lysine side groups.

```
    |                                                          |
 H—N                                                           N—H
    |                                                          |
 H—C—(CH2)2—CH2—CH2—+NH2    H3N+—CH2—CH2—(CH2)2—C—H
    |                                                          |
 O=C                                                           C=O
    |                                                          |
```

Lysine residues

Two adjacent lysine residues undergo, with the catalytic help of lysyl oxidase, aldehyde conversion. The structures thereby become

```
    |                                                   |
 H—N                                                    N—H
    |               O        O                          |
 H—C—(CH2)2—CH2—C//          \\C—CH2—(CH2)2—C—H
    |               \        /                          |
 O=C                 H      H                           C=O
    |                                                   |
```

Aldehyde derivatives

and these are then linked by an aldol condensation, thus:

```
    |                            |
 H—N                             N—H
    |               H            |
 H—C—(CH2)2—CH2—C=C—(CH2)2—C—H
    |                 |          |
 O=C                  C          C=O
    |               //  \        |
                   O     H
```

Aldol cross-link

As is usual in a condensation reaction, the by product is water. The aldol condensation was discovered by Aleksandr Borodin (1833–1887), better known as the composer of the opera *Prince Igor*.

only a limited number of different odour substances. The receptor cells in the olfactory system are therefore rather specialized, and their axons take equally specialized routes to tidily distinct regions of the brain's primary olfactory receiving area, that is to say the olfactory bulb. Only thereafter is there mingling of nerve signals from a broader range of receptors, and the degree of signal mixing increases the deeper into the brain's hierarchy one penetrates. Ultimately, of course, that mixing is total because the nerve cells in the motor-activating part of the brain can respond to any incoming odour. The senses of taste and smell are not independent. The diminution of the ability to smell, suffered by a patient with a head cold, is accompanied by a well-known impairment of taste.

The results obtained by Axel and Buck are reminiscent of those found with other sensory modes. In the auditory system, for example, there is similar channelling of different information into different signalling routes, the information of interest being primarily frequency. As is well known, sound waves funnelled through the outer ear impinge upon the eardrum, causing it to vibrate. These vibrations are transmitted across the air-filled middle ear via a set of three small bones collectively known as the *ossicle chain*, the final member of which – the stirrup – makes contact with the *oval window* of the fluid-filled inner ear. The key component of this latter region, the ear's inner sanctum, is the *basilar membrane* that divides the spiral-shaped *cochlea* in its longitudinal direction. Through a series of delicate experiments, around the middle of the twentieth century, Georg von Békésy established that the vibrations of the oval window are supplanted to the basilar membrane, which in turn activates the underlying receptor cells, better known as the *hair cells* because of their hairlike appearance. He was able to confirm Hermann von Helmholtz'sassumption that the frequency of the incident sound waves dictates the position along the basilar membrane at which the stimulation reaches its maximum, and he also demonstrated that the information implicit in this position is encoded in the frequency of the nerve impulses generated by the hair cells. We thus see that the system converts frequency to position, and then position to frequency, but one must bear in mind that the frequency of the nerve impulses is much lower than that of the incident sound.

The best-studied of the sensory receptors is that involved in *sight*. The visual nerve endings in the human are of two types, both named for their physical appearance: *cones*, which function only at high light intensity and also provide *colour vision*, and *rods*, in which a single very sensitive pigment provides low-illumination vision with negligible colour discrimination. The adult human retina has about three million cones and a billion rods. The former are confined to the centrally located *fovea* while the latter are spread uniformly over the rest of the retina. Both types of cell are elongated, with a region at the receptor end comprising a series of disc-shaped membranes stacked pancake-wise and lying transverse to the long axis. It is these inner membranes which contain the visual pigment. Further along the cell we find a region rich in mitochondria and ribosomes, which respectively generate ATP and protein at the high rate consistent with a large turnover of energy and material; the average disc lifetime is a mere 10 days. Finally comes a region containing the nucleus, and the terminal synaptic connection with the next neuron in the retina. The *optic nerve* consists of about 1200 bundles of myelinated nerve fibres. Optical sensitivity derives from the interaction of light energy with the visual pigment, and in the case of the rods, as shown by Selig Hecht in 1938, a cell can be triggered by a single photon.

The photosensitive molecule in rod cells is rhodopsin, which consists of the protein *opsin* and the attached prosthetic group *11-cis-retinal*. The latter cannot be produced by the body, and its ingested precursor is

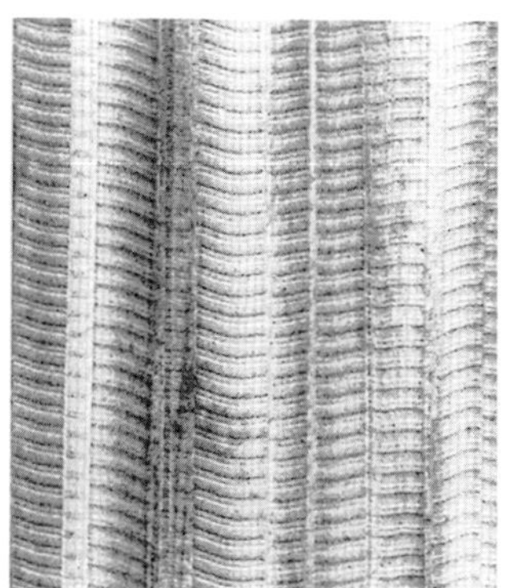

The basic structural unit of collagen, which provides the basis for the body's fibrous connective tissue, is the tropocollagen molecule. This is about 300 nanometres long and 1.5 nanometres wide, and it comprises three linear polypeptide chains. Their relative positions are shown in the diagram, with the hydrogen bonds and the side groups indicated by the usual dotted lines and green spheres. Every third side group is glycine, and this allows the chains to lie close enough to permit the hydrogen bonding. In collagen, the tropocollagen molecules are arranged in parallel, but with a stagger that produces the characteristic 67-nanometre repeated band structure visible in the electron photomicrograph of tissue from a rat-tail tendon. The gap bands are believed to mark the sites at which bone formation can be nucleated.

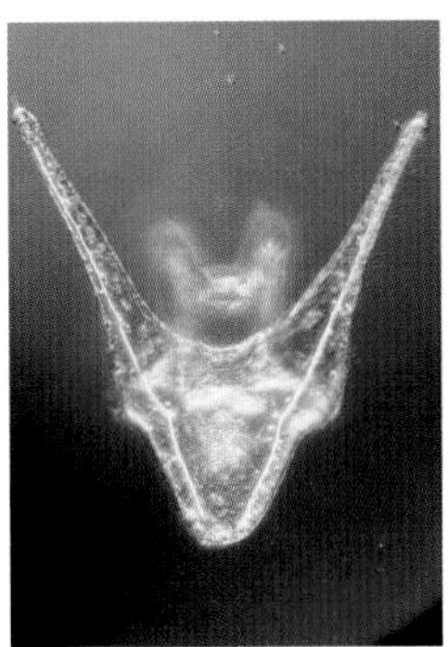
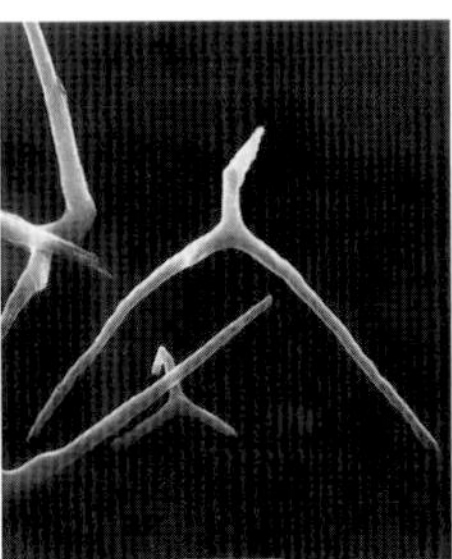

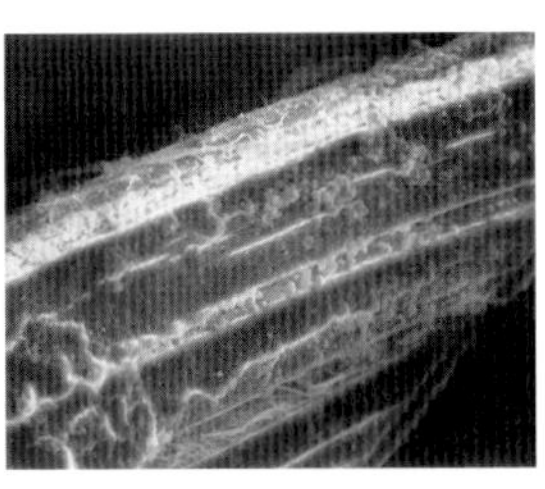
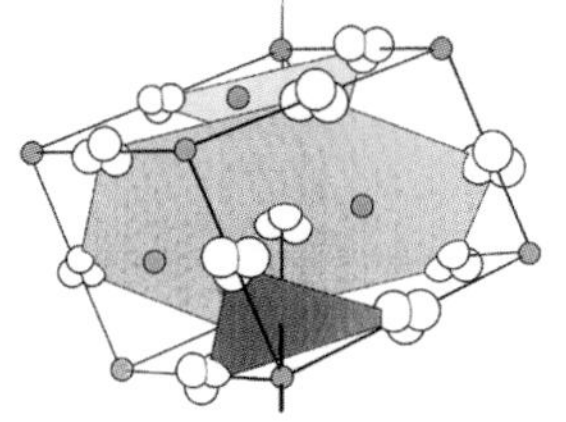
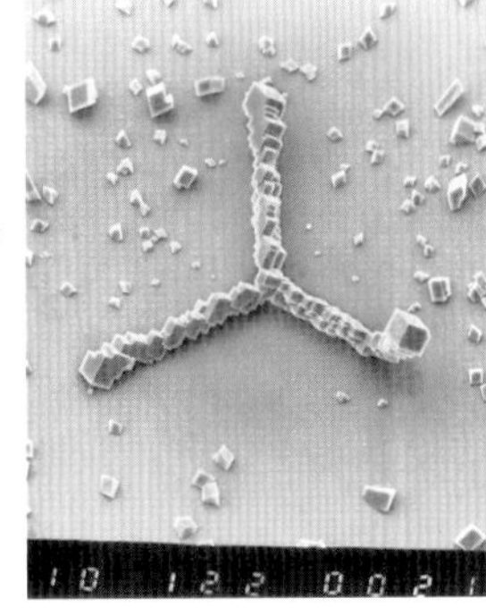
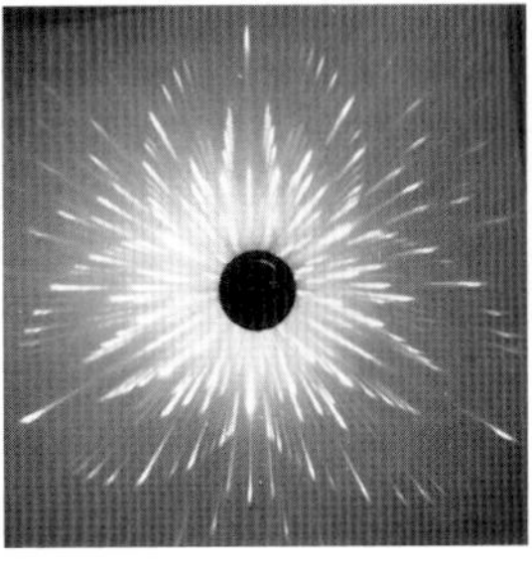

One of the best-studied biological crystal formations is the skeleton of the sea-urchin. In its embryonic stage, this creature grows two three-pronged structures known as spicules, which eventually fuse to create the supporting skeleton for the four-horned larval form. These beautiful structures are clearly visible by polarized-light microscopy (top left), and they are composed essentially of calcite with a few per cent of magnesium impurities. The crystal structure of calcite is shown bottom left, with its alternating planes (grey) of calcium ions (purple) and carbonate groups (white) perpendicular to a crystallographic axis which is here shown lying vertically. The development of smooth surfaces on the spicules appears to depend upon maintenance of the correct internal calcium and magnesium concentrations. The two scanning electron photomicrographs show normal spicule growth (top row, second from left) and abnormal sharp-edged growth (bottom row, second from left), the latter having been induced by lowering the calcium and magnesium levels in the seawater to only one tenth their normal values. There is an amusing similarity between the outer appearance of the mature sea-urchin (top row, third from left) and a diffraction pattern (bottom row, third from left) taken of one of its spikes. The streaks on the latter are a result of the elongated shape of the spike, which is found to be an almost perfect single crystal. The calcification of collagen, to form bone in animals, is believed to be under enzymatic control. The bottom left picture shows a scanning electron photomicrograph of collagen encrusted with an apatite deposit. Apatite is the animal counterpart of the sea-urchin's calcite, and it has a considerably more complicated composition and structure. Many of the oldest known archaeological artifacts were carved from bone or shell. The Sumerian bull pendant (bottom right) dates from about 2500 BC.

all-trans-retinol, commonly known as *vitamin A*. As was discovered by George Wald, the initial event in the visual process is the photon-induced *isomerization* of 11-*cis*-retinal to *all-trans-retinal*. In terms of molecular geometry this is simply a swivelling of one of the carbon–carbon bonds. Light energy is thereby converted to mechanical energy. The configurational change in

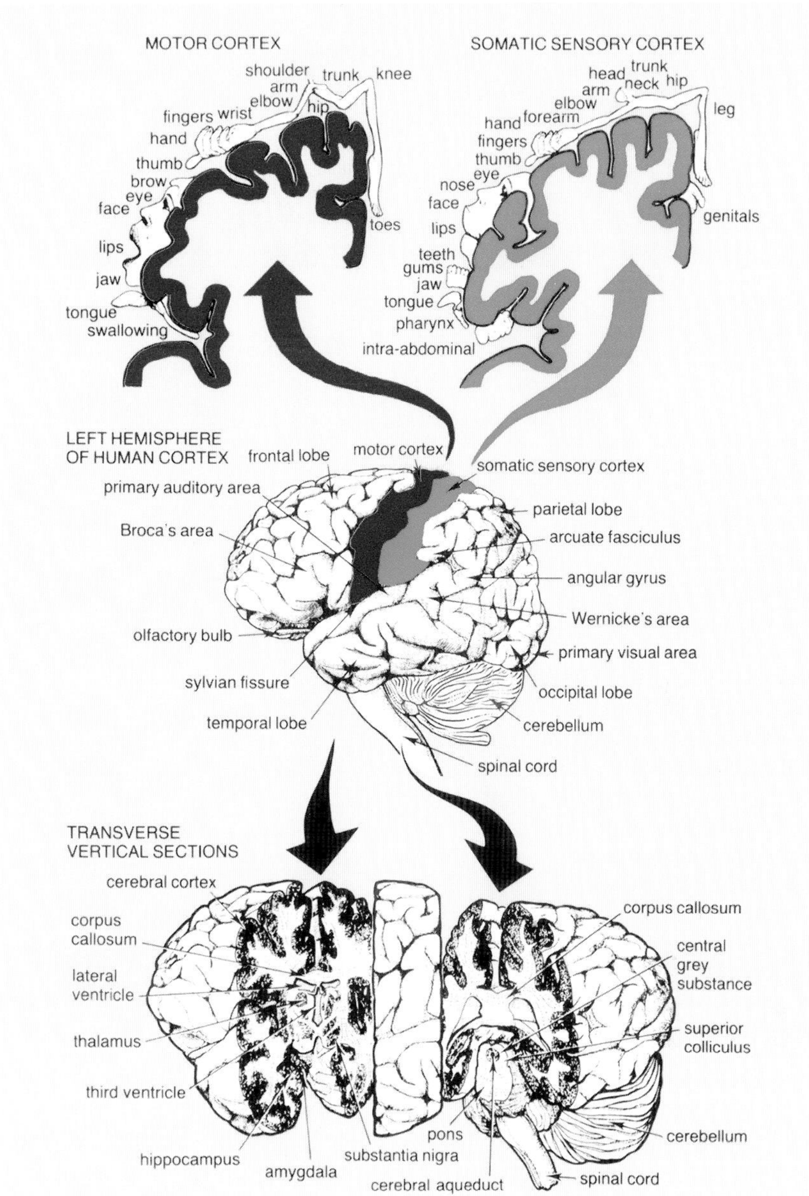

Many of the human brain's structural features have now been related to specific physiological functions. This is particularly true of the motor and sensory regions of the cerebral cortex; it has been found that there is actually a spatial mapping of the body's components, as shown in the upper part of this composite diagram, the motor and sensory maps lying roughly parallel. The lower sectional views show the location of the corpus callosum, which forms a bridge between the brain's two major lobes. Severance of this region appears to deprive the body's two halves of mutual knowledge.

the retinal also causes it to dissociate from the opsin. Just how these light-induced changes, occurring as they do at the disc membrane, induce cation permeability changes and hence a nerve signal in the outer cell membrane, which lies several hundred nanometres away, has not been fully elucidated. One possibility is that calcium ions once again serve to mediate transfer of information from one site to another, by simple aqueous diffusion.

Thomas Young's theory of *colour vision*, dating from 1802, hypothesized three types of colour receptor, sensitive to the three *primary hues*: blue, green and red. It has now been established that the *chromophore*, or *light-sensitive unit*, in all three types of cone is 11-*cis*-retinal, but that it is bound to a different form of opsin in each case. It is generally the case that the physical

Scientists are invariably influenced by current wisdom. When Leonardo da Vinci (1452–1519) was active, it was believed that the brain carries out its functions through the mediating agency of its animal spirits, which were taken to reside in the four ventricles. Leonardo displayed great technical resource when he determined the form of these chambers by pouring molten wax into them. When the wax had set, he was able to observe the ventricles' shape by cutting away the surrounding tissue. Unfortunately, he was thereby discarding the material which really held the mind's secrets.

properties of a prosthetic group depend on the nature of the protein to which it is attached, and the retinal–opsin complexes are a case in point. The three forms have peak sensitivities at 455 nm (blue), 530 nm (green), and 625 nm (red), nicely endorsing Young's original idea. *Colour blindness* results from genetically determined lack of one or more of the types of opsin. Deficiency of vitamin A produces *night blindness* and, if prolonged, deterioration of the outer segments of the rods. Instead of passing on to the brain, which might seem a logical step at this juncture, discussion of that complicated structure will be deferred until later in this chapter. And the way in which it functions, to produce what we call the mind, will be discussed in the final chapter.

The energy required to drive the various physiological functions of the animal is produced by the *respiratory processes*. Because of their high energy

The brain's layers

In popular depictions, the brain is usually represented only by its external surface, the underlying structures being far less familiar. Broadly speaking, the entire structure can be regarded as consisting of three layers of tissue, the oldest in terms of evolution lying innermost. The largest part of the brain is the cerebral cortex (left picture), also known as the telencephalon. It is encased by the dura matter and the pia-arachnoid meninges, just under the skull. It is the main part of the forebrain, and it comprises over fifty contiguous areas. Groups of these are known as lobes, prominent examples being the posteriorly located occipital lobe, which is concerned with vision; the laterally located temporal lobe, which comprises the auditory cortex, the limbic cortex and the temporal association cortex; the parietal cortex, which is located near the centre of the top of the head, and which is also instrumental in forming associations; and the frontal lobe, which appears to direct certain aspects of thought, and participate in the planning of actions. Also clearly visible in this figure are the motor cortex, which governs muscular movement, and the somatosensory cortex, which receives tactile signals from the various regions of the body. The diencephalon (middle picture) lies inside the telencephalon, and together with the midbrain it directs the more overt aspects of behaviour and the emotions. One of its most prominent members is the thalamus. This structure receives nerve signals from the various sensory-input regions (except the one serving smell), and it sends signals onward to the corresponding areas of the cerebral cortex. But it is more than just a relay station, because it is also the target of counter-running signals that convey information *from* the cerebral cortex. This figure could be misleading in that it makes the hippocampus appear like an almost independent worm-like structure. In reality, it is merely an internal edge of the cortex. Like the thalamus, it both receives signals from and returns signals to the cortex. The amygdala is the structure that appears to be most directly involved in the orchestration of emotions. The cerebellum and the pons are two of the most prominent members of the metencephalon of the hindbrain (right picture). They wrap around the top of the brainstem on the back and front sides, respectively. The cerebellum is implicated in the fine tuning of motor functions, and there is additional evidence that it plays a role in cognition. The pons is a distribution point for nerve fibres involved in the control of movement and posture. The reticular activating system exercises control over attention mechanisms, and it is also involved in regulation of the sleep–wake cycle, as is the locus coeruleus. The superior colliculus, which is located on the rear side of the brainstem, quite close to the locus coeruleus, receives signals from the retinas and also from the cerebral cortex. It controls the direction of gaze.

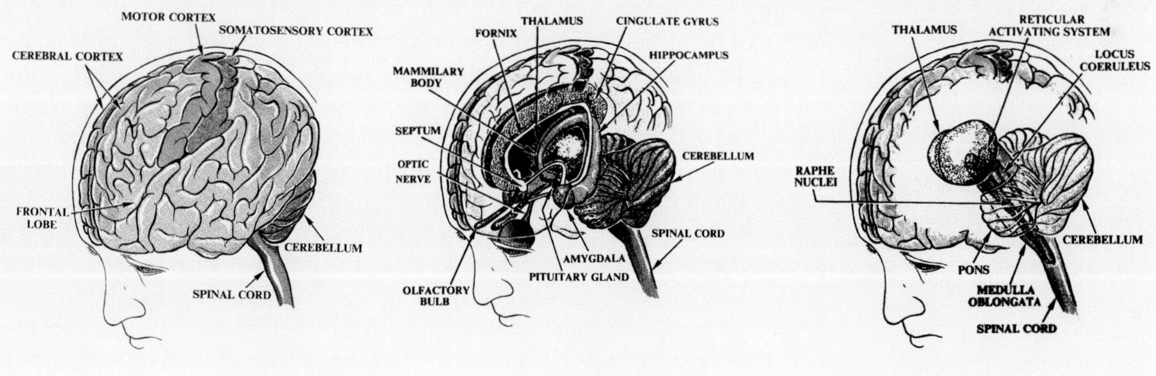

needs, all but the very smallest of living organisms have evolved *aerobic respiration*. Eighteen times as much energy is liberated from glucose when oxygen is involved as in the *anaerobic* process. To meet the increased demand for oxygen, those species for which simple gaseous diffusion would be too slow have evolved *circulatory systems*, and the most advanced enjoy the added boost of oxygen-transporting molecules. Without its oxygen carrier, haemoglobin, 1 l of human *arterial blood* at body temperature could dissolve

Studies of various animals, including the chicken, rat, guinea pig, cat and several species of monkey, have shown that the same parts of the nervous system control attack and defence. The cat in the left picture, photographed by Walter Hess (1881–1973) and M. Brügger, is displaying typical defensive behaviour against a threatening dog. The cat in the larger picture had electrical leads implanted into its midbrain, and Richard Bandler used it to demonstrate that electrical stimulation of that region evokes a similar response (note the white rat that is partly visible behind one of the cat's front paws). This is the case even if the midbrain has been surgically separated from the cerebral cortex, so the former is the real seat of such behaviour. In the most spectacular demonstration of the same principles, José Delgado implanted electrodes in the midbrain of a bull and controlled its behaviour by transmitted radio signals; he even managed to stop the animal in mid-charge. In 1939, Heinrich Klüver (1897–1979) and Paul Bucy showed that removal of the temporal lobe, including the underlying amygdala and hippocampus, produced a dramatic change in monkeys. They became much tamer, and yet reacted more vigorously to various stimuli. Their memory function was clearly impaired, and their level of sexual arousal had risen so much that they even mounted inanimate objects and members of other species.

and transport a mere 3 ml of oxygen. In the presence of that important substance the oxygen-carrying capacity increases seventy-fold. A single *red blood cell*, properly known as an *erythrocyte*, contains almost three hundred million haemoglobin molecules. It is thus appropriate that we consider the structure and mode of action of haemoglobin, the paradigm of a transport molecule. And it transpires that there is a related molecule, myoglobin, which serves as a temporary storage for oxygen in the tissues.

Myoglobin has a molecular weight of about 17 000 da, and 1260 of its atoms are not hydrogen. The molecule comprises two parts: *globin*, which is a protein consisting of a single polypeptide chain, having 153 amino-acid residues, and a prosthetic *haem group*, a ring structure which is the key to the oxygen-carrying function. *Haemoglobin* is almost exactly four times larger, the reason for this simple ratio lying in its relationship to myoglobin, as will be discussed shortly. Let us consider haem in the context of the larger of the two carriers. The reaction between haem and oxygen is reversible: in an abundance of oxygen, as occurs in the lining of the lung, oxygen is absorbed, whereas oxygen is released when this is scarce, as in the tissues. The active centre of the haem group is an iron atom. Only if this is in the divalent ferrous (Fe^{2+}) state, when the compound is called *ferrohaemoglobin*, can oxygen be bound. Thus although the optical properties of the haemoglobin molecule depend upon whether or not it is carrying oxygen, making arterial blood red and *venous blood* purple, the iron atoms are always in the same charge state. This was demonstrated by James Conant in 1923. The trivalent ferric state of the iron atom can be achieved, if the haemoglobin is removed from the blood cell and exposed to air for a prolonged period. In that *ferrihaemoglobin*, or *methaemoglobin*, state, the colour is brown. Reversibility is of course vital, so the oxygen must not be held too tenaciously: the iron must not become rust. The desired function is provided by cooperation between the globin and the haem. Indeed, the latter is not capable of binding oxygen if the globin is absent, so we must consider this largest part of the carrier molecule. And it is convenient to start with myoglobin.

As with other proteins, the way myoglobin works becomes apparent only through knowledge of its three-dimensional, tertiary structure. (Having but a single polypeptide chain, myoglobin has no quaternary configuration.) The structure was determined by John Kendrew and his colleagues, atomic resolution being achieved by 1959, using X-ray diffraction. This technique normally yields only half the desired information, producing scattered intensity but not the phases of the various diffraction beams. The *phase*

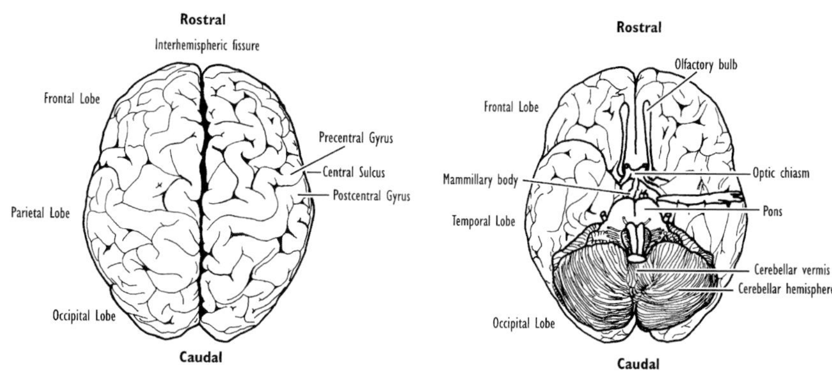

The cerebral cortex is a slab of neuronal tissue, about three millimetres thick, with a surface area too large to be smoothly accommodated within the skull. It is consequently convoluted into the pattern of elevations (gyri) and depressions (sulci) clearly visible in this plan view seen from above (left picture). Its tissue is commonly known as the grey matter, a name that derives from the darkening caused by the presence of the nerve cell bodies. The interhemispheric fissure, which runs from the front (rostral direction) to the back (caudal direction) would totally divide the brain into two separate halves, were it not for the bridging structure known as the corpus callosum. The right picture shows the view from below. The rostral part of the left temporal lobe (which would have lain on the right in this view) has been cut away to reveal the underlying (actually the *overlying*, when its owner is standing errect) left parietal lobe. The medial (centrally located) portion of this same temporal lobe was removed from *both* sides of the brain of the famous patient H. M., in a successful attempt to allay his epilepsy. Unfortunately, and as discovered by Brenda Milner in the mid 1960s, he thereafter suffered from anterograde amnesia, that is to say an inability to store new memories.

problem was solved by *heavy atom isomorphous replacement*. Corresponding diffraction pictures are taken of crystals of the compound with and without a heavy foreign atom substituting one of the normal atoms. The heavy atom, having a high scattering power, shifts all the phases, and the absolute values of the latter are found from the differences between the two X-ray pictures. A prerequisite is that addition of the heavy atom does not deform the molecular configuration. Lawrence Bragg compared this with the ruby on the forehead of the emperor's elephant; it doesn't know it's there.

Neither myoglobin nor haemoglobin has any disulphide bonds to help stabilize it. The required rigidity is supplied partly by the fact that the hydrophobic side-groups tend to lie at the inside of the molecule, while the haem group itself also makes a contribution. Kendrew and his colleagues found that myoglobin's secondary structure comprises eight distinct alpha-helical regions, and that the tertiary structure is a tight folding of these into an asymmetrical but compact globule having the approximate dimensions $4.5 \times 3.5 \times 2.5$ nm. The haem group resides in a fissure lying between the third, fifth, sixth and seventh helices from the amino end of the protein. This tertiary structure holds the secret of the reversible operation of the molecule. The iron is bound on one side of the haem plane to the histidine side group of the eighth residue in the sixth helical region. On the other side of the plane it is in close proximity to another histidine side group, on the seventh residue of the fifth helical region. It is not actually bound to the latter, however, and this allows an oxygen molecule to approach close enough to the iron atom to become loosely bound. The role of the other surrounding helical portions, which are conveniently non-polar, is to protect the iron of the haem group from the surrounding water, which would cause it to lose an electron, and be converted to the ferric state, which cannot bind oxygen. Sitting in its cleft, the haem group is like a tongue between jaws, the latter permitting it a taste of the oxygen, but not of the water.

The structure of haemoglobin was determined, using the same approach, by Max Perutz and his colleagues, this work reaching fruition in 1962. It turns out that haemoglobin consists of four linear polypeptides, which are divided into two pairs: two identical alpha chains having 141 amino-acid residues each; and two identical beta chains, each with 146 residues. Although mutually different in their amino-acid sequences, these two types of chain, and the single chain of myoglobin, resemble one another in their tertiary structures. All comprise eight helical regions, and for some positions of the sequence all three have the same amino-acid residue. In the sixth helical region, for instance, this is true of the fourth and eighth residues.

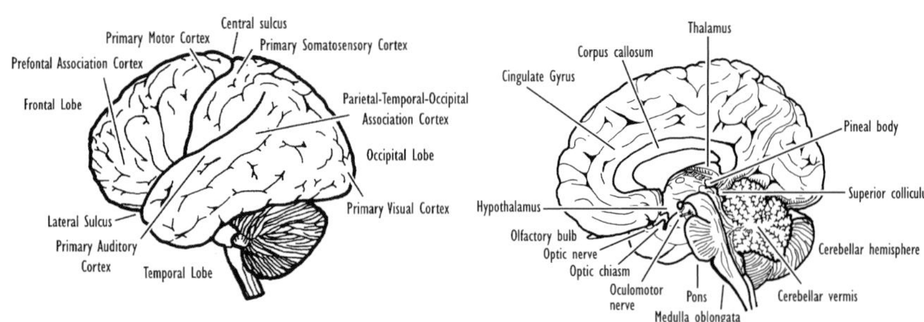

The spatial relationship between the pons, cerebellum and medulla oblongata are clearly revealed in this medial view of the brain's right-hand side (right picture). This view also shows the position of the superior colliculus, which controls the direction of gaze. It receives nerve signals from both the retinas, and also from the primary visual area of the cerebral cortex. Two of its layers of nerve cells comprise maps of the visual and tactile (touch) space, respectively, and these are in mutual registry. Hence our ability to coordinate these senses when we manipulate objects simultaneously seen and touched. The left picture shows a lateral view of the brain's left side.

This interesting fact, that the three structures are physically similar although only some of the corresponding residues are identical, exposes a complexity of *protein folding*: some amino-acid residues are more important than others. Each of the four units carries a haem group, so a haemoglobin molecule is capable of transporting up to four molecules of oxygen.

We might speculate as to why Nature should use two different molecules for its oxygen-bearing task, myoglobin and haemoglobin, when one might suffice. The answer seems to lie in an intriguing cooperativity shown by the larger molecule. Although myoglobin actually has a higher natural affinity for oxygen, haemoglobin is the more efficient carrier. The oxygen saturation of haemoglobin changes more rapidly with variation in the oxygen pressure than would be the case if its four sub-units acted independently, and the outcome of this is a doubling of the amount of oxygen that can be delivered. When one of the four units, an alpha or a beta, captures an oxygen atom, the probability that another unit will accomplish the same thing is thereby increased. (This is reminiscent of a familiar biblical adage, from St Matthew's gospel: to him who hath shall be given.) The cooperativity derives from *allosteric structural changes*, and the fully oxygenated form, *oxyhaemoglobin*, is actually the more compact. *Deoxyhaemoglobin* is, conversely and surprisingly, more rigid, owing to the reversible formation of certain inter-unit and intra-unit electrostatic bonds known as *salt bridges*. There had already been a hint of structural changes in 1937, when Felix Haurowitz noticed that oxygenation of deoxyhaemoglobin crystals causes them to shatter. Haemoglobin has another important property not shared by its smaller cousin: its ability to bind oxygen depends both on pH and on the local concentration of carbon dioxide. A decrease in pH (i.e. an increase in the concentration of hydrogen ions), and also an increase in the carbon dioxide concentration, both cause a drop in the oxygen affinity. This is another example of something that has been described earlier: a conformational change of a protein due to changes in the aqueous environment, the alteration in structure having a pronounced influence on function. These effects are collectively known as the *Bohr effect*, named after their discoverer, Christian Bohr. The biological advantage they confer is obvious; metabolically active tissue produces carbon dioxide, so the oxyhaemoglobin is induced to shed its oxygen, this being a prerequisite for the respiratory function.

The particular importance of some amino-acid residues in haemoglobin can be appreciated by contemplation of its mutant forms. The best-known of these is carried by patients suffering from *sickle-cell anaemia*. This disease, which, because of a chromosome linkage, selectively afflicts black-skinned people, was discovered in 1904 by James Herrick. He named it for the sickle-like appearance of the carrier's red blood cells. Vernon Ingram established, in 1954, that the fault lies in the beta chains. At the sixth residue from the amino terminal, the normal form has glutamate while the mutant sickle variety has a valine. The trouble arises because whereas glutamate is polar, and thus hydrophilic, valine is non-polar and hence hydrophobic. The mutant molecule can hide its water-shunning patches by linking these with the corresponding patches on other molecules and, as Max Perutz and his colleagues, J. T. Finch, J. F. Bertles and J. Döbler, have established, this produces rod-like superstructures which give the diseased cells their characteristically elongated shape. Impairment of the oxygen-carrying function results partly from the actual distortion of the red cells, which tend to get stuck in the smallest capillary vessels, and partly from the decreased affinity for oxygen suffered by the clustered haemoglobin molecules. Upwards of a hundred mutant varieties of haemoglobin have now been identified and characterized, most of the associated diseases being relatively mild. An amino-acid substitution on the outside of the molecule, for instance, usually has a negligible influence on function, the notable exception being the sickle-cell case itself. The most disastrous mutation is found in *haemoglobin M*, the letter denoting *methaemoglobin*. Substitution of tyrosine for histidine at either the haem-bound or haem-associated residue positions causes the iron to adopt the ferric state, to which oxygen cannot become bound.

The systems described thus far could be called the fancy bits of the body; their leading roles involve dynamic changes at the molecular level. But the relatively static part played by the *structural tissue* does not imply dullness on the microscopic scale. In the higher animals the main structural molecule is *collagen*. The *chitin* that provides the basis for the horny parts of members of the *Insecta* and *Crustacea* classes of *Arthropoda* is a polysaccharide, and it is thus more closely allied to the *cellulose* found in plants. In *mammals*, collagen is ubiquitous. It is the main constituent of tendon, skin, bone, teeth, ligament, cartilage, blood vessels and various other ducts. It also forms the basis of such specialized tissue as the cornea of the eye. Collagen holds the organs in their correct positions and dictates the external shape of the body. Its importance is revealed when disease causes either a deficiency or surplus of this structural polymer, as occurs, for example, with *arthritis* and *atherosclerosis*, respectively.

The basic unit of collagen is the *fibril*, an elongated structure with diameter usually between 100 and 300 nm. In the *cornea* the diameter is only 16 nm, however, and this is the reason for that tissue's transparency: the characteristic dimension is well below the approximately 500 nm wavelength of visible radiation. The arrangement of the fibrils varies from one tissue to another; they lie parallel in the *tendon*, whereas they are woven into a criss-cross pattern in *cartilage*. Each fibril consists of a large number of *tropocollagen* molecules, which are long and thin, measuring 300 nm $\times$ 1.5 nm. These filaments are secreted by connective tissue cells called *fibroblasts*, or *fibrocytes*. Each tropocollagen molecule is a triple-stranded rope-like arrangement of polypeptide molecules, the helical structure of which was elucidated by Francis Crick, G. N. Ramachandran and Alexander Rich, in 1960. The three polypeptides are identical and they are unusual in two important respects: they have an uncharacteristically high proportion of glycine residues, about one third, and their amino-acid sequence is surprisingly regular for a protein. The structure of the tropocollagen molecule is intimately connected with both these features. The triple helix arrangement causes every third side group to lie inside the molecule, close to the symmetry axis where there is relatively little space. Glycine has the smallest side

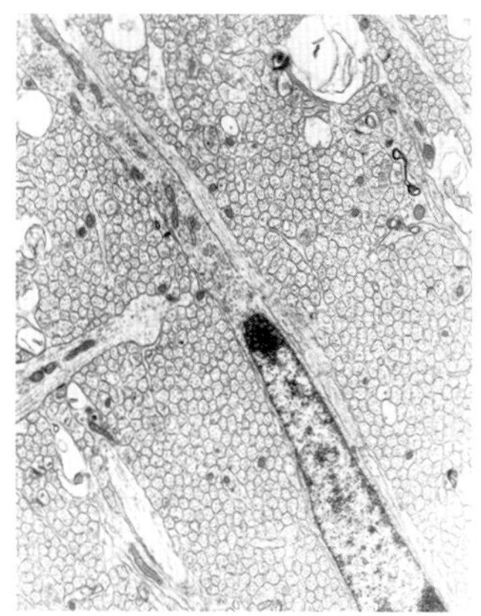

Studies of newts with surgically rotated eyes, carried out by Roger Sperry (1913–1994), showed that growing neurons follow a predetermined pattern of development, migrating and establishing contacts with other neurons in a manner suggestive of precise programming. Rita Levi-Montalcini subsequently demonstrated that the presence of a complex of proteins, collectively called the *nerve growth factor*, is vital to the development of several types of neuron. This electron photomicrograph shows a small region of the cerebellum of a foetal monkey. A roughly 40-micrometre-diameter neuron is seen migrating in a diagonal direction from the lower right corner, guided by a fibrous cord of glial cell processes, or extensions. Although this migrating cell is in intimate contact with thousands of other cells, including the numerous axons here seen as roughly circular cross-sections, it nevertheless finds its way from the cerebellar surface to the interior of the cerebellum.

group of all the amino acids, a single hydrogen atom, and a glycine residue occurs at every third position of the sequence.

The vital elastic qualities of collagen derive from extensive cross-linking of the molecules in the mature tissue. The intra-molecular bridges within each triple helix are provided by lysine side groups, which form *aldol cross-links*. The nature of the intermolecular cross-links between adjacent triple helices is still rather obscure. A prerequisite for the formation of these strengthening ties appears to be the conversion of proline and lysine residues to hydroxyproline and hydroxylysine, respectively. This is mediated by *hydroxylase enzymes*, in the presence of ferrous iron and molecular oxygen. A reducing agent such as *ascorbic acid*, more commonly known as *vitamin C*, is also required. The disease *scurvy*, the dread of mariners in earlier centuries, involved the breakdown of collagen cross-links, and manifested itself in skin lesions, bleeding gums, and a general debility. It was caused by defective hydroxylation caused by dietary deficiency of vitamin C, and its cure was arrived at empirically in 1753, by James Lind. Redistribution of connective tissue is a routine occurrence at some stages of the life cycle in many species. This happens in the *metamorphosis* of the tadpole into the adult frog when the tail rapidly disappears. Another example is the reorganization of the uterus following pregnancy. In each case the breakdown of existing collagen is mediated by a *collagenase enzyme*. Breakdown by simple hydrolysis is accomplished during the production of *gelatin*, which is simply fragmented collagen, by boiling bones and related tissue in water. (One of the present author's more disagreeable childhood memories is the pungent smell that used to emanate from the glue factory near his home, where the bones of worn-out horses provided the raw material.)

Regularity in the amino-acid sequence of collagen is not limited to the positions of the glycine residues. There is also a natural periodicity of 67 nm in the occurrence of charged and uncharged amino acids, similar residues being grouped into fairly regular clusters. This has a strong influence on the assembly of collagen molecules in fibrils which, observed in an electron microscope after suitable chemical staining, have a characteristic banded appearance. There are just four and a half intervals of 67 nm along the 300 nm length of the tropocollagen molecule, and the association of complementary side groups produces a crystalline arrangement in which the molecules lie with their long axes parallel. In cross-section, the pattern is pseudo-hexagonal. Most importantly, gaps of 67 nm width occur regularly along the fibre length, and it is these that give rise to the banding.

Collagen is one of the two main components in the important biological composite, *bone*, the other being the mineral *hydroxyapatite*. The latter gives bone its hardness, while collagen supplies the valuable flexibility without which the skeletal framework would be unacceptably brittle. Bone can be compared with that other well-known composite, *reinforced concrete*. Both concrete and hydroxyapatite perform poorly under tension, and collagen plays a role similar to that of the reinforcing steel rods in providing tensile strength. Hydroxyapatite, produced by specialized cells known as *osteoblasts*, has the basic composition $Ca_{10}(PO_4)_6(OH)_2$, but there is evidence of non-stoichiometry, and considerable variation in the calcium–phosphorus ratio appears to be tolerated. Moreover, a certain amount of fluorine is present, this replacing hydroxyl ions. Complete replacement would produce the hexagonal crystals of *fluorapatite*, $Ca_{10}(PO_4)_6F_2$, but this extreme is never

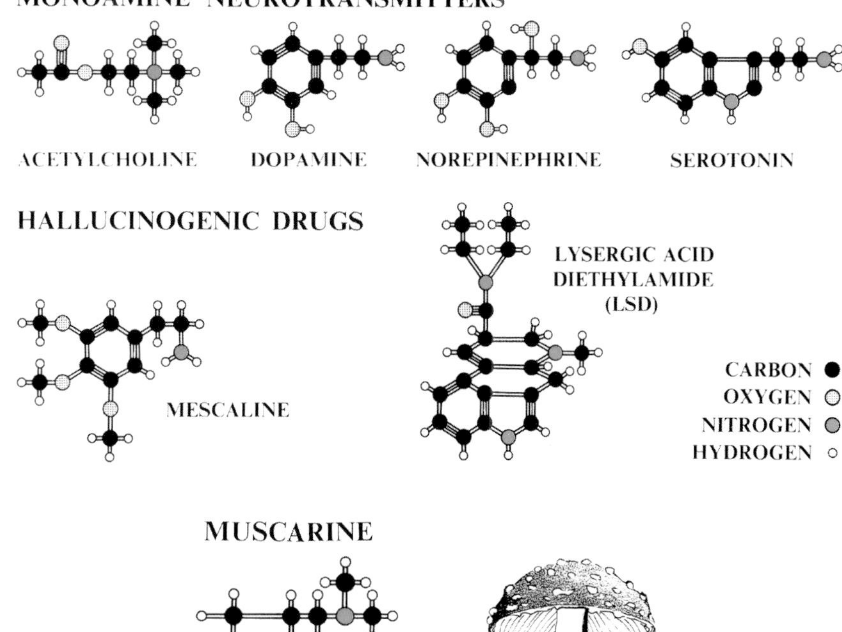

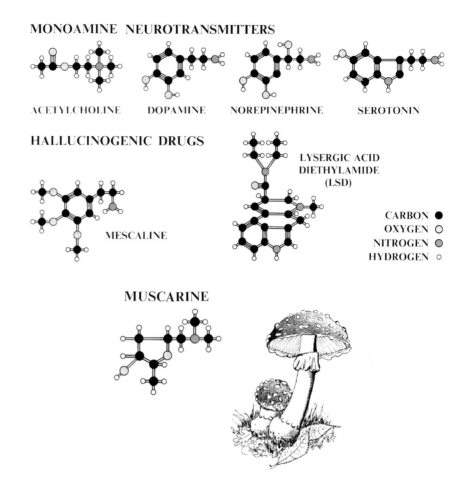

Neurotransmitters are synthesized close to the presynaptic terminal of a nerve axon, packaged into phospholipid vesicles known as synaptosomes, and guided towards the presynaptic membrane by microtubules. The arrival of a nerve impulse releases approximately 10 000-molecule bursts of these chemical messengers, which thereupon diffuse across the roughly 50-nanometre synaptic gap, that trip taking a mere 10^{-7} seconds or so. They then dock with receptor proteins in the postsynaptic membrane, causing a change in its sodium and potassium permeabilities, and provoking a new nerve impulse. Many neurotransmitters are small molecules having a single amine end group. Some of them, such as norepinephrine (noradrenaline) also serve elsewhere in the body as hormones. Several hallucinogenic drugs owe their potency to a structural similarity with certain neurotransmitters. Examples are mescaline, which can mimic dopamine and norepinephrine, and lysergic acid diethylamide, otherwise known as the infamous LSD, which can masquerade as serotonin. Muscarine is found in the toadstool known as fly agaric (*Amanita muscaria*), sketched here. Chewing the poisonous, but seldom lethal, tissue of this scarlet fungus was popular among the Assyrian priesthood, about four thousand years ago. They gnawed on the fungus in such large doses that a high concentration of the drug was present in their urine. Quite how they discovered this remains a mystery, but the fact is that they treated the liquor with great reverence, and even regarded it as being fit for consumption. As with most human institutions, that priesthood was hierarchical, so the senior members passed the drug on to their subordinates in more senses than one.

Glutamate and GABA

Neurotransmitters can be broadly divided into two classes: excitatory and inhibitory. An excitatory neurotransmitter promotes activity in a nerve cell it impinges upon, by tending to depolarize the post-synaptic membrane of that neuron. But it should be stressed that the receiving neuron requires input from many (usually hundreds) of other neurons in order that its threshold depolarization be reached. Inhibitory neurotransmitters have the opposite effect, hyperpolarizing the post-synaptic membrane to a more negative voltage (inside, with respect to that of the extracellular fluid), and thus driving it farther from the threshold. This neat distinction fails in the case of at least one neurotransmitter, however, because the influence – positive or negative – depends upon the type of receptor (in the post-synaptic membrane) for dopamine. A multiplicity of receptor types is indeed the rule for most neurotransmitters, though in most cases they all lead to excitation or all to inhibition, when the neurotransmitter molecule docks with them. Considerable neurotransmitter variety is observed in the basal ganglia, this probably reflecting the subtleties of reaction required of those sub-cortical nuclei. In the cerebral cortex, on the other hand, excitation is left almost exclusively to the neurotransmitter glutamate, and inhibition just as exclusively to GABA. Glutamate is an amino acid, of course, and we encountered it in Chapter 15. Its molecular structure is:

$$\begin{array}{c} \mathrm{H} \\ | \\ \mathrm{H_3\overset{+}{N}-C-CH_2-CH_2-COO^-} \\ | \\ \mathrm{COO^-} \end{array}$$

As we have noted elsewhere in this book, the interior of the body is a closed system, and it has to husband its resources in such a way that there are no net surfeits or deficits of its working substances. It accomplishes this by using one type of molecule as the raw material for another, glutamate being a case in point. By letting glutamic acid decarboxylase act as a catalyst upon that molecule, it converts it to gamma-aminobutyrate, usually referred to by its acronym GABA:

$$\mathrm{H_3\overset{+}{N}-CH_2-CH_2-CH_2-COO^-}$$

It is by this deft piece of biochemical manipulation that the system gets two for the price of one, as it were, and the really impressive thing is that the conversion changes an exciter into an inhibitor.

reached. The hydroxyapatite crystals, which have a monoclinic structure, occur as platelets up to about 40 nm in length and width and 5 nm in thickness. In the mid 1960s it became clear that not all bone mineral is in the crystalline form. As much as half exists as amorphous calcium phosphate, and this is particularly significant because the glassy state is less susceptible to cleavage. Structural studies have revealed that the crystalline flakes are nucleated at specific sites in the 67 nm gaps on the collagen fibrils, while the amorphous component occurs preferentially between the fibrils. The delicate chemical balance required for proper bone growth is underlined by experiments in which the concentration of one element is deliberately altered. Studies of skeletal growth in the *sea-urchin*, for instance, reveal that the smooth surface of that creature's backbone develops a crystalline angularity in a magnesium-depleted environment. The role of magnesium might be to poison the surface of the growing hydroxyapatite, preventing the growth of large regions of perfect crystal. Of the many known bone diseases, mention should be made of *rickets*, which is caused by a

Caffeine

The *methylxanthines* are *alkaloids*, three familiar examples being caffeine (1,3,7-trimethylxanthine), theophylline (1,3-dimethylxanthine) and theobromine (3,7-dimethylxanthine). Caffeine is the best known, it being a constituent of one of the widest-used beverages, namely coffee. It is also present in tea, which additionally contains considerably smaller amounts of theophylline and theobromine, neither of which are present in coffee. The chief source of theobromine, however, is the cocoa bean. The caffeine molecule consists of a six-membered ring and a five-membered ring, these sharing a double-bond, as shown in the following two-dimensional diagram:

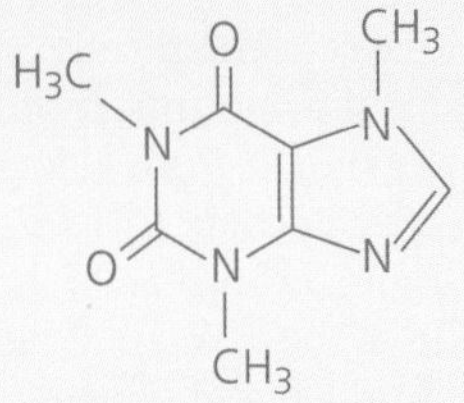

Caffeine acts as a stimulant on the nervous system, on cardiac muscle, and on the respiratory system. It also functions as a diuretic (that is to say, it promotes the passing of urine) and it delays fatigue in the short term. Theophylline similarly acts as a cardiac stimulant and diuretic, as does theobromine. Both the latter compounds also relax the smooth muscles (as distinct from the skeletal muscles). The potency of these compounds stems from their inhibiting cyclic adenosine monophosphate (cAMP) phosphodiesterase, this thereby prolonging the influence of cAMP on certain hormones. Moderate amounts of caffeine – the equivalent of three or four cups a day for an adult – are harmless, but it becomes toxic when consumed in excess, the lethal dose being about 150 mg per kg of body mass. This corresponds to very roughly 100 cups of coffee for the average adult, which would be rather difficult to consume sufficiently rapidly in any case, but it is worth bearing in mind that accidental ingestion of 10 g of caffeine could prove to be fatal. Chronic caffeine poisoning causes irritability, anxiety, muscle spasms and palpitations, while an acute attack additionally produces nausea, vomiting, tachycardia (a racing pulse) and ultimately delirium, a pronounced irregularity of the heartbeat and muscular seizures. Human consumption of caffeine has been said to date from around 850 AD, when an Egyptian goatherd (named Khaldi) noticed that his animals became pleasantly agitated after consuming the red berries of a shrub. These were coffee beans, and Khaldi passed on the information to the local monks, who put the substance into production.

dietary deficiency of *vitamin D*, a lipid related to *cholesterol*. This impairs the absorption of calcium from the gastrointestinal tract, and the calcium depletion manifests itself in stunted and deformed growth. The improved dental health that results from *fluoridation* of drinking water and toothpaste is connected with the replacement of hydroxyl ions by fluoride ions, as mentioned earlier, but it is not yet clear why this substitution is advantageous.

The *catabolic processes* by which the animal extracts energy, chemical compounds, and individual elements from ingested food, although being specifically tailored to its own needs and environment, are nevertheless simply modifications of the general mechanisms outlined in Chapter 16. As such they require little elaboration. Large molecules are split up into smaller units. Thus *proteins* are hydrolyzed to produce the various amino acids, *polysaccharides* give the simple sugars such as glucose, and *fats* supply glycerol and fatty acids. These processes actually consume more free energy than they

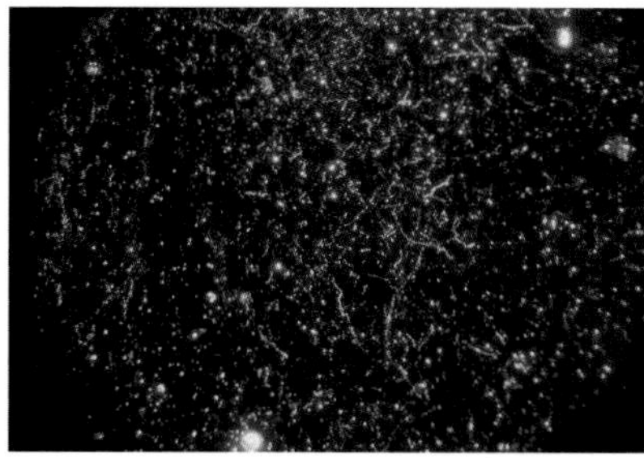
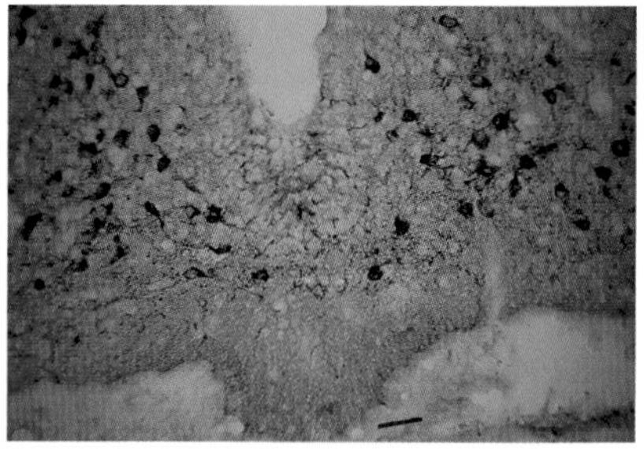

It was earlier believed that neurotransmitters and hormones are chemically and functionally distinct, the former providing fast and localized communication between nerve cells while the latter act as the much slower signals from glands to specific target cells. More recently, however, substances have been discovered which participate in both processes. Adrenaline, for example, is both a neurotransmitter and a hormone released by the medulla of the adrenal gland. It is for this reason that a sudden discharge of adrenaline from the gland produces effects similar to those caused by massive stimulation of the sympathetic nervous system: increased blood pressure, dilation of blood vessels, widening of the pupils, and erection of the hair. Many of these dual agents are amino-acid polymers, and are thus termed neuropeptides: the small enkephalins and the larger endorphins. These pictures demonstrate detection of beta-endorphin, a 31 amino-acid residue neuropeptide, in the arcuate nucleus of a rat hypothalamus. The technique employs antibodies to label with the enzyme horseradish peroxidase those cells which contain the neuropeptide. The enzyme catalyzes polymerization of the brown pigment diaminobenzidine, which colours the endorphin-rich nerve cells (right-hand picture). The cells have diametres around 25 micrometres, and the light patch at the upper middle is the third ventricle, a fluid-filled cavity imbedded in the brain. The left-hand picture shows the same tissue under dark-field illumination, with the light incident from the side. The nerve cell axons and dendrites show up as gold threads, while the cell bodies appear as elongated triangles. Some of the bluish spots are tissue surface irregularities, whereas others are capillaries seen in cross-section.

liberate, but the further breakdown of some of these intermediate units, via the citric-acid cycle and oxidative phosphorylation, as described earlier, provide the animal with its primary energy source: molecules of energy-rich ATP. Control of these processes is the task of the numerous *enzymes*, and the *feedback mechanisms* whereby the reaction rates are adjusted, are of various types. Some involve regulation of the rate of protein synthesis, more of the enzyme giving a speed-up of the catabolic rate. Others involve environment-induced allosteric modification of the enzyme's structure and hence of its catalytic capability. Similar principles are in operation in the *anabolic processes* through which smaller units are assembled into larger molecules, at the expense of free energy gained from ATP. Most of the individual elements are obtained by direct aqueous solution and diffusion. In the properly fed individual, a balanced diet supplies the raw materials in the correct proportions and amounts, and this minimizes the burden on the control mechanisms. The result is a body structure best suited to permit those activities which guarantee the further acquisition of its needs.

The final item to be discussed in this chapter is the most complicated structure in not only the animal body but in the entire material world:

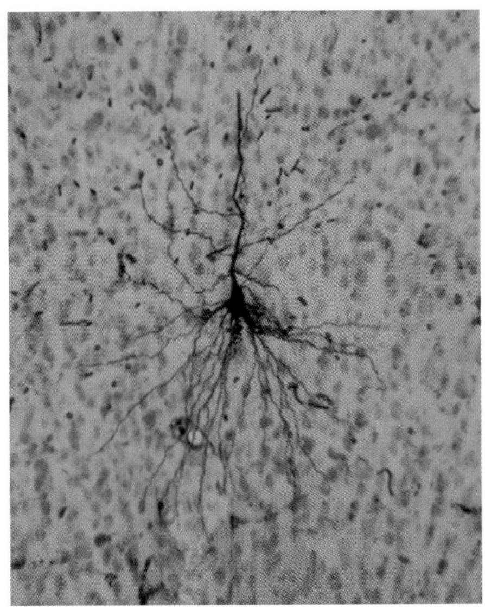

Some of the most extensively studied brain cells are those involved in vision. The classical experiments of David Hubel and Torsten Wiesel established that certain cells in the visual cortex respond only to specific patterns of light stimulation, such as line segments of given orientation, length, and direction of motion. Cortical neurons display great diversity of shape, size, and function and they are segregated into layers in which the cells have distinct forms and densities of packing. This lamination is believed to be important in cortical function, the layers determining such factors as location in the visual field and type of input received, electrical measurements on individual cells play an important role in studies of neural function, as do direct observations of cell morphology. This optical photomicrograph shows a single cell in a cat's visual cortex, made visible by injection with horseradish peroxidase and then counterstaining the background cells with cresyl violet. Several dendrites, through which the cell receives electrical signals, are seen radiating outwards from the cell body. The cell sends out signals through its single axon, which is not readily visible at this magnification. The thickest fibre is known as the apical dendrite, and it gives the cell the pyramidal shape after which it is named. The fibre appears to terminate abruptly but this is an artifact of the sectional slicing. Short sections of other cells are also visible. The way in which the interconnections between the neurons determine physiological function remains a mystery.

the *brain*. It is the clearing-house for the information passed on from the senses, and it is the origin of the commands transmitted to muscles, organs and glands. It also appears to be the site of an individual's sense of identity, its feeling of self, and in the higher animals it is the seat of the mind. In the case of *Homo sapiens*, there has been an inclination to take things one step further and associate the brain with that nebulous concept, the soul, thus linking it to the supernatural and the religious. In recent decades, this tendency has been reversed, and in some cases, as with comparison of the brain to an electronic computer, one is probably erring towards the other, too simplistic, extreme. If the task of understanding the brain seems formidable, it is because it must grapple with such concepts as thought, perception, emotion, rationalization, understanding and imagination, and indeed conception itself. Finally, and most dauntingly, because this introduces the dimension of time, a proper understanding of the brain must reveal the nature of memory.

The brain was not always accorded the prominence it now enjoys. The Sumerians, Assyrians and early Israelites believed the soul to reside in the liver. At the time, this was a reasonable guess; the liver is a focal point in the circulatory system, and if a sufficient amount of blood flowed out of the body, the soul also seemed to depart. The Ancient Egyptians favoured the bowels and the heart, regions still loosely associated with such qualities and emotions as courage and affection. Although René Descartes appeared to perceive the essential unity of mind and body, when he uttered his immortal phrase *Cogito ergo sum*, he actually advocated total divorce of these two attributes. He saw the body as nothing more than a sophisticated device, while the mind was regarded as something akin to a *Ghost in the Machine*, or *homunculus*. Descartes even identified the *pineal gland* as the terminal with which the body communicated with the mind. (He fared better with his speculations on the nervous system, correctly guessing this to be a system of conduits, some 150 years before Luigi Galvani's famous observation of the twitching of a frog's leg.) One's 'me-ness' appears to lie at the centre

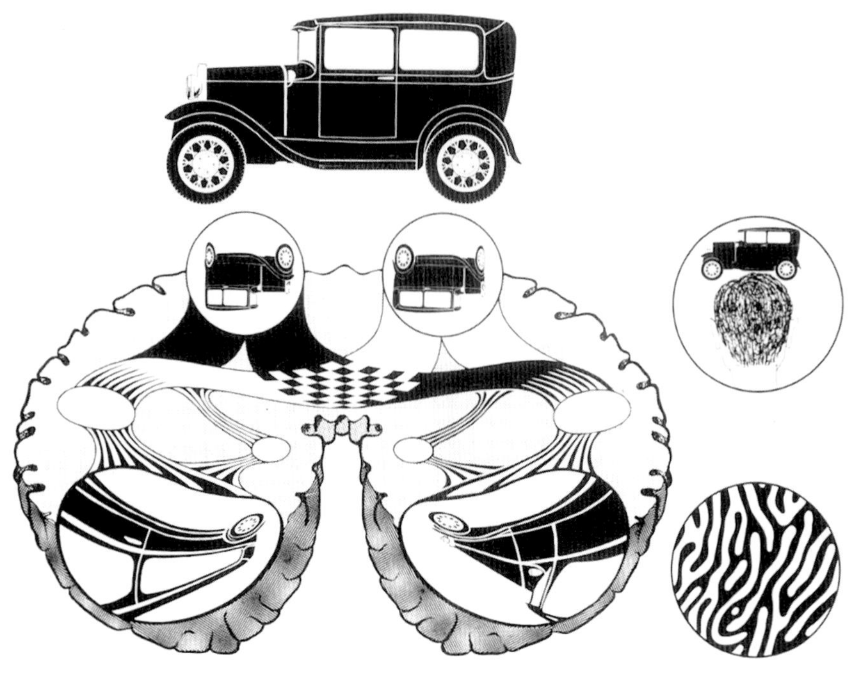

The brain's left side controls the body's right, and vice-versa, so the visual system has a problem because there is an additional left–right inversion as the light rays pass through each eye lens. (There is also an up–down inversion, as indicated in this *highly* schematic view of a horizontal section through the brain.) The system's solution is division of each optic nerve into two halves, one passing signals to the occipital area of one cerebral hemisphere and the other to the other. These signals actually employ two different routes, the major path passing through the lateral geniculate nucleus and the minor one passing through the superior colliculus (see the centrally located large and small white ovals, respectively). These relay stations also receive counter-running signals from the occipital cortices, the return signals being vital to the visual system's correct functioning; the return signals in the major route help to maintain focus, while those in the minor route help to hold the gaze in the appropriate direction despite any simultaneous head movements. An antique car was chosen so as to better illustrate the distortions introduced by the system. The central part of the visual field is better represented in the primary visual region than is the periphery, and it is remarkable that we see the two halves of the observed object joined seamlessly, despite the fact that one half is handled by each hemisphere. Also noteworthy is the interleaving of signals from the left and right retinas, as indicated by the black–white alternations. This arrangement was discovered in a cat brain by David Hubel and Torsten Wiesel (see lower right inset), using radioactive tracer techniques. It provides the basis of depth perception.

of the head, between the ears and just behind the eyes. But if the brain therefore seems the obvious location of our own consciousness, we should contemplate whether a person congenitally deaf and blind would feel the same way.

Observed macroscopically, the brain gives little hint as to how it performs. In the human, it has a familiar walnut-like appearance with its convoluted surface and clear division into two lobes. This major portion is the *cerebral cortex*. Closer inspection reveals such distinct minor compartments as the *cerebellum*, located at the lower rear; the *pons* and *thalamus*, positioned centrally near the junction with the spinal cord; and smaller structures like the *basal ganglia, hypothalamus, pituitary, hippocampus, septum* and *mammillary bodies*. The pituitary gland, which controls other glands, and through them the body's *endocrine system*, is part of what was formerly known as the *limbic system*, a term that is now falling out of favour. Endocrine imbalance is a recognized source of behavioural aberration, and surgical destruction of the *amygdala*, located near the hippocampus, has been used to quell extreme aggression. There has been an understandable temptation to associate different parts of the brain with specific functions, an approach which has clear limitations. Such localization is certainly not justifiable to the extent advocated by Gregor Reisch, in the sixteenth century, who identified regions purported to be responsible for imagination, estimation, recognition, memory and fantasy. His cerebral chart, which also assigned one area to a sense of the whole, was at least superior to the phrenological map, the adherents of which analyzed character by studying the external shape of the skull. Phrenology, which originated in 1812, and a famous accident that occurred thirty-six years later, revealing much about brain function, will be discussed in the final chapter.

Experimental work on animals was in progress at this time, and Marie-Jean-Pierre Flourens established that the cerebral cortex is the seat of the

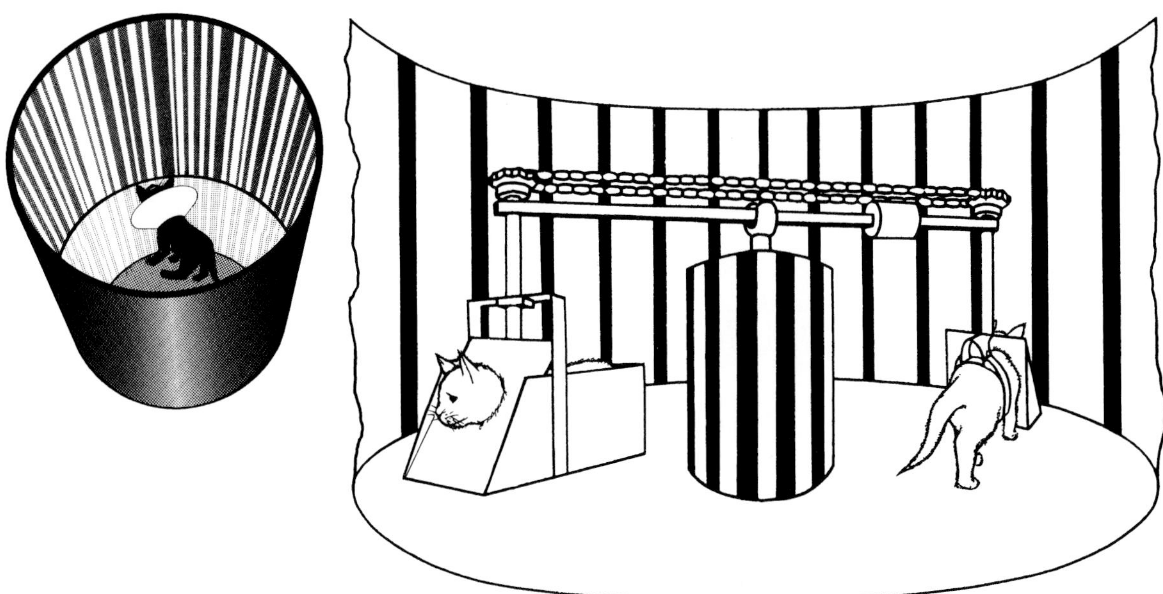

David Hubel and Torsten Wiesel discovered the key feature of signal processing in the primary visual cortex when their investigations revealed the presence of nerve cells each responding preferentially to observed lines lying in particular direction; here was the manner in which the system reduces a complex scene into its constituent parts. Donald Hebb (1904–1985) had suggested, two decades earlier, that learning takes place by synaptic modification, and this was subsequently confirmed by Eric Kandel and his colleagues. The coarser consequences of these facts were investigated with experiments on newborn kittens. Richard Held and Alan Hein (right) restricted head movements by neck-yokes, their animals being confined to a cylindrical room decorated exclusively with vertical stripes, the active subject being free to walk about whereas the passive animal rode in a gondola. The active kitten's grosser movements were imposed upon its partner by a rod and chain mechanism; both therefore had the same visual experiences. Subsequent testing revealed that the active kitten had acquired normal visual-motor associations whereas its now-freed passive partner was clumsily naive in its attempts to negotiate its physical environment. It clearly lacked the synaptic connections underlying such expertise. In Colin Blakemore and Grahame Cooper's experiment (left) a newborn kitten was similarly confined to a vertically decorated environment, and a fitted collar even prevented it from seeing its own body. Within a few months, it had become quite clumsy, bumping into objects as it moved about. Direct electrical recording from nerve cells in its primary visual cortex revealed the reason: cells sensitive to horizontal lines were sparse, whereas their vertically tuned counterparts were over-represented. This demonstrated, in addition, that nerve cells are able to change their preferred direction, through modification of the synapses via which they receive input signals.

functions usually associated with the *mind*. Animals deprived of their cerebral lobes lost their ability to perceive, judge, and remember. In the 1860s, Jules Dejerine and Hugo Liepmann made independent investigations of stroke-related damage to the *corpus callosum*, a thick bundle of nerve fibres that is the main connection between the two sides of the brain, and found evidence that the various faculties are associated with different sides of the cortex. It was found that the right side of the body is governed by the left side of the brain, and vice versa. Around the same time, Paul Broca discovered

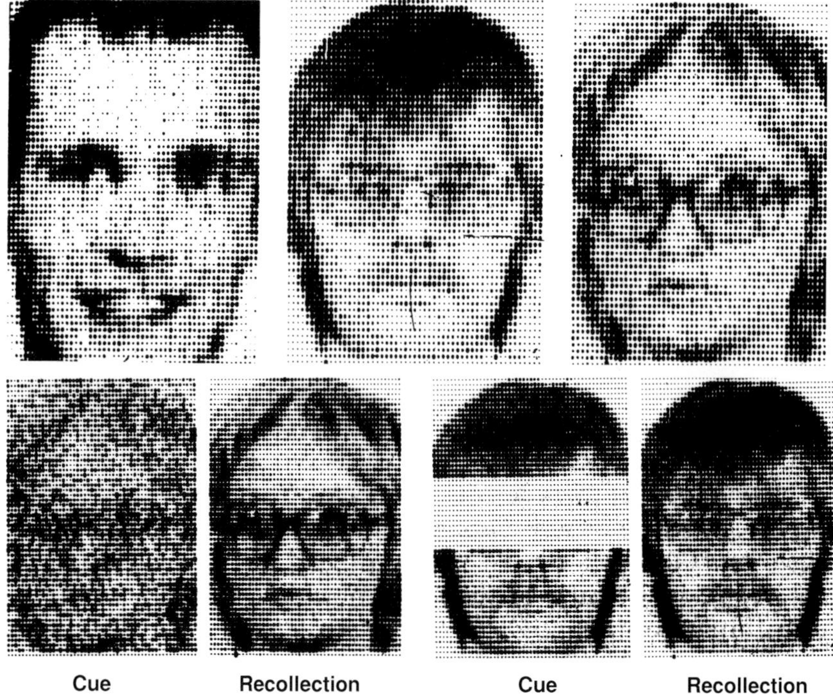

The brain's vast mesh of interconnected cells, the neural network, stores its memories by distributing each of them over many synaptic contacts. The latter are slightly modified each time a new item is deposited, and the network can thereby hold many memories superimposed on one another. The stored memories are robust in that they can survive the loss of a fraction of the synapses, and because the network functions in a parallel manner, recall is very rapid. The network can also associate; presented with a sufficiently large part of one of its stored memories, it will reconstruct the entire item. In the computer simulation of this process shown here, Teuvo Kohonen has stored digitalized versions of the pictures of fifty different faces, three of which appear above. The pictures demonstrate the ability of his 5120-element network to recall perfectly despite being presented with incomplete cues.

an area at the left side of the cortex which, if damaged, produced speech disorder. Thirteen years later, in 1874, Karl Wernicke found another region, somewhat to the rear of *Broca's area*, which controls understanding of both the spoken and written word but which is not directly involved in the speech mechanism. Since then the areas related to various physiological functions have been located with increasing precision. Thus hearing is associated with an area at the side, on each half of the brain, just above the ear, and the sense of time with a region just below this, while vision, oddly enough, stems from the rear of the brain, in a region known as the *occipital lobe*. Hermann Munk's studies of mild damage to the cortex revealed that animals such as dogs are able to rally from impairment of sight and hearing, but human casualties of the First World War examined by Gordon Holmes showed no similar recovery. The potential for repair seems to diminish with increasing sophistication.

To a certain extent, it appears that we can think in terms of mapping of function in some regions of the cortex. In the 1860s, John Hughlings Jackson correlated the regions of the body affected by *epileptic convulsions* with the injured areas in the corresponding patients' brains, and found that control of movement is located along a well defined strip now called the *motor cortex*. Intriguingly, it transpired that adjacent parts of the body are governed by adjacent parts of the cortex. The same was found to be true of a second band, close to and parallel with the motor cortex, from which skin sensation emanates. But this cortical mapping involves distortion. Edgar Adrian found that a disproportionately large amount of a pig's *somatosensory cortex* – that is to say the area which receives tactile information – is devoted to the snout; in a mouse it is the whiskers that are most generously represented. David Hubel and Torsten Wiesel have shown that the visual cortex also faithfully maps the visual field, but the arrangement is more complex, some cells

Positron emission tomography provides pictures of the brain in action. A mutant form of glucose, which cannot be fully metabolized, is preferentially absorbed by the most active regions and tends to accumulate in them. The glucose molecules are radioactively labelled with an emitter of positrons, and these mutually annihilate with electrons. The resulting release of energy produces two gamma rays, which can easily penetrate the surrounding tissue and bone. Because these rays travel in straight lines, the direction from which they emerge from the head can be used to determine their point of origin, and the appropriate circuitry yields a picture showing which parts of the brain are most active in a particular situation. In the four examples shown here, which are reproduced from investigations carried out by Niels Lassen (1926–1997), David Ingvar and Erik Skinhøj, the white colour indicates maximum intensity, while successively lower values are displayed in orange, red, yellow, green and blue. The pictures correspond to situations in which the subject was following a moving object with his eyes (upper left); listening to spoken words (upper right); moving his fingers on the side of his body opposite the hemisphere being examined (lower left); and reading aloud (lower right). It is instructive to compare these pictures with the diagram shown below, which indicates the areas responsible for several of the faculties that have been localized through studies of patient with brain injuries.

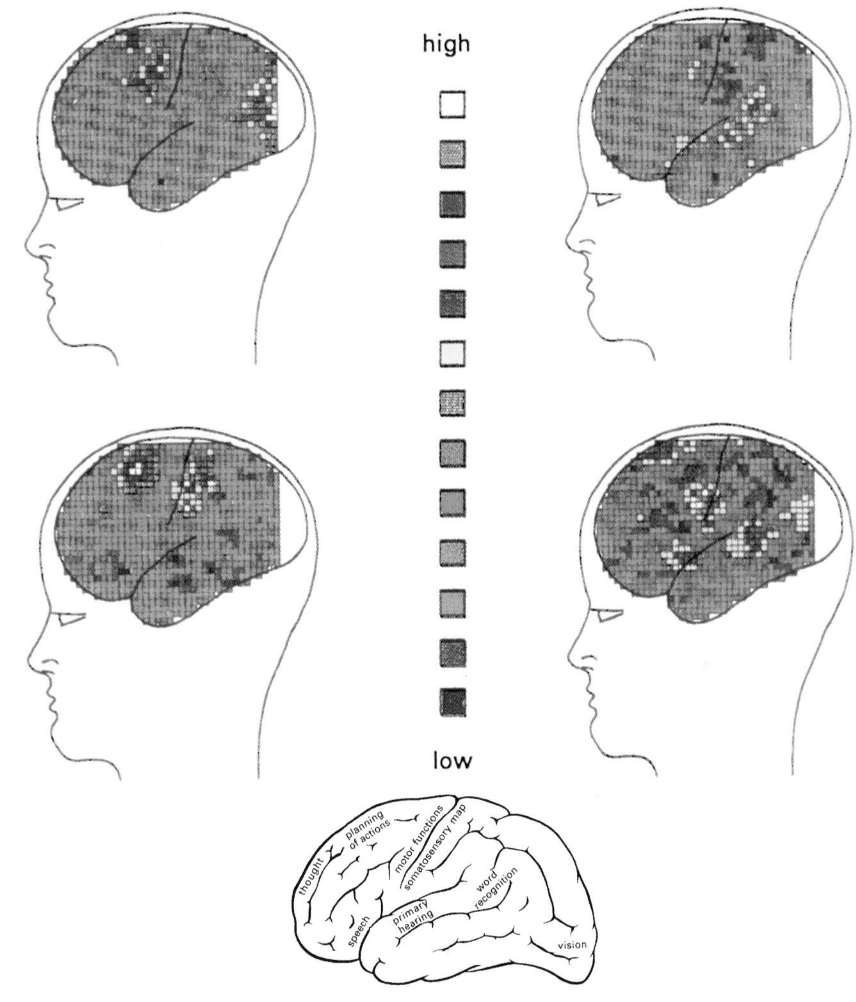

responding only to lines within a particular range of orientation. Indeed, there is even discrimination between different directions of motion, which suggests that the brain might use such information to locate the edges of objects and thereby identify shapes and patterns. Immediately posterior to the somatosensory cortex lies the *parietal lobe*, and amongst other things this appears to be related to a sense of actual existence. A patient injured in this region of one lobe ignores the opposite side of the body: half the hair goes uncombed, and half the beard unshaved.

Recent advances promise further pinpointing of specialized parts. Louis Sokoloff developed a technique in which radioactively labelled deoxyglucose is used to identify areas of the brain which have been particularly active under a given clinical situation. The deoxyglucose is taken up by the cells as if it were ordinary glucose, but its breakdown is arrested after the first catabolic stage because of its peculiar chemical structure. Radioactivity thus accumulates in the most active cells, which are subsequently located during biopsy. A significant improvement, because it permits study in the live state, was achieved by Niels Lassen and David Ingvar. *Positron-emission tomography*, as the technique is called, permits detection of the deoxyglucose, or indeed other chemicals, by labelling with positron-emitting radioactive isotopes.

Gamma-ray emission from a radioactive technetium isotope was exploited in these studies of blood flow to the brain. Red blood cells labelled with this isotope were injected intravenously while the patient was seated in front of a gamma camera. A sequence of pictures, each lasting about one second, then permitted the monitoring of blood flow. The first two images are anterior views, with the patient's left side on the right side of the image. Blood flow in a normal patient is shown on the left, while the right-hand picture reveals evidence of left-sided carotid artery stenosis, a narrowing of the arteries probably caused by atherosclerosis. In the third picture, the obstructed blood flow to the lobe of another patient's brain is revealed by the same tomographic technique, using the same isotope. In this case, the brain is viewed from above. Studies of patients with brain injuries have established many of the connections between cerebral location and physiological function.

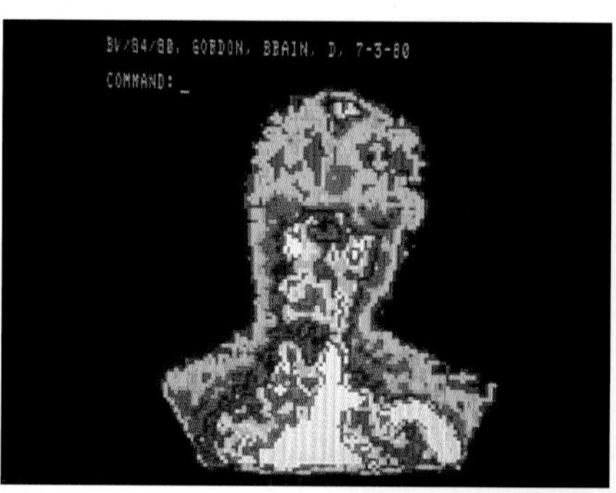

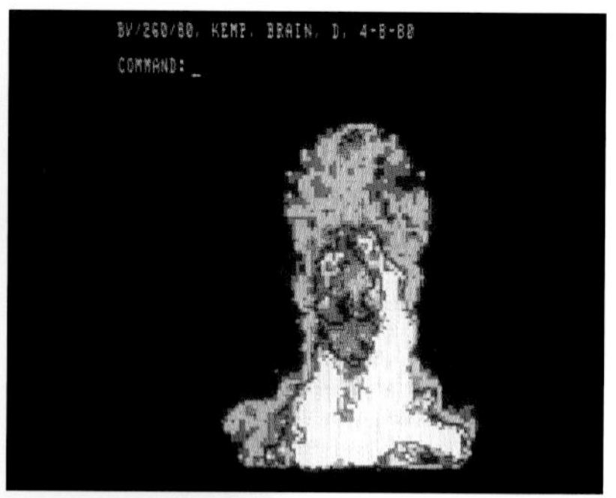

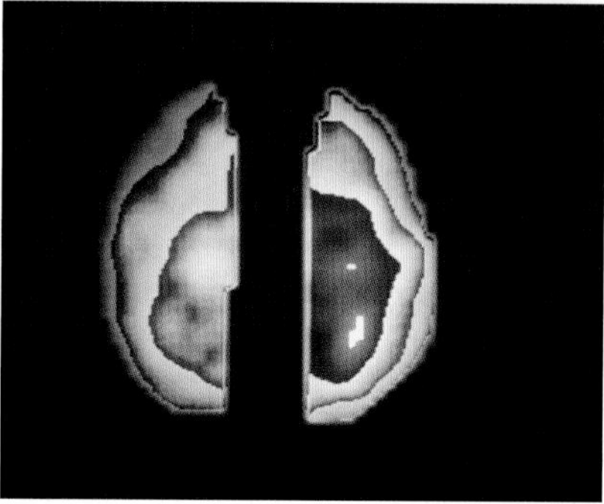

The emitted positrons mutually annihilate with electrons to produce gamma rays which pass out through the skull, and angle-sensitive detection reveals their point of origin. A picture of the brain obtained with this technique reveals which areas are most active, under a given set of conditions. Similar information is provided by magnetic resonance imaging, pioneered by Paul Lauterbur and Peter Mansfield in the 1980s, and this technique has the advantage of not requiring radioactive material to be introduced into the body.

Although increasingly precise mapping of the cerebral cortex is obviously a commendable goal, the grosser divisions are not without interest. In the 1950s and early 1960s, many epileptic patients were successfully treated by severance of the corpus callosum, but Roger Sperry discovered bizarre qualities in such *split-brain individuals*. Blindfolded, and allowed to identify an object with the left hand, such a person can subsequently retrieve it from a collection of different items, but only if the left hand is again used. Initial familiarization with the right hand produces the opposite bias. But asked to name the object, the patient reveals a clear one-sidedness, succeeding only if it is in the right band. This is consistent with the location of Broca's area, mentioned earlier, and shows that information is unable to traverse the surgically imposed gap. It is intriguing that the right side of each retina feeds visual information to the right side of the brain, thus compensating for the inversion caused by the lens of the eye. This facilitates coordination of the visual and tactile senses, and Sperry demonstrated analogous left–right discrimination between objects seen and selected without recourse to speech. His subsequent work with both split-brain patients and normal subjects allowed Sperry to further catalogue the specializations of the two lobes. The right side appears to be primarily concerned with emotion, intuition, creation, spatial relationships, and appreciation of art and music. The left side, apart from looking after speech, is dominant for analysis and logic,

This shows how one side of the macaque monkey cortex would look if it were unfolded and laid out flat on a laboratory bench. The various areas are identified by numbers and letters, the picture having been compiled from numerous sources by Daniel Felleman and David Van Essen. The areas serving the five senses are indicated, as are those involved in sending (motor) signals to the muscles. It is noteworthy that vision lays claim to more of the cortex than does any other sense. This fact, and the existence of multiple visual areas, was established by Semir Zeki. The dissected retinas of both eyes are included, in order to show how they project to the cerebral cortex via the lateral geniculate nucleus (which is part of the thalamus), the superior colliculus and the pulvinar. The frontal eye field (FEF) is of particular significance for what is known as working memory, because it appears to cooperate with several distantly positioned areas in providing the system with a sort of erasable sketch-pad or blackboard.

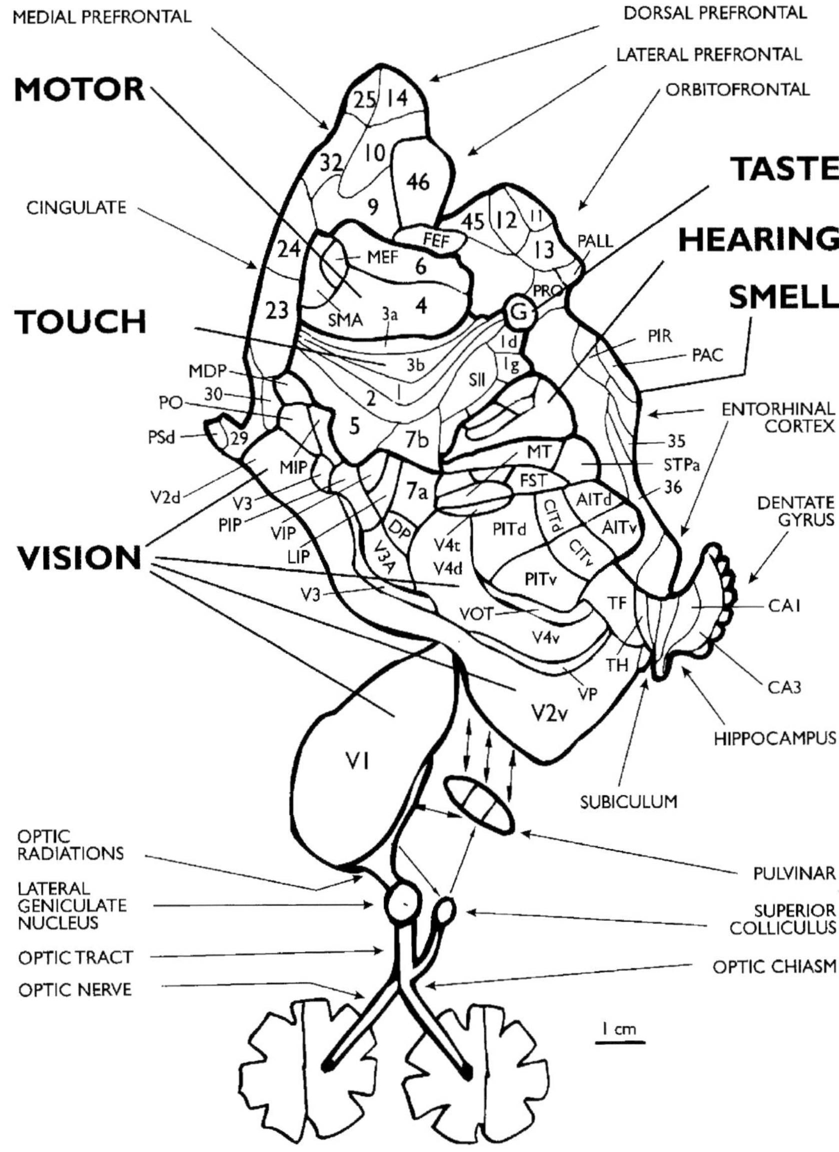

abstraction, planning and coordination, and the sense of time. There is indeed a hint that these specializations reflect C. P. Snow's emphasis of the schism between the two cultures, but we must remember that most of us enjoy the benefits of an intact corpus callosum. As well as the identification of faculties with left and right, one can also distinguish between inside and outside. We earlier noted that the human foetus at three weeks resembles that of several other animals earlier in the gestation period. This is particularly true of the embryonic brain. During the latter stages of development, however, the human brain clearly possesses parts not seen in those of lower species.

The brain and the *electronic computer* are often conceptually linked, and considered in the light of the functions they perform, information processing, storage, and retrieval, the two do bear a superficial resemblance to each

The positions of the various components of the basal ganglia are revealed in this *coronal section* of the human brain (that is to say a vertical section made at right angles to the brain's long axis), the words *dorsal* and *ventral* indicating top and bottom, respectively. Also indicated are the positions of the thalamus and some of the ventricles. The influence of the basal ganglia on muscular movement is clearly more subtle than is the case with the cerebellum because the former have no direct connections with the spinal cord. In the view of the present author, these groups of nerve cells cooperate to provide what could be called a differential clutch mechanism, permitting muscular movement only when the overall system has deemed this to be desirable. Given that the system must be able to draw on various options – move, think, move and think about the movement, move and think about something else – it is not surprising that this region possesses a greater multiplicity of neurotransmitters than is the case anywhere else in the brain.

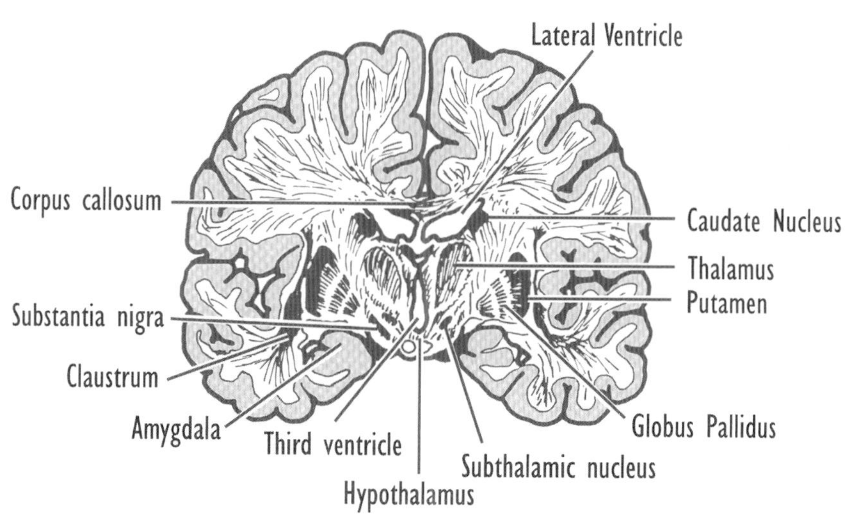

other. There are also important differences, however. Computer operation is based on the binary system, with its two options: yes or no. In terms of circuitry, or hardware as it is called, these are provided by the two possible directions of magnetization of a ferrite core. The brain is not so limited; its neurons can respond in a more graduated fashion to incoming signals. Another difference lies in the relationship between stimulus and response. Although other modes of operation are possible, electronic circuits are usually designed to function linearly, giving simple proportionality between input and output. The brain appears to be nonlinear, a consequence both of the graded response of its neurons and of the fact that both *excitatory synapses* and *inhibitory synapses* are present in its circuitry. Then again, the computer performs its functions in a consecutive manner, following the steps encoded in the appropriate program and in the sequence thereby dictated. The brain seems to work in another manner, with numerous steps being carried out simultaneously. This is a result of the more complicated wiring of the living structure, with each neuron being in synaptic contact with upwards of tens of thousands of other neurons. Finally, and probably most significantly, there is a huge disparity between the storage sizes of the brain and the computer. At the start of the third millennium, a good personal computer had a memory capable of holding about 1000 million individual pieces of information – that is to say, about a gigabyte. The human brain comprises some 10^{11} neurons, and the total number of synaptic contacts must lie in the region of 10^{14}. In such a vast system, we might anticipate a fundamentally different mode of operation, with the whole being far more than just the sum of its parts; the brain might possess capabilities, or *emergent properties* as they have been called, that are consequences of its sheer size.

Although a complete understanding of every detail of brain function remains a remote prospect, there are grounds for optimism that the general principles will be elucidated in the reasonably near future. The unique

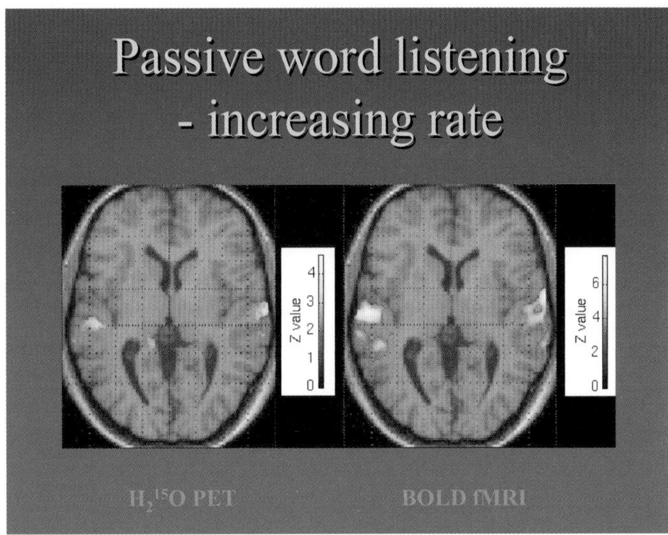

Positron emission tomography (PET), invented by Niels Lassen (1926–1997), and functional magnetic resonance imaging (fMRI), invented by Paul Lautebur and Peter Mansfield, are here revealed to be capable of producing quite similar information. The same human subject was passively listening to words spoken by someone else while his brain activity was monitored first by PET, using water containing a ^{15}O positron-emitting isotope, and then by blood oxygenation level dependent (BOLD) fMRI. As can be seen from the activity levels in these horizontal sections of the brain (with the nose pointing upward in both cases), the primary auditory area in the lateral region of the temporal lobe is being provoked in each case. Of the two techniques, fMRI is now preferred because it does not involve introducing radioactive material into the body, even though the doses used in PET are admittedly very small. It must be borne in mind, moreover, that high magnetic fields are used in fMRI, so the technique could be lethal for people with pace-makers.

qualities of the brain are provided by the three peculiar properties of neurons: their shape, their excitability, and the specialized nature of their mutual contacts. That neurons are unique was first proposed by Richard Goldschmidt, in 1912, following observations on the intestinal parasite *Ascaris*. He found that the solid masses of nervous tissue known as the *ganglia* in the brains of these small worms always contain 162 cells, and that these have a standard arrangement. More recently Sydney Brenner and his colleagues have established that the brain of the nematode worm, *Caenorhabditis elegans*, has exactly 279 neurons, which are connected to one another in essentially the same fashion in every individual. To what extent this same invariance applies to more advanced species remains to be investigated. Even though the brain's wiring, to borrow a term from electronics, appears to be rather precise, this does not preclude modification of the contacts and their ability to transmit signals. How this is accomplished will be discussed shortly.

Small clusters of cells ultimately destined to develop into the brain's major parts are identifiable early in embryonic growth. The precursors are located near the medial cleft in the *blastula*, which was discussed in Chapter 17. In the balance of the prenatal period, and during the months following birth, the developing brain gains neurons at the rate of hundreds of thousands per minute. The frenetic activity during this period includes not only growth and proliferation, but also differentiation of immature neurons, migration of entire cells, the establishment of synaptic connections, and the selective elimination of certain cells that play only a formative role. The *glial cells*, which provide mature neurons with both physical support and nourishment, appear also to serve as structural guides for the developing cells. Retention of its increasingly complex structure during growth puts certain limitations on the plasticity of the brain, and at birth the head is closer to its final adult size than is the rest of the body. Major discoveries regarding neural interconnections were made by Roger Sperry in the 1940s. He surgically rotated the eye of a newt through 180° so that relative to the retina, up would appear to be down and vice versa. After a period of recovery, the creature thereafter lunged downwards when prey flew overhead and leaped upward when a potential meal passed under its nose. This indicated specificity of connections between the retina and the relevant region of the newt brain, known as the *tectum*. It also suggested a correspondence between areas of the tectum and parts of the brain controlling motor responses: a spatial mapping of both structures, similar to that discussed above. Repair of the surgical damage had been demonstrated, and the regenerating nerve fibres even seemed to know where to go. Finally, and most remarkably, incision and eye rotation at the embryonic stage still produced an adult newt displaying the bizarre behaviour; the developing neural fibres and their connections to the tectum are apparently predestined – they seem to know just which direction to grow in and which other neurons to link up with. The nature of the biochemical mechanism that permits this astonishing feat is, of course, part of the broader mystery of developmental biology discussed earlier. Rita Levi-Montalcini discovered the vital chemical orchestrator in the case of the nervous system, and this is now known as *nerve growth factor*. As impressive as the specificity of neural connections undoubtedly is, we must remember that the mere circuitry of the nervous system is not the whole story. Consider, for instance, the *blind spot*. This is the place where the *optic nerve* enters the retina, and it is devoid of rods and cones. But even when we see with one eye closed, and therefore lack the benefit of overlapping visual fields, we are unconscious of the blind region. There must be some mechanism whereby the brain compensates for the missing information.

The restricted description of neurotransmission given earlier was intentionally illustrative; it ignored the fact that several dozen transmitter substances have now been identified. Such chemical multiplicity gives the brain a diversity of modulation and response. The absolute quantities involved are minute, and the background purity requirement for correct functioning is consequently stringent. The brain is, indeed, isolated from the rest of the body by the *blood–brain barrier*, in which the sheathing capacity of the glial cells and the selective permeability of blood-vessel walls are both crucial. The various transmitters are not ubiquitous throughout the brain, being instead localized in specific sites and conduits as mentioned earlier. Well-documented examples include monoamines and certain amino acids. Among the former are *dopamine*, found in the mid-brain basal ganglia structures known as the *substantia nigra* and the *ventral tegmentum*, which regulates such emotions as arousal and pleasure; *norepinephrine*, located in the part of the brain stem called the *locus coeruleus*, which governs mood, dreaming and the sense of satisfaction; and *serotonin*, in the nearby *raphe nuclei*, which controls sensory perception, the onset of sleep and body temperature. Impairment of the delivery system of a neurotransmitter underlies many nervous diseases. Some dopamine-transporting fibres, for instance, extend into a central region called the *corpus striatum*, which regulates complex movements, and their degeneration leads to *Parkinson's disease*, characterized by muscular tremors and rigidity. There is also evidence that *schizophrenia* is caused by malfunction of the dopamine pathway, either through overproduction of the transmitter itself or by oversensitive receptors in certain parts of the brain. Some symptoms of this serious disease are mimicked by excessive doses of *amphetamines*.

The potency of many *hallucinogenic drugs* has now been linked to structural similarities between their molecules and those of transmitter chemicals. Thus *lysergic acid diethylamide*, better known as *LSD*, incorporates the two-ring indole structure, as does serotonin. Similarly, *mescaline*, the basis of the Mexican Indian drug *peyotl*, bears a molecular resemblance to dopamine and norepinephrine. The efficacy of numerous *pain-relieving drugs* – also known as *analgesics* – has led to speculation as to why the brain developed *opiate receptors*. It was conjectured that the body might produce its own pain killers. Receptors were discovered by Candace Pert, Eric Simon, Solomon Snyder and Lars Terenius, who found them to be located in regions of the brain and spinal cord known to be involved in emotions and pain perception. The hypothesized internal opiates were discovered in the mid 1970s, by Choh Hao Li, in the pituitary glands of camels, animals whose insensitivity to pain has long intrigued scientists, and by John Hughes and Hans Kosterlitz, in the brains of pigs. The latter named their substances *enkephalins*, invoking the Greek word for head. There were two of them, both linear polypeptides consisting of five residues. In Met-enkephalin the sequence of amino-acid residues is Tyr–Gly–Gly–Phe–Met, while in Leu-enkephalin the methionine is replaced by leucine. Other related polypeptides, discovered subsequently, were dubbed *endorphins*, meaning the morphine within, a reflection of the fact that they too resemble morphine chemically. Several neuropeptides have been found in neural terminals, where their release is mediated by calcium. The release can be triggered by electrical stimulation, hypnosis, heat, and insertion of needle points, the last two effects being suggestive of Indian fakirs walking over hot coals and lying on beds of nails. Endorphins may also hold the key to *acupuncture*.

Apart from the variety of transmitter substances, diversity arises from a multiplicity of neural mechanisms. Henry Dale first suggested that each neuron emits a single transmitter, but this might not be inviolable. Excitatory synapses are characterized by round vesicles while the inhibitory vesicles are flatter. However, even this seemingly reliable guide is compromised. John Eccles showed that some transmitters are excitatory in one part of the brain and inhibitory at another. There is evidence, moreover, of some transmitters acting presynaptically. Not all transmitters act directly. In the mid 1950s, Earl Sutherland discovered a mechanism in which the primary transmitter modulates the concentration of a *second messenger*, which proved to be *cyclic adenosine monophosphate*, usually abbreviated to *cyclic-AMP*. The transmitter activates the membrane-bound enzyme *adenylate cyclase*, which catalyzes the conversion of ATP to cyclic-AMP. The latter then provokes a particular chemical reaction, producing the physiological response. One transmitter molecule can stimulate production of thousands of cyclic-AMP molecules, giving large chemical amplification.

Having considered its layout, and the associated chemistry, we now turn to the brain's electrical phenomena. In 1924, Hans Berger detected electrical signals with the aid of metal electrodes attached to the head of his son Klaus. The pattern of such signals is now called an *electroencephalogram*, or *EEG*. When Klaus was relaxed, his EEG consisted of regular waves with a frequency of around 10 cycles/s; Berger called this the *alpha rhythm*. Paradoxically, the alert brain seemed to produce a diminished EEG, but subsequent probing by Edgar Adrian showed that this had a more complex structure in spite of its relatively small amplitude. It is now clear that the EEG of the active brain consists of waves with many different frequencies. These cannot all be mutually in step, so periodic cancellation of their effects is inevitable. In 1952, Eugene Aserinsky and William Dement discovered that the EEG of a sleeping subject displays a period of the above desynchronization about once every 90 min. Because this is accompanied by a frenzied motion of the eyes beneath their closed lids, as detected by electrodes taped to the latter, these episodes became known as *rapid eye movement sleep*, or *REM sleep*. A person awakened during the REM period can usually remember having dreamt, and will frequently be able to recall the dream's content. Although the distinction between normal sleep and REM sleep is not ideally sharp, it does seem that the latter, and its peculiar EEG, can be identified with the *dreaming state*. And this state appears to be essential. Subjects continually deprived of their REM sleep, through being disturbed at the onset of the characteristic EEG, show impaired mental capacity when awake. As if in an attempt to catch up on lost dreaming time, the sleeping brain then initiates REM periods more often. Prolonged deprivation of REM sleep can even provoke *hallucinations*, which are essentially waking dreams.

If the brain's processing of incoming signals is impressive, then its ability to store experience and regurgitate it on demand is astounding. And this, together with the way in which commands are initiated, was a challenging mystery. An important clue is provided by the fact that only a fraction of the input is permanently preserved; *memory* is selective. This indicates that permanence is not achieved spontaneously, because the immediate response can subsequently be influenced, constructively or destructively. The importance of *forgetting* is easily overlooked. The brain could become intolerably burdened if it could not rid itself of information having only transient value. *Long-term memory*, at least, appears to be associated with the cerebral cortex. In 1892, Friedrich Goltz showed that when this is damaged, a dog is unable to learn from past experience. As Goltz remarked, a decerebrated dog is essentially nothing but a child of the moment. It is now well established that memory involves structural modification at the neuronal level. J. Eschner and P. Glees trained monkeys for a specific task and then cooled them to a body temperature of 18 °C, at which point all EEG activity had ceased. Restored to normal temperature 45 min later, the creatures' performance was undiminished, indicating a permanent structural basis for their acquired facility rather than a dynamic pattern of electrical signals. But at what level do the structural changes occur? We saw earlier that genetic information is stored and mediated by the nucleic acids. Suggestions that memory too is recorded through modification of nucleotide base sequences have generally met with scepticism. The reticence is understandable; memory having such biochemical specificity would be capable of synthesis in the test tube. Similarly, George Ungar's claim to have discovered a substance that induces fear of the dark when injected into rats, the controversial scotophobin peptide, has won little support. But modification of larger structures by molecular intermediates is plausible, and is supported by pharmacological evidence. The pituitary hormones *lysine vasopressin* and *adrenocorticotropic hormone*, both small polypeptides, stimulate learning in animals.

Although it is doubtful that nucleic acids serve directly as memory molecules, they do appear to play a role in the retentive mechanism. Victor Shashoua detected RNA with modified base sequences in the brain of the goldfish *Carassius auratus*, following 5 h of intense training. By radioactively labelling the amino acid valine, Shashoua was also able to demonstrate that the learning procedure led to an increase in three proteins, each with molecular mass around 30 000 da, in the brains of his subjects. A number of models have been proposed for modifications to neural networks that could provide the various types of memory. For the short-term variety, Eric Kandel and Marc Klein have suggested that sensitization could arise from a rise in the level of *adenylate cyclase* near the synapse. This would increase the concentration of *cyclic-AMP*, which in turn was conjectured to open more calcium channels and ultimately lead to increased synaptic transmitter release. The opposite to such reinforcement, namely repression by *habituation*, was envisaged as resulting from a decrease in the number of open calcium channels, because of repeated terminal impulses. For long-term memory, Paul Greengard has invoked the discovery by Thomas Langan, that *histones* are susceptible to phosphorylation by a cyclic-AMP dependent *protein kinase*. As we saw in Chapter 15, histones are positively charged proteins that bind to the negatively charged ribose phosphate backbone of DNA. When histones are bound to DNA, they block expression of that molecule's information. Phosphorylation by protein kinase makes the charge on the histones less positive, reducing their ability to bind to DNA. Thus the chain of events might lead from adenylate cyclase, through cyclic-AMP, protein kinase, histone and DNA, to a protein that is inserted into the synaptic membrane, thereby modifying its functioning. L. Bass and W. J. Moore suggested that the laying down of the memory record might involve protein degradation rather than synthesis. The latter process is, after all, usually remote from the membrane at which the electrical activity occurs. The electric field across the membrane is about 10^7 V/m, and this is sufficient to segregate proteins, pushing the negatively charged ones to the outside and those bearing positive charges to the inside. A nerve signal lasts about 1 ms, as we saw earlier, and this provides ample time for redistribution of the membrane proteins during the voltage reversal. Termination of the pulse would then leave certain proteins on the wrong side of the membrane, where they might be attacked by enzymes. Their loss would presumably alter the electric characteristics of the membrane, thereby producing a record of the event. Bass and Moore noted that whereas most proteins normally survive for several weeks, there are some, such as *acetylcholinesterase*, which have appreciably higher turnover rates, making them suitable candidates for such a mechanism.

These memory models are based on modification of synapses through use, a mechanism first postulated by Donald Hebb in 1949, but must not be taken to imply localization of conception to specific synapses. Estimates of memory capacity based on mental performance put this at about 10^{10} individual bits of information. This is no higher than the total number of neurons, and it is far less than the total number of synaptic contacts. So a memory storage system based on binary circuit elements (i.e. switching units with only two switch conditions) is a possibility. But retention of memory by mildly damaged brains, albeit with diminished resolution, suggests delocalization. Evidence supporting this view was provided by Karl Lashley's experiments with rats in the second quarter of the twentieth century. He found that ability to perform previously learned tasks diminished in rough proportion to the amount of damage to the cortex. Similarly, our mental capacities suffer a general decline, because of the inevitable daily loss of about 50 000 brain cells, but individual memories are not switched off like the lights in a building. H. Julez, Christopher Longuet-Higgins and Karl Pribram have independently drawn an analogy to holography, in which the record of each detail of an object is distributed across the whole of the recording medium, an image of the original being regenerated in response to a specific stimulating input. As we saw in Chapter 10, optical holography uses the coherent light provided by a laser, and a three-dimensional image is reconstructed even though the recording emulsion is only two-dimensional.

If part of the record, called a *hologram*, is removed, the entire object can still be reconstructed, but with less resolution. The brain would presumably provide a three-dimensional recording medium and might thus, as Longuet-Higgins has suggested, serve as a hologram which also encompasses the time domain. The optical oscillations that provide the basis for normal holography would of course be replaced by the electrical oscillations that produce the EEG.

Initiation of memory retrieval was studied by Wilder Penfield, who performed operations in which the temporal lobe of the brain was exposed while the patient was fully conscious. Mild activation of certain regions of the cortex by a small electrode caused the patient to relive a previous experience in great circumstantial detail, as if a particular memory file was being opened and a tape recording from it read out. The access key to the file, which Penfield called the *experiential centre*, appeared to lie in the hippocampus. As promising as these analogies might seem, the warning must be reiterated that the essence of brain function probably stems from the enormous number of its individual units. To endeavour to understand this organ solely on the basis of molecular neurobiology is like studying the transistors and capacitors of a tape recorder with the hope of thereby appreciating the intricacies of a Beethoven symphony. Many years will pass before the microscopic realm is united with the macroscopic approach of the psychologist. But science continues to seek the holy grail of explaining the mechanism underlying the mind. The final chapter will chronicle the progress that has been achieved with this ultimate of all challenges in the material world.

In a previous chapter, we noted that the tenacious theory of vitalism in organic chemistry gave way only in the face of overwhelming evidence that there is no mystical extra ingredient in chemical compounds having a biological origin. It could be that attitudes towards the brain will one day undergo a similar overhaul. The complete elucidation of the molecular processes underlying the nerve action potential, the transmission of signals across synaptic junctions, and the storage of experience, will probably fall short of explaining the emergent properties of a system comprising over a hundred thousand million individual units. There is a school of opinion which lays considerable emphasis on the role of hereditary molecules in the functioning of the brain, and results such as those of Shashoua mentioned earlier could be cited in support. But it still requires a large extrapolation of this idea to then maintain that individuals are intimately controlled by the strands of nucleic acid in their brain cells. Since human culture and ethics are founded on the belief that every person acts as a free agent, such a suggestion would have an ominous ring. Surely we cannot be prisoners of our own genes. This is the nature versus nurture debate in starkest relief, and opposition to the reductionist view must contend, among other things, with the nagging mystery of sleep. Sixteen hours do not make a day, and the question remains as to what is happening when we are unconscious. Until recently, sleep was looked upon as being little more than a period of rest and recuperation from the day's efforts, but the EEG shows that the brain is involved in certain activities during dreaming, at least. These could involve a processing of the day's experiences and, as a consequence, an updating of the brain's circuits. If memory does have a hereditary link, and if it is established that the programming of future decisions is even merely conditioned during sleep, we might be forced to re-evaluate our attitude towards the concept of free will. The issue is of course still quite open, but it would be difficult to imagine a more traumatic discovery than this: that the mind and soul are nothing more than manifestations, albeit complex, of the material world. It is to these deeply serious issues that we must turn in the final chapter.

Summary

Because animals possess no chloroplast-bearing cells, they have no means of directly using solar energy. They are forced to obtain energy by consuming materials in which it is stored: plants or the tissue of plant-eating animals, or both. This requires a means of locomotion and feeding. Larger organisms move by ATP-driven muscular action which, at the molecular level, involves the relative sliding of filaments of two polymeric molecules: actin and myosin. This mechanism might be universal, providing the basis for movement of the cilia by which some simple organisms travel, and underlying the process of cell division. Muscle contraction is triggered by electrochemical impulses in nerve fibres, which are elongated electrical conduits. These support a concentration imbalance of sodium and potassium ions, maintained by ATP-driven molecular pumps. The nerve impulse, lasting about 1 ms, involves an influx of sodium ions, across the nerve cell membrane, and a subsequent potassium outflow. The senses convert optical, sound, mechanical and chemical stimuli into successions of nerve signals, which are transmitted to the control centre, the brain, the latter directing muscle contraction. In advanced species, the oxygen required for respiration – conversion of chemical energy at the molecular level – is provided by the lungs and vascular system. Blood cells in the latter contain the carrier of oxygen and carbon dioxide, haemoglobin, an iron-bearing protein. In vertebrates, mechanical support of the body is provided by the connective tissue, primarily the triple-stranded polymer collagen, and skeletal bone, a composite of collagen and apatite. In the brain, the synaptic connections between nerve cells – neurons – are reasonably specific and predetermined, but their efficiencies are modifiable by use. In primitive species, there is even evidence that every individual has an identical set of neural connections. In higher animals, specific regions are connected to the different senses by bundles of nerve fibres, and in the motor and tactile sensory areas of the cerebral cortex there is a direct spatial relationship with the corresponding parts of the body. But this cortical mapping is distorted, disproportionately large areas representing particularly sensitive parts, such as the pig's snout and the mouse's whiskers. The number of neurons in the entire brain, about a hundred thousand million in the adult human for example, could have been sufficient for a memory system based on binary circuit elements, as in a digital computer. But loss of brain cells leads to a general blurring rather than elimination of specific memories so it is more likely that past experience is stored in a parallel and distributed manner among the synaptic connections. The brain's electrical activity, recorded as an electroencephalogram, consists of trains of waves which are simple in the resting state and more complicated when the animal is alert. The pattern is also complex during dreaming.

20

The sum of all the parts of Such —
Of each laboratory scene —
Is such. While science means this much
And means no more, why, let it mean!
But were the science-men to find
Some animating principle
Which gave synthetic Such a mind
Vital, though metaphysical —
To Such, such an event, I think,
Would cause unscientific pain:
Science, appalled by thought, would shrink
To its component parts again.

Robert Graves (1895–1985)
Synthetic Such

Enchanted loom: *the mind*

The first nineteen chapters of this book were written in the third person, with the author assuming the role of the proverbial fly on the wall. This was appropriate because the things described in those chapters are accessible to any observer, directly or indirectly, through his or her senses. In that respect, they exemplify what John Ziman said of science in general – they are *public knowledge*. The human mind has long been regarded as basically different because of its subjective dimension. We take a third-person stance when considering the workings of minds, admittedly, but our own thoughts are perforce first person and *private*. In this last chapter, therefore, I shall feel justified in making occasional first-person excursions, and there will even be expressions of my own subjective beliefs about such issues as consciousness and intelligence. But that doesn't imply acceptance of the view that the mind is fundamentally different from anything else in Nature. On the contrary, I believe the much-vaunted *mind–body problem* will be solved, and that the existence of thoughts will be given a physical explanation.

The materialist view of the mind was being espoused even among the Ancient Greeks. As Hippocrates put it: *From the brain, and the brain only, arise our pleasures, joys, laughter and jests, as well as our sorrows, pains, grief and tears.* He appears to have been the first to realize that the right side of the body is controlled by the left side of the brain, and vice versa. And the Egyptians were documenting relevant evidence at least as early as 1850 BC. A papyrus found by Edwin Smith in a junk shop in Luxor dates from that time, and there are indications that it is actually a copy of much earlier records. Smith's serendipitous discovery, in 1862, gave us access to detailed case histories of 48 injuries, including lists of symptoms, diagnoses and treatments. The ancient medical practitioners recorded their surprise that injuries to the head can have repercussions for remote parts of the body, as in paralysis of the lower limbs caused by lesions of what we now realize was the brain's motor cortex. One of the giants of early Greek anatomy, Galen of Bergama, was particularly concerned with the brain's ventricles, and saw them as the ultimate destination of a series of transformations that envisaged nutrients in the gut producing natural spirits in the liver, vital spirits in the heart and animal spirits in the *rete mirabile* – a system of blood vessels at the base of the brain – those latter spirits, also called psychic pneuma, finally winding up in the ventricles.

Anatomists were still preoccupied with the ventricles a thousand years later, when the Carthusian monk Gregor Reisch apportioned the main mental faculties among those cavities. The *sensus communis*, which sounds like a reference to prudence but which was merely perceived as the confluence

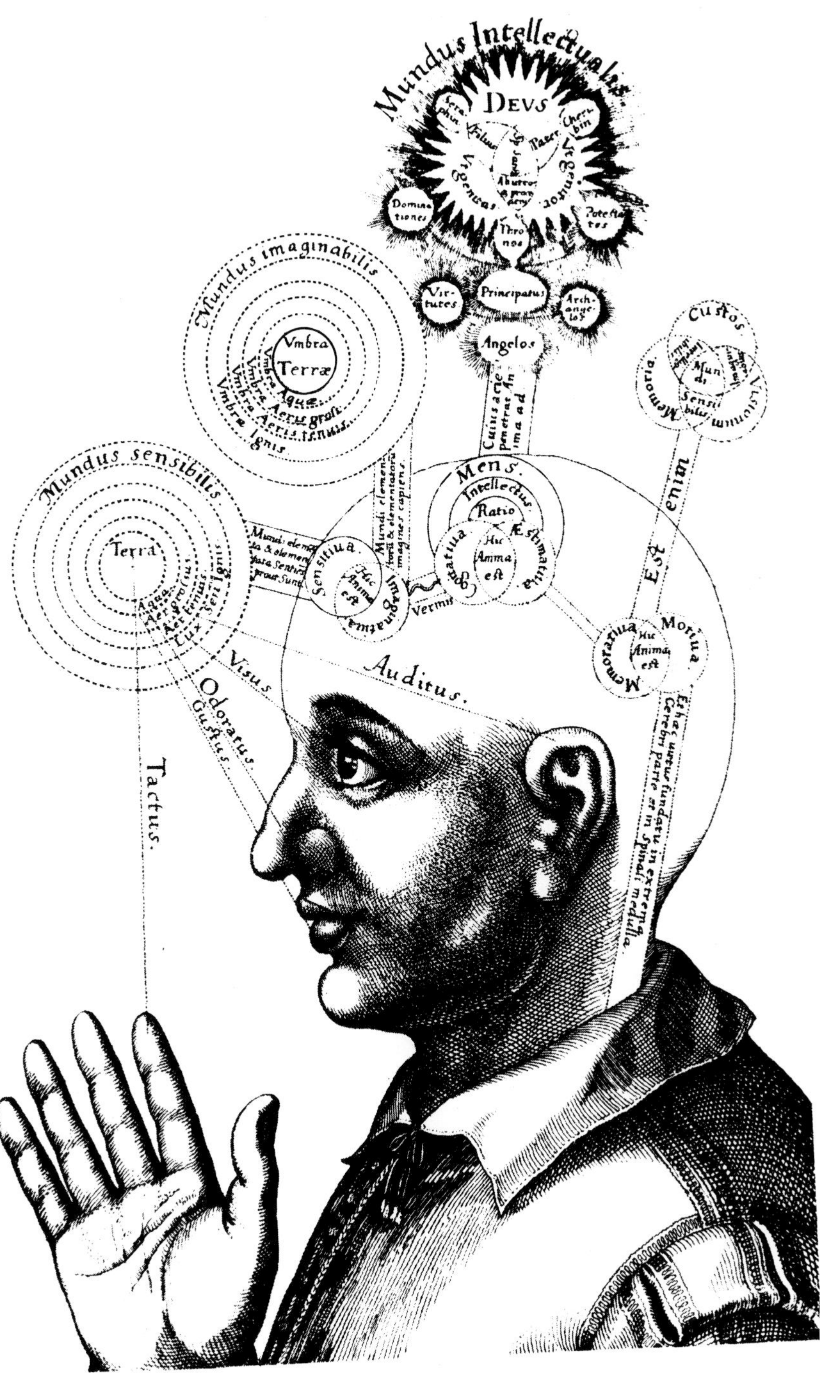

Until the late sixteenth century, those who studied the brain's anatomy were preoccupied with the ventricles, rather than the surrounding grey-matter tissue. This reflected the primacy accorded the animal spirits, which were taken to reside in the ventricles, and which were believed to be the mind's working substance. It was assumed that these spirits would need the series of interconnecting tunnels shown in this fanciful circuit diagram by the Paracelsian alchemist, philosopher and mystic Robert Fludd (1574–1637), and such connecting passages do indeed exist. There are four ventricles, but because the two at the front are symmetrically arranged, the brain theorists of that time believed that there were just three chambers, amongst which all the faculties were to be apportioned. The *sensus communis* was attributed to the front ventricles, which were perceived as a rendezvous point for messages arriving from the various senses, rather than as the seat of common sense. The central ventricle on typical anatomical drawings of that era was usually labelled with words such as *ratio* (reason), *cognatio* (thought) and *aestimato* (judgement). Memory (*memoria*) was invariably attributed to the aftermost ventricle. The dualism of brain and mind, a concept later championed by René Descartes (1596–1650), is depicted by the links between the ventricles and the externally situated centres of thought, these latter tunnels being perceived to be just as tangible as their inter-ventricle conduit counterparts.
It is at least commendable that Fludd included a connection with the spinal cord.

of information from the eyes, nose and tongue, was consigned to the symmetrically arranged pair of ventricles in the front of the brain. The more posterior ventricles were postulated to be the repositories of judgement, memory, fantasy, thought and imagination. Philosophers were no less fascinated by the interaction between the mind and the animal spirits, and the Paracelsian philosopher Robert Fludd produced a circuit diagram embodying the wisdom of that time. It showed the familiar arrangement of ventricles

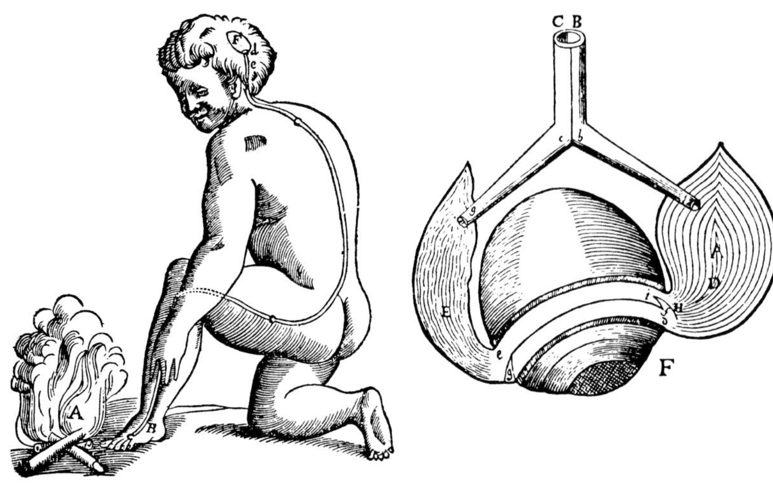

L'HOMME
DE RENE
DESCARTES
ET VN TRAITTÉ
DE LA FORMATION DV FOETVS
DV MESME AVTHEVR.
Auec les Remarques de LOVYS DE LA FORGE, Docteur en Medecine, demeurant à la Fleche, Sur le Traitté de l'Homme de RENE' DESCARTES; & sur les Figures par luy inuentées.

A PARIS,
Chez CHARLES ANGOT, Libraire Iuré, ruë S. Iacques, au Lion d'Or.
M. DC. LXIV.
AVEC PRIVILEGE DV ROY.

Although René Descartes (1596–1650) is best remembered for his dualist view of body and mind, and for the phrase *cogito ergo sum* (I think, therefore I am), he was also responsible for introducing the concept of the reflex. Because the current wisdom of his era saw all control as ultimately emanating from the animal spirits, which were believed to reside in the brain's ventricles, it is not surprising that his conjectured reflex mechanism was based on the principles of hydraulics. The tubes in his system were dual structures that served to carry messages both to the brain from the skin and from the brain to the muscles. In this reproduction of one of his illustrations, the fire **A** displaces the skin of the foot, thereby tugging at the tube **C** and opening the pore **D** of the ventricle **F**. This releases animal spirits into the core of the tube, and they flow down to the foot **B**, to inflate the muscles and cause the foot to be removed from the heat. The other figure shows his resourceful system of valves (such as the one marked **H**) which allows one muscle to relax as the other contracts when the spirits flow through the tubes **B** and **C**.

and interconnecting tunnels, one of the latter continuing down toward the spinal cord. And in keeping with the view that the mind hovered nebulously in the vicinity of the body, Fludd depicted the various faculties in a series of orbs, connected to the ventricles by a system of corridors no less tangible than those inter-ventricle tunnels. It is hardly surprising, therefore, that when Leonardo da Vinci clandestinely investigated the brain, the ventricles were the target of his attention. He poured molten wax into them, let it set, and then revealed their three-dimensional form by scraping away the surrounding tissue. As we now know, he had thus discarded the material that provides the basis of the mind, and focussed on components that merely serve the brain's housekeeping; they contain the *cerebrospinal fluid*.

There is an understandable tendency for scientists to base their theories on the cutting edge of current technology. Around the turn of the previous century, therefore, it is not surprising that researchers were comparing the brain to a telephone exchange, and more recently there has been the popular comparison of the brain with an electronic computer. When René Descartes speculated about brain function, engineers were proudly putting hydraulic devices through their paces, and this caught his imagination. He saw the reflexes as being mediated by a series of hydraulic conduits, linking the body's surfaces to the ventricles, the working substance being the animal spirits. Descartes perceived the muscles as being activated through inflation by these spirits, just as we would inflate a balloon by expelling air into it, and he conceived an ingenious system of valves to govern differential expansion and contraction. But in contrast to Fludd's model, his mind–body interaction was not based on further tubes. Instead, he identified the brain's pineal gland as the vital location, because of its lying on the brain's medial plane, and because it was thus believed to be unique among the brain's centres in not consisting of a left–right pair. Such symmetry considerations weighed heavily with philosophers of earlier times. It is fascinating to note that Niels Steensen (Nicolaus Steno), who lived in the same period as Descartes, was far surpassing his better-known contemporary in appreciating what was important. When other anatomists were barely getting to grips

Niels Steensen (1638–1686), also known as Nicolaus Steno, must be one of the strangest figures in the entire history of brain study. Remarkably broad in his intellectual pursuits, he is credited with pioneering the sciences of geology and crystallography, and he was also one of the founding fathers of anatomy and physiology. He turned his attention to the brain in 1667, and soon established himself as a leading expert in its anatomical exploration. At a time when most other people working in the field were barely getting to grips with the grey matter, Steensen quickly realized that the underlying white matter must be equally important in giving the brain the power of thought. This picture shows one of the remarkably detailed illustrations from his most famous publication on the subject. Six years after it appeared in print, Steensen suddenly turned his back on all scientific investigation, and devoted the rest of his short life to religion. He had risen to the rank of bishop by the time he died. He was beatified in the late 1980s, and now seems well on the way to actual sainthood.

DISCOVRS
DE
MONSIEVR STENON,
SVR
L'ANATOMIE
DV CERVEAV.
A
MESSIEVRS DE
l'Aſſemblée, qui ſe fait chez
Monſieur Theuenot.

A PARIS,
Chez ROBERT DE NINVILLE, au bout du Pont
S. Michel, au coin de la ruë de la Huchette,
à l'Eſcu de France & de Nauarre.

M DC. LXIX.
AVEC PRIVILEGE DV ROY.

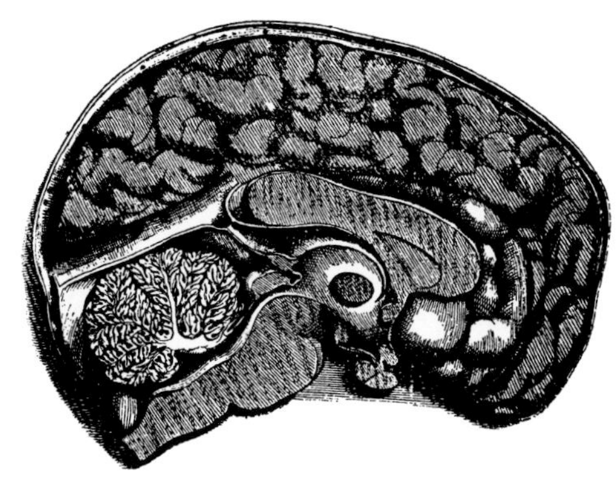

with the brain's *grey matter*, Steensen correctly perceived the relevance of the underlying *white matter*, via which the various cerebral regions interact. But Steensen was the author of no catchy phrase, whereas the immortality of Descartes was assured when he enunciated his *Cogito ergo sum* (I think, therefore I am). The question remains, however, as to what thought actually is, and how it arises in the brain.

I shall now exercise my author's prerogative and give this narrative a bias that will enable us to avoid much wasted effort. This seems desirable because many regard elucidation of the mind's mechanism as a monumentally daunting task – one of forbidding complexity. There are even those who, like Colin McGinn, believe that it will lie forever beyond our grasp, that there will never be a scientific explanation of the phenomenon. But *never* is a dangerous word in science. Let me start by noting that it is often said of elephants that they seldom forget. But one could ask what an elephant would actually have to remember. Rather little, one might suppose, given that elephants have not been able to match our culture and technology. Evolution is only indirectly concerned with such things, however. They make an impact only insofar as they influence the chance of survival until the age of reproduction. So what *would* an elephant have to remember? If one has never contemplated that question, the answer may come as a surprise. *An elephant has to remember no less, and no more, than every other creature that ever existed, namely how to move under the prevailing circumstances in the environment and within its own body.* On the output side, movement is all there has ever been. All we can ever hope to do is move. At first encounter, this will sound both wild and obviously incorrect. What about thought, one could ask. This strikes at the heart of the issue, and I shall be arguing that thought is not the passive activity one might take it to be.

In coming to this surprising conclusion, I have drawn on the speculations of others during much of the twentieth century. But George Bernard Shaw is not one of them. In his *Man and Superman*, he asks: *What made this brain of*

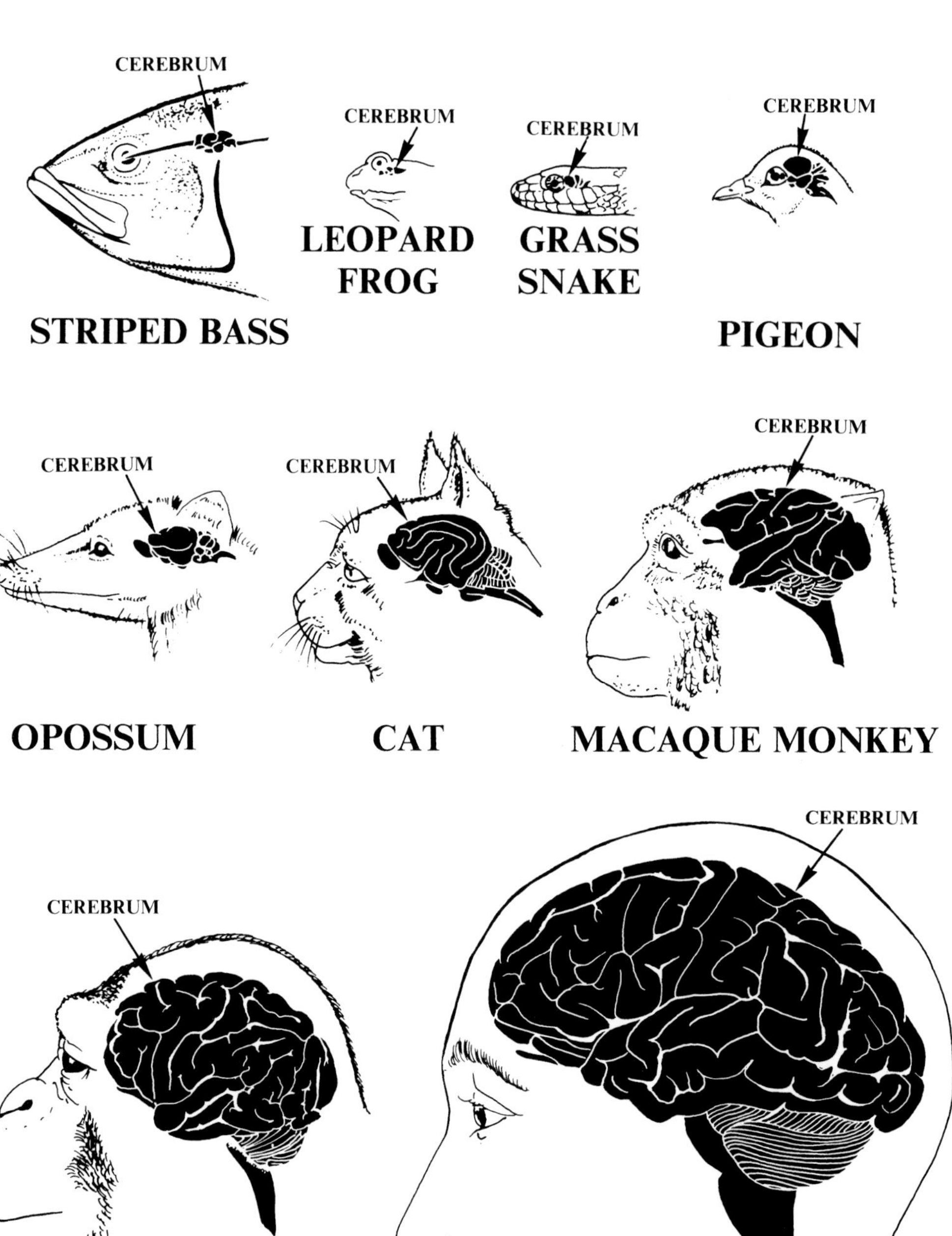

The evolutionary succession from bony fish, through amphibians, reptiles, birds and finally mammals was accompanied by a gradual increase in the ratio of brain weight to body weight. The increase in the relative size of the cerebral cortex was even more pronounced, as can be seen in these schematic diagrams. Within the mammals, there has been a systematic increase in the relative size of the frontal region of the skull and brain, and also of the cerebellar hemispheres, both of these features reaching their current maxima in humans.

Homo sapiens

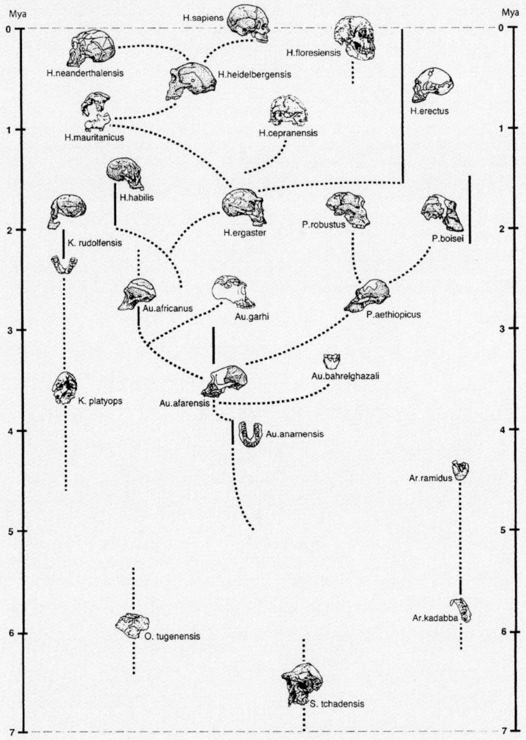

The human species belongs to the genus *Homo,* of the family *Hominidae*, of the suborder *Anthropoidea* (also known by the names *Pithecoidea* and *Simae*), of the order *Primates,* of the subclass *Eutheria,* of the class *Mammalia,* of the subphylum *Vertebrata,* of the phylum *Chordata,* of the subkingdom *Metasoa,* of the animal kingdom. The *Anthropoidea,* or higher primates, appeared as a distinct group about forty million years ago, in the Early Oligocene epoch. This group is typified by large forward-facing eyes and pronounced use of the hands for manipulation. It split into the two branches *Platyrrhini* and *Catarrhini,* the former being characterized by a broad nasal septum and including such New World monkeys as the marmoset and the howler. The catarrhines are the Old World anthropoids, which are characterized by a narrow nasal septum and the female menstrual cycle, and they include baboons, chimpanzees, gibbons, orangutans, and gorillas, which are all anthropoid apes, and humans. The anthropoid apes constitute the family *Pongidae* while we are the only living members of the family *Hominidae*. The difficulties inherent in taxonomic classification are epitomized in the search for the ancestral relationships between humans and their forebears, popularly known as the hunt for the missing link. Like us, our predecessors were notable for having the largest ratio of brain weight to body weight in the entire animal kingdom, and because body weight has shown only minor variation, endocranial volume is taken as a reliable measure of evolutionary rank. The last common ancestor of ape and human, before the catarrhine suborder split into its two families, was found in Kenya and given the name *Proconsul.* Following the separation, about twenty-five million years ago, the main representative of the branch that would ultimately lead to the anthropoid apes was *Dryopithecus.* Our own branch is now regarded as having passed through *Ramapithecus,* who, about fifteen million years ago, was walking on two legs and using bits of wood and stones as implements. During the period stretching from about five million to one million years ago, there were at least three – and possibly several more – different types of primitive human, all of whom used such rudimentary tools as shaped stones, pointed sticks, and animal bones. *Australopithecus,* discovered in 1924 by Raymond Dart, had a brain volume of about 0.5 litres; there appear to have been two species: *Australopithecus africanus* and *Australopithecus robustus,* both of which turned out to be evolutionary dead-ends. *Homo habilis,* discovered in 1960 by Louis Leakey, had an even larger brain, with a capacity of about 0.6 litres. It now seems that the member of this trio most likely linked to ourselves was 1470-man, this name indicating a catalogue number in the National Museum of Kenya. Discovered in 1972 by Richard Leakey, 1470-man had an endocranial volume of 0.8 litres, which is double that of the modern chimpanzee. The next development, around half a million years ago, was the emergence of *Homo erectus,* the earliest generally-accepted link with the present human. The brain volume had now reached 1.0 litres, and this species knew how to use fire and hunt large game. It is possible that speech began with *Homo erectus,* who existed in several varieties, including *Homo erectus erectus,* also known as Java man or *Pithecanthropus,* and *Homo erectus pekinensis,* alias Peking man or *Sinanthropus.* Our own species, *Homo sapiens,* appears to have been established around 250 000 years ago, fossilized examples having been found at Swanscombe in England and Steinheim in Germany. There was another branching about 100 000 years later, into the two subspecies *Homo sapiens neanderthalensis* and *Homo sapiens sapiens.* The former had heavy brow ridges, a low forehead and a somewhat crouched stance. But Neanderthal culture, deriving as it did from an endocranial volume of about 1.5 litres, was surprisingly well advanced and included use of animal skins for clothing and ritual burial of the dead. The earliest example of *Homo sapiens sapiens* is also known as Cro-Magnon man, after the place in France where fossilized remains were discovered. His brain volume was about the same as that of the Neanderthals, and it is not clear whether the latter became extinct or were reabsorbed by our more upright and high-browed subspecies. The current view is that there was not peaceful coexistence, however, and that the Neaderthals lost a competition that may have been rather violent, their last stand taking place in what is now southern Spain. The present version of the human being was already in evidence about 30 000 years ago, and further differentiation is well under way. There are four main varieties: Negroid, Australoid, Caucasoid, and Mongoloid, and the latter has further divides into Classic Mongoloids, Arctic Mongoloids, and American Indians. It is in the nature of this branch of science that the consensus changes as more fossilized evidence becomes available. And as indicated in the schematic evolutionary tree compiled by Ian Tattersall, and reproduced here, the consensus is tending to move away from a single, well-defined evolutionary route to our species; on the contrary, it seems that such features as walking on the two hind limbs, manual dexterity and a larger endocranial volume evolved more than once. Tattersall's scheme includes the diminutive species *Homo floresiensis*, bones of which were discovered in 2004 on the island of Flores in Indonesia. The dating of these bones, and of objects found buried near them, indicate that these metre-high beings were living as recently as 13 000 years ago. Their forebears have not yet been identified, so their connection to the rest of the scheme remains speculative. The changing evolutionary paradigm is particularly well illustrated by the finding of six hominid specimens in the Toros-Menalla area of the Djurab Desert in northern Tchad, around the turn of the millennium, by Michel Brunet, Patrick Vignaud and their colleagues. One fragment is a remarkably complete cranium, reproduced here, which has been named *Sahelanthropus tchadensis*. It dates from about 6–7 million years ago, which is astonishing, given that many of its features were previously believed to have emerged only about 2 million years ago.

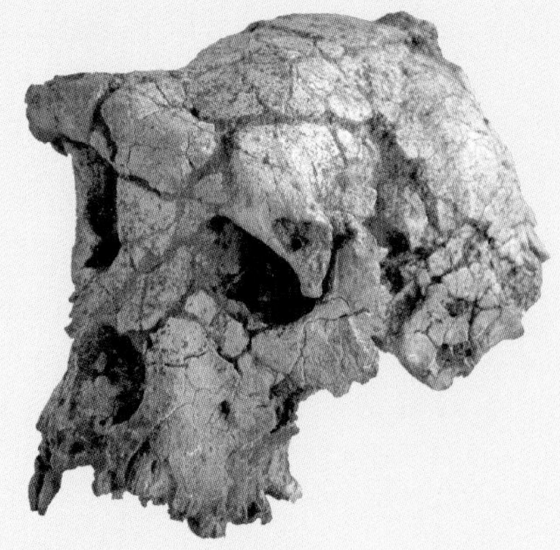

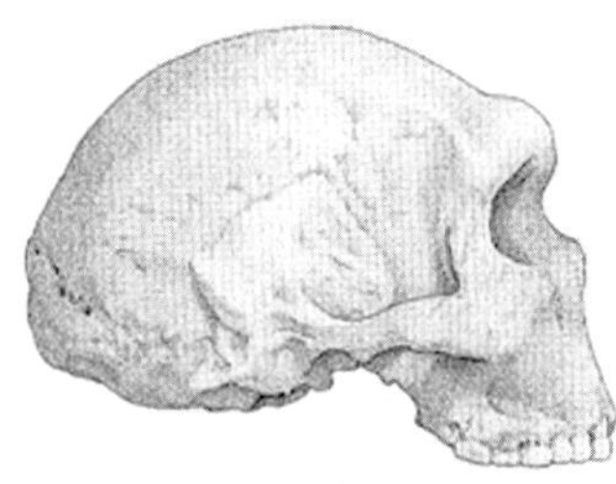

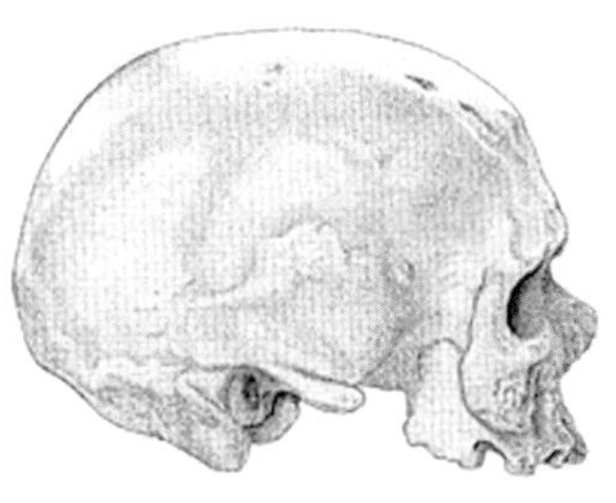

The evolution of what Alex Marshack and Ian Tattersall have referred to as the *human capacity*, appears to have been brought about in particular by the expansion of the frontal part of the cranium, and thus of the underlying brain tissue. The most significant part of the latter is the prefrontal cortex, which is known to mediate what is referred to as working memory. The expansion is clearly visible in these two drawings by Tattersall, the one above showing the cranium of *Homo heidelbergensis*, who lived about 250 000 years ago, while the one below shows *Homo sapiens sapiens*, also known as Cro-magnon man, who emerged about 30 000 years ago. John Skoyles has suggested that the evolution of the human capacity owes much to the emergence of a large, functionally uncommited prefrontal cortex.

mine, do you think? And he supplies an answer with which I cannot identify, namely: *Not the need to move my limbs; for a rat with half my brain moves as well as I.* I prefer the view of Shaw's contemporary, Charles Sherrington, author of the seminal *Man on his Nature*. As he noted: *To move is all mankind can do, and for such the sole executant is muscle, whether in whispering a syllable or felling a forest.* In a similar vein, Edgar Adrian's *The Mechanism of Nervous Action* includes the statement that: *the chief function of the central nervous system is to send messages to the muscles which will make the body move effectively as a whole.* Adrian's use of the word *chief* reminds us that although activation of muscles is the system's sole concern on the output side, it also has the job of controlling such internal functions as the beating of the heart and the working of the glandular apparatus.

Even if one readily endorses the importance of movement, however, it might nevertheless seem subordinate to cognition. Why do we have senses, after all? By way of making the plot thicken, let me invoke the words of another physiologist, Roger Sperry. They are worth quoting in full:

> *An analysis of our current thinking will show that it tends to suffer from a failure to view mental activities in their proper relation, or even in any relation, to motor behaviour. The remedy lies in further insight into the relationship between the sensory-associative functions of the brain on the one hand and its motor activity on the other. In order to achieve this insight, our present one-sided preoccupation with the sensory avenues to the study of mental processes will need to be supplemented by increased attention to the motor patterns, and especially to what can be inferred from these regarding the nature of the associative and sensory functions. In a machine, the output is usually more revealing of the internal organization than is the input.*

The bias that I confessed to above takes the words of Sperry as a lead, but it goes much further. I shall be suggesting that the detection of sensory input unavoidably implicates a vital component of the brain's motor-directing apparatus. By this I mean that the latter is not merely a *possible* recipient of signals emanating from already completed sensory processing, earlier in the nervous system; sensory processing, in my view, necessarily *involves* activation of part of the motor-directing apparatus.

Sperry's ideas were put forth many decades ago, and one could speculate as to why they failed to gain a lasting foothold in the mind–body debate. Why, in Sperry's own words, have they become *the old motor theory of thought, now largely abandoned*? Arthur Ritchie played the Devil's advocate by suggesting that the theory failed because of its inability to counter simple rebuttals. He cited the case of a man who uttered the continuous sound '-E-E-E-E-' while successfully doing mental arithmetic, and noted that the man therefore could not have been making 'implicit' mouthed movements while thinking about the numerical task. Ritchie's conclusion was premature. That mental arithmetic was possible because the vocalized '-E-E-E-E-' was surely calling upon different portions of the motor repertoire, so simultaneous execution of both tasks was possible. If the man had been required to calculate mentally while reciting the seven times table, say, he would have found the dual challenge far more difficult, and perhaps even impossible. (It should be emphasized that the articulation of words and the making of other vocal sounds is just as muscular as flexing one's biceps.)

The case of Phineas Gage: sudden change of personality

The location of the brain's specialized areas has been facilitated by studies of individuals with head injuries. This approach can be traced to as far back as the ancient Egyptians, the modern era starting on September 13, 1848. On that day, an appalling accident befell a much-liked construction foreman of the Rutland and Burlington Railroad of New England, Phineas Gage (1823–1860). Lightly tapping down a charge of gunpowder with a tamping iron, prior to blasting a rock, and failing to note that his assistant had not yet added the customary plug of sand, for safety, he inadvertently caused a spark. The ensuing explosion shot the 3 cm thick and 109 cm long rod into his skull, just above the left eye, and it emerged from the top of his head, ultimately to land some fifty metres away. Miraculously, he survived this horrible injury, and he was attended to by a local doctor, John Harlow (1819–1907), who managed to stop the profuse bleeding and limit the ensuing infection. But the irreparable damage to the frontal lobes of his brain caused a profound change of character; the mild manner was gone, and for the rest of his days, Gage was obstinate, violent, mercurial and profane. It was the clear linking of tissue damage to the sudden change of personality which made Gage's case so important to medical history. As can be seen from the life mask and skull now preserved at the Harvard Medical School (together with the tamping iron – reproduced here at a much smaller magnification than for the skull), the injury lifted away about ten square centimetres of the cranium, which subsequently grew back in place. The colour picture was produced by Hanna Damasio and her colleagues, and shows a computer fitting of an average male brain into the actual skull, the brain image having been obtained by modern techniques (see the preceding chapter). This computer-aided reconstruction was important because it revealed precisely which brain sites had been destroyed and which had been spared. Damasio and her colleagues were

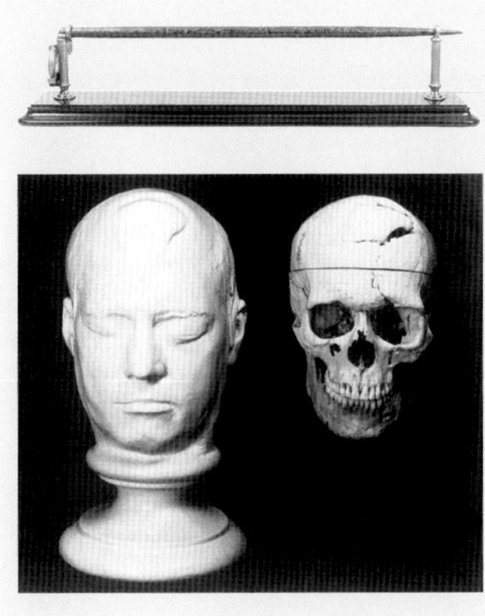

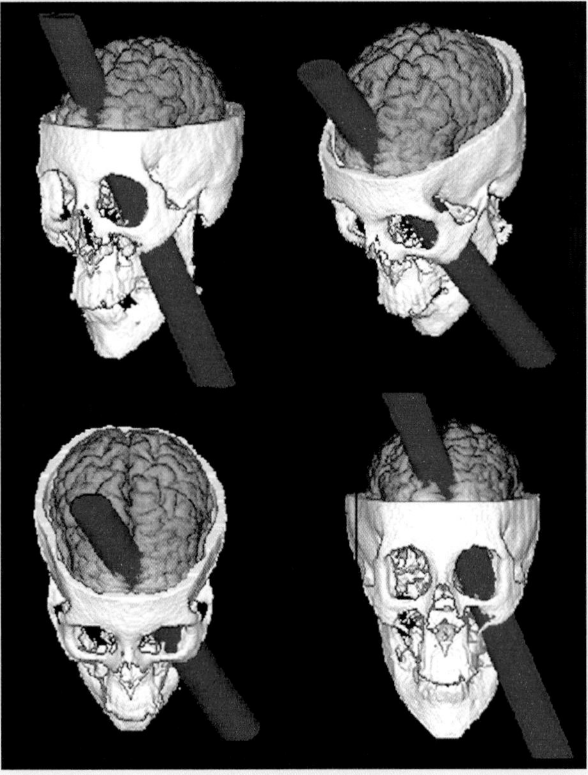

thereby able to demonstrate that the prefrontal cortex in both cerebral hemispheres had sustained widespread damage, whereas the premotor cortices had escaped, and remained fully intact. In effect, Gage's accident had thus destroyed the part of the brain responsible for the most recent increase in the human capacity, cancelling out the gains that had been achieved during the 220 000 years of evolution from *Homo heidelbergensis* to *Homo sapiens sapiens*. According to the views being expressed in this chapter, the extent of Gage's injuries indicate that he would have retained the faculty of consciousness, but that his intelligence would have been severely impaired. Harlow's clinical notes, reproduced in the second volume of the *Massachusetts Medical Society Publications*, eight years after Gage died, are consistent with this finding.

Given that the *motor theory of thought* has such an indifferent track record, one might guess that it lacked something vital, despite its promise. I believe that its poor showing in the scientific arena stems from its implicit acceptance of something which has been part of the psychological canon since the time of Aristotle. That something is now referred to as the *stimulus–response paradigm*, and it as just as firmly entrenched as *dualism* – the idea of mind–body separation advocated by Descartes. The stimulus–response paradigm sees consciousness as arising at some time between the sensing of a stimulus and the decision to respond. Indeed, one could say that it believes the task of consciousness to be mediation of that decision. This idea seems to enjoy the support of common sense. For example, when the reader scans these words – which are a visual stimulus – he or she can contemplate their meaning and respond accordingly, by continuing to read or by stopping and doing something else. It seems logical to assume that the contemplation is the business of consciousness, and its lack of any obligatory connection with movement seems equally obvious.

I suspect that unhesitating acceptance of the stimulus–response paradigm owes much to our knowledge of simple reflexes. The doctor taps me with his little rubber hammer, just below the knee, and up jumps my lower leg. Similarly, with the help of other senses, I instinctively turn my head toward a sudden loud noise or bright flash of light. In such cases, the stimulus–response paradigm just *has* to be correct. But these are examples of the unconscious reflexes we are all born with. The reflex tensing of our muscles when we hear the name Adolf Hitler, on the other hand, is something we acquired when we read the history of the twentieth century, or experienced that history in the making; we did not inherit that reaction through our genes. There seems to be something different about such reflexes, and one has the feeling that they involve consciousness. The stimulus–response paradigm might not be applicable to such situations. If that is the case, a concept accepted without question for two and a half millennia is ripe for revision. I believe that it is indeed, and I shall show in the following pages how this change of direction can lead to a remarkably full picture of consciousness and its ramifications.

Howard Berg's studies of bacterial locomotion are relevant to this story, as unlikely as that may seem. He wondered why bacteria swim, given that they meet as much resistance when moving through water as we would if swimming through molasses. The answer seems obvious: they swim because they have to obtain food. But the bacterium does not know where potential food is located, because it possesses no senses. Why, then, *would* it swim? A bacterium moves because of the propulsion provided by its flagellum,

Mad cow disease: gradual disappearance of mind

About 10 per cent of large proteins require the help of another protein molecule to nudge them into their final native conformation, the mediating molecule being referred to as a *chaperone*. Large proteins require proportionately more time to attain their final conformation, and in the absence of a suitable chaperone, there might be the possibility of infelicitous aggregation with other molecules in the intervening period, with possibly dire consequences for the host cell. Something akin to the chaperoning mechanism appears to underlie the virulence of one form of the molecule known as the *prion*, this name being acronymic for *proteinaceous infectious particle*. The existence of the prion was postulated by Stanley Prusiner, who boldly predicted the occurrence of hereditary molecules other than the nucleic acids. The prion has a single primary sequence of amino acid residues, but it comes in two different conformations, one innocuous and the other highly dangerous (see the colour diagram, in which the helices are merely shown as cylinders, while the beta-sheet is indicated by arrows, in the usual labour-saving manner). The unfortunate fact is that the latter can induce the former to adopt its pernicious conformation, the process obviously having the potential for touching off a chain reaction. Prions occur naturally in the brain, and their virulent conformation has been shown to have the capability of producing the devastating brain disease in cattle known as *bovine spongiform encephalopathy* (*BSE*), also known as *mad cow disease*. The sponge-like appearance of the brain's surface that is characteristic of this affliction has been linked to the formation of so-called plaques, which are aggregates of prion molecules in their virulent conformation. It is not yet clear whether the same type of molecule underlies the sheep disease scrapie, or indeed the rare – but terrible – human affliction Creutzfeldt–Jakob disease (CJD). The significance of a proved connection to CJD, in the context of the present chapter, would lie in its demonstration that the mind can gradually disappear because of imperfections at the molecular level. As described elsewhere in this chapter, Phineas Gage's mind was changed suddenly because of gross structural damage to the brain. In the case of CJD, the putative root cause would lie at a far more microscopic level, and thus be much more subtle, but the end result would be just as devastating.

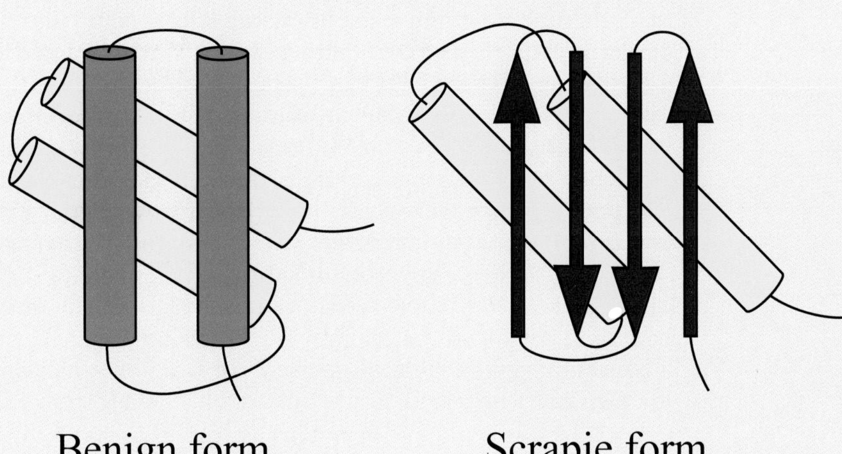

which resembles a ship's propeller. Anticlockwise rotation of this device, looking in the forward direction, produces forward movement. Clockwise rotation does not cause the creature to swim backward, as might be expected; it provokes tumbling, and hence a change of direction. Berg showed that the bacterium's direction of movement is dictated by the local availability of food. When this is adequate, a primitive memory mechanism based on a chain of chemical reactions produces anticlockwise rotation of the flagellum, and hence forward motion. If the amount of food impinged upon and ingested during the previous few seconds is insufficient, however, there is a brief period of clockwise rotation and the bacterium moves off in a different direction. This does not guarantee that the creature will now be moving toward a plentiful food supply, but it is a better strategy than doggedly staying on course. The bacterium behaves in this fashion because it has evolved to survive. But the mechanism is unconscious and mechanical; there is no thought involved. There could not possibly be, given that the bacterium possesses no nervous system.

The important thing about this primitive scenario is that it is diametrically opposite to the traditional stimulus–response paradigm. It is the bacterium's motion that is the relevant stimulus, in this case, while the relevant response is that of the feedback from the surroundings, in the form of impinged-upon food. This is just the reverse of the reflexes we were considering earlier. But what relevance can be attached to behaviour at the other end of the evolutionary ladder – behaviour that has served these primitive creatures for almost four thousand million years? As extraordinary as it may seem, I believe that it holds the key to understanding consciousness. I am going to suggest that the bacterium's probing of its surroundings, and its unconscious recording of the outcome of that probing, has been the basic behavioural strategy used by all creatures, great and small, throughout the entire history of the animal kingdom. But we have much ground to cover before the link to consciousness can be demonstrated.

There is not room here to consider the behaviour of many other species, but let us take the case of the honeybee. As is well known, this insect informs its hive-mates of a successful forage through its famous waggle dance, which indicates both the distance and the direction to a discovered supply of nectar. Karl von Frisch, pioneer investigator of bee behaviour, believed that these insects' distance-gauging abilities stem from an unconscious mechanism that keeps tally of the energy expended in the hunt for food, but in 2000 Mandyan Srinivasan and his colleagues Shaowu Zhang, Monika Altwein and Jürgen Tautz showed that this is not the case. They cunningly forced bees to fly through a tube whose inner surface was decorated with a sort of distorted chess-board pattern, on their way to candidate flowers less than 1 m from the hive. Reading the bees' subsequent dance, the researchers noted that the insects were telling their mates that the food lay tens of metres away. The bees had been unconsciously duped into interpreting the black–white–black–white alternations in the tube's decoration as the similar changes of shading encountered in their normal environment, those latter alternations usually being on the scale of several metres rather than several centimetres. The ingenious experiment had demonstrated that a bee unwittingly records the amount of 'visual flow' for a given expenditure of energy. This is simply a more sophisticated version of the bacterial mechanism described above.

The most bizarre chapter in the history of brain science started in 1790, when Franz Gall (1758–1828) put forward the idea that mental propensities are reflected in the contours of the skull. A bump in a given area was taken as an indication of strength in the corresponding aspect of personality. Several different systems subsequently appeared, one by Gall's leading disciple Johann Spurzheim (1776–1832), all claiming to most accurately chart the various phrenological regions. This pseudoscience, the popularity of which can be attributed to its tempting scope for quackery, reached its grotesque zenith with the publication of a particularly elaborate scheme by James Redfield, M.D., of New York City (left figure); it listed 160 numbered faculties. The fact that Republicanism (region 149) lies adjacent to Faithful Love (area 148), in his system, probably reveals more about Redfield's political sympathies than about his medical qualifications. The most advanced scientific knowledge of any period takes time to filter through to the public awareness; the *American Phrenological Journal* (right figure) was still thriving in the 1930s, and the British Phrenological Society was not disbanded until 1967. Porcelain phrenological heads, such as the beautiful examples produced in the mid-nineteenth century by Lorenzo Fowler (1811–1896) and his brother Orson (1809–1887) of Ludgate Circus in London (colour picture), and marketed by their brother-in-law business associate Samuel Wells (1820–1875), are now expensive collectors' items.

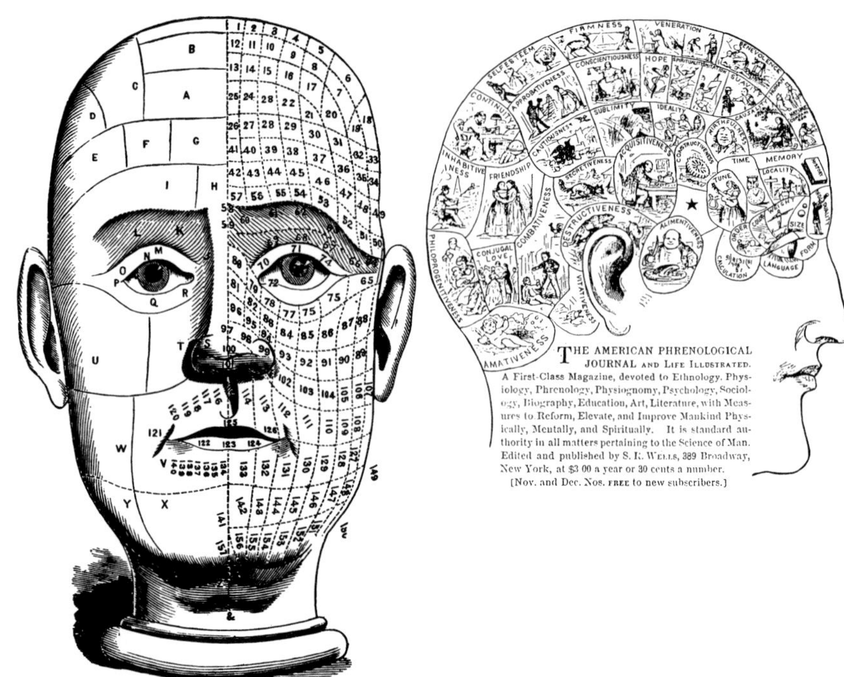

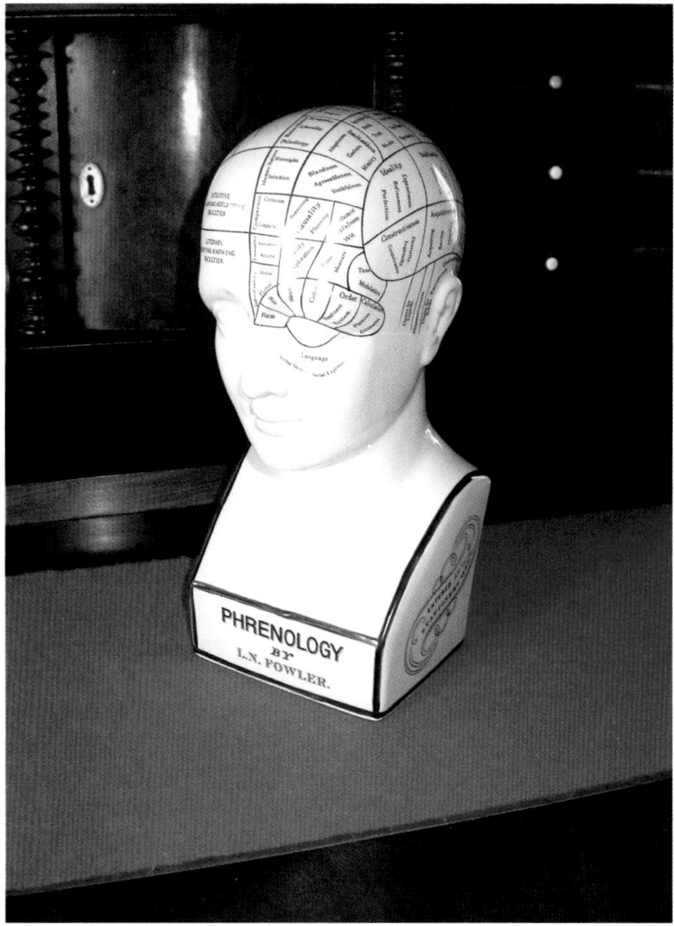

A sound case can be made for using the word *cognition* when considering these mechanisms. It is unconscious cognition, admittedly, but it is cognition nevertheless. In effect, the twirling of the bacterium's flagellum asks *Is there food in this direction?* – and bacteria have been putting such questions to their surroundings for the past four thousand million years. Humans possess a counterpart of such unconscious cognition. For example, every step one takes while walking asks unconsciously *Is the ground still there?*, and our nervous systems just as unconsciously record the answer *Yes, the ground is still there.* Only if our unconscious probing encounters something unexpected – a hole or an overlooked tree root, for example – does the event force itself into our conscious awareness. But we are not born with a capacity for automated walking. An infant must concentrate on every attempted step, its effort to maintain balance lying at the focus of attention. This led me to suggest, in 1994, that the chief goal of consciousness is acquisition of new reflexes – reflexes that are more sophisticated than those we inherit, because these novel variants depend upon context. (It ought to be added, here, that although the newborn infant is unable to support its own weight with its legs, and thus lacks the vital prerequisite for walking, its legs will automatically mimic locomotion if its feet are permitted to touch the ground while its weight is being supported by an adult. Interestingly, the infant loses this automatic reflex when it starts to crawl, the nervous system then having begun to modify its motor repertoire.)

This theory of active acquisition of information has been touched upon in various forms by others, including Jean Piaget, who noted that *'all knowledge is tied to action, and one can only know an object or an event by using them through assimilation into schemes of action'*. More recently, one has been able to encounter variations on this idea in the writings of Alain Berthoz, Andy Clark, Walter Freeman, Susan Hurley, Marc Jeannerod and Maxine Sheets-Johnson. And James Gibson was, in the present author's opinion, moving in the right direction when he observed that *'the causal link is from response to stimulus as much as it is from stimulus to response'*.

But Gibson's testimony seems to admit of a duality that I feel is misleading, because it leaves the door open for a school of thought that many regard as defunct, namely *behaviourism*. The school's founder, John Watson, believed that the mental processes underlying feelings and intentions do not determine what a person does. And his fellow enthusiast, Clark Hull, saw psychology as resembling physics before the discovery of atoms. Just as knowledge of those entities was not required when experimentally determining the relationship between pressure and volume, for example, Hull believed that the response caused by a given stimulus could be gauged just as adequately by exclusively external observations of behaviour. The behaviourist movement could find no place for consciousness in its scheme of things, and its adherents were encouraged to expunge the word from their vocabularies. Behaviourism's most ardent champion was Burrhus (B. F.) Skinner, who believed that the way we respond is determined solely by the history of reinforcements to which we have been subjected. Like his fellow behaviourists, he regarded feelings as mere consequences of actions, and consciousness as just an epiphenomenon.

Behaviourism enjoys at least implicit support even in this twenty first century, though its modern invokers seek to use it as a stepping-stone to consciousness. Several writers have sketched plots in which the stimulus–response mechanism ultimately spawned an internal counterpart that permitted thought, the later being essentially an exclusively private stimulus–response duplicate. This is tantamount to saying that the mind evolved from reflexes. One sees such a suggestion in the ideas of Nicholas Humphrey, for example. (I shall later describe a thought mechanism mediated by internal simulation of external events that does not fall into the behaviourist trap.) I am strongly opposed to this line of reasoning because it admits, *inter alia*, of the possibility of pre-conscious language. Two extreme examples of this faulty logic are seen in the writings of Daniel Dennett and Julian Jaynes. Dennett describes an imagined pre-historic scene in which one of our forebears attempted to call to her mates one day, not realizing that she was alone, and found herself listening to her own verbalizations. This got her talking to herself and – hey presto – consciousness followed! The Jaynes story, put forward in his *The Origin of Consciousness in the Breakdown of the Bicameral Mind*, is even more extreme. He sees consciousness as having arisen from social pressures, the triggering events being catastrophes. According to Jaynes, pre-conscious Man communicated more directly with his gods than with his peers, and consciousness developed only as the authority of the deities started to wane. What a bizarre idea! How extraordinary to suppose that cities could have been built by our supposedly pre-conscious forebears, without questioning those ancestors' ability to accomplish such impressive feats of engineering without the benefit of consciousness. The application of mortar to the space between two bricks was not an inherited reflex. And neither was the joining of several syllables to make a word.

Indirect endorsement of the idea that perception requires activity in the muscle-sequencing region of the brain is provided by studies of an auditory illusion carried out by Diana Deutsch. She found that certain series of composite tones are perceived by some subjects to follow a descending scale whereas others hear them as ascending. Her fascinating discovery was that whether a particular person heard the notes rise or fall depended upon the spectral composition of that person's own voice. So here again was evidence that perception is not the passive process many take it to be. Taken to its logical limit, such evidence suggests that cognition is dependent upon covert activation of the brain's relevant muscle-sequencing region. I shall now submit that *all* cognition is of this type, though unlike the bacterium and the bee, we are capable of both overt and covert variants of the faculty – the covert variant being thought. One consequence of this theory is that we will be able to repeat what another person has been saying only if we have been covertly setting up the relevant sequences of articulatory muscle movements required to mimic his or her words. It is not often that such repetition is demanded of us, of course, but those sequences of muscular movements are nevertheless being set up, *if we are attending to the speaker's words*. If this theory is tenable, it would have to imply that analogous muscular-sequencing preparations apply to things perceived by the other four senses. This could seem to have little relevance to the visual domain, but we must bear in mind that the eyes are not stationary when we observe something that is not moving. On the contrary, the eyes abruptly change their direction of gaze about four times every second, as they search for visual clues.

Let us now return to the idea that all creatures gain information about their surroundings by dint of their own movements (though members of advanced species such as our own can make do with movements merely carried out in the imagination). This is all too familiar to the blind person, who has to grope about when reaching for something, and await the feedback signals from the surroundings which announce that the item has been successfully located. Similarly, a blind person reads by moving the fingers across the pattern of raised dots in a Braille text. Passive reading of such a text is impossible if the pattern is merely moved by someone else across the blind person's stationary finger-tip; there has to be active exploration.

As interesting as these aspects of the issue might seem, there are those who argue that they are only scratching the surface of the real problem of consciousness. In 1994, David Chalmers divided aspects of consciousness into two classes. What could be called the mechanics of the phenomenon – identification of the vital brain circuitry and explanation of how it produces conscious perception – belong to what Chalmers calls the *easy problem of consciousness*. The *hard problem of consciousness*, according to Chalmers, concerns the question of why there should even *be* such a thing as consciousness, and the actual nature of raw (conscious) sensation – the painfulness of pain, for example, rather than the mere fact of there being pain. Chalmers believes that the easy problem should be solvable within the coming centuries, whereas the hard problem may never be solved. Others had expressed similar reservations. Joseph Levine referred to the *explanatory gap*, in 1983, when asking why consciousness has to exist in Nature, given that so many creatures manage perfectly well without it. And already in 1974, Robert Kirk had asked why an unthinking machine should not be capable of everything humans can manage, despite being bereft of consciousness. This is now known as *the zombie argument*, a zombie in this case being such an unthinking automaton rather than a product of voodoo. These questions raised by philosophers are both basic and important, given that no method has yet emerged by which one could unambiguously demonstrate the presence of consciousness in another human being, let alone in a machine such as a computer. In a similar vein, Thomas Nagel famously wondered *What is it like to be a bat?* After a little contemplation, most of us would probably conclude that we could not possibly know.

I believe that there *is* an answer to the question of why there has to be consciousness, if a certain level of behaviour is to be achieved. The argument – an admittedly protracted one – runs like this. Movement, of either the overt or covert type, is the only way in which a creature can achieve anything, and this extends to the gaining of information from the surroundings. In this respect, as has already been stressed, the relevant stimulus is the creature's own self-paced movement (actual or *actively* planned for – see later), while the relevant response is that of the sensory feedback from the surroundings (actual or *actively* anticipated – again, see later). So the creature must possess a mechanism that tells the system that a given response from the surroundings is associated with the recently undertaken movement-mediated probing of those surroundings. As we saw earlier, even the lowly bacterium possesses a short-term memory system that keeps it informed of that movement-plus-feedback association. But the bacterium is unable to augment its behavioural repertoire. The plot of its humble life is written in its genes. In its case, there is only nature, no nurture.

If a creature is to acquire new reflexes it must analogously associate the pattern of its recent movements with the resulting feedback from the surroundings. But it is confronted with the complicating factor that there might simultaneously be sensory input that it has not itself provoked. It must therefore have a mechanism that permits it to distinguish between the two types of input, that is to say between the sensory feedback it has provoked and the sensory input over which it had no control; it must be able to distinguish between touching and being touched, for example. And it has to be able to make this vital distinction because it has dominion only over the former type of input. Failing that, the creature will not be aware that it is the agent of its own actions. To take a simple example, a child will not be able to control its voice if it cannot distinguish between the sounds it makes with its vocal apparatus and the sounds it simultaneously picks up from its parents' voices.

What would be required of a nervous system that such distinction be possible? For a start, there would have to be a mechanism that labels the self-provoked feedback from the surroundings, and simultaneously evaluates its consequences for the creature. Such labelling, I suggest, is part of the basis of raw sensation, and raw sensations are what awareness is made of; in a sense, they *are* consciousness. Ned Block recounts a story, apparently ancient among philosophers, of a biographer researching the life of Mark Twain. He stumbles across the surprising fact that there is a mysterious second person named Samuel Clemens who led a remarkably similar life to Twain, almost as if he had been shadowing the famous writer. Everywhere that Twain went, Clemens went also, and everything that Twain did, Clemens did too. When the biographer discovers that Twain's royalty checks always wound up in Clemens's bank account, it finally dawns on him that Clemens *was* Twain, and that Mark Twain was just a pseudonym. I relate this tale because it might well be analogous to the issue we've just been discussing. The simultaneous labelling and evaluating mechanism, and the raw sensations it generates may simply *be* consciousness.

But something more is required. As we just noted, the mechanism purported to be carrying out these multiple tasks would be available only if the brain possessed the appropriate anatomy, and the relevant circuits would have to function at the appropriate times. Moreover, there would be the requirement of suitable physiology. Clearly, it would be an advantage if the sensory receiving areas were put on the alert every time the brain's muscle-sequencing region had initiated a series of movements. It transpires that there is indeed such an internal flagging of planned movement, and it is known as *efference copy*. We should pause to consider this technical term. Inwardly travelling signals within the brain are said to be *afferent*, while outward going signals are referred to by the term *efferent*. Thus the arrival of light rays at the eye's retina gives rise to afferent signals in the brain's *primary visual area*, located in the occipital lobe at the rear of the head. These afferent signals travel toward the brain's movement-sequencing region, gradually being modified *en route*. Ultimately, the suitably modified signals provoke the movement-sequencing region to dispatch efferent signals to what is known as the primary motor region, which, in turn, sends efferent signals to the muscles. We can extend this terminology to describe what is going on in the surroundings, partly because of the brain's activity. Imagine that a person is observing his or her own act of writing. The light rays from the paper to the retina are said to constitute *exafference* signals, and the marks being made on the paper by the writer are, logically, *exefference* signals. However, it is vital to note that the exafferance signals will be of two varieties: there will be the signals emanating from the just-drawn lines, but there will also be signals from other marks that were not of the writer's doing. (One could imagine that the person was writing responses into the various pre-drawn boxes on a questionnaire, for example.) One thus says that the light rays coming from the writer's just-drawn lines constitute *reafference*, that is to say feedback signals from the surroundings that the writer has provoked. Although there is more to this story, we might already find ourselves wondering whether consciousness arises at the watershed between afference and efference. (It doesn't; things are not that simple.)

Referring back to our earlier discussion, an animal will be aware that it is the agent of its own actions if, and only if, it can successfully distinguish between (self-provoked) reafferent signals and other (non-self-provoked) exafferent signals. As noted above, members of sufficiently advanced species have the neural equipment to do precisely that, the vital component being internal feedback routes that convey the efference copy signals. It is useful to think of those latter signals as *internal reafference*. They are made available to the sensory receiving areas, where they can be compared with the signals arriving via the external route, that is to say compared with the (external) reafference signals. The classical example of such a comparative mechanism is seen in the monitoring of – and correction for – what is known as *retinal slip*. This mechanism comes into play if the head is turning while one tries to maintain one's focus on a stationary object. That situation was analyzed by Erich von Holst, who showed how comparison of the efference copy of instructions sent to the eye muscles with the resulting afference signals being returned from the retina could prevent such slip. It is in this manner that the image of the object is held steady on the retina. The reader can test this concept by gently poking the corner of one eye and noting that this makes a viewed stationary object appear to move, the reason for the illusion being that no efference copy signals are available to serve compensation because no movement commands have been sent to the eye muscles.

This chapter is emphasizing my personal view that movement – in either its overt or covert form – underlies all cognition, irrespective of which sensory mode is being called into play. So one might expect counterparts of that eye-poking illusion to occur with other senses. A particularly serious one has been suggested by Christopher Frith to underlie the alien voices heard by sufferers of *schizophrenia*, the idea being that their efference copy system is somehow impaired. But there is more to the mechanism than the mere provision of efference copy; the system also needs a means of comparison. Indeed, it needs a means of *anticipation*, and I shall later be describing a mechanism – provided by the cerebellum – which I believe serves that purpose.

William James, widely regarded as the doyen of psychologists around the end of the nineteenth century, was much concerned with the *stream of consciousness*, and particularly with its continuity. Our experiences and thoughts seem to enjoy a brief period centre stage in our attention, and then give way to other thoughts and experiences usually generated by what went before. Imagination, in the absence of outside stimuli, represents a pure form of the single-file traffic of our thoughts. Cheves Perky's investigation reported in 1910 deserves to be more widely known. She positioned a subject in front of a translucent screen, onto which was projected – from behind – a picture of a tomato, and she controlled that object's intensity. The subject was asked to say when the picture disappeared, as the intensity gradually diminished. Perky then switched the projector off and asked the subject to merely imagine seeing the same tomato in its original location on the screen. Without revealing the fact, she switched the projector

As depicted in this picture of her at Domrémy, painted by Lenepveu in 1889, Joan of Arc was prone to hearing voices, a typical symptom of schizophrenia. This delusion, and other bizarre behavioural traits characteristic of the affliction, ultimately led to her execution. But it is now clear that her plight was attributable to an organically determined overproduction of the neurotransmitter dopamine, which is particularly prominent in the brain's basal ganglia, and which exercise a strong influence on the muscular activation. Christopher Frith has suggested that schizophrenia additionally arises because of an impairment of the efference-copy mechanism. According to the present author's view, this mechanism alerts the auditory receiving region to the imminent arrival of vocal sounds generated by the (articulatory) muscle-sequencing region, and its impairment could prevent the victim from realizing that she was hearing her own voice. According to the view expressed in the present chapter, this would even apply to the covert form of speech – so-called inner speech – that is a common aspect of thought.

on again and gradually increased the intensity, carefully noting when the subject became aware of what was happening. In every repetition of this test, the intensity had reached a level greater than the one at which the subject had originally reported disappearance. The subjects clearly experienced difficulty distinguishing between observed and imagined objects.

That finding was endorsed by the resourceful – and equally inexpensive – investigations carried out by Stephen Kosslyn. Before contemplating these tests, we should note that the *visual field* of an eye is simply the angle subtended by the full width of the scene it can take in without moving. (The visual fields of the two eyes overlap in the central part of our vision, of course, as can be checked by closing each eye in turn, without moving the head or the eyes.) Moreover, an eye's visual field is constant; what we can take in with a single eye, without moving eye or head, depends upon the distance to the scene we are inspecting. We can see only one tree in a forest if it lies just a few centimetres from our nose, but we see many trees lying more distant. And the entire forest may be visible from a distance of several kilometres. Kosslyn started by noting that the same constraints apply to mental imagery. For example, the angular spread of an imagined object lying at a particular distance is about the same as that observed with the eyes open. And when imagining specific scenes, he found that the amount of available detail depended upon distance. Imagining observing a person holding up a newspaper, for example, he felt that he could read the print only when the paper was close at hand.

Those preliminaries undertaken, Kosslyn tested subjects by having them draw a map of simple terrain – the sort of thing one finds in illustrated editions of Robert Louis Stevenson's *Treasure Island*. Essential features to be included in the sketch were the beach, a hut, a coconut tree, a rock and buried treasure (suitably indicated with an ×). He then asked each subject to give a running commentary on his or her imagined exploration of the territory via a specified route, for example beach–tree–rock–hut–treasure, timing that subject with a stop-watch as the path was followed. It transpired that the times taken to reach the various landmarks during these mental expeditions were a reliable indicator of distance on the subjects' physical maps. Imagination is clearly a spatially accurate faculty. We should not be surprised, therefore, that Marc Jeannerod detected activity in the same brain region – the motor-sequencing area – irrespective of whether an object was being directly observed or merely imagined.

We must now identify a technical term first used by Frederic Bartlett, following Immanuel Kant's introduction of the concept. This is the *schema* (plural: *schemata*). Kant saw it as the link between *percept* and *concept* – between cognition and comprehension – and he saw it as a procedure for constructing an image. There is an obvious connection with the ideas of Jean Piaget, mentioned above. Ulric Neisser improved on Kant's rather vague notion, and more recently I suggested a particularly brief definition: *a schema is a reproducible cognitive element comprising a pattern of actual or planned motor movements and a concurrent pattern of actual or anticipated feedback from the surroundings, the reproducibility stemming from the fact that schemata are laid down in the available form of memory*. (The words *planned* and *anticipated* will later allow us to include the case of thought in the absence of movement.) This concise and more specific definition has the advantage that it is applicable even to organisms devoid of a nervous system. In the bacterium, as we have seen, the available form of memory is primitive but adequate for the creature's purposes. The definition also reminds us that a schema can be brought into play by purely internal events, as when hunger provokes foraging. In all creatures, that urge to consume food derives from reactions that monitor the internal level of certain chemical compounds, such as the glucose level in the blood. Even in advanced creatures such as ourselves, some schemata are inherited – one could call them *basic instincts* – while others may be acquired by mere happenstance, rather than the more customary deliberate exploration.

Samuel Johnson once stated that '*What we hope ever to do with ease, we must first learn to do with diligence*'. A novice walker needs to be conscious of every step, and so does an experienced walker negotiating unfamiliar terrain. Diligence is essential in such situations. Returning to our definition of a schema, and to our walking scenario, each successful step invokes the schema *Is the ground still there? Yes, the ground's still there!* A schema is thus a question that comes equipped with its own answer, and the concept provides a natural basis for all cognition; we recognise things by fitting them to the relevant schemata, and our knowledge is simply the totality of our remembered schemata. The advantage of possessing a given schema is that it can be used automatically and unconsciously, as a *context-specific reflex*, walking being an obvious example.

Turning again to language, whose articulation naturally involves muscular movement, we have no difficulty in reading correctly the two sentences '*The dog's lead was on the hook*' and '*The molten lead was in the crucible*', despite our giving two different intonations – and meanings – to the four-letter sequence l-e-a-d. Indeed, we read such sentences so quickly that there is no time to concentrate on the individual words. Experienced readers manage such reflex reading because context provides clues sufficient for correct pronunciation. But consciousness is vital in the overall mechanism of reading; it acts at the higher level of meaning, providing a sentinel for the unexpected. (Good writing eases the task of that sentinel.) Similarly, experienced walkers need concentrate only on the overall goals of many-step expeditions, consciousness providing a lookout for obstacles. Reflexes, both innate and learned, are *ballistic* whereas conscious behaviour is a special type of *guided* mode – what the engineer refers to as a *servo mechanism*, the gist of the word *servo* being that it implies the availability of feedback information.

If schemata *are* so basic, their acquisition and use must be fundamental in behaviour. How are new ones learned? Primarily through consciousness, I believe, though subtle augmenting mechanisms may exist. And before we continue, we ought to consider briefly the concept of drive. The driving force for (possible) muscular movement originates externally in the traditional stimulus–response view of consciousness, and the same always applies to the reflex. But the force stems from *within* the animal in the new paradigm I am proposing here, the drive ultimately serving *homeostasis* – the body's aim to maintain the internal *status quo*. Drive chooses the sequence of muscular movements best suited to achieving the system's immediate goal, the link between goal and sequence being stored in memory through prior experience (or through inheritance). But because the surroundings may have changed since that experienced was gained, a check needs to be kept that the feedback from those surroundings is as expected; *an activated schema is a hypothesis seeking confirmation or revision*. And as we have seen, the system does this by alerting the sensory receiving areas to the imminent arrival of feedback, the apprising mechanism involving dispatch to those areas of signals that are duplicates – *efference copies* – of the ones sent to the muscles. Other signals arising from any mismatch between the actual and anticipated feedback are available for correcting the ongoing muscular movements, in the standard servo manner. Such corrections do not require consciousness, however, a typical unconscious example being that mechanism we considered earlier for holding the image of an object at a fixed position on the retina, despite movement of the head.

What is additionally required for consciousness? The vital extra ingredients, I believe, are what Alan Baddeley and Graham Hitch referred to as *working memory* and a mechanism to control *attention* – working memory because an answer is useless if the question has been forgotten, and attention because a question may be useless if misdirected. The latter statement reflects my belief that attention is more related to the question than to the answer. When conscious, we can (a) think without moving, (b) move while thinking about that movement, and (c) move while thinking about something else. In mode (c), the thinking requires more of our attention than the movement; *conscious thoughts are never unattended*. But what is happening in the brain when we are in mode (b), attending to our movement? Given that the movement is provoked by the (internal) drive, it must involve activation of the corresponding schema, so efference-copy signals will have alerted the appropriate receptors to the imminent arrival of sensory feedback. But what if there is simultaneously sensory input that is not a consequence of our movement? Its detection could lead to acquisition of a false schema, having no subsequent utility for the system. (If we tapped an infant on the forehead every time its hand touched the side of its cot, for instance, we could fool its nervous system into concluding that the forehead was located at the side of the cot.)

How can the system distinguish between concurrent provoked and unprovoked inputs – between touching and being touched, for example? It does so by subjecting the just-activated schema to attention that stems from the drive, and by briefly holding a record of that attended-to schema in working memory. Unprovoked sensory input will then be perceived as such because it will not have been associated with efference copy – it will not be able to satisfy the sensory expectations of the already-activated schema. And the resulting mismatch may then cause a shift of attention. In essence, our consciously performed overt actions are immediately preceded by covert (imagined) versions, the latter guiding the former.

How do we recognise sensory input we have not provoked, however? According to the new paradigm I'm proposing here, such input would have to generate a cascade of signals penetrating through to the brain's movement-sequencing region, this then dispatching efference-copy signals back to the sensory receptors, the upshot being looping signals that permit the system to lock on to an appropriate schema. (The looping process inevitably consumes time, and I offer this as an explanation of the experimentally observed delay-to-consciousness, as will be expanded upon below.) But how do we perceive sensory input we are *unable* to provoke, such as sound that lies beyond our vocal range? It too will give rise to looping signals, and when the system learns to recognise them, the corresponding schema permits us to say *high pitch*, for example. Any perceptual limitation is thus in the sensory range of the receptors, not in the movement-sequencing region. In his *An Essay Concerning Human Understanding*, published in 1690, John Locke introduced the concept of the *tabula rasa*, a sort of metaphorical blank slate on which life's experiences are written. The looped circuit I am invoking here, and not the least its modifiable synapses, would constitute such a *tabula rasa*.

When drive-provoked movement produces the desired feedback from the surroundings, there must be a mechanism whereby the drive is

Consciousness: two opposing paradigms

The traditional stimulus–response paradigm for consciousness (top left diagram) regards consciousness as arising solely from the processing of sensory input (full red arrow), a *possible* consequence (open red arrows) being activation of a drive mechanism, which selects and activates an appropriate sequence of muscular movements the response. In this paradigm, therefore, the drive is provoked by the surroundings, as in the simple reflex (full green arrow). In the new paradigm favoured by the present author (top right diagram), the muscular movements are the stimulus, as the animal probes its surroundings, and the response is the resulting sensory feedback from those surroundings. The drive in this case comes from within, and it controls attention. When an animal consciously probes its surroundings, an internal simulation guides its muscular movements (full red arrows). This simulation precedes the actual movements by a fraction of a second, and it mimics those movements and the anticipated feedback from the surroundings. It therefore invokes the appropriate drive-activated schema (upper full green arrows), and any mismatch between the anticipated and actual sensory feedback is available for correcting the movement, and revizing the schema. The animal's nervous system is thereby able to update its knowledge of the surroundings. In the new paradigm for conscious perception (bottom diagram), consciousness will not prevail unless nerve signals have reached the movement-sequencing region of the brain, and activated the appropriate schema (full red arrows), a possible consequence being actual muscular movement (open red arrow). The theory thus predicts that even apparently passive conscious perception has to activate the movement-sequencing region, which in humans is the premotor cortex (together with the supplementary motor cortex and frontal eye field).

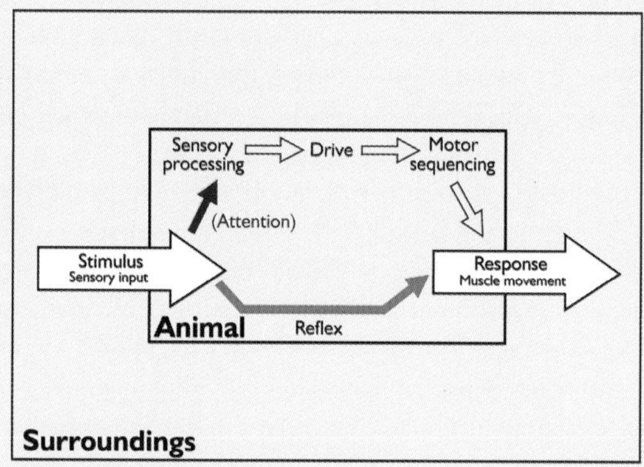

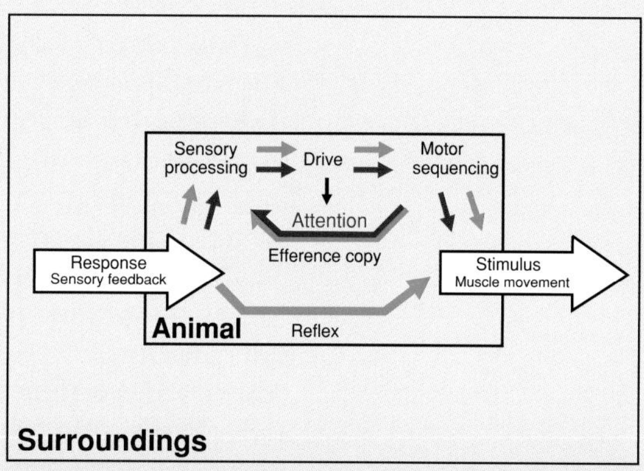

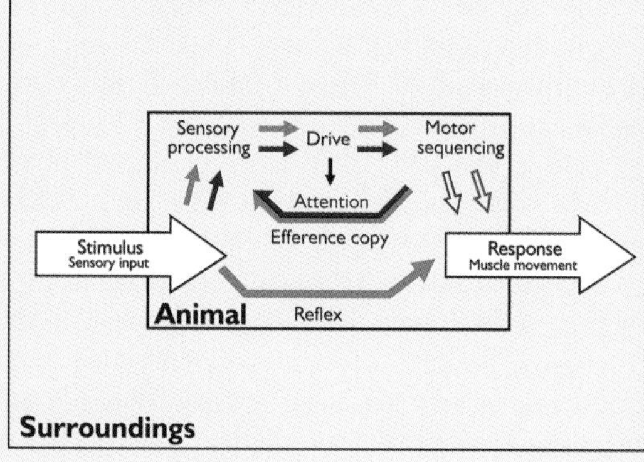

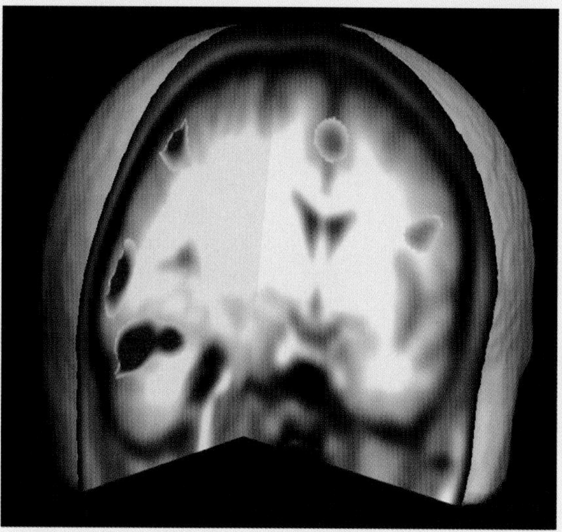

The additional time compared with the shorter time course suggested by the upper diagram required for the incoming sensory signals to penetrate through to the movement-sequencing region, set up signals returning via the efference-copy route, and thereby permit the signals to self-organize and home in on an appropriate schema, is believed to be the origin of the delay-to-consciousness reported by Benjamin Libet and his colleagues. Those self-organizing neural signals circulating around the route described by the full red arrows will display a certain resistance to instantaneous change, and this inertial effect is believed to underlie the sense of effort, even in the absence of overt muscular movement. The colour photograph, taken by Mark Christensen and his colleagues, shows the brain regions activated when a subject passively observes an abstract visual pattern. It reveals that the movement-sequencing region of the brain the premotor cortex is indeed activated under these circumstances, in agreement with the lower diagram. The picture was obtained by blood oxygenation level dependent functional magnetic resonance imaging (BOLD fMRI), and it shows a cut-away view of the brain from the right rear of the head, the red and yellow colours indicating the areas maximally activated. Progressing from right to left, the areas shown being activated are the right premotor cortex; the right and left anterior cingulates (which lie medially, and which are known to be implicated in attention); the left parietal cortex (which has been linked to visual processes serving action); and finally, various visual areas.

discontinued. Mismatch between the anticipated and actual feedback, conversely, must somehow cause the drive to be maintained, and the discordant upshot may be the source of emotion. Because detection of such mismatch provides consciousness with its justification, this suggests that emotion is indispensable to the phenomenon. Deposition in working memory of the record of a movement-embodied question – actual or merely imagined – must perforce precede the arrival of the corresponding answer – actual or merely anticipated – so provided that working memory somehow labels time of arrival, this will give the system its sense of cause and effect; it will serve as time's arrow, and underlie recognition that one is the agent of one's own actions. The overall sense of self, on the other hand, must stem from the system's gradually acquired ability to distinguish between self-provoked and externally arising sensory input.

Until now, little has been said about the anatomy believed to underlie the functions required by the proposed new paradigm. If there were no appropriate anatomical underpinnings, the ideas I've been expressing could be worth rather little. And if the ideas make no contact with the compendious detail given in the preceding chapter, that detail might seem to have been described to no purpose. Let us enquire, therefore, as to how these ideas might draw on the known features of the brain's anatomy.

One learns little about unseen objects by merely touching them – much more when exploring them through sequences of movements. Preconscious creatures such as the honeybee commonly use sequenced movements, their nervous systems comprising a *sequence-orchestrating component* (counterpart of the premotor, supplementary motor and frontal eye areas in mammals) in addition to the *muscle-activating component* (counterpart of the primary motor area in mammals). The former clearly controls the latter, when the system uses all its options. The sequence-orchestrating component is thus better suited as the source of efference-copy paths, and in animals capable of consciousness these internal feedback routes are known to provide links to all the sensory receptor areas, those routes becoming subject to a global attention mechanism that acts close to the sequence orchestrator.

In mammals, the (composite) *thalamic intralaminar nucleus* is a strong candidate for attention's focal point. It receives signals from the premotor, supplementary motor and frontal eye areas, and projects to all sensory input areas, as well as to the basal ganglia. It is known to be indispensable to consciousness, and its output is controlled by the *nucleus reticularis thalami*, which has been conjectured by Charles Yingling and James Skinner to serve attention. Moreover, that latter nucleus is controlled by the *brainstem reticular formation*, which governs the sleep–wake cycle. The *basal ganglia* are strong candidates as primary mediators of drive; like the intralaminar nuclei, they have widespread connections to the cerebral cortex, and they are under the influence of the homeostasis-serving hypothalamus.

The sequence orchestrator must select patterns of muscular movements that do not threaten the overall integrity of the body's skeletal musculature: the muscles must not be directed to tear the body apart. For example, a sequence of muscular movements will position a finger between the upper and lower sets of teeth, and sufficiently vigorous contraction of the jaw muscles will then cause the finger to be bitten through, with attendant pain. Putting it mathematically, the sequence orchestrator guides the system through *muscular hyperspace* – *hyper*space because we must consider *all* the muscles simultaneously – and the tracing of a trajectory through that hyperspace produces the familiar serial nature of consciousness. So the crux of the theory is that consciousness arises through the tracing of the hyperspace trajectory and the *rapid* detection of sensory feedback signals from the surroundings, the animal thereby being aware that it is the agent of its own actions. And just as the more familiar form of navigation would be impossible if no record were kept of the path recently followed, so is viable navigation through muscular hyperspace dependent upon brief (working) memory – albeit unconscious because of its brevity – of what the muscles have been recently doing, in the service of self-paced probing of the surroundings.

These arguments are beginning to sound rather abstrusely mathematical, and some readers may feel that they are consequently becoming too removed from biological reality. So let us get our feet back on the ground by considering a specific example. A juggler astounds us by accomplishing feats that we would find impossible, keeping an impressively large number of items simultaneously flying through the air, while using just the same complement of arms and legs that we have. So it is not just possession of the appropriate number of limbs that is important; it is a question of moving those limbs, and their attached hands and feet, at precisely the right moments. And the ease with which the juggler performs is surely a consequence of training that involved the word I highlighted earlier: *diligence*. That word conveys a sense of deliberation and laborious effort, the latter stemming from the fact that the muscles seem to offer resistance when they are being required to move at times and in directions that do not come naturally. With sufficient training, of course, they *do* come naturally, and the juggler can then impress us with consummate ease. We who hold driving licences manage our own juggling feats, as we simultaneously work the various pedals, turn the steering wheel, manipulate the levers that activate or deactivate the windscreen wipers, and so on. And if we are transporting passengers, we may also be engaged in conversation, conversation involving the muscles that serve articulation, as we have noted earlier. Navigation through muscular hyperspace, therefore, is just a shorthand way of saying that the muscles are performing a series of appropriately sequenced and properly timed contractions.

So this mechanism acts globally but is nevertheless centrally – and thus *dictatorially*, not democratically – controlled by the brain's muscle-sequencing area; hence the unity of conscious experience. And it precludes what is known as a *binding problem* because it forces the senses to function cooperatively, thereby permitting *cross-modal perception*; the animal is able to recognize feedback received exclusively through one type of sensory receptor on the basis of what has previously been received only through a different type of receptor.

Here too, we have technical terms that must be expanded upon. In 1965, John Eccles published a monograph based on some lectures in which he addressed the problem of explaining the unity of conscious experience. Like the majority of people researching in this area, Eccles endorsed the stimulus–response paradigm, and he was also convinced that the brain's various regions function as a sort of democracy, each having its fair share of influence on the body's muscular actions. In holding these views, he was obviously in good company; given our knowledge of history, most of us shun the idea of dictatorship. But why should what is undeniably good for society be relevant for an individual organism? The circulation of the blood could have been managed by a series of pumping stations, strategically deployed around the body, though there would then have been the problem of making them execute their power strokes at the correct times. Evolution sidestepped that problem by giving the system a single *dictatorial* heart. Eccles, and several others who have followed his lead, had to tackle the conundrum of having the various democratic senses get their act together – of identifying the nervous system's equivalent of the ballot box, one could say. It is this that is known as the *binding problem*, and once again we ought to pause and contemplate things at greater length.

Let us consider the case of a lower animal in which the binding problem has not been solved, namely the snake. One would think that such a creature would have no difficulty in feeding itself, but we are about to discover that the way it tackles this menial task is decidedly ham-fisted. The North-American corn snake, *Elaphe guttata*, studied in impressive detail by Sverre Sjölander, is a typical example. It feeds on mice. Snake eyesight is not particularly acute, but it is augmented by a set of sensitive heat detectors, arranged around the eyes. So when a mouse is in front of the snake, its presence is detected by way of the thermal radiation from its body. Radiation travels in straight lines; its rays are blocked by opaque objects. If the mouse passes behind a sufficiently large stone or plant, therefore, it will disappear from the snake's view. But the remarkable fact is that it will also have ceased to exist as far as the snake is concerned. It's not just a case of out of sight being out of mind; we are going to discover that a snake *has* no mind.

Every now and then, a mouse fails to avoid the snake's clutches, and it is squeezed to death in the reptile's coils. This process of strangulation is a reflex, automatically triggered by detection of the mouse's body heat. One might think that the snake's job would now be quite straightforward, but here is the first surprise. It doesn't know where the mouse is! Lying lifeless in the predator's coils, the little mammal is no longer in view of the latter's eyes or heat detectors. So after making its kill, the snake is likely to crawl around as if looking for further potential victims. It will seem to have forgotten its successful forage, and indeed it has. So what was the point of the exercise? Has Nature made the snake embark on a fruitless expedition? The answer is *no*, of course. If the snake's behaviour provided it with no benefit, that behaviour would have been eliminated by evolution. But how does the snake reap the reward of its efforts? The answer is that it uses its nostrils. It is not long before a dead mouse will begin to smell, and the snake is able to locate it by sniffing the resulting odour. And because the mouse's body will have cooled off by that time, the snake's sense of smell must act unaided in this part of the drama. Rather remarkably, the eyes and the heat detectors play no further role.

Our assumption would be that the job of consuming the mouse is now as good as accomplished, but there is another surprise. It transpires that a snake always swallows a mouse front-end-first, presumably because the mouse's body presents a narrower challenge from that direction. Here, surely, is another task for the snake's odour detectors. Aren't they just what is required if the creature has to distinguish between the prey's nose and its backside? But again we encounter something unexpected. The sense of smell goes off duty once the rotting victim has been found, and the sense of touch takes over. In other words, the snake feels its way to the mouse's sharp end, and that's the end it pushes into its mouth. Now the job really is done, and the meal is soon being digested.

The important – and surprising – message of this story is that a snake's sensory systems do not work in unison; they function independently. But is this of any great consequence? If we believe that consciousness merely provides an animal with an internal movie show of life's events, it might seem unimportant that the snake never gets its sensory act together. We could imagine that the creature is nevertheless able to think about the division of labour that provides work first for its eyes and heat detectors, then for its odour detectors, and finally for its sense of touch. The snake might even be impressed by the democratic nature of the arrangement. If one believes that consciousness is a passive attribute, and that it serves no biological function, the snake's method of obtaining food seems eminently logical. After all, the system does work perfectly well. This argument is invalid, however; consciousness simply isn't possible if the senses are fragmented. Serpents are incapable of thought. Not only do they make heavy weather of the job of eating; they have no inner life. They are to all intents and purposes just slippery machines. We tend to be frightened of them, admittedly, but the threat they pose has nothing to do with their possession of devious mental capacities. In that respect, they are not more menacing than a meat grinder.

The term for the inadequacy we have just been discussing is the one we encountered earlier: the snake's sensory systems are unable to solve the *binding problem*. Let's try a small experiment, to clarify this unfamiliar technical term. Hold your hands out in front of you and clap them together. If you actually did this, rather than just imagined yourself doing it, you will have demonstrated that at least some of your senses *are* able to solve the binding problem. You *saw* your hands come together, you *heard* your hands come together, and you *felt* your hands come together, and there was no doubt in your mind that the signals generated in all three senses belonged to one and the same event. (We could easily bring in another sense. When I depress the top of my can of shaving cream, for example, I simultaneously hear a squirting sound and smell the perfumed foam.) And if you merely imagined the handclap, you will similarly have imagined seeing, hearing and feeling the sensory inputs simultaneously. So our senses properly bind events together in both the external and imagined domains.

We can now return to that second term introduced earlier; animals unable to bind their sensory inputs are said to lack *cross-modal perception*. The word *mode* refers to a given type of sensory input, and as was discussed in the preceding chapter, humans have sensory apparatus for handling the *visual* (sight) mode, the *auditory* (hearing) mode, the *olfactory* (smell) and *gustatory* (taste) modes, and finally the *tactile* (touch) mode. An animal with cross-modal perception can recognize something with one sense even though it has only ever experienced the object with another sense. It is this ability that allows us to recognize objects previously seen but not touched, even when we are blindfolded, by using our sense of touch. Such an achievement is quite beyond a snake.

Here, then, is the advantage of being able to solve the binding problem, and why it is beneficial for the nervous system to be properly informed about what goes with what. Let's briefly consider an amphibian, namely the frog. There are aspects of its behaviour that could fool us into believing that it is rather sophisticated. If something irritating is placed on its back, for example, a frog will try to remove it with one of its flippers. This reaction looks so deliberate that we might be tempted to take it as evidence of consciousness. But it is actually just a reflex that has been retained through evolution because of its usefulness. Another behavioural trait exposes the frog's serious limitations. The creature feeds on flies, which it catches by shooting out its long tongue. But it does not recognize flies as such. If a fly-sized bead is swung in front of it on a piece of thread, it will lunge at this too, and subsequently spit it out when it proves to have the wrong taste. Its fly-catching routine is a reflex based on visual receptors that detect moving objects having a given size. This sensory apparatus cannot detect stationary objects. So we have the extraordinary situation that a frog will starve to death in the absence of moving flies even though it might be knee deep in a freshly killed supply of the insects. The same would be true of a snake surrounded by freshly killed – and thus not-yet-smelling – mice.

So inability to handle the binding problem, and the associated lack of cross-modal perception, puts an animal at great disadvantage. Such a creature is nothing more than a complex organic machine. It is unable to achieve things we find so simple that we take them for granted. And above all, its nervous system is incapable of generating imagined events. We hear the screech of a car's brakes and leap out of harm's way, even though we may not have seen the vehicle. Our cross-modal perception has allowed us to sense the impending impact before it has actually happened, and to take evasive action. And such conjured-up scenarios are, of course, the very stuff of consciousness. Creatures unable to bind their sensory inputs, such as reptiles and more primitive species, cannot possess this wonderful attribute.

In conscious animals, as has already been stressed, schemata must be stored in a distributed and superimposed manner. But what can we say about the dynamics of their selection? It is at this point that we must introduce the quantitative aspects of the story, without which any theory risks being just so much hand waving. By definition, this selection must take place within the time span of a *cognitive event*, but how brief may this be? We have already noted that the eyes unconsciously change their direction of gaze about four times every second. It has been experimentally established that the actual

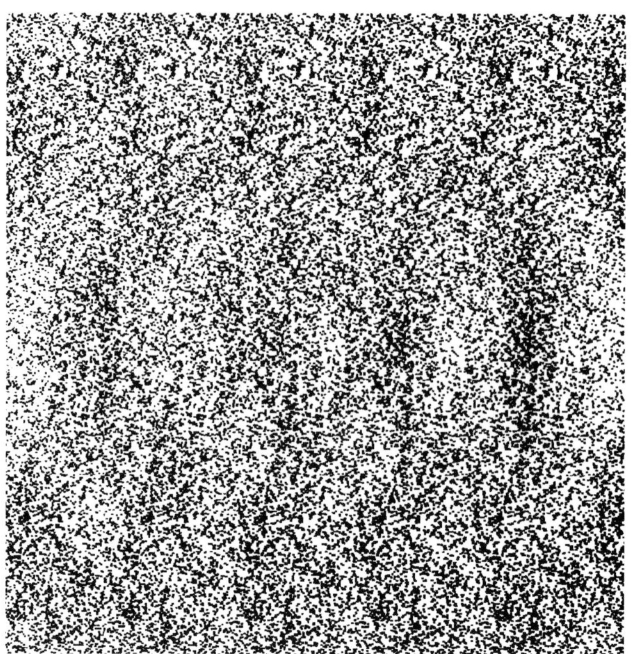

If one is not experienced in viewing *autostereograms*, such as the one reproduced here, it can be quite difficult to discover what is being conveyed. One trick is to position the picture right in front of the nose, so that focussing on the dots is impossible, and then gradually move it away, without changing the position of gaze. When the picture is about 40 centimetres from the tip of the nose, a beautiful three-dimensional shape should suddenly be perceived. The final result is well worth the effort. It is theoretically significant that the entire three-dimensional shape emerges simultaneously, this being a consequence of a self-organizing process of the type described by Hermann Haken. In the present author's view, this process has to involve movement-preparation events in the premotor cortex.

eye movement takes about 50 ms, so the eyes remain gazing in a given direction for roughly 200 ms. And as Kirk Thompson has demonstrated, from electrode measurements of nerve cell activities in a monkey's frontal eye field (which is the sequence-dictating area for the eye muscles), there is no clear differentiation between responses to targets and distractors (i.e. false targets) until around the 200 ms mark. But what about other muscles, as they mediate probing of the surroundings? Top male sprinters take 45–50 steps when covering 100 m in about 10 s, their movements being guided because they can avoid an obstacle – such as a falling competitor – suddenly appearing in their path; this too indicates around 200 ms for a cognitive event. (The far higher cycling rate of a humming bird's wing indicates that these innate movements are automatic and ballistic.) These various pieces of evidence support the view that selection in the sequence-orchestrating area is the governing and rate-limiting process. And they suggest that even apparently passive conscious perception must involve activation of the premotor, supplementary or frontal eye areas. It should be noted, also, that we are finding the same basic time period of 200 ms for both external events involving actual muscular movement and purely internal events involving the looping of nerve signals. This is interesting because it suggests that the internal and external events have to be matched, if the system is to function properly.

While on this important subject of time, we ought to note corroborating evidence from language, not the least because of its central role in our lives. How quickly can English words be uttered and still be intelligible? The 1963 edition of the *Guinness Book of World Records* accorded the record to President John F. Kennedy, who in 1961 reached the astounding rate of 327 words per minute, during one of his political speeches. That same source noted that tobacco auctioneers can reach a rate of 400 per minute, but that no one outside the trade can understand what they are saying. Kennedy's rate corresponds to just over five words per second, or one word in about 200 ms.

Once again, therefore, we arrive at our standard period for a cognitive event, and here too we see a matching of the rate at which the muscles can perform and the rate at which the product of muscular movement can be reliably perceived. We can go one step further, and note that if the average number of syllables per word is around two, while the typical syllable comprises two phonemes, the average time taken to utter a phoneme is around 50 ms. Phonemes are intoned ballistically, therefore, whereas the speaking of a word must involve their being strung together in the guided manner we are claiming to be the province of consciousness. We get tongue-tied when that stringing together has produced uncommon combinations, and when we are being required to intone them rapidly. This is why we dissolve in laughter when trying to repeat sentences such as A BLOKE'S BACK BRAKE-BLOCK BROKE, ten times in quick succession.

By way of invoking further quantitative evidence, we could hardly make a better choice than the work of Benjamin Libet and his colleagues Bertram Feinstein, Dennis Pearl and Elwood Wright Jr. They carried out experiments on patients undergoing brain surgery, exploiting the fact that this involves exposing the surface of the cerebral cortex. The patients' permission had naturally been obtained beforehand, and we should bear in mind that the cortical tissue itself has no pain receptors; the patient is not under duress. The other advantage of this situation is that the patient usually has to be awake, in order to give the doctor feedback during the surgery. Each patient was thus able also to give Libet and his colleagues feedback during their investigations. The experimenters first located the region in the tactile cortex corresponding to the back of one hand, and monitored any nerve cell activity there with the help of a suitable electrode. They then stimulated the back of that hand with an electrical pulse lasting about 0.5 ms, and measured the time it took for this to elicit a signal in the corresponding cortical area. That delay turned out to be about 40 ms, which is roughly what one would expect, given the distance the signals would have to travel, and their known speed of propagation.

Libet and his colleagues then stimulated the cortex's back-of-hand area, and the patient reported feeling a tingle on the back of the actual hand. But it was discovered that this direct cortical stimulation had to be continued for at least 350 ms for anything to be felt at all. Libet and his colleagues thereafter tried various combinations of hand and cortical stimulations, varying their order and the time lapse between them. They observed what is known as a *masking effect*; if the direct cortical stimulation followed on too soon after stimulation of the actual hand, the patient simply did not experience the hand stimulation. By using this type of approach, and more complicated ones involving stimulation of both hands with various time delays, they were able to arrive at a remarkable conclusion: it takes about 500 ms for a person to become consciously aware of an external stimulation to a tactile sensory receptor. That numerical value was subsequently revised downward to about 350 ms, but it is nevertheless surprising, given that nerve signals can travel from any point in the brain to any other point in about 100 ms, taking the most direct route. The conclusion to be drawn from this famous result must be that the development of conscious awareness requires nerve signals to make circuitous – and probably looping – journeys within the brain. Why should this be so? The signals cannot be undertaking these travels just to use up time. They must be achieving something, and it seems very likely that they would be modified *en route*. I believe the answer lies in what the physicist calls *self-organization*. The delay-to-consciousness observed by Libet and his colleagues must be a consequence of the incoming signals having to lock on to an appropriate schema, this process clearly requiring time. In fact, comparing the 350 ms with the 200 ms we were discussing earlier, schemata activated externally, as in the case of the experiments carried out by Libet and his colleagues, appear to require more time to establish themselves than those which are the exclusive product of the internal drive.

One should not underestimate the importance of the 200 ms period to the conflict between the two paradigms we have been considering. To the stimulus–response paradigm, that 200 ms just happens to be the time it takes for a sensory item to be recognized; the period cannot be accounted for in terms of the underlying anatomy and physiology. But to the paradigm I have been advocating – what could be called the *inside-out paradigm for consciousness* – it represents the natural time period both for the fastest muscular movements and for the looping nerve signals that underlie their selection. Neither the muscles nor the self-organizing signals can be induced to change their patterns at a more rapid rate; this is the maximal rate at which the muscular hyperspace can be traversed, and therefore the maximal rate at which the surroundings can be explored in the self-paced manner we have been discussing. Putting it in terms more familiar in the type of physical phenomena explored by Isaac Newton, Robert Hooke and Galileo Galilei – and for that matter by Albert Einstein – the 200 ms is a measure of the *inertia* of the system. This is what the incoming sensory signals have to negotiate with in their attempts to guide the muscular movements – actual or merely imagined – in the feedback-based servo manner we have been discussing.

I am tempted to take the argument one step further, and suggest that this could hint at a solution of the hard problem identified by David Chalmers, provide the refutation of Robert Kirk's zombie argument and show the way in which Joseph Levine's explanatory gap could be closed. The point is that despite one's best efforts, there always seems to be a level lying below that of one's analysis – a sort of infinite regress that keeps on saying '*Yes, but what is steering **that** process?*'. This is the familiar problem of the homunculus needing an inner homunculus, which in turn is in need of a still-more-internal homunculus, and so on. The significance of the 200 ms, I suggest, is that it places a limit on how much we can regard time as being divisible, when considering the workings of the mind. It introduces an indeterminacy principle, not of the type one associates with Werner Heisenberg, as discussed in the first chapter, but an indeterminacy principle nonetheless. And it supplies an answer to what could be called *the what-is-being-reported-to problem*. In effect, the incoming sensory signals report to those self-organizing looping signals, either confirming or attempting to revise the corresponding schema that has just been – or is in the process of being – activated. And because there is that matching of natural time periods between the internally simulated and externally executed muscular events, we feel the same sense of effort irrespective of whether we are actually moving or merely imagining that we are doing so. This is surely why those experiments carried out by Stephen Kosslyn produced such suspiciously accurate results. And this, I believe, provides a description of the physics underlying the mind–body problem; it shows why there has to *be* sensation. A zombie would need access to the same mechanism in order to move in the way we do, and it would therefore require access to what in us produces sensation. Its zombie status would thereby be revoked.

But what does activation of a schema entail, at the level of the nerve cells in the movement-sequencing premotor cortex (and the related supplementary cortex and frontal eye field)? It might be imagined that this location is the repository of complete patterns of movements lasting several seconds, rather in the manner of a tape-recording, but this picture would be misleading. It is rather the case that the movement-sequencing region merely dictates which elementary movements are to follow which, on a much briefer time scale. Taking again the example of language, we saw earlier that articulation of a single phoneme requires a minimum of about 50 ms, and we have just noted that a nerve signal could travel half-way across the brain in that period. The job of the sequencing area must therefore be to select what is to follow – by way of muscular movement – during the next 50 ms, and so on. And the prescription for this particular succession of movements must be laid down in the synapses. We now begin to glimpse the beauty of this system; a given set of phonemes can be strung together in different sequences, so as to permit articulation of different words, once they have been *diligently* acquired, the corresponding prescriptions being stored in the synapses.

Results obtained by Reza Shadmehr and Henry Holcomb are of relevance here. They used the brain-activity imaging technique known as *positron emission tomography* to verify that acquisition of motor skill involves learning an internal model of the task's dynamics. They found that within 6 h after the appropriate practice, while performance remains unchanged, the system recruits new brain regions to perform the job. They observed a shift of activity from the prefrontal region – known to play a major role in working memory – to the (muscle-sequencing) premotor area. Similarly, Gregory Berns, Jonathan Cohen and Mark Mintun, using the same experimental technique, required their subjects to perform simple reaction-time tasks, various stimuli being equally probable and apparently occurring at random. But in reality the stimuli followed a complex pattern, which subsequent behavioural tests showed had been nevertheless learned. And the acquisition proved to have been accompanied by increased blood flow – that is to say, increased energy expenditure – in the premotor area (and in an area called the *anterior cingulate*, which is known to be implicated in attention).

This brings us to the important issue of dreaming, which I am going to suggest mediates incorporation of recently acquired temporary (or candidate) schemata into the brain's permanent repertoire of schemata. In 1983, Francis Crick and Graeme Mitchison conjectured that dreaming serves to eliminate accidentally formed – and thus unwanted – memories, but it is difficult to believe that a system prone to such erroneous storage would have survived in the competitive evolutionary arena. Dreaming sleep is now more commonly referred to as rapid-eye-movement (REM) sleep, because the eyes have been observed to move about erratically under their closed lids during this phase. It was previously believed that the eyes were somehow involved in a dreaming counterpart of wakeful vision during such episodes. But perhaps the real explanation is more mundane. The point is that all the other skeletal muscles are immobilized during REM sleep, presumably because whatever is going on in the brain might otherwise cause the limbs to move (something which does in fact occur in certain pathological conditions). But with the eyelids closed, the eyeballs would be free to move without causing any sensory input. There is more to this story, however, because the muscle-immobilized REM period is limited to a few minutes, in order to ensure adequate blood circulation (the recent cases of deep vein thrombosis during long air journeys serving to illustrate that point). But the corneas of the eyes also contain blood vessels, and the rapid movement of the eyes against their lids might be nothing more sophisticated than a periodic bout of massage.

In 1998, Robert Stickgold found evidence of consolidation and integration of memories during sleep, including the REM phase, and four years later he and his research group demonstrated that short bouts of sleep can prevent perceptual deterioration. Around the same time, J. Allan Hobson, Amir Mazur and Edward Pace-Schott made observations which suggest that the differences between the self-awareness typical of waking and its diminution during dreaming may be the result of biochemical deactivation of the working-memory-mediating prefrontal cortex in REM sleep. Another facet of the story emerged in 2004, when Stickgold and his colleagues found that memory reinforcement during sleep applies particularly to what is known as *procedural memory*, that is to say memory for simple muscular sequences such as those involved in playing the piano. This is undoubtedly interesting, but it acts as a reminder that the whole question of memory is complicated by the number of different forms identified in the literature of the subject. In contrast to procedural memory, for example, there is what Larry Squire calls *declarative memory*, that is to say memory related to something specific, such as the name of a friend or a place. And Endel Tulving has identified what he calls *episodic memory*, which stores a record of an event having a specific sequence of contexts – what one could call a plot. Confronted with such multiplicity, Steven Rose has speculated as to how many different forms of memory there really are. If the mode of acquiring information that I have been describing here is as universal as I take it to be, one cannot rule out that there is really only one type of memory, at the most fundamental level. What appear to be different forms would then be a consequence of there being different types of context, when the memories are laid down.

Before enlarging on that unexpected possibility, let us return to the question of acquisition and incorporation of schemata, and consider another surprise. This process could be of considerable potential danger to an animal. When the surroundings are being explored, by the self-paced probing that we have been discussing, the schema or schemata being provisionally stored are related only to the currently prevailing context. The accompanying evaluation mechanism will have guaranteed that there is no *immediate* conflict. But this does not rule out the possibility of conflict with one or other schema already stored in long-term memory, and such schemata represent the great majority, of course. Some of these schemata must relate to muscular movements that are vital to survival, in fact, such as those involved in breathing. It would be a logical strategy, therefore, for the system to ensure compatibility with these well-entrenched schemata while the body's skeletal musculature is otherwise off-duty, during sleep. There is an observation that supports this idea, and it concerns the porpoise. It has been discovered that the two cerebral hemispheres of this creature take turns in sleeping, one or the other hemisphere always being awake. It seems possible that this is related to the fact that the porpoise lives in water, and that it would risk drowning if possible conflict with its breathing mechanism arose during a schema consolidation that involved the entire brain.

I cannot resist recounting a dream – a nightmare – I had a few years ago. I woke shaking like a leaf, having dreamt of a number of mice forming a living chain, like a hanging bridge spanning a gorge. The scene was horrible because each mouse held on to the mouse in front of it by digging its claws into that mouse's haunches. Blood was oozing. I staggered to the kitchen to make some tea, wondering why I'd been dreaming about something unrelated to the preceding day's events, this being unusual for my dreams. Then my eyes fell on a memo I had written to myself before retiring, reminding me to take a certain adaptor to my laboratory the following day. The mouse attached to my computer at work had malfunctioned, and the replacement could not be connected because it had an incompatible plug. Let's say that it had a B-type plug whereas the computer had an A-type port. One of my resourceful students had a number of adaptors, and we built up a bizarre looking chain – B-to-C, C-to-D and D-to-E – that nevertheless promised to do the job. We still lacked the E-to-A adaptor that would complete the chain, and I recalled having precisely such a thing at home. Hence the memo. The computer mouse is now so common that we do not associate it with the animal; it has acquired a totally separate identity. But the brain apparently nevertheless bases cognition on the muscular sequencing involved in intoning the five-letter string M-O-U-S-E. That is just what the inside-out paradigm would suggest is the case. In fact, this and similar examples may be hinting at the existence of what could be called *metalinks* in and between the premotor, supplementary motor and frontal eye fields of the cortex. They would no doubt be mediated by synapses, an example being linking of the articulation of the word *ship* with visual recognition of a ship, when the brain is sufficiently trained. But we are not always able to recall the names of things visually observed, so the putative metalinks would have to be vulnerable to impaired functioning.

Which animals are conscious? It seems logical to conclude that only those species sufficiently high on the evolutionary scale could be candidates, but where is the dividing line? Many assume that only humans are conscious, though most who have mammalian pets would probably disagree. I would like to suggest that the watershed species are the interesting creatures known as the *monotremes*. There are just three members in this subclass of Mammalia, namely the duck-billed platypus and the two genera of *spiny anteater*. They are very primitive, and only distantly related to all other mammals. Unlike the latter, they lay eggs, and they display many other reptilian characteristics, but they do have hair and they do secrete milk. Zoo keepers have frequently said of spiny anteaters that they never learn from experience; they seem to be rather dumb creatures. One fascinating thing about the spiny anteater is its brain anatomy. It lacks a properly developed nucleus reticularis thalami, the thing we earlier noted is a key component of the attentional apparatus. Without the latter, it is difficult to believe that the creature possesses consciousness. And possibly decisive evidence on the issue is provided by the following fact: *unlike all proper mammals, the spiny anteater does not dream*! (It does appear to have REM sleep, however, but this merely suggests that REM sleep and actual dreaming did not evolve at the same time, the former preceding the latter.)

That said, I am prepared to hazard a guess as to which creatures can be conscious, because I believe there are two reliable indicators, even though we cannot interview non-humans and enquire as to their inner life. Those indicators are the presence of dreaming and the presence of *play*. When animals play, they are exploring their surroundings and interacting with their peers, challenging both to react in novel ways. This is just what I have

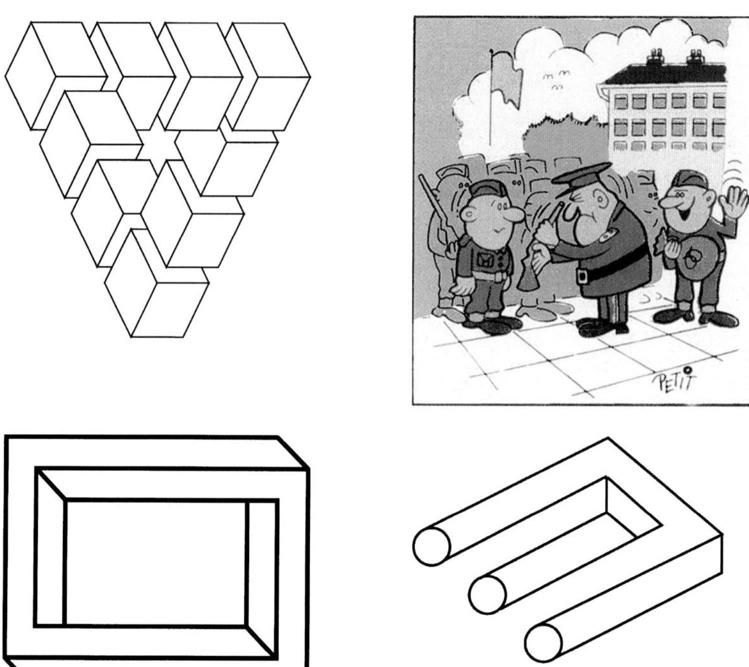

This famous arrangement of cubes to form an impossible triangle was devised by the Swedish artist Oscar Reutersvard in 1934, and a continuous form of the triangle was subsequently studied by Lionel Penrose (1898–1972) and his son Roger. The impossibility of the corresponding object emerges only through careful study. The same is true of the other two black-and-white pictures. The visual system instantly accepts their local features, because they violate none of the rules of geometry one has learned through experience. The underlying conflict is revealed only through sequential (conscious) inspection of the various parts, together with brief retention of their mutual juxtapositions. The importance of this type of picture lies in its exposure of possible discord between percept and concept: the figure is not impossible, because it can be drawn, but the object it conveys is impossible. The first recorded drawing of an impossible object, by the French artist Marcel Duchamp (1887–1968), depicted an impossible bed frame. The ability to hold a number of facts in short-term memory, and grasp their relative consequences, might be the most satisfactory measure of intelligence. As discussed in the present chapter, this ability probably also underlies the related faculty of acquiring new schemata, as well as the possibility of possessing what is known as a *theory of mind* – the ability to realize that other people think. A normal person has little difficulty in getting the point of the piece of humour in the coloured picture by the cartoonist Petit. When the soldier on the right pops the paper bag, the officer is going to think that the rifle he is inspecting was loaded, and that it has gone off in his face. But such understanding is beyond the capability of the majority of autistic people. They appear to be unable to impute thought processes to others, and they may even be unaware of their own consciousness.

been saying provides consciousness with its *raison d'etre*. And in doing so, they acquire new schemata that are candidates for permanent storage in their behavioural repertoires – hence the need for dreaming.

If tenable, these views could suggest that *intelligence* is a measure of the ability to acquire and manipulate schemata. And as noted, a contributing factor is the span of working memory. It is thus important that *functional magnetic resonance imaging* has revealed a direct correlation between intelligence, as measured by the English National Adult Reading Test, and the degree of activation of the prefrontal cortex, which we noted above is involved in working memory. This result was obtained in 2003 by John Geake, Peter Hansen and their colleagues. At a higher level, although autism and low intelligence often appear together, they are nevertheless distinct, and the former may have a broader aetiology (thus explaining why it occurs less frequently – about once in every 2000 births). That there is a strong genetic factor in autism has been demonstrated by Susan Folstein and Michael Rutter, who compared the incidence of the syndrome in identical (monozygotic) and non-identical (dizygotic) twins with what is seen in the general population. If one monozygotic twin is autistic, it is almost (though not quite) certain that the other twin will also be autistic. The other twin of a dizygotic autistic twin is, on the other hand, not more likely to be autistic than is any of its non-twin siblings. For all the children in such an autism-affected family, the incidence is nevertheless considerably higher than in the population as a whole. (An eight-nation collaboration involving upwards of a hundred scientists, led by Anthony Bailey, Kathy Lord, Anthony Monaco and Michael Rutter, has now located autism-related sites on a number of chromosomes.)

The guided mode that epitomizes consciousness may be undermined by its sheer complexity, compared with the ballistic mechanism. Several brain components must cooperate for optimal behaviour, and the possibility of autism may be the price the system has to pay for its sophistication. Working memory, the attention mechanism, the drive and the efference-copy route are all vulnerable to malfunction. Difficulty with controlling attention has been observed in autistic people, by Eric Courchesne and his colleagues, and diminution of (drive-mediating) basal ganglia function has more generally been shown to impair learning. The inevitable link between movement and cognition, envisaged in the new paradigm, is endorsed by studies of the cerebellum, which has indeed been found to be underdeveloped in many autistic people. Then again, Edward Ornitz has noted that a number of autistic symptoms suggest dysfunction of the attentional gating mechanism controlled by the above-mentioned nucleus reticularis thalami. Moreover, he has found evidence in the autistic nervous system of an inability to differentiate between exafference – the impinging signals *not* provoked by the person – and reafference – those impinging self-provoked signals. The devastating upshot could be non-receipt of the instructions, via the efference copy route, to ignore the sensory input associated with the autistic person's own routine movements. He or she might be perpetually bombarded with nerve signals that normally do not need to be attended to, and this could lead to the overlooking of signals that are, contrariwise, important.

Finally, a study of movement impairment in few-month-old infants later diagnosed with autism supports the conjectured link to schema acquisition, the victims being found to encounter difficulty in learning sequences of movements. (Reliable diagnosis of autism is not possible before the end of a child's second year, because the paediatrician needs to gauge whether language acquisition is normal or otherwise. But we must bear in mind that language too involves muscular movement.) As Philp Teitelbaum and his colleagues discovered, they apparently *do* find it hard to acquire new schemata. This could also be the cause for their apparent preoccupation with repetitive movement, as first reported in the 1940s by Leo Kanner, and the underlying factor in their resistance to change, which he also stressed. It

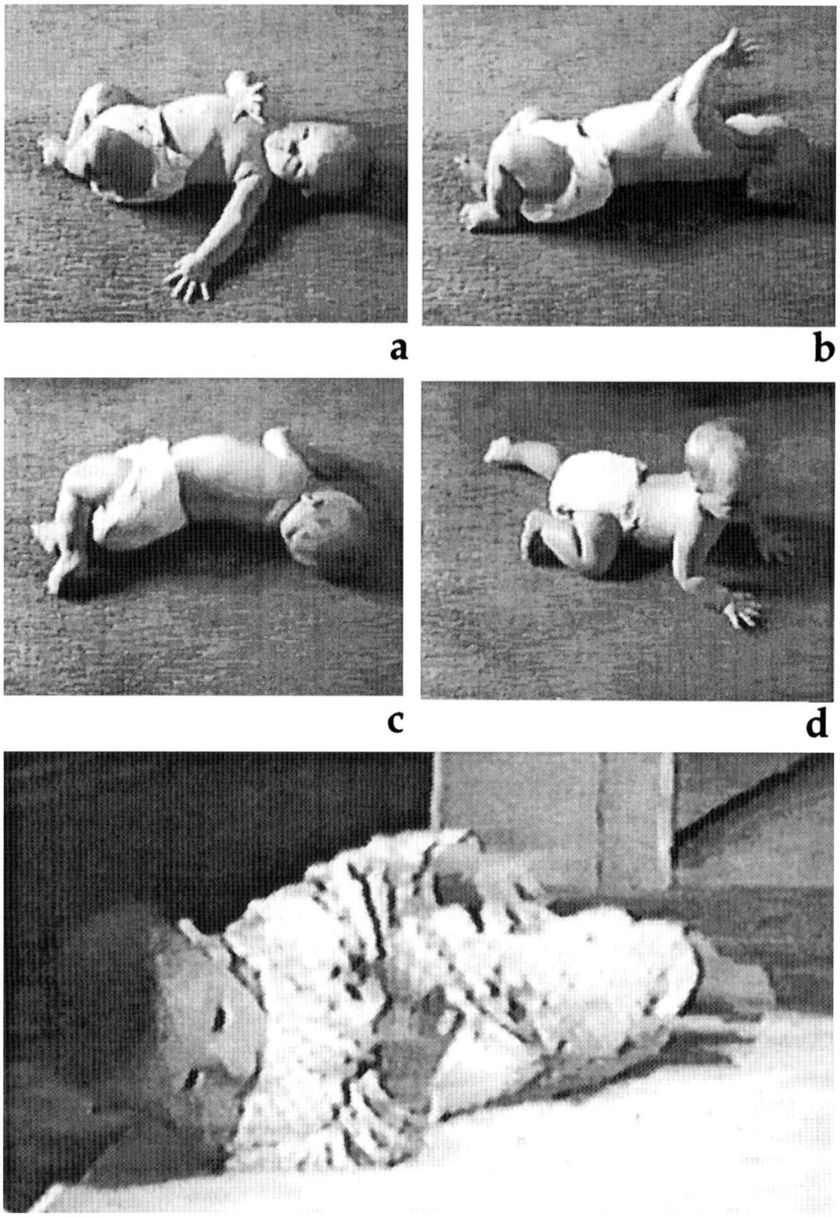

a

b

c

d

When a normal few-month-old infant wishes to turn from its back to its front (upper set of pictures), it soon learns that a sequential turning of the head, then the shoulders, and finally the hips produces the supine–prone transition with the minimum of effort. But as discovered by Philip Teitelbaum and his colleagues, such motor sequencing is impaired in autistic infants. The few-month-old autistic person encounters great difficulty in smoothly executing the supine–prone transition. Instead, he/she awkwardly arches the back, and ultimately falls onto his/her front (lower picture, from low resolution amatear video). Such an autistic child clearly has difficulty in acquiring the schemata that govern whole-body movement, just as he/she will later encounter problems in learning the schemata involved in language, during the second year of life.

could indeed be the reason why even high-functioning autistic people tend to take language too literally.

Once again I will recount one of my own observations. An autistic female was informed that she had forgotten to roll up the car window, after she had left the vehicle and was making her way into a building. Without further ado, she returned to the car, opened the door, sat in her seat, closed the door and locked it, fastened her seatbelt (!), and rolled up the window. She then reversed the entire process, unfastening her seatbelt, unlocking the door, and so on, eventually retracing her steps to the building. So here too, we have an example of the difficulty of acquiring a new schema. This was an isolated observation, admittedly, and it thus carries no statistical weight. But that doesn't matter because it harmonizes with what has been firmly established by the Wisconsin Card Sorting Test. Playing cards are shown to the subject, one by one, and he or she is asked to categorize them according to (say) number (ace, two, three, etc.). At a certain point, the tester specifies that the basis of categorization is to be switched to (say) suit. A normal person will immediately conform with the new sorting rule, but the typical autistic person is found to encounter great difficulty in making such an adjustment, and simply continues to follow the original rule.

As was noted earlier, the new paradigm being advocated here regards thought as internal simulation of the body's interactions with its surroundings. But this is not a particularly profound idea; indeed, one could ask what else it could possibly be. The important thing is to elucidate the underlying mechanism. The inside-out paradigm sees that mechanism involving the travelling of nerve signals around a closed loop that has the premotor area of the cortex and the relevant sensory-receiving areas of the cortex as the two terminal stations. By using the same analogy, one could say that the backward-directed efference-copy and forward-directed routes are the rails along which the signals run, so as to produce thoughts. What about the other brain regions? The important thing to bear in mind is that consciousness is a product of the brain as a whole; no one region bears the full responsibility for its generation. The same can be said of a car, for example; the engine, carburettor, distributor head, coupling and so on all contribute to the mechanism, but none of them individually can claim credit for propelling the vehicle.

What, then, are the contributions of some of the components that have been mentioned in this and the preceding chapter? Our guiding principle must be that they are present not by design but by evolutionary accident, they being retained if they proved to be useful. We must not regard them as being too sophisticated, therefore, that adjective being appropriate only to the overall system. Consider the cerebellum, for example. Its anatomy and physiology make it eminently well suited for letting the system ignore prosaic feedback, such as the tactile variety produced when we sit in a chair. It would be an intolerable burden for the system if it had to attend to the reactive force of the chair on the buttocks, and the same can be said of all the feedback that routinely results when we touch the various parts of our own bodies. One refers to body schemata, in fact, and the manner in which they come into play unconsciously is attributable to the role played by the cerebellum. In effect, it provides the system with a prediction of the feedback that is just about to arise when a hand approaches an ear, for example. This does not make the cerebellum a brilliantly engineered device; it is a facility that merely arises from the cerebellum's ability to record and store oft-repeated coincidences – it simply informs the system of what usually happens. A theoretical model showing how the cerebellum could function in this manner has been described by Christopher Miall, Donald Weir, Daniel Wolpert and John Stein.

Likewise, the basal ganglia are well suited for mediating the link between the internally generated drive and the appropriate course of muscular action. But here we encounter an interesting complication, in those animals capable of consciousness at least. As was stressed earlier, such animals can draw on a variety of possible options. They can think without moving, move without thinking, move while thinking about that movement, and move while thinking about something else. This indicates that the system's anatomy would have to possess a certain degree of heterogeneity, and this is precisely what was found by Ann Graybiel and Clifton Ragsdale Jr. They discovered that a key component of the basal ganglia in mammals, namely the striatum, has tandem sets of nerve cells, both of which seem to be serving roughly the same function. The above car analogy is relevant to the basal ganglia, in that they appear to act collectively as a clutch. But in their case the clutch is apparently able to act in a differential manner, simultaneously handling the start–stop instructions for both (covert) thoughts and (overt) muscular contractions. In fact, one can take that analogy one step further because Roger Albin, Anne Young and John Penney have traced a number of different muscular-control malfunctions to faulty physiology in one or other of the basal ganglia components. There is, for example, a sad similarity between the jerky movements of the victim of *Parkinson's disease* and the jerky movements of a car when it is being operated by a novice driver who fails to control the clutch adequately.

It is high time, here toward the end of this final chapter, that something be said about other current theories of consciousness. Those aspiring to explanations at the level of anatomy and physiology have invariable assumed without question that the traditional stimulus–response paradigm is correct, the exceptions being those I mentioned at the start of my analysis. I have also mentioned the theories that attempt to link the emergence of consciousness to the development of language, and I have explained why I disagree with them. I feel that they are aiming at the wrong level in the hierarchy discussed near the end of this book's Prologue. Finally, there is a type of theory that operates at the other extreme of that hierarchy, namely the type that tries to link consciousness to things occurring at the sub-cellular and near-atomic level. One of these theories, championed by Roger Penrose and Stuart Hameroff, sees consciousness as arising because of quantum coherence associated with the microtubules that were discussed in Chapter 16 of the present book. In a sense, the underlying argument is that since both quantum mechanics and consciousness are mysterious, they might be fundamentally linked. But if that were the case, one might wonder why there are contributing factors that operate at a level manifestly higher than that of single molecules: processes implicating synapses, nerve impulses, and events strikingly suggestive of the inertial effects that the microtubule picture is incapable of explaining.

It has wisely been said that what all theories of consciousness lack is an operational definition of awareness. It is indeed frustrating that something so familiar and obvious, namely our ability to observe and contemplate the world around us, should be so difficult to pin down. Is it possible that the processes I have described here have permitted such a definition to slip in unobtrusively, through the back door as it were? That indeterminacy principle appears valid, at least, because it can explain the fascinating illusion reported in 1972 by Frank Geldard and Carl Sherrick. They constructed a three-component apparatus through which a series of brief taps – two or three in each case – could be administered first to the wrist, then to the elbow, and finally to the shoulder. If the intervals between the taps is suitably adjusted, the subject observes something that is clearly illusory, namely a series of equally spaced taps along the length of the arm, as if a little rabbit were running along it! The nervous system – drawing on past experience – is clearly making its best guess as to what is going on, given the inherent uncertainty we have been discussing. And it has produced awareness of something that does not exist, namely the taps felt at positions intermediate to wrist and elbow, and the others felt between the elbow and shoulder.

When the eyes inspect an object, they move position about four times every second, the small shifts of gaze direction being known as saccades. Their scanning of the object is not haphazard. The eyes tend to spend proportionately more time on features of greater potential interest. In the examples shown in the black-and-white part of this illustration, the straight line segments trace the route taken by the direction of gaze. There is a mechanism which suppresses the formation of a visual image during the saccadic movements, and evidence suggests that the system pieces the various quarter-second snapshots together, jigsaw fashion. In the two examples shown here, the system clearly knows that the regions of a human face having particular potential interest are the eyes and the mouth, so the gaze is tending to linger in those places. This is even the case for the eyes' scrutiny of the right-hand object, despite the fact that it is merely a statue. Given that consciousness of a perceived object requires about half a second to develop, as discussed in the present chapter, it is clear that the saccades are undertaken unconsciously. They are nevertheless steered by what is clearly a well-informed mechanism. A cursory glance at the upside-down coloured part of the illustration is sufficient to identify it as Leonardo's famous Mona Lisa, replete with her elusive smile. If the page is rotated to recover the more familiar orientation, however, the picture is found to illustrate something else, namely that the individual elements of an image are handled separately by the early part of the visual system, and that they can thus have a life of their own; the jigsaw process obviously has its limitations.

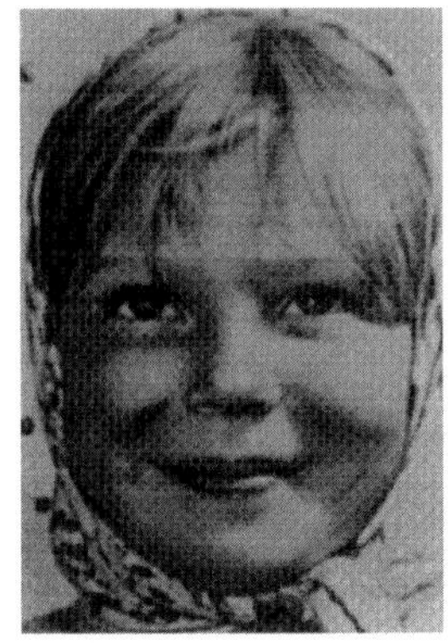

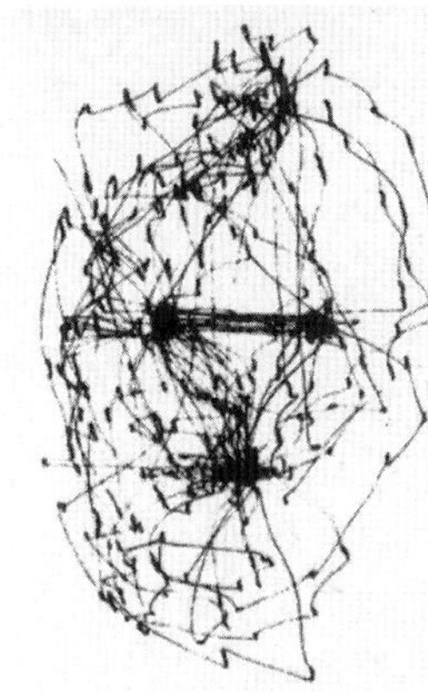

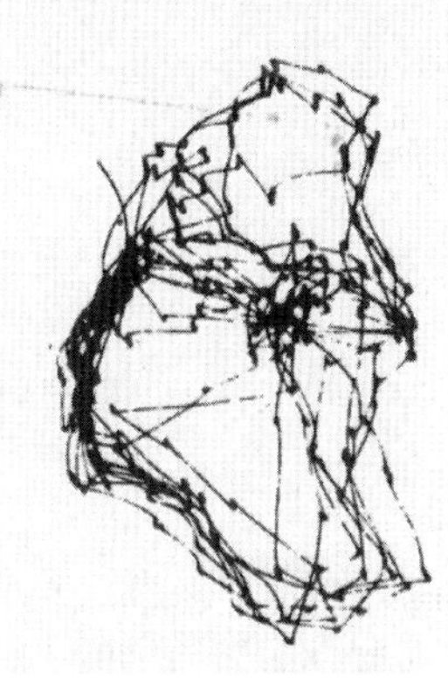

Stroop interference and ambiguity

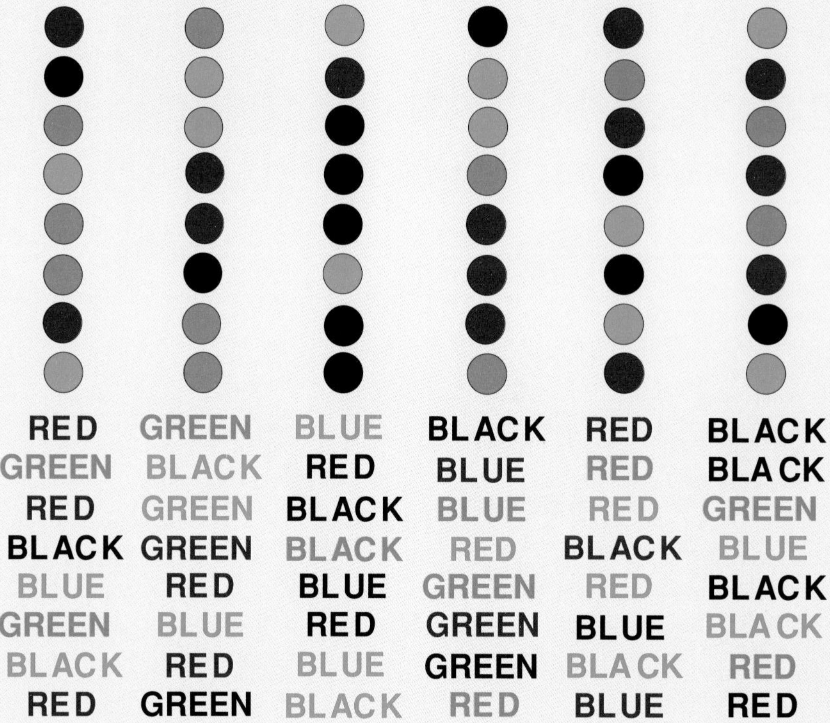

To a certain extent, the brain can simultaneously direct two or more different patterns of muscular movements, and this ability extends to cases in which some of the movements are merely covert, as in thought. But as first demonstrated by J. Ridley Stroop in 1935, interference between two such sets of movements can arise if these relate to things displaying a high degree of mutual similarity. Such *Stroop interference*, as it is called, can be demonstrated by comparing the times taken to read ***aloud*** the ***actual colours*** in the two sequences of objects (circles and words) shown in the colour diagram. According to the view expressed in the current chapter, the brain's basal ganglia are probably responsible for permitting performance of simultaneous sets of movements (including the covert variety), but that brain component clearly has its limitations. The black-and-white figures are all examples of the type of ambiguity related to Stroop interference, this phenomenon preventing perception of both alternatives simultaneously. The duck-rabbit was originally produced by Ludwig Wittgenstein (1889–1951), while the reversing cube was first described in 1832 by the Swiss naturalist and crystallographer L. A. Necker, the reversing staircase being a variation on that theme. The celebrated picture published by Edgar Rubin (1886–1951) in 1921 can be interpreted either as a white vase against a black background or two black counter-disposed faces against a white background. Finally, and most sophisticatedly, the famous picture drawn by the cartoonist W. E. Hill, and reproduced in the 6 November number of the magazine *Puck*, was cited by Edwin Boring (1886–1968) in a brief article published in 1930. It can be taken to show either an elderly woman, seen at almost full face, or a younger woman seen in profile from her left side. The old woman's lips serve as the younger woman's neckband; her left eye is the younger woman's left ear; her right eyelash is the younger woman's left eyelash; and her nose is the other's jaw. It is significant that with practice one can switch from one interpretation to the other, in any of these ambiguous figures, in as little as about half a second, but not much more rapidly. This is consistent with the result obtained by Benjamin Libet and his colleagues that conscious awareness takes about that long to develop, as discussed in the present chapter.

This *cutaneous rabbit*, as it is called, has it in common with all illusions that top-down processes based on what usually happens are influencing bottom-up processes occurring close to the sensory input. Concept is influencing percept. In essence, the problem arises because the nervous system usually does not have the luxury of being able to take its time; perceptions invariably have to be generated very rapidly, before making way for the new input already impinging on the senses. So an operational definition could be as follows: *Conscious awareness is possible if temporal changes in sensory input (or in internal patterns of signals) can be detected while the signals resulting from that input are still reverberating in the system, in a fully functional – though no doubt modified – form, such detection permitting the system to capture correlations between cause and effect so rapidly that their significance for the organism can be immediately appreciated, and also immediately used to modify the organism's repertoire of responses*. But as we have seen, there is a limit on just how rapidly the system can function in this manner.

In his beautifully written *Man on his Nature*, Charles Sherrington famously likened the neural networks in the brain to an enchanted loom. But this did not mean that he regarded the brain's functioning as in any way magic. I feel no misgivings in embracing a theory of consciousness and intelligence rooted in the brain's anatomy and physiology, one devoid of anything mysterious. And I will continue to cherish a strong belief that the individual brain will turn out to be rather simpler – in principle, at least – than has been generally expected. On the other hand, and as mentioned in the Prologue, I am equally convinced that understanding the interactions *between*

Top-down influences in brain function

In his thirteenth sonnet, William Shakespeare wrote *My most true mind thus makes mine eye untrue*, and Roger Carpenter's interesting remark added clarification when he wrote *Much artistic endeavour consists in trying to recapture the novelty of primary perceptions which our internal model continually discounts*. Both statements emphasize the influence of top-down process on the way we interpret sensory input. And that influence is captured in a particularly dramatic fashion by the triangle illusion created by Gaetano Kanizsa; one sees a complete white triangle even though most it has not been drawn. A remarkable fact emerges if one measures the activity in cells in the primary visual cortex corresponding to the position of the middle of any of the three sides of the white triangle. If a black line had been drawn at that position, with the appropriate orientation, activity would be expected, and is indeed observed. The astonishing thing is that the same activity level is observed even when only the white triangle is 'present'. But that activity drops gradually to zero as more and more of the partial black circles and V shapes are removed, even though these objects are not in the cells' field of view. This exposes the top-down determinants of the illusion, and the importance of input signals from nerve cells further up in the cortical hierarchy. Semir Zeki and Stewart Shipp discovered that these inputs have larger catchment angles than is the case for the forward projections lower in the hierarchy. The principles of top-down learning were explored by Claus Nielsen and the present author in 1991, and more completely by Wolf Singer four years later. The same top-down influences permit detection of the well-camouflaged Dalmatian dog, drawn by Ronald James, and they underlie our easy reading of Oliver Selfridge's famous two words – THE CAT – despite the fact that there is no difference between the sloppily drawn H and the equally careless A. Top-down influences are also at central stage when the psychiatrist probes the unconscious mind of his or her patients by using the test devised by Hermann Rorschach (1884–1922). As is well known, the patient is asked to identify the shapes suggested by the various ink blobs, immediately and without contemplation. Because this test accesses memory through unconscious processes, it is capable of revealing some of the patient's personality traits, although such interpretation requires skill on the part of the tested.

brains and their associated minds will continue to be a scientific challenge for a very long time.

So let us return to the Prologue, and contemplate the actual course of events. It was inevitable that mankind's early view of things would be self-centred, and that all the things we see around us would be perceived as having been put there for our convenience. It is a sobering thought that, on the contrary, we are mere latecomers on the scene, and that, for example, the black hole at the centre of our galaxy was already ancient before the first primitive forms of life appeared on Earth. As we have seen, the gradual evolution that followed ultimately produced creatures possessing consciousness,

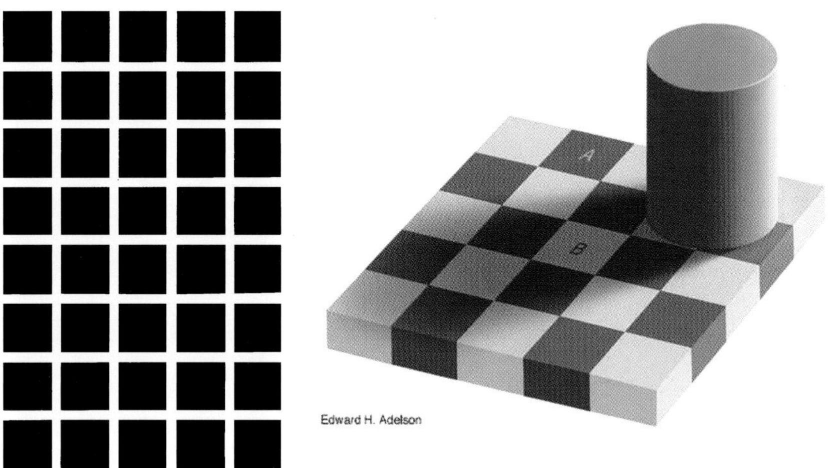

There are many well-known visual illusions, but they do not all arise in the same manner; some are attributable to processes occurring early in the visual system while others have a marked top-down character. These two extremes are illustrated here. In the grid illusion, discovered by L. Hermann in 1887, nebulous shadows are seen at the intersections of the white strips, but they seem to evaporate if one stares directly at them. The effect is traceable to the ganglion cells that are the retina's output neurons. These were first studied systematically by Haldan Hartline (1903–1983), and they were subsequently demonstrated by Horace Barlow and Stephen Kuffler (1913–1980) to respond maximally either to a light central circle surrounded by a dark ring or the opposite, there being roughly equal numbers of these two classes. The intersections in the grid's image on the retina come closer to the former of these optimal stimuli, and the slight departure from the ideal gives rise to the shadows. But it is still not known why they disappear when one looks directly at them. The other extreme is here represented by the astonishing illusion created by Edward Adelson. It is very difficult indeed to convince oneself that the two squares marked A and B really have been drawn with the same shade of grey. Top-down influences play a major role in this ingenious illusion.

and I have suggested that this faculty emerged with the mammals (and possibly with some bird species). But the self-centredness persists. Many regard what went before as just a preliminary to the agenda's main point of business: the arrival of mankind. And people holding that view tend also to accord consciousness exclusively to us. They see *Homo sapiens* as different in kind from other mammals, rather than just different in degree. In that scenario, consciousness becomes the great releaser, permitting us to be the only exercisers of free will.

Well, let's take a closer look at our much-vaunted freedom. The pivotal experiment, in my view at least, was carried out by Roger Carpenter and Matthew Williams, and it involved the tracking of eye movements. Some might regard these as too specialized to have any bearing on such a major issue, but we should bear in mind that vision lays claim to almost half the entire cerebral cortex; sight is our senior sense. Carpenter and Williams required their subjects to fix their gaze on a centrally located cross, and subsequently to let their eyes swivel toward a suddenly appearing spot of light. The latter could appear either to the left or to the right of the cross, in a random manner, and the subjects were not told exactly when the spot would become visible. Neither were they informed that the experimenters could control the left/right statistics – for example 25% to the left and 75% to the right. By remotely tracking the eye movements, Carpenter and Williams could measure the time delay before each swivelling, left or right, and their subsequent statistical analysis demonstrated that the movements were being made in a totally mechanical fashion. Furthermore, it transpired that the subjects' nervous systems were *unconsciously* adapting themselves to the prevailing left/right statistical distribution.

Sometimes, a subject's eyes would swivel before the spot became visible, as if anticipating its appearance, and such jumping the gun, as it could be called, sometimes shifted the eyes in the wrong direction, on average half the time of course. But the remarkable thing was that these express shifts of gaze were also found to occur in a totally mechanical manner. Bearing in mind that all the gaze shifts recorded in these experiments were being made in a time shorter than the half second measured by Benjamin Libet and his colleagues, we see that they were all made unconsciously. And yet they were extracting useful information from the surroundings, and acting on it accordingly. Free will simply didn't come into the process.

Bird-brains or brainy birds?

There is increasing evidence that birds have impressive cognitive abilities. Most spectacularly, New Caledonian crows – *Corvus moneduloides* – have been observed fabricating tools in order to get at food that would otherwise have been inaccessible. Alex Weir, Jackie Chappell and Alex Kacelnik filmed individuals of this species and documented that they were bending pieces of wire into hooks, which they then used to lift small food-laden baskets from the bottom of a glass tube, the latter being much longer than the bird's beaks and necks (see the colour photograph). Other individuals observed these performances, and clearly learned from them. Pigeons have been found to be capable of what is known as *transitive inference*, which is regarded as cognitively challenging. Individuals were exposed to situations involving unequal quantities of things – let's denote the things and their quantities by the letters A, B, C, D and E – in which they would be interested. And if they were inspecting these items adequately, and drawing the appropriate conclusions, they would be able to see that A was greater than B, B was greater than C, C was greater than D, and D was greater than E. Such adequate inspection and inference resulted in them being given a positive reward. As reported by Lorenzo von Fersen and his colleagues, subsequent testing revealed that the pigeons were able to appreciate that B was greater than D, for example, with better than 85 per cent reliability. Günter Vetter and his colleagues have discovered that individuals of this same species also display an aptitude for handling the type of visual ambiguity illustrated elsewhere in the present chapter; they can adjust their perception of movement in order to differentiate between horizontal and vertical motion, for example. Robert Lanza has observed what could be called lying by pigeons. Trained to differentiate between a number of different colours, and then given a bigger reward when correctly identify one particular colour, they will then deliberately report seeing that colour, regardless of what they are actually confronted with, just to get the bigger reward. Some birds also show an ability to remember sequence, and to associate it with time, these aspects collectively constituting what is known as episodic memory. Nicola Clayton and Anthony Dickinson have discovered that scrub jays can apply this faculty to location, content and the relative time since specific foods items were cached. Impressive avian abilities are also observed in other sensory modes. Peter Marler has documented remarkable examples of flexibility in the acquisition of novel song patterns in several species, and the verbal ability of parrots is of course legendary. Irene Pepperberg has observed high levels of communicative prowess in African grey parrots, for example, and these birds appear to understand numbers up to about seven, including the number zero. Ann Butler and the present author have constructed the overall circuit diagrams for mammalian and avian brains, in an effort to compare their similarities and differences. They found that essentially all the main regions of the mammalian brain have their counterparts in the brains of birds, and that the same is true of the way in which the various regions are interconnected. The actual arrangements of the nerve cells in some of the regions showed marked differences between the two taxa, however. If the numerous observations described above can be taken as reliable indicators of consciousness, therefore, the conclusion would

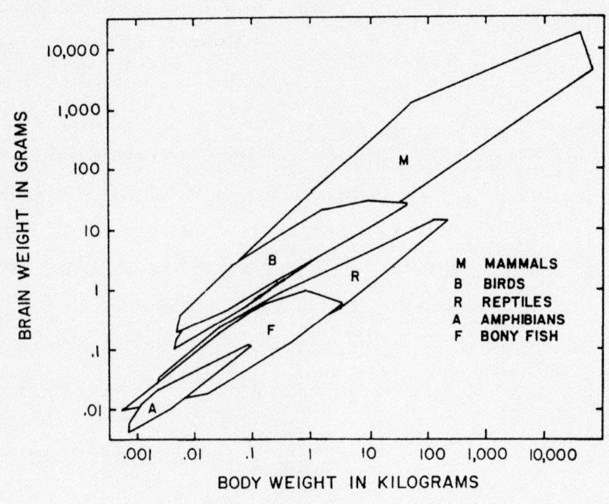

be that it is the circuitry rather than the local pattern of nerve cells that is the critical determinant for that faculty. It is well known that the ratio of brain weight to body weight is remarkably similar in mammals and birds (see the graph), whereas the ratios for reptiles, amphibians and fish are much less favourable. This suggests, in turn, that it is the brain's gross pattern of components and its circuitry that are chiefly influenced by that ratio. Referring back to the evolutionary tree reproduced in Chapter 17, and bearing in mind the evidence against consciousness in those borderline creatures, the monotremes (that is to say, the duck-billed platypus and the two species of spiny anteater), as discussed elsewhere in the present chapter, one sees that consciousness would have to have evolved independently in the birds and in the proper mammals. One possibility is that the vital genes were present in creatures more primitive than either the birds or the mammals, and that they were merely waiting to be activated. The conclusion would then have to be that such activation has occurred twice, in separate branches of the evolutionary tree.

A widely held view is that the evolutionary event that produced us must have been cataclysmic, given that it created a species totally unlike any that had existed before, that is to say conscious beings endowed with free will. Many would maintain that this is self-evident. The different view I will now put forward will no doubt shock many readers. I myself find it shocking. But science has its rules, to which all its practitioners must adhere. One cannot turn one's back on a theory just because it leads to uncomfortable conclusions.

The thing which I believe exposes the chink in the traditional view's armour looks innocent enough. It is the frequently asked question: Why do we sleep? It has long since been demonstrated that sleep is not strictly necessary for restoration of energy; wakeful resting is just as effective. Could it be that the question 'Why do we sleep?' is inappropriately posed? What if the emergence of consciousness simply *had* to be accompanied by the emergence of sleep? What if an important function of the latter is processing of the new information acquired during waking, and its possible incorporation into the brain's permanent store of schemata, in a manner that precludes their mutual interference? If this were the case, one would expect the system to have evolved in such a way that the wakeful period occurs when there is most new information in the offing, that is to say during the daylight hours for our highly visual species. And this is just what we find, of course. There is a centre in the forebrain, the *suprachiasmatic nucleus*, located adjacent to a region where the nerve paths carrying visual information are particularly closely packed. It is thus well positioned to tell the sleeping system when daylight returns, and therefore when it is time to wake and start probing the surroundings for new information. (It is this brain centre that makes it hard to sleep during the summer months.) There are of course nocturnal species who burn the midnight oil, as it were, but they have evolved to do so as a result of prey–predator exigencies, which themselves involve the acquisition and use of information; they do not invalidate the thesis.

So the shocking possibility has to be entertained that the ultimate control is nevertheless exercised by the unconscious, even in creatures who possess the power of consciousness – and thus must also spend part of their time sleeping. Consciousness would then merely serve to make the creature's behavioural repertoire more sophisticated. But that word *merely* should not be taken to denigrate the richness of conscious experience; it simply refers to the widened operational scope enjoyed by the owner of the faculty. The richness stems from the enormous variety of things and scenarios in our

The ability to deceive is often taken as a sign of relatively high mental sophistication, and there is evidence of it in the behaviour of both mammals and birds. With a resolve suggestive of thought, this nesting killdeer (colour photograph) is deliberately leading a potential predator away from its nest and its young. It is acting as if it were injured, to make itself appear more vulnerable, and thus more attractive to the intruder. The bird shown here, which was photographed by Noble Proctor, is feigning a broken wing. The cartoon, based on actual observations of hamadryas baboons, demonstrates that members of at least one non-human mammalian species are capable of attributing thought to their peers. The picture shows a situation in which a female is grooming a young male that is hidden from her harem master's view by a rock. The female took twenty minutes to work her way casually over to the rock, in order not to arouse suspicion, and it was clear that she could calculate that the young male would be invisible to the senior baboon.

surroundings, rather than from the underlying mechanism. The enchanted looms of our nervous systems are wondrous things, to be sure, but there is no ghost in the machine. Thinking is a bodily function. The emergence of consciousness (and its accompanying sleep) has been, I feel, evolution's boldest trick – to date, at least. It is deeply impressive that the material world should have admitted of such a possibility.

Summary

There are two opposing paradigms for the generation of consciousness and the mind. One, the so-called stimulus–response paradigm, regards consciousness as arising exclusively near the brain's sensory-input regions, a *possible* consequence being the dispatch of nerve signals to the muscle-activating region. The alternative theory, which could be called the inside-out paradigm, maintains that consciousness will not be generated unless the sensory-input signals penetrate through to the muscle-sequencing region, and activate an appropriate schema, the latter being a cognitive element reproducibly linking motor-planning activity to anticipated sensory feedback activity. This mechanism adds a servo-guided form of movement to the more primitive ballistic mode. A covert version of the mechanism underlies thought, which is internal simulation of the body's interactions with its surroundings. Movement, of either the overt or covert type, is the only way an animal can gain information. The relevant stimulus is thus the creature's own self-paced movement – actual or *actively* planned for – while the relevant response is that of the sensory feedback from the surroundings – actual or *actively* anticipated. So the animal must possess a mechanism which tells the nervous system that a given response from the surroundings is associated with the recently undertaken movement-mediated probing of those surroundings. The apprising mechanism involves efference-copy signalling from the muscle-sequencing region to the sensory-input regions, and there are the additional requirements of an attention-focussing mechanism and adequate working memory. The evolution of the latter, which depends on the brain's frontal-lobe tissue, accompanied the increasing size of the front of the skull. It also plays a key role in intelligence, which is the ability to link existing schemata into more complicated schemata. Play and dreaming appear to be reliable signs of consciousness in other species, and this indicates that all true mammals – as opposed

Patients who have been close to death and whose vital systems have been resuscitated frequently report having seen a tunnel with a bright light at its end. This is not new. A fifteenth-century painting by Hieronymus Bosch (1450–1516) shows departing souls being helped toward such a tunnel by attendant angels. The idea of posthumous survival of the mind belongs to the domain of religious faith, and belief in it clearly brings comfort to millions of people. But there is no scientific evidence supporting such a thing. If one looks for a scientific explanation of such near-death visual experiences, one possibility is that the illusion may be caused by perturbations of the nervous system as its normal functioning breaks down, the central bright patch possibly relating to the rich endowment of receptor cells in the retina's centrally located fovea. Another possibility is that the illusion is attributable to top-down influences stemming from the fact that the idea of passing through a tunnel is familiar in human culture's view of the end of life.

to the borderline monotremes – are conscious. The same is possibly true of at least some bird species. But despite the wider behavioural options available to possessors of consciousness, there is evidence that ultimate control is nevertheless retained by the unconscious. There is also evidence that free will is an illusion.

Index

Page numbers in italic refer to figures, those in bold refer to boxes

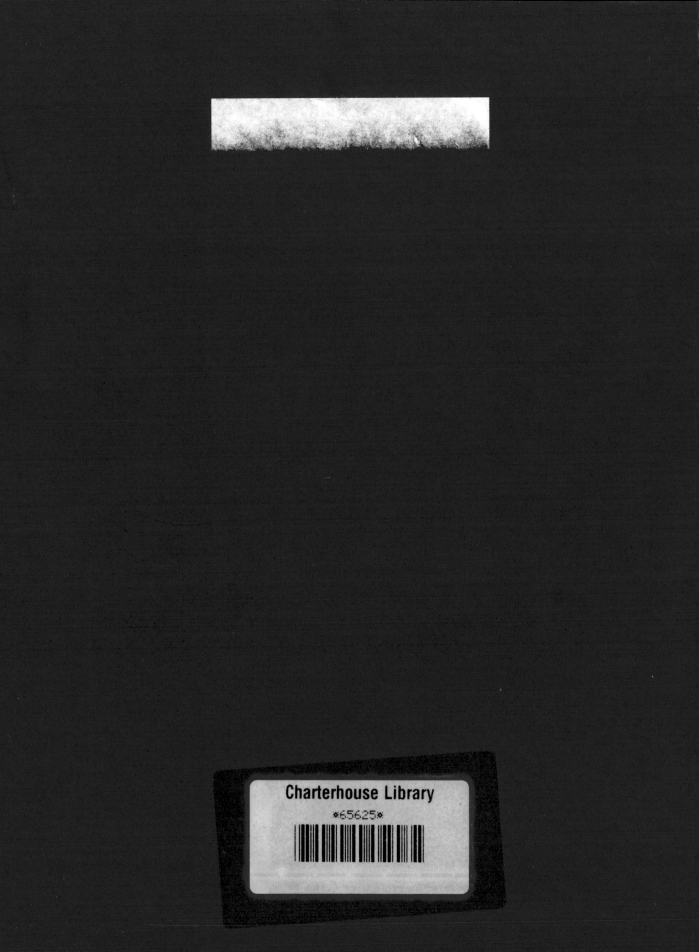